PHYSICS

PHYSICS OF
NUCLEI AND PARTICLES

VOLUME I

ACADEMIC PRESS, INC.
111 Fifth Avenue, New York, New York 10003

United Kingdom Edition published by
ACADEMIC PRESS, INC. (LONDON) LTD.
Berkeley Square House, London W.1

LIBRARY OF CONGRESS CATALOG CARD NUMBER: 68-14644

PRINTED IN THE UNITED STATES OF AMERICA

PHYSICS OF
NUCLEI AND PARTICLES

PIERRE MARMIER

ERIC SHELDON

Laboratory of Nuclear Physics
Federal Institute of Technology
Zürich, Switzerland

VOLUME I

ACADEMIC PRESS

NEW YORK AND LONDON

FOREWORD

The last decade or so has witnessed an increasing rate of progress in the attack on problems concerned with nuclei. Not only have new and sophisticated techniques been responsible for a vast improvement in the quality of experimental data but also new and penetrating theoretical ideas, supported by rapid electronic computational facilities, have gone far toward consolidating the experimental gains. Models of the static and dynamic properties of nuclei have been refined and polished to account for the improved data, but we still lack a complete and general model that incorporates all of the essential features of the individual models. Significant progress has been made in dealing with the many-body aspects of nuclei, but a practical general approach has not yet been devised. Great effort has been expended in the interpretation of proton-proton and neutron-proton scattering data, but we still do not have precise knowledge of the details of the fundamental nucleon-nucleon interaction. Although there has been much progress, there are still many outstanding unsolved problems; nuclear physics is far from being a closed subject.

With the current intensity of interest in nuclear problems, it is particularly appropriate that there should appear, at this time, a new textbook devoted to a comprehensive review of the status of the entire field. Professors Marmier and Sheldon have succeeded in this staggering task by producing, in these three volumes, a thorough and incisive discussion of nuclear physics today.

These volumes, a prerequisite for which is a one-year course in quantum mechanics, were inspired by the success of the lecture notes which the authors developed for the course at E.T.H., Zurich. The text is unique in that the presentation attempts to unify the concepts of particle physics and nuclear physics. Each chapter starts from an elementary point of view and the subject matter is developed to a high degree of sophistication.

The reader will find here, in this collaboration of experimentalist and theorist, an approach to the subject that reveals the authors' deep and active involvement in current research problems. Details have not been spared but neither have the authors lost sight of the unified structure of the subject. This is indeed a welcome and needed addition to the literature of the field.

JERRY B. MARION

University of Maryland

March, 1969

CONTENTS

Foreword v

Summary of Contents, Volumes II and III xvii

1. Historical Development of Nuclear Physics. The Size and Constitution of the Atomic Nucleus

A SURVEY OF NUCLEAR AND PARTICLE PHYSICS WITHIN THE
CONTEXT OF PHYSICS AS A WHOLE

1.1. The Present Status of Nuclear Physics 1
1.2. Brief History of the Development of Atomic, Nuclear, and
Particle Physics 2
1.3. The Domain of Nuclear Physics 7
1.4. The Size of the Nucleus and Nuclear Constitution . . . 8
Exercises 19

2. Nuclear Radii and the Liquid Drop Model of the Nucleus

STABILITY AND RADIUS OF NUCLEI

2.1. Energy Considerations 22
2.2. The Radius of Nuclei and the Liquid Drop Model . . . 28

vii

2.3. The Liquid Drop Model and the Semiempirical Mass Formula   31
    2.3.1. Von Weizsäcker's Approach to Binding Energy   31
    2.3.2. Volume Energy   32
    2.3.3. Surface Energy   32
    2.3.4. Coulomb Energy   33
    2.3.5. Asymmetry Energy (Neutron Excess)   34
    2.3.6. Pairing Energy   36
    2.3.7. Summary   37
    2.3.8. Other Mass Formulae   39

2.4. Applications of the Mass Formula to Considerations of Stability   42
    2.4.1. Coulomb Radius Constant   42
    2.4.2. Radius Constant from Binding Energy for Mirror Nuclei   .   43
    2.4.3. Uranium Fission   44
    2.4.4. The Behavior of Isobars in β Decay   .   .   .   45
    Exercises   51

3. Interactions and Nuclear Cross Sections

INTERACTIONS, TRANSITION PROBABILITY, AND REACTION CROSS SECTION

3.1. Nuclear Force Characteristics   53
3.2. Classification of Interactions   55
    3.2.1. Strong Interactions   55
    3.2.2. Electromagnetic Interactions   .   .   .   55
    3.2.3. Weak Interactions   55
    3.2.4. Gravitational Interactions   57

3.3. Response of Particles to Strong, Electromagnetic, and Weak
    Interactions   60
    3.3.1. Strong Interactions   61
    3.3.2. Electromagnetic Interactions   .   .   .   61
    3.3.3. Weak Interactions   62

3.4. Transition Probability   64
    3.4.1. Time-Dependent Perturbation Theory .   .   .   64
    3.4.2. Transition Probability per Unit Time   .   .   67

3.5. Reaction Probability and Cross Section .   .   .   68
    3.5.1. Definition of Reaction Cross Section   .   .   71
    3.5.2. Partial and Total Cross Sections .   .   .   77
    3.5.3. Differential and Total Cross Section (Angular Distribution)   .   77
    3.5.4. Double-Differential Cross Section (Angular Correlation)   .   78
    3.5.5. Geometrical and Absolute Cross Section .   .   .   79
    3.5.6. Cross Sections Referred to Electrons and Atoms .   .   82
    3.5.7. Classical Elastic Coulomb Scattering Cross Section (Rutherford
        Formula)   83
    3.5.8. Coulomb Scattering Cross Section for Like Particles (Mott
        Formula)   88

3.6. Transition Probability and Cross Section 89

    3.6.1. Box Normalization and the Relationship of Transition Probability per Unit Time to the Differential Cross Section for Elastic Scattering 89

    3.6.2. Elastic Scattering Matrix Element. 91

    3.6.3. Level Density 92

    3.6.4. Elastic Scattering Cross Section (Born Collision Formula) . . 93

    Exercises 95

4. Passage of Ionizing Radiation through Matter

IONIZING EFFECTS OF ELECTROMAGNETIC RADIATION AND CHARGED PARTICLES

4.1. Survey of Electromagnetic Interaction Processes . . . 98

4.2. Thomson and Compton Scattering of Gamma Radiation . . 98

    4.2.1. Thomson Cross Section 100

    4.2.2. Compton Effect 103

    4.2.3. Compton Cross Section (Klein-Nishina Formula) . . . 106

    4.2.4. Atomic Compton Cross Section 110

4.3. Rayleigh Scattering 112

4.4. Photoelectric Effect 113

    4.4.1. Energy and Atomic-Number Dependence 117

    4.4.2. Angular Distributions 119

    4.4.3. Attenuation Coefficients 120

4.5. Auger Effect 120

4.6. Pair Production 122

    4.6.1. Dirac Electron Theory 123

    4.6.2. Electron-Positron Conjugation 125

    4.6.3. Feynman Graphs 127

    4.6.4. Differential and Total Pair Cross Sections 138

    4.6.5. Inverse Pair Production: Annihilation Radiation and Bremsstrahlung 144

4.7. Nuclear Scattering of Gamma Rays 147

4.8. Total Attenuation Coefficient for Electromagnetic Radiation Passing through Matter 148

4.9. Interaction of Charged Particles with Matter 155

    4.9.1. Energy Loss of Heavy Charged Particles (Stopping Power, Range, and Straggling) 156

    4.9.2. Energy Loss of Electrons (Stopping Power, Range, and Straggling) 170

    4.9.3. Bremsstrahlung 176

    4.9.4. Čerenkov Radiation 190

4.10. Energy Loss of Heavy Ions 199

    Exercises 204

5. Nuclei and Particles as Quantum-Mechanical Systems

QUANTUM PROPERTIES OF NUCLEI AND PARTICLES

5.1. The Need to Treat Nuclei and Particles Quantum-Mechanically   208
5.2. Quantization of Angular Momentum   210
5.3. Quantum Numbers of Individual Particles   212
   5.3.1. Orbital Quantum Number l 212
   5.3.2. Magnetic Orbital Quantum Number m_l 212
   5.3.3. Spin Quantum Number s 212
   5.3.4. Magnetic Spin Quantum Number m_s 213
   5.3.5. Total Angular Momentum Quantum Number j 213
   5.3.6. Magnetic Total Angular Momentum Quantum Number m_j 214
   5.3.7. Radial (n_r) and Principal Quantum Number n 214
   5.3.8. Isospin Quantum Numbers T and T_z 215
   5.3.9. Strangeness S 216
   5.3.10. Parity π 220
5.4. Quantum Properties of Nuclear States   224
   5.4.1. Nuclear Energy Levels   224
   5.4.2. Nuclear Angular Momentum (Spin and Coupling Schemes)   225
   5.4.3. Nuclear Parity   227
   5.4.4. Isospin   227
   5.4.5. Magnetic and Electric Moments of Particles and Nuclei.   240
   5.4.6. Anomalous Nucleon Spin Magnetic Dipole Moments   247
   5.4.7. Magnetic Dipole Moments of Nuclei   250
   5.4.8. Electric Moments   251
5.5. Symmetries, Invariances, and Conservation Laws   257
Exercises   261

6. Radioactivity

RADIOACTIVE DECAY

6.1. Mean Lifetime toward Radioactive Decay   263
   6.1.1. Level Width and Decay Probability   265
   6.1.2. Half-Life and Specific Activity   269
6.2. Branching Ratios (Partial Widths)   270
6.3. Radioactive Decay: Daughter Activity   272
   6.3.1. Daughter Activity in Special Cases   275
   6.3.2. Production of Radioactive Sources (Induced Radioactivity)   281
   6.3.3. Mixture of Activities   283
6.4. Decay Schemes of Widely Used Radioactive Sources   286
6.5. Parent-Daughter Relationships in Radioactive Dating   290
6.6. Nuclear Stability Limits according to the Liquid Drop Model   294
Exercises   295

7. Alpha Decay

ALPHA DECAY

7.1. Introduction	298
7.2. Semiempirical Mass Formula Applied to α Decay	299
7.3. Relation between α Energy and Decay Half-Life	301
7.4. Penetration of Potential Barriers	303
7.4.1. Rectangular Barrier.	306
7.4.2. Barrier of Arbitrary Shape	309
7.4.3. Nuclear Potential Barrier	310
7.5. Short- and Long-Range α Radiation	320
7.6. Application of the Gamow Formula to α Decay	321
7.6.1. α Energy and Intensity	321
7.6.2. Nuclear Radius Constant	322
7.6.3. Spontaneous Nuclear Disintegration	323
Exercises	324

8. Beta Decay

THE WEAK BETA-DECAY INTERACTION

8.1. Introduction	327
8.1.1. Decay Modes	328
8.1.2. Mass-Energy Balance	329
8.1.3. Beta-Energy Spectrum	331
8.2. The Neutrino	334
8.2.1. Neutrino Properties	334
8.2.2. Neutrino Hunting	336
8.3. Beta-Decay Theory	351
8.3.1. Formulation	351
8.3.2. Probability Function and the Beta-Momentum Spectrum	352
8.3.3. Statistical Factor (Final-State Density)	353
8.3.4. Interaction Matrix Element	355
8.3.5. Coulomb Correction Factor	357
8.3.6. Kurie Plot	360
8.3.7. Neutrino Rest Mass	361
8.3.8. Neutrinos and Cosmology	367
8.3.9. Beta-Decay Lifetime	368
8.4. Classification of Beta Transitions	371
8.4.1. Degrees of Forbiddenness	373
8.4.2. Superallowed Transitions	374
8.4.3. Allowed Transitions.	375
8.4.4. Forbidden Transitions	376

8.5. Electron Capture 377
    8.5.1. Discrete Neutrino-Energy Spectrum 377
    8.5.2. Capture Lifetime and *ft* Value 378
    8.5.3. Electron Capture Ratios 381

8.6. Forms of Beta Interaction 388
    8.6.1. Restrictions upon the Coupling Strengths. . . . 391
    8.6.2. Fermi and Gamow-Teller Transitions 391
    8.6.3. *ft* Values and Nuclear Matrix Elements 393
    8.6.4. Electron-Neutrino Angular Correlation 396

8.7. Parity Nonconservation in Beta Decay 397
    8.7.1. Test of Parity Violation 399
    8.7.2. Neutrino Helicity 404

8.8. Beta Decay Coupling Strengths and Interaction Characteristics 408
    Exercises 410

9. Radiative Transitions in Nuclei

GAMMA DECAY

9.1. Multipole Character of Gamma Radiation 414
9.2. Multipole Transition Probability 418
    9.2.1. Reduced Transition Probability 425
    9.2.2. "Forbidden" Transitions 427
    9.2.3. Nuclear Isomerism 427
    9.2.4. Multipole Mixing 428
9.3. Nuclear Level Scheme Compilation 433
9.4. Angular Distributions and Correlations 440
9.5. Recoil-Free Gamma Spectroscopy 446
    9.5.1. Nuclear Resonance Absorption and Fluorescence . . 446
    9.5.2. Mössbauer Effect 450
    Exercises 460

10. Internal Conversion

INTERNAL CONVERSION AND INTERNAL PAIR FORMATION

10.1. Conversion Coefficients 467
    10.1.1. Partial and Total Conversion Coefficients . . . 467
    10.1.2. Experimental Study of Conversion 468
    10.1.3. Evaluation of Internal Conversion Coefficients . . 473
    10.1.4. Approximate Analytic Expressions for Conversion Coefficients
        and Mean Lifetime 478

10.2. Selection Rules 481
    10.2.1. Mixed Multipolarity 482

10.3. Conversion Distributions and Correlations (Particle Parameters) 482
    Exercises. 483

11. Fundamental Characteristics of Nuclear Reactions

NUCLEAR REACTION CHARACTERISTICS

11.1. Reaction Energetics 485
    11.1.1. Energy and Momentum Conservation in Nuclear Reactions . . 485
    11.1.2. Nonrelativistic Q Equation 487
    11.1.3. Relativistic Q Equation 488
    11.1.4. Threshold Energetics 488
    11.1.5. Energy-Correlation Analysis 490

11.2. General Features of Reaction Cross Sections 505
    11.2.1. Probability Considerations 505
    11.2.2. General Cross-Sectional Trend for Elastic Neutron Scattering . . 507
    11.2.3. Characteristic Cross Section for Exothermic Reactions
        Induced by Low-Energy Neutrons 507
    11.2.4. Characteristic Cross Section for Inelastic Neutron Scattering . . 509
    11.2.5. Characteristic Cross Section for Endothermic Neutron-
        Induced Reactions Leading to Emission of Charged Particles . . 509
    11.2.6. Cross-Sectional Trend for Exothermic Reactions involving
        Charged Incoming and Uncharged Outgoing Particles . . 510
    11.2.7. Characteristic Cross Section for Exothermic Reactions with
        Charged Incident and Emergent Particles . . . 511

11.3. Detailed Balance Predictions for Inverse Reaction Cross Sections 511
    11.3.1. Experimental Investigation of Inverse Reactions . . . 514

11.4. Resonance Reactions 519
    11.4.1. Resonance Anomalies in Excitation Functions. . . 519
    11.4.2. Breit-Wigner Formula 520
    11.4.3. Resonance Cross-Section Nomenclature . . . 523
    11.4.4. Modified Breit-Wigner Theory for Elastic Scattering . . 524
    11.4.5. Statistical Spin Factor 530

11.5. Formal Reaction Theory 531
    11.5.1. Partial-Wave Approach to Scattering of Spinless Particles . . 532
    11.5.2. Phase Shifts 538
    11.5.3. Resonance Cross Sections 539
    11.5.4. Reaction Theory in Matrix Formalism. . . . 541
    11.5.5. Transition Amplitude in S-Matrix Theory . . . 541
    11.5.6. Basic Properties of the S Matrix (Probability Conservation
        and Time-Reversal Invariance). 542
    11.5.7. Cross Sections in Matrix Formalism 544
    Exercises. 555

APPENDIX A. **Kinematics of Relativistic Particles**

A.1. Lorentz Transformation 560
  A.1.1. Lorentz Contraction and Time Dilation 562
  A.1.2. Geometrical Representation of the Lorentz Transformation . . 563
  A.1.3. Composition of Collinear Velocities 565
  A.1.4. Relativistic Addition of Noncollinear Velocities . . . 565
  A.1.5. Doppler Effects (Frequency Shifts) 567

A.2. Relativistic Mass, Momentum, and Energy . . . 569
A.3. "Relativistic" Particles 572
A.4. Lifetimes of Relativistic Particles 574
A.5. Speeds of Relativistic Charged Particles 574
  Exercises 578

APPENDIX B. **Transformation Relations between the Laboratory and Center-of-Mass Systems for Elastic Collisions**

B.1. Characteristics of the Center-of-Mass System 580
B.2. Nonrelativistic Elastic Collision of a Moving Particle with a Stationary Target 581
  B.2.1. Velocity Relations 582
  B.2.2. Kinetic Energy Relations 583
  B.2.3. Angular Relations 583
  B.2.4. Relations between Velocities, Energies, and Scattering Angles in the Laboratory System 585
  B.2.5. Solid Angle Relations 586
  B.2.6. Angular Intensity Relations 586

B.3. Relativistic Elastic Collision of a Fast-Moving Particle with a Stationary Target 590
  B.3.1. Velocity Relations 591
  B.3.2. Expressions for Energy in the Center-of-Mass System . . 592
  B.3.3. Angular Relations 595
  B.3.4. Solid Angle Relation 597
  Exercises 599

APPENDIX C. **The Dynamics of Decay and Reaction Processes**

C.1. Decay and Reaction Kinematics 601
C.2. Energetics and Kinematics for Two-Particle Decay . . . 601
C.3. Scattering Kinematics 604
  C.3.1. Nonrelativistic Elastic Scattering Kinematics . . . 605
  C.3.2. Relativistic Elastic Scattering Kinematics . . . 605

C.3.3. Graphical Treatment of Elastic Scattering 607
C.3.4. Graphical Treatment of Inelastic Scattering 609
C.3.5. Inelastic Scattering at High Incident Energies Leading to the
Threshold of Particle Creation 610

C.4. Nonrelativistic Reaction Kinematic Formulae . . . 612
Exercises 614

APPENDIX D. **Wave Mechanics**

D.1. Schrödinger Equations 617
D.2. Probability Density and Electron Probability Distribution . 620
D.3. Heisenberg Uncertainty Relations 624
D.4. Klein-Gordon Equation for Spin-0 Particles 626
D.4.1. Relativistically Invariant Notation 626

D.5. Dirac Equation for Spin-$\frac{1}{2}$ Relativistic Particles . . . 628
D.5.1. Covariant Form of the Dirac Equation (Gamma Matrices) . . 631
D.5.2. Properties of the Gamma Matrices 633

D.6. Dirac Electron-Positron Theory 637
D.7. Weyl Equation for Massless Particles (Two-Component Neutrino
Theory) 639
D.8. Wave Equations for Bosons. 641
Exercises 645

APPENDIX E. **Angular Momentum in Quantum Mechanics
(Racah Algebra)**

E.1. Angular Momentum Operators 646
E.2. Composition of Angular Momentum Wave Functions (Clebsch-
Gordan Coefficients) 647
E.3. Properties of Clebsch-Gordan Coefficients and Wigner 3-j
Symbols 649
E.4. Values of Simple 3-j Symbols 651
E.5. Examples of Wave-Function Coupling 651
E.6. Recoupling of Angular Momenta (Racah Coefficients and
Wigner 6-j Symbols) 658
E.7. Coupling of Four Angular Momenta (Wigner 9-j Symbols) . 661
E.8. Racah Functions in Angular Distribution and Correlation
Theory 664
E.8.1. Composition of Angular Distribution Functions. . . . 666
E.8.2. Composition of Angular Correlation Functions 670
Exercises 674

APPENDIX F. **Feynman Interaction Theory**

F.1. The Underlying Motivation behind a Field-Interaction
     Approach.   675
F.2. Interaction Matrix Elements   676
F.3. Feynman Graphs   680
    F.3.1. Relation to Matrix Elements   680
    F.3.2. Feynman Graphs in Momentum Space   686
    F.3.3. Second Quantization   687
    F.3.4. Formation of the S-Matrix Element   .   .   .   693

APPENDIX G. **Some Measurement Techniques in Nuclear Physics**

G.1. Introduction   698
G.2. Beta Spectrometry   698
    G.2.1. Principle of Magnetic Spectrometers   .   .   .   699
    G.2.2. Focusing Arrangements   703
G.3. Scintillation Counters   706
    G.3.1. Principle   706
    G.3.2. Energy Resolution.   709
    G.3.3. Pile-Up   713
G.4. Semiconductor Detectors   713
G.5. Energy Scale in Low-Energy Nuclear Spectroscopy   .   721
G.6. Coincidence Techniques   723
    Exercises   727

APPENDIX H. **Radiation Dosimetry**

H.1. Biological Effects of Radiation   731
H.2. Dosimetry Units   732
    Exercises   734

APPENDIX I. **Constants and Conversion Factors in Atomic,
            Nuclear, and Particle Physics**

Text and Tables   738

References   745
Solutions to Exercises   776
Subject Index   791

SUMMARY OF CONTENTS
Volumes II and III

12. Nuclear Particles and Their Interactions

Neutrons; antinucleons; deuterons and two-body nuclear forces; three-nucleon systems [^{3}H, ^{3}He]; four-nucleon systems [^{4}He]; heavy ion physics

13. Nuclear Forces and Potentials, as Deduced from Nuclear Dynamics (Scattering and Polarization)

Nuclear force characteristics: central, tensor, spin-orbit components; velocity-dependent forces; scattering formalism and effective-range theory; Wolfenstein parameters; phase-shift analysis

14. Scattering and Reaction Models in Nuclear Physics

Optical model; compound-nucleus formalism; direct-interaction formalism; unified formalism; statistical fluctuations and strength functions

15. Nuclear Models

Statistical models; liquid-drop model; Fermi-gas model; shell model; collective model; unified model; cluster model; nuclear matter

16. Certain Specialized Reaction Processes

Spallation; fission; fusion; nuclear astrophysics; high-energy physics

17. Fundamental Particle Physics

Physics of leptons, mesons, baryons, and resonances; conservation properties

18. Group-Theoretical Methods in Nuclear and Particle Physics

Elements of group theory and unitary groups; group-theoretical treatment of angular momentum; the $SU(3)$ group: algebra and irreducible representations; particle classification in the $SU(3)$ scheme; quark states, mass formulae and broken symmetries; electromagnetic interactions and dynamic predictions of the $SU(3)$ scheme; $SU(6)$ and higher symmetry groups; principal linear groups; relativistic group theory for hadrons; group-theoretical classification of nuclear states

Appendix J. Rotation and Angular-Momentum Calculus

Rotation group, rotation matrices and application to the quantum mechanics of angular-momentum coupling

PHYSICS OF
NUCLEI AND PARTICLES

VOLUME I

Chapter 1

HISTORICAL DEVELOPMENT OF NUCLEAR PHYSICS. THE SIZE AND CONSTITUTION OF THE ATOMIC NUCLEUS

A Survey of Nuclear and Particle Physics within the Context of Physics as a Whole

1.1. The Present Status of Nuclear Physics

Man's basic knowledge of the ultimate constitution and physical properties of matter has been derived from studies in atomic, nuclear, and particle physics. As a fledgeling of the well-founded quantum theory of molecules and atoms, nuclear physics is as yet a comparatively young discipline, still in its formative stages and not completely disciplined. But in its unruliness lies its very fascination and diversity, for in its various facets it displays not only the power but also the elegance of methods, concepts and models which are often of a purely empirical nature. It is well to bear in mind that it represents no more than a visualization of processes manifested through their effects upon our sense impressions: "All our knowledge has its origin in our perceptions" [Le 91]. Behind the questions we ask and the interpretation of the results which emerge lies our own imagery. We are therefore led to build conceptual models of "particles" and their "orbits," "potential barriers," "interaction modes," etc., in order to account for observed phenomena. While tacitly acknowledging their empirical nature, we some-times run the risk of imputing to such concepts a specious absolute "reality"

1

and erecting too rigid a descriptive framework around them. As we extend our knowledge—to a large degree simply by defining the problem more appositely and learning to ask the right questions—we refine and unify the visualization of inherent physical properties and interactions. This can bring with it fundamental changes in thought (such as the relinquishing of the assumption of parity invariance in certain processes), the introduction of new approaches (such as those in dispersion theory), or new representations (such as those encountered in unified reaction theories).

Perhaps the main contribution of nuclear physics to other sciences is to be found in the manifold applications of its techniques and in the influence of its methods. Developments in its own theoretical, experimental, and technological fields have set up widespread repercussions in associated disciplines, and in several instances have given rise to new branches of science. Much still remains to be clarified in nuclear physics, and requisite mathematical extensions have yet to be established. Nevertheless, great strides in our understanding of nuclear matter, nuclear structure, interactions, and reactions have been made, particularly in recent years, and in a sense we now stand at the threshold of comprehending many aspects of a complex and hitherto diffusely understood subject.

For a survey and appraisal in simple terms of the current situation, attention may be drawn to two articles, namely, that by Zucker and Bromley [Zu 65] concerning nuclear physics and that by Weisskopf [Wei 65] concerning fundamental particle physics, which might with advantage be perused at this stage and reread later when more of the material in this book has been assimilated.

We proceed now to give a brief review of historical developments in the rise of nuclear and particle physics in order to place this subject in perspective and to serve as a preliminary to the consideration of nuclear composition, interactions, and reaction phenomena.

1.2. Brief History of the Development of Atomic, Nuclear, and Particle Physics

The problem of the chemical constitution of matter was clarified at the start of the nineteenth century by Dalton's atomic hypothesis [Da 08], which also served to explain the course of chemical reaction processes and the structure of molecules. On the other hand, the physical elucidation of the fundamental structure of matter stems basically from discoveries which took place at the end of the century, namely, Röntgen's discovery of x rays [Rö 95], Becquerel's discovery of natural radioactivity [Be 96] (the emitted α, β, and γ rays were later shown to be helium nuclei, electrons, and electromagnetic radiation, i.e., photons, respectively), and Thomson's discovery of electrons [Th 97].

By the start of the twentieth century, Thomson had shown that all atoms contained electrons, and in 1904 he propounded the theory that their negative charge was counterbalanced by an equal charge of opposite sign distributed uniformly throughout the atom [Th 04]—a rival model to that proposed by Nagaoka [Na 04] in which atoms were conceived to be composed of a core surrounded by rings of rotating electrons, i.e., essentially the modern picture of the atom.

Barkla [Ba 11] in 1911 showed that the energies of x rays scattered by light elements could, on the basis of Thomson's model, be explained by taking the number of scattering electrons to be approximately equal to half the atomic weight (up to $A = 32$). The significance of this finding can be appreciated when viewed in relationship to the argument put forward by Rutherford [Ru 11] in the same year, that aside from the electronic charge there must be a large charge at the center of the atom, since otherwise it would be impossible (e.g., on the basis of a homogeneous charge distribution as in the Thomson model) to account numerically for the incidence of large-angle scattering of Becquerel's α particles at a single encounter with atoms such as those in a gold target. The underlying set of experimental data was that of Geiger and Marsden [Gei 09] who, appreciating its importance, thereupon reinvestigated the scattering process and in 1913 published their conclusion [Gei 13] that the number of neutralizing positive charges on the atomic NUCLEUS, as it was termed by Rutherford—hence also the number of planetary electrons—was approximately half the atomic weight. This introduced the concept of ATOMIC NUMBER Z as the positive electric charge on the nucleus and resolved many of the apparent inconsistencies in the arrangement of Mendeléev's periodic table of the elements, since, as van den Broek [Br 13] had pointed out, an average difference of 2 units found to prevail between the atomic weights of adjacent elements betokened a difference of exactly 1 unit in their charges (i.e., their atomic numbers) and thus revealed Z to be the arbiter of position in the periodic table. The assignment of atomic numbers from measurements of x-ray spectra constituted the major achievement of Moseley's work [Mo 13, Mo 14].

Bohr immediately adopted van den Broek's conjecture and used it in his early papers on atomic structure [Bo 13b]. Thus the Rutherford-Bohr model of the atom came into being. It pictured the atom as consisting of a small positively charged core, of diameter $\approx 10^{-12}$ cm, surrounded by electrons moving in circular orbits of radius $\approx 10^{-8}$ cm. The total negative charge of the planetary electrons compensates the equal but opposite charge of the nucleus in a neutral (un-ionized) atom. Bohr's concept of discrete, stationary, *nonradiating* orbits, while at variance with classical electrodynamic theory, represented a major step in the application of Planck's quantum hypothesis [Pl 00] and opened the way to a detailed quantitative treatment of atomic and nuclear physical problems. Successive refinements in the model then followed, such as the generalization from circular to elliptical orbits for the electrons [So 16] and the quantum-mechanical reinterpretation of such orbits in terms of cloud-like probability distributions of location [Bo 26, Bo 27]. The development of quantum theory and of new basic concepts, such as the intrinsic spin of electrons and other particles [Uh 25, Uh 26], led to detailed theories of atomic structure and atomic interactions, which in turn paved the way for study of the atomic nucleus itself and of fundamental particles.

The first fundamental particle, the electron, was identified in 1897, and by 1903 its properties such as mass and charge had been established to a fair degree of accuracy. In 1903, α particles were shown to be helium nuclei, but it was not until 1919 that the first fundamental nuclear particle, the proton,[†] was definitely identified. It was discovered in a series of investigations by Rutherford as a product of the first induced nuclear transmutation, namely, the conversion of nitrogen nuclei under bombardment by naturally emitted α particles into an isotope of oxygen and a hydrogen nucleus (proton), according to the reaction scheme

$$^{14}N + {}^4He \rightarrow {}^{17}O + {}^1H$$

Thereupon, it was (erroneously) contended that the nucleus was made up of protons and electrons—a model which, though plausible at the time, ran into difficulties when examined from the standpoint of quantum theory. The dilemma was resolved in 1932 when the neutron was discovered by Chadwick [Ch 32] as an uncharged particle of practically the same mass as a proton, with the right properties to be a nuclear constituent. Independently, Iwanenko [Iw 32] and Heisenberg [He 32] then proposed a model of the nucleus as composed of protons and neutrons—but not electrons—and this remains the basic nuclear model today.

The year 1932 was further distinguished by the discovery [An 33] of another fundamental particle, the positron (or "positon"), a positively charged electron, predicted by Dirac [Di 31] (see also [Op 30]) on quantum-theoretical grounds. As "antiparticle" to the electron, it interacts annihilatively with the latter, releasing energy in the form of electromagnetic γ radiation. In addition to this, deuterium, the heavy isotope of hydrogen (D $= {}^2$H), was discovered

---

† The name "proton" ($\pi\rho\tilde{\omega}\tau o\nu$, first) harks back to a suggestion by Prout, first proposed anonymously in the Annals of Philosophy (1815) and reiterated in the same journal the following year, that the $\pi\rho\tilde{\omega}\tau\eta$ $\ddot{\upsilon}\lambda\eta$ of the ancients (i.e., the primordial substance from which the atoms of all elements are built) is realized in hydrogen.

[Ur 32a, Ur 32b], and the first successful nuclear transmutation induced by artificially accelerated particles. Cockcroft and Walton [Co 32] observed that protons accelerated to energies around 150 keV caused lithium nuclei to break up and yield a pair of α particles as reaction products, according to the reaction scheme

$$^7\text{Li} + {}^1\text{H} \rightarrow {}^4\text{He} + {}^4\text{He}$$

In 1934 this was complemented by the finding that stable light nuclei could be rendered radioactive by α-particle irradiation [Cu 34] and that nuclei could be transmuted by capture of *neutrons* as well as protons. This latter observation by Fermi [Fe 34a] was followed later that year by the publication of his theory of β decay [Fe 34b]. Thereafter, the study of nuclear reactions and interactions acquired even greater impetus and significance, aided by advances in the production, acceleration and detection of particles used in nuclear research, and in the physico-chemical methods of radioisotope extraction and separation. Widespread recognition of the potentialities of nuclear physics research followed upon Hahn and Strassmann's observations of nuclear fission [Ha 39] and Bohr and Wheeler's theoretical explanation of the process [Bo 39], based upon the simple liquid drop model of the nucleus. This raised problems of nuclear stability having a close bearing upon the properties of nuclear forces, which in turn directly involve consideration of particle physics. The conjecture by Yukawa [Yu 35a] in 1935 that the strong nuclear binding forces and interactions can be regarded as occasioned by the exchange of a particle whose mass lies midway between that of the electron and that of the proton or neutron gave rise to meson physics ($\mu\acute{\epsilon}\sigma\sigma\varsigma$, middle) which, although the *weakly* interacting μ meson or "heavy electron" was observed very shortly afterward [An 36], did not really get under way until the discovery of the *strongly* interacting π meson in 1947 [La 47b] and its artificial production in 1948 [Ga 48]. Thereafter, a whole range of heavier mesons, as well as hyperons (namely, particles whose mass exceeds that of the proton or neutron) have been brought to light. More recently, particle physics has been further extended by the investigation of particle "resonances," which represent short-lived quantum-mechanical bound states of one or more fundamental particles. The classification of the "elementary" particles has recently been systematized by the consideration of group-theoretical quantum methods concerned with basic symmetry properties, but there is as yet no fundamental theory of particles, just as there is no conclusive theory of nuclear matter and many-body interactions. There has, however, especially of late, been ever-increasing progress in the elucidation of basic problems, in the evolution of methods to treat an ever-larger range of phenomena in ever-greater detail, and in fundamental insight into the characteristics of the microworld comprised of nuclear and particle physics.

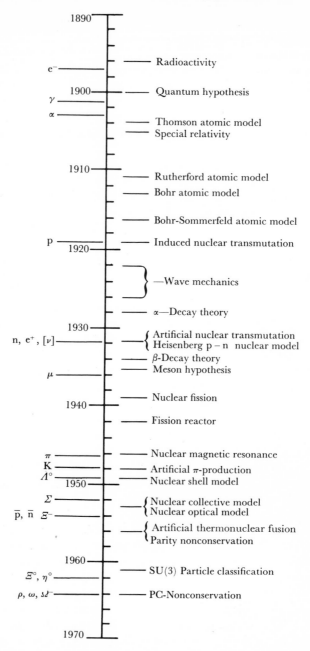

Fig. 1-1. Time-scale representation of the development of discoveries and theories in atomic, nuclear, and particle physics.

1.3. The Domain of Nuclear Physics

The preceding outline of the development of our knowledge concerning atomic, nuclear and particle physics can be represented by a time-scale diagram (Fig. 1-1). To get some feeling of where the microworld of nuclear

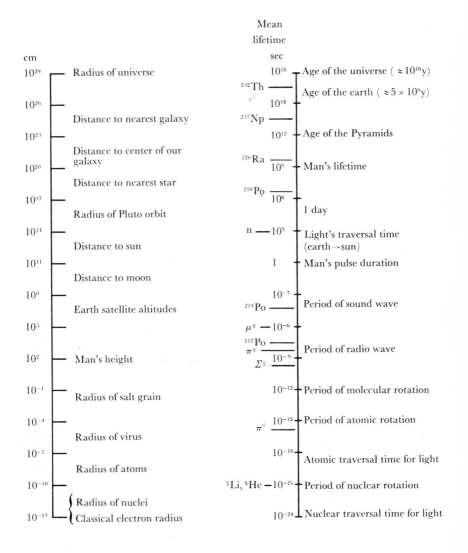

(a) Length intervals   (b) Time intervals

physics lies in relation to other physical systems and phenomena, similar diagrams can be set up, using logarithmic scales for length and time intervals. The two diagrams in Fig. 1-2 are of this type; their range extends over 42 powers of 10.

1.4. The Size of the Nucleus and Nuclear Constitution

Whereas *atomic* radii are of the order of Ångström (Å) units (1 Å $= 10^{-8}$ cm), those of *nuclei* range from but 1 to 10 fm (1 fm = 1 femtometer $= 10^{-13}$ cm, also termed 1 Fermi). Hence the time taken by a body moving with the speed of light ($c = 3 \times 10^{10}$ cm/sec) to traverse the diameter of one of the smaller nuclei is around 10^{-23} sec, a figure which represents a lower limit to nuclear interaction times in direct reaction processes. This is indicated in Fig. 1-2, which also shows nuclear and particle dimensions to lie at the extreme lower end of the scale of lengths. Some impression of the minute size of nuclei may be derived from the evaluation of Exercise 1-1; there are, indeed, intimations that quantum field theory may break down at separations appreciably below 1 fm [McK 60, Me 64, Me 68], and the circumstance that quantum theory developed for treatment of atomic phenomena remains applicable to nuclear systems having dimensions 10^5 times smaller is both fortunate and remarkable.

As might be expected, the investigation of the size of nuclei requires special techniques, foremost among those employed at present being the analysis of elastic scattering experiments with high-energy electrons. The examination of increasingly small structures entails the use of progressively more complex techniques, ranging from optical, ultraviolet and phase-contrast microscopes, through electron microscopes able to resolve single molecules and field-emission microscopes capable of resolving individual atoms, to the somewhat indirect means of determining nuclear size from scattering distributions of high-energy electrons. *High* energies are needed in order to get sufficiently *small* wavelengths for details of *nuclear* structure to be resolved by a beam of electrons. The inverse relationship between wavelength and energy is embodied in the de Broglie formula for the wavelength of "matter waves" in accordance with the quantum-mechanical principle of complementarity.

In referring to quanta of electromagnetic radiation as quasi-corpuscular "photons," we have already implicitly made use of the so-called wave-particle duality. In electrodynamics, one has the relation

$$p = \frac{E}{c} \qquad (1\text{-}1)$$

between the linear momentum p and the total energy E of electromagnetic radiation propagated with velocity c. The energy E is quantum-mechanically related to the frequency ν of the radiation via Planck's constant h:

$$E = h\nu = \frac{hc}{\lambda} \tag{1-2}$$

where λ is the wavelength of the radiation. Combining these expressions, one gets the relation

$$p = h/\lambda \tag{1-3}$$

or, in the rationalized units commonly employed in quantum mechanics, $\hbar \equiv h/2\pi$ and $\lambdabar \equiv \lambda/2\pi$,

$$\lambdabar = \hbar/p \tag{1-4}$$

In 1924, de Broglie [de B 24] conjectured that a similar relation applies to matter in motion, that is to say, to moving particles of rest mass m_0 which, unlike that of photons,† is *nonzero*. Einstein's special theory of relativity [Ei 05a] gives the relativistic mass of particles moving with velocity v as

$$m = \gamma m_0 \tag{1-5}$$

where

$$\gamma \equiv \left(1 - \frac{v^2}{c^2}\right)^{-1/2} \tag{1-6}$$

The expression it provides for the linear momentum is

$$p = mv = \gamma m_0 v \tag{1-7}$$

The total energy E of the moving particles, composed of their kinetic energy E_{kin} and their rest energy $E_0 \equiv m_0 c^2$,

$$E = E_{\text{kin}} + E_0 \tag{1-8}$$

is related to the momentum p by the important relativistic expression

$$E^2 = p^2 c^2 + E_0^2 \tag{1-9}$$

whence

$$p = \frac{(E^2 - E_0^2)^{1/2}}{c} = \frac{[E_{\text{kin}}(E_{\text{kin}} + 2E_0)]^{1/2}}{c} \tag{1-10}$$

so that

$$\lambdabar = \frac{\hbar}{p} = \frac{\hbar c}{[E_{\text{kin}}(E_{\text{kin}} + 2E_0)]^{1/2}} \tag{1-11}$$

which, in the extreme-relativistic limit ($E \approx E_{\text{kin}} \gg E_0$) becomes

$$\lambdabar = \hbar c/E \tag{1-12}$$

† A recently determined [Go 68] geomagnetic upper limit on the possible mass of a photon is $m_\gamma < 4.0 \times 10^{-48}$ g.

and in the nonrelativistic limit $(E_0 \gg E_{kin})$ becomes

$$\lambda = \frac{\hbar}{(2m_0 E_{kin})^{1/2}} = \frac{\hbar}{m_0 v} \tag{1-13}$$

To derive convenient numerical formulae for electron wavelengths, we substitute numerical values in the general de Broglie relation (1-11). The energy units conventionally used in nuclear physics are eV (electron-volts), where

$$1 \text{ eV} \equiv 10^{-3} \text{ keV} \equiv 10^{-6} \text{ MeV} \tag{1-14}$$

$$= (e^{[C]} \times 1^{[V]}) \text{ J} \tag{1-15}$$

$$= 10^7 (e^{[C]} \times 1^{[V]}) \text{ erg} \tag{1-16}$$

i.e.,

$$1 \text{ eV} = 1.6021 \times 10^{-12} \text{ erg} \tag{1-17}$$

since the elementary charge is $e = 1.6021 \times 10^{-19}$ C. In these units, the REST MASS OF THE ELECTRON,

$$m_e = 9.1091 \times 10^{-28} \text{ g} \tag{1-18}$$

is equivalent to an ELECTRON REST ENERGY

$$E_0 \equiv m_e c^2 = 0.511\ 006 \text{ MeV} \tag{1-19}$$

The rationalized Planck constant has the value

$$\hbar \equiv \frac{h}{2\pi} = \frac{6.6256 \times 10^{-27}}{2\pi} = 1.0545 \times 10^{-27} \text{ erg sec} \tag{1-20}$$

whence the DE BROGLIE WAVELENGTH OF THE ELECTRON takes the value

$$\lambda_e = 1.226\ 378 \times 10^{-7} \left[E_{kin}^{[eV]} \left(1 + \frac{E_{kin}^{[eV]}}{1.022 \times 10^6} \right) \right]^{-1/2} \text{ cm} \tag{1-21}$$

and

$$\lambda_e \equiv \frac{\lambda_e}{2\pi} = 1.951\ 841 \times 10^{-8} \left[E_{kin}^{[eV]} \left(1 + \frac{E_{kin}^{[eV]}}{1.022 \times 10^6} \right) \right]^{-1/2} \text{ cm} \tag{1-22}$$

In the nonrelativistic limit (e.g., for $E_{kin} \lesssim 50$ keV), the term in square brackets reduces to simply $[E_{kin}^{[eV]}]^{-1/2}$.

The electron wavelength λ_e over the energy range from $E_{kin} = 0.2$ eV to $E_{kin} = 3 \times 10^{12}$ eV has been tabulated numerically [Ma 56b] and is depicted graphically over this range in Fig. 1-3, which also displays the de Broglie wavelengths for some other particles frequently encountered in nuclear and particle physics. Clearly, accelerating voltages of the order of keV as used in electron microscopes suffice to resolve details of *atomic* structure ($\lambda_e \simeq 0.4$ Å when $E_{kin} = 1$ keV), but they are inadequate for the study of nuclear structure. In than that of the electron:

$$m_p = 1.6725 \times 10^{-24} \text{ g} = 1836\, m_e \qquad (1\text{-}23)$$

Hence 9-MeV protons have a rationalized de Broglie wavelength $\lambda_p = 1.5$ fm, and 1-MeV protons have $\lambda_p = 4.6$ fm.

There is an alternative way of looking at this situation, namely by employing Heisenberg's uncertainty relations (see Appendix D.3), which can be written in the form

$$\Delta p\, \Delta q \geqq \hbar \qquad (1\text{-}24)$$

connecting a momentum uncertainty Δp with a position uncertainty Δq in the direction of the momentum vector, or

$$\Delta E\, \Delta t \geqq \hbar \qquad (1\text{-}25)$$

connecting an energy uncertainty ΔE with a time interval Δt available for defining the energy. If we apply the first of these two relations to the proton-*electron* model of nuclear composition and assume that at the very most the mean binding energy of an electron conceived to be in the nucleus is 8 MeV, it follows that the maximum possible uncertainty in the electron's energy is $\Delta E = 8$ MeV, which, according to Eq. (1-1), corresponds to an upper limit for the momentum uncertainty of

$$\Delta p \cong \frac{\Delta E}{c} = \frac{8 \times 1.6021 \times 10^{-6}}{3 \times 10^{10}} = 4.27 \times 10^{-16} \text{ g cm sec}^{-1} \qquad (1\text{-}26)$$

and thus, from Eq. (1-24) to a lower limit for uncertainty of location of

$$\Delta q \geqq \frac{\hbar}{\Delta p} = \frac{1.055 \times 10^{-27}}{4.27 \times 10^{-16}} \cong 2.5 \times 10^{-12} \text{ cm} = 25 \text{ fm} \qquad (1\text{-}27)$$

This value exceeds the diameter of even the largest nuclei. Conversely, if an electron were to be confined to a region having a spatial extension $\Delta q = 10$ fm, its corresponding energy uncertainty of $\Delta E = 20$ MeV would considerably exceed the maximum energy available on an average for binding a particle within a nucleus. Of course, a *proton* confined to a region with $\Delta q = 10$ fm would have exactly the same momentum uncertainty, but because of the larger mass this would correspond to a much smaller energy uncertainty.

We note in passing that if we set ΔE equal to E in the classical expression $E = \frac{1}{2}mv^2$, we find that for $\Delta E = E = 8$ MeV, the numerical value of the velocity is $v = 1.7 \times 10^{11}$ cm/sec for electrons, which indicates that these have to be treated as extreme-relativistic particles at the above energy, whereas $v = 4 \times 10^9$ cm/sec for protons, which is sufficiently small against $c = 3 \times 10^{10}$ cm/sec for *nuclear protons* to be treated as *nonrelativistic* particles.

Other considerations, such as the spin of nuclei in relation to that of the

order to derive information on the distributions of mass and charge in nuclei, linear accelerators have had to be designed which are capable of providing high-energy electron beams around 1000 MeV (10^9 eV $= 10^3$ MeV $= 1$ GeV,

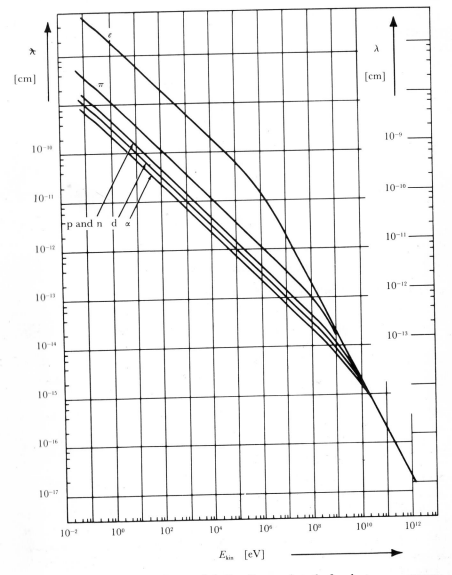

Fig. 1-3. Rationalized and normal de Broglie wavelengths for electrons, π mesons, nucleons, deuterons, and α particles in terms of kinetic energy from 10^{-1} to 10^{12} eV (log-log scale).

or, in American terminology, 1 BeV: Note that 1 billion = 10^9 in the United States, but 10^{12} elsewhere). The principal such facility has hitherto been the 1-GeV electron accelerator at Stanford, extensively used by the Hofstadter group to investigate nuclear structure [Ho 57, El 61, Ho 63, Ho 64]. With the recent completion of another Stanford linear accelerator (SLAC) capable of furnishing a 20-GeV electron beam [Da 66b] after acceleration along a 2-mile path, fresh strides in our knowledge and understanding of nuclear

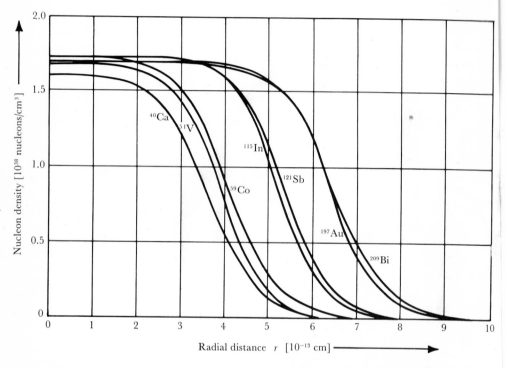

Fig. 1-4. Charge-density distribution in medium-heavy and heavy nuclei as a function of radial distance from the center of the nucleus, as indicated by scattering of 1-GeV electrons (after [Ho 57]).

and particle structure are anticipated. Some of the results obtained at 1 GeV are collected in Fig. 1-4, which displays the charge-density distribution in terms of the nuclear radius, as derived from the analysis of electron scattering experiments undertaken with the Stanford accelerator.

As inspection of Fig. 1-3 indicates, electron wavelengths below $\lambda_e = 1$ fm are attained only with energies above about 0.3 GeV. The 1-GeV machine, yielding an electron beam of wavelength $\lambda_e = 0.195$ fm has therefore been

admirably suited to the investigation of nuclear structure, but even 20-GeV SLAC ($\lambda_e \cong 0.01$ fm) should provide still more detailed and data on particle structure. Of course, the fineness of the information matter of wavelength alone (cf., Exercise 1-2). Although heavier p have a still smaller wavelength—e.g., the largest *proton* accelerators operation, the 28-GeV proton synchrotron at CERN and the 33-GeV at Brookhaven, provide proton beams of wavelength $\lambda_p \cong 0.007$ fm, w new accelerator at Serpukhov provides a 74-GeV proton beam, of wav $\lambda_p \cong 0.003$ fm—a beam of high-energy *electrons* is basically preferable of their purely *electromagnetic* interaction with nuclei, conjoined wi small mass and the small inherent size of the particles, which offer t bility of high resolution. Although it is not possible to define the "siz electron in other than somewhat artificial classical terms (the "c ELECTRON RADIUS" $r_e \equiv e^2/m_e c^2 = 2.818$ fm is introduced in Section electron clearly has smaller dimensions than nuclei. Indeed, the latte be conceived to have a clear-cut bounding surface either, as is evid Fig. 1-4. Instead of a step function (giving a sharp radius), the s distributions evince a gradual decline to zero within a diffuse surfac and thereby necessitate some convention for defining the radius (half-height).

From this information on the size of nuclei, it is possible to infer th are too small to harbor electrons. For if electrons were to be confin nuclear dimensions, their de Broglie wavelength would have to about 20 fm, for example, which means that their energy woul exceed about 9.4 MeV. Energies of this magnitude have, howev been observed experimentally for electrons emitted as β rays fro In practice, one finds typically that $E_e \approx 1$ MeV, and for this e rationalized de Broglie wavelength $\lambda_e \approx 100$ fm far exceeds even t nuclear diameters.

Further evidence is provided by a consideration of the mean er which particles are bound within a nucleus. As discussed in mor Section 2.1, the MEAN NUCLEAR BINDING ENERGY PER PARTICLE ne 9 MeV. (Indeed, for the present consideration this figure represen generous upper limit, since it has been based upon differences in of nuclei pictured as composed of protons and neutrons only, whe model of a nucleus as constituted of protons and electrons wou about half as many particles again, so that the appropriate me energy per particle would have to be reduced to about two-th above value, viz. 6 MeV.) We see that electrons confined to nuclear would have an energy higher than the requisite mean binding en an unacceptable situation. It should be noted that no such diffi in the case of protons, because the REST MASS OF THE PROTON is so n

constituent particles, or the fact that the magnetic dipole moment of an
electron is very much larger than that of nuclei, together with evidence that
apart from the purely electrostatic interaction, electrons have practically no
residual interaction with nuclear particles, all provide further confirmation
that the proton-*electron* hypothesis of nuclear structure is untenable. It has
been replaced, following the discovery of the neutron in 1932, by the PROTON-
NEUTRON MODEL of nuclear constitution which, by excluding bound electrons
from the nucleus, resolves the above difficulties and renders a good account of
all observed nuclear properties. This pictures the nucleus as comprised of
Z protons (the ATOMIC NUMBER Z identifies the specific chemical element El,
and is equal to the number of orbital electrons in the neutral atom) and N
neutrons (the NEUTRON NUMBER N characterizes the particular physical
ISOTOPE of the given element). The sum

$$Z + N = A \tag{1-28}$$

is termed the MASS NUMBER A of the particular nuclear species, i.e., of the
NUCLIDE or NUCLEIDE. By recent convention, the nomenclature $^A_Z\text{El}_N$, or
simply ^AEl, has been adopted to denote each individual nuclide—a change
from the old form El^A which, though still in common use, does not permit the
charge-states of ionized atoms to be expressed in a simple way.

Convention in nuclear physics:

$$^1\text{H} = {}^1_1\text{H}_0 = \text{p} \qquad ^2\text{H} = {}^2_1\text{H}_1 = \text{d} \qquad ^3\text{H} = {}^3_1\text{H}_2 = \text{t}$$

$$^3\text{He} = {}^3_2\text{He}_1 \qquad ^4\text{He} = {}^4_2\text{He}_2 = \alpha$$

$$^{235}\text{U} = {}^{235}_{92}\text{U}_{143} \qquad ^{238}\text{U} = {}^{238}_{92}\text{U}_{146}$$

Convention in ion physics:

$$^4\text{He}^+ \text{ (singly ionized)} \qquad ^4\text{He}^{2+} \text{ (doubly ionized)} = \alpha$$

There is no basic relation between A and Z, but the empirical formula
(substantiated in Sections 2.3.5 and 2.4.4),

$$Z = \frac{A}{1.98 + 0.0155\, A^{2/3}} \tag{1-29}$$

furnishes a good approximation in the case of stable nuclei. This immediately
prompts the question, what makes a nucleus stable? For very light nuclei
($A \lesssim 40$) the formula shows that stability obtains when Z and N are equal,
whereas for the heavier nuclei, stability prevails only if N somewhat exceeds
Z (for the heaviest nuclei, N has to exceed Z by about 50 percent). So it
would seem that neutrons play the role of a binding agent when present in
the right number to preserve the balance, as evinced for example by the

lifetimes of the lithium isotopes:

$$^{5}_{3}\text{Li}_2 : 10^{-21} \text{ sec}$$

$$^{6}_{3}\text{Li}_3 \text{ and } ^{7}_{3}\text{Li}_4 : \infty \text{ (stable)}$$

$$^{8}_{3}\text{Li}_5 : 1.24 \text{ sec}$$

$$^{9}_{3}\text{Li}_6 : 0.25 \text{ sec}$$

Yet in their *free* state, neutrons are themselves unstable. Being slightly heavier than protons,† namely,

$$m_\text{n} = 1.674\ 82 \times 10^{-24} \text{ g} = 1.008\ 665 \text{ u} \simeq 939.550 \text{ MeV} \qquad (1\text{-}30)$$

whereas

$$m_\text{p} = 1.672\ 52 \times 10^{-24} \text{ g} = 1.007\ 277 \text{ u} \simeq 938.256 \text{ MeV} \qquad (1\text{-}31)$$

they decay into the latter (plus other products) within a mean lifetime [So 58, So 59a, So 59b]

$$\tau_\text{n} = 16.9 \text{ min} = 1.01 \times 10^3 \text{ sec} \qquad (1\text{-}32)$$

(The value obtained in a recent remeasurement [Chr 67], $\tau_\text{n} = 15.58 \pm 0.23$ min, is at variance with this; as it has not yet been confirmed, we retain the accepted value of 16.9 ± 0.4 min throughout this book.)

Clearly, the condition of neutrons within a nucleus is not the same as in the free state. Soon after the discovery of the neutron, Heisenberg [He 32] suggested that the proton and neutron could be regarded as different charge states of the same fundamental particle, the NUCLEON. In the free state, the proton carries a positive charge whose magnitude is (to better than 2 parts in 10^{20}) the same as that carried by the electron,

$$e = 4.802\ 98 \times 10^{-10} \text{ esu} = 1.6021 \times 10^{-19} \text{ C} = 1.6021 \times 10^{-20} \text{ emu}$$

whereas that of the free neutron is zero (or at any rate, below 2×10^{-15} e). We now think of nucleons within a nucleus or interacting with other nucleons as composed of a core surrounded by at least one "cloud" of charged mesons and accordingly conceive of the binding force between nucleons as due to

---

† It is convenient to express atomic and nuclear masses in ATOMIC MASS UNITS u (or, in terms of the mass-energy relation $E = mc^2$, in energy units):

$$1 \text{ u} \equiv 1 \text{ amu} = 1.660\ 43 \times 10^{-24} \text{ g} \simeq 931.478 \text{ MeV}$$

Whereas up to 1961 a "physical amu" had been defined as $\frac{1}{16}$th of the mass of the neutral ^{16}O atom, and the definition of a "chemical amu" had been based upon the natural mixture of oxygen isotopes, the current unified definition, as adopted by the International Union of Physics and Chemistry, takes the atomic mass unit u to be $\frac{1}{12}$th of the mass of the neutral ^{12}C atom, as determined by atomic mass spectrography.

order to derive information on the distributions of mass and charge in nuclei, linear accelerators have had to be designed which are capable of providing high-energy electron beams around 1000 MeV (10^9 eV $= 10^3$ MeV $= 1$ GeV,

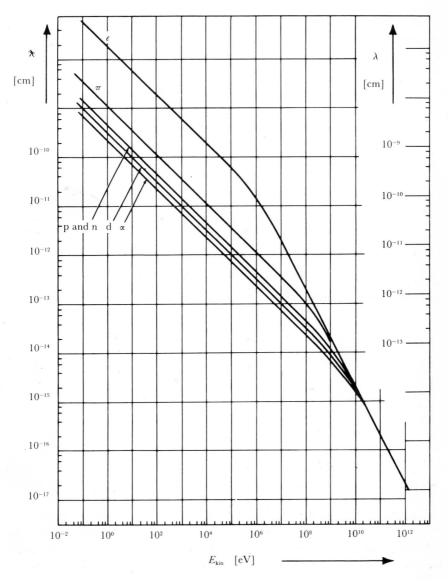

Fig. 1-3. Rationalized and normal de Broglie wavelengths for electrons, π mesons, nucleons, deuterons, and α particles in terms of kinetic energy from 10^{-1} to 10^{12} eV (log-log scale).

or, in American terminology, 1 BeV: Note that 1 billion $= 10^9$ in the United States, but 10^{12} elsewhere). The principal such facility has hitherto been the 1-GeV electron accelerator at Stanford, extensively used by the Hofstadter group to investigate nuclear structure [Ho 57, El 61, Ho 63, Ho 64]. With the recent completion of another Stanford linear accelerator (SLAC) capable of furnishing a 20-GeV electron beam [Da 66b] after acceleration along a 2-mile path, fresh strides in our knowledge and understanding of nuclear

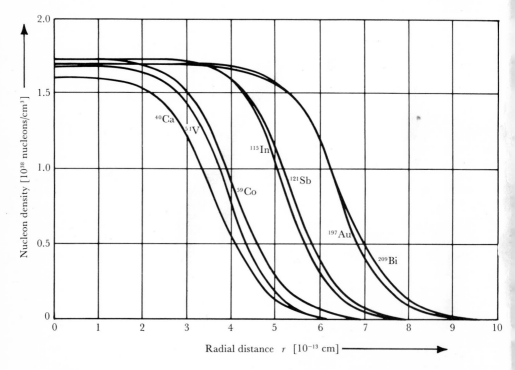

Fig. 1-4. Charge-density distribution in medium-heavy and heavy nuclei as a function of radial distance from the center of the nucleus, as indicated by scattering of 1-GeV electrons (after [Ho 57]).

and particle structure are anticipated. Some of the results obtained at 1 GeV are collected in Fig. 1-4, which displays the charge-density distribution in terms of the nuclear radius, as derived from the analysis of electron scattering experiments undertaken with the Stanford accelerator.

As inspection of Fig. 1-3 indicates, electron wavelengths below $\lambda_e = 1$ fm are attained only with energies above about 0.3 GeV. The 1-GeV machine, yielding an electron beam of wavelength $\lambda_e = 0.195$ fm has therefore been meson exchange between them (just as in chemistry the covalent linkage can be regarded as due to electron exchange between atoms, in ion physics the H_2^+ ion as held together by electron exchange between two protons, and in general physics electromagnetic forces as due to photon exchange). Heisenberg's conception of "exchange forces" was used by Yukawa [Yu 35a] to account for nuclear forces and nuclear stability, since the purely gravitational attraction between nucleons is far too weak to overcome the electrostatic repulsion of the individual like charges which make up the nuclear charge Ze. Nowadays, the force between neutrons in a nucleus is envisaged to be due to the exchange of neutral mesons.

The second of the two uncertainty relations (1-25) can be used to derive an estimate of the mass of the meson through whose agency the nuclear force acts. To this end, we consider the agent as a particle traveling with the maximum possible velocity, namely c, over a distance equal to the range of the strong nuclear binding force. A considerable body of evidence established this to be about $r_0 = 1.4$ fm, whence the time interval for the transfer of a meson from one nucleon to the other is

$$\Delta t = \frac{r_0}{c} = \frac{1.4 \times 10^{-13}}{3 \times 10^{10}} = 4.7 \times 10^{-24} \text{ sec} \qquad (1\text{-}33)$$

and the corresponding energy uncertainty according to the uncertainty principle is

$$\Delta E \approx \frac{\hbar}{\Delta t} = \frac{1.05 \times 10^{-27}}{4.7 \times 10^{-24}} = 2.23 \times 10^{-4} \text{ erg} = 139 \text{ MeV} \qquad (1\text{-}34)$$

This energy uncertainty is none other than the *mass* of the meson expressed in energy units. In terms of the electron rest mass

$$m_e = 9.1091 \times 10^{-28} \text{ g} = 5.485\ 97 \times 10^{-4} \text{ u} \triangleq 0.511\ 006 \text{ MeV} \qquad (1\text{-}35)$$

this estimate furnishes the value

$$m = 270\ m_e \qquad (1\text{-}36)$$

in excellent agreement with the mass of the π meson (or "pion") of mass $m_\pi = 273\ m_e$, the particle now regarded as in the main responsible for the occurrence of the nuclear force. The exchange of such a particle, without noticeably violating the all-important physical principle of mass-energy conservation, provides a model for the linkage between the protons and neutrons which constitute the atomic nucleus and an explanation for the enhanced stability of bound neutrons.

The actual structure of protons and neutrons is indicated by electron-scattering experiments to be rather more complicated. By analogy with procedures in electron diffraction and x-ray diffraction, it is customary to

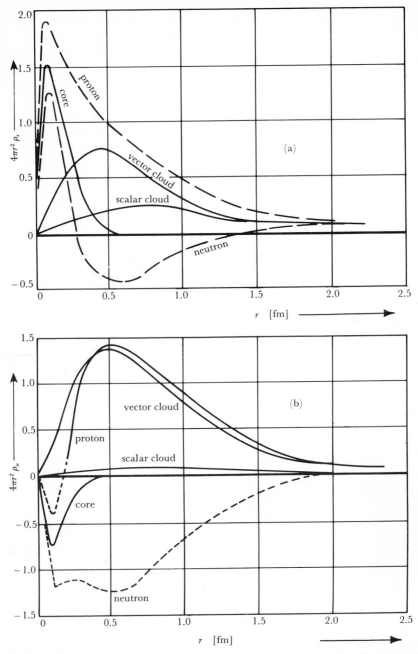

Fig. 1-5. Nucleon structure as indicated by form-factor analysis based upon high-energy electron scattering data. The curves show (a) normalized charge distributions and (b) normalized magnetic-moment distributions for the bare nucleon core, two π-mesic clouds (scalar and vector), proton, and neutron as functions of radial distance from the nucleon center (after [Li 61b]).

present the structural findings in terms of "form factors," which are numerical functions expressing the radial distribution of the charge and magnetic moment densities. A set of such numerical values is determined empirically on the basis of a given structural self-consistent model and adjusted to give the best measure of agreement with the entire range of experimental data. An analysis of this type undertaken by Littauer *et al.* [Li 61b] appears to indicate that the proton and neutron in common have a core, of positive charge around 0.35*e*, and of probable radius 0.2 fm, surrounded by two distinct clouds of mesons. The inner cloud, of radius 0.8 fm, has a charge whose value is +0.5*e* in the case of a proton and −0.5*e* in the case of a neutron. The outer cloud appears to have a radius of about 1.4 fm and carries a positive charge of 0.15*e* for both protons and neutrons, thus conferring a net charge of *e* upon the proton and 0 upon the neutron. Because of their different properties as regards the mathematical formalism, the inner and outer clouds of mesons have been termed the "vector" and "scalar" clouds respectively; they appear to be associated with two-pion and three-pion aggregates, though this is somewhat uncertain. The radial distributions of charge density and anomalous magnetic moment suggested by this analysis have been depicted in Fig. 1-5.

More recent results [Ak 66] obtained for proton-proton elastic scattering at 90° in the energy range 5.0 to 13.4 GeV would seem to indicate an internal structure to the proton along the lines of "onion shells" with an outer pion cloud at a radius 0.92 fm, an inner heavy cloud of radius 0.50 fm and a core of radius 0.32 fm, but these measurements and their interpretation are still in a preliminary stage. A further discussion has been given by Schopper [Scho 67].

EXERCISES

1-1 On a scale in which a drop of water (radius $r = 0.1$ cm) is magnified to the size of the Earth (radius $R = 6.38 \times 10^8$ cm), what is the radius of (a) a typical atom, (b) a typical nucleus?

1-2 What is your de Broglie wavelength when you move with a velocity $v = 1$ m/sec?

1-3 Taking the wavelength of blue light to be 5000 Å and of red to be 7000 Å, compare the mass of a blue photon with that of a red.

1-4 At what angle θ do 0.01-eV neutrons experience first-order Bragg reflection on a crystal of sodium chloride? (Bragg condition: $n\lambda = 2d \sin \theta$. Lattice separation in NaCl: $d = 2.8$ Å.) What is the energy of x rays which undergo reflection at the same angle?

1-5   If a 500-μA beam of 20-MeV protons impinges on a target in which it is completely absorbed (a) how many protons are captured per second; (b) how much heat is generated if there is perfect conversion of kinetic energy into heat; (c) what force does the proton beam exert; (d) by how many percent does the relativistic mass of the protons exceed their rest mass?

1-6   A reactor provides an external beam of slow neutrons ($E_n^{(kin)} = 0.025$ eV) whose flux is $\Phi_n = 5 \times 10^6$ cm^{-2} sec^{-1}. Taking the mean lifetime of free neutrons to be $\tau_n = 1.01 \times 10^3$ sec, calculate the number of neutrons which decay per cm^3 in 1 minute as they emerge from the reactor.

1-7   By allowing fast protons from an accelerator to impinge upon a suitable target, a 170-MeV beam of π mesons can be produced. Assuming these particles to have a mean decay lifetime of $\tau_{\pi^\pm} = 2.60 \times 10^{-8}$ sec, calculate the path length s from the target corresponding to a 50 percent diminution in intensity for a parallel pion beam.

1-8   The following particles are accelerated from rest to such speeds that they respectively have a kinetic energy equal to twice their rest energy: (a) electron ($m_e = 9.109 \times 10^{-28}$ g), (b) π meson ($m_\pi = 273 \ m_e$), (c) proton ($m_p = 1836 \ m_e$).

   What further accelerating voltage would have to be applied in a second stage of acceleration in order to double the *total* energy so attained by each particle species, and what are the respective values of $\beta \equiv v/c$ before and after this second stage?

1-9   Chadwick [Ch 32] discovered neutrons in the course of experiments in which α particles (from a natural polonium source) were directed on to a beryllium target. Highly penetrating uncharged radiation was thereby produced, which could yield recoil ions whose maximal energy in the case of hydrogen was measured as $E_H = 5.7$ MeV and in the case of nitrogen as $E_N = 1.5$ MeV. Assuming the collision to be elastic, calculate the mass of the neutral particles and compare it with the modern value of the neutron mass.

1-10 Calculate the radius a_0 and momentum p of an electron in the innermost orbit of a Bohr hydrogen atom, and show that on setting these equal to the maximum uncertainty in each case, their product satisfies the Heisenberg uncertainty relation.

1-11 A nuclear power station is designed to provide 1000 MW power. Assuming that each fission event releases 200 MeV energy and that the steam engine operating between temperatures $T_1 = 500°C$ and $T_2 = 100°C$ has an efficiency equal to 40 percent of that which an ideal Carnot engine would have, calculate (a) how many grams of ^{235}U are used up per day; (b) how many ^{235}U nuclei undergo fission per sec; (c) how many metric tons (1 tonne $= 1000$ kg) of coal having a heat of combustion $L = 7000$ cal/g would be needed per day in a conventional power station with the same efficiency to produce the same amount of power?

Chapter 2

NUCLEAR RADII AND THE
LIQUID DROP MODEL
OF THE NUCLEUS

Stability and Radius of Nuclei

2.1. Energy Considerations

The simplest nucleus is that of hydrogen, $^1_1H_0$. This is a stable nucleus, consisting as it does of a single stable proton. The foregoing discussion leads us to expect that the next-simplest nucleus, $^2_1H_1$, the deuteron, will, in view of the strength of the proton-neutron linking force, also be stable—and so it is, but the binding energy is quite low. The nucleus formed by addition of a second neutron, the triton $^3_1H_2$, is no longer stable, its mean lifetime being 17.7 y, for the unpaired neutron renders the nucleus too massive to be energetically in a state of equilibrium. The emission of a β^- particle (or electron, originating from decay of a neutron) converts the nucleus into the slightly lighter nuclide $^3_2He_1$, which is stable. So also, to a very marked degree, is the nuclide $^4_2He_2$ (or α particle), constituted as it is of a couple of linked proton-neutron pairs. It is striking that the nuclei $^5_2He_3$ and $^5_3Li_2$, respectively formed by addition of a neutron or of a proton to the stable $^4_2He_2$ system, are outstandingly unstable, having lifetimes around 10^{-21} sec and decaying by neutron or proton emission to $^4_2He_2$ (see Fig. 2-1). Clearly, the nuclear or atomic mass is one of the most decisive factors governing nuclear stability, and considerations of mass-energy balance can shed much light upon the stability properties of various nuclear species such as those classified in Table 2-1.

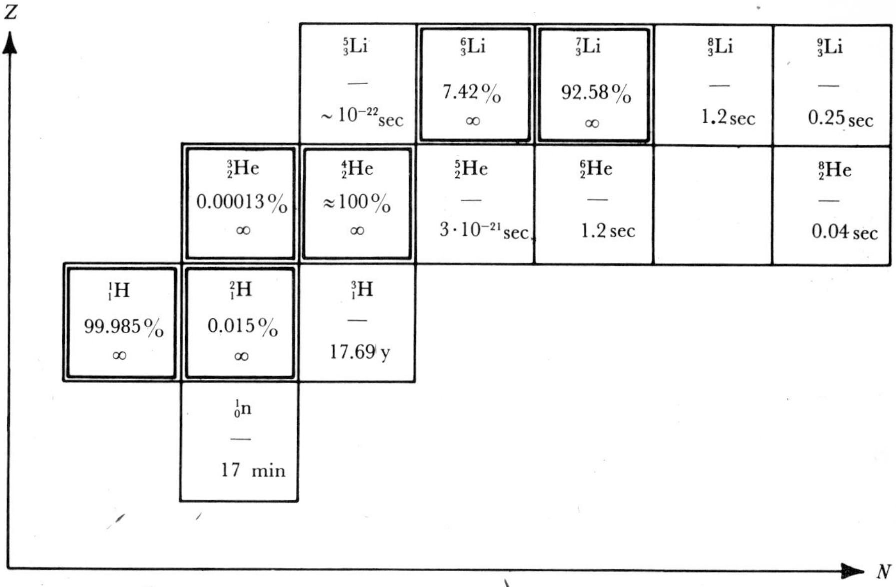

Z

	$^5_3$Li	$^6_3$Li	$^7_3$Li	$^8_3$Li	$^9_3$Li
	—	7.42%	92.58%	—	—
	$\sim 10^{-22}$sec	∞	∞	1.2 sec	0.25 sec
$^3_2$He	$^4_2$He	$^5_2$He	$^6_2$He		$^8_2$He
0.00013%	$\approx 100\%$	—	—		—
∞	∞	$3\cdot10^{-21}$sec	1.2 sec		0.04 sec
$^1_1$H	$^2_1$H	$^3_1$H			
99.985%	0.015%	—			
∞	∞	17.69 y			
$^1_0$n					
—					
17 min					

$\longrightarrow N$

Fig. 2-1. Representative chart of the lightest nuclides, showing how stability changes with constitution. The nuclides are arranged according to atomic number Z and neutron number N, and characterized by their mass number A and Z. The stable nuclides are indicated by framing. Natural abundances and mean lifetimes are also cited. Comprehensive nuclide charts of this type have been prepared by D. T. Goldman [Go 65a] and W. H. Sullivan [Su 57] based upon a schematic by E. Segrè, and are accordingly also known as "Segrè charts." (Used by permission of McGraw-Hill Book Co.)

Table 2-1. NOMENCLATURE OF NUCLEAR SPECIES

Nomenclature	Characteristic	Examples	Comment
Nucleus (Nuclide)	(A,Z)	$^1_1$H, $^{63}_{29}$Cu, $^{235}_{92}$U	At present, over 1250 nuclear species are known
Isotope	Z = constant	$^{126}_{56}$Ba to $^{144}_{56}$Ba	Up to 23 isotopes per element are known
Isobar	A = constant	$^{14}_6$C, $^{14}_7$N, $^{14}_8$O	A = constant in β decay
Isotone	N = constant	$^{14}_6$C$_8$, $^{15}_7$N$_8$, $^{16}_8$O$_8$	
Isomer	^{A}El* or ^{A}El^m	$^{90}_{40}$Zr$^m_{50}$ (2.315-MeV state)	Metastable excited states of nuclei having a mean lifetime which exceeds 10^{-9} sec
Mirror nuclei	$N_1 = Z_2,\ N_2 = Z_1$	$^3_1$H$_2$, $^3_2$He$_1$	Only among light nuclei do mirror pairs exist with mutually interchanged Z and N

The very accurate methods of atomic mass spectrography show that the experimentally determined atomic masses are smaller than the sum of the individual masses of the constituent particles. Clearly, some of the latter overall mass goes toward providing the intrinsic energy associated with binding the particles within the nucleus. The measured mass of a neutral atom in an isotopically pure specimen, termed the ATOMIC WEIGHT, is close to being an integer multiple A of the atomic mass unit u, the numerical difference when expressed in u being termed the MASS DEFECT (or MASS DECREMENT) ΔM, where

$$\Delta M \equiv M - A \qquad (2\text{-}1)$$

and the specific mass difference, the mass defect per nucleon, being termed the PACKING FRACTION f, with

$$f \equiv \frac{\Delta M}{A} \equiv \frac{M - A}{A} \qquad (2\text{-}2)$$

The value of f varies with the atomic number A as shown in Fig. 2-2, which indicates that there is a variation in the intrinsic energy of nuclear configurations over the range of the periodic table. It has become customary to quote the masses in u of single *atoms* of a given isotope. The NUCLEAR MASS $m(Z,A)$ can be derived therefrom by subtracting the sum of the atomic electron masses (Zm_e). Strictly speaking, since the electrons are bound within the atom, their atomic binding energy should be added, but this is so small that it can usually be neglected:

$$m(Z,A) = M - Zm_e\,(\,+\,B_e) \qquad (2\text{-}3)$$

where $m(Z,A)$ is the nuclear mass, M is the atomic mass, Zm_e is the electronic mass, and B_e is the electronic binding energy expressed in u.

The nuclear mass can analogously be decomposed further into its component terms:

$$m(Z,A) = Zm_p + Nm_n - B \qquad (2\text{-}4)$$

where B is the NUCLEAR BINDING ENERGY, here expressed in mass units u. It is the energy which would have to be supplied before the nucleus could be disassembled into separate free nucleons. Since it applies only to stable nuclei, it is necessarily always positive. A rather more meaningful quantity is the MEAN BINDING ENERGY PER NUCLEON, B/A, which bears a close resemblance to the atomic packing fraction f. Figure 2-3 shows the manner in which B/A varies with A, and in particular indicates that its behavior mirrors that of f. Over most of its range, i.e., for all but the very light nuclides below $A \approx 20$, the binding energy per nucleon (expressed in energy rather than mass units for convenience) can be seen to remain roughly constant around 8 MeV/nucleon,

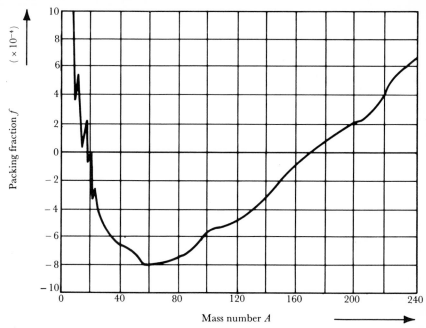

Fig. 2-2. Packing fraction $f \equiv (M - A)/A$ represented in terms of the mass number A (linear scale). It is zero around $A = 20$ (neon) and $A = 170$ (erbium-rare earth region), maximal positive at the extremes, and maximal negative at $A = 60$ (cobalt, nickel). (Used by permission of McGraw-Hill Book Co.)

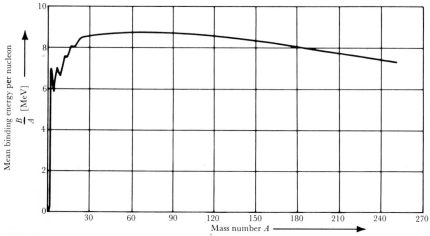

Fig. 2-3. Mean binding energy per nucleon B/A *vs* mass number A. Energy release accompanies transmutations which lead to higher values of B/A, i.e., to the peak of the curve. Above about $A \cong 30$, the value of B/A remains essentially constant at around 8 MeV per nucleon, which indicates that nuclear forces are saturable.

with a maximum of about 8.8 MeV/nucleon around $A \simeq 60$, and falling monotonically to about 7.5 MeV/nucleon at $A \simeq 240$. This behavior of B/A immediately provides an insight into the magnitude of the ENERGY RELEASE IN FISSION of heavy nuclides or FUSION of light nuclides, for in both instances there is a *gain* in binding energy per nucleon, e.g., in the symmetric fission of a nucleus around $A = 240$, the increase in B/A is equal to $8.5 - 7.5 = 1.0$ MeV/nucleon, which corresponds to a total energy release $B \approx 1 \times 240 \sim 200$ MeV. In the hypothetical case of fusion of two ^{2}H nuclei (deuterons), for which $B/A = 1.113$ MeV/nucleon, to form an α particle, for which $B/A = 7.075$ MeV/nucleon, the gain in energy is almost 6 MeV *per nucleon*, and hence very considerably greater than even that in fission. A gain in binding energy of course represents a diminution in mass, and this in turn is manifested as an external energy release.

Other nuclear reactions also entail changes in the binding energy and therefore either release or take up external energy as they proceed. The reaction equation for the process

$$I(i,f)F$$

in which an incident particle i impinging on a target nucleus I brings about a change into a residual nucleus F together with an emitted particle f, is written as

$$I + i \rightarrow f + F + Q \qquad (2\text{-}5)$$

where Q represents an energy (whose value in MeV can be positive or negative, depending upon the type of reaction) included in order to preserve mass-energy balance. The equation thus has to fulfil *conservation rules* for total mass-energy, as well as for charge and total number of nucleons. The Q-VALUE is, accordingly,

$$Q = [(m_I + m_i) - (m_F + m_f)]c^2 = [(M_I + M_i) - (M_F + M_f)]c^2 \qquad (2\text{-}6)$$

and specifies the change in the binding energies of the reaction partners in the course of the reaction. The total energy of the initial or final system comprises the sum of the rest energies of the partners plus the sum of their kinetic energies, and since there is a change in the total rest mass, this is manifested as a REACTION ENERGY Q, which usually appears in the form of kinetic energy of the outgoing particle f, together with the recoil energy of F. Every reaction possesses a characteristic Q-value, and is termed EXOTHERMIC when $Q > 0$ (liberation of energy), or ENDOTHERMIC when $Q < 0$ (absorption of energy). An exothermic reaction can accordingly, at least in principle, occur spontaneously, but an endothermic reaction cannot take place unless the incident particle has a kinetic energy exceeding the THRESHOLD ENERGY, whose value in the center-of-mass system is equal to the absolute Q-value. At threshold, the reaction can just commence, but the reaction products have zero kinetic energy.

To illustrate this by a numerical example, we consider the process

$$^{10}_{5}\text{B} + {}^{1}\text{n} \rightarrow {}^{7}_{3}\text{Li} + {}^{4}_{2}\text{He} + Q \tag{2-7}$$

On the one hand, the Q-value can be derived from a consideration of the respective binding energies:

$$Q = B_{\text{Li}} + B_{\alpha} - B_{\text{B}} = 39.245 + 28.296 - 64.750 = 2.791 \text{ MeV} \tag{2-8}$$

Alternatively, it can be evaluated from the atomic masses:

$$Q = (M_{\text{B}} + m_{\text{n}} - M_{\text{Li}} - M_{\text{He}})c^2 \tag{2-9}$$

Inserting the values

$$M_{\text{B}} = 10.012\ 939 \text{ u} \qquad m_{\text{n}} = 1.008\ 665 \text{ u}$$
$$M_{\text{Li}} = 7.016\ 004 \text{ u} \qquad M_{\text{He}} = 4.002\ 603 \text{ u} \tag{2-10}$$

we get

$$Q \triangleq 0.002\ 997 \text{ u} \triangleq 2.791 \text{ MeV} \tag{2-11}$$

Note that if the binding energies of the electrons in the respective atoms are regarded as negligible, it is immaterial whether atomic masses or nuclear masses are throughout used to evaluate Q, since the total number of electrons, and therefore their total rest mass, is the same in the initial and final atomic systems.

The fact that Q is *positive* in the above case shows that the reaction $^{10}\text{B}(n,\alpha)^7\text{Li}$ is *exothermic* and can be caused to take place even with *slow* neutrons. It would, however, be a mistake to assume that the *inverse, endothermic*, reaction $^7\text{Li}(\alpha,n)^{10}\text{B}$ could be brought about by α particles having just the energy $E_{\alpha} = Q = 2.79$ MeV, since this would be overlooking the fact that some of the kinetic energy of the impinging particle is taken over by the struck nucleus in the form of kinetic recoil energy. For the above forward (n,α) reaction, this recoil energy is vanishingly small when slow neutrons are employed, and even with neutrons of a higher energy, remains small because of the 10:1 mass ratio between target and projectile. However, for the reverse endothermic reaction $^7\text{Li}(\alpha,n)^{10}\text{B}$, the mass ratio is only 7:4, the incident energy is appreciable, and the recoil energy is in consequence far from negligible. Hence the α-particle energy in the laboratory system must considerably exceed 2.79 MeV before a reaction will commence. For the occurrence of a nuclear reaction, it is the total kinetic energy in the center-of-mass system E_{CM} which is the decisive factor, composed of the sum of the kinetic energies of the partners I and i in the center-of-mass system:

$$E_{\text{CM}} = E'_{\text{I}} + E'_{\text{i}} \tag{2-12}$$

(quantities in the center-of-mass system are primed, and those in the laboratory system unprimed). A consideration of reaction kinematics, as in Appendix C.4, shows that

$$E'_{\text{i}} < E_{\text{i}} \tag{2-13}$$

when the target nucleus I is at rest initially $(E_{\text{I}} = 0)$. The total kinetic energy in the center-of-mass system is

$$E_{\text{CM}} = \tfrac{1}{2}\mu v_{\text{rel}}^2 \tag{2-14}$$

where

$$\mu \equiv \frac{m_{\mathrm{I}} m_{\mathrm{i}}}{m_{\mathrm{I}} + m_{\mathrm{i}}} \tag{2-15}$$

is called the REDUCED MASS of the collision partners, and v_{rel} is the relative velocity of i with respect to I (independent of the choice of system). In the laboratory system the total kinetic energy E_{lab} is just E_{i}, since $E_{\mathrm{I}} = 0$, and therefore

$$E_{\mathrm{lab}} = E_{\mathrm{i}} = \tfrac{1}{2} m_{\mathrm{i}} v^2 = \tfrac{1}{2} m_{\mathrm{i}} v_{\mathrm{rel}}^2 = \frac{m_{\mathrm{i}}}{\mu} E_{\mathrm{CM}} = \frac{m_{\mathrm{I}} + m_{\mathrm{i}}}{m_{\mathrm{I}}} E_{\mathrm{CM}} \tag{2-16}$$

The *endothermic reaction* proceeds from threshold when $E_{\mathrm{CM}} \gtrless Q$, i.e., when

$$E_{\mathrm{i}} = E_{\mathrm{lab}} \gtrless \frac{m_{\mathrm{I}} + m_{\mathrm{i}}}{m_{\mathrm{I}}} Q \tag{2-17}$$

For the $^7\mathrm{Li}(\alpha,\mathrm{n})^{10}\mathrm{B}$ reaction, the minimum energy of the α particle in the laboratory system therefore takes on the value

$$E_\alpha = \frac{7+4}{7} Q = \frac{11}{7} \times 2.79 = 4.38 \text{ MeV} \tag{2-18}$$

The difference between the threshold energy in the laboratory system and the Q-value for an endothermic reaction is particularly large when the target mass m_{I} is small, and can run to very appreciable values in the region of relativistic energies. For example, although protons can be accelerated to energies of 34 GeV in the laboratory system, the effective reaction energy when they impinge upon stationary protons in a target is only that in the center-of-mass system, namely 6.5 GeV. The recoil effect of the struck nucleus can be obviated only by using target nuclei which move against the incident particles at the instant of impact. This represents the advantages inherent in the use of colliding-beam assemblies, since for these the total energies in the laboratory system are equal to those in the center-of-mass system.

In this connection, it should perhaps be emphasized that because a Q-value represents an energy *difference*, it is independent of the system of reference, but its numerical value expressed in u or MeV depends upon the mass values of nuclei and the mass convention. A consistent set of Q-values based upon the present $^{12}\mathrm{C} = 12$ u convention has been published by Everling *et al.* [Ev 60] up to $A = 200$. Mattauch *et al.* [Ma 65d] have tabulated a consistent set of atomic masses and binding energies up to $A = 257$, and list [Ma 65e] a commensurate set of consistent Q-values, also up to $A = 257$.

2.2. The Radius of Nuclei and the Liquid Drop Model

Scattering investigations with electrons, nucleons, deuterons, and α particles confirm, as one might conjecture from minimal-energy considerations, that to a first approximation nuclei can be regarded as spherical in shape. Right from the earliest scattering studies by Rutherford and by Chadwick it has been concluded that the radius of all but the very lightest nuclei can fairly accurately be described by a relation of the form

$$R = r_0 A^{1/3} \tag{2-19}$$

where A is the mass number of the nucleus (and thus specifies the number of nucleons) and r_0 is a universal constant related to the nucleon radius. The numerical value of r_0 varies slightly with the method employed for the determination of nuclear radii but can in general be taken as

$$r_0 = 1.4 \text{ fm} \tag{2-20}$$

in close agreement with the range of nuclear forces employed in the Yukawa theory, and approximately equal to half the value of the CLASSICAL ELECTRON RADIUS,†

$$r_e = \frac{e^2}{m_e c^2} = 2.82 \text{ fm} \tag{2-21}$$

† A name not to be interpreted too literally. Our only present source of information as to the structure, and therefore the radius, of an electron lies in collision studies, but in these the electron interacts solely via its Coulomb field. This accordingly does not lend itself to any meaningful definition of collision radius. In its stead, one arbitrarily derives a definition of the "classical electron radius" by setting the electron's rest energy $m_e c^2$ equal to the electrostatic potential energy e^2/r_e associated with a charge e distributed over the surface of a sphere having a radius r_e. Thus

$$r_e = e^2/m_e c^2 = 2.817\ 77 \times 10^{-13} \text{ cm}$$

However, the assumption of equality is open to considerable argument. Furthermore, if one instead carries out a classical calculation treating the net charge as brought piecemeal in infinitesimal amounts on to the surface of the sphere, a factor 2 enters, which reduces r_e to one half of the above value. Alternatively, a more sophisticated classical calculation which considers the electric and magnetic field energies of a rotating electron yields a value of r_e which is two thirds of the value cited. At the other extreme, when, as in electrodynamics, the electron is treated as a point charge, one is faced with the problem that its self-energy e^2/r_e becomes infinite. It is interesting to note a rather striking relationship between r_e and $\lambdabar_C$, the rationalized COMPTON WAVELENGTH of the electron, namely

$$r_e = \alpha \lambdabar_C$$

where

$$\lambdabar_C = \hbar/m_e c = 386 \text{ fm}$$

and

$$\alpha = e^2/\hbar c \simeq 1/137 = \text{FINE STRUCTURE CONSTANT}$$

Another such relation connects $\lambdabar_C$ with the BOHR RADIUS of the electron in the hydrogen atom,

$$a_0 = \hbar^2/m_e e^2 = 0.529 \text{ Å}$$

namely

$$\lambdabar_C = \alpha a_0$$

Attention may also be drawn to the simple relation connecting the de Broglie wavelength with the Compton wavelength of a particle:

$$\lambdabar_D = \lambdabar_C [\epsilon(\epsilon + 2)]^{-1/2},$$

where $\epsilon \equiv E_{\text{kin}}/m_0 c^2$ is the kinetic energy measured in units of the rest energy $m_0 c^2$.

This striking correlation between r_0 and A directly points to a universal *constant* MEAN DENSITY ρ_0 OF NUCLEAR MATTER, since A cancels out when the expression for ρ_0 is built:

$$\rho_0 \cong \frac{A m_p}{\frac{4}{3}\pi R^3} = \frac{m_p}{\frac{4}{3}\pi r_0^3} \approx \text{constant} \cong 10^{14} \text{ g cm}^{-3} \qquad (2\text{-}22)$$

Even though the realization that over 99.9 percent of the mass of an atom is lodged within its nucleus would lead one to expect a high value for nuclear

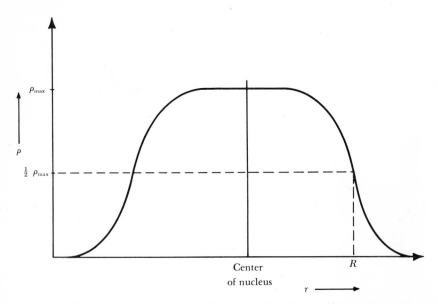

Fig. 2-4. Schematic representation of the mass distribution in a nucleus, showing how the density gradually diminishes in the surface region and indicating a definition of the nuclear radius R as the radial distance corresponding to half-maximum mass density.

density, the above is quite outstandingly large. From ρ_0 we can derive some measure of the closeness of packing of nucleons within a nucleus, for it represents a nucleon density of around 10^{38} cm^{-3}, many orders of magnitude higher than that of electrons in an atom.

The property of constant density possessed by nuclei prompts an analogy with the physics of an incompressible fluid and suggests that certain aspects of nuclear structure and stability might be treated by a model in which the nucleus is regarded as a liquid drop. Further support for this seemingly crude visualization is provided by the constancy of B/A, which by analogy would correspond to a constant heat of vaporization, independent of the size of the droplet. The liquid drop model has, indeed, proved of great value for the consideration of nuclear mass systematics and thereby of nuclear stability, as also for the related approach to the treatment of nuclear fission. (A rather clear description of such application to the problem of fission has been given by Nix and Swiatecki [Ni 65].) As a model of nuclear structure, it is not only one of the first to have been put forward, but has the merit of being particularly simple to picture in physical terms. We therefore examine it here in some detail with a view to the elucidation of nuclear masses and stability. But first, it might perhaps be stressed that a certain discretion has to be exercised when drawing analogies with the physical system of a liquid drop. For instance, as we have seen, it would be erroneous to conceive of the nucleus as a sharply bounded entity. Even the undeformed nuclei have no sharply defined surface: the structure of the nucleus is described by distributions of nuclear mass and charge whose density in the surface region dwindles continuously rather than abruptly (Fig. 2-4). In consequence, meaningful definitions of the nuclear radius R can be constructed either from setting

(a) $r = R$ at the point in the surface region where $\rho = \frac{1}{2}\rho_{max}$ or

(b) $r = R$ at the point in the surface region where ρ drops most steeply, i.e., $r = R$ when $d^2\rho/dr^2 = 0$ at the surface.

2.3. The Liquid Drop Model and the Semiempirical Mass Formula

2.3.1. Von Weizsäcker's Approach to Binding Energy

The mass of a nucleus is given by the formula

$$m(Z,A) = Zm_p + Nm_n - B \qquad (2\text{-}23)$$

where B is the binding energy expressed in mass units. If it were possible to calculate B from a general formula, all nuclear masses could be evaluated theoretically. To this end, von Weizsäcker [von W 35] employed the liquid drop model of the nucleus, regarding B as akin to a latent energy of condensation. This enabled him to propose a semiempirical formula for B in which various contributions making up the overall binding energy B are deduced theoretically in turn,

$$B = B_1 + B_2 + B_3 + \cdots \qquad (2\text{-}24)$$

but their relative magnitudes are determined empirically by adjustment of the weighting parameters in the formula to obtain the best fit to actual measured nuclear masses. We now retrace this approach in outline, to some extent introducing slight modifications which have been indicated by information that has since come to hand. We also base the treatment upon a somewhat simplified form of expression due to Bethe and Bacher [Be 36], published shortly after von Weizsäcker had initiated the approach. For this reason, the modern form of the semiempirical mass formula is often referred to as the BETHE-WEIZSÄCKER FORMULA.

2.3.2. VOLUME ENERGY

The major contribution to B, namely B_1, stems from the mutual interactions of the nucleons under the influence of nuclear forces. Some important characteristics of the strong internucleon forces, as distinct from the much weaker long-ranged Coulomb interaction between protons, can be deduced from evidence which has already been presented. For instance, the constancy of these forces is indicated by the constancy of B/A over almost the entire A-range. (The departure from the value $B/A \simeq 8$ MeV/nucleon at the very light and very heavy extremes of the mass spectrum occurs because the contribution B_1/A is overlaid by other effects there.) Furthermore, the constancy of the matter density ρ_0 indicates the nuclear forces to be *short-ranged*, so that interactions between nucleons are essentially confined to just those between nearest neighbors, for if the binding were influenced by those nucleons lying beyond the nearest neighbors its strength would increase as the total number of nucleons increases, and B_1 would be proportional to A^2 rather than to A. The density ρ_0 would thereby also have to increase with A instead of remaining constant.

For the VOLUME ENERGY B_1 we may accordingly write

$$\frac{B_1}{A} = \text{constant} \tag{2-25}$$

or

$$B_1 = a_v A \tag{2-26}$$

with a_v a constant to be determined empirically.

2.3.3. SURFACE ENERGY

Strict proportionality between B_1 and A implicitly assumes overall constancy in the strength of the interaction of each nucleon with its immediate surroundings. However, those nucleons which are situated in the surface region of the nucleus are necessarily more weakly bound than those in the nuclear interior because they have fewer immediate neighbors. The number of such nucleons is proportional to the surface area of the nucleus, and therefore to $R^2 \sim A^{2/3}$. In expressing this proportionality,

$$B_2 = -a_s A^{2/3} \tag{2-27}$$

the sign of the SURFACE ENERGY B_2 must be opposite to that of B_1 since this effect, which corresponds to the surface tension of a liquid drop, represents a *weakening* in the binding energy. Carrying the analogy further, we see that this weakening is least, and therefore the stability is greatest, when the droplet is spherical in shape, since then the surface area is minimal for a given volume. This gives us reason to expect that spherically symmetric nuclei will also be the most stable—though we shall find that this argument does not apply to very light nuclei below $A = 30$ and to certain heavier nuclei which are permanently deformed but nevertheless stable.

2.3.4. COULOMB ENERGY

The electrostatic repulsion between like charges in the nucleus has a long-range character, and the resultant COULOMB ENERGY B_3 is therefore proportional to Z^2. It furthermore depends upon the charge distribution: the expression given by electrostatic theory for the energy due to a net charge Ze uniformly distributed over a sphere of radius R is

$$E = -a \frac{(Ze)^2}{R} \tag{2-28}$$

where the constant a takes the value $a = \frac{3}{5}$ for a homogeneous distribution of charge throughout the volume of the sphere (for a *surface* distribution of charge, the constant would have the value $a = \frac{1}{2}$). It has been found that to a good approximation the value $a = \frac{3}{5}$ accounts for the Coulomb energy of nuclei. Replacing R by $r_0 A^{1/3}$ and noting the presence of the minus sign to indicate the diminution in binding energy due to repulsive effects, we obtain the Coulomb energy term as

$$B_3 = -a_c \frac{Z^2}{A^{1/3}} \tag{2-29}$$

where

$$a_c = \frac{3}{5} \frac{e^2}{r_0} \tag{2-30}$$

Sometimes, and particularly in the case of light nuclei having comparatively small values of Z, the above is modified slightly to read

$$B_3 = -a_c \frac{Z(Z-1)}{A^{1/3}} \tag{2-31}$$

This modification is effected in order to rid B_3 of a self-energy contribution by each of the Z protons. It can be conceived as though each proton were spread throughout the entire volume of the nucleus and interacted with itself. For each of the Z protons, the self-energy is $\frac{3}{5}(e^2/r_0)$ and hence Z times this amount is deducted from B_3 to yield the modified expression (2-31). This "granularity" modification has, however, been shown [Pe 54] to be unjustified quantum-theoretically.

2.3.5. ASYMMETRY ENERGY (NEUTRON EXCESS)

A reasonable question in view of the existence of strong nuclear forces between neutrons is to ask why there are not stable nuclei composed exclusively of neutrons. There has, indeed, of late been a reawakening of interest in the question of neutron matter and the possibility of existence of bound multineutron states, such as ^{2}n, ^{3}n, ^{4}n, and ^{6}n. Thus, Bösch *et al.* [Bö 63a] disprove an experiment [Sa 61] which claimed to have detected the dineutron ^{2}n. (Further evidence pointing to the nonexistence of the dineutron and a list of references concerning relevant experiments has been presented by Willard *et al.* [Wi 64].) Ajdačić *et al.* [Aj 65] tentatively offer evidence in support of the possible existence of the trineutron ^{3}n, as does also [Mi 66], although a later search [Th 66, De 68] yielded negative results. Tang and Bayman [Ta 65] demonstrate the nonexistence of the tetraneutron ^{4}n (for which there had been vain experimental searches, e.g. [Ci 65, Co 65b]).

Leaving such rather speculative considerations aside, however, and concentrating attention upon stable atomic nuclei, we see from a plot of N against Z as in Fig. 2-5 that up to about $Z = 20$ ($A \simeq 40$), nuclei tend to be stable if $N = Z$, whereas for heavier nuclei stability ensues only if there is a neutron excess—indeed, the heaviest nuclei have a ratio $N/Z \simeq 1.5$. With increasing Z, the magnitude of the repulsive Coulomb force rises so rapidly that heavy nuclei with equal numbers of protons and neutrons could not remain stably bound. A NEUTRON EXCESS ($N - Z$) must be present in order to provide a sufficiently powerful nuclear attractive force to counter the longer-range electrostatic disruptive force.

For example, if in the mass equation no further "correction" terms beyond B_3 were inserted, so that the mass formula would read

$$m(Z, A) = Zm_p + Nm_n - a_v A + a_s A^{2/3} + a_c \frac{Z^2}{A^{1/3}} \qquad (2\text{-}32)$$

we would, on setting the differential to zero, i.e.,

$$\left.\frac{\partial m}{\partial Z}\right|_A = 0$$

obtain the condition for maximum stability as

$$Z = 0.66\, A^{1/3} \qquad (2\text{-}33)$$

This is in marked conflict with experience, since it would for example predict that the mass number of the stable nucleus with $Z = 20$ is nearly $A = 28,000$. Obviously, at least one more correction term is needed.

Whereas the previously considered effects influencing B are of a classical nature, the origin of the decrease in the binding energy due to the N, Z

asymmetry lies in a purely quantum-mechanical effect, and its evaluation is somewhat complicated [von W 35, Fe 50]. Merely to get some idea of the form that the correction term takes, we note that since for a liquid of given composition the energy is proportional to the amount of liquid, we would by analogy expect the term to be proportional to A. We also want it to be dependent upon N and Z in such a way that it is minimal at $N = Z$, and that

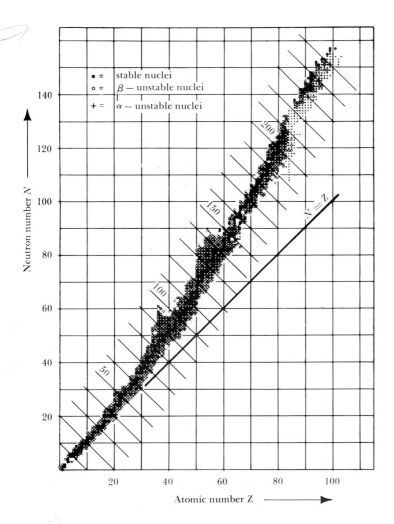

Fig. 2-5. Representation of N vs Z for stable and radioactive nuclides, showing the onset of a neutron excess $(N > Z)$ beyond the very light nuclides and indicating the trend of the "valley of β stability."

furthermore it would apply in just the same way if it were to describe a proton excess, i.e., it should be symmetric in N and Z. These requirements point to a term containing a numerator of the form $(N - Z)^2$ which, together with its first derivative with respect to Z, vanishes when $N = Z$ and thus corresponds to a maximum value of the binding energy in the absence of Coulomb interaction. This is "normalized" by a denominator $(N + Z)^2$, so that

$$B_4 = -a_a A \frac{(N - Z)^2}{(N + Z)^2} = -a_a \frac{(N - Z)^2}{A} = -a_a \frac{(A - 2Z)^2}{A} \qquad (2\text{-}34)$$

with a_a an empirical constant and $N \equiv A - Z$.

Again, B_4 carries a minus sign because it expresses the weakening in the binding occasioned by an asymmetry in N and Z, since beyond a certain stage the neutrons cease to act as "binding agents" within the nucleus.

On incorporating B_4 within the mass formula and setting $\partial m/\partial Z|_A$ to zero, one obtains the aforementioned relation

$$Z_{\text{stable}} = \frac{A}{1.98 + 0.0155\, A^{2/3}} \qquad (2\text{-}35)$$

which connects Z with A for the most stable nuclei.

2.3.6. PAIRING ENERGY

Nuclei are found to display a systematic trend in that those having *even* numbers of protons and neutrons ("even-even," or e-e, nuclei) tend to be very stable, those with even-Z, odd-N ("e-o" nuclei) or odd-Z, even-N ("o-e" nuclei) somewhat less stable, and those with odd Z and N (o-o nuclei) in the main unstable. This is illustrated by the data in Table 2-2.

Table 2-2. SYSTEMATICS OF STABILITY TRENDS IN NUCLEI

A	Z	N	Type	Alternative designation	Number of stable + long-lived nuclides	Degree of stability	Usual number of stable isotopes per element
Even	Even	Even	e-e	Even mass, even N	$166 + 11 = 177$	Very pronounced	Several (2 and 3)
Odd	Even	Odd	e-o	Odd mass, odd N	$55 + 3 = 58$	Fair	1
Odd	Odd	Even	o-e	Odd mass, even N	$51 + 3 = 54$	Fair	1
Even	Odd	Odd	o-o	Even mass, odd N	$6 + 4 = 10$	Low	0
					$278 + 21 = 299$		

The first four stable o-o nuclei are $^2_1$H, $^6_3$Li, $^{10}_5$B, $^{14}_7$N (for a mnemonic, note $Z = 1, 3, 5, 7$); being so light, they lie outside the range of validity of the liquid drop model. There is a fifth stable o-o nucleus, $^{180}_{73}$Ta, which constitutes a special case.

To take account of this PAIRING EFFECT, an additional term is incorporated into the mass formula,

$$B_5 = \begin{cases} +\delta & \text{for e-e nuclei} \\ 0 & \text{for e-o and o-e nuclei} \\ -\delta & \text{for o-o nuclei} \end{cases} \tag{2-36}$$

in which either δ itself is to be established empirically or, on the basis of slightly more detailed analysis [Fe 50], is set to

$$\delta = a_p A^{-3/4} \tag{2-37}$$

and a_p is determined empirically. A rather more sophisticated approach to the derivation of the pairing energy through consideration of the proton-neutron residual interaction has been presented by Hebach and Kümmel [He 65].

2.3.7. SUMMARY

Incorporating all the terms B_i into the mass equation, we get the SEMI-EMPIRICAL MASS FORMULA

$$M(Z,A) = ZM_H + (A - Z)m_n - a_v A + a_s A^{2/3} + a_c \frac{Z^2}{A^{1/3}} + a_a \frac{(A - 2Z)^2}{A}$$

$$+ \begin{cases} +a_p A^{-3/4} & \text{for o-o nuclei} \\ 0 & \text{for e-o and o-e nuclei} \\ -a_p A^{-3/4} & \text{for e-e nuclei} \end{cases} \tag{2-38}$$

It should be noted that as it has become the convention to quote *atomic* rather than nuclear masses, e.g., $M(Z,A)$ rather than $m(Z,A)$, the formula has been rewritten to give the mass M of the neutral atom, and therefore in the first term the mass m_p has been replaced by M_H, the mass of the neutral ^{1}H atom,

$$M_H = 1.007\ 825\ \text{u} \triangleq 938.767\ \text{MeV} \tag{2-39}$$

in order to take into account the mass of the orbital electrons (neglecting the electronic binding energies). This change in no way affects the use of the mass formula for calculation of nuclear transformation energies when the number of electrons cancels out on both sides of the reaction equation.

Different approaches to the evaluation of the weighting parameters a_v, a_s, a_c, a_a, and a_p furnishing the best fit to observed masses when inserted in the mass formula expressed in the above form have yielded different sets of results (some of these are compared on p. 383 of [Ev 55] and p. 31 of [Wa 58a]), of which we quote the ensemble

$$\left.\begin{array}{l} a_v = 1.51 \times 10^{-2} \text{ u} \triangleq 14.1 \quad \text{MeV} \\ a_s = 1.40 \times 10^{-2} \text{ u} \triangleq 13 \quad \text{MeV} \\ a_c = 6.39 \times 10^{-4} \text{ u} \triangleq 0.595 \text{ MeV} \\ a_a = 2.04 \times 10^{-2} \text{ u} \triangleq 19 \quad \text{MeV} \\ a_p = 3.60 \times 10^{-2} \text{ u} \triangleq 33.5 \quad \text{MeV} \end{array}\right\} \qquad (2\text{-}40)$$

as a good representative set. These represent contributions to the total binding energy per nucleon whose relative magnitudes are depicted in Fig. 2-6. There have, however, been elaborations upon the derivation of a mass formula and the choice of correction terms, which we discuss briefly in the next section.

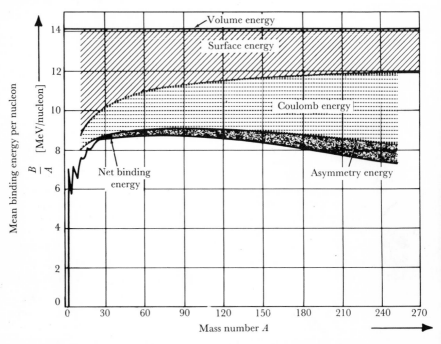

Fig. 2-6. Relative contributions to the net mean binding energy per nucleon in function of the mass number A, showing the relative importance of various terms in the semiempirical Weizsäcker expression for binding energy (after [Ev 55]). (Used by permission of McGraw-Hill Book Co.)

2.3.8. OTHER MASS FORMULAE

Except at the extreme ends of the mass spectrum, the Bethe-Weizsäcker formula generally furnishes values of the binding energy which are within 1 percent of the values determined experimentally. It thus gives atomic masses which are correct to roughly 1 part in 10^4. Conspicuous discrepancies occur, however, in the neighborhood of so-called magic nuclei having Z or N

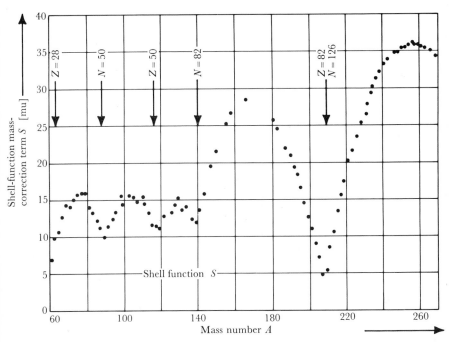

Fig. 2-7. Variation of the mass-correlation function S as used by Mozer [Mo 59b] to take shell-model effects into account in a semiempirical mass formula. The data refer to stable odd-A nuclides and are based upon a square-well potential. The dips occur at "magic" numbers corresponding to shell closure, viz. $Z = 28$, $N = 50$, $Z = 50$, $N = 82$, ($Z = 82$ and $N = 126$), respectively.

equal to 28, 50, 82, or 126. Here we encounter a strength of binding which transcends that of ordinary nuclei, for reasons which will later be discussed as part of the characteristics of the nuclear shell model. Attempts to take account of this effect empirically have led to generalizations of the mass formula, such as those introduced by Levy [Le 55], Cameron [Ca 57], Mozer [Mo 59b], Seeger [Se 59a, Se 61a], Ayres *et al.* [Ay 62], and Kümmel *et al.* [Kü 66]. These lead beyond the detailed survey of atomic masses and semiempirical formulae undertaken by Wapstra [Wa 58a] (in this connection, mention may also be made of a review of the masses of atoms with $A > 40$ by Duckworth [Du 57]).

To pick out but one of these treatments, we select that of Mozer, who having undertaken a least-squares analysis of atomic masses for odd-A nuclei (for which $\delta = a_{\mathrm{p}} = 0$) to obtain optimum values of the parameters in the Bethe-Weizsäcker formula, found the values to vary so markedly with the

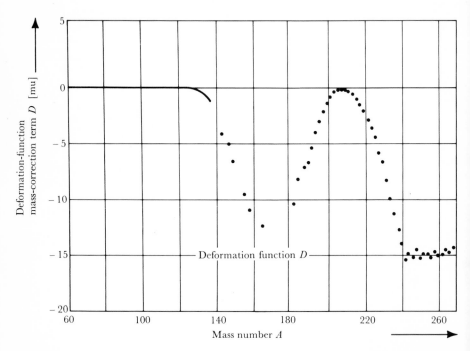

Fig. 2-8. Variation of the mass-correction function D introduced by Mozer [Mo 59b] to take cognisance of the influence of nuclear deformation for odd-A nuclides upon a semiempirical mass formula.

optimization procedure, and the deviations from measured masses to remain quite large even under the best conditions, in consequence proposed a generalized formula in which a shell-effect energy term and a nuclear deformation energy were appended to the existing binding energy terms. He expressed his mass equation in the form

$$M(Z,A) = ZM_{\mathrm{H}} + (A - Z)m_{\mathrm{n}} - a_{\mathrm{v}}A + a_{\mathrm{s}}A^{2/3} + a_{\mathrm{c}}\frac{Z(Z-1)}{A^{1/3}}$$

$$+ a_{\mathrm{a}}\frac{(A - 2Z)^2}{A} + S(Z,A) + D(Z,A) \qquad (2\text{-}41)$$

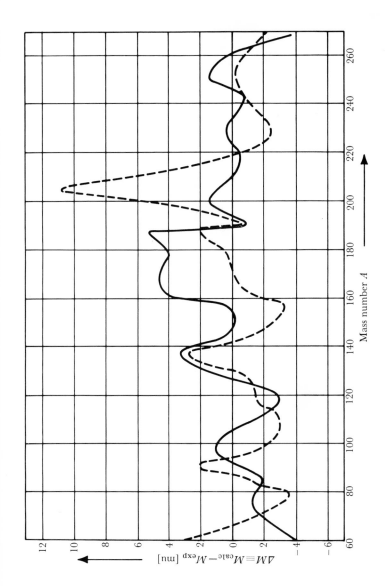

Fig. 2-9. Comparison of the deviation between calculated and measured atomic masses for stable odd-*A* nuclides. The broken curve refers to masses as calculated from a semiempirical formula of the Weizsäcker type with constants adjusted to give a least-squares fit, while the solid curve refers to masses calculated from Mozer's formula [Mo 59b].

with the shell and deformation corrections S and D determined semiempirically to be smooth functions of A, as shown in Figs. 2-7 and 2-8. The modified form $[Z(Z-1)]$ was chosen for the numerator of the Coulomb term. With S and D established, Mozer found new parameter weightings by least-squares analysis, their values being

$$\left.\begin{array}{l} a_v = 1.6412 \times 10^{-2}\ u \,\triangleq\, 15.29\quad \text{MeV} \\ a_s = 1.6922 \times 10^{-2}\ u \,\triangleq\, 15.76\quad \text{MeV} \\ a_c = 7.4120 \times 10^{-4}\ u \,\triangleq\, 0.6904\ \text{MeV} \\ a_a = 2.4292 \times 10^{-2}\ u \,\triangleq\, 22.628\ \text{MeV} \end{array}\right\} \qquad (2\text{-}42)$$

The overall fit to experimental atomic masses was thereby greatly improved, as is evident from Fig. 2-9.

Except for [Kü 66], no such analysis has so far been performed using the more recent values of atomic masses [Ma 65d], but one may presume that a still better degree of fit could be attained, since these latter mass values have been adjusted to be mutually consistent and carry appreciably smaller error limits than those employed in the analyses mentioned above.

By rearrangement of the expressions furnished by another model of the nucleus, namely the superfluid model, Bauer and Canuto [Ba 65c] have shown that the appropriate binding energy can be rewritten in a form which exactly reproduces the Bethe-Weizsäcker semiempirical formula, and therefore lends considerable support to this representation.

2.4. Applications of the
Mass Formula to Considerations of Stability

The mass formula not only lends itself to the evaluation of atomic masses and nuclear binding energies *per se*, but furnishes theoretical predictions concerning a number of features of nuclei and their behavior. By discussing some of these predictions, we illustrate the use of a nuclear model—here the simple liquid drop model—to elucidate nuclear phenomena.

2.4.1. Coulomb Radius Constant

One of the most directly derivable items of information provided by the mass formula is the magnitude of the nuclear radius parameter r_0 contained within the expression for the Coulomb energy constant $a_c = \frac{3}{5}(e^2/r_0)$, assuming a uniform volume distribution of charge. By establishing the value of a_c, together with its probable error limits, on the basis of a least-squares analysis of atomic masses, the value of r_0 and its probable error can be determined uniquely. A detailed investigation of this method of finding the value of r_0 has been carried out by Green [Gr 54], who used a special method to derive the optimal value of a_c and thence found

$$a_c = 7.6278 \times 10^{-4} \text{ u} \triangleq 0.710\,51 \text{ MeV} \qquad (2\text{-}43)$$

whence

$$r_0 = 1.216 \text{ fm}$$

The more recent value obtained by Mozer [Mo 59b],

$$a_c = 7.4120 \times 10^{-4} \text{ u} \triangleq 0.6904 \text{ MeV} \qquad (2\text{-}44)$$

yields

$$r_0 = 1.251 \text{ fm} \qquad (2\text{-}45)$$

and this is in excellent agreement with values derived by other methods.

2.4.2. RADIUS CONSTANT FROM BINDING ENERGY FOR MIRROR NUCLEI

Another means of determining r_0 with the aid of the mass formula involves the consideration of mirror nuclei, i.e., nuclides having the same A but mutually interchanged Z and N. The respective atomic masses are therefore $M(Z,A)$ and $M(A - Z, A)$. On building the mass difference, we find that many terms cancel out, leaving simply

$$\Delta M \equiv M(Z,A) - M(A - Z, A) = [A - 2Z][(m_n - M_H) - a_c A^{2/3}] \qquad (2\text{-}46)$$

where $(A - 2Z)$ is the neutron excess of nuclide (Z,A). The difference in respective binding energies is still simpler,

$$\Delta B \equiv B(Z,A) - B(A - Z, A) = (A - 2Z)(m_n - M_H) - \Delta M = a_c A^{2/3}(A - 2Z) \qquad (2\text{-}47)$$

Thus ΔB for mirror nuclei furnishes a value for a_c, and from this r_0 can again be determined. Table 2-3 collates some results for mirror nuclei with $A - 2Z = 1$ which are not too light to permit the application of the liquid drop model.

Table 2-3. COULOMB RADIUS CONSTANT, AS DETERMINED FROM THE BINDING ENERGY DIFFERENCES FOR MIRROR NUCLEI WITH $A - 2Z = 1$

Nuclide (Z,A)	Nuclide $(A - Z, A)$	$B(Z,A)$ MeV	$B(A - Z, A)$ MeV	ΔB MeV	$a_c = A^{-2/3} \Delta B$ MeV	r_0 fm
$^{37}_{18}\text{Ar}$	$^{37}_{19}\text{K}$	315.510	308.587	6.923	0.6235	1.39
$^{31}_{15}\text{P}$	$^{31}_{16}\text{S}$	262.916	256.688	6.228	0.6311	1.37
$^{23}_{11}\text{Na}$	$^{23}_{12}\text{Mg}$	186.565	181.726	4.839	0.5983	1.44
$^{15}_{7}\text{N}$	$^{15}_{8}\text{O}$	115.494	111.952	3.542	0.5824	1.48

With decrease in A, the nuclei become ever less suited to application of the liquid drop model, but even though the value of the radius constant r_0 tends to diminish to a reasonable magnitude for the heavier nuclides in Table 2-3 it nevertheless remains rather high in absolute terms. It should not be thought, however, that mirror-nucleus radii are for some reason conspicuously larger than those for other nuclei; indeed, as Cooper and Henley [Co 53] have shown, the radius constant takes the value $r_0 = 1.2$ fm when a more refined calculation is undertaken, in which provision is made for a Coulomb exchange energy, as well as for angular momentum effects. An extensive critical survey of mirror nuclei [Ko 58a] has yielded the value $r_0 = 1.28$ fm.

In the selection of pairs of mirror nuclei for analysis, only those should be used for which the atomic masses and binding energies have not been derived by adjustment from values for neighboring nuclides. This excludes some of the heavier mirror nuclei such as ^{33}Cl, ^{35}Ar, ^{39}Ca, ^{41}Sc, and ^{43}Sc, rendering this method of determining r_0 less reliable than that described previously, because the set of data is much more restricted.

2.4.3. Uranium Fission

When a ^{235}U nucleus captures a neutron, an unstable ^{236}U nucleus is formed, which decays by a fission process into a pair of nuclides X,Y whose masses we assume for the present consideration to be approximately equal:

$$^{235}\text{U} + \text{n} \rightarrow {}^{236}\text{U}^* \rightarrow \text{X} + \text{Y} + x\text{n} \tag{2-48}$$

As we have already seen in Section 2.1, this process liberates energy amounting to about 1 MeV per nucleon, or 200 MeV per uranium nucleus.

For the special case of SYMMETRICAL FISSION which we consider here, the mass formula, neglecting the small pairing-energy contribution δ, gives

$$Q = M(Z,A) - 2M(\tfrac{1}{2}Z,\tfrac{1}{2}A) = a_s A^{2/3}(1 - 2^{1/3}) + a_c Z^2 A^{-1/3}(1 - 2^{-2/3}) \tag{2-49}$$

i.e.,

$$Q = -0.2599 a_s A^{2/3} + 0.3700 a_c Z^2 A^{-1/3} \tag{2-50}$$

Hence for the nucleus $^{235}_{92}$U, substituting $a_s = 13$ MeV and $a_c = 0.595$ MeV, we obtain

$$Q \simeq -129 + 302 = 173 \text{ MeV} \tag{2-51}$$

Even though this value for the energy liberated per nucleus undergoing fission is, just as the value of 200 MeV above, merely an estimate, the discrepancy is too large to be ignored. The explanation is not hard to provide. The gross overall energy liberated by fission and subsequent processes indeed amounts to around 200 MeV, but the values of B/A read off for the initial and final nuclei were those for stable nuclei. (In Section 2.1 we considered the case of ^{238}U fission, but the numerical estimate applies just as well to the more pertinent case of ^{235}U symmetric fission treated here.) On the other hand, the energy of 173 MeV is set free during just the one initial fission step. The fission products are, however, extremely unstable because they have a large neutron excess (the neutron emission is indicated by the term xn in the reaction equation) and consequently undergo a series of decay stages involving n, β^-, and γ emission before finishing up as stable nuclei. It is these subsequent stages which are associated with the evolution of the remaining 30 to 40 MeV of energy.

The two terms which make up the value of Q originate from the Coulomb energy which is set free and the surface energy which has to be supplied. The energy balance in fission is accordingly a tug-of-war between electrostatic and surface tension forces from the viewpoint of the liquid drop model.

From a close examination of the behavior of B/A depicted in Fig. 2-3, one can infer that the fission process becomes exothermic for nuclei which are much lighter than thorium or uranium; in fact, it is already exothermic below about $A \simeq 150$. Why, then, does the process not set in spontaneously for nuclei which are much lighter than thorium or uranium? As we shall see in Section 7.6.3, the reason is to be found in the existence of nuclear potential barriers whose inhibiting effect has to be overcome before fission can occur. Only for the heaviest nuclides is there sufficient energy available to render the potential barrier ineffectual so far as fission is concerned.

2.4.4. The Behavior of Isobars in β Decay

We have already considered a special case of nuclei with $A = $ constant, namely pairs of mirror nuclei. A more general situation, however, is represented by the parent and daughter nuclides in β decay, for these also are isobaric pairs.

Most of the unstable nuclides depicted in Fig. 2-5 are natural β emitters, typical examples being ^{40}K, ^{87}Rb, ^{115}In, etc. The stability of β-emitting isobars can be predicted from the mass formula and from an examination of the above plot of N vs Z, for when the latter is viewed diagonally—that is to say, along a series of normals to the line $N = Z$—each set of isobars is located on a straight line. Closer scrutiny reveals that whereas each set of *odd-A isobars* normally includes just *one stable nuclide*, the sets of even-A isobars often include two, and sometimes even three, stable members. This characteristic property becomes even more evident if one conceives of a third coordinate axis as perpendicular to the (N,Z)-plane, to represent the values of atomic mass $M(Z,A)$. The ensuing three-dimensional figure has been termed a NUCLEAR ENERGY SURFACE, for the stable region lies along the trough to such a surface. The stable members lie in the valley, and the unstable along the sides and rim of the slopes, not unlike a class of beginners in skiing. The

line joining the most stable nuclides traces the bottom of the valley, as characterized by Z_{stable}, and is called the LINE OF β-STABILITY. The perpendicular
sections along sets of isobars through this valley give the roughly parabolic
profile of $M(Z,A)$ vs Z, with the unstable isobars straddling the most stable
member or members, which lie at the base of the valley. There is, moreover,
an important distinction between the mass-energy profiles for odd-A isobars
and those for even-A isobars inasmuch as the former have $\delta = 0$ and are
therefore represented by just a single parabola, as in Fig. 2-10, whereas the

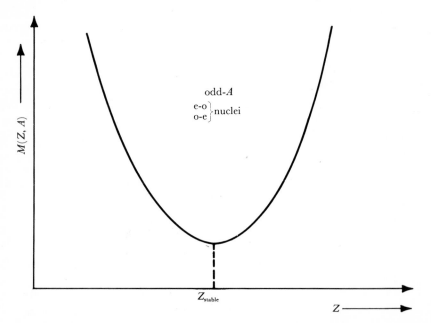

· **Fig. 2-10.** Mass parabola for odd-A nuclei. The trough at Z_{stable} represents the
most stable isobar. (Used by permission of McGraw-Hill Book Co.)

latter have a pairing energy of $+\delta$ for e-e nuclei and $-\delta$ for o-o nuclei, so that
the masses of e-e isobars fall on a separate lower parabola than that for the
o-o isobars (Fig. 2-11).

To examine this briefly with the aid of the mass formula, it is convenient
to introduce the abbreviated notation

$$M(Z,A) = k_1 A + k_2 Z + k_3 Z^2 \pm \delta \qquad (2\text{-}52)$$

with

$$k_1 = m_{\text{n}} - (a_{\text{v}} - a_{\text{a}} - a_{\text{s}} A^{-1/3}) \qquad (2\text{-}53)$$

$$k_2 = -[4a_{\text{a}} + (m_{\text{n}} - M_{\text{H}})] \qquad (2\text{-}54)$$

$$k_3 = \frac{4a_a}{A}\left(1 + \frac{A^{2/3}}{4a_a/a_c}\right) \qquad (2\text{-}55)$$

Then in all the above cases the base of the parabola lies at the point Z_{stable} given by the condition

$$\left.\frac{\partial M}{\partial Z}\right|_A = 0 = k_2 + 2k_3 Z_{\text{stable}} \qquad (2\text{-}56)$$

namely

$$Z_{\text{stable}} = -\frac{k_2}{2k_3} \qquad (2\text{-}57)$$

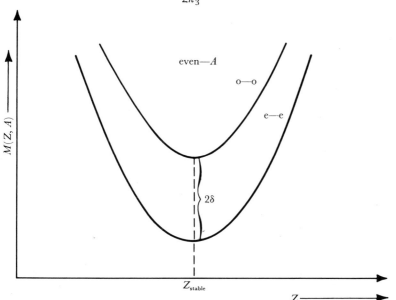

Fig. 2-11. Mass parabolas for even-A nuclei, having a separation 2δ, where δ is the pairing energy, at the valley of maximum stability. (Used by permission of McGraw-Hill Book Co.)

which is identical with the numerical result (1-29), (2-35) quoted in Sections 1.4 and 2.3.5. The isobar or isobars with (integer) Z lying closest to the noninteger Z_{stable} are then the most stable. Every isobar whose mass exceeds that of the most stable member of the isobaric multiplet is converted by β emission to the adjacent, next lower member of the isobaric set. In β^- decay Z increases by 1 unit, and in β^+ decay it decreases by 1 unit. Hence in the case of *odd-A isobars*, only β^- decay can take place for the nuclides which lie along the left arm of the parabola, and β^+ decay for those lying along the right arm, the decay chain in each case finishing up with just *one* stable isobar, as shown in Fig. 2-12.

In the case of *even-A isobars*, β decay changes an e-e nucleus into an o-o nuclide and *vice versa*, so that the decay steps zigzag between the two parabolas. The decay chain invariably terminates on the lower of these parabolas, since this represents a state of greater stability (lower mass); the only exceptions to the rule that there are no stable o-o nuclei are the very light nuclides ^{2}H, ^{6}Li, ^{10}B, ^{14}N for which the liquid drop model is inapplicable, and the special

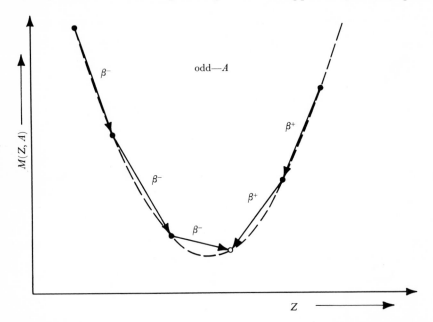

Fig. 2-12. Mass parabola for odd-A nuclei, showing how β^--decay steps on the left branch and β^+-decay steps on the right branch both terminate in a unique stable isobar (designated by an open circle). Electron-capture transitions may also appear on the right-hand branch of the parabola.

case of ^{180}Ta. However, there is nothing to prevent the occurrence of more than one stable e-e isobar for a given mass number A, since neighboring isobars on the e-e parabola are separated by an interval $\Delta Z = 2$ and cannot normally transmute into each other (i.e., not by simple single-stage β decay). Thus it can happen that a single o-o parent can have two (or more) stable e-e daughters, the one formed by β^+ decay and the other by β^- decay. A frequently cited instance is that of $^{102}_{45}$Rh, which has a 54 percent probability of decaying into $^{102}_{46}$Pd by β^- emission and a 46 percent probability of forming $^{102}_{44}$Ru by β^+ emission, with a mean lifetime of 303 d. Figure 2-13 illustrates another such case, namely the decay of the parent ^{108}Ag into stable ^{108}Cd by β^- emission and into stable ^{108}Pd by β^+ emission.

The mass formula in abbreviated notation can be used to trace out the form of the mass-energy parabolas and to evaluate the β-decay energy. Substituting $Z_{\text{stable}} = -k_2/2k_3$ in the mass formula for an odd-A nucleus, we

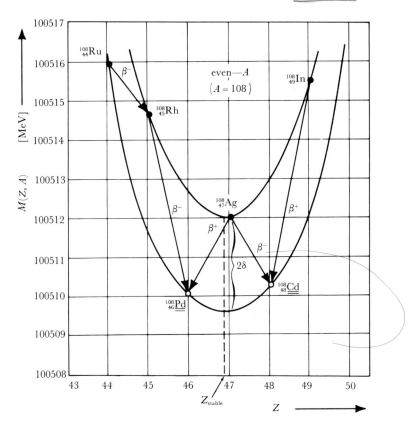

Fig. 2-13. Mass parabolas for the even-A isobars $A = 108$, demonstrating the existence of *two* stable isobars (open circles). The separation between the parabolas is $2\delta = 2.35$ MeV, which may be compared with the semiempirical prediction $2\delta_{\text{th}} = 2a_pA^{-3/4} = 2$ MeV. The value of Z_{stable}, deduced from β systematics [Ev 55, p. 379], is $Z_{\text{stable}} = 46.8$, as drawn in, while with Mozer's parameters [Mo 59b] one obtains $Z_{\text{stable}} = 46.4$.

get the corresponding hypothetical mass

$$M(Z_{\text{stable}}, A) = k_1 A - (2k_3 Z_{\text{stable}}) Z_{\text{stable}} + k_3 Z_{\text{stable}}^2 \qquad (2\text{-}58)$$

$$= k_1 A - k_3 Z_{\text{stable}}^2 \qquad (2\text{-}59)$$

This represents the lowest point of the parabola. The distribution of masses about this is given by

$$M = M(Z,A) - M(Z_{\text{stable}},A) = k_1 A + k_2 Z + k_3 Z^2 - k_1 A + k_3 Z^2_{\text{stable}}$$
(2-60)

$$= k_3 (Z - Z_{\text{stable}})^2 \qquad (2\text{-}61)$$

for odd A

which exhibits the parabolic form. Analogously, for even-A, odd-Z nuclei, we have

$$M(Z_{\text{stable}},A) = k_1 A - k_3 Z^2_{\text{stable}} + \delta \qquad (2\text{-}62)$$

and for even-A, even-Z isobars,

$$M(Z_{\text{stable}},A) = k_1 A - k_3 Z^2_{\text{stable}} - \delta \qquad (2\text{-}63)$$

showing that the vertical separation between the two parabolas is 2δ.

The β-decay energy can be calculated immediately, noting that for β^- decay $Z \to Z + 1$, so that

$$Q_{\beta^-} = M(Z,A) - M(Z+1, A) = 2k_3(Z_{\text{stable}} - Z - \tfrac{1}{2}) \qquad (2\text{-}64)$$

and for β^+ decay $Z \to Z - 1$, whence

$$Q_{\beta^+} = M(Z,A) - M(Z-1, A) = -2k_3(Z_{\text{stable}} - Z + \tfrac{1}{2}) \qquad (2\text{-}65)$$

Both cases can be collected within the formula

$$Q_{\beta^\pm} = 2k_3[\pm(Z - Z_{\text{stable}}) - \tfrac{1}{2}] \qquad (2\text{-}66)$$

For even-A nuclei the hypothetical "most stable isobar" is that lying on the (lower) e-e parabola. When referred to this point, the mass differences of the individual isobars are

$$\delta M \equiv M(Z,A) - M(Z_{\text{stable}}, \quad \text{even } A)$$

$$= \begin{cases} k_3(Z - Z_{\text{stable}})^2 & \text{for e-e nuclei} \\ k_3(Z - Z_{\text{stable}})^2 + 2\delta & \text{for o-o nuclei} \end{cases} \qquad (2\text{-}67)$$

In every case, the decay energies increase as the separation from the "most stable isobar" grows. Since the lifetimes diminish as the decay energy increases, those isobars lying high at the extreme ends of the arms of the parabolas have characteristically short lifetimes, and consequently the number of observable isobars is restricted.

The fission of heavy nuclei gives rise to products having a large neutron excess. These lie high on the left arm of the appropriate parabola, and accordingly lead to a long β^--decay chain involving a markedly large number of isobars. An example of such a decay chain is the sequence

$$^{139}\text{I} \xrightarrow[2.7 \text{ sec}]{\beta^-} {}^{139}\text{Xe} \xrightarrow[41 \text{ sec}]{\beta^-} {}^{139}\text{Cs} \xrightarrow[9.5 \text{ min}]{\beta^-} {}^{139}\text{Ba} \xrightarrow[85 \text{ min}]{\beta^-} {}^{139}\text{La (stable)}$$

EXERCISES

2-1 Does an accelerator which can produce a 20-MeV beam of ^{12}C ions in the laboratory system suffice for an investigation of the reaction

$$^{12}\text{C} + {}^{16}\text{O} \rightarrow {}^{17}\text{F} + {}^{11}\text{B} - 15.368 \text{ MeV}?$$

2-2 At what kinetic energy E_{kin} does the de Broglie wavelength of an electron become equal to (a) its Compton wavelength, (b) the classical electron radius?

2-3 If a spherical drop of water having a radius $R = 0.15$ cm were to have the density of nuclear matter, what would its mass be?

2-4 Show that for a homogeneous distribution of electric charge (Ze) throughout the volume of a sphere of radius R the constant a in Eq. (2-28) takes the value $\frac{3}{5}$, as against the value $a = \frac{1}{2}$ applicable to a uniform *surface* distribution. What is the percentage difference between the corresponding Coulomb energies?

2-5 What value is obtained for the radius of an electron if its Coulomb energy E_C is set equal to its rest energy $E_0 = m_e c^2$, on the assumption of a homogeneous (a) volume, (b) surface distribution of charge.

2-6 The semiempirical mass formula with a Coulomb term applicable to homogeneous *volume* distribution of charge yields values of the mass which in the region of heavy nuclei differ from measured values by at most 0.01 percent. Using ^{238}U as an example (atomic mass $M = 238.12$ u), show that this error limit absolutely rules out the extreme alternative of a uniform *surface* distribution of charge, even if the mass discrepancy were due exclusively to the Coulomb term.

One could nevertheless conceive of other forms of charge distribution; for instance, a homogeneous distribution throughout the volume of a hollow spherical shell having an inner radius R_1 and an outer radius R_2 (equal to the nuclear radius). What is the largest value of the ratio R_1/R_2 commensurate with the above mass discrepancy limit of 0.01 percent?

2-7 With the aid of the parameters (2-40) confirm the numerical result (2-35) deduced from the semiempirical mass formula along the lines suggested in Section 2.3.5.

2-8 Use Eq. (2-57) to establish whether the radioactive nuclide $^{142}_{54}$Xe is β^- unstable or β^+ unstable.

2-9 The transition energy for the β^+ transition between the mirror nuclei ^{13}N and ^{13}C is $E_0 = 1.19$ MeV. What value does this indicate for the radius parameter r_0?

2-10   From the β^--transition energies of the two $A = 89$ isobars,

$$^{89}_{37}\text{Rb}_{52} \ (E_\beta = 3.92 \text{ MeV})$$
$$^{89}_{38}\text{Sc}_{51} \ (E_\beta = 1.46 \text{ MeV})$$

use (2-66) to calculate the energy of the β^+ transition

$$^{89}_{41}\text{Nb}_{48} \xrightarrow{\ \beta^+\ } {}^{89}_{40}\text{Zr}_{49}$$

and compare this with the experimental value

$$E_{\beta^+_{\text{max}}} = 2.86 \text{ MeV}$$

2-11   Show with the aid of the semiempirical mass formula that only for heavy nuclei can nuclear fission be an exothermic process. Around which region of A and Z does the Q-value vanish?

(One way of solving this problem is to set

$$\left. \begin{array}{l} A = A_1 + A_2 = xA + (1 - x)A \\[2mm] Z = Z_1 + Z_2 = xZ + (1 - x)Z \end{array} \right\} \qquad (0 \leqslant x \leqslant 1)$$

where (A_1, Z_1) and (A_2, Z_2) specify the two product nuclei from fission; symmetric fission corresponds to $x = \frac{1}{2}$.)

INTERACTIONS AND NUCLEAR CROSS SECTIONS

Interactions, Transition Probability, and Reaction Cross Section

3.1. Nuclear Force Characteristics

The preceding sections have been devoted to the presentation of ideas upon which a simple and readily visualizable model has been based. The model itself, because of its uncomplicated nature, provides a clear example of the way one often proceeds in nuclear physics studies: on combining classical with quantum-mechanical considerations, one derives theoretical expressions which are then so weighted empirically that good overall agreement with experimental observations ensues. The fact that in its field, which we concede to be restricted, such a model is able to provide good qualitative and quantitative agreement with experimental results—as regards both *static* properties, such as those concerning ground-state nuclear masses and relative energies which are not subject to external influences and, on the other hand, *dynamic* features, such as those involved in β decay and fission—provides strong support for the validity of its underlying concepts and assumptions. In particular, it suggests that the foregoing picture of nuclear forces has been constructed along reasonable lines. It therefore provides us with a starting point for the consideration of the interactions which play an essential role in nuclear binding or disintegration.

Unlike other types of forces such as gravitational or electrostatic attractions which, because of their *long-range* character as evinced by their inverse-square dependence upon distance, preclude any unique definition of interaction radius, the basic *internucleon forces* show evidence of being *short-ranged* and having a strictly restricted interaction distance. This follows from the constancy of the mean binding energy per nucleon over virtually the entire range of nuclides, for this suggests a "nearest neighbor" type of interparticle force such as one encounters in the case of exchange forces which involve the interchange of a relatively massive particle. In putting forward this hypothesis to account for the "saturation property" of nuclear forces, Heisenberg [He 32] in his original paper ascribed this to interchange of an electron-neutrino pair, but this was subsequently seen to be insufficiently strong as an interaction to account for the *very intense* forces between nucleons and was superseded by Yukawa's conjecture [Yu 35a] of (π) *meson exchange*. To take account of the short-range character of the nucleon-nucleon interaction Yukawa modified the mathematical expression for the potential between two nuclear particles from the classical form, as typified by the Coulomb potential $-e^2/r$, to

$$\frac{g^2}{r} \exp\left(-\frac{r}{r_0}\right) \tag{3-1}$$

where g represents the "charge of the exchange field" and r the interparticle separation. By assigning a definite range $r_0 \simeq 1.4$ fm to the interaction, the calculation fixes the mass of the exchange particle at $\simeq 270 m_e$ and accounts for the strong internucleon binding energy of about 8 MeV. This strength can be contrasted with that of classical forces operating between two particles which have a separation of the order of 2 fm (assuming Coulomb's law and Newton's law to hold at such small distances): the *electrostatic* energy between a proton and a neutron is identically zero and that between two protons is -0.72 MeV; the relative *magnetic* potential energy (due to magnetic moments) amounts to about 0.03 MeV and the *gravitational* energy to roughly 6×10^{-37} MeV. The influence of classical forces therefore represents essentially a perturbation on the specifically nuclear forces, and can in most instances be neglected in a quantitative treatment of nuclear interactions—though in certain circumstances it must be kept in mind that the Coulomb interaction may have to be taken into consideration.

We note that when the exponential factor in Yukawa's potential is set equal to unity, the expression reduces to the same kind of potential as that in atoms, but with g replacing e. It is accordingly interesting to examine the numerical values of some nuclear analogs to atomic quantities. For example, the nuclear analog to the Bohr radius (Section 2.2.2) might be written as

$$(a_0)_N = \frac{\hbar^2}{m_n g^2} \tag{3-2}$$

If we anticipate the result $g^2/\hbar c = 0.08$, to be given in the next section, and set g^2 to 2.5×10^{-18} erg-cm, this yields a numerical value

$$(a_0)_N \simeq 2.6 \times 10^{-13} \text{ cm} = 2.6 \text{ fm} \tag{3-3}$$

which is in rather good agreement with the order of size of light nuclei and the range of the strong interaction. Along the same line of reasoning, we might expect a relationship between the mean binding energy of 8 MeV/nucleon and the nuclear analog to atomic binding energies, which are of the order

$$Ry \equiv Rhc \equiv \frac{2\pi^2 m_e e^4}{h^3 c} hc = \frac{m_e e^4}{2\hbar^2} \triangleq 13.61 \text{ eV} \tag{3-4}$$

namely

$$(Ry)_N \cong \frac{m_n g^4}{2\hbar^2} \cong 3 \text{ MeV} \tag{3-5}$$

3.2. Classification of Interactions

In classifying the various possible types of interactions, their relative strengths can best be expressed through values of the respective dimensionless COUPLING CONSTANTS which are used to describe the energy of a particle in the mathematical formalism for an appropriate field of force.

3.2.1. STRONG INTERACTIONS

Yukawa-type forces associated with a *π-exchange field charge g* have the characteristic coupling constant $g^2/\hbar c$. Its numerical value far exceeds that of other interactions: from the observed strength of the pion-nucleon interaction when described in the formalism of pseudovector meson theory, the value has been estimated to be 0.08. (In pseudoscalar theory it is around 15, which explains why sometimes a compromise value of 1 is quoted for the strong-interaction coupling constant.)

3.2.2. ELECTROMAGNETIC INTERACTIONS

Inverse-square Coulomb forces may be conceived as due to photon exchange. In terms of the elementary charge e, the dimensionless coupling constant is $\alpha \equiv e^2/\hbar c = 7.297\ 20 \times 10^{-3}$ which, because of its significance in atomic spectroscopy, has been termed the FINE-STRUCTURE CONSTANT.

3.2.3. WEAK INTERACTIONS

Many of the interactions in particle physics have been found to be very much weaker than the above. To cite one example, the μ meson (muon), first detected by Anderson and Neddermeyer [An 36] in the year following Yulawa's meson theory was subsequently found to have the wrong properties in regard to interaction strength to be the "Yukawa particle," even though its mass of 207 m_e agreed well with Yukawa's prediction. Whereas the Yukawa meson necessarily had to interact *strongly* with matter, the fact that muons penetrated through the entire atmosphere to the Earth's surface indicated that they interacted but weakly with the nuclei they encountered. Also, their comparatively long lifetime of 2.2 μsec is a reflection of their stability against decay and indicates that the decay process, too, is due to a weak interaction. The free muons decay as follows:

$$\left.\begin{array}{l} \mu^- \rightarrow e^- + \nu_\mu + \bar{\nu}_e \\ \mu^+ \rightarrow e^+ + \bar{\nu}_\mu + \nu_e \end{array}\right\} \tag{3-6}$$

into an electron e^- (or positron, e^+) together with a neutrino ν_μ (or *anti*-neutrino, $\bar{\nu}_\mu$: the bar by convention symbolizes an ANTIPARTICLE, or CON-JUGATE PARTICLE) characteristic of μ decay and an *anti*neutrino $\bar{\nu}_e$ (or neutrino, ν_e) associated with the β decay of the neutron:

$$n \rightarrow p^+ + e^- + \bar{\nu}_e \tag{3-7}$$

again a process involving the weak interaction. (The neutron's mean lifetime is 1.01×10^3 sec.)

From Fermi's theory of β decay in its up-to-date form, one obtains a numerical constant which is characteristic of the weak interaction and whose value is

$$g_F = 1.41 \times 10^{-49} \text{ erg cm}^3 \tag{3-8}$$

To construct a dimensionless constant from this, it is reasonable in analogy with the expressions for the other coupling constants to set g_F^2 in the numerator and to balance this by a denominator whose dimensions must accordingly be $[(\hbar c)^2 \times (\text{length})^4]$. A suitable quantity to take for the length might appear to be the range of the strong nuclear force, but this choice could be criticized in that it is not known whether the weak interaction is of the same form or range as the strong force. Instead, the quantity taken as fundamental length is the rationalized Compton wavelength of the pion, namely

$$(\lambda_C)_\pi \equiv \frac{\hbar}{m_\pi c} = 1.413 \text{ fm} \tag{3-9}$$

since this is of the same order of magnitude as the rationalized Compton wavelengths of other fundamental particles and, indeed, corresponds to the range of the nuclear force. Insertion of this value yields a dimensionless WEAK-INTERACTION COUPLING CONSTANT of magnitude

$$\frac{g_F^2}{(\hbar c)^2} \left(\frac{m_\pi c}{\hbar}\right)^4 \cong 5 \times 10^{-14} \tag{3-10}$$

It may be pointed out that we neither know the range of the weak interaction nor have experimentally identified the agent responsible for its occurrence. If we draw an analogy with the other types of interaction and conjecture the existence of an "exchange particle", we are led to the concept of an INTER-MEDIATE CHARGED VECTOR BOSON W [Gl 60, Le 60], whose lifetime would be less than 10^{-17} sec and whose rest mass would most probably exceed 1500 m_e. Taking further the analogy with the structure of the other dimensionless constants, we would infer a "field charge" q such that $q^2/\hbar c \simeq 5 \times 10^{-14}$ and thus $q \simeq 1.3 \times 10^{-15}$ erg$^{1/2}$ cm$^{1/2}$, but this does not lend itself to any simple connection with g_F or with models of a weakly interacting force. Even though an intensive effort has been devoted to the detection of a W particle through such reactions as

$$\nu_\mu + p \rightarrow W^+ + p + \mu^-$$
$$\downarrow$$
$$e^+ + \nu_e$$
$$(\text{or } \mu^+ + \nu_\mu) \qquad (3\text{-}11)$$

no convincing evidence attesting to its existence has yet come to light (see, e.g., [Be 64, Bu 65b]). Some of the mounting evidence for the possible existence of an INTERMEDIATE NEUTRAL VECTOR BOSON, W^0, has recently been discussed by Neito [Nei 68].

3.2.4. GRAVITATIONAL INTERACTIONS

The strength of the inverse-square attractive forces due to gravity† is expressed by the Newtonian gravitational constant $G = 6.670 \times 10^{-8}$ dyn cm^2 g^{-2}. In an analogous form to the above, we can construct a dimensionless coupling constant from this as $Gm^2/\hbar c$ and, taking m to be the mass of a nucleon, namely $m_\mathcal{N} = 1.67 \times 10^{-24}$ g, we can express this as

$$\frac{Gm_\mathcal{N}^2}{\hbar c} \approx 2 \times 10^{-39} \qquad (3\text{-}12)$$

a quantity so small that it can be neglected in most considerations of nuclear effects.

Nevertheless, the most sensitive of nuclear techniques have proved sufficiently delicate to respond even to gravitational interaction effects. In particular, the specific energy shift (or, what amounts to the same thing, frequency shift, since in quantum theory an energy E is related to a frequency ν through the expression $E = h\nu$) experienced by electromagnetic radiation passing through a gravitational field at the Earth's surface has been measured, and found to be in accord with the theoretical prediction. The numerical value can be derived most simply by proceeding from the corpuscular viewpoint and calculating the specific energy gain of a photon "falling" through a

† It is possible to construct a mathematical formalism to describe the quantization of the gravitational field and to express the propagation of such a field in terms of a conjectured particle called a GRAVITON, but this particle has so far eluded observation. Its properties can be predicted to some extent: It would be a spin-2 particle in the Einstein gravitational theory of general relativity (in all other gravitational theories, there would be a spin-0 field component, as has been pointed out by Sexl [Se 66]) and, assuming its rest mass to be non-zero, it would have a Compton wavelength (Section 4.2.2) which has been estimated [Sa 60a] to be about 3×10^{26} cm.

height difference H cm which corresponds to a difference in gravitational potential that, to the first order of approximation, is given by

$$\Delta\phi \simeq gH \qquad (3\text{-}13)$$

where $g = 981$ cm sec^{-2} is the acceleration due to gravity at the point on the Earth's surface where the experiment is being undertaken. By analogy with the energy change $e\,\Delta V$ experienced by a particle which carries a charge e when it traverses an electric potential difference ΔV, we can write the increase in energy of a photon whose energy is originally $E(= mc^2$ where m is the initial relativistic photon mass) when it passes through a gravitational potential $\Delta\phi$ as

$$\Delta E = m\,\Delta\phi \qquad (3\text{-}14)$$

and thence derive the specific energy change as

$$\frac{\Delta E}{E} = \frac{m\,\Delta\phi}{mc^2} = \frac{\Delta\phi}{c^2} \simeq \frac{gH}{c^2} \approx 10^{-18}\,H \qquad (3\text{-}15)$$

Experiments which were able to employ a height difference of $22\frac{1}{2}$ m [Po 60, Po 64, Po 65] (see also [Cr 60b, Cr 64]) accordingly had to be designed to detect and measure a specific energy shift of about 2 parts in 10^{15}. Expressed in another way, since they investigated the gravitational shift in energy of 14.4-keV γ radiation, they had to respond quantitatively to an energy change amounting to no more than about 3×10^{-11} eV. The experimental series of measurements provided very satisfactory confirmation of the above theoretical prediction, the mean result being expressed by the ratio

$$\frac{\Delta E_{\text{exp}}}{\Delta E_{\text{th}}} = \frac{\Delta\nu_{\text{exp}}}{\Delta\nu_{\text{th}}} = 0.9990 \pm 0.0076 \qquad (3\text{-}16)$$

In its turn, this finding can be used as evidence indicating that *anti*particles must be subject to the same *attractive* gravitational force as that felt by their particle conjugates, and thus that the gravitational mass for either particles species is *positive*, a point which has been under discussion for some time [Ma 58c, Mo 58b, Schi 58, Schi 59]. Since a "falling" photon experiences exactly the theoretically predicted energy shift, it follows that if we elaborate the physical situation in our mind from that depicted schematically in Fig. 3-1 (a), and conceive of the photon changing in the course of its progress from emitter to detector into a particle-antiparticle pair which subsequently reunites annihilatively to re-form a photon (the overall process can involve one or more such stages *ad libitum*) according to the scheme indicated in Fig. 3-1(b), then the net mass-energy gain in this modified situation must be just the same as that in the simpler situation where the photon retained its identity unchanged throughout the entire journey. But we know that a particle falling in a gravitational field experiences a positive relativistic gain in energy, and hence, in order to avoid violating the principle of energy conservation, we are obliged to ascribe to the antiparticle twin the same positive energy gain and to regard the antiparticle as subject to the same *attractive* gravitational force.

A second, alternative, line of reasoning might be to think of the work necessary to raise a newly created particle-antiparticle pair from the site of the γ detector to that of the γ emitter against a gravitational potential difference $\Delta\phi$. It is clear that if the energy balance is to be preserved, the work done against gravity in raising the particle must in magnitude and sign be equal to that done in raising the twin antiparticle, and the net total work must be equal in magnitude to the extra energy which a photon formed by recombination at the emitter site would acquire in falling back down through the field $\Delta\phi$ to the detector, as depicted schematically in Fig. 3-1(c).

Of course, a direct way to check the sign of the gravitational mass of antimatter would be to study the gravitational deflection of a horizontal beam of positrons (beams of more massive antiparticles cannot as yet be produced with sufficient intensity). However, such an experiment is simpler in conception than in execution, for the elimination of extraneous electromagnetic fields which would perturb the positron motion in a highly evacuated tube of considerable length presents very pronounced technical difficulties.

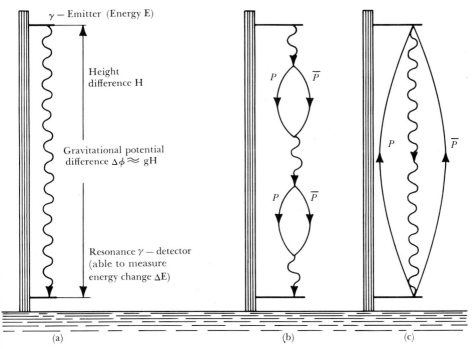

Fig. 3-1. *Gedankenexperiments* designed to demonstrate that antiparticles ($\bar{P}$) are subject to the same attractive gravitational forces as particles (P). (a) Assembly to detect a gravitational energy shift using a Mössbauer technique. (b) The same with intermediate formation of two P-$\bar{P}$ pairs which recombine annihilatively to reconstitute the γ quanta. (c) Elaboration of the experiment as discussed in the text.

In conclusion, it may be mentioned that the effect of gravity upon matter has been tested not only with massive bodies but also with beams of neutral free cesium and potassium atoms [Es 47], as well as with beams of high- and low-energy neutrons [McR 51, Da 65a], and the results conclusively indicate that *all* matter is *at all times* subject to the same gravitational interaction.

3.3. Response of Particles to Strong, Electromagnetic, and Weak Interactions

As regards the nongravitational interactions, e.g., the strong, electromagnetic, and weak modes, there is no such uniformity in particle behavior. Some particles respond exclusively to but one interaction mode (for example, neutrinos are associated solely with weak interactions), whereas others may under various conditions avail themselves of two or even three of the above interaction modes, though they tend to evince a preference as regards the type of interaction to which they are sensitive. For instance, nucleons and pions have a predilection for the strong interaction, electrons and photons for the electromagnetic, and muons for the weak interaction. Hyperons (particles of rest mass exceeding that of the neutron, such as the sigma, xi, lambda, and omega hyperons Σ, Ξ, Λ, Ω) interact strongly with other particles (such as nucleons), but in general decay through a weak interaction.

The interaction may serve to effect the binding of particles, or it may promote reaction and decay processes. Indeed, the sharp distinction so often drawn between a reaction and a decay process is essentially artificial, although for convenience we shall also distinguish between them in the following sections. It will be observed that a decay process, for instance the β decay of the neutron, to take a specific example,

$$n \rightarrow p^+ + e^- + \bar{\nu}_e \tag{3-17}$$

can be matched by a corresponding reaction, such as would in this case be represented by INVERSE β DECAY,

$$p^+ + \bar{\nu}_e \rightarrow n + e^+ \tag{3-18}$$

or ELECTRON CAPTURE,

$$p^+ + e^- \rightarrow n + \nu_e \tag{3-19}$$

Moreover, the decay can be regarded as being none other than a reaction outcome which ensues when a nucleus has been "energized" in some way in the course of the initiating step of the reaction. This initiating step need not necessarily be apparent: in the case of naturally radioactive nuclides it would correspond to the formation phase of the parent nuclei. An "energized" nucleus (or particle) will thereafter seek ways and means to get rid of the surplus mass-energy by emission of particles and radiation in order to attain to progressively lower states of inherent energy, and the manner in which it does this is governed by conservation rules which may exclude an otherwise feasible decay mode. Thus, unless there is something to prevent it, the strong —and fast—interaction serves to effect the decay, but when this is excluded the nucleus or particle avails itself of the slower electromagnetic or, failing that, of the appreciably more sluggish weak interaction in order to reduce its energy content. But the actual process is always the one which, without contravening various selection and conservation rules, is fastest.

At the risk of anticipating some points discussed more fully in later sections, we present some examples to illustrate the various types of interaction, pointing out meanwhile that there may in fact be further as yet undiscovered interaction modes (such as "superweak" interaction, currently being studied in connection with K-meson ["kaon"] phenomena), the detection of which constitutes one of the aims of particle physics.

3.3.1. STRONG INTERACTIONS

Clearly, internucleon *binding forces* accounted for by the exchange of pions furnish an instance of a strong interaction. The pion-nucleon interaction that operates in pion *scattering* is also strong, as is that in the scattering of kaons by nucleons. A strong interaction is, moreover, responsible for kaon *production*: it operates in *reactions* such as

$$\pi^- + p^+ \rightarrow \Sigma^- + K^+ \tag{3-20}$$

However, the *decay* of mesons, nucleons, and hyperons proceeds by an electromagnetic or weak interaction (or both) except when, as we shall see in examples which follow, it involves as intermediate step the (virtual) formation of a nucleon-antinucleon pair, this being an instance of a strong interaction.

3.3.2. ELECTROMAGNETIC INTERACTIONS

In most instances, processes which proceed by the electromagnetic interaction involve one or more photons. Photon exchange is, of course, responsible for electromagnetic *binding forces*, and photon capture can effect the *production* of mesons or hyperons by an electromagnetic interaction, for example:

$$\gamma + p \begin{cases} \nearrow \pi^0 + p & \text{(photopion production)} \\ \searrow \Lambda^0 + K^+ & \text{(associated photoproduction of a} \\ & \Lambda \text{ hyperon and K meson)} \end{cases} \tag{3-21}$$

An example of a *radiative capture reaction* is provided by the process

$$\pi^- + p \rightarrow n + \gamma \tag{3-22}$$

Decay processes illustrate a particularly important point which at first sight may appear paradoxical, namely that *uncharged* particles can also seem to be subject to electromagnetic interactions, even though this would appear to be fundamentally impossible. For example, π^0 decay into two photons,

$$\pi^0 \rightarrow \gamma + \gamma \tag{3-23}$$

occurs via an electromagnetic interaction, with a mean lifetime of about

5×10^{-17} sec. The paradox is resolved by introducing as an intermediate step in the overall reaction the virtual production of a nucleon-antinucleon pair (by a strong interaction), the energy for which is "borrowed" for a time so short that no contravention of the uncertainty relation $\Delta E \, \Delta t \approx \hbar$ occurs. This $\mathcal{N} - \bar{\mathcal{N}}$ pair then decays into two photons by an electromagnetic process, so that the overall decay scheme is

$$\pi^0 \xrightarrow[\text{strong}]{} \underset{\text{virtual}}{(\mathcal{N} + \bar{\mathcal{N}})} \xrightarrow[\text{em}]{} \gamma + \gamma \tag{3-24}$$

The process of mutual annihilation of particles and antiparticles to γ radiation possesses, as might be expected, an (electromagnetic) inverse, e.g., $\gamma \rightarrow e^+ + e^-$, and accordingly a competing reaction to the above π^0 decay mode is $\pi^0 \rightarrow \gamma + e^+ + e^-$ (occurring once in 80 cases) and the still rarer decay mode into two electron pairs, $\pi^0 \rightarrow e^+ + e^- + e^+ + e^-$ (occurring once in 29,000 cases). In the latter we have an instance of an electromagnetic interaction in which no γ quantum is explicitly featured. Another such example is the fast electromagnetic decay of the η^0 meson,

$$\eta^0 \rightarrow \pi^+ + \pi^- + \pi^0 \tag{3-25}$$

with a mean lifetime below 10^{-16} sec.

Mention should also be made here of the decay process

$$\Sigma^0 \rightarrow \Lambda^0 + \gamma \tag{3-26}$$

in which an uncharged particle again decays via the electromagnetic mode. Once more, this is explained by the intermediate formation of a virtual nucleon-antinucleon pair, according to the sequence

$$\Sigma^0 \xrightarrow[\text{strong}]{} \Lambda^0 + \underset{\text{virtual}}{(\mathcal{N} + \bar{\mathcal{N}})} \xrightarrow[\text{em}]{} \Lambda^0 + \gamma \tag{3-27}$$

What renders this particularly interesting is the fact that it represents the only known instance of hyperon decay occurring by a process other than via a weak interaction.

3.3.3. Weak Interactions

Weak interactions do not appear to have a capacity for binding (they may, indeed, even have repulsive character) but are responsible for a number of *capture reactions*, such as

$$\mu^- + p \rightarrow n + \nu_\mu \tag{3-28}$$

and, in particular, for many *decay processes*, e.g., of mesons:

$$\mu^+ \rightarrow e^+ + \nu_e + \bar{\nu}_\mu \tag{3-29}$$

$$\pi^+ \to \mu^+ + \nu_\mu \qquad \text{(or, once in 10,000 cases,} \qquad \pi^+ \to e^+ + \nu_e) \quad (3\text{-}30)$$

$$\pi^- \to \mu^- + \bar{\nu}_\mu \qquad \text{(or, once in 10,000 cases,} \qquad \pi^- \to e^- + \bar{\nu}_e) \quad (3\text{-}31)$$

$$K^+ \to \pi^+ + \pi^0 \quad (3\text{-}32)$$

$$K^0 \nearrow \pi^+ + \pi^- \quad (3\text{-}33)$$
$$\searrow \pi^0 + \pi^0 \quad (3\text{-}34)$$

of the neutron:

$$n \to p + e^- + \bar{\nu}_e \quad (3\text{-}35)$$

and of all hyperon decays (except $\Sigma^0 \xrightarrow[\text{em}]{} \Lambda^0 + \gamma$):

$$\Sigma^+ \nearrow p + \pi^0 \quad (3\text{-}36)$$
$$\searrow n + \pi^+ \quad (3\text{-}37)$$

$$\Sigma^- \to n + \pi^- \quad (3\text{-}38)$$

$$\Xi^0 \to \Lambda^0 + \pi^0 \quad (3\text{-}39)$$

$$\Omega^- \nearrow \Xi^0 + \pi^- \quad (3\text{-}40)$$
$$\searrow \Lambda^0 + K^- \quad (3\text{-}41)$$

It will be observed that throughout a distinction has been made between neutrinos and antineutrinos associated with the muon, ν_μ and $\bar{\nu}_\mu$, and those associated with the electron, ν_e and $\bar{\nu}_e$, since considerable evidence exists showing that they are not identical. For example, if they were the same, one would expect to observe a muon decay mode of the type $\mu^\pm \to e^\pm + \gamma$ as a result of ν-$\bar{\nu}$ annihilation, but though extensively sought, it has never been seen. It might at this stage be pointed out, too, that whereas $\bar{\nu}_e$'s are readily produced through the β decay of neutrons set free, for example, in fission (a reactor accordingly represents a prolific source of $\bar{\nu}_e$'s), the conjugate particle ν_e originates primarily from solar and stellar thermonuclear reactions such as

$$p + p \to d + e^+ + \nu_e \quad (3\text{-}42)$$

One associates ν and $\bar{\nu}$ exclusively with the weak interaction, just as one associates photons with the electromagnetic interaction. The other particles respond less exclusively to the various different types of interaction: for example, pion scattering is effected by a strong force, pion radiative capture by an electromagnetic interaction, and pion decay by a weak interaction.

3.4. Transition Probability

After this brief excursion into particle physics, we turn our attention to the consideration of the probability for a transition (such as that represented by a decay process) to take place. We accordingly sidestep the binding aspects of interparticle interactions in order to examine quantum-mechanically those features concerned with the occurrence of transitions that change the state of a system, e.g., by β decay or γ decay, to which the present considerations have special relevance.

Admittedly, the mathematical procedures so introduced demand an appreciably wider and more refined knowledge of wave mechanics than has been invoked up to the present stage of our treatment, but the consideration of transition probabilities is so fundamental to the study of nuclear processes that it should neither be neglected nor shirked. The reason for introducing such details here is that they stem naturally from the consideration of interactions and provide a logical bridge to the (less demanding) subject of nuclear cross sections. An understanding of the material which follows will greatly aid in the comprehension, appreciation, and application of probability considerations in nuclear physics and provide a firm foundation for the subsequent treatment of nuclear decay processes. The detailed derivation and applications, printed in smaller type, may, however, be omitted in a preliminary study of this section.

3.4.1. Time-Dependent Perturbation Theory

The behavior of a quantum-mechanical system described by a wave function ψ in terms of the time t is expressed by the time-dependent Schrödinger equation (Appendix D.1)

$$i\hbar \frac{\partial \psi}{\partial t} = H\psi \tag{3-43}$$

where H is the Hamiltonian describing the energy of the system at any given time t. The wave function ψ depends not only upon the time but also upon the relative coordinates of the interacting system, which we collect within the variable q, writing $\psi \equiv \psi(q,t)$. Since, in general, the Schrödinger equation containing a time-dependent Hamiltonian cannot be solved with the current mathematical methods, the procedure for arriving at a solution which describes the behavior of the system in course of time, and thus gives an expression for the probability that it undergoes a transition from an initial state i to a final state f, consists of splitting the Hamiltonian into a time-independent part H_0, for which the exact stationary solution is known, and a time-dependent part H', for which an approximate solution is sought. When H' is small compared with H_0, a *perturbation method* (the variation of constants) introduced by Dirac [Di 26, Di 27] (see also [Schi 55, Se 64]) can be used to derive a suitable approximate solution to the problem.

We start by separating the wave function ψ into spatial and time-dependent factors,

$$\psi \equiv \psi(q,t) = u(q) \exp(-iEt/\hbar) \qquad (3\text{-}44)$$

and assume that up to the time $t = 0$ the system remains in an unperturbed initial state i of energy E_i, described by the stationary Schrödinger equation

$$H_0 u_i = E_i u_i \qquad (3\text{-}45)$$

in terms of the orthonormal energy eigenfunctions u_i which have the property

$$\left.\begin{array}{l} \displaystyle\int |u_i|^2 \, dq = 1 \\[3mm] \displaystyle\int u_f^* \, u_i \, dq = \delta_{if} \equiv \begin{cases} 1 & \text{if } i = f \\ 0 & \text{otherwise} \end{cases} \end{array}\right\} \qquad (3\text{-}46)$$

where the asterisk denotes complex conjugation, and $|u_i|^2 \equiv u_i^* u_i$.

As soon as the perturbation which induces the transition (e.g., a variable electromagnetic field causing the emission of a γ quantum, or the elastic collision of two particles, etc.) starts to act at the time $t = 0$, the wave function ψ has to be expressed as an expansion in the eigenfunctions $u_i \exp(-iE_i t/\hbar)$ of the unperturbed problem, viz.

$$\psi = \sum_i a_i(t) \, u_i \exp\left(\frac{-iE_i t}{\hbar}\right) \qquad (3\text{-}47)$$

where the u_i are time-independent, but the expansion coefficients a_i depend on the time and express the likelihood of the system's being in a state i at any given instant t. Thus by the relation

$$a_f(0) = \delta_{if} \qquad (3\text{-}48)$$

we express our certainty of locating the system in state i at the moment $t = 0$ and the impossibility of its being in any other state $f(\neq i)$ at that instant. When the perturbation has had a time t to act, however, the probability of the system's having effected a transition to state f, of energy E_f, is no longer zero, but is given by the absolute square of the transition amplitude (i.e., the amplitude of the unperturbed wave function), in conformity with Born's probability interpretation of a wave function's amplitude:

$$\text{transition probability} = |a_f(t)|^2 \qquad (3\text{-}49)$$

The index f characterizes any one of a number of final states in a discrete or continuous spectrum (for the latter, summation over a discrete set of eigenfunctions must be replaced by integration over a continuous set, and the Kronecker δ_{if} replaced by the Dirac δ function $\delta(i - f)$) within an energy band of width $\Delta E = \hbar/t$.

To evaluate the transition probability, we accordingly have to derive $a_f(t)$ by solving the perturbed Schrödinger equation

$$i\hbar \frac{\partial \psi}{\partial t} = H\psi = (H_0 + H')\psi \qquad (3\text{-}50)$$

Replacing ψ by its perturbed eigenfunction expansion and noting that $H_0 u_i = E_i u_i$ we find that

$$i\hbar \sum_i \frac{\partial a_i}{\partial t} u_i \exp\left(-\frac{iE_i t}{\hbar}\right) = \sum_i H' a_i u_i \exp\left(-\frac{iE_i t}{\hbar}\right) \tag{3-51}$$

Multiplying through on the left by u_f^* and integrating over the volume, while observing the orthogonality property $\int u_f^* u_i \, dq = \delta_{if}$, we reduce this to

$$i\hbar \frac{\partial a_f}{\partial t} \exp\left(-\frac{iE_f t}{\hbar}\right) = \sum_i a_i \exp\left(-\frac{iE_i t}{\hbar}\right) \int u_f^* H' u_i \, dq \tag{3-52}$$

i.e.,

$$\frac{\partial a_f}{\partial t} = \frac{1}{i\hbar} \sum_i a_i H'_{fi} \exp(i\omega_{fi} t) \tag{3-53}$$

written in the usual abbreviated notation for the matrix element of the perturbation H' which takes the system from state i to f,

$$\int u_f^* H' u_i \, dq \equiv H'_{fi} \tag{3-54}$$

and introducing the Bohr frequency

$$\omega_{fi} \equiv \frac{E_f - E_i}{\hbar} \tag{3-55}$$

with matrix indices read by convention from right to left.

The treatment so far is exact; the perturbation approximation is introduced at this stage in order to perform the integration of $\partial a_f / \partial t$. This consists in replacing H' by $\epsilon H'$ and expressing the a's as a power series in ϵ,

$$a_s = a_s^{(0)} + \epsilon a_s^{(1)} + \epsilon^2 a_s^{(2)} + \cdots \tag{3-56}$$

substituting the series expansions in Eq. (3-53), equating equal powers of ϵ, and finally setting ϵ to unity. We note the (unperturbed) starting condition

$$a_i^{(0)} = a_i(0) = 1 \tag{3-57}$$

Substituting in (3-53), we get

$$\frac{\partial a_f^{(0)}}{\partial t} + \epsilon \frac{\partial a_f^{(1)}}{\partial t} + \epsilon^2 \frac{\partial a_f^{(2)}}{\partial t} + \cdots = \frac{1}{i\hbar} \sum_i [\epsilon a_i^{(0)} + \epsilon^2 a_i^{(1)} + \cdots] H'_{fi} \exp(i\omega_{fi} t) \tag{3-58}$$

whence, in zeroth order,

$$\frac{\partial a_f^{(0)}}{\partial t} = 0, \qquad \text{so that } a_f^{(0)} = \text{constant} = \delta_{if} \tag{3-59}$$

In first order of approximation we have

$$\frac{\partial a_f^{(1)}}{\partial t} = \frac{1}{i\hbar} \sum_i a_i^{(0)} H'_{fi} \exp(i\omega_{fi} t) = \frac{1}{i\hbar} H'_{fi} \exp(i\omega_{fi} t) \tag{3-60}$$

This can be integrated to obtain $a_f^{(1)}$ if we assume the perturbation to be describable by a step function which has a constant finite value between times 0 and t, viz.

$$a_f^{(1)} = \frac{1}{i\hbar} H_{fi}' \int_0^t \exp(i\omega_{fi} t)\, dt = \frac{1}{\hbar \omega_{fi}} H_{fi}'[1 - \exp(i\omega_{fi} t)] \tag{3-61}$$

If we neglect higher orders of approximation, we obtain the probability that a transition from state i to f has taken place within a time t under the influence of a step-function perturbation,

$$|a_f|^2 \approx |a_f^{(1)}|^2 = 4|H_{fi}'|^2 (\hbar \omega_{fi})^{-2} \sin^2(\tfrac{1}{2}\omega_{fi} t) \tag{3-62}$$

3.4.2. Transition Probability per Unit Time

If we have a group of final states f whose energy is nearly the same as that of the initial state i, we can treat them as a continuum and express the number of such states per unit energy interval around E_f in the form of an energy LEVEL DENSITY $\rho(E_f)$, whose explicit value we shall derive in Section 3.6.3. Here, we avail ourselves of this concept which represents a measurable quantity in order to derive the final expression for the transition probability per unit time in a simple widely used form.

If we assume that H_{fi}' varies but slowly with f within the group of states in the region around E_f, the quantity $|H_{fi}'|^2$ can be taken to denote a mean value, independent of the actual energy E_f and treated as essentially constant in the calculation which follows. The evaluation of the transition probability per unit time to all accessible states f according to first-order perturbation theory then consists of replacing the summation in the expression

$$W_{fi} = \frac{1}{t} \sum_f |a_f|^2 \tag{3-63}$$

by an integral over all energy intervals dE_f,

$$W_{fi} = \frac{1}{t} \int_{-\infty}^{\infty} |a_f|^2 \, \rho(E_f) \, dE_f \tag{3-64}$$

noting that $dE_f = \hbar \, d\omega_{fi}$. Thus, if we confine the consideration to final states within an energy band of width $\Delta E \ll E_f$, we may bring $|H_{fi}'|^2 \rho(E_f)$ outside the integral and write

$$W_{fi} = \frac{4|H_{fi}'|^2 \rho(E_f)}{\hbar t} \int_{-\infty}^{\infty} \frac{\sin^2 (\tfrac{1}{2}\omega_{fi} t)}{\omega_{fi}^2} \, d\omega_{fi} \tag{3-65}$$

in which the integral is simple to evaluate since it is of the form

$$\int_{-\infty}^{\infty} x^{-2} \sin^2 (kx) \, dx = \pi k \tag{3-66}$$

We thus obtain the TRANSITION PROBABILITY PER UNIT TIME for a change in the system from an initial state i to all accessible final states f:

$$W_{\text{fi}} = \frac{2\pi}{\hbar}\, \rho(E_{\text{f}})\, |H'_{\text{fi}}|^2 \tag{3-67}$$

We note that W_{fi} has the dimensions $\sec^{-1}$, since H'_{fi} has the dimensions of an energy and ρ that of a reciprocal energy. This expression forms the second of Fermi's "Golden Rules" [Fe 50] and is of wide-ranging importance in nuclear physics since it relates an interaction strength H' which can in many cases be determined explicitly with the probability for the occurrence of a given process, such as the statistical likelihood of β or γ emission, or scattering, etc. Before going on to illustrate its use, we introduce and explain the concept of a nuclear "cross section," relating this first to reaction probability and subsequently take up once more the threads of the present considerations in order to relate it to transition probability.

3.5. Reaction Probability and Cross Section

Although in principle an approach similar to that employed in deriving the transition probability could be used for expressing the likelihood that a *reaction* will take place under given circumstances, a more direct means of expressing this probability is offered by the elementary concept of a REACTION CROSS SECTION, which has the dimension cm^2 rather than $\sec^{-1}$ and thus is represented by an *area*. The connection between transition probability and cross section will in due course be elucidated in Section 3.6.

The concept of cross section is based upon a consideration of the physical situation encountered in determining reaction probability in the laboratory. An incident collimated beam of particles impinges upon target nuclei in a suitable specimen of material and interacts with these through the processes of scattering, absorption and/or reaction, thereby becoming attenuated —it may be in intensity or energy, or both—by an amount which can be determined by measurements effected on the emergent beam. The most naive way of picturing the likelihood of interaction is to visualize the incident beam as made up of point particles which if they directly strike some part of a target nucleus, set up an interaction, whereas if they miss the target nucleus, they proceed unaffected. However, this naive conception overlooks both the finite extension of the impinging particles and the finite interaction radius which may be presumed to extend quite some way beyond the immediate confines of the target nucleus. Hence rather than treating the *geometrical* cross-sectional area of a nucleus (πR^2) as a measure of interaction probability, it is meaningful to ascribe to each nucleus an effective area σ perpendicular to the incident beam such that if a bombarding particle impinges upon any part of such an imaginary disk, a reaction will occur, but otherwise no interaction takes place if the particle's path falls outside the target zone (Fig. 3-2). The magnitude of the disk's CROSS SECTION σ depends upon the reaction and upon the energy; its size is suitably expressed in units of

$$1 \text{ b} = 1 \text{ barn} = 10^{-24} \text{ cm}^2 \tag{3-68}$$

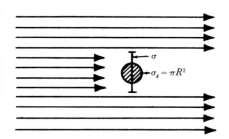

Fig. 3-2. Pictorial representation of a cross section as an imaginary target disk whose area can be appreciably different from the geometrical cross section σ_g presented to an incident flux of particles.

To take an example, if boron is bombarded with neutrons having a speed of about 10^5 cm/sec, the boron nuclear cross section for neutron capture is around 1200 b, whereas for neutrons traveling at 10^6 cm/sec, the absorption cross section drops to about 120 b. On the other hand, the reaction cross section for the photodisintegration of deuterium, viz. the $^2\text{H}(\gamma,\text{n})^1\text{H}$ reaction at $E_\gamma = 2.6$ MeV, is $\sigma = 1.2 \times 10^{-3}$ b $= 1.2$ mb (millibarns).

By way of another example to illustrate the wide range of numerical values of measured cross sections in nuclear physics at present, it may be mentioned that current values extend from $\sim 10^6$ b down to $\sim 10^{-19}$ b. As shown in Fig. 3-3, the total cross section for neutrons of energy around 0.07 eV incident upon $^{135}_{54}\text{Xe}$ has been measured to be $\sigma = 3.6 \times 10^6$ b. This is so large a magnitude that if a collimated beam of slow neutrons were sent through a pure gaseous ^{135}Xe target at STP, the beam intensity would be reduced by a factor of roughly one million for each millimeter of target traversed. Expressed in another way, a high-intensity beam containing 10^{13} slow neutrons cm^{-2} sec^{-1} would be reduced to the negligible magnitude of 1 neutron cm^{-2} sec^{-1} by passage through 0.31 cm of such a target.

At the very opposite extreme lie reaction cross sections for weak-interaction processes such as those used for neutrino detection in experiments described more fully in Section 8.2.2. Thus the cross section for an antineutrino-initiated reaction,

$$\text{p}^+ + \bar{\nu}_e \rightarrow \text{n} + \text{e}^+ \tag{3-69}$$

which formed the basis of experiments (e.g., [Re 60, Ne 66]) to detect antineutrinos produced by the β decay of reactor neutrons is $\sigma = (0.94 \pm 0.13) \times 10^{-43}$ cm^2, while a chemical radioactivity neutrino experiment [Da 55] was designed to be responsive to a cross section almost 3 orders of magnitude lower. An indication of the minuteness of such a cross section is provided by the fact that quite intense beams of terrestrial antineutrinos or solar neutrinos could pass through targets the size of the Earth without perceptible diminution through interaction. The fact that the mean penetration depth

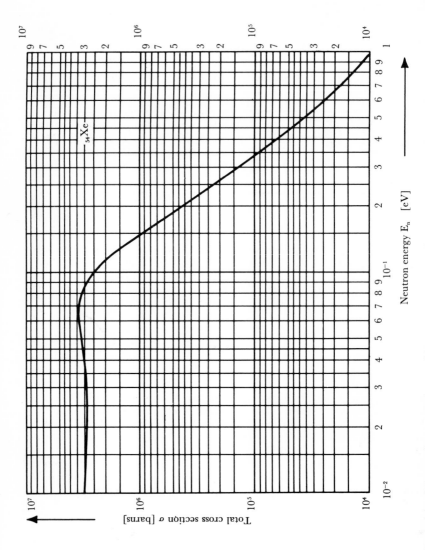

Fig. 3-3. Variation of the total cross section σ with neutron energy E_n for xenon, illustrating the high magnitude maintained in the low-energy region [Hu 58]. (Courtesy of Brookhaven National Laboratory.)

of neutrinos in solid matter can be expressed in light-years may give some impression of the minuteness of the associated cross sections and the weakness of the weak interaction. The small magnitude of the neutrino capture cross section also serves to illustrate the fallacy of picturing the cross section as some sort of actual disk area, for a disk having an area of 0.94×10^{-43} cm² would have a radius $R = 1.73 \times 10^{-22}$ cm, a dimension so small that it is unlikely to be physically meaningful.

Regarded as a *hypothetical* target area, the reaction cross section offers a means of picturing a reaction probability as a numerical hit-or-miss probability, a statistical concept which can be expressed quantitatively.

3.5.1. DEFINITION OF REACTION CROSS SECTION

We consider the situation depicted in Fig. 3-4, in which a collimated beam of monoenergetic particles impinges upon a specimen having the shape of a

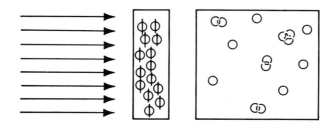

Fig. 3-4. Representation of overlapping disk cross sections in side view and front view as "perceived" by a collimated incident particle beam.

rectangular prism of area F normal to the beam and thickness d. If this contains n target nuclei, the nuclear density, i.e., the number of target nuclei per unit volume, is

$$n^{\square} \equiv \frac{n}{Fd} \tag{3-70}$$

For example, the nuclear density of a monatomic gas at STP is $n^{\square} = 2.687 \times 10^{19}$ cm⁻³.

For a *thin target*, we assume d to be sufficiently small to prevent any appreciable overlapping or masking of the individual effective (hypothetical) target disks, each of area σ. The overall area then presented as target is $n\sigma$ and the probability for an impinging particle to give rise to a reaction is thus

$$P = \frac{n\sigma}{F} = n^{\square}\sigma d \tag{3-71}$$

expressed in dimensionless units. If the number of incident particles per
second is N_0, the reaction rate, or the number of reactions per second, is

$$R = PN_0 = N_0 n^\square \sigma d \qquad (3\text{-}72)$$

In the case of a *thick target*, however, there can be considerable masking of
the individual disks, and the overall effective target area is no longer obtained
simply by direct summation as $n\sigma$. Such linear summation is valid only for

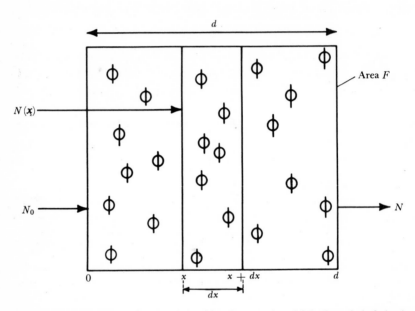

Fig. 3-5. Segmentation of target into thin elements in which there is infinitesimal
overlapping permits the evaluation of the outgoing flux N for an incident flux N_0
through integration over x from 0 to d.

a thin section (having coordinates $x \rightarrow x + dx$) within which the overlapping
is so infinitesimal that direct summation is permissible (Fig. 3-5). By the
time the incident beam of particles reaches the layer at x, it has become
attenuated as a result of interaction with target particles in preceding layers
and its intensity per second is $N(x)$, which is smaller than N_0. The number of
reactions per second in the layer under consideration is equal to the rate of
diminution in the number of particles which, impinging upon the next layer,
can initiate further reactions, and this is

$$-dN(x) = N(x)\, n^\square \sigma\, dx \qquad (3\text{-}73)$$

On integration over all such sections, this gives N, the number of particles per second which traverse the entire thickness of the target specimen without having encountered any interaction:

$$\int_{N_0}^{N} \frac{dN(x)}{N(x)} = \int_{0}^{d} -n^{\square}\sigma \, dx \tag{3-74}$$

i.e.,

$$N = N_0 \exp(-n^{\square}\sigma d) \tag{3-75}$$

It should be noted that any energy-dependence of σ has tacitly been suppressed in deriving the above result, which shows that there is an *exponential* weakening in the particle flux upon passage through a thick target. The attenuation that a beam experiences as it goes through a target may take the form of a reduction in the actual number of particles traversing the target without change in energy—this is the case with a γ-ray beam—or a diminution in the mean energy of the particles due to scattering, ionization, or absorption processes.

From Eq. (3-75), we see that the number of reactions taking place in the target per second is

$$R = N_0 - N = N_0[1 - \exp(-n^{\square}\sigma d)] \tag{3-76}$$

When d is small, this reduces to the thin-target expression (3-72).

To illustrate the application of this reaction-rate formula, we present a solution to the following problem:

The cross section for the $^{113}\text{Cd}(n,\gamma)^{114}\text{Cd}$ reaction at the resonance energy $E_n = 0.178$ eV has been measured to be $\sigma = 63{,}600$ b (Fig. 3-6). A target of natural cadmium (which contains the natural isotopic abundance $C = 12.26$ percent of the isotope ^{113}Cd), 10^{-4} cm thick is irradiated with a parallel beam of mono-energetic 0.178-eV neutrons whose intensity is $N_0 = 10^8$ n/sec. Assuming the distribution of the emitted γ radiation to be isotropic, what is the counting rate registered by a γ detector having an effective area $S = 5$ cm^2 and a response probability $\eta = 0.25$, situated at a distance $d = 1$ m from the target?

Taking account of the isotopic abundance C, we write the nuclear density of ^{113}Cd nuclei as

$$n^{\square} = N_A \, C \, (\rho/A) \tag{3-77}$$

where $N_A = 6.022\,52 \times 10^{23}$ mol^{-1} is Avogadro's constant, $\rho = 8.65$ g/cm^3 is the density of cadmium at room temperature, and $A = 113$ g is the atomic weight. This yields a value

$$n^{\square} \sigma d = N_A \, C \, \rho \, \sigma \, (d/A) = 0.035\,95 \tag{3-78}$$

on numerical substitution, and since this is very much smaller than unity, the exponential term in the reaction-rate formula (3-76) need be expanded to first order only, giving

$$R = N_0 \, n^{\square}\sigma d = 3.595 \times 10^6 \text{ sec}^{-1} \tag{3-79}$$

a reaction rate for which the commensurate counting rate is

$$r = R \, \eta \, (\Omega/4\pi) = 35.8 \text{ sec}^{-1} = 2145 \text{ min}^{-1} \tag{3-80}$$

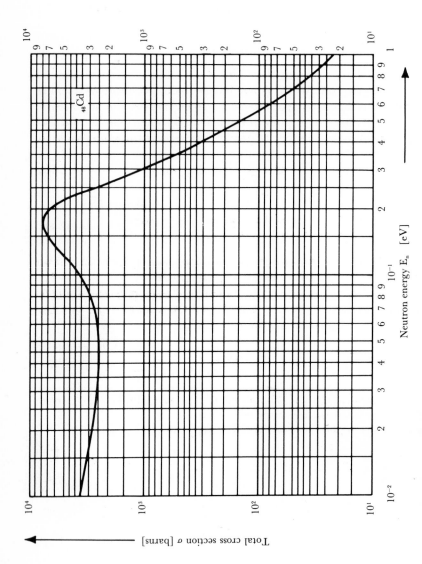

Fig. 3-6. Energy dependence of the slow-neutron cross section for natural cadmium, showing the resonance at $E_n = 0.178$ MeV evinced by ^{113}Cd (isotopic abundance $= 12.26$ percent). On dividing the peak cross section for natural cadmium, σ_0 (Cd) $= 7800$ b, by the ^{113}Cd abundance, one obtains the resonance cross section referred to ^{113}Cd, namely σ_0 (^{113}Cd) $\simeq 63{,}600$ b [Hu 58]. (Courtesy of Brookhaven National Laboratory.)

where $\Omega = S/d^2 = 5 \times 10^{-4}$ is the solid angle subtended by the counter at the center of the target.

The cross section can be related directly to parameters which express the experimental conditions:

$$\sigma = \frac{R}{N_0} \frac{1}{n/F} = \frac{\text{reactions/sec}}{(\text{incident particles/sec}) \times (\text{target nuclei/cm}^2)} \qquad (3\text{-}81)$$

This has been based upon a *corpuscular* representation of the situation, but we might equally well have adopted a *wave picture* of the process and obtained an analogous expression for the cross section. For the interaction between an incident (plane) wave with a target, the reduction dI in intensity from an initial value I_0 is

$$-dI = I_0\, n^\square \sigma\, dx \qquad (3\text{-}82)$$

and the definition of cross section now reads

$$\sigma = \frac{\Delta I}{I_0} \frac{1}{n/F} = \frac{\text{diminution in incident wave intensity}}{(\text{incident intensity}) \times (\text{target nuclei/cm}^2)} \qquad (3\text{-}83)$$

with σ pictured now as a cross-sectional area within the incident wave front impinging upon target nuclei conceived as point particles.

In order to reconcile these alternative but analogous descriptions of a cross section, a definition of σ can be adopted which applies equally well to either representation, being based upon a relative flux diminution:

$$\sigma = \frac{\Delta Q/t}{Q_0/t} \frac{1}{n/F} = \frac{\text{diminution in energy flux}}{(\text{incident energy flux}) \times (\text{target nuclei/cm}^2)} \qquad (3\text{-}84)$$

where Q_0/t constitutes the incident beam power and $\Delta Q/t$ is the power absorbed from the beam as it traverses the target.

Generally, σ is a function of the kinetic energy E of the incident flux,

$$\sigma = \sigma(E) \qquad (3\text{-}85)$$

that is, a function of the wavelength of the incident wave,

$$\sigma = \sigma(\lambda) \qquad (3\text{-}86)$$

where λ stands for the de Broglie wavelength of the incident wave in the laboratory system.

In carrying out the integration above, we have tacitly assumed σ to be independent of x, which is equivalent to assuming σ to be independent of E, and E independent of x. This is valid if each incident particle experiences no more than just a *single* reaction in traversing the target, as would for instance be the case when the chance of a second interaction is precluded, e.g., by absorption of the incident particle at the first reaction, or by the prevalence of so large a mean free path between successive reactions that its value greatly exceeds the thickness of the target. Under these circumstances the exponential law of attenuation holds rigorously. It is accordingly *not* valid for the passage of charged particles through matter, since these experience a progressive decrease in energy as a result of multiple collisions, but it *is* valid in the case of γ radiation passing through matter, since γ rays, unless they undergo Compton scattering, suffer no decrease in energy, but merely in intensity, as they penetrate a target.

Hence for γ *rays traversing a target* of thickness d one can write

$$I_{\text{out}} = I_{\text{in}} \exp(-n^{\square}\sigma d) = I_{\text{in}} \exp(-\mu d) \tag{3-87}$$

where $\mu = n^{\square}\sigma$ is called the ATTENUATION COEFFICIENT and is equal to the reciprocal of the mean free path. The MEAN FREE PATH itself represents the target thickness in which the beam is reduced to $1/e$-th of its incident intensity. Instead of the linear attenuation coefficient μ, many numerical compilations of data list values of the MASS-ATTENUATION COEFFICIENT μ/ρ (in g^{-1} cm^2), where ρ is the density of the target material (in g cm^{-3}). Noting, as before, that the number of target *nuclei* per unit volume is identical to the number of *atoms* per unit volume of the material, we write

$$\mu \equiv n^{\square}\sigma = \rho N_A \sigma / A \tag{3-88}$$

with N_A the Avogadro constant (or "Loschmidt number" in Continental usage) and A the atomic weight of the target material. To eliminate the dependence of μ via ρ upon such physical factors as temperature, pressure, and phase, it is convenient to use the mass-attenuation coefficient

$$\frac{\mu}{\rho} = \frac{N_A \sigma}{A} = \text{constant} \times \sigma \tag{3-89}$$

and at the same time to express the target thickness as ρd in grams per square centimeter (rather than directly as d in centimeters) in the attenuation expression

$$I_{\text{out}} = I_{\text{in}} \exp\left[-\left(\frac{\mu}{\rho}\right)(\rho d)\right] \tag{3-90}$$

A typical value for the thickness of a metal target in nuclear physics experiments might be 1 mg/cm²: this corresponds to an actual thickness ranging from 3.7×10^{-4} cm for the light element Al ($\rho = 2.70$ g cm^{-3}) to 8.8×10^{-5} cm for the heavy element Pb ($\rho = 11.35$ g cm^{-3}). It is sometimes the practice in a description of experimental work involving incident particles of a given type and energy to quote target thicknesses in keV, this being the energy diminution which the given beam experiences on traversing the target. Thus, for example, a Cr target of thickness 3.83 mg/cm² (equivalent to $d = 5.4 \times 10^{-4}$ cm) traversed by 5.8-MeV protons could be described as "169 keV thick."

3.5.2. PARTIAL AND TOTAL CROSS SECTIONS

The treatment so far has in all cases assumed that merely *one* reaction process takes place as a result of the interaction, whereas in practice usually several reactions of different types occur simultaneously. Their separate probabilities, expressed as individual cross sections, can simply be added (since areas are additive) to give an overall probability expressed as a *total cross section*

$$\sigma_{total} = \sum \sigma_{partial} \qquad (3\text{-}91)$$

where the summation extends over each of the individual PARTIAL CROSS SECTIONS $\sigma_{partial}$ referring to the individual simultaneous reaction processes, e.g., scattering, absorption, emission, etc.

For example, the total cross section of neutrons having an energy of 0.0253 eV (and thus a speed of 2.2×10^5 cm/sec—so-called "thermal neutrons") incident upon ^{135}Xe is, according to Fig. 3-3, equal to $\sigma_{total} = 3 \times 10^6$ b, whereas the measured absorption cross section at that energy is $\sigma_{absorption} = 2.72 \times 10^6$ b.

3.5.3. DIFFERENTIAL AND TOTAL CROSS SECTION (ANGULAR DISTRIBUTION)

Since any reaction invariably results in the emission of a reaction product, the cross section expressing reaction probability at the same time expresses emission probability. However, the detection of reaction products is accomplished by setting up detectors which "view" the target at a given angle to the incident beam direction, and it is therefore desirable to define a *differential* cross section to express the emission probability in that particular direction. Thus the DIFFERENTIAL CROSS SECTION is introduced to describe the angular distribution of particles or radiation emitted by a target when a reaction occurs, and this is written as

$$\sigma(\theta) \equiv \frac{d\sigma}{d\Omega} \qquad (3\text{-}92)$$

The differential quantity $d\sigma = \sigma(\theta)\, d\Omega$ represents the probability for radiation to be emitted into an element of solid angle $d\Omega$ lying at a mean angle θ to the incident beam direction. Hence the units of $\sigma(\theta)$ are in cm^2/steradian or, more usually, in millibarns per steradian (mb/sr). Sometimes, for experimental reasons, angular distributions are determined in relative rather than absolute terms; for example, the probability of emission into a cone at a mean angle θ to the incident direction is measured relative to that into the same cone but perpendicular to the incident direction. The quantity so determined is a dimensionless ANGULAR DISTRIBUTION FUNCTION $W(\theta)$, normalized, in general, to $W(90°) = 1$.

Clearly, the TOTAL CROSS SECTION σ is simply the integral of the differential cross section over all space,

$$\sigma = \int_{\phi=0}^{2\pi} \int_{\theta=0}^{\pi} \sigma(\theta) \sin\theta \, d\theta \, d\phi = 2\pi \int_0^{\pi} \sigma(\theta) \sin\theta \, d\theta \qquad (3\text{-}93)$$

In the special case of *isotropic emission* of radiation,

$$\sigma = 4\pi \, \sigma(\theta) \qquad (3\text{-}94)$$

since the integration over a sphere gives 4π sr. It will be noted that an ambiguity exists with regard to the term *total* cross section: As used here, it implies the integral over all directions of a differential cross section for a single type of reaction, whereas the sense used previously implied the sum of partial cross sections for a number of concurrent types of interaction. The meaning is generally clear from the context.

3.5.4. DOUBLE-DIFFERENTIAL CROSS SECTION (ANGULAR CORRELATION)

In a nuclear reaction of the type $A(a,b\ c)B$, detector-counter combinations can not only be set up to measure the emission rate of radiation b alone, or c alone, in function of the respective emission angle θ_b, or θ_c, referred to a specified direction such as the incident direction of a (which amounts to a determination of the *angular distribution*), but can be set up as in Fig. 3-7 to measure *both b and c* in coincidence, the usual procedure being to leave one detector fixed (at θ_b, say) while the other is rotated through a series of angles θ_c. Since the three radiations a, b, c need not necessarily be coplanar, the emission probability also depends upon the azimuth ϕ, which is the angle between the (a,b) and (a,c) planes. Measurement of the coincidence rate then betokens a determination of the ANGULAR CORRELATION, which, if expressed in absolute units (mb/sr^2) represents the DOUBLE-DIFFERENTIAL CROSS SECTION

$$\sigma(\theta_b, \theta_c, \phi) \equiv \frac{d^2\sigma}{d\Omega_b \, d\Omega_c} \qquad (3\text{-}95)$$

or if expressed in relative units is a measure of the CORRELATION FUNCTION $W(\theta_b,\theta_c,\phi)$. The quantity $\sigma(\theta_b,\theta_c,\phi) \, d\Omega_b \, d\Omega_c$ expresses the probability that when particle b is emitted into an element of solid angle $d\Omega_b$ at a mean angle θ_b to the incident direction, a particle c is concurrently emitted into the element of solid angle $d\Omega_c$ at an angle θ_c.

Clearly, by integrating the double-differential cross section over all the possible emission directions of one of the radiations (b, say)—thus, essentially, integrating over the surface of a sphere—one obtains the differential cross section for emission of the other radiation partner c:

$$\int \frac{d^2\sigma}{d\Omega_b \, d\Omega_c} \, d\Omega_b = \frac{d\sigma}{d\Omega_c} \qquad (3\text{-}96)$$

In nuclear physics the term "double-differential cross section" is sometimes employed in a different sense, namely to describe the change of the differential cross section in terms of the incident particle energy. It is then written as

$$\frac{d^2 \sigma}{d\Omega \, dE} \qquad (3\text{-}97)$$

and expressed in units of mb sr^{-1} MeV^{-1}. It is usually straightforward to establish from the context which of these alternative possibilities is meant.

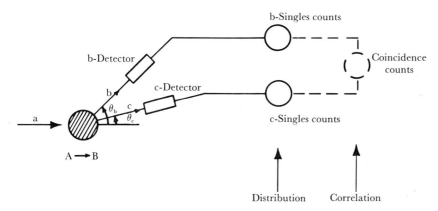

Fig. 3-7. Experimental arrangement for measuring the distribution of *b* or *c* radiation, or the *b-c* correlation for a reaction of the type (*a*, *b c*).

3.5.5. GEOMETRICAL AND ABSOLUTE CROSS SECTION

The GEOMETRICAL CROSS SECTION σ_g is simply the cross-sectional area presented to the incident particle flux by the actual target nucleus of radius R, namely

$$\sigma_g = \pi R^2 = \pi r_0^2 A^{2/3} \qquad (3\text{-}98)$$

expressed in barns. It usually bears no numerical resemblance to the actual reaction cross section: for example, the geometrical cross section for ^{198}Au is $\sigma_g = 2.1$ b, whereas the measured cross section for the process ^{198}Au(n,γ)^{199}Au, representing the radiative capture of slow neutrons, is $\sigma = 35{,}000$ b. Indeed, whereas geometrical cross sections range over a fairly restricted set of values, from about 0.1 b to 2.7 b, reaction cross sections cover, as we have seen, a far more extensive range. Even if we confine ourselves to the radiative capture of slow neutrons, the measured cross section for ^{13}C is $\sigma = 9 \times 10^{-4}$ b, which one may compare with a geometrical cross section of $\sigma_g = 0.3$ b.

The (ABSOLUTE) CROSS SECTION is a concept introduced in the treatment of collision processes, in which the incident particle can enter into a *noncentral* collision with a nucleus. The IMPACT PARAMETER D, namely the separation at which the collision partners (i.e., their centers) would pass one another in the center-of-mass system if there were no interaction between them, is then non-zero. Thus the incident particle, of mass m and relative velocity v, possesses an orbital angular momentum mvD with respect to the target nucleus. Quantum theory requires the quantization of momenta as well as of energies: the orbital angular momentum can therefore assume only such values as are integer multiples of the rationalized Planck constant $\hbar$, e.g.,

$$mvD = l\hbar \tag{3-99}$$

where l is an integer which takes on the value 0 for a central collision, or 1, 2, 3, ... for noncentral collisions having progressively larger relative orbital angular momenta. The integer l is termed the ORBITAL QUANTUM NUMBER describing the momentum state of the system before collision. Strictly, the quantum-theoretical expression for the numerical orbital momentum corresponding to the quantum number l is not simply $l\hbar$, but rather $[l(l+1)]^{1/2}\hbar$. However, the approximation $l\hbar$ suffices for our purposes. This yields a value for the impact parameter

$$D = l\frac{\hbar}{mv} \equiv l\lambda \tag{3-100}$$

where λ is the rationalized de Broglie wavelength for the nonrelativistic particle. Thus one could also conceive of D as quantized, in the way shown in Fig. 3-8. Only certain discrete values of the impact parameter D could be assumed by the incident particle with respect to the nucleus, e.g.,

particle waves with $l = 0$ (S waves†) have impact parameter $D = 0$,

particle waves with $l = 1$ (P waves) have impact parameter $D = \lambda$,

particle waves with $l = 2$ (D waves) have impact parameter $D = 2\lambda$,

particle waves with $l = 3$ (F waves) have impact parameter $D = 3\lambda$,

particle waves with $l = 4$ (G waves) have impact parameter $D = 4\lambda$,

The beam can be conceived as arranged in concentric cylindrical zones, each having a discrete value of orbital momentum. Thus particles representing

---

† The partial-wave nomenclature is historically based upon the characteristics of spectral series in atomic spectroscopy. Radiation with orbital momentum $l = 0$ is classed as an S (sharp) wave, that with $l = 1$ as a P (principal) wave, that with $l = 2$ as a D (diffuse) wave, that with $l = 3$ as an F (fundamental) wave, and thereafter the progression is alphabetical (leaving out J).

a partial D wave with $l = 2$ would be visualized to proceed along the zone having an impact parameter $D = 2\lambdabar$, and if they interact with the target, the effective cross section would have to be equal to the corresponding disk area and thus, in other words, *approximately* equal to the area of the annulus between $D = 2\lambdabar$ and $D = 3\lambdabar$. Hence, generalizing, particles with orbital momentum l effectively enter into collision with the target if they impinge

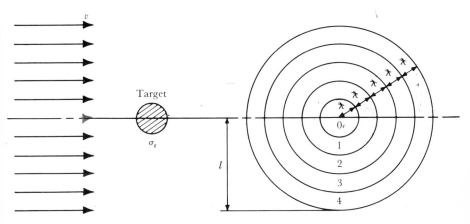

Fig. 3-8. Diagram illustrating the quantization of incident-particle orbital momenta (concept of partial waves).

upon the annular area between radii $l\lambdabar$ and $(l + 1)\lambdabar$. We can accordingly define an ABSOLUTE CROSS SECTION as

$$\sigma_l = \pi (l + 1)^2 \lambdabar^2 - \pi l^2 \lambdabar^2 = (2l + 1)\pi \lambdabar^2 \qquad (3\text{-}101)$$

which is, as usual, expressed in barns.

Unlike σ_g, which depends upon the actual dimensions of the target nuclei, the absolute cross section σ_l depends only on the orbital momentum l and the rationalized de Broglie wavelength $\lambdabar$ of the incident radiation. Particles of low energy have low velocity and therefore a large wavelength $\lambdabar$, in consequence of which it is not surprising that σ_l can assume large values considerably exceeding σ_g. The order of magnitude of σ_l is in fact determined by $\lambdabar$ rather than by l, and hence even though a wave of low-energy particles may enter into an essentially S-wave interaction, the absence of higher partial waves will not preclude σ_l from being very much larger than σ_g, as is, indeed, the case with the $^{198}\text{Au}(n,\gamma)^{199}\text{Au}$ reaction at about $E_n = \frac{1}{40}$ eV. From the relation

$$E_n = \frac{p^2}{2m_n} = \frac{l(l + 1)\hbar^2}{2m_n D^2} \qquad (3\text{-}102)$$

we can deduce that the impact parameter D takes on large values as soon as l departs from zero. For instance, $D = 40,000$ fm for $l = 1$ and $E_n = 0.025$ eV, so that in a sense the P-neutron trajectory is already so far off the actual nucleus that no interaction can take place. Thus for slow neutrons, only S-wave scattering or absorption occurs; as the energy increases, higher partial waves progressively enter into interaction. At energies of the order of MeV, the particles have a wavelength of comparable magnitude to R, and, as a result, σ and σ_g can become roughly equal in value. This general argument should not, however, be regarded to apply rigorously to all cases: the low measured cross section of the $^{13}C(n,\gamma)^{14}C$ reaction is but one of many exceptions to the above prediction.

3.5.6. Cross Sections Referred to Electrons and Atoms

Although we have so far been concerned exclusively with cross sections which express the probability of interaction with a nucleus, the realm of nuclear physics can also embrace the consideration of certain *extranuclear* processes, such as collisions and interactions with the electrons surrounding the nucleus in an atom. For instance, in Section 4.1.2 we shall deal with the Compton effect, in which the energy of a γ-ray beam is degraded as a result of collisions with atomic electrons. In such cases, it is expedient to base the calculation upon a consideration of the interaction with a single electron and to derive the reaction probability in terms of a CROSS SECTION PER ELECTRON, $_e\sigma$, a quantity independent of the target material. In the absence of interference effects and neglecting binding-energy effects of the electrons in different shells, the CROSS SECTION PER ATOM, $_a\sigma$, is derived by direct summation over the Z electrons in the atom:

$$_a\sigma = Z \, _e\sigma \qquad (3\text{-}103)$$

The distinguishing suffixes are written on the left to avoid confusion with the usual nomenclature in nuclear physics. It will be appreciated that in such cases as those above it is logically preferable to speak of a cross section per *atom* rather than per nucleus, although the numerical values in either case are identical.

Also, when there is no interference and $_a\sigma = Z \, _e\sigma$, the mass-attenuation coefficient can be written as

$$\frac{\mu}{\rho} = N_A \frac{Z}{A} \, _e\sigma \qquad (3\text{-}104)$$

This has a roughly constant value for any particular reaction throughout the entire range of nuclides, since to a rough approximation $Z/A \simeq$ constant.

3.5.7. CLASSICAL ELASTIC COULOMB SCATTERING CROSS SECTION (RUTHERFORD FORMULA)

In connection with his proposal of a new model of the atom as constituted of a very small nucleus in which mass and charge are concentrated, surrounded by a much more diffuse electron distribution, Rutherford [Ru 11] derived the classical differential cross section for the elastic scattering of charged particles under the influence of a Coulomb force. The theory was prompted by the implications of experimental results obtained by Geiger and Marsden [Gei 09] who, at Rutherford's instigation, extended the range of a previous set of investigations to include the determination of the *large*-angle scattering probability for α particles incident upon heavy-element targets. On the basis of the Thomson model then extant or the consideration of multiple scattering,† this probability for back-scattering was expected to be vanishingly small, yet the measurements showed the incidence to be surprisingly high—a finding which could be explained only by supposing that atoms do not have a fairly homogeneous charge distribution but rather that a massive concentration of charge at the center of the atom is responsible for an interaction powerful enough to cause back-scattering of 7.68-MeV RaC′($= {}^{214}$Po) α particles when these impinge upon atoms of gold, silver, copper, and aluminum. The fact that the cross section for Coulomb scattering derived from Rutherford's model was found to give an excellent qualitative and quantitative description of the observed results provided strong support for replacing Thomson's atomic theory by that of Rutherford, especially when subsequent careful measurements by Geiger and Marsden [Gei 13] reinforced other corroborative evidence pointing to the validity of this model, which was then further developed by Bohr.

In its essentials, the Rutherford derivation of the Coulomb cross section ran as follows:

Under the action of a repulsive Coulomb interaction, a particle of charge Z_1e, mass m_1 and velocity u approaching a stationary target of charge Z_2e and mass m_2 (which is assumed to be very large, so that no recoil effect needs to be considered), will follow a hyperbolic path, as shown in Fig. 3-9. The scattering center m_2 will lie at the outer focus S of the hyperbola whose apex A lies at a distance x from S and whose asymptotes each make an angle α with

---

† Statistical multiple-scattering theory indicates that the probability for an α particle to be deflected through an angle greater than ϑ in traversing a thin foil of thickness d is $\exp(-\vartheta^2/\overline{\vartheta^2})$, where $(\overline{\vartheta^2})^{1/2}$ is the root-mean-square (rms) angle of deflection and is proportional to d. Thus in the case of 8-MeV α particles incident upon a gold target 4×10^{-5} cm thick, the rms deflection is $(\overline{\vartheta^2})^{1/2} = 1°$, and hence the probability of deflection beyond $\vartheta \geq 10°$ by multiple scattering is only about e^{-100}, a value far below the observed incidence.

the line SA. The scattering angle ϑ is accordingly

$$\vartheta = \pi - 2\alpha \tag{3-105}$$

and the velocity v of m_1 at the apex A is given by the equation for energy conservation,

$$\frac{1}{2} m_1 u^2 = \frac{1}{2} m_1 v^2 + \frac{Z_1 Z_2 e^2}{x} \tag{3-106}$$

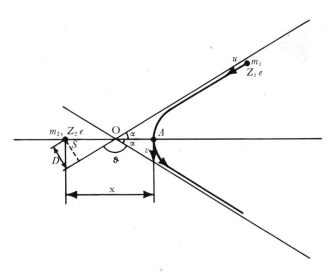

Fig. 3-9. Geometrical relationships for a classical Rutherford trajectory pursued by an incident particle m_1 upon interaction with a scattering center m_2.

In terms of the COLLISION DIAMETER b, which is the distance of closest approach in a head-on collision and has a value

$$b = \frac{Z_1 Z_2 e^2}{\frac{1}{2} m_1 u^2} \tag{3-107}$$

(equivalent to x when v is set to zero in (3-106)), this can be rewritten as

$$u^2 = v^2 + \frac{b}{x} u^2 \tag{3-108}$$

whence

$$\frac{v^2}{u^2} = 1 - \frac{b}{x} \tag{3-109}$$

Strictly speaking, the collision diameter is defined in terms of the REDUCED MASS

$$\mu = \frac{m_1 m_2}{m_1 + m_2} \tag{3-110}$$

as

$$b = \frac{|Z_1 Z_2| e^2}{\frac{1}{2} \mu u^2} \tag{3-111}$$

but when $m_1 \ll m_2$ the reduced mass μ approaches m_1 and b assumes the value (3-107).

The IMPACT PARAMETER D follows from a consideration of the angular momentum about S:

$$m_1 u D = m_1 v x \tag{3-112}$$

and therefore

$$D^2 = x^2 \frac{v^2}{u^2} = x(x - b) \tag{3-113}$$

By making use of the analytical properties of a hyperbola, we can eliminate x from this. The eccentricity of a hyperbola is

$$\epsilon = \frac{1}{\cos \alpha} = \frac{\overline{SO}}{\overline{OA}} \tag{3-114}$$

whence

$$x \equiv \overline{SA} = \overline{SO} + \overline{OA} = \overline{SO}(1 + \cos \alpha) \tag{3-115}$$

i.e.,

$$x = \frac{D(1 + \cos \alpha)}{\sin \alpha} = D \cot \frac{\alpha}{2} \tag{3-116}$$

Substituting in (3-113), we find

$$D^2 = D^2 \cot^2 \frac{\alpha}{2} - Db \cot \frac{\alpha}{2} \tag{3-117}$$

i.e.,

$$b = 2D \cot \alpha = 2D \tan \frac{\vartheta}{2} \tag{3-118}$$

since

$$\alpha = \frac{\pi}{2} - \frac{\vartheta}{2} \tag{3-119}$$

An interesting feature of the calculation is the fact that formulae such as (3-118), which stem from a simplified derivation in which the target recoil is excluded, can straightway be transcribed into the rigorous expressions which take account of target recoil. One need merely replace the laboratory scattering angle ϑ by the corresponding center-of-mass angle ϑ' throughout. Of course, when $m_1 \ll m_2$ the distinction between these systems disappears and $\vartheta \approx \vartheta'$; we shall, however, retain the distinction and rewrite (3-118) in its general form

$$b = 2D \tan \frac{\vartheta'}{2} \tag{3-120}$$

It is obvious from this that when the impact parameter D is equal to the collision radius $b/2$, the center-of-mass scattering angle is $\vartheta' = 90°$.

The differential cross section for scattering into a solid angle $d\Omega' = 2\pi \sin \vartheta' \, d\vartheta'$ corresponding to an angular aperture between ϑ' and $\vartheta' + d\vartheta'$ (equivalent to impact parameters lying between D and $D - dD$, and assuming the azimuthal distribution to be isotropic) is the area of a ring having a mean radius D and width $-dD$,

$$d\sigma' = -2\pi D \, dD \tag{3-121}$$

and from (3-120) we have

$$D = \frac{1}{2} b \cot \frac{\vartheta'}{2} \quad \text{and} \quad dD = -\frac{1}{4} b \, \mathrm{cosec}^2 \frac{\vartheta'}{2} \, d\vartheta' \tag{3-122}$$

which, upon substitution in the above yields

$$d\sigma' = \frac{1}{4} \pi b^2 \cot \frac{\vartheta'}{2} \, \mathrm{cosec}^2 \frac{\vartheta'}{2} \, d\vartheta' = \left(4\pi \sin \frac{\vartheta'}{2} \cos \frac{\vartheta'}{2} \, d\vartheta' \right) \left(\frac{b^2}{16 \sin^4 (\vartheta'/2)} \right) \tag{3-123}$$

Thus the RUTHERFORD SCATTERING CROSS SECTION (also termed the COULOMB CROSS SECTION or CLASSICAL CROSS SECTION) is

$$\sigma_R(\vartheta') \equiv \sigma_C(\vartheta') \equiv \frac{d\sigma'}{d\Omega'} = \frac{b^2}{16 \sin^4 (\vartheta'/2)} \equiv \left(\frac{Z_1 Z_2 e^2}{\frac{1}{2}\mu u^2} \frac{1}{4 \sin^2 (\vartheta'/2)} \right)^2 \tag{3-124}$$

This classical result, illustrated graphically in Fig. 3-10 holds rigorously when the interaction is purely electrostatic, and this is effectively the case when the minimum distance of approach b (which is equal to x only when $D = 0$ and $v = 0$) significantly exceeds the target radius or the rationalized de Broglie wavelength of the projectile, i.e., when

$$b \gg \frac{\hbar}{m_1 u} \tag{3-125}$$

or

$$\left| \frac{2Z_1 Z_2 e^2}{\hbar u} \right| \gg 1 \tag{3-126}$$

This condition may also be expressed in terms of the fine-structure constant

$e^2/\hbar c \simeq 1/137$ and the reduced velocity $\beta \equiv u/c$ as

$$\frac{b}{\lambda} = \frac{2Z_1 Z_2}{137\beta} \gg 1 \qquad (3\text{-}127)$$

For 7.8-MeV α particles on gold, the above critical parameter takes the

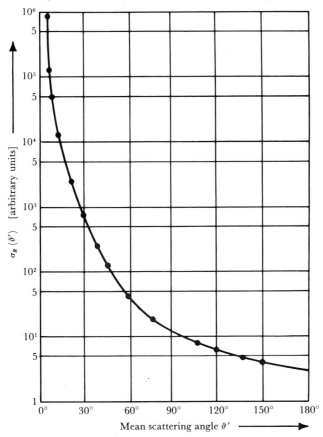

Fig. 3-10. Differential Rutherford scattering cross section for the single scattering of α radiation by a thin gold foil. The theoretical curve, having a $1/\sin^4(\vartheta/2)$ form, provides a perfect fit to the experimental data of Geiger and Marsden [Gei 13]. (Adapted from [Ev 55]. Used by permission of McGraw-Hill Book Co.)

value 36, which vindicates Rutherford's application of the above theory to the experimental data of Geiger and Marsden.

As it stands, the formula (3-124) predicts an infinite differential cross section at very small (forward) angles (corresponding to very large impact parameters) and, in its integrated form, an infinite total cross section, but this is in practice obviated by the fact that it needs to be modified when applied to very small angles because the effect of the screening of the nuclear charge by the atomic electrons then becomes appreciable.

It can be shown that the Rutherford formula holds for pure Coulomb scattering not only classically, but also quantum-mechanically. Modifications arise in a rigorous quantum-mechanical treatment only in the case of collisions between *indistinguishable like particles*, e.g., the scattering of α particles on helium nuclei.

3.5.8. Coulomb Scattering Cross Section for Like Particles (Mott Formula)

As first derived by Mott [Mo 30] (see also [Mo 49], pp. 90 ff., and [La 58a], pp. 423 ff.), the differential cross section in the center-of-mass system for the Coulomb scattering of two like particles having an initial relative velocity u and a reduced mass $\mu = \frac{1}{2}m$ is

$$\sigma_M(\vartheta') \equiv \left(\frac{d\sigma'}{d\Omega'}\right)_{\text{Mott}}$$

$$= \left(\frac{Z^2 e^2}{\frac{1}{2}\mu u^2}\right)^2 \left[\frac{1}{16 \sin^4(\vartheta'/2)} + \frac{1}{16 \cos^4(\vartheta'/2)} + \frac{\Phi}{\sin^2(\vartheta'/2)\cos(\vartheta'/2)}\right] \tag{3-128}$$

where

$$\Phi \equiv K \cos\left[\frac{Z^2 e^2}{\hbar u} \ln\left(\tan^2 \frac{\vartheta'}{2}\right)\right] \tag{3-129}$$

with $K = \frac{1}{8}$ in the collision of spin-0 particles (α particles), or $K = -\frac{1}{16}$ in the collision of spin-$\frac{1}{2}$ particles (electrons, protons).

In Eq. (3-128), which is termed the Mott scattering formula, it is evident that the first term, identical to the Rutherford expression (3-124), refers to the scattered particle, and the second term, of the same form, to the indistinguishable recoil particle which is emitted at 90 degrees to the scattered particle in the laboratory system. It is the third term, dependent upon Φ and thence upon $\hbar$, which represents a wave-mechanical quantum interference term. Thus the Mott formula reduces to the Rutherford formula when $\Phi \to 0$. Strictly, the Mott formula is a nonrelativistic expression applicable only to *low-energy* Coulomb scattering of like particles (e.g., to protons of energy below about 0.4 MeV), particularly since at higher energies the interaction is unlikely to be of a pure Coulomb type since strong interparticle forces of short range begin to play a role. At low energies, $e^2 \gg \hbar u$, and Φ is a rapidly oscillating function which averages out to zero when meaned over even a small range of angles ϑ'.

A good approximation, applicable to electrons of energy greater than 1 keV (but nonrelativistic), is, with $K = -\frac{1}{16}$,

$$\frac{d\sigma'}{d\Omega'}_{\text{Mott}} \cong \left(\frac{e^2}{\frac{1}{2}\mu u^2}\right)^2 \frac{1}{4}\left[\frac{1}{\sin^4(\vartheta'/2)} + \frac{1}{\cos^4(\vartheta'/2)} - \frac{1}{\sin^2(\vartheta'/2)\cos^2(\vartheta'/2)}\right]\cos\frac{\vartheta'}{2}$$

(3-130)

Furthermore, in the high-energy nonrelativistic limit, one has $e^2 \ll \hbar u$ and $\Phi \approx K$, so that for spin-$\frac{1}{2}$ particles,

$$\frac{d\sigma'}{d\Omega'}_{\text{Mott}} \approx \left(\frac{e^2}{\frac{1}{2}\mu u^2}\right)^2\left(\frac{4 - 3\sin^2\vartheta'}{4\sin^4\vartheta'}\right)$$

(3-131)

The foregoing formulae can be expressed in terms of ϑ, the angle of scattering in the laboratory system, on noting that

$$\vartheta' = 2\vartheta$$

(3-132)

and

$$d\Omega' = 4\cos\vartheta\ d\Omega$$

(3-133)

3.6. Transition Probability and Cross Section

3.6.1. Box Normalization and the Relationship of Transition Probability per Unit Time to the Differential Cross Section for Elastic Scattering

A simple, direct relationship exists between the transition probability per unit time W_{fi} and the differential cross section $d\sigma$ for emission of radiation into an element of solid angle $d\Omega$. Before we can express this algebraically, however, we must take a closer look at certain features of the calculations presented in Section 3.4. In particular, we need to consider the sense in which the eigenfunctions u_{i} and u_{f} are regarded as normalized in the condition (3-46).

In the wave-mechanical treatment of energy eigenfunctions it is appropriate to employ "box normalization," as described, for example, by Schiff [Schi 55, pp. 43ff.] The system described by the wave function u_{i} or u_{f} is deemed to be confined within a cube of volume L^3 having perfectly rigid sides at which the functions obey periodic boundary conditions (the amplitude of the wave function vanishes at the walls of the enclosure). The integral $\int |u(q)|^2\ dq$ then converges for all eigenfunctions $u(q)$, and the $u(q)$ can be normalized by choosing their coefficients to be such as to make the volume integral unity. This mode of normalization applies equally to MOMENTUM EIGENFUNCTIONS, which are solutions of equations of the type

$$i\hbar\frac{\partial}{\partial x}u_{\text{p}}(q) = p_x u_{\text{p}}(q)$$

(3-134)

and all cyclic permutations of this form. The eigenfunctions take the form

$$u = L^{-3/2} \exp[i(\mathbf{k} \cdot \mathbf{q})] \tag{3-135}$$

which corresponds to a plane wave and is well suited to describe the motion of particles in collision problems. In the normalization of energy or momentum eigenfunctions, the finite box volume L^3 can be made arbitrarily large by allowing L to expand at will, and in the limit one can conceive of L as tending to infinity, with the origin of the wave function located at the center of the infinite periodicity cube.

The periodicity dimension L enters directly into a consideration of the relationship between transition probability per unit time and differential cross section, as we shall see by focusing attention on the case of elastic scattering, which we have here selected as a straightforward illustrative example, postponing until later the consideration of decay processes. As a result of an interaction between projectile and target, elastic scattering can occur, which represents a transition in the momentum state of the projectile. The probability of such an occurrence as evaluated by a perturbation approximation method is simply the transition probability per unit time as given by the "Golden Rule No. 2" [Eq. (3-67)], and is none other than the ratio of the effective interaction volume swept through by the incident particle to L^3, the total volume available for the system to occupy. A particle initially moving with a velocity v_i in the center-of-mass system sweeps out an effective interaction volume $v_i \, d\sigma$ in unit time, and the ratio of this to L^3 gives W_{fi},

$$W_{fi} = \frac{v_i}{L^3} \, d\sigma \tag{3-136}$$

expressing the likelihood of elastic scattering into a solid angle $d\Omega$ per unit time. It should be noted that all quantities are referred to the center-of-mass system, but for convenience they have been left unprimed to conform with customary usage.

It may also be mentioned that elastic scattering is distinguished from inelastic scattering by the fact that in the center-of-mass system there is *no transfer of energy* from projectile to target although there *is* a transfer of momentum, as evinced by the change in the flight direction of the particle even though in *magnitude* its velocity remains unchanged. This suggests the use of momentum eigenfunctions to determine the transition probability rather than energy eigenfunctions, since there is basically no "energy transition." It need hardly be pointed out that in the laboratory system there is both energy and momentum transfer in elastic collisions, just as in inelastic, for the target originally at rest acquires a recoil energy. If it acquires additional energy over and above this in the form of an "excitation energy," the collision is an inelastic one.

3.6.2. ELASTIC SCATTERING MATRIX ELEMENT

To calculate the differential cross section which expresses the scattering probability in absolute terms, we make use of the relation (3-136) to obtain $d\sigma$ in terms of W_{fi} and derive the latter from the "Golden Rule", Eq. (3-67), substituting therein appropriate expressions for the perturbation matrix element and the final-state level density, the derivation of which we now present.

In the formulation of the treatment, it is convenient to introduce WAVE-NUMBER (or PROPAGATION) VECTORS $\mathbf{k}_i$ and $\mathbf{k}_f$, defined as

$$\mathbf{k} \equiv \frac{1}{\lambda} = \frac{\mathbf{p}}{\hbar} = \frac{m\mathbf{v}}{\hbar} \tag{3-137}$$

where all quantities (including λ) are throughout this section referred to the center-of-mass system. It is also useful to introduce a symbol for the vector difference

$$\mathbf{k}_i - \mathbf{k}_f \equiv \mathbf{K} \tag{3-138}$$

We note that for elastic scattering, $|\mathbf{v}_i| = |\mathbf{v}_f| = v$ and therefore

$$|\mathbf{k}_i| = |\mathbf{k}_f| = k \tag{3-139}$$

Introducing orthonormal momentum eigenfunctions (representing incoming and outgoing plane waves) to serve as unperturbed eigenfunctions,

$$u_i = \frac{1}{L^{3/2}} \exp[i(\mathbf{k}_i \cdot \mathbf{q})], \qquad u_f = \frac{1}{L^{3/2}} \exp[i(\mathbf{k}_f \cdot \mathbf{q})] \tag{3-140}$$

and assuming the perturbation H' to be produced by a spherically symmetric scattering force we can write the transition matrix element as

$$H'_{fi} \equiv \int u_f^* H' u_i \, d\mathbf{q} = \frac{1}{L^3} \int H' \exp[i(\mathbf{K} \cdot \mathbf{q})] \, d\mathbf{q} \tag{3-141}$$

or, in terms of polar coordinates,

$$H'_{fi} = \frac{1}{L^3} \int_0^\infty \int_0^\pi \int_0^{2\pi} H'(r) \exp[i(\mathbf{K} \cdot \mathbf{r})] \, r^2 \sin\theta \, dr \, d\theta \, d\phi \tag{3-142}$$

The integration over ϕ and θ is straightforward, and yields the result

$$H'_{fi} = \frac{2\pi}{L^3} \int_0^\infty \int_1^{-1} H'(r) \exp(iKr\cos\theta) \, r^2 \, dr \, d(-\cos\theta) \tag{3-143}$$

$$= \frac{4\pi}{L^3} \int_0^\infty H'(r) \, r^2 \, \frac{\sin Kr}{Kr} \, dr \tag{3-144}$$

which is a suitable form for substitution into the Golden Rule formula, together with the final-state level density $\rho(E_f)$ which we derive next.

3.6.3. Level Density

The final states whose density we seek to determine are distinguished by the orientation of the wave-number vector $\mathbf{k}_f$. The level density is defined by the expression

$$\rho(E_f) \equiv \frac{dN_f}{dE_f} \tag{3-145}$$

where dN_f is the number of final states per energy interval dE_f lying within an energy band of width $\Delta E = \hbar/t$ about the energy E_f. When u_f is normalized within a cube whose sides, of length L, lie along the axes of a Cartesian coordinate system, it follows that L must be an integer multiple of $\lambda = 2\pi\lambdabar$, and hence the vector

$$\mathbf{n} = \frac{L}{2\pi} \mathbf{k} \tag{3-146}$$

must have integer components—in $\mathbf{n}$-space the possible states of $\mathbf{n}$ are those for which the tip of the $\mathbf{n}$-vector coincides with integer lattice points. We need to write this in differential form

$$dn = \frac{L}{2\pi} dk \tag{3-147}$$

to see that by analogy, the number of end-points of the $\mathbf{k}$-vector in $\mathbf{k}$-space (having orthogonal coordinates k_x, k_y, k_z) is

$$dN = \left(\frac{L}{2\pi}\right)^3 d^3k \tag{3-148}$$

within a volume element of magnitude $d^3k = dk_x \, dk_y \, dk_z$ or, in polar form

$$d^3k = k^2 \, dk \, d\Omega \tag{3-149}$$

The quantity dN is none other than the number of possible final states dN_f whose wave vector has a magnitude which lies in the range k_f to $k_f + dk_f$ and a direction which lies within the solid angle $d\Omega$.

Also, differentiating the final-state energy

$$E_f = \frac{p_f^2}{2m} = \frac{(\hbar k_f)^2}{2m} \tag{3-150}$$

we get

$$dE_f = \frac{\hbar^2 k_f}{m} dk_f \tag{3-151}$$

which can, together with (3-148) and (3-149), be substituted in (3-145) to yield the desired level density:

$$\frac{dN_f}{dE_f} = \rho(E_f) = \left(\frac{L}{2\pi}\right)^3 k_f^2 \, d\Omega \frac{m}{\hbar^2 k_f} = \left(\frac{L}{2\pi}\right)^3 \frac{m}{\hbar^2} k_f \, d\Omega \tag{3-152}$$

3.6.4. ELASTIC SCATTERING CROSS SECTION (BORN COLLISION FORMULA)

We are now in a position to substitute explicit expressions in the Golden Rule formula (3-67). Noting that $|\mathbf{k}_i| = |\mathbf{k}_f| = k$, we can write the transition probability per unit time as

$$W_{fi} = \frac{2\pi}{\hbar} \rho(E_f) |H'_{fi}|^2 = \frac{2\pi}{\hbar} \left(\frac{L}{2\pi}\right)^3 \frac{m}{\hbar^2} k \, d\Omega \left| \frac{4\pi}{L^3} \int H'(r) r^2 \frac{\sin Kr}{Kr} \, dr \right|^2$$

(3-153)

We then substitute

$$v_i = |\mathbf{v}_i| = |\mathbf{v}_f| = \frac{\hbar k}{m}$$

(3-154)

in the expression (3-136) relating W_{fi} to the cross section $d\sigma$ to get

$$\frac{d\sigma}{d\Omega} = \frac{L^3}{v_i} \frac{W_{fi}}{d\Omega} = \frac{L^3 m}{\hbar k} \frac{W_{fi}}{d\Omega}$$

(3-155)

$$= \frac{L^3 m}{\hbar k} \frac{2\pi}{\hbar} \left(\frac{L}{2\pi}\right)^3 \frac{m}{\hbar^2} k \left(\frac{4\pi}{L^3}\right)^2 \left| \int H'(r) r^2 \frac{\sin Kr}{Kr} \, dr \right|^2$$

(3-156)

i.e.

$$\frac{d\sigma}{d\Omega} = \left| \frac{2m}{\hbar^2} \int_0^\infty H'(r) r^2 \frac{\sin Kr}{Kr} \, dr \right|^2$$

(3-157)

This formula, termed the BORN COLLISION FORMULA, gives the result of applying the quantum-mechanical Born approximation to the derivation of the collision cross section for the case of elastic scattering brought about by an interaction whose strength is $H'(r)$. We have derived the result by treating the spherically symmetric force as a perturbation, and left it general inasmuch as we have not stipulated the form taken on by $H'(r)$.

Criteria for the validity of applying the Born approximation and using the Born scattering formula have been discussed by Schiff [Schi 55, pp. 169 and 170]. For our purposes it suffices to note the comment of Williams [Wi 45] that the domain of validity corresponds to the extreme opposite of the domain of validity of classical scattering theory as expressed by the condition (3-127), namely

$$\frac{b}{\lambda} \equiv \frac{2 Z_1 Z_2}{137 \beta} \ll 1$$

(3-158)

Accordingly, the Born formula cannot be used to treat the scattering of low-energy particles for which $\beta \equiv v/c$ is small—it cannot, thus, be used for evaluating the cross section for the scattering of low-energy electrons by the Coulomb field of an atom. But in very many other cases of practical interest it can be used to advantage when appropriate expressions are substituted for H' and the integration is carried out analytically or numerically. The perturbation H' is usually described by a scattering potential having a definite depth, range, and shape. We illustrate this by two examples.

The simplest type of potential is presented by a square well of depth V_0 and width (range) a. In this case

$$H'(r) = \begin{cases} -V_0 & \text{if } r \leq a \\ 0 & \text{if } r > a \end{cases} \qquad (3\text{-}159)$$

and the integration limits in Eq. (3-157) after substituting the above for $H'(r)$ can accordingly be reduced to just 0 to a.

For *elastic scattering,*

$$K \equiv |\mathbf{k_i} - \mathbf{k_t}| = 2k \sin \frac{\vartheta'}{2} \qquad (3\text{-}160)$$

where ϑ' is the scattering angle in the center-of-mass system.

Hence we derive the differential cross section for elastic scattering by a square well explicitly from Eq. (3-157) as

$$\frac{d\sigma'}{d\Omega'} = \left| -\frac{2mV_0}{K\hbar^2} \int_0^a r \sin Kr \, dr \right|^2 = \left| \frac{2mV_0}{K^3 \hbar^2} (Ka \cos Ka - \sin Ka) \right|^2 \qquad (3\text{-}161)$$

with K given by Eq. (3-160). At high energies $(ka \gg 1)$ the scattering distribution peaks strongly in the forward direction: most of the particles are scattered into a cone whose opening angle is of the order of $1/ka$.

By way of a second illustration, we consider the case of a Coulomb potential,

$$H'(r) = \frac{Z_1 Z_2 e^2}{r} \qquad (3\text{-}162)$$

which, despite the fact that it has no definite range, nevertheless proves amenable in the majority of instances to treatment in terms of the Born approximation. This example is instructive in that it provides a very short and elegant derivation of the Rutherford scattering cross section (see Section 3.5.7). Inserting (3-162) into the Born scattering formula, we obtain

$$\frac{d\sigma'}{d\Omega'} = \frac{4m_1^2 \, Z_1^2 Z_2^2 e^4}{K^2 \hbar^4} \left| \int_0^\infty \sin Kr \, dr \right|^2 = \frac{b^2 m_1^4 u^4}{K^2 \hbar^4} \left| \int_0^\infty \sin Kr \, dr \right|^2 \qquad (3\text{-}163)$$

where

$$b \equiv \frac{Z_1 Z_2 e^2}{\frac{1}{2}m_1 u^2} \qquad (3\text{-}164)$$

is the collision diameter and the notation of Section 3.5.7 has been employed. The value of the integral in Eq. (3-163) poses a problem at its upper limit, since it oscillates about zero there; it is, in fact, convenient to take the integral to vanish at its upper limit (this can be substantiated by a more rigorous treatment in which screening effects are considered) and to write

$$\frac{d\sigma'}{d\Omega'} = \frac{b^2}{K^2} \left(\frac{m_1 u}{\hbar} \right)^4 \left| \left[-\frac{\cos Kr}{K} \right]_{r=0} \right|^2 = \frac{b^2}{K^2} \left(\frac{m_1 u}{\hbar} \right)^4 \frac{1}{K^2} = b^2 \left(\frac{m_1 u}{K\hbar} \right)^4 \qquad (3\text{-}165)$$

If now we substitute the relation

$$K = 2k \sin \left(\frac{\vartheta'}{2} \right) = 2 \frac{m_1 u}{\hbar} \left(\sin \frac{\vartheta'}{2} \right) \qquad (3\text{-}166)$$

we obtain the Rutherford formula in an identical form to Eq. (3-124):

$$\frac{d\sigma'}{d\Omega'} = \frac{1}{16} b^2 \operatorname{cosec}^4\left(\frac{\vartheta'}{2}\right) \tag{3-167}$$

An approach similar in principle to the above can be used in the consideration of inelastic scattering, but the mathematics becomes considerably more complicated. Examples of such calculations may be found in Schiff [Schi 55, pp. 205 ff.] and Flügge [Fl 64, pp. 327 ff.]. Instead of pursuing these considerations further, we will next turn our attention to the evaluation and discussion of cross sections for the coherent and incoherent scattering of electromagnetic radiation by the electrons in an atom, since this will provide us with some further distinctive characteristics of cross sections and at the same time furnish an insight into processes which prevail when ionizing radiations pass through matter.

EXERCISES

3-1 Calculate the ratio of the pure gravitational attraction to the Coulomb attraction between proton and electron in a hydrogen atom, assuming the electron to be in the innermost Bohr orbit. How does this value compare with the relative magnitude of the gravitational and electromagnetic interaction as deduced from coupling constants?

3-2 Compare the Coulomb energy of a ^{238}U nucleus with its corresponding gravitational energy, $E_G = \frac{3}{5}\gamma(m^2/R)$, where $\gamma = 6.67 \times 10^{-8}$ dyn cm^2 g^{-2}.

3-3 What is the minimum value of the kinetic energy of a proton in electron-volts if, emitted normally from the Sun's surface, it is to escape from the solar system? What temperature corresponds to this value of energy? (Assume that only gravitational forces act. Sun's radius $R = 6.96 \times 10^{10}$ cm; the Sun's mass M is to be calculated from the Earth-Sun separation $D = 1.5 \times 10^{13}$ cm; other constants in Appendix I.)

3-4 Would an electron situated on the surface of a sphere whose mass is 10^{10} times that of the electron and which carries the same amount of negative charge as the electron experience a net attraction or repulsion? How large would the mass of the sphere be at equilibrium?

3-5 On irradiating a thick aluminum target with 7.8-MeV α particles having a mean range $\bar{R} = 2.5$ mg/cm^2, one finds that 8 protons are produced in an (α,p) process for every million incident α particles. What is the mean cross section for this process? (Density of aluminum: $\rho = 2.69$ g cm^{-3}.)

3-6   The cross section for reactor antineutrinos to interact with protons is
of the order $\sigma \approx 10^{-43}$ cm^2. What is the probability for a neutrino to
be captured in an interaction as it passes diametrically through the
Earth? (Mean density of Earth: $\bar{\rho} \simeq 5$ g cm^{-3}; radius of Earth:
$R = 6.38 \times 10^8$ cm. Assume a one-to-one neutron-to-proton ratio in
terrestrial matter.)

3-7   A thin target film consisting of an element X which has two isotopes
X_1 and X_2 of natural isotopic abundance x_1 and x_2, respectively, is
subjected to bombardment by a beam of thermal neutrons whose
flux is Φ neutrons cm^{-2} sec^{-1}. Determine the isotopic composition
(x_1', x_2') of X after a time t if the isotope X_1 can undergo an (n,α) reaction
with thermal neutrons (the cross section being σ) whereas X_2 cannot
react with them at all.

3-8   The fission cross section for thermal neutrons on natural uranium
(isotopic composition: 0.72 percent ^{235}U, 99.27 percent ^{238}U, <0.01
percent others) is $\sigma_f = 4.22$ b. Calculate the fission cross section of
^{235}U for thermal neutrons, taking account of the fact that ^{238}U does
not enter into fission with thermal neutrons. If a thermal neutron flux
$\Phi = 10^{10}$ neutrons cm^{-2} sec^{-1} impinges upon a thin target of natural
uranium, how many fission events occur per second when the target
is 0.01 cm thick? (Density of uranium: $\rho = 18.68$ g cm^{-3}.)

3-9   A 6-MeV α particle makes a head-on approach toward a stationary
Au atom. Plot the force as a function of separation, assuming a homogeneous distribution of positive electric charge throughout (a) the
entire atom, regarded as a sphere of radius $R_A = 10^{-8}$ cm; (b) the
atomic *nucleus* $^{197}_{79}$Au, whose radius is $R = r_0 A^{1/3} = 1.4 \ A^{1/3}$ fm.

3-10   The Rutherford formula (3-111) is able to account satisfactorily for
the scattering of protons on a thin $^{232}_{90}$Th target up to an incident kinetic
energy $E^{(\text{kin})} = 4.3$ MeV. Use this observation to estimate a value for
the range of nuclear forces which bring about the deviation.

Chapter 4

PASSAGE OF IONIZING RADIATION
THROUGH MATTER

Ionizing Effects of Electromagnetic Radiation and
Charged Particles

Ionizing processes embrace a wide field within nuclear, atomic, and solid state physics. The role they play in nuclear and particle physics is of especial significance, however, since on the one hand all experimental investigations involve the passage of ionizing radiations through matter, and on the other the operation of particle detectors depends essentially upon ionization effects.

It is expedient to subdivide the consideration of these effects into a section concerned with electromagnetic radiation, e.g., x rays and γ rays passing through matter, and one concerned with the passage of heavy and light charged particles, e.g., protons and electrons, respectively. A qualitative difference should be noted from the outset, namely, the ATTENUATION OF γ RADIATION as it passes through a material is evinced as an *intensity diminution,* and not as an energy change, according to the exponential law

$$I(x) = I_0 e^{-\mu x} \tag{4-1}$$

which gives the intensity (i.e., number of quanta per square centimeter per second) of an electromagnetic beam of original intensity I_0 after traversing a thickness x of a homogeneous material whose ATTENUATION COEFFICIENT is μ for γ radiation of the particular constant energy under consideration. As against this, *charged particles* passing through matter are *slowed down* by an amount determined by the STOPPING POWER of the material for the given

97

particles at an energy which diminishes as the particles traverse the target. The change in the particle flux can thus involve a change in energy as well as a change in the number of particles.

4.1. Survey of Electromagnetic Interaction Processes

We start by focusing attention upon *electromagnetic radiation* and noting that a γ ray is confronted by a prolific choice of interaction possibilities as it impinges upon matter. The actual process or processes in which it participates is decided by the relative magnitudes of the partial cross sections for each of the feasible "competing reactions"—a situation not unlike that in Section 3.3 in which a particle avails itself of interaction modes according to the available possibilities between which it discriminates. This wide-ranging choice requires that we restrict ourselves almost exclusively to considering those effects which are particularly prominent at low energies, namely:

(a) THOMSON AND COMPTON SCATTERING on electrons at such incident energies that the binding energy of the scattering electrons in the target atoms may by comparison with the γ-ray energy be neglected and the electrons treated as if "free," yielding fairly simple but only rather approximate results for the probability of processes in which *some* of the γ-ray energy is transferred to the scattering particles;

(b) THE PHOTOELECTRIC EFFECT, in which *all* the γ-ray energy is transferred to bound atomic electrons, causing them to be ejected from the atom;

(c) PAIR PRODUCTION, in which the γ-ray energy is high enough to give rise to the creation of an electron-positron pair $(\gamma \to e^- + e^+)$ in the vicinity of a collision partner such as a nucleus or an electron.

These and other γ-interaction processes are discussed in detail by Davisson [Da 52b], [Si 65, pp. 37 ff.], Heitler [He 54, pp. 175 ff.], Evans [Ev 55, pp. 672 ff. and 819 ff.], Fano *et al.* [Fa 59], and Leipunskii *et al.* [Le 65]; some impression of their scope and characteristics may be derived from Table 4-1, in which the principal interactions have briefly been summarized.

4.2. Thomson and Compton Scattering of Gamma-Radiation

It is appropriate to commence with these processes which at first sight appear altogether dissimilar but inherently are related by the fact that in the limit of low γ-ray energies the Compton scattering formulae go over into the Thomson expressions. Not only are both processes important from a historical and practical standpoint, but their treatment follows on from the considerations presented in Chapter 3, and consolidates the approach outlined there. Whereas Thomson scattering involves essentially the wave features of electromagnetic radiation, leading to a coherent scattering process, the Compton

Table 4-1. PROCESSES WHEREBY ELECTROMAGNETIC RADIATION INTERACTS WITH MATTER[a]

Process	Kind of interaction	Significant energy region	Approximate Z dependence
Scattering from electrons coherent scattering			
—Rayleigh scattering	with bound atomic electrons	<1 MeV (mainly at forward angles)	Z^2 (small angles) Z^3 (large angles)
—Thomson scattering	with "free electrons"	Independent of E	Z^2 (small angles) Z^3 (large angles)
incoherent scattering	with bound atomic electrons	<1 MeV (mainly at large angles)	Z
—Compton scattering	with "free electrons"	around 1 MeV ($\downarrow$ as $E\uparrow$)	Z
Photoelectric effect	with bound atomic electrons, causing ejection of electrons	0 to 0.5 MeV ($\downarrow$ as $E\uparrow$)	Z^5
Nuclear photoeffect	with nucleus as a whole, causing emission of photons, particles, and (above threshold) mesons	$\gtrsim 10$ MeV	
Interaction with a Coulomb field			
—pair production	with nuclear Coulomb field (elastic pair production)	>1 MeV, especially at 5–10 MeV	Z^2
	with electron Coulomb field (inelastic pair production and triplet production)	>2 MeV ($\uparrow$ as $E\uparrow$)	Z
—Delbrück scattering	in nuclear Coulomb field		Z^4
Nuclear scattering			
—coherent scattering	Mössbauer, resonance, Thomson		
—incoherent scattering	Compton, with individual nucleons	$\gtrsim 100$ MeV	

[a] A more extensive tabulation is given by Siegbahn [Si 65, pp. 38 and 39].

effect relates to the corpuscular features of the incident radiation, of which it provided one of the first vivid demonstrations, and is an incoherent scattering process: it is classed as *inelastic* although the kinematics of the interaction with electrons which are taken to be essentially unbound are those of an *elastic* collision. Thomson scattering, regarded as the low-energy (high-wavelength) limit of Compton scattering yields cross sections which approximate to those obtained experimentally at small scattering angles, but discrepancies arise when measurements are taken at larger angles. The latter data evince a diminution in the intensity of the coherently scattered radiation and a corresponding increase in the incoherent-scattering intensity characterized by the fact that its wavelength invariably exceeds that of the incident beam.

4.2.1. THOMSON CROSS SECTION

Much of the early information concerning atomic number and atomic structure was derived from x-ray scattering investigations upon atoms whose orbital electrons were considered to be essentially unbound, i.e., "free." Thomson assumed that the incident beam set each quasi-free electron into forced resonant oscillation. He then used nonrelativistic electrodynamic theory to calculate the magnitude of the cross section for the re-emission of electromagnetic radiation as a result of the induced oscillation of particles carrying a charge e. The process whereby energy is taken from the incident beam in setting the electrons into forced vibration and re-emitted as from an electric dipole corresponds to the elastic coherent scattering of electromagnetic radiation in which, in the limit $E \to 0$, there is no change in wavelength.

The cross section for this process can be derived concisely and elegantly through the use of the second "Golden Rule" (as has been done, e.g., by Fowler [Ba 62b, pp. 58ff.]) or, even more simply, deduced by taking the limiting case of the appropriate Compton cross section as the incident wavelength tends to infinity. The latter procedure will be demonstrated in Section 4.2.3; we here present a classical electrodynamic derivation.

The electric field $\mathbf{E}$ of the incident wave is, in terms of the field amplitude $\mathbf{E}_0$, given by the harmonic relation

$$\mathbf{E} = \mathbf{E}_0 \sin \omega t \tag{4-2}$$

where $\omega = c/\lambda$ is the frequency. Under the action of this field the electron is brought to vibrate about its mean position, and in terms of a displacement $\mathbf{x}$ from the latter, which corresponds to an instantaneous electric dipole moment $\mathbf{M} = e\mathbf{x}$, the equation of motion is

$$m_e \ddot{\mathbf{x}} = e\mathbf{E} = e\mathbf{E}_0 \sin \omega t \tag{4-3}$$

The rate at which energy is taken from the incident plane wave per unit perpendicular area as electrons are set into induced oscillation is equal to the incident wave intensity, expressed by the time-average of the Poynting vector, whose magnitude is

$$|\mathbf{S}_{\text{in}}| = \frac{c}{4\pi}\,\overline{|\mathbf{E}|^2} = \frac{c}{4\pi}\,\overline{|\mathbf{E}_0|^2 \sin^2 \omega t} = \frac{c}{8\pi}\,\mathbf{E}_0^2 \quad \text{erg cm}^{-2}\ \text{sec}^{-1} \quad (4\text{-}4)$$

since the average value of a $\sin^2$ fluctuation is $\frac{1}{2}$.

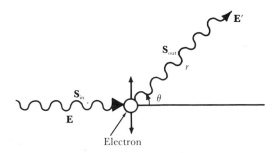

Fig. 4-1. Electromagnetic energy flux emission in the form of electric dipole radiation through induced electron oscillation (Thomson scattering). The energy flux is denoted by $\mathbf{S}_{\text{in}}$ and $\mathbf{S}_{\text{out}}$, while $\mathbf{E}$ and $\mathbf{E}'$ represent electric field intensities.

The energy flux emitted as electromagnetic dipole radiation in the direction given by the scattering angle θ in Fig. 4-1 is

$$|\mathbf{S}_{\text{out}}| = \frac{c}{4\pi}\,\overline{|\mathbf{E}'|^2} \qquad (4\text{-}5)$$

where $\mathbf{E}'$, the field strength due to an electric dipole at a distance r, is given by the classical electrodynamic theory of Hertz as

$$|\mathbf{E}'| = \frac{|\ddot{\mathbf{M}}|}{c^2 r}\,\sin\Theta = \frac{e|\ddot{\mathbf{x}}|}{c^2 r}\,\sin\Theta \qquad \checkmark (4\text{-}6)$$

The angle Θ is that between the direction of emission $\mathbf{r}$ and the polarization vector of the incident wave $\mathbf{E}$. When we substitute for $|\ddot{\mathbf{x}}|$ in (4-6) from (4-3) and insert the resultant expression into Eq. (4-5), we must again form the time-average of $|\mathbf{E}'|^2$, which in a similar manner to the above introduces a factor $\frac{1}{2}$. Furthermore, if the incident wave is unpolarized, we must build an average over all angles Θ of the form

$$\overline{\sin^2 \Theta} = \tfrac{1}{2}(1 + \cos^2 \theta) \qquad (4\text{-}7)$$

in terms of the scattering angle θ. Thus for unpolarized radiation,

$$|\mathbf{S}_{\text{out}}| = \frac{e^4 E_0^2}{16\pi m_e^2 c^3 r^2} (1 + \cos^2 \theta) \qquad \text{erg cm}^2 \text{ sec}^{-1} \qquad (4\text{-}8)$$

The energy per unit time passing through a segment of area dA perpendicular to $\mathbf{r}$ which is bounded by a solid angle element $d\Omega = dA/r^2$ is therefore

$$|\mathbf{S}_{\text{out}}| \, dA = \frac{1}{r^2} \frac{e^4 E_0^2}{16\pi m_e^2 c^3} (1 + \cos^2 \theta) \, dA = \frac{e^4 E_0^2}{16\pi m_e^2 c^3} (1 + \cos^2 \theta) \, d\Omega$$

$$(4\text{-}9)$$

The differential cross section for re-emission of radiation (Thomson scattering) into a solid angle $d\Omega$ is, by definition, the ratio of this quantity to the incident intensity $|\mathbf{S}_{\text{in}}|$,

$$\sigma_{\text{Th}}(\theta) \, d\Omega = \frac{|\mathbf{S}_{\text{out}}| \, dA}{|\mathbf{S}_{\text{in}}|} = \frac{1}{2} \left(\frac{e^2}{m_e c^2} \right)^2 (1 + \cos^2 \theta) \, d\Omega \qquad (4\text{-}10)$$

Thus in terms of the classical electron radius $r_e = e^2/m_e c^2$, the Thomson differential cross section is

$$\sigma_{\text{Th}}(\theta) = \tfrac{1}{2} r_e^2 (1 + \cos^2 \theta) \qquad \text{cm}^2/\text{electron} \qquad (4\text{-}11)$$

The structure of this differential scattering cross section per electron as a function of the scattering angle θ is shown in Fig. 4-2.

The total scattering cross section follows from integration over all scattering angles θ:

$$\sigma_{\text{Th}} = \int \sigma_{\text{Th}}(\theta) \, d\Omega = \int_0^\pi \tfrac{1}{2} r_e^2 (1 + \cos^2 \theta) \, 2\pi \sin \theta \, d\theta = \pi r_e^2 [-\cos \theta - \tfrac{1}{3} \cos^3 \theta]_0^\pi$$

$$(4\text{-}12)$$

i.e.,

$$\sigma_{\text{Th}} = \frac{8}{3} \pi r_e^2 \equiv \frac{8}{3} \pi \left(\frac{e^2}{m_e c^2} \right)^2 = 0.665 \text{ b} \qquad (4\text{-}13)$$

It will be noted that the result is a function of r_e even though in the theory the electron has implicitly been treated as a point charge. On the basis of a naive geometrical interpretation of σ_{Th} one might consider $(8/3)^{1/2} r_e$ to represent an "effective" electron radius, but this must not be taken too literally.

The result is noteworthy in that it contains no energy-dependent term and predicts no change in wavelength upon re-emission of electromagnetic radiation. Since it stands for a cross section *per electron* it should, strictly, be written as $_e\sigma_{\text{Th}}$ and the corresponding *atomic* cross section as $_a\sigma_{\text{Th}}$ for which, assuming there to be an independent interaction with each of the Z quasi-free electrons in the atom, we may write

$$_a\sigma_{Th} = Z_e\sigma_{Th} = \tfrac{8}{3}\pi r_e^2 Z \tag{4-14}$$

This linear dependence upon Z provided the basis for the elucidation of atomic numbers of light elements by Barkla [Ba 11] and others. The cross section for scattering on particles heavier than electrons, e.g., on protons, is

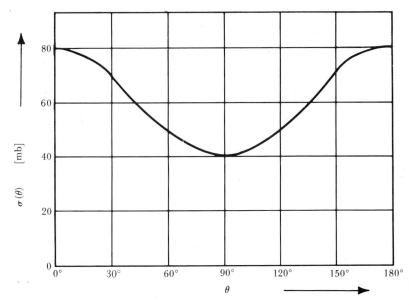

Fig. 4-2. Basic form of the differential cross section for Thomson scattering, as given by Eq. (4-11).

vanishingly small because of the inverse-square dependence upon the mass of the scattering body.

It is found that for γ-ray energies exceeding the electron binding energy but small in comparison with $m_e c^2 = 0.511$ MeV, the cross section measured at small angles approaches Thomson's value, but when the scattering is observed at progressively larger angles there is a diminution in the intensity of *coherently* scattered radiation, offset by an increase in the incoherent portion whose wavelength is larger than that of the incident radiation.

4.2.2. COMPTON EFFECT

The effect in which there is a wavelength shift in incoherent scattering of electromagnetic radiation on quasi-free atomic electrons bears Compton's name since he undertook the first experimental investigations [Co 22] and provided† a theoretical explanation [Co 23a, Co 23b, Co 24] which constituted early evidence for the corpuscular photon conception of electromagnetic radiation according to quantum ideas. The increase of wavelength as shown in Fig. 4-3, which corresponds to a decrease in energy upon scattering, can be derived from a consideration of relativistic momentum and energy conservation in a collision which is treated as though it were elastic (although

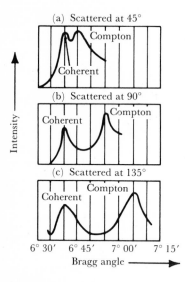

Fig. 4-3. Reproduction of Compton's measurements on x-ray lines (the $K\alpha$ lines from molybdenum) scattered from graphite, the left-hand peak in each case being the coherent line, of wavelength $\lambda = 0.71$ Å and the right-hand peak the Compton line, displaced by an amount $\Delta\lambda$, which increases as $(1 - \cos\theta_\gamma)$ (from [Sm 65]). (Reprinted with permission of Pergamon Press Ltd.)

incoherent scattering on bound electrons constitutes an *inelastic* process). The Compton effect assumes importance when the γ energy becomes comparable with, or higher than, $m_e c^2 = 0.511$ MeV. It is maximal around 1 MeV and is especially pronounced at fairly large scattering angles.

Figure 4-4 depicts the situation in which a photon of energy $h\nu$ and momentum $h\nu/c$ incident upon a stationary free electron is scattered through an angle θ_γ and emerges with an energy $h\nu'$ and momentum $h\nu'/c$. The electron recoils at an angle θ_e with an energy E_{kin} and momentum **p** which may be sufficiently large to be treated relativistically. Conservation of momentum perpendicular to the scattering plane requires that all momentum vectors be coplanar. In the scattering plane, momentum conservation along the incident direction requires that   x - direction

$$\frac{h\nu}{c} = \frac{h\nu'}{c} \cos\theta_\gamma + p \cos\theta_e \qquad (4\text{-}15)$$

† Debye [De 23] independently gave a quantitative theoretical explanation. Bartlett [Ba 64b] gives a good description of the historical background.

and perpendicular thereto, γ - *direction*

$$0 = \frac{h\nu'}{c} \sin \theta_\gamma - p \sin \theta_e \qquad (4\text{-}16)$$

Energy conservation furnishes a third equation,

$$h\nu = h\nu' + E_{\text{kin}} \qquad (4\text{-}17)$$

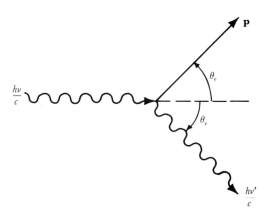

Fig. 4-4. Compton scattering of an incident photon (momentum $h\nu/c$) by a quasi-free electron through an angle θ_γ. The momentum of the scattered photon is $h\nu'/c$ ($< h\nu/c$); the electron recoils at an angle θ_e with a momentum $\mathbf{p}$.

which with the above and the relativistic energy-momentum relation [Eq. (1-10), Appendix A.2]

$$pc = [E_{\text{kin}}(E_{\text{kin}} + 2m_ec^2)]^{1/2} \qquad (4\text{-}18)$$

can be manipulated to yield the following general results:

(a) For all incident photon energies the COMPTON SHIFT of wavelength is

$$\varDelta\lambda \equiv \lambda' - \lambda \equiv \frac{c}{\nu'} - \frac{c}{\nu} = \lambda_C(1 - \cos\theta_\gamma) \qquad (4\text{-}19)$$

and therefore increases with the scattering angle θ_γ. The maximum shift occurs when $\theta_\gamma = 180°$ and takes the value $\varDelta\lambda_{\text{max}} = 2\lambda_C$. The quantity

$$\lambda_C \equiv \frac{h}{m_ec} = 2.426\ 21 \times 10^{-10}\ \text{cm} \qquad (4\text{-}20)$$

is termed the COMPTON WAVELENGTH of the electron and is numerically equal to the wavelength of a photon whose energy is $m_ec^2 = 0.511$ MeV. Excellent agreement has been obtained between measured and predicted Compton shifts.

(b) The outgoing photon energy in terms of the scattering angle is

$$\frac{E_{\gamma(\text{out})}}{E_{\gamma(\text{in})}} \equiv \frac{h\nu'}{h\nu} = \frac{1}{1 + \mathscr{E}(1 - \cos \theta_\gamma)} \tag{4-21}$$

where $\mathscr{E} \equiv h\nu/m_e c^2$ represents the "reduced" incident γ energy. When $\theta_\gamma = 0$ it follows that $\nu' = \nu$ for all incident energies $\mathscr{E}$, so that $\Delta\lambda = 0$ in the forward direction. The relation (4-21) represents an ellipse whose eccentricity increases with $\mathscr{E}$. For very low incident γ energies, when $\mathscr{E} \ll 1$, it reduces to a circle and indicates that $\nu' \approx \nu$ for all scattering angles θ_γ.

(c) The kinetic energy of the Compton recoil electron,

$$E_{\text{kin}} = h\nu - h\nu' \tag{4-22}$$

can be expressed in terms of θ_γ or θ_e as

$$E_{\text{kin}} = h\nu \, \frac{\mathscr{E}(1 - \cos \theta_\gamma)}{1 + \mathscr{E}(1 - \cos \theta_\gamma)} \tag{4-23}$$

$$E_{\text{kin}} = h\nu \, \frac{2\mathscr{E} \cos^2 \theta_e}{(1 + \mathscr{E})^2 - \mathscr{E}^2 \cos^2 \theta_e} \tag{4-24}$$

(d) The angles θ_γ and θ_e can be related to each other through the formula

$$\cot \theta_e = (1 + \mathscr{E}) \tan \frac{\theta_\gamma}{2} \tag{4-25}$$

which shows that the electron recoil angle θ_e cannot exceed 90 degrees.

(e) The maximum energy transfer to the electron, which occurs when the photon is back-scattered, namely when $\theta_\gamma = 180°$ and $\theta_e = 0°$, is

$$(E_{\text{kin}})_{\max} = \frac{h\nu}{1 + \dfrac{1}{2\mathscr{E}}} \tag{4-26}$$

and in this case the Compton shift of wavelength is maximal.

4.2.3. Compton Cross Section (Klein-Nishina Formula)

The evaluation of the differential cross section is rather complicated, since it involves the use of relativistic wave mechanics (see Heitler [He 54, pp. 217 ff.] and Appendix F), and we therefore confine ourselves to citing results first derived by Klein and Nishina [Kl 29]. The differential collision cross section per electron in the case of a *linearly polarized* incident plane electromagnetic wave is, in the notation employed hitherto, and in terms of the angle Θ between the polarization directions of the incident and emergent quanta, given by the KLEIN-NISHINA FORMULA:

$$_e\sigma_C(\theta_\gamma)_{pol} \equiv \left.\frac{d\,_e\sigma_C}{d\Omega_\gamma}\right|_{pol} = \frac{1}{4}r_e^2\left(\frac{\nu'}{\nu}\right)^2\left(\frac{\nu}{\nu'}+\frac{\nu'}{\nu}+4\cos^2\Theta-2\right) \text{ cm}^2 \text{ sr}^{-1}/\text{electron}$$

(4-27)

When the incident wave is *unpolarized*, the differential collision cross section per electron is obtained by calculating the mean value of the above expression averaged over all angles Θ, and this in terms of the scattering angle θ_γ yields the result

$$_e\sigma_C(\theta_\gamma)_{unpol} \equiv \left.\frac{d\,_e\sigma_C}{d\Omega_\gamma}\right|_{unpol}$$

$$= \frac{1}{2}r_e^2\left(\frac{\nu'}{\nu}\right)^2\left(\frac{\nu}{\nu'}+\frac{\nu'}{\nu}-\sin^2\theta_\gamma\right) \qquad \text{cm}^2 \text{ sr}^{-1}/\text{electron} \qquad (4\text{-}28)$$

or, expressed in terms of $\mathscr{E}$ on making use of the relationship (4-21),

$$\left.\frac{d\,_e\sigma_C}{d\Omega_\gamma}\right|_{unpol} = \tfrac{1}{2}r_e^2\{[1+\mathscr{E}(1-\cos\theta_\gamma)]^{-3}[-\mathscr{E}\cos^3\theta_\gamma$$

$$+ (\mathscr{E}^2+\mathscr{E}+1)(1+\cos^2\theta_\gamma)-\mathscr{E}(2\mathscr{E}+1)\cos\theta_\gamma]\}$$
$$\text{cm}^2 \text{ sr}^{-1}/\text{electron} \qquad (4\text{-}29)$$

In the case of low energies, $\mathscr{E}$ is negligibly small, and this reduces to the Thomson differential cross section,

$$\left.\frac{d\,_e\sigma_C}{d\Omega_\gamma}\right|_{unpol} \xrightarrow{\;\mathscr{E}\to 0\;} \frac{d\,_e\sigma_{Th}}{d\Omega_\gamma} = \tfrac{1}{2}r_e^2(1+\cos^2\theta_\gamma) \qquad \text{cm}^2 \text{ sr}^{-1}/\text{electron} \quad (4\text{-}30)$$

The expression (4-29) is depicted graphically in Fig. 4-5 in polar representation for various values of $\mathscr{E}$, and the angular distribution for various values of $\mathscr{E}$ is shown in the usual representation in Fig. 4-6.

The TOTAL COLLISION CROSS SECTION is obtained by integration over all scattering angles,

$$_e\sigma_C = \int\,_e\sigma_C(\theta_\gamma)\,d\Omega_\gamma = \int_0^\pi\,_e\sigma_C(\theta_\gamma)\,2\pi\sin\theta_\gamma\,d\theta_\gamma \qquad (4\text{-}31)$$

the end result being

$$_e\sigma_C = 2\pi r_e^2\left\{\frac{1+\mathscr{E}}{\mathscr{E}^2}\left[\frac{2(1+\mathscr{E})}{1+2\mathscr{E}}-\frac{1}{\mathscr{E}}\ln(1+2\mathscr{E})\right]\right.$$

$$\left.+\frac{1}{2\mathscr{E}}\ln(1+2\mathscr{E})-\frac{1+3\mathscr{E}}{(1+2\mathscr{E})^2}\right\} \qquad \text{cm}^2/\text{electron}$$

(4-32)

which, when $\mathcal{E}$ is small, can to a good approximation be reduced [Le 47] to

$$_e\sigma_C \cong \tfrac{8}{3}\pi r_e^2(1 - 2\mathcal{E} + 5.2\mathcal{E}^2 - 13.3\mathcal{E}^3 + \cdots) \qquad \text{cm}^2/\text{electron} \qquad (4\text{-}33)$$

In the limit $\mathcal{E} \to 0$ this is equal to the Thomson value

$$_e\sigma_C \xrightarrow{\;\mathcal{E}\to 0\;} {}_e\sigma_{Th} = \tfrac{8}{3}\pi r_e^2 \qquad \text{cm}^2/\text{electron} \qquad (4\text{-}34)$$

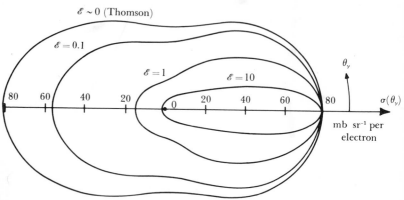

Fig. 4-5. Polar representation of the angular dependence of the differential cross section for Compton scattering of unpolarized radiation, as given by Eq. (4-29) for various values of the reduced incident energy $\mathcal{E} \equiv h\nu/m_e c^2$ (adapted from [Da 52b]).

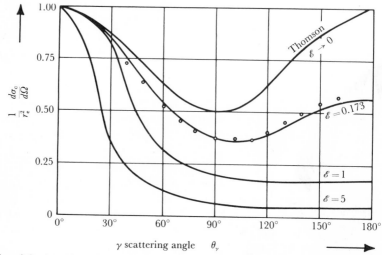

Fig. 4-6. Angular distribution of Compton-scattered γ quanta expressed as a normalized differential cross section $\sigma(\theta_\gamma)/r_e^2$ for various reduced incident energies $\mathcal{E} \equiv h\nu/m_e c^2$. At $\mathcal{E} = 0.173$ the curve is fitted to experimental data of Friedrich and Goldhaber [Fr 27] for scattering of 0.14-Å x rays in carbon (from [He 54]). (Used by permission of The Clarendon Press, Oxford.)

It is to be emphasized that the differential cross section (4-29) gives the number of photons scattered into unit solid angle at a mean scattering angle θ_γ, and that the total COLLISION CROSS SECTION (4-32) expresses the probability that a collision will occur in which a photon of energy $h\nu$ is removed from the collimated incident beam. It is therefore concerned with the energy taken out of the primary beam, and if the incident radiation is monochromatic, then this extracted energy is simply proportional to the number of Compton collisions. Not *all* of this energy is, however, scattered, since some of it is imparted to the electrons as recoil energy. The scattered energy is smaller than the incident energy by a factor $h\nu'/h\nu$, and in consequence the so-called ENERGY-SCATTERING CROSS SECTION σ_s which deals with the *scattered* γ energy, as against the extracted γ energy, is smaller than σ_C by this amount:

$$\sigma_s = \frac{h\nu'}{h\nu}\,\sigma_C \tag{4-35}$$

The transformation from the Klein-Nishina and derived formulae which feature the collision cross section per electron, $_e\sigma_C$ to the corresponding expressions for $_e\sigma_s$ is trivial therefore. A third concept which is sometimes encountered is the ENERGY-ABSORPTION CROSS SECTION σ_a which is simply the difference between the above two entities:

$$\sigma_a = \sigma_C - \sigma_s \tag{4-36}$$

and which represents the probability for the recoil kinetic energy

$$E_{kin} = h\nu - h\nu' \tag{4-37}$$

to be imparted to the electron in course of a Compton collision. It therefore concerns itself with the *transferred* energy which is subsequently dissipated within the target specimen through inelastic collisions and ionizing processes which the electron initiates.

Note that for clarity we have purposely referred to *energy*-scattering and *energy*-absorption cross sections, writing these as σ_s and σ_a respectively, in order to prevent confusion with the more usual *particle* scattering and absorption cross sections, σ_{sc} and σ_{abs} which we shall treat later.

The variation with γ energy of the various total cross sections defined above is indicated in Fig. 4-7 on a logarithmic energy scale.

One can also concern oneself with the differential and total cross sections for the Compton recoil electrons. No ambiguity exists in this case; the differential cross section can be derived from the photon collision cross section by multiplying the latter by an appropriate solid-angle factor:

$$\frac{d\,_e\sigma}{d\Omega_e} = \frac{d\,_e\sigma_C}{d\Omega_\gamma}\frac{d\Omega_\gamma}{d\Omega_e} = \frac{(1+\mathscr{E})^2(1-\cos\theta_\gamma)^2}{\cos^3\theta_e}\frac{d\,_e\sigma_C}{d\Omega_\gamma} \tag{4-38}$$

and this is shown in polar representation in Fig. 4-8.

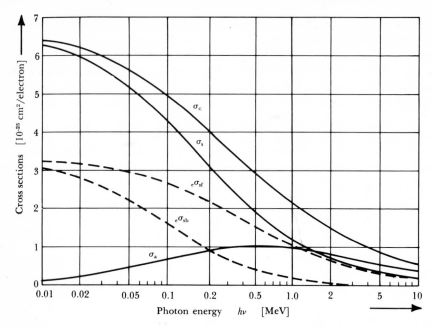

Fig. 4-7. Energy dependence of cross sections for the Compton effect. The full curves show the total collision cross section per electron σ_C together with its components σ_a, the absorption cross section and σ_s, the scattering cross section as given by (4-32), (4-35), and (4-36). Also depicted as dashed curves are the scattering cross sections per electron for the forward direction $({}_\mathrm{e}\sigma_\mathrm{s\,f})$ and the backward direction $({}_\mathrm{e}\sigma_\mathrm{s\,b})$ (from [Da 65c]). (Used by permission of North-Holland Publishing Co.)

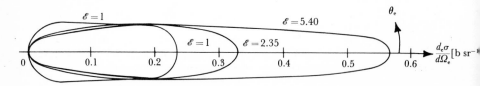

Fig. 4-8. Polar representation of the variation of the differential *electron* cross section for the Compton recoil electron emerging at an angle θ_e (from [Da 65c]).

4.2.4. Atomic Compton Cross Section

Only at high incident photon energies is the binding energy of the electrons in an atom negligible and the linear formula

$$_\mathrm{a}\sigma = Z\,_\mathrm{e}\sigma \tag{4-39}$$

applicable. At lower energies it is necessary to take account of the distributions and momenta of the electrons in the atom, which involves a cognizance of the phase relations that exist.

The scattering is *coherent* when $h\nu = h\nu'$ and it is then appropriate to add up the individual *amplitudes*, rather than intensities, for each of the atomic electrons involved, whereas it is *incoherent* when $h\nu' < h\nu$ and one then sums the intensities. The atomic differential Compton cross section can accordingly

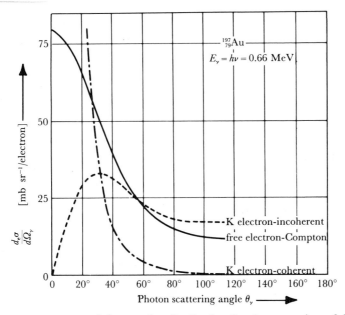

Fig. 4-9. Comparison of the angular distribution for the scattering of 0.66-MeV photons on gold, as calculated for Compton scattering on free electrons using the Klein-Nishina formula (4-28) for unpolarized radiation (solid curve), or for coherent scattering from electrons in the *K* shell [Br 57] (dot-dash curve) with the experimental results for *incoherent K*-electron scattering (dashed curve) (from [Mo 61]).

be represented as the sum of a coherent and an incoherent partial cross section:

$$_a\sigma(\theta_\gamma) = {}_a\sigma_{\text{coh}}(\theta_\gamma) + {}_a\sigma_{\text{incoh}}(\theta_\gamma) \tag{4-40}$$

The calculation of the constituent terms is still in process of refinement. Figure 4-9 shows a comparison between the results furnished for the angular distribution of 0.66-MeV photons on gold by the Klein-Nishina formula and the calculated coherent differential cross section for scattering on bound electrons of the *K* shell, which shows more pronounced forward peaking. They may be contrasted with the measured incoherent cross section for *K*-shell electrons, whose angular dependence is of altogether different form.

4.3. Rayleigh Scattering

Scattering by free electrons as discussed so far can be regarded as a high-energy limit to the more general situation of scattering by *bound* atomic electrons, whose motion and distribution within the atom, as well as the binding energy itself, need to be considered in a unified treatment. The quantum-mechanical calculations prove to be very involved, and even in the case of coherent, elastic scattering, i.e., RAYLEIGH SCATTERING, for which the theory is simplified by the fact that the bound electron reverts to the same state after scattering as it was in originally, the theoretical (and experimental) data are sparse (see, e.g., [Ma 56a, Br 57, Co 59]).

When the atom as a whole is considered to absorb the transferred momentum, it follows that the energies of incident and scattered photons are the same and there is a phase relation between the scattering amplitudes for different electrons of the atom. The total scattered intensity is accordingly a coherent sum of the individual contributions; i.e., the amplitudes of the radiation scattered by each electron must be added together first, and the sum squared to obtain the resultant intensity. Binding-energy effects render this a more involved situation than merely a coherent summation of Thomson scattering, but one can nevertheless derive such properties as the well-known $1/\lambda^4$-dependence of the average radiated power, which is a characteristic of dipole radiation, by applying the formulae of Section 4.2.1.

Combining Eqs. (4-3) and (4-6) we can write

$$\mathbf{M} = \mathbf{M}_0 \sin \omega t \tag{4-41}$$

where $\mathbf{M}_0$ is the maximum dipole moment, and therefore

$$\dot{\mathbf{M}} = \omega \mathbf{M}_0 \cos \omega t \quad \text{and} \quad \ddot{\mathbf{M}} = -\omega^2 \mathbf{M}_0 \sin \omega t \tag{4-42}$$

whence as before, the time-average $\overline{|\ddot{\mathbf{M}}|^2}$ is

$$\overline{|\ddot{\mathbf{M}}|^2} = \tfrac{1}{2}\omega^4 M_0^2 \tag{4-43}$$

Substitution of this in Eqs. (4-6) and (4-5) yields

$$|\mathbf{S}_{\text{out}}| = \frac{\overline{|\ddot{\mathbf{M}}|^2}}{4\pi c^3 r^2} \sin^2 \Theta = \frac{cM_0^2}{8\pi r^2} \frac{1}{\lambda^4} \sin^2 \Theta \tag{4-44}$$

when use is made of the relation $\lambda\omega = \lambda\nu = c$. This indicates the $1/\lambda^4$-dependence, familiar in general physics as the basis of explanations of the blue color of the atmosphere and redness of sunsets.†

---

† As Rayleigh showed in 1871, even dust-free air will, due to molecular density fluctuations, scatter light sufficiently to account for the blue color of the sky. "Dust dipole" scattering enhances this effect and provides at least a partial explanation of red sunsets.

Of course, the above representation can be retained to yield an expression for the differential Thomson cross section of the form

$$\sigma_{\mathrm{Th}}(\theta) = \frac{1}{2}\left(\frac{M_0}{E_0}\right)^2 \frac{1}{\lambda^4}(1 + \cos^2\theta) \qquad (4\text{-}45)$$

for unpolarized radiation on using the relation (4-7) and building

$$\sigma_{\mathrm{Th}}(\theta)\, d\Omega = \frac{|\mathbf{S}_{\mathrm{out}}|\, dA}{|\mathbf{S}_{\mathrm{in}}|} \qquad (4\text{-}46)$$

as before. Despite the appearance of $1/\lambda^4$ in the result, the cross section is not dependent upon the incident γ energy, since it can readily be reduced to the standard formula (4-11) on noting that

$$|\dot{\mathbf{M}}|^2 = \frac{1}{2}\left(\frac{c}{\lambda}\right)^4 M_0^2 = e\overline{|\ddot{\mathbf{x}}|^2} = \frac{1}{2}\left(\frac{eE_0}{m_{\mathrm e}}\right)^2 \qquad (4\text{-}47)$$

The Rayleigh cross section *per atom* as derived by *coherent* summation is proportional to Z^2 for that part of the radiation which experiences small-angle scattering, namely

$$\theta < \theta_{\mathrm c} = 2 \text{ arc sin } (0.026 Z^{1/3}/\mathscr{E}) \qquad (4\text{-}48)$$

This condition covers over three-quarters of the scattered intensity, and for this forward fraction, the differential cross section as given by Debye theory [De 30] is

$$_{\mathrm a}\sigma_{\mathrm R}(\theta \leqslant \theta_{\mathrm c}) \sim r_{\mathrm e}^2 Z^2(1 + \cos^2\theta) \qquad \mathrm{cm}^2\ \mathrm{sr}^{-1}/\mathrm{atom} \qquad (4\text{-}49)$$

whereas the remaining fraction of the scattered radiation emitted at larger angles has a distribution which has a Z^3-dependence.† Relativistic calculations to which references are given in Bel'skii *et al.* [Be 59] indicate an even more pronounced dependence upon atomic number, namely as Z^5 or Z^8–Z^{10}.

4.4. Photoelectric Effect

The energy of incident γ quanta can also be completely absorbed by electrons in atoms, particularly by those in the innermost shells (momentum conservation precludes absorption by unbound electrons). Thus about 80 percent of the photoelectric absorption occurs with electrons in the K shell

† The theory of Franz [Fr 35, Fr 36, Mo 50; Si 65, p. 58] gives the atomic differential cross section as

$$_{\mathrm a}\sigma_{\mathrm R}(\theta > \theta_{\mathrm c}) \simeq \frac{64\pi^2 e^{10}}{9m_{\mathrm e}^2 h^3 c^7}\left(\frac{Z}{\mathscr{E}}\right)^3 \frac{\frac{1}{2}(1 + \cos^2\theta)}{\sin^3\frac{1}{2}\theta} = 8.73 \times 10^{-33}\left(\frac{Z}{\mathscr{E}}\right)^3 \times$$

$$\times \frac{\frac{1}{2}(1 + \cos^2\theta)}{\sin^3\frac{1}{2}\theta} \qquad \mathrm{cm}^2\ \mathrm{sr}^{-1}/\mathrm{atom}$$

if the γ energy exceeds the energy with which the electrons are bound. The interaction essentially occurs with the entire atomic electron cloud rather than with individual "corpuscular" electrons; when the incident energy $h\nu$ exceeds the electronic binding energy B_e an electron is ejected with a kinetic energy

$$E_{\text{kin}} = h\nu - B_e \tag{4-50}$$

and the residual atom as a whole takes up the (quite small) recoil energy, so that the momentum and energy balance is preserved. The energy uptake may be insufficient to bring about ejection as such (i.e., ionization), but may simply raise an electron from one of the innermost shells to an "outer" bound state (if one exists). The binding energy not only depends upon the atomic number Z but also upon the orbital electron shell: it diminishes as one proceeds to "outer" shells according to the approximate formulae (which take account of the progressive screening of the attractive nuclear charge by inner electron shells):

$$(B_e)_K = Ry(Z-1)^2 \qquad (K\text{-shell binding}) \tag{4-51}$$

$$(B_e)_L = \tfrac{1}{4}Ry\,(Z-5)^2 \qquad (L\text{-shell binding}) \tag{4-52}$$

$$(B_e)_M = \tfrac{1}{9}Ry(Z-13)^2 \qquad (M\text{-shell binding}) \tag{4-53}$$

The Rydberg constant Ry (equivalent in first order to the ionization energy of a normal hydrogen atom) is here expressed in energy units,

$$Ry = hcR = hc\,\frac{2\pi^2 m_e e^4}{ch^3} = \frac{m_e e^4}{2\hbar^2} \triangleq 13.61 \text{ eV} \tag{4-54}$$

where $R = 1.097\,373 \times 10^5$ cm^{-1} is the spectroscopic Rydberg constant (for infinite mass). It will be noted that an electron raised to a *higher* energy state in an atom is more loosely bound.

The total binding energy of an atom for its normal complement of electrons is given by the semiempirical expression

$$(B_e)_{\text{tot}} = 15.73\,Z^{7/3} \text{ eV} \tag{4-55}$$

which is plotted in function of Z in Fig. 4-10.

Depending upon the atomic number of the absorber, the photoelectric effect is the predominant mode by which electromagnetic radiation of energy up to roughly 0.5 MeV interacts with matter. The absorption curve as a function of incident energy displays the characteristic sawtooth structure shown in Fig. 4-11 in which the sharp discontinuities known as ABSORPTION EDGES arise whenever the incident energy coincides with the ionization energy of electrons in the K, L, M, ... shell. The sawtooth shape evinces fine structure because the electron shells, with the exception of the K shell, each possess subshells having slight variations in binding energy owing to the slightly different spatial probability distribution of electrons in respective subshells, which gives rise to slight differences in the screening of the nuclear charge. Thus there are $(2 \times 1) + 1 = 3$ subshells in the L shell responsible for the appearance of 3 nicks in the L-absorption edge, and $(2 \times 2) + 1 = 5$ subshells in the M shell, etc., the MULTIPLICITY being of the form $(2l + 1)$ with $l = 0, 1, 2, ...$ for the K, L, M, ... shells.

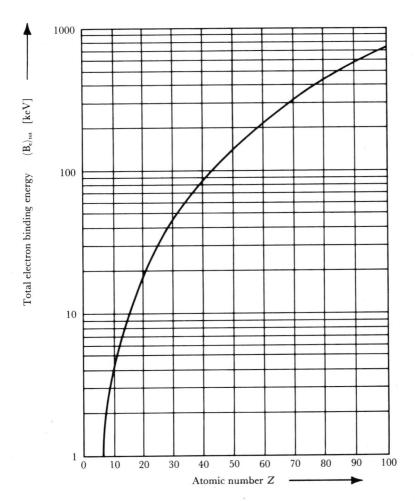

Fig. 4-10. Increase of the total electron binding energy $(B_e)_{tot}$ with atomic number Z.

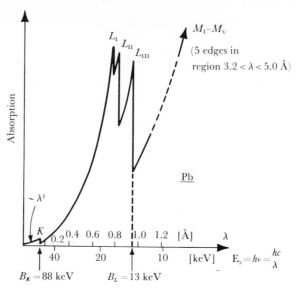

Fig. 4-11. Absorption curve for x rays in Pb as a function of incident γ wavelength and energy, showing the characteristic absorption edges. The K-edge lies at $\lambda_K = 0.1405$ Å ($\hat{=} 88.3$ keV), the three L-edges at $\lambda_L = 0.780$ Å ($\hat{=} 15.91$ keV), 0.813 Å ($\hat{=} 15.26$ keV) and 0.950 Å ($\hat{=} 13.06$ keV) respectively, and the five M-edges would lie in the region $\lambda_M = 3.2$–5.0 Å ($E_\gamma = 3.88$–2.48 keV), whilst the group of seven N-edges would commence at $\lambda_N \cong 14$ Å ($\hat{=} 0.89$ keV).

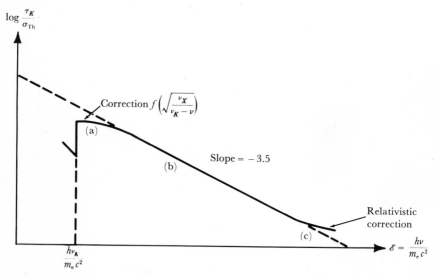

Fig. 4-12. Semilogarithmic representation of the appearance of the total photoelectric cross section in function of the reduced incident γ energy $\mathscr{E}$. The corrections which tend to flatten the curve are discussed in the text.

4.4.1. ENERGY AND ATOMIC-NUMBER DEPENDENCE

The exact theoretical evaluation of the photoelectric cross section presents very great difficulty and in general it is now customary to employ numerical computation based upon appropriate semiempirical formulae in order to derive values for the cross section per atom τ. A survey of the various theoretical methods and approximations has been given by Pratt [Pr 60]; for our purposes it suffices to note that the basic elements of the sawtooth structure can be divided into three regions, viz.

(a) In the vicinity of the absorption edge;

(b) At some distance from the edge;

(c) In the relativistic region ($\mathscr{E} \gg 1$).

The TOTAL PHOTOELECTRIC CROSS SECTION in the intermediate region (b) for *both* K electrons together, according to the nonrelativistic Born-approximation treatment [He 54, pp. 207 ff.] depends upon the reduced incident energy in a nonsimple way:

$$\tau_K = (32/\mathscr{E}^7)^{1/2} \alpha^4 Z^5 \sigma_{Th} \qquad \text{cm}^2/\text{atom} \qquad (4\text{-}56)$$

where α is the fine-structure constant and σ_{Th} is the Thomson cross section per electron. In the logarithmic representation of Fig. 4-12 this corresponds to a straight line of slope -3.5 modified at either end by corrections which tend to flatten the graph. Thus in the neighborhood of the absorption edge, the γ energy is comparable with the K-electron binding energy and it becomes necessary to use rather more exact electron wave functions, whose introduction gives rise to a multiplicative correction function of the form $f[\nu_K/(\nu_K - \nu)]^{1/2}$ in the region (a), where $h\nu_K = (B_e)_K$. At the high-energy end of the sawtooth region (c), a relativistic correction again causes flattening with the result that, as shown in Fig. 4-13, an entire range of elements displays the energy dependence depicted in Fig. 4-12. Currently, theoretical work has been extended to a consideration of higher electron shells, and results tend to show that τ for other shells also varies as Z^5, but that its energy dependence ranges from $\mathscr{E}^{-7/2}$ at comparatively low energies to $\mathscr{E}^{-1}$ at high, as evinced by the formula of Sauter [Sa 31] for the case $\mathscr{E} \gg 1$:

$$\tau_K = \frac{1.5}{\mathscr{E}} \alpha^4 Z^5 \sigma_{Th} \qquad (4\text{-}57)$$

An indication of the Z-dependence, which is rendered complicated by the introduction of the above edge-correction function, is given by setting $\tau_K \sim Z^n$ and using empirical data to arrive at the value of the power n, which is found to vary with energy in the manner shown in Fig. 4-14.

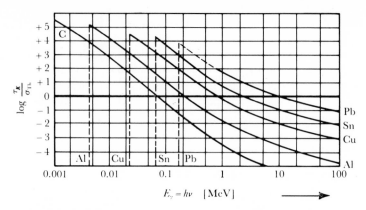

Fig. 4-13. Log-log plots of the ratio of the photoelectric absorption cross section for the K shell to the Thomson cross section in the case of carbon, aluminum, copper, tin, and lead. The K-absorption edges are shown (that for carbon lies just outside the diagram) (adapted from [He 54]). (Used by permission of The Clarendon Press, Oxford.)

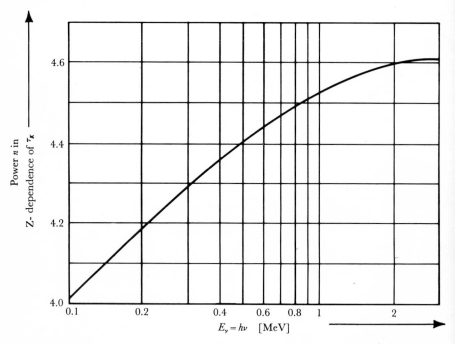

Fig. 4-14. Semilogarithmic representation of the approximate energy dependence of the power n in the proportionality $\tau_K \sim Z^n$ between the atomic photoelectric cross section τ_K and the atomic number Z (from [Ev 55], based upon data by N. C. Rasmussen). (Used by permission of McGraw-Hill Book Co.)

4.4.2. ANGULAR DISTRIBUTIONS

At low incident γ-ray energies, photoelectrons are to a very marked extent emitted in a direction perpendicular to the incident beam, but as the γ energy increases the peaking moves progressively to more forward electron emission angles θ_e, as shown in Fig. 4-15. In polar representation, the normalized distributions are observed to shift with $\mathscr{E}$ in the manner shown in Fig. 4-16.

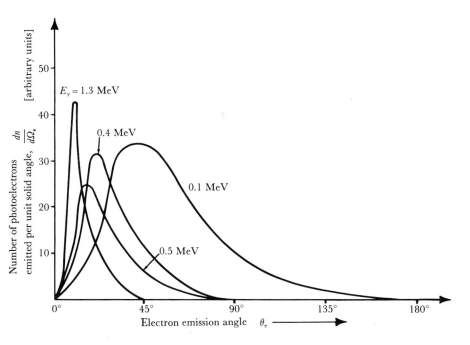

Fig. 4-15. Angular distribution of photoelectrons at various incident γ energies, as calculated relativistically by Sauter [Sa 31] (adapted from [Da 52b]).

The angular distribution for *linearly polarized* incident radiation is of the form

$$W(\theta_e) \sim \sin^2 \theta_e \cos^2 \phi (1 - \beta \cos \theta_e)^{-4} \qquad (4\text{-}58)$$

with ϕ the angle between the reaction plane and the polarization plane of the incident photons, and $\beta = v_e/c$ the reduced velocity of the electron emitted in the direction θ_e.

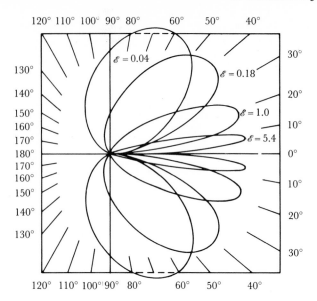

Fig. 4-16. Polar representation of the angular distribution of photoelectrons for various reduced incident γ energies $\mathscr{E} \equiv h\nu/m_e c^2$. The major axes of the curves have been normalized to the same magnitude for ease of comparison: however, this suppresses the marked variation of the differential cross section with energy $\mathscr{E}$ (from [Da 52b]).

4.4.3. ATTENUATION COEFFICIENTS

The LINEAR ATTENUATION COEFFICIENT is, by definition,

$$\tau = n^{\square}\ _a\tau \qquad \mathrm{cm}^{-1} \tag{4-59}$$

where $n^{\square}$ denotes the number of *atoms* per unit volume of the attenuator and $_a\tau$ the atomic photoelectric cross section. The corresponding MASS-ATTENUA-TION COEFFICIENT is τ/ρ. It should be noted that its value, unlike that for the other processes, is *not* roughly constant for all elements at a given energy.

4.5. Auger Effect

For atoms in their ground state, the various electron shells from the inner-most shell in turn have their full complement of electrons. The ejection of an electron from a given shell through a photoelectric process then creates a vacancy in that shell and leaves the atom in an energetically excited state. It reverts to a lower state when an electron from a higher shell makes a transition to fill the gap. Thereby an amount of energy equal to the difference in binding energy, for example $(B_e)_K - (B_e)_L$, is set free as a quantum of

electromagnetic radiation, and this can be emitted in the form of CHARAC-
TERISTIC X RAYS, or alternatively be imparted to an electron in a higher shell
and cause it to be ejected from the atom itself, since for instance

$$(B_e)_L < (B_e)_K - (B_e)_L \qquad (4\text{-}60)$$

whence the emitted electron or AUGER ELECTRON, leaves with a discrete energy
$(B_e)_K - 2(B_e)_L$, as shown in Fig. 4-17. The AUGER EFFECT, so named after

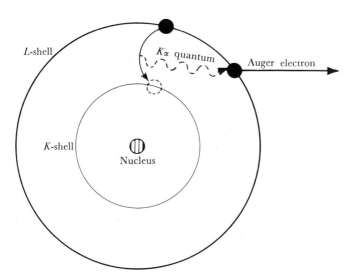

Fig. 4-17. Schematic diagram to illustrate the ejection of an Auger electron with discrete energy as a result of an atomic electron transition, associated with the emission of characteristic x radiation.

Auger's discovery [Au 25] of electron tracks of nonradiative atomic origin in photographic emulsions following x-ray irradiation, can be classified as an internal collision of the second type, namely a collision in which one of the collision partners is originally in an excited state. Because of the comparatively low energy of the Auger electrons, the effect has in the past not been studied very extensively; but, because of its bearing upon internal conversion, electron capture, and nuclear spectroscopy, it has recently received renewed attention, especially as new designs of β-ray spectrometers and improvements in experimental techniques now permit measurements to be effected at quite low energies, e.g., to $E_e \approx 1$ keV or even less. A survey of this subject has been presented by Burhop [Bu 52] and, more recently, by Bergström and Nordling [Be 65] (see Siegbahn [Si 65, pp. 1523 ff.]).

The influence of the Auger effect is especially marked in the case of light elements, as is evident from the empirical plot against Z of the K-FLUORESCENCE YIELD, the number of emitted electromagnetic quanta N_K per vacancy in the K shell, shown in Fig. 4-18. The corresponding K-Auger yield is $1 - N_K$. The Z-dependence of the ratio K-fluorescence/K-Auger has on semi-empirical grounds been suggested [Bu 55b, Ha 60a] to be of the form

$$\frac{N_K}{1 - N_K} = (-6.4 + 3.4Z - 0.000\ 103Z^3)^4 \times 10^{-8} \qquad (4\text{-}61)$$

The value of N_K, accurate to about $\frac{1}{2}$ percent, from this relation has been plotted in Fig. 4-18.

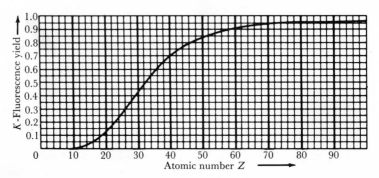

Fig. 4-18. Plot of the K-fluorescence yield *vs* atomic number Z, as determined from the semiempirical formula (4-61) (from [Be 65]). (Used by permission of North-Holland Publishing Co.)

4.6. Pair Production

If the energy of the incident electromagnetic radiation exceeds $2\ m_ec^2 = 1.02$ MeV, the production of a positive and negative electron pair is energetically possible—but only in the vicinity of a collision partner which can take up a suitable amount of momentum to preserve the linear momentum balance. In pair creation, any excess incident energy over $2m_ec^2$ is imparted as kinetic energy to the pair of particles, whose motion has to be treated relativistically, so that

$$(E_{\text{pair}})_{\text{total}} = E_{\text{kin}} + 2m_ec^2 = 2\gamma m_ec^2 \qquad (4\text{-}62)$$

where $\gamma \equiv (1 - \beta^2)^{-1/2}$, with $\beta \equiv v/c$ the reduced velocity of each electron. The total linear momentum of the pair of particles is, according to relativistic kinematics,

$$p_{\text{pair}} = 2\gamma m_ev = \frac{v}{c^2}\,(E_{\text{pair}})_{\text{total}} \qquad (4\text{-}63)$$

The commensurate quantities for the incident γ radiation are

$$E_\gamma = h\nu, \qquad p_\gamma = \frac{h\nu}{c} = \frac{E_\gamma}{c} \tag{4-64}$$

and energy conservation requires that

$$(E_{\text{pair}})_{\text{total}} = E_\gamma \tag{4-65}$$

i.e.,

$$\frac{c^2}{v} p_{\text{pair}} = p_\gamma c \tag{4-66}$$

viz.

$$p_{\text{pair}} = \frac{v}{c} p_\gamma < p_\gamma \tag{4-67}$$

which shows that the photon possesses a momentum excess that has to be absorbed by a collision partner. The latter can be a nucleus or an atomic electron: in the case of a nucleus whose mass is large, the recoil energy can be vanishingly small, but in the case of an electron there is a large transfer of energy and in consequence the threshold energy for pair production then lies† at $4m_ec^2 = 2.04$ MeV. Pair production in the field of a nucleus can be distinguished from that in the field of an electron not only by the different threshold energy and kinematics but by the characteristic tracks observable in a cloud chamber or bubble chamber—the former produces a pair of tracks whereas the latter produces three tracks, the extra trace being produced by the high-energy recoil electron. Although small at comparatively low incident energies, pair production becomes the dominant process above $\mathscr{E} \approx 10$ and essentially accounts thereafter for all γ-ray absorption in matter. The early investigations [Ch 30a, Ch 30b, Ta 30, Ta 32] of scattering and absorption of 2.62-MeV γ rays showed that the interaction with high-Z elements exceeded that predicted by the Compton and photoelectric effect alone; in 1932/3 Anderson [An 33] observed the ejection of e^+-e^- pairs from lead foil subjected to cosmic ray irradiation, which provided evidence for Oppenheimer and Plesset's interpretation [Op 33] of pair production on the basis of Dirac's electron theory.

4.6.1. DIRAC ELECTRON THEORY

The Dirac relativistic wave equation has energy eigenvalues for the free electron which can be negative as well as positive, viz.

$$E = \pm(p^2c^2 + m_e^2c^4)^{1/2} \tag{4-68}$$

---

† The general expression for the threshold energy is $E_{\text{thresh}} = 2m_ec^2[1 + (m_e/M)]$, where M is the mass of the recoiling particle.

These values range from $-\infty$ to $-m_e c^2$ and then from $+m_e c^2$ to $+\infty$, with a gap between $-m_e c^2$ and $+m_e c^2$, of width $2m_e c^2$, as shown in Fig. 4-19. Although from a classical viewpoint, an electron could not traverse this barrier region, quantum mechanics yields a nonvanishing probability for such a transition. The properties associated with an electron which is located in the "sea" of possible negative-energy states were elucidated by Dirac in the following way: For a system in its ground state, the electrons occupy the lowest energy states compatible with the Pauli exclusion principle, which forbids the presence of more than two electrons in any one energy state. The region of negative energy eigenvalues is conceived to be normally replete with the full complement of

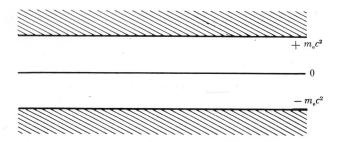

Fig. 4-19. Energy gap of magnitude $2\ m_e c^2 = 1.02$ MeV between the "sea" of negative-energy electron states and positive-energy levels.

electrons. The presence of the latter, although not susceptible to direct physical observation, nevertheless plays a role in physical phenomena such as vacuum polarization and the anomalous magnetic moment of the electron. Electrons in positive energy states constitute those which can be observed experimentally: they cannot make the transition to the negative region unless a vacancy in the otherwise completely filled "sea" occurs, but a transition in the other direction can readily take place when sufficient energy is supplied to the system. Thus a photon of energy $2m_e c^2$ can confer the requisite energy upon an electron in the negative "sea" to raise it to an unoccupied permitted level in the positive-energy region, a process of "pair creation" which is manifested as a "hole" in the negative sea and a normal observable electron in the positive region. The "hole" or vacancy is evinced as an observable particle too, namely, as a POSITRON e^+, the conjugate particle to the e^-. It has a positive charge and mass since it corresponds to the *absence* of a particle, having, as it were, negative charge and mass (charge and mass being universally normalized to zero when all negative states are filled and all positive states are empty).

4.6.2. ELECTRON-POSITRON CONJUGATION

Whereas Dirac had based his theory of positrons upon a visualization of these as holes in an otherwise full negative-energy sea of electrons, Feynman [Fe 49a, Fe 49b, reprinted in Schw 58]—and, before him, Stueckelberg

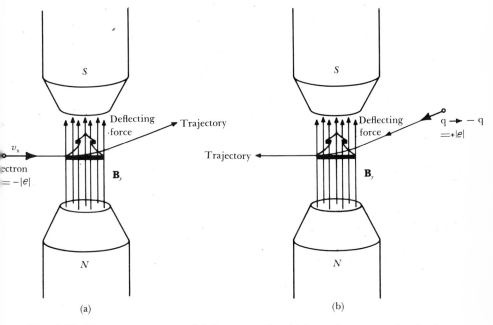

Fig. 4-20. Diagrams to contrast (a) the process in which an electron moving in the positive x direction is deflected in the negative z direction as it passes through a magnetic field $\mathbf{B_y}$ with (b) the time-inverted situation in which the trajectory is reversed while the field direction (and the Lorentz force direction) is left unchanged. Consistency demands a change in the sign of the particle's charge; e.g., a positron may be regarded as a time-inverted electron. (This constitutes merely an illustrative argument, since the time-reversal operation is actually more complicated than that of time-inversion depicted here: for example, the field $\mathbf{B}$ reverses its direction upon time reversal. One might picture this as a reversal of the direction of current flow in an electromagnet, or a reversal of the spin sense of atomic electrons in a permanent magnet.)

[St 41a]—adopted a radically different approach, which had far-reaching consequences. In this alternative visualization, a positron is pictured as a particle which behaves exactly like an electron in a time-inverted world.

To demonstrate this in a simple, albeit not rigorously accurate, way we imagine an electron to pass between the pole pieces of a permanent magnet which produces a perpendicular field of induction $\mathbf{B}$ [Fig. 4-20(a)]. The motion of the electron, whose charge is $q = -|e|$, under the action of the deflecting Lorentz force is expressed by the equation

$$m_e \frac{d^2z}{dt^2} = qv_xB_y = -|e|\frac{dx}{dt}B_y \qquad (4\text{-}69)$$

with v_x perpendicular to B_y in a right-handed Cartesian system. In the time-inverted situation depicted in Fig. 4-20(b), this equation of motion is left invariant when t is replaced by $-t$ provided that the sign of the charge

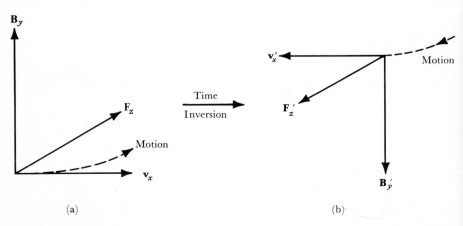

Fig. 4-21. Vector diagram to indicate relationships between the variables which are featured in the Lorentz-force expression (4-70).

is also reversed, $q \to -q = +|e|$, as would correspond to a *positron* traveling in the reverse direction. Thus, electrodynamic relations which describe electrons go over into those describing positrons when subjected to the operation $t \to -t$ of time-inversion.

In the interests of accuracy, it should be realized that a strict application of the time-reversal operation causes not only v_x to change sign, but also B_y. (One can picture the magnetic field as set up by a flow of current through the coils of an electromagnet—thus in the time-inverted world, the current would flow in the reverse direction and the sign of the induction would change.) The above argument then takes the following form: The Lorentz force upon the electron,

$$\mathbf{F}_z = q[\mathbf{v}_x \times \mathbf{B}_y] = -|e|[\mathbf{v}_x \times \mathbf{B}_y] \qquad (4\text{-}70)$$

acts in the *negative* z direction [Fig. 4-21(a)]. In the time-inverted world [Fig. 4-21(b)] there is a change of sign ($\mathbf{v}'_x = -\mathbf{v}_x$ and $\mathbf{B}'_y = -\mathbf{B}_y$) and a reversal of the trajectory which corresponds to the action of a force $\mathbf{F}'_z$ in the *positive* z direction, e.g., a Lorentz force upon a particle which carries a *positive* charge

$q = +|e|$. Hence in this situation the effect of time inversion is akin to that of charge conjugation (particle $\rightarrow$ antiparticle).

4.6.3. FEYNMAN GRAPHS

The considerations upon which Feynman based his theory of positrons led additionally to improvements in the description and treatment of inter-action processes such as those with which we are dealing in the present section. Feynman proposed a method of depicting interaction processes in a simple pictorial way which not only brings out the underlying physical features but which can be used to ease considerably the task of formulating

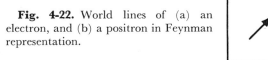

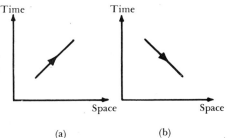

Fig. 4-22. World lines of (a) an electron, and (b) a positron in Feynman representation.

(a)   (b)

the interaction probability in quantitative mathematical terms. Thus even quite complicated problems in quantum electrodynamics can be treated directly by linking together expressions appropriate to each segment of the requisite FEYNMAN GRAPH or DIAGRAM [Fe 49b, Fe 49c, Dy 49]. We approach this by first examining the construction of such diagrams, proceeding from examples of interactions which we have already encountered and going on to preview some of those which assume importance in later parts of this section.

For our purposes, we can view Feynman graphs as space-time diagrams. We accordingly depict the motion of an electron by a WORLD LINE as in Fig. 4-22(a), with the abscissa representing a general spatial coordinate r and the ordinate a time t. Then the corresponding world line for a positron, which Feynman regards as the time-inverse of an electron, will be a line implying motion along *increasing r* but *decreasing t*, as in Fig. 4-22(b). Feynman's device, though novel, in no way stands at variance with the particle-hole viewpoint of Dirac: the backward progression in the time, introducing as it does negative energy components in the transition term of the integral which describes the motion, corresponds identically to the propagation of a hole in the negative-energy Dirac sea. A change in the motion, as occurs during inelastic scattering for example, is represented by a change in slope of the electron's or positron's world line, conjoined with a world line which depicts the motion of the field quantum, be it a photon (wavy line) or a meson (broken line), through whose mediation the change is brought about. Thus the impinging of a photon upon an electron which is thereupon scattered inelastically is depicted by the Feynman graph in Fig. 4-23(a), in which by convention the space-time axes are omitted and the VERTEX of the diagram represents the point of interaction. When a photon is *emitted*, rather than absorbed, on electron scattering, the

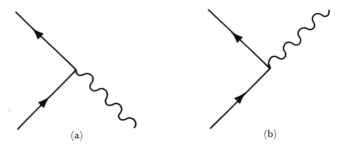

(a)   (b)

Fig. 4-23. Feynman diagrams for (a) the scattering of an electron upon absorption of a γ quantum, and (b) the emission of a γ quantum upon electron scattering.

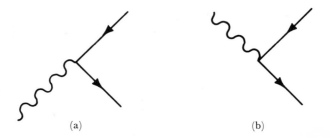

(a)   (b)

Fig. 4-24. The same processes as depicted in Fig. 4-23, but for *positron* scattering.

diagram takes the form of Fig. 4-23(b). The photon line carries no arrow because a photon is its own antiparticle, but the arrows on the electron lines essentially distinguish the graphs from the corresponding diagrams for inelastic *positron* scattering, Fig. 4-24(a, b), in which the arrows have been reversed.

The Coulomb interaction in which electron-electron scattering is brought about by exchange of a (virtual) photon is shown in Fig. 4-25, whereas the diagram for positron-positron scattering takes the same form but with reversed arrows. In the case of electron-positron scattering we must distinguish between two situations which involve different intermediate states: The straightforward exchange of a virtual photon is depicted in Fig. 4-26(a) and the more complicated process ("Møller scattering") of annihilation of an electron-positron pair followed by its re-creation from the virtual photon in Fig. 4-26(b) (vertices *A* and *B*, respectively). The Feynman graph in Fig.

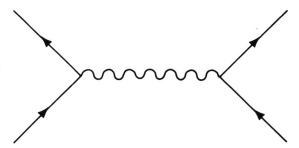

Fig. 4-25. Electron-electron scattering mediated by the exchange of a (virtual) photon in Feynman representation. The same diagram with reversed arrows would represent positron-positron scattering.

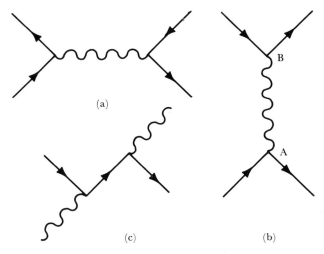

Fig. 4-26. Feynman graphs for electron-positron scattering (a) through exchange of a (virtual) photon, (b) through a Møller-scattering process involving pair annihilation and re-creation, (c) through a similar process, but in a different sequence.

4-26(c) shows a process made up of the same elements, but in a different sequence, namely pair creation followed by pair annihilation, a basically different situation which might in the Feynman description be interpreted as the motion of but a single particle which can run in both time directions and is on two occasions scattered by photons. The basic elements are reproduced in Fig. 4-27(a,b), a representation of pair creation and annihilation which is strikingly similar to its anticipation by Stueckelberg [St 41b], but which has to be modified to express the physical situation in which one-quantum absorption or emission is precluded by the dictates of energy and momentum conservation for particles in the free state. As discussed in Section

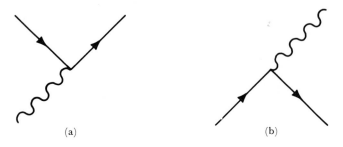

Fig. 4-27. Elements of Feynman graphs corresponding to (a) pair creation, (b) pair annihilation.

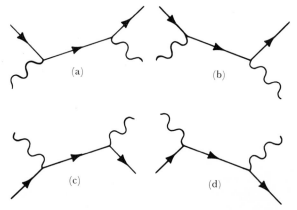

Fig. 4-28. Feynman graphs for two-quantum pair creation (a) and (b) and pair annihilation (c) and (d).

4.6.5, each of the actual processes thus entails the participation of at least two photons: *Two-quantum pair creation* is accordingly represented by a Feynman graph of the type shown in Fig. 4-28(a) or, since the particle in the intermediate state can equally well be a (virtual) positron, that in Fig. 4-28(b), and the corresponding diagrams for two-quantum pair annihilation by Fig. 4-28(c, d). On the other hand, pair annihilation in the Coulomb field of an electron is described by the Feynman diagram in Fig. 4-29.

The scattering of γ rays can be illustrated by progressively more complicated Feynman graphs, which enable the interaction probability to be derived with increasing numerical accuracy. The basic diagram for the *Compton scattering* by an electron is shown in Fig. 4-30(a) but this must be complemented by the graph in Fig. 4-30(b) when the calculation is carried out in the lowest order of perturbation theory (as in Appendix F.3).

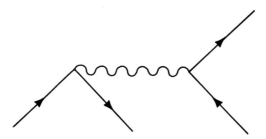

Fig. 4-29. Feynman graph depicting pair annihilation in the Coulomb field of an electron.

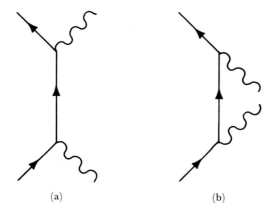

    (a)   (b)

Fig. 4-30. Compton scattering of a photon by an electron represented by two complementary Feynman graphs.

  The actual mathematics which leads to the Klein-Nishina formula is presented in detail by Schweber *et al.* [Schw 56, pp. 248ff.] and by Mandl [Ma 59b, pp. 105ff.], whereas a list of possible processes associated with lowest-order Compton and double Compton scattering has been given by Kockel [Ko 49]. An outline of the basic steps in the calculation has been presented in Appendix F. In applying the Feynman method to physical processes between initially free particles, a more satisfactory description of the particle states is provided by going over to momentum space and specifying momenta p (or wave-numbers $k = 2\pi/\lambda = p/\hbar$ in the case of photons) rather than depicting the process in coordinate space. The lowest-order Feynman graphs for Compton scattering by an electron in this representation are shown in Fig. 4-31, and

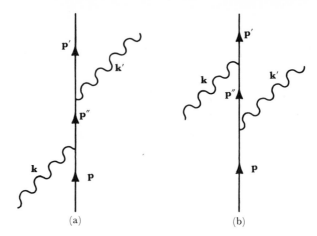

Fig. 4-31. Feynman representation in *momentum space* of the Compton scattering of a photon (momentum $\hbar k \rightarrow \hbar k'$) by an electron (momentum $p \rightarrow p'$).

higher-order radiative correction diagrams in Fig. 4-32. These latter involve closed-loop segments, such as arise in a consideration of self-energies of particles. When one

regards a physical electron as a "bare" electron surrounded by a photon cloud, the interaction of such an electron with its radiation field, responsible for a difference in energy (mass) of a physical electron as against that of a bare electron, is termed the SELF-ENERGY of the electron, and in Feynman's representation is characterized by a graph of the type shown in Fig. 4-33 corresponding to a perturbation term of second order which, because it diverges, makes it necessary to introduce renormalization procedures into quantum electrodynamics. An example of a fourth-order self-energy part inserted into an internal electron line of a Feynman diagram for Compton scattering is given in Fig. 4-34. On the other hand, a PHOTON SELF-ENERGY diagram, Fig. 4-35, features a closed loop corresponding to a (virtual intermediate) electron-positron pair (the self-interaction of the electromagnetic field has no simple classical analog and is not incorporated within the classical Maxwell equations. It is describable as a VACUUM POLARIZATION, since it entails modification of the electron-positron distribution and is thus akin to a polarization effect.

Fig. 4-33. Feynman graph representing the self-energy of the electron.

Closed-loop diagrams with incident and emergent photons can also take other forms, as is evident from Fig. 4-36, which depicts several types. None of these involves free electron functions, but only electron propagation functions. The square diagram depicted in Fig. 4-36(d) represents the scattering of light by light: two incident photons scatter each other through a process of (virtual) pair production and give rise to two outgoing photons. Thus the process of DELBRÜCK SCATTERING which is featured in Section 4.7 is described by the Feynman graph of Fig. 4-37, where

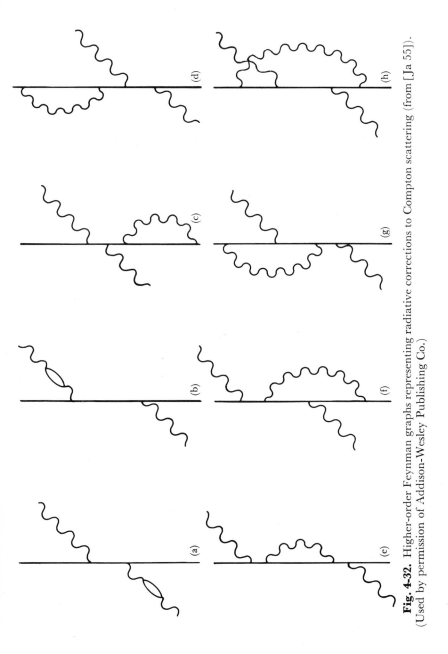

Fig. 4-32. Higher-order Feynman graphs representing radiative corrections to Compton scattering (from [Ja 55]). (Used by permission of Addison-Wesley Publishing Co.)

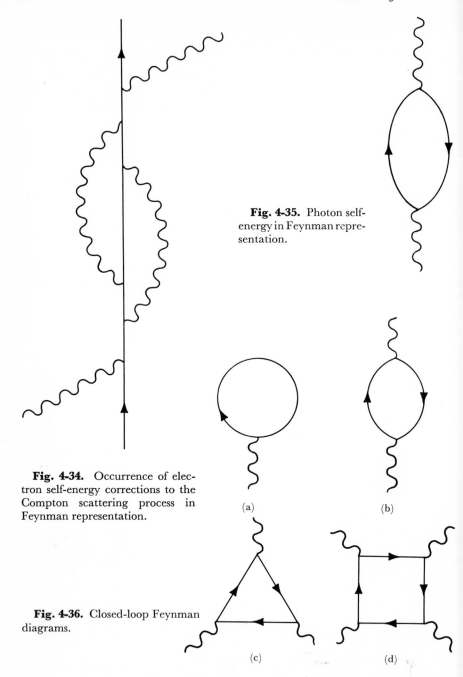

Fig. 4-35. Photon self-energy in Feynman representation.

Fig. 4-34. Occurrence of electron self-energy corrections to the Compton scattering process in Feynman representation.

(a)

(b)

Fig. 4-36. Closed-loop Feynman diagrams.

(c)

(d)

the double vertical line represents the world line of the nucleus in whose electromagnetic field an incident γ ray is scattered. Mathematically, a closed loop indicates that a trace enters into the perturbation formalism. An important mathematical property possessed by diagrams containing closed loops was found by Furry [Fu 37], whose name has been associated with the theorem that diagrams which contain a closed loop with an *odd* number of corners make a null contribution to the matrix elements describing the process. In form of a GENERALIZED FURRY THEOREM this may

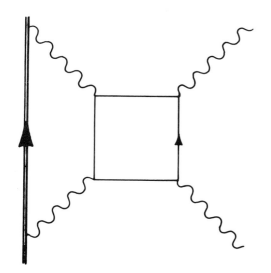

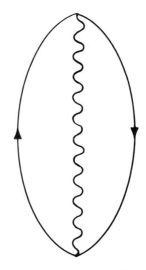

Fig. 4-37. Feynman graph for the Delbrück scattering of photons by photons. The double line represents a world-line of a nucleon or a nucleus.

Fig. 4-38. Vacuum fluctuation Feynman diagram.

be expressed as follows: *Any diagram or part of a diagram from which only photon lines emerge does not contribute to the matrix element if the number of these photon lines is odd.* A graph of the form shown in Fig. 4-38 has no external lines at all and therefore does not cause any transitions; a *vacuum fluctuation diagram* of this type can (at any rate, in elementary applications) be omitted altogether.

In Fig. 4-37 we have introduced the world line of a nucleus; equally permissible is the incorporation of the world lines of other particles or antiparticles within a Feynman graph. For example, the diagram which represents the *chemical binding* between two protons by an electron-exchange force (covalent linkage) is shown in Fig. 4-39, and that for the *nuclear binding force*

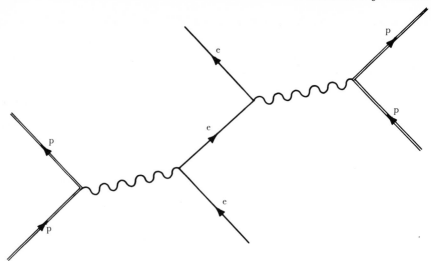

Fig. 4-39. Representation of chemical binding forces in terms of Feynman-like diagrams.

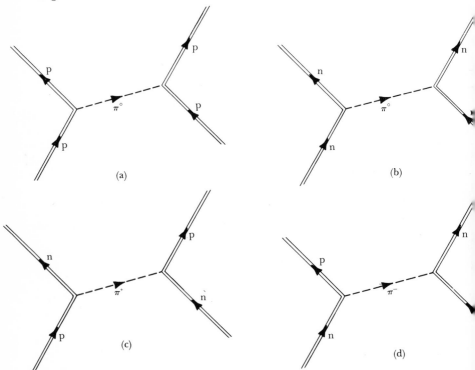

Fig. 4-40. Feynman representation of nuclear forces due to π-meson exchange: (a) p-p force, mediated by π^0 mesons; (b) n-n force, mediated by π^0 mesons; (c) p-n force, mediated by π^+ mesons; (d) n-p force, mediated by π^- mesons. Nucleons are conventionally represented by double lines, mesons by single broken lines.

due to π-*meson exchange* by Fig. 4-40(a)–(d). The (virtual) annihilation of a π^- meson (an antiparticle) to form a neutron-antiproton pair is depicted in Fig. 4-41 (cf. the radiative capture reaction $\pi^- + p \rightarrow n + \gamma$, shown in Fig. 4-42), and the qualitative representation of the neutron decay process $n \rightarrow p + e^- + \bar{\nu}_e$ illustrative of the Fermi weak interaction is provided by Fig. 4-43. The reversed arrow indicating the world line of the emergent

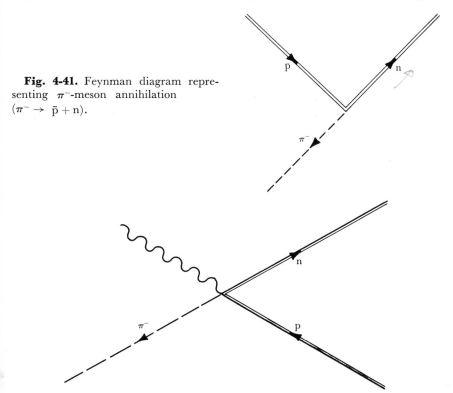

Fig. 4-41. Feynman diagram representing π^--meson annihilation $(\pi^- \rightarrow \bar{p} + n)$.

Fig. 4-42. Feynman diagram depicting π^- capture by a proton $(\pi^- + p \rightarrow n + \gamma)$.

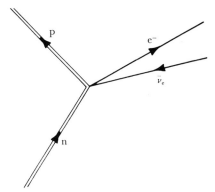

Fig. 4-43. Feynman representation of the β decay of a neutron $(n \rightarrow p^+ + e^- + \bar{\nu}_e)$.

*anti*neutrino $\bar{\nu}_e$ becomes even more meaningful in the case of the graphs for decay of the μ^- meson,

$$\mu^- \rightarrow e^- + \bar{\nu}_e + \nu_\mu \qquad (4\text{-}71)$$

and of its antiparticle,

$$\mu^+ \rightarrow e^+ + \nu_e + \bar{\nu}_\mu \qquad (4\text{-}72)$$

as is evident from Fig. 4-44(a, b). The latter of these represents the final stage of the π-μ-e decay process shown in Fig. 4-45, which can be modified as in Fig. 4-46 to depict the rare decay mode $\pi^+ \rightarrow e^+ + \nu_e$ by linking the (virtual) ν_μ and $\bar{\nu}_\mu$ lines into a loop. Since the Feynman diagram for this and the other elementary particle processes is to be regarded as merely a qualitative, rather

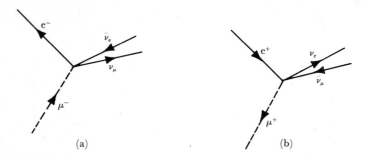

Fig. 4-44. Feynman graphs contrasting (a) the decay of a μ^- meson ($\mu^- \rightarrow$ $e^- + \bar{\nu}_e + \nu_\mu$) and (b) of its conjugate particle, the μ^+ meson ($\mu^+ \rightarrow e^+ + \nu_e + \bar{\nu}_\mu$).

than a strictly quantitative, way of representing the interaction, such a deformation is permissible. In this way the graphs can provide a pictorial description of an extensive range of fundamental processes, especially since by reversing the arrows one can derive the diagram for the time-inverse process. Indeed, the diagrams in momentum space can, according to the sophisticated approach associated with Regge theory in high-energy physics be "read" in a horizontal as well as a vertical sense and thereby be used to extract even further inherent quantitative information. Rather than to expatiate on the mathematical aspects at this stage, we relegate them to Appendix F and take up once more the consideration of pair creation, together with further interaction processes which occur when radiation or particles move through matter.

4.6.4. DIFFERENTIAL AND TOTAL PAIR CROSS SECTIONS

The quantum-theoretical treatment of pair production from which the angular and energy dependence of the cross section may be derived is unfortunately extremely involved and leads to complicated expressions which lend themselves to simplification only under very special conditions. The basic evaluation by Heitler *et al.* [He 33, Be 34a, He 54, pp. 256 ff.] using the Born approximation for the production of an e^+-e^- pair with energy E_+, E_- and (relativistic) velocity v_+, v_- in the Coulomb field of a nucleus of charge Ze

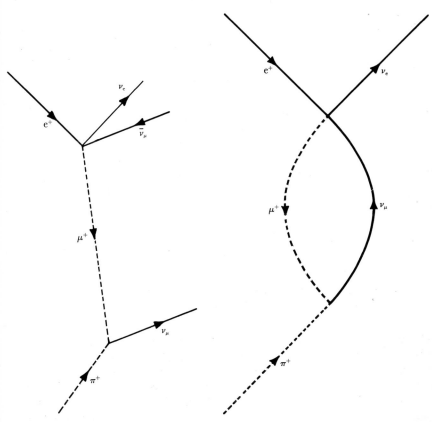

Fig. 4-45. The $\pi^+ - \mu^+ - e^+$ decay mode involving the intermediate formation and decay of a μ^+ meson, shown in Feynman representation.

Fig. 4-46. As Fig 4-45, but with a closed loop formed on linking the ν_μ and $\bar{\nu}_\mu$ lines, thereby representing the rare decay mode $\pi^+ \rightarrow e^+ + \nu_e$.

applies when

$$\frac{2\pi Ze^2}{\hbar v_\pm} = \frac{2\pi Z\alpha}{\beta_\pm} \ll 1, \qquad 2\alpha Z^{1/3} \frac{E_+ E_-}{(h\nu)(m_e c^2)} \ll 1 \qquad (4\text{-}73)$$

with α the fine-structure constant. Its results prove to be correct for light elements, but about 10 percent too high for heavy elements, and the calculation has since been refined by Maximon *et al.* [Ma 52, Be 54, Da 54] avoiding the Born approximation, but concluding with formulae bearing a close resemblance to those of Heitler.

Qualitatively, the *angular distribution* is found to peak increasingly in the forward direction with increase in the incident photon energy $h\nu$, a trend akin to that of previously discussed rival processes. In the limit of very high energies ($\mathscr{E} \equiv h\nu/m_e c^2 \gg 1$) the mean angle of positron or electron emission is of the order

$$\bar{\theta} \approx \frac{1}{\mathscr{E}} \qquad (4\text{-}74)$$

In place of the involved formula for the *energy distribution* of pairs expressed by the differential cross section for creation of a positron of total energy E_+ to $E_+ + dE_+$ with a negative electron of total energy E_-, we write the latter in the form

$$_a\sigma_p(E_+) \equiv \frac{d\,_a\sigma_p}{dE_+} = \frac{\bar{\sigma}}{h\nu - 2m_e c^2} f(\mathscr{E},Z) \qquad \text{cm}^2 \ \text{MeV}^{-1}/\text{atom} \quad (4\text{-}75)$$

where $(h\nu - 2m_e c^2)$ is the combined kinetic energy of the particles, $\bar{\sigma} \equiv \alpha r_e^2 = 5.794 \times 10^{-28} \ \text{cm}^2$, and $f(\mathscr{E},Z)$ is a dimensionless nonsimple function of $\mathscr{E} \equiv h\nu/m_e c^2$ and Z, whose numerical value can be derived from Fig. 4-47, which depicts f in terms of the fraction

$$x = \frac{E_+ - m_e c^2}{h\nu - 2m_e c^2} \qquad (4\text{-}76)$$

which the positron receives of the overall kinetic energy. This representation is convenient inasmuch as the TOTAL PAIR CROSS SECTION in Born approximation is given in terms of σ as simply the area under the appropriate curve, since

$$_a\sigma_p \equiv \int \,_a\sigma_p(E_+) \, dE_+ = \bar{\sigma} \int f \, dx \qquad \text{cm}^2/\text{atom} \quad (4\text{-}77)$$

This integrated cross section can also be expressed analytically in the extreme relativistic limit when one assumes either (a) no screening ($1 \ll \mathscr{E} \ll 1/\alpha Z^{1/3}$):

$$_a\sigma_p = \bar{\sigma} Z^2 \left(\frac{28}{9} \log 2\mathscr{E} - \frac{218}{27} \right) \qquad \text{cm}^2/\text{atom} \quad (4\text{-}78)$$

or (b) complete screening of the nuclear Coulomb field by the field of the atomic electrons ($\mathscr{E} \gg 1/\alpha Z^{1/3}$):

$$_a\sigma_p = \bar{\sigma} Z^2 \left(\frac{28}{9} \log \frac{183}{Z^{1/3}} - \frac{2}{27} \right) \qquad \text{cm}^2/\text{atom} \quad (4\text{-}79)$$

a result independent of the reduced photon energy $\mathscr{E}$. With the help of numerical integration for situations to which the limiting expressions cannot be applied, the total pair cross section in terms of σ varies with incident energy $\mathscr{E}$ in the way shown in Fig. 4-48. The results differ but little from those obtained from Bethe-Maximon theory which avoids the Born approximation. For instance, at high energies in the absence of screening the more exact formula

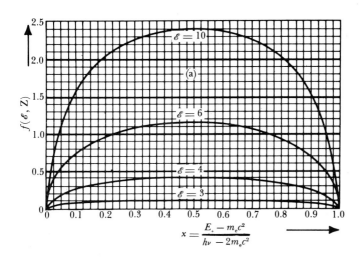

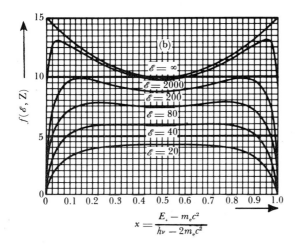

Fig. 4-47. Curves showing the energy distribution of electrons in an electron pair, plotted as a function $f(\mathscr{E}, Z)$ *vs* x for various reduced incident γ energies $\mathscr{E}$. The units are explained in the text: The total pair cross section is given by the area under each graph (a) at low energies, (b) at high energies [in units $Z(Z+1)\,\bar{\sigma} \equiv Z(Z+1)\,\alpha r_e^2$] (from [Be 53]).

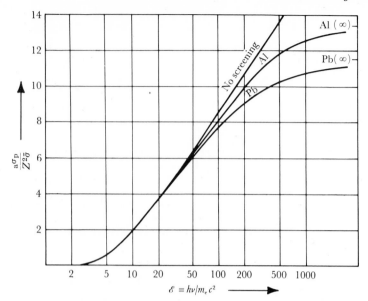

Fig. 4-48. Semilogarithmic representation of the energy dependence of the pair cross section in which the ratio $_a\sigma_p/Z^2\bar\sigma$ calculated using the Born approximation is plotted against $\mathscr{E}$ for various elements. The asymptotic values are shown on the right-hand ordinate (adapted from [He 54]). (Used by permission of The Clarendon Press, Oxford.)

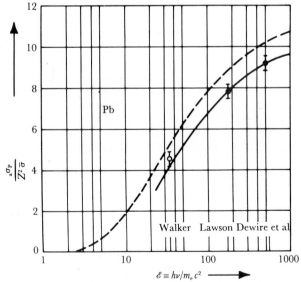

Fig. 4-49. Comparison of experimental results with theoretical curves derived from the Born approximation (dashed curve) and the Bethe-Maximon theory (solid curve) for the pair-production cross section on lead (from [Da 54]).

can be approximated to

$$_a\sigma_p = \bar\sigma Z^2 \left(\frac{28}{9} \log 2\mathscr{E} - \frac{218}{27} - 1.027 \right) \qquad \mathrm{cm}^2/\mathrm{atom} \qquad (4\text{-}80)$$

and the correction amounts to an 11.8 percent reduction to the cross section in lead at 88 MeV. A comparison of the results furnished for lead with and without the Born approximation is presented in Fig. 4-49.

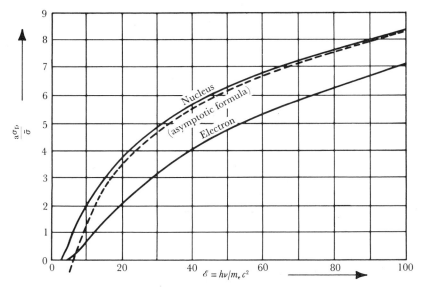

Fig. 4-50. The total cross section in normalized form $_a\sigma_p/\bar\sigma$ *vs* reduced energy $\mathscr{E}$ for pair production in the field of a nucleus and in the field of an electron, neglecting screening. The solid curves have been calculated from the theory of Borsellino [Bo 47]; the dashed curve is derived from the asymptotic formula (4-81), and lies higher than that for the electron-field cross section at the energies here considered (from [Be 53]).

When pair creation occurs in the field of an electron, at an energy exceeding the threshold value $4m_ec^2$, the total cross section per electron is of similar form, namely (without screening),

$$_e\sigma_p = \bar\sigma \left(\frac{28}{9} \log 2\mathscr{E} - 11.3 \right) \qquad \mathrm{cm}^2/\mathrm{electron} \qquad (4\text{-}81)$$

the absolute value of the second numerical term, while somewhat uncertain, being appreciably larger than 218/27 [cf. Eq. (4-80)]. Thus the electron-field cross section is lower than that for pair production in a nuclear field, as is evident from Fig. 4-50, and at low γ-ray energies the contribution to σ_p due to pair production in the electron field is small.

The LINEAR ATTENUATION COEFFICIENT for pair production is, of course,

$$\kappa = n^\square {}_a\sigma_p \qquad cm^{-1} \tag{4-82}$$

where $n^\square$ is the number of atoms per cubic centimeter of the target.

4.6.5. INVERSE PAIR PRODUCTION: ANNIHILATION RADIATION AND BREMSSTRAHLUNG

On the basis of Dirac's electron theory the inverse process to the photo-production of a particle-hole pair is the filling of a negative-energy vacancy by an electron from a positive energy state. Taking place in the vicinity of a nucleus or an electron, this transition would be accompanied by the emission of energy in the form of electromagnetic radiation. The process of radiative ANNIHILATION of a particle-antiparticle pair in the Coulomb field, as the exact converse of pair production, is matched by an extreme case of BREMSSTRAH-LUNG. The latter "braking radiation" is emitted whenever a charged particle experiences a change in its velocity $\mathbf{v}$ as, for instance, when an electron's motion is altered under the influence of the Coulomb field of a nucleus.

Assuming the acceleration $\mathbf{b} = \dot{\mathbf{v}} = \ddot{\mathbf{x}}$ to be uniform in time, the radiated energy S can easily be derived from the formulae of Section 4.2.1, since

$$|\mathbf{S}_{out}| = \frac{c}{4\pi} \frac{e^2 |\ddot{\mathbf{x}}|^2}{c^4 r^2} \sin^2 \Theta \rightarrow \frac{e^2}{c^3} |\ddot{\mathbf{x}}|^2 \frac{(1 + \cos^2 \theta)}{8\pi r^2} \tag{4-83}$$

and therefore

$$S = \int |\mathbf{S}_{out}| \, dA = \frac{e^2}{c^3} |\ddot{\mathbf{x}}|^2 \int \frac{1 + \cos^2 \theta}{8\pi} \, d\Omega = \frac{e^2}{c^3} |\ddot{\mathbf{x}}|^2 \int_0^\pi \tfrac{1}{4}(1 + \cos^2 \theta) \sin \theta \, d\theta \tag{4-84}$$

i.e.,

$$S = \frac{2}{3} \frac{e^2}{c^3} |\ddot{\mathbf{x}}|^2 \tag{4-85}$$

Since the only difference between the normal bremsstrahlung transition and that corresponding to inverse pair production lies in the sign of the final-state energy level of the electron, it is not surprising that there is a close similarity between the pair creation and bremsstrahlung cross sections, as will be seen when they are compared in due course.

When positrons annihilatively interact with electrons which are bound to a nucleus capable of absorbing the recoil momentum, a *single* γ quantum can be emitted. This process of 1-QUANTUM ANNIHILATION yields essentially mono-energetic γ radiation and occurs with a small, but finite, probability when ANNIHILATION IN FLIGHT takes place. A Born-approximation calculation [He 54, pp. 272 ff.] for positrons of energy up to 50 MeV falling on stationary electrons has yielded the results shown in Fig. 4-51 for the relative cross sections for 1-quantum and 2-quantum annihilation. When referred to a heavy element, such as lead, the results indicate the most favorable ratio of 1-quantum to 2-quantum annihilation to be about 20 percent and to occur around 5 MeV. When 2-quantum annihilation takes place, there can be a continuous mutual distribution in the emergent energies and angles.

When, however, annihilation takes place in the absence of a "buffer" capable of taking up recoil momentum, 1-quantum annihilation is rigorously

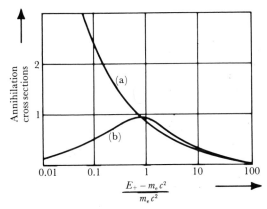

Fig. 4-51. Cross sections for annihilation of a positron with a stationary electron as a function of reduced positron energy: (a) 2-quantum annihilation, in units $Z\pi r_e^2$ per atom; (b) 1-quantum annihilation, in units $Z^5 \pi r_e^2 \alpha^4$, (from [He 54]). (Used by permission of The Clarendon Press, Oxford.)

forbidden by kinematic considerations, and two or more photons are simultaneously emitted at an angle to one another. The probability is largest for an electron at the base of the positive energy continuum to make a transition which fills a vacancy near the surface of the negative energy "sea"; the kinetic energies of the particle and antiparticle are then vanishingly small and the ANNIHILATION AT REST results in the emission of two γ quanta essentially of energy 0.51 MeV each at an angle of 180 degrees to one another.

Positrons which have been slowed down sufficiently by passage through matter can form an intermediate metastable system with electrons in which both conjugate particles revolve about their mutual center of gravity in a hydrogen-like "atom" called POSITRONIUM (see [De Be 54, De 58, Si 58]). Its energy states are akin to those of a hydrogen atom, and its lowest state has a binding energy of $\frac{1}{2}Ry = 6.8$ eV (it therefore decays readily by annihilation in which an energy of at least $2m_ec^2 = 1.02$ MeV is liberated). The conjugate particles in a positronium atom in its ground state may have *antiparallel* spins, in which case they form the singlet system called PARAPOSITRONIUM whose mean lifetime against annihilation is $\tau_{\text{para}} \cong 1.25 \times 10^{-10}$ sec, or they may have parallel spins and form the triplet ORTHOPOSITRONIUM system, whose mean lifetime is appreciably longer, $\tau_{\text{ortho}} \cong 1.4 \times 10^{-7}$ sec. A simple consideration of angular-momentum conservation indicates that *para*positronium is subject to *two*-quantum annihilation [Fig. 4-52(a)] whereas annihilation

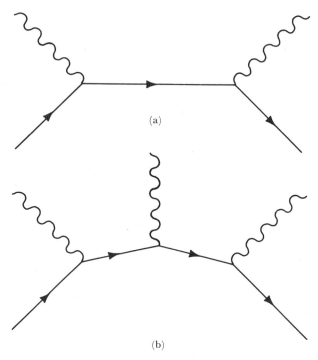

Fig. 4-52. Feynman graphs illustrating annihilative decay of (a) *para*positronium, by 2-quantum annihilation; (b) *ortho*positronium, by 3-quantum annihilation.

from the triplet configuration entails three-quantum emission [Fig. 4-52(b)] and is therefore much less probable, as is evinced by the lifetime, which is longer by a factor of about $8\hbar c/e^2$. In the case of free collisions, the probability of formation of singlet *para*positronium stands in a ratio of $1:3$ to that for formation of triplet *ortho*positronium, and accordingly the ratio of 2-quantum to 3-quantum annihilation is simply

$$\frac{1}{3} \frac{8\hbar c}{e^2} = 372 \qquad (4\text{-}86)$$

on the assumption that there is no conversion from the *ortho* to the *para* form.

4.7. Nuclear Scattering of Gamma Rays

The scattering of electromagnetic radiation by the Coulomb field of the nucleus is less of practical than of theoretical importance, since the magnitudes of the effects due to nuclear Thomson and resonance scattering, the nuclear photoeffect and Delbrück scattering are all small, but their significance as regards quantum electrodynamics and the characteristics of nuclear inter-actions is considerable. For example, elastic scattering of photons by the nuclear Coulomb field can take place not only through nuclear Thomson scattering (for which the cross section is given by replacing e by Ze and m_e by the nuclear mass $Am_{\mathcal{N}}$ in the formulae of Section 4.2.1) but also through more involved processes such as DELBRÜCK SCATTERING, in which virtual or real creation and subsequent annihilation of an e^+-e^- pair occurs as an intermediate step in the course of the photon's interaction with the nuclear field. The effect, named after Delbrück, who proposed this explanation in a note correcting an experimental paper by Meitner and Kösters [Me 33], is of interest because it is predicted by quantum electrodynamics but not by linear classical theory, and thus serves as an additional check of vacuum polarization and associated phenomena. Its detection is fraught with difficulty because it cannot readily be distinguished experimentally from the other elastic nuclear scattering processes, i.e., nuclear Thomson, nuclear resonance, and Rayleigh scattering; moreover, as it combines coherently with the latter, its amplitude (made up of a real and an imaginary part) must be added to the Thomson amplitude and the other (complex) amplitudes before squaring in order to derive the overall differential cross section:

$$\frac{d\sigma}{d\Omega} = |a_{\mathrm{Th}} + (a_{\mathrm{r}} + ia_{\mathrm{i}})_{\mathrm{D}} + (a_{\mathrm{r}} + ia_{\mathrm{i}})_{\mathrm{R}} + (a_{\mathrm{r}} + ia_{\mathrm{i}})_{\mathrm{Res}}|^2 \qquad (4\text{-}87)$$

The imaginary part $(a_{\mathrm{i}})_{\mathrm{D}}$ of the Delbrück scattering amplitude has now been established fairly reliably by measurements at 17 MeV by Stierlin *et al.* [St 62] as also at 87 MeV by Moffat and Stringfellow [Mo 60a], and particularly by the experi-ments of Bösch *et al.* [Bö 63b] at 9 MeV complemented by the theoretical analysis undertaken by Ehlotzky and Sheppey [Eh 64]. Figure 4-53 shows the measure of agreement between theory (correct to better than 10 percent) and experiment, while at the same time indicating tentative evidence for the existence of a real Delbrück amplitude $(a_{\mathrm{r}})_{\mathrm{D}}$. Because the Delbrück amplitude for small scattering angles is roughly proportional to the energy while the amplitudes for the other effects prove to be prac-tically independent of $\mathscr{E}$ and appreciably larger than Delbrück amplitudes at low energies it is unlikely that such investigations can easily be pursued below about 5 MeV (while on the other hand, investigations at large angles are vitiated by a lack of theoretical knowledge of the requisite nuclear resonance contribution).

Just as Delbrück scattering can be regarded as pair production in the nuclear field followed by pair annihilation, and its imaginary (absorptive) amplitude $(a_{\mathrm{i}})_{\mathrm{D}}$ related to the pair-production cross section, so can the other effects be related to processes already considered: For instance, the Rayleigh imaginary amplitude $(a_i)_R$ can be related to the photoelectric absorption cross section, and nuclear Thomson and resonance amplitudes to the nuclear photoeffect cross sections. The nuclear photoeffect itself, also termed NUCLEAR PHOTODISINTEGRATION, which embraces such processes as (γ,n) and (γ,p), has been subjected to much theoretical and experimental study since it yields information on nuclear energy levels as well as on nuclear models; its cross section ranges from millibarns for light elements to hundreds of millibarns for heavy elements and it begins to contribute to γ attenuation above around 10 to 20 MeV.

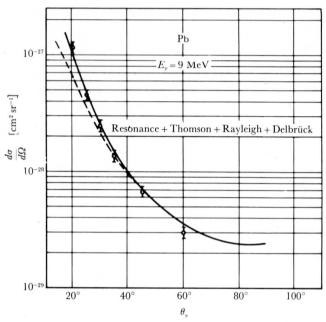

Fig. 4-53. Differential cross section for scattering of 9-MeV γ rays through an angle θ_γ. The experimental data of [Bö 63b] are compared with theoretical curves [Eh 64]: the solid curve represents a coherent superposition of nuclear resonance, nuclear Thomson, Rayleigh, and (real and imaginary) Delbrück scattering components, while the dashed curve corresponds to the omission of the real part of the Delbrück scattering amplitude (from [Eh 64]).

4.8. Total Attenuation Coefficient for Electromagnetic Radiation Passing through Matter

To summarize and contrast the effect of the various processes which contribute to the attenuation of a beam of electromagnetic radiation passing through matter according to the relation

$$I(x) = I_0 \exp(-\mu x) \qquad (4\text{-}88)$$

we examine the characteristics of the principal terms which make up the
TOTAL LINEAR ATTENUATION COEFFICIENT μ. It is formed of a superposition of
the following three main effects:

(a) Compton scattering, represented by the atomic cross section $_a\sigma_C$ and
the corresponding linear attenuation coefficient $\sigma \equiv n^\square {}_a\sigma_C$;

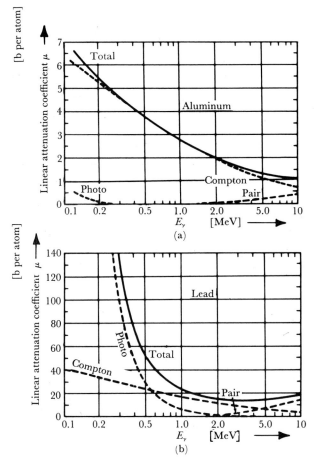

Fig. 4-54. Total linear attenuation coefficients μ in function of the γ energy E_γ for
γ rays passing through (a) aluminum, (b) lead (from [Da 65c]). (Used by permission
of North-Holland Publishing Co.)

(b) Photoelectric effect, represented by the linear attenuation coefficient
τ; and

(c) Pair production, represented by the linear attenuation coefficient κ,
each of which have a different $\mathscr{E}$- and Z-dependence.

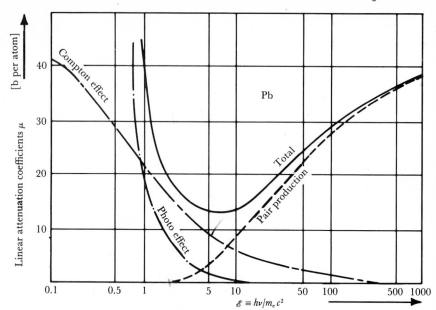

Fig. 4-55. Total linear attenuation coefficients μ in function of the reduced γ energy $\mathscr{E} \equiv E_\gamma / m_e c^2$ for lead in semilogarithmic representation extending to higher energies than Fig. 4-54(b). There is a slight, unimportant, difference between the older numerical data used in this figure and that set employed in Fig. 4-54(b); the latter is more reliable (adapted from [He 54]).

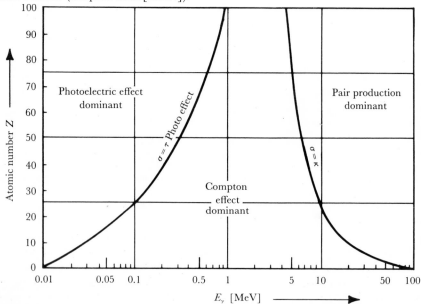

Fig. 4-56. Representation of the relative predominance of the three main forms of interaction according to γ energy and atomic number Z (from [Ev 55]). (Used by permission of McGraw-Hill Book Co.)

Writing

$$\mu = \sigma + \tau + \kappa \qquad cm^{-1} \qquad (4\text{-}89)$$

the energy-dependence of μ and its constituent terms is shown in Fig. 4-54 for a low-Z and a high-Z element. Although dominant at very low energies, the Compton attenuation decreases monotonically toward medium energies,

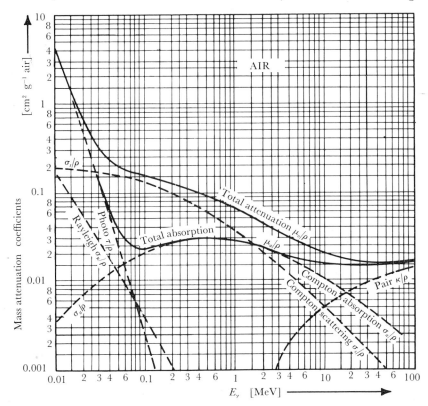

Fig. 4-57(a). Mass attenuation coefficients for photons in air at STP (density $\rho = 0.001\ 293$ g cm^{-3}), of volume composition 78.04 percent N$_2$, 21.02 percent O$_2$, 0.94 percent Ar, calculated and plotted by Evans [Ev 55] from tables of atomic cross sections compiled by White [Whi 52]. (Used by permission of McGraw-Hill Book Co.)

and the photo-effect attenuation also decreases—very much more sharply in the case of a high-Z element, with a roughly $\mathscr{E}^{-3}$ energy-dependence. On the other hand, pair production sets in only above $2m_ec^2$ but increases with increasing energy, roughly as $\log \mathscr{E}$. In the case of a heavy element its rise can overcompensate the fall of the Compton attenuation, with the result that instead of decreasing monotonically with energy as it does for light elements [(Fig. 4-54(a)], the total attenuation coefficient μ can for heavy elements display a dip and thereafter rise appreciably with energy. This is illustrated for lead at medium energies in Fig. 4-54(b) and at high energies in Fig. 4-55. In the medium-Z region around copper ($Z = 29$) the two above effects compensate almost perfectly and μ displays a fairly constant value beyond 6 MeV over a considerable energy range.

The Z-dependence is rendered more clearly evident by using the representation of Fig. 4-56, which shows that:

(a) Pair production dominates at high energies $\mathcal{E}$ and high Z;

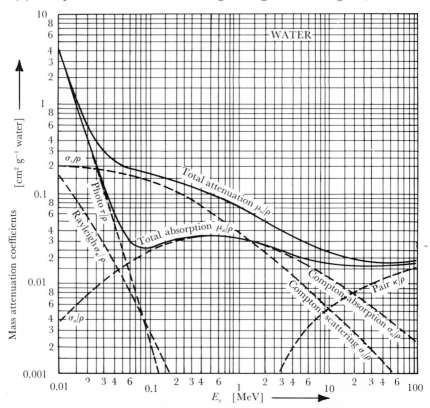

Fig. 4-57(b). Mass attenuation coefficients for photons in water or biological tissue (*cf.* Appendix H), calculated and plotted by Evans [Ev 55] from tables of atomic cross sections compiled by White [Whi 52]. (Used by permission of McGraw-Hill Book Co.)

(b) The Compton effect dominates at medium energies $\mathcal{E}$ and low Z;

(c) The photoelectric effect dominates at low energies $\mathcal{E}$ and high Z, since the Compton attenuation at all but the lowest energies is roughly proportional to Z, while that for the photoelectric effect goes as Z^4 or Z^5 and that for pair production as Z^2.

Since the MASS-ATTENUATION COEFFICIENTS μ/ρ and σ/ρ, τ/ρ, κ/ρ have the property that they do not depend upon the density and the physical state of the absorber, numerical data are often expressed in terms of these. The attenuation is given by the formula

$$I(x) = I_0 \exp\left[-\left(\frac{\mu}{\rho}\right)(\rho x)\right]$$

with the target thickness conveniently expressed as ρx (in mg cm^{-2}). Figures 4-57(a) through 4-57(d) show some attenuation data as calculated and plotted by Evans [Ev 55] from theoretical tables given by White [Whi 52] (see also Davisson [Si 65, pp. 827 ff.]). The "total absorption" curve represents the

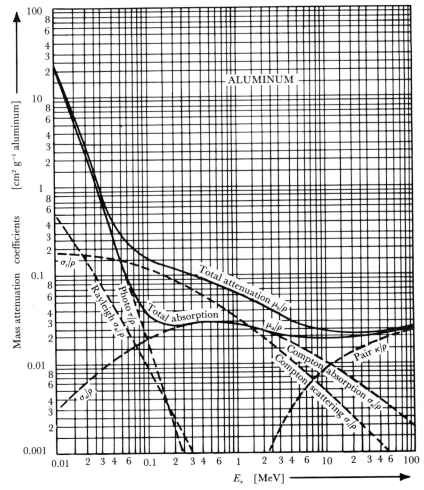

Fig. 4-57(c). Mass attenuation coefficients for photons in aluminum (density $\rho = 2.70$ g cm^{-3}), calculated and plotted by Evans [Ev 55] from tables of atomic cross sections compiled by White [Whi 52]. (Used by permission of McGraw-Hill Book Co.)

sum $(\mu_a/\rho) = (\sigma_a/\rho) + (\tau/\rho) + (\kappa/\rho)$, whereas the "total attenuation" curve also includes the contribution of Compton scattering, $(\mu_0/\rho) \equiv (\mu_a/\rho) + (\sigma_s/\rho) = (\sigma_C/\rho) + (\tau/\rho) + (\kappa/\rho)$. Attenuation due to Rayleigh scattering is shown separately as (σ_R/ρ). In Fig. 4-57(d), the characteristic K and L edges are to be seen at low energies.

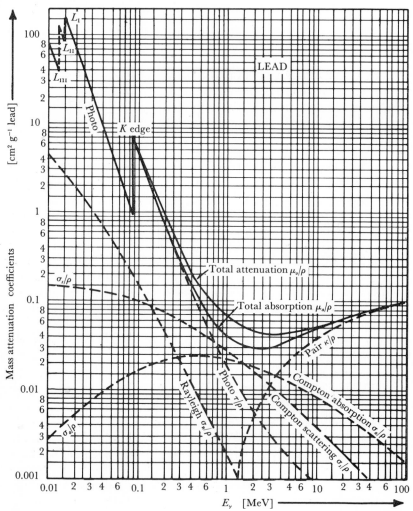

Fig. 4-57(d). Mass attenuation coefficients for photons in lead (density $\rho = 11.35$ g cm^{-3}), of isotopic composition 1.5 percent $^{204}_{82}$Pb, 23.6 percent $^{206}_{82}$Pb, 22.6 percent $^{207}_{82}$Pb, 52.3 percent $^{208}_{82}$Pb, calculated and plotted by Evans [Ev 55] from tables of atomic cross sections compiled by White [Whi 52]. The characteristic K- and L-edges associated with the photoelectric effect can be observed at low energies. (Used by permission of McGraw-Hill Book Co.)

4.9. Interaction of Charged Particles with Matter

Rutherford scattering, i.e., the classical elastic scattering of charged particles by a Coulomb field already discussed in Section 3.5.7 constitutes just one of several interaction modes by which particles experience a change in energy on passing through matter. Any or all of the following collision processes may take place—not to mention nuclear reactions and other strong interactions—and in their course bring about energy diminution and deflection of the particles' motion:

(a) Elastic collisions with atomic electrons;
(b) Inelastic collisions with atomic electrons;
(c) Elastic collisions with atomic nuclei; and
(d) Inelastic collisions with atomic nuclei.

Particles having kinetic energies of the order of MeV experience some 10,000 such collisions as they traverse a typical target; consequently their behavior is described by a statistical theory of MULTIPLE SCATTERING embracing the individual processes (a) to (d).

The process (a) represents the RAMSAUER EFFECT [Al 31] and is essentially negligible at all but the lowest particle energies, below about 100 eV. As against this, inelastic collisions of type (b) with electrons betoken the principal mode of kinetic energy diminution of charged particles passing through an absorber; the electrons are thereby raised to higher energy levels in the atom and may, as indeed happens in the majority of instances, have sufficient energy transferred to them to cause them to be ejected from the atom altogether. Thus the preponderant mode of energy loss is *ionization of absorber atoms*. Compared with (b) the elastic *nuclear* collisions of type (c) are *less* frequent in the case of heavy projectiles such as mesons, protons, deuterons, α particles, and heavy ions, but much more frequent in the case of light incident particles ($e^{\pm}$). Inelastic nuclear collisions of type (d) are altogether less probable than (c) and can therefore be neglected in considering the energy diminution of heavy particles, although in the case of light projectiles having high energies they may need to be taken into consideration. We have already mentioned one process which falls into this last category, namely bremsstrahlung, the emission of energy in the form of electromagnetic radiation. Another process by which a nucleus can be raised into an excited state by an inelastic interaction with a heavy charged particle is COULOMB EXCITATION (heavy ions can effect MULTIPLE COULOMB EXCITATION), but the relative probability of occurrence is low for this mode, which is described in more detail in Section 12.7.2.

There are accordingly characteristic differences between the processes which cause the slowing down of heavy particles as against those applying to light projectiles; instead of presenting a detailed derivation of the requisite theory, which may be found in several sources, e.g., [Be 53, Ev 55, Wha 58, NRC 64], we restrict ourselves to a qualitative discussion and presentation of some empirical formulae for various situations. The practical importance of a consideration of these effects lies not only in the fact that energy loss on traversal of a target can influence the effective reaction cross section or could be used for the energy degradation of a charged-particle beam, but that the ionizing action of charged particles on the one hand constitutes a means of detecting them with the aid of suitable detectors whereas on the other it gives rise to detrimental biological radiation effects in which the energy loss to cells furnishes a measure of the radiation damage.

4.9.1. ENERGY LOSS OF HEAVY CHARGED PARTICLES (STOPPING POWER, RANGE, AND STRAGGLING)

The classical calculation of the energy loss in collisions with electrons which stems from Bohr [Bo 13a, Bo 48] and Bethe [Be 30, Be 32] is approximated by the following simplified considerations:

When an ion of charge ze and velocity v passes within a distance b (b is the impact parameter) of an electron which, for the moment, is assumed to be free and at rest, it imparts a momentum Δp to the electron, where

$$\Delta p = \int_{-\infty}^{\infty} F_C \, dt \equiv \int_{-\infty}^{\infty} \frac{ze^2}{(x^2 + b^2)} \, dt \qquad (4\text{-}90)$$

with F_C the Coulomb force between the negative electron of charge e and the positive ion which, at a given instant, is at a position specified by the coordinate x. If b does not change appreciably during the encounter, the component of F_C parallel to v cancels out in the integration, so that only the component perpendicular to v, namely,

$$(F_C)_\perp = F_C \frac{b}{(x^2 + b^2)^{1/2}} \qquad (4\text{-}91)$$

need be considered for the evaluation of the integral, which may most readily be effected by setting dt equal to dx/v and treating v as constant, since it is practically unaltered by an encounter with one single electron:

$$\Delta p \cong \int_{-\infty}^{\infty} (F_C)_\perp \frac{dx}{v} = \frac{ze^2 b}{v} \int_{-\infty}^{\infty} \frac{dx}{(x^2 + b^2)^{3/2}} = \frac{2ze^2}{vb} \qquad (4\text{-}92)$$

Then the classical kinetic energy lost by the ion and imparted to the electron at an impact parameter b is

$$\Delta E_b = \frac{(\Delta p)^2}{2m_e} \cong \frac{2z^2 e^4}{m_e v^2 b^2} \qquad (4\text{-}93)$$

This now has to be summed over all the electrons at the impact parameter b. Supposing there are $n^\square$ atoms of atomic number Z in unit volume of the absorber, the number of electrons per unit volume will be $n^\square Z$. Thus in traveling a distance dx, the incident ion will encounter $2\pi b \, db \, dx \, n^\square Z$ electrons and will thereby lose energy

$$dE_{b\to b+db} = \Delta E_b \, 2\pi b n^\square Z \, db \, dx \tag{4-94}$$

Accordingly, the rate of energy loss per unit path length, $-dE/dx$, is given by integrating this expression over b from a minimum to a maximum value:

$$-\frac{dE}{dx} = \frac{4\pi z^2 e^4}{m_e v^2} n^\square Z \int_{b_{\min}}^{b_{\max}} \frac{db}{b} = \frac{4\pi z^2 e^4}{m_e v^2} n^\square Z \ln \frac{b_{\max}}{b_{\min}} \tag{4-95}$$

It will be noted that the natural choice of integration limits, namely $b_{\min} = 0$, $b_{\max} = \infty$, must be avoided since it leads to an infinite rate of loss of energy per unit path. The appropriate choice of limits, as indeed the entire calculation, must be based upon quantum-mechanical considerations, as in the treatments by Bethe [Be 30, Be 32] and Bloch [Bl 33b] but a rough indication of the result can be derived classically as follows: The lower limit $b_{\min}$ ensues from Eq. (4-92) on noting that the maximum momentum transfer to an electron in a classical central collision is $2m_e v$:

$$b_{\min} = \frac{ze^2}{m_e v^2} \tag{4-96}$$

The upper limit $b_{\max}$ is determined by the finite binding energies of electrons in an atom. A *bound* electron cannot accept arbitrarily small amounts of energy, but must absorb sufficient energy to be raised to an unfilled energy state. We may introduce an effective excitation energy I, averaged over all the electrons in an atom of the absorber, and contend that there is no energy transfer unless

$$\Delta E_b \geqslant I \tag{4-97}$$

i.e.,

$$\frac{2z^2 e^4}{m_e v^2 b^2} \geqslant I \tag{4-98}$$

whence

$$b \leqslant b_{\max} = \frac{ze^2}{v} \left(\frac{2}{m_e I} \right)^{1/2} \tag{4-99}$$

Substitution of these limits (4-96), (4-99) in (4-95) yields the approximate expression

$$-\frac{dE}{dx} \approx \frac{4\pi z^2 e^4}{m_e v^2} n^\square Z \ln \left(\frac{2m_e v^2}{I} \right)^{1/2} \tag{4-100}$$

which differs from the corresponding quantum-mechanical expression only in the occurrence of the power $\frac{1}{2}$ in the logarithmic term.

The quantum-theoretical STOPPING POWER for heavy particles is thus

$$-\frac{dE}{dx} = \frac{4\pi z^2 e^4}{m_e v^2} n^\square Z \ln \left(\frac{2m_e v^2}{I} \right) \tag{4-101}$$

in the absence of relativistic considerations and other corrections. When these are incorporated, the BETHE EQUATION reads

$$-\frac{dE}{dx} = \frac{4\pi z^2 e^4}{m_e v^2}\, n^{\square} Z \left[\ln\left(\frac{2m_e v^2}{I}\right) - \ln\left(1 - \beta^2\right) - \beta^2 - \frac{C_\mathrm{K}}{Z}\right] \qquad \text{erg cm}^{-1}$$

(4-102)

where $\beta \equiv v/c$, and the MEAN IONIZATION (EXCITATION) POTENTIAL is given by

$$I \approx 11.5Z \qquad \text{eV}$$

(4-103)

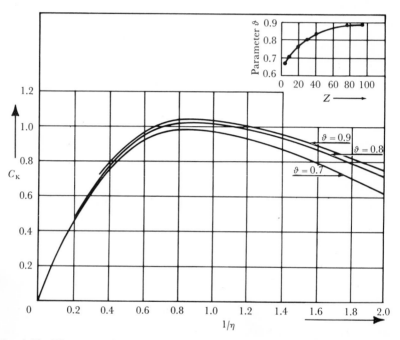

Fig. 4-58. The correction term C_K in the stopping-power formula (4-102), due to non-participation of bound K electrons in the slowing-down process. The theory, stemming from Bethe and others [Li 37, Wa 52] (see also [Be 53]), is formulated in terms of a parameter $1/\eta$, here plotted as abscissa,

$$\frac{1}{\eta} \equiv \left(\frac{u_k}{v}\right)^2 = \left(\frac{\alpha c Z_\mathrm{eff}}{v}\right)^2 \simeq \frac{m}{m_e}\left(\frac{Z - 0.3}{137}\right)^2 \frac{m_e c^2}{2E}$$

where $u_k = \alpha c Z_\mathrm{eff} \simeq \alpha c(Z - 0.3)$ is the speed of the K electron in an atom in which screening by one K electron is considered, but all other screening ignored, while m, v, and E are the mass, velocity, and kinetic energy of the incident heavy particle. Unless $1/\eta \ll 1$, Eq. (4-101) needs correction by including C_K which can assume appreciable values. The parameter ϑ is a function of Z only, as in the inset graph; and is the ratio of the actual K-shell binding energy to the "ideal ionization potential" $\frac{1}{2}m_e c^2 \alpha^2 (Z - 0.3)^2$, neglecting screening by L, M, N, ... electrons (from [Wa 52]).

or, to a better approximation,

$$I \simeq 9.1Z(1 + 1.9Z^{-2/3}) \qquad \text{eV} \qquad (4\text{-}104)$$

The quantity C_K is a term dependent upon E and Z which needs to be included only at comparatively low energies, e.g., below 4 MeV for protons passing through an aluminum absorber. Its value, which ranges between 0 and about 1, is shown in a suitable representation in Fig. 4-58.

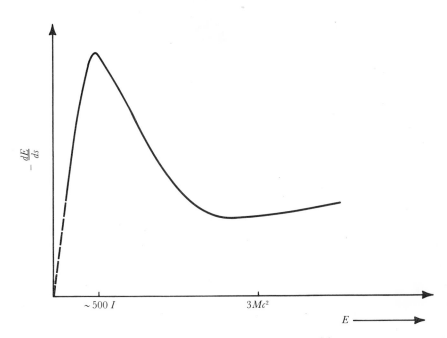

Fig. 4-59. Variation of atomic stopping power with energy.

The stopping power for heavy particles as given by the Bethe equation (4-102) varies with energy E in the manner shown in Fig. 4-59. Qualitatively, this form of energy dependence follows from the fact that in the medium energy range ($I \ll E \ll Mc^2$, where M is the mass of an absorber atom) the first of the terms in the square brackets varies but slowly with energy and the others are negligibly small, so that very roughly

$$-\frac{dE}{dx} \sim \frac{1}{E} \qquad (4\text{-}105)$$

On the other hand, at high energies the relativistic terms in conjunction with the initial term cause the numerical value of the stopping power to rise very gently, so that a broad minimum is set up in the neighborhood of $E \approx 3\ Mc^2$, whereas at low energies the initial term is dominant, causing the curve to drop rapidly back to the origin at energies below about $E \approx 500\ I$.

The z^2-dependence, confirmed by experiments using protons and α particles of the same velocity incident upon the same target material, shows that the ionizing power of charged particles, which essentially determines the

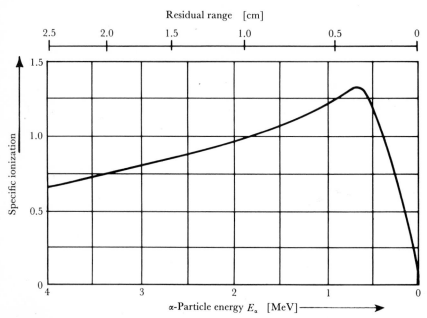

Fig. 4-60. Specific ionization of a beam of α particles in air at 15°C and 1 atm, plotted as a function of energy and residual range. The peak corresponds to 6600 ion pairs per mm (adapted from [Rie 55]).

stopping power of the absorber, increases quadratically with the charge state and assumes very high values for multiply charged heavy ions. The SPECIFIC IONIZATION, or number of charge pairs created per millimeter of the charged particle's trajectory depends upon the projectile's charge and energy. It rises to a maximum when the projectile has lost nearly all its energy at the end of its range, as may be seen in Fig. 4-60, which shows the ionization in terms of energy (or residual range) for an α particle in air at 15°C and 760 mm Hg. The peak corresponds to 6600 ion pairs/mm, and at that point the *effective* charge of the α particle is not 2 but $\bar{z} \cong 1.5$, due to the charge-exchange process

$$He^{2+} + e^- \rightleftharpoons He^+ \tag{4-106}$$

which begins to be appreciable at energies below about $E_\alpha = 2\ \text{MeV}\ (\bar{z} = 1.883$

at 1.7 MeV, and 1.500 at 0.65 MeV, thereafter dropping linearly with *velocity* to zero). The stopping powers for various charged projectiles in air are plotted as a function of particle energy in Fig. 4-61. The quantity $-(1/n^{\square})\,(dE/dx)$ is called, somewhat loosely, the ATOMIC STOPPING CROSS SECTION and expressed in eV cm²/atom. For protons in various materials, the variation with energy is shown in Fig. 4-62. It is independent of the physical state of the stopping

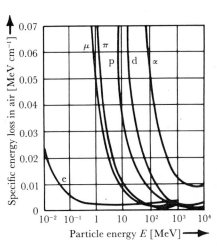

Fig. 4-61. Semilogarithmic plot of the stopping power, or specific energy loss, of various charged particles in air at 15°C and 1 atm, as a function of particle energy E (from [Bei 52]).

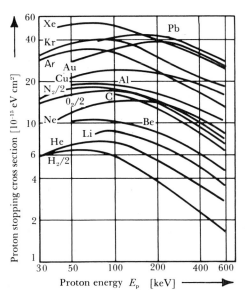

Fig. 4-62. Log-log plot of experimental stopping cross sections for protons in various materials, expressed in terms of energy (from [Wha 60]).

material and of physical conditions such as pressure, but varies considerably from one stopping material to another. A quantity which, analogously to the mass-attenuation coefficient, is, within roughly a factor of 2, constant for all absorbers is the MASS STOPPING POWER,

$$-\frac{1}{\rho}\frac{dE}{dx} = -\frac{dE}{d(\rho x)} \equiv -\frac{dE}{d\xi} \qquad (4\text{-}107)$$

where $\xi \equiv \rho x$ (with ρ = density) is the penetration coordinate measured in

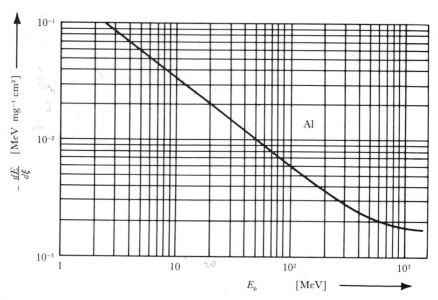

Fig. 4-63. Log-log plot of stopping power $-dE/d\xi$ vs energy E_p for protons in aluminum.

g/cm² rather than cm. This follows from the rough constancy of Z/A for all elements, since

$$-\frac{dE}{d\xi} \equiv -\frac{1}{\rho}\frac{dE}{dx} \sim n^{\square}\frac{Z}{\rho} = \frac{\rho N_A}{A}\frac{Z}{\rho} \approx \text{const.} \qquad (4\text{-}108)$$

If the stopping power for given charged particles at a given energy is known for a standard absorber S, that for any other absorber A under the same conditions can be derived from the calculable ratio of absorber thickness corresponding to equal energy loss:

$$Q \equiv \frac{(-dE/d\xi)_A}{(-dE/d\xi)_S} = \frac{-(1/\rho_A)(dE/dx)_A}{-(1/\rho_S)(dE/dx)_S} \approx \frac{A_S}{A_A}\frac{Z_A}{Z_S}\frac{\ln(2m_e v^2/I_A)}{\ln(2m_e v^2/I_S)} \qquad (4\text{-}109)$$

which is termed the RELATIVE STOPPING POWER. Although theoretically hydrogen would represent the most suitable choice for standard element, aluminum is in fact employed in practice. The stopping power of aluminum in terms of *proton* energy is shown in Fig. 4-63, and experimental values of the relative stopping power Q for a range of absorbers are depicted in terms of a logarithmic Z-scale in Fig. 4-64. The conversion from $-dE/d\xi$ for protons to that for particles of charge ze of the same energy is effected by simply multiplying the former by z^2.

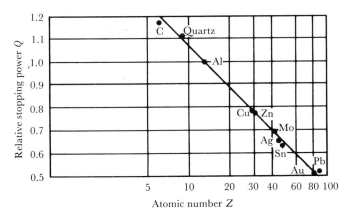

Fig. 4-64. Semilogarithmic plot of relative stopping power, defined by Eq. (4-109), for 6.3-MeV protons as a function of atomic number Z (from [Ma 51b]).

As the path of a heavy particle through an absorber is practically rectilinear, the RANGE R of the projectile can be taken to be equal to the total path length, which can be obtained by integration over the net energy loss from an initial energy E_0 to an end energy of zero:

$$R = \int_{E_0}^{0} \frac{dE}{(dE/dx)} \qquad (4\text{-}110)$$

Since the stopping power is unknown at very low energies this relation can be used only for the high-energy section of the range, and the remaining part has to be determined empirically, e.g., from measured ranges. Obviously a calculation of this sort gives merely a MEAN RANGE $\bar{R}$ in the absorber. Since the stopping of particles is brought about by a very large number of statistically distributed collisions, the actual ranges of individual particles differ slightly from one another and straggle around the mean value. The range-energy relations used in practice always refer to the mean range. When the mean range $\bar{R}_S$ in a standard absorber S is known, that for another absorber A is given by the integral

$$R_A = \int_0^{\bar{R}_s} \frac{dx}{Q} \tag{4-111}$$

in terms of the relative stopping power Q, which is a function of energy.

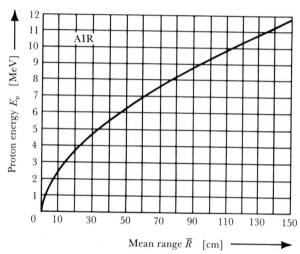

Fig. 4-65. Mean range of protons in dry air at STP (adapted from a plot by [Ev 55], based upon data by H. A. Bethe in Brookhaven National Laboratory Report BNL-T-7, 1949). (Used by permission of McGraw-Hill Book Co.)

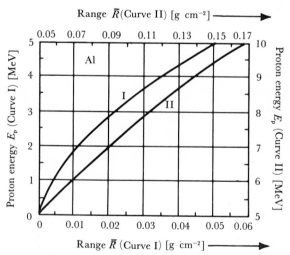

Fig. 4-66. Mean range of protons in aluminum, for proton energies from 0 to 10 MeV (adapted from [Be 53]).

Thus, although mean ranges can be depicted graphically in function of energy as in Figs. 4-65 to 4-70, one cannot give an analytic expression for the integral from which $\bar{R}$ is constructed. A rough estimate of the range of protons of energy between some MeV and 200 MeV in dry air can be made from the approximate formula proposed by Wilson and Brobeck [Wi 47b]:

$$\bar{R} \approx \left(\frac{E^{[\text{MeV}]}}{9.3} \right)^{1.8} \times 10^2 \qquad \text{cm} \qquad (4\text{-}112)$$

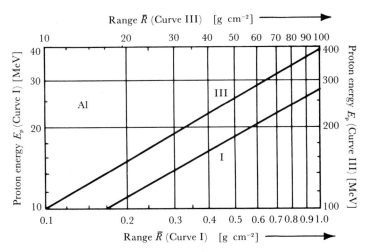

Fig. 4-67. As Fig. 4-66, but for energies from 10 to 400 MeV (adapted from [Be 53]).

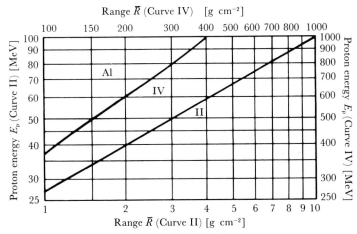

Fig. 4-68. As Fig. 4-66, but for energies from 25 to 1000 MeV (adapted from [Be 53]).

The above reference also gives more accurate analytical approximations. A fairly rough relation, accurate to about ±15 percent, for the range in centimeters is provided by the BRAGG-KLEEMANN RULE:

$$\frac{\bar{R}_A}{\bar{R}_S} \cong \frac{\rho_S}{\rho_A}\left(\frac{A_A}{A_S}\right)^{1/2} \tag{4-113}$$

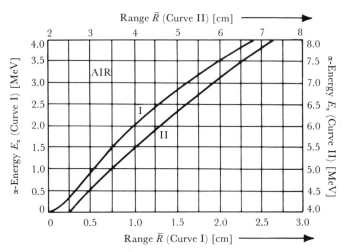

Fig. 4-69. Mean range of α particles (expressed in cm) in dry air at STP for α energies from 0 to 8 MeV (adapted from [Be 53]).

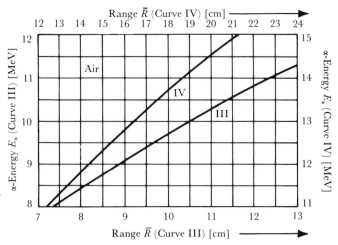

Fig. 4-70. As Fig. 4-69, but for α energies from 8 to 15 MeV (adapted from [Be 53]).

In the case of a mixture of various elements with atomic weights A_1, A_2, A_3, ... and relative mass abundances γ_1, γ_2, γ_3, ..., the term $A_A^{1/2}$ in the above must be replaced by the expression

$$A_A^{1/2} = \gamma_1 A_1^{1/2} + \gamma_2 A_2^{1/2} + \gamma_3 A_3^{1/2} + \cdots \tag{4-114}$$

When referred to air, for which

$$A_{\text{Air}}^{1/2} = 3.81 \quad \text{and} \quad \rho_{\text{Air}} = 1.226 \times 10^{-3} \quad \text{g cm}^{-3} \tag{4-115}$$

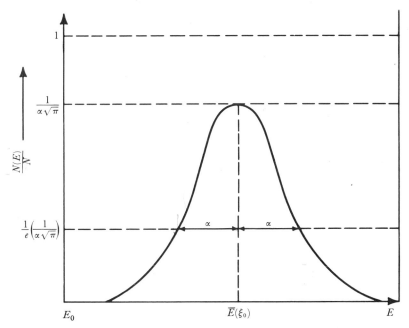

Fig. 4-71(a). Schematic diagram depicting energy straggling as the specific number of particles whose mean energy is $\bar{E}$ after traversing a thickness ξ_0 of absorber. The straggling parameter α specifies the half-width at $(1/e)$th height.

the mean range $\bar{R}_A$ of an absorber A can be calculated from that in air under the same conditions from the relation

$$\bar{R}_A = 3.2 \times 10^{-4} \left(\frac{A_A^{1/2}}{\rho_A} \right) \bar{R}_{\text{Air}} \tag{4-116}$$

Not only is there a STRAGGLING of *range*, but also of *energy* and *angle* as an originally collimated, monoenergetic beam of heavy charged particles is slowed down and stopped in a medium. The statistical distribution of energy-depleting collisions leads to a statistical distribution of end energies, for instance, after traversal of an absorber thickness $\xi_0 \equiv \rho x_0$. Since a heavy particle loses its energy in multiple-scattering collisions in each of which there is but small energy transfer without appreciable path deflection, the ENERGY

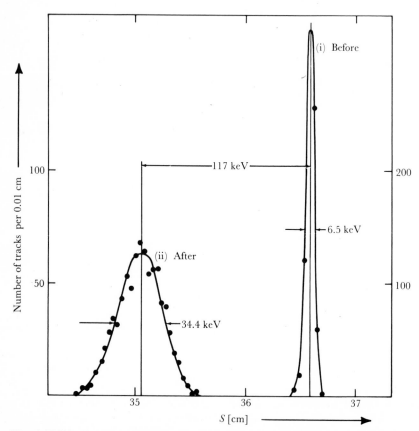

Fig. 4-71(b). Energy straggling as evinced by the broadening and flattening of a line in a proton spectrum due to 3-MeV protons (i) before and (ii) after passage through a 3.3-mg cm^{-2} gold foil, as measured by a magnetic spectrograph (from [Nie 61]).

STRAGGLING can be described in terms of a Gaussian distribution:

$$\frac{N(E)\,dE}{N} = \frac{1}{\alpha \pi^{1/2}} \exp\left[-\frac{(E - \bar{E})^2}{\alpha^2}\right] \qquad (4\text{-}117)$$

depicted in Fig. 4-71(a). This presents schematically the specific number $N(E)$ of particles having energies in the range E to $E + dE$ referred to the total number of particles N, whose mean energy is $\bar{E}$ after traversing a thickness ξ_0 of absorber. The distribution parameter, or STRAGGLING PARAMETER α, which expressed the half-width at $(1/e)$th height, is given by the expression

$$\alpha^2 = 4\pi z^2 e^4\, n^\square Z\, x_0 \left[1 + \frac{kI}{m_e v^2} \ln\left(\frac{2m_e v^2}{I}\right)\right] \qquad (4\text{-}118)$$

with k a constant which depends upon the electron shells of the absorber

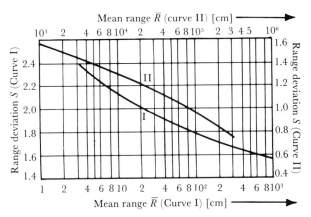

Fig. 4-72. Semilogarithmic plot of the range straggling of protons in air, expressed through the percentage range deviation S in function of the mean range $\bar{R}$ (see Eq. (4-120)) (from [Be 53]).

atoms and which varies between about $\frac{2}{3}$ and $\frac{4}{3}$. The value $k \simeq \frac{4}{3}$ is suitable in the majority of instances. A practical illustration of energy straggling is shown in Fig. 4-71(b) for protons of initial energy 3.0 MeV upon passage through a 3.3-mg cm^{-2} gold foil, which reduces their energy by 0.117 MeV and broadens the line width at half maximum by more than a factor of 5.

Along analogous lines, the RANGE STRAGGLING, expressing the specific number of particles with ranges R to $R + dR$ as related to the total number of particles of the same initial energy, is given by the Gaussian distribution

$$\frac{N(R)\, dR}{N} = \frac{1}{\alpha \pi^{1/2}} \exp\left[-\frac{(R - \bar{R})^2}{\alpha^2}\right] \qquad (4\text{-}119)$$

with $\bar{R}$ the mean range.

The straggling can also be expressed as a percentage in terms of the mean range:

$$S = \frac{R - \bar{R}}{\bar{R}} \times 100 \quad \text{percent} \qquad (4\text{-}120)$$

For protons in air, this has the form shown in Fig. 4-72.

The multiple collisions do, nevertheless, broaden out the originally colli-mated beam, as shown in Fig. 4-73, and cause the emergent beam to evince some ANGLE STRAGGLING. This is described by the mean angle of divergence $\bar{\theta}$ whose value can be deduced from the expression [Fe 50, p. 37]:

$$\overline{\theta^2} = \frac{2\pi z^2 e^4}{\bar{E}^2}\, n^\square Z^2 x_0 \ln\left(\frac{\bar{E}a_0}{zZ^{4/3}e^2}\right) \tag{4-121}$$

where $a_0 \equiv \hbar^2/m_e e^2$ is the Bohr radius.

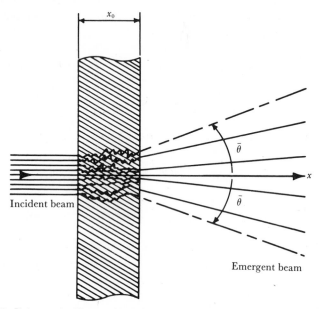

Fig. 4-73. Schematic illustration of angle straggling through a mean angle of divergence $\bar{\theta}$ as heavy particles pass through an absorber.

For example, on passing through an aluminum foil 55 mg/cm² thick (corres-ponding to 205×10^{-4} cm), protons whose initial energy is 7 MeV are slowed to 4 MeV, in course of which the energy straggling parameter becomes $\alpha = 0.07$ MeV and the angle straggling $\bar{\theta} = 5°$.

4.9.2. Energy Loss of Electrons (Stopping Power, Range, and Straggling)

Essentially, three effects contribute to the slowing down of electrons as they pass through an absorber, namely,

(a) inelastic scattering on atomic electrons,

(b) elastic scattering on atoms, and

(c) inelastic nuclear scattering with emission of bremsstrahlung.

The scattering of incident electrons on atomic electrons, leading to a diminution of the incident beam energy, provides an instance of collisions between identical particles for which the Mott scattering formula, discussed in Section 3.5.8, might be expected to hold, at least at low energies. At higher energies, we need to employ an appropriate *relativistic* extension to Mott's approach, as originally put forward by Møller [Mø 32a]. For this reason, electron-electron scattering in the relativistic region is termed MØLLER SCATTERING, and the stopping power is expressed in terms of the relativistic factor $\gamma \equiv (1 - \beta^2)^{-1/2}$, where $\beta \equiv v/c$, as

$$-\frac{dE}{dx} = \frac{4\pi e^4}{m_e v^2} n^\square Z \left[\ln\left(\frac{2m_e c^2}{I}\right) + \ln(\gamma - 1) + \tfrac{1}{2}\ln(\gamma + 1) \right.$$

$$\left. -\left(3 + \frac{2}{\gamma} - \frac{1}{\gamma^2}\right)\ln(2^{1/2}) + \frac{1}{16} - \frac{1}{8\gamma} + \frac{9}{16\gamma^2} \right] \qquad (4\text{-}122)$$

The low-energy limit of this, as obtained by setting γ to 1, is

$$-\frac{dE}{dx} = \frac{4\pi e^4}{m_e v^2} n^\square Z \left[\ln\left(\frac{2m_e v^2}{I}\right) - \frac{5}{2}\ln 2 + \frac{1}{2} \right] \qquad (4\text{-}123)$$

i.e.,

$$-\frac{dE}{dx} = \frac{4\pi e^4}{m_e v^2} n^\square Z \left[\ln\left(\frac{2m_e v^2}{I}\right) - 1.2329 \right] \qquad \text{erg cm}^{-1} \qquad (4\text{-}124)$$

which differs from the low-energy form of the Bethe formula (4-101) for *heavy*-particle stopping only in the appearance of the numerical term (-1.2329). On the other hand, at very high energies where $E_{\text{kin}} \approx E_{\text{tot}} = (\gamma - 1)mc^2 \approx (\gamma + 1)mc^2$ since $\gamma \gg 1$, Eq. (4-122) may be written as

$$-\frac{dE}{dx} = \frac{4\pi e^4}{m_e v^2} n^\square Z \left[\ln\left(\frac{2m_e}{I}^2\right) + \frac{1}{2}\ln\frac{(\gamma - 1)^2(\gamma + 1)}{8} + \frac{1}{16} \right] \qquad (4\text{-}125)$$

i.e.,

$$-\frac{dE}{dx} \cong \frac{4\pi e^4}{m_e v^2} n^\square Z \left[\frac{1}{2}\ln\left(\frac{E^3}{2m_e c^2 I^2}\right) + \frac{1}{16} \right] \qquad (4\text{-}126)$$

For comparison, the extreme-relativistic form of the *proton* stopping power may, on using the expansion

$$\ln y = (y - 1) - \tfrac{1}{2}(y - 1)^2 + \tfrac{1}{3}(y - 1)^3 - \cdots \qquad (0 < y < 2) \quad (4\text{-}127)$$

to first order, be derived from Eq. (4-102) on setting C_K to zero as

$$-\frac{dE}{dx} = \frac{4\pi e^4}{m_e v^2} \left[\ln\left(\frac{2m_e c^2}{I}\right) + \ln(\gamma^2) - 1 \right] \qquad \text{erg cm}^{-1} \qquad (4\text{-}128)$$

The difference between Eqs. (4-125) and (4-128) amounts in general to only about 10 percent.

The *elastic scattering of electrons by the Coulomb field of an atom* is in itself a more probable process than the electron-electron encounters which have just been considered, and leads to a diminution in the beam energy through the cumulative action of MULTIPLE SCATTERING encounters in each of which the light incident particle loses a small amount of energy, namely just that sufficient to effect a deflection of the electron's path. Since large-angle deflection of the light projectile can occur even without the latter's passing very close to the nucleus, the screening of the nuclear Coulomb field by that of the surrounding electrons in their atomic shells plays a significant role which invalidates the use of the simple Rutherford scattering theory of Section 3.5.7. A nonrelativistic theory which takes screening into account without using the Born approximation has been proposed by Molière [Mo 47] and a relativistic theory, neglecting screening since its effect is confined to the region of small angles, by Mott [Mo 29, Mo 32b], whose resultant expression for the differential cross section can, according to McKinley and Feshbach [McK 48], be related to the Rutherford cross section. The ratio is, for SINGLE SCATTERING,

$$\mathcal{R} \equiv \frac{d\sigma'_{\text{Mott}}}{d\sigma'_{\text{Rutherford}}} = \left[1 - \beta^2 \sin^2 \frac{\vartheta'}{2} + \pi\alpha\beta Z \left(1 - \sin \frac{\vartheta'}{2} \right) \sin \frac{\vartheta'}{2} \right] \quad (4\text{-}129)$$

where α is the fine-structure constant and $d\sigma'_{\text{Mott}}$ is not to be confused with that for collisions between *like* particles, considered in Section 3.5.8 (an additional source of possible confusion is the term MOTT SCATTERING as applied in some descriptions to electron-*atom* encounters). The ratio $\mathcal{R}$ is a function of angle, energy, and atomic number Z. It should be noted that for relativistic electrons the quantity γm_e has to be inserted for the reduced mass μ in the Rutherford formula (3-124) for scattering on a massive atom.

As the absorber thickness increases beyond about $1/\sigma n^\square$, the likelihood that an incident particle will undergo several successive single-scattering encounters becomes appreciable. This represents a complicated situation described as PLURAL SCATTERING. With further increase in absorber thickness the mean number of successive encounters rises to a value which permits of statistical treatment of the process; e.g., when the number of collisions rises above about 20, the angular distribution of the scattered electrons approximately follows a Gaussian probability law. We then speak of MULTIPLE SCATTERING, for which the probability that an electron emerges at an angle θ to $\theta + d\theta$ is

$$P(\theta) \, d\theta = \frac{2\theta}{\overline{\theta^2}} \exp \left(-\frac{\theta^2}{\overline{\theta^2}} \right) d\theta \quad (4\text{-}130)$$

where the mean square angle of scattering, $\overline{\theta^2}$ can be calculated from $d\sigma'_{\text{Mott}}$

to obtain, for example, the numerical approximations for electrons of kinetic energy $\bar{E}$ (in MeV) passing through x_0 cm ($= \xi_0$ g cm^{-2}) of

(a) air:

$$\overline{\theta^2} \approx 9 \times 10^6 \frac{x_0}{(\bar{E})^2} \approx 7 \times 10^9 \frac{\xi_0}{(\bar{E})^2} \qquad (4\text{-}131)$$

(b) lead:

$$\overline{\theta^2} \approx 7 \times 10^{15} \frac{x_0}{(\bar{E})^2} \approx 6 \times 10^{14} \frac{\xi_0}{(\bar{E})^2} \qquad (4\text{-}132)$$

The actual expression for $\overline{\theta^2}$ in the case of particles having a mass m, charge ze, linear momentum p, and velocity v, which traverse a foil of atomic weight A, atomic number Z, and thickness x_0 is

$$\overline{\theta^2} = \frac{4\pi z^2 Z(Z+1)e^4 n^\square}{p^2 v^2} \ln \left[4\pi z^2 Z^{4/3} n^\square \left(\frac{\hbar}{mv} \right)^2 x_0 \right] \qquad (4\text{-}133)$$

The traversal of matter by electrons differs from the passage by heavier particles in that the latter can at most lose merely a small fraction ($\sim m_e/M$) of their kinetic energy in an encounter with an atomic electron, whereas the former can, in a Møller collision, lose as much as one-half of their original energy and, in the emission of bremsstrahlung, lose the whole of their kinetic energy. The large amount of energy straggling in the motion of electrons through an absorber stands in direct relation to the large deviations which the actual path lengths can evince about their mean,

$$\bar{S} = \int ds = \int \left(\frac{dE}{ds} \right)^{-1} dE \qquad (4\text{-}134)$$

A calculation of the above MEAN PATH LENGTH $\bar{S}$ is accordingly of little utility in the case of electrons, especially as the "drunken man's path" of the electrons as they traverse an absorber is associated with a RANGE R which is far less than the path length S, and which can differ appreciably from a mean value $\bar{R}$, as is evident from Fig. 4-74. The concept of range as the penetration depth in the incident direction furnishes a definition of the MEAN RANGE $\bar{R}$, equal in value to the thickness of a layer of absorber which reduces the number of emergent electrons to one-half of the incident intensity. For an initially mono-energetic beam of electrons, the mean range $\bar{R}$ is roughly one-half of $\bar{S}$, as may be seen from Fig. 4-75, which shows the statistically distributed ranges R and actual path lengths S for 0.02-MeV electrons passing through a gaseous oxygen target at STP. From the intersections of the tangents at the respective points of inflection with the abscissa, the values of the EXTRAPOLATED RANGE R_{ex} and EXTRAPOLATED PATH LENGTH S_{ex} can be determined. It will be noted that only about 3 percent of the electrons have sufficiently straight paths to enable them to traverse an absorber whose thickness is equal to the mean path length $\bar{S}$.

It has so far not proved possible to set up basic analytic expressions for the mean range or the extrapolated range of electrons, but the following EMPIRICAL

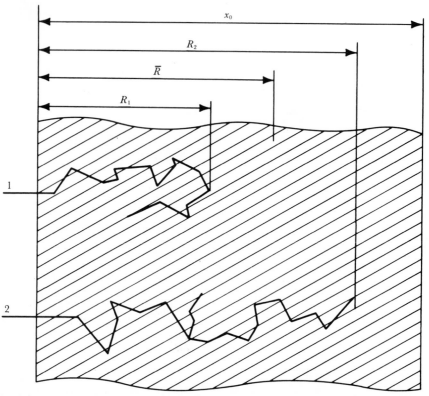

Fig. 4-74. Diagram indicating the possible paths of two light particles having the same initial energy in a thick absorber of thickness x_0. The actual ranges R_1 and R_2 follow a statistical distribution; the mean range R is a well-defined quantity, roughly one-half of the actual mean path length $\bar{S}$, which is determined by the energy.

EXPRESSIONS FOR MEAN RANGE (in g cm^{-2}) as functions of kinetic energy E (in MeV) have been proposed by various sources, such as [Fe 38]

$$\bar{R} = 0.543E - 0.160 \qquad (E > 0.7 \text{ MeV}) \tag{4-135}$$

to be compared with more recent range-energy relations:
[Bl 46c]

$$\bar{R} = 0.571E - 0.161 \qquad (1.2 < E < 2.3 \text{ MeV}) \tag{4-136}$$

[Gl 48]

$$\begin{cases} \bar{R} = 0.407E^{1.38} & (0.15 < E < 0.8 \text{ MeV}) & (4\text{-}137) \\ \bar{R} = 0.542E - 0.133 & (0.8 < E < 3 \text{ MeV}) & (4\text{-}138) \end{cases}$$

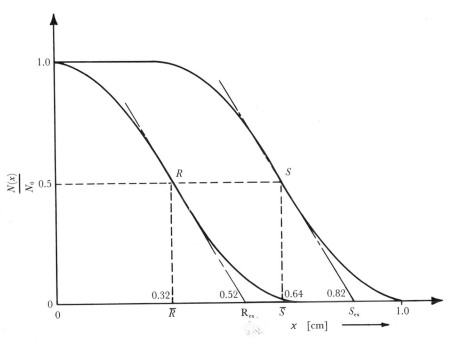

Fig. 4-75. The distribution of ranges R and path lengths S for 0.02-MeV electrons in a gaseous oxygen absorber at STP. The specific number of electrons having $R > x$ or $S > x$, where x is the oxygen target thickness in cm is plotted against x. The mean and extrapolated ranges and path lengths are shown. Under these conditions, $\bar{S}$ is about $1.24\,R_{\text{ex}}$ (from [Wi 31a]).

[Ka 52a]

$$\begin{cases} \bar{R} = 0.412E^{n} & \text{with } n = 1.265 - 0.0954 \ln E \quad (0.01 < E < 3 \text{ MeV}) \\ & \hspace{7.5cm} (4\text{-}139) \\ \bar{R} = 0.530E - 0.106 & (2.5 < E < 20 \text{ MeV}) \hspace{2.9cm} (4\text{-}140) \end{cases}$$

Useful numerical tabulations of mean ranges and stopping power for negative and positive electrons have been compiled by Nelms [Ne 56, Ne 58]. Figure 4-76 shows the excellent agreement between measured values and those furnished by the empirical Katz-Penfold relations (4-139) and (4-140).

The ENERGY STRAGGLING of electrons has been treated by various authors, e.g., [La 44, Bl 50b], and can be described in terms of an energy spread at half height about the mean,

$$\Gamma = 1.220\rho \left(\frac{Z}{A}\right)\left(\frac{c}{v}\right)^2 x_0 \qquad \text{MeV} \qquad (4\text{-}141)$$

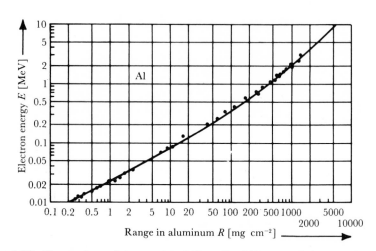

Fig. 4-76. Comparison of the empirical Katz-Penfold curve giving range in terms of electron energy according to the relations (4-139) and (4-140), and experimental points for electrons in aluminum as measured by several independent groups. In the case of monoenergetic electrons, the extrapolated range R_{ex} was taken for R, while in the case of a continuous spectrum, the maximal energy was taken for E and the maximal range for R (from [Ev 55]). (Used by permission of McGraw-Hill Book Co.)

4.9.3. BREMSSTRAHLUNG

As discussed in Section 4.6.5, emission of electromagnetic radiation always accompanies the acceleration or deceleration of charged particles. The emitted radiation, accordingly termed BREMSSTRAHLUNG, represents an energy dissipation per unit time of

$$\frac{dE}{dt} = \frac{2}{3}\frac{e^2}{c^3}|\ddot{\mathbf{x}}|^2 \qquad (4\text{-}142)$$

according to the classical equation (4-85), and extends over a continuous range of frequencies.

An instructive application of this expression lies in evaluating the energy loss per second due to bremsstrahlung when an electron accelerates rectilinearly from infinity to the K shell (whose radius is a_0) of an atom of nuclear charge $+ Ze$ (see Fig. 4-77).

In this idealized case, the total bremsstrahlung energy $E_B = E'$ is the integral from $x = \infty$ to $x = a_0$ of the dissipation rate:

$$E_B = \frac{2}{3}\frac{e^2}{c^3} \int |\ddot{\mathbf{x}}|^2 \, dt \qquad (4\text{-}143)$$

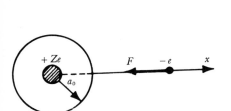

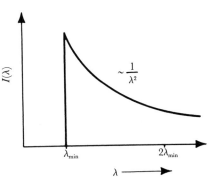

Fig. 4-77. Diagram to illustrate the approach of an electron from infinity toward an atomic K orbit (around a nucleus of charge $+Ze$) along the negative x direction under the influence of an attractive Coulomb force **F**.

Fig. 4-78. Bremsstrahlung intensity from a thin target as a function of wavelength λ. Corresponding to energies from 0 to E_{max}, the wavelength spectrum ranges continuously from ∞ to $\lambda_{min} = c/\nu_{max}$, with a sharp cut-off at the latter limit, subject to Duane and Hunt's law [Du 15]. The form of the spectrum follows from the relation $I(\lambda)\,d\lambda = I(\nu)\,d\nu$ as $I(\lambda) = I(\nu)\,d\nu/d\lambda \sim 1/\lambda^2$ (from [Ev 55]). (Used by permission of McGraw-Hill Book Co.)

A change in the integration variable can be effected with the aid of the equation of motion,

$$F = -m_e \ddot{\mathbf{x}} = \frac{Ze^2}{x^2} \qquad (4\text{-}144)$$

(in which the minus sign indicates that the acceleration $\ddot{\mathbf{x}}$ is in the *negative x* direction) and with the expression equating the kinetic energy with the potential energy,

$$\frac{1}{2} m_e v^2 = \frac{Ze^2}{x} \qquad (4\text{-}145)$$

(this follows immediately from (4-144) on setting $\ddot{x} = (d\dot{x}/dx)\dot{x}$ and integrating):

$$dt = \frac{dx}{v} = \left(\frac{m_e x}{2 Ze^2}\right)^{1/2} dx \qquad (4\text{-}146)$$

Thus

$$E_B = \frac{2}{3}\frac{e^2}{c^3} \int_{\infty}^{a_0} \left(-\frac{Ze^2}{m_e x^2}\right)^2 \left(\frac{m_e x}{2 Ze^2}\right)^{1/2} dx = \frac{2\sqrt{2}}{15}\frac{Z^{3/2}e^5}{m_e^{3/2}c^3 a_0^{5/2}} \qquad \text{erg} \qquad (4\text{-}147)$$

which, substituting

$$a_0 = \frac{\hbar^2}{Zm_e e^2} \tag{4-148}$$

for the Bohr radius, reduces to

$$E_B = \frac{2\sqrt{2}}{15} \frac{m_e Z^4 e^{10}}{\hbar^5 c^3} = 0.189 \frac{m_e Z^4 e^{10}}{\hbar^5 c^3} \text{ erg} \tag{4-149}$$

a result which is in good agreement with the more rigorous expression obtained by taking the electron's motion to lie along a hyperbolic path:

$$E_B = \frac{\pi}{4\sqrt{2}} \frac{m_e Z^4 e^{10}}{\hbar^5 c^3} = 0.556 \frac{m_e Z^4 e^{10}}{\hbar^5 c^3} \text{ erg} \tag{4-150}$$

When screening effects are neglected, the kinetic energy of a K electron is

$$(E_{kin})_K = \frac{m_e Z^2 e^4}{2\hbar^2} = -\tfrac{1}{2}(E_{pot})_K \tag{4-151}$$

and hence classically the total energy of a K electron is

$$(E_{tot})_K = (E_{pot})_K + (E_{kin})_K = -(E_{kin})_K \tag{4-152}$$

so that the ratio of bremsstrahlung energy to total energy is, for the K shell,

$$\left| \frac{E_B}{E_{tot}} \right|_K = \frac{\pi}{2\sqrt{2}} \frac{Z^2 e^6}{\hbar^3 c^3} = \frac{\pi}{2\sqrt{2}} Z^2 \alpha^3 = 4.316 \times 10^{-7} \, Z^2 \tag{4-153}$$

Since the rate of energy dissipation due to bremsstrahlung is proportional to $|\ddot{\mathbf{x}}|^2$, where $|\ddot{\mathbf{x}}|$ in the case of radiating particles of mass m and charge ze passing through a target of atomic number Z is in turn proportional to zZe^2/m, it follows that

$$\frac{dE_B}{dt} \sim \frac{z^2 Z^2}{m^2} \tag{4-154}$$

Thus the bremsstrahlung intensity not only varies as z^2 and Z^2, but is *inversely proportional to the square of the mass of the incident radiating particle*, for which reason it is very much higher for electrons than it is for mesons, protons, α particles, and heavy ions, and, indeed for the latter is in practice insignificant. As established in 1915 by Duane and Hunt [Du 15], the incident light particle can radiate an amount of energy which ranges from zero to the incident kinetic energy, whence

$$(h\nu)_{max} = E_{kin} \tag{4-155}$$

with a sharp cutoff at the short-wavelength end of the continuous x-ray spectrum (Duane and Hunt's law), as shown in Fig. 4-78.

It is interesting to note that the cross section for emission of bremsstrahlung takes the same form in classical and in quantum theory, namely

$$\sigma_B \sim \alpha \, Z^2 \left(\frac{e^2}{m_e c^2} \right)^2 \equiv \alpha r_e^2 Z^2 \qquad \text{cm}^2/\text{nucleus} \tag{4-156}$$

where r_e is the classical electron radius, even though there is a distinct qualitative difference between the detailed microscopic picture which the two theories furnish. The classical theory of bremsstrahlung provides a visualization of the emission process which is fundamentally in error inasmuch as it pictures every change in a charged particle's motion to be accompanied by a *large* probability (indeed, a *certainty*) of *low-energy* photon emission, whereas the corresponding prediction from the quantum-theoretical standpoint is a *low* probability of *higher-energy* photon emission. The statistically averaged bremsstrahlung intensity for an ensemble of charged particles which experience changes in motion thus proves to be roughly the same according to either approach, even though there is a marked distinction in the two theories between the actual spectral distributions of the bremsstrahlung photons.

The quantum-mechanical calculation of cross sections for radiative collisions of electrons with nuclei is along lines closely similar to those for the evaluation of pair-production cross sections, as summarized in Section 4.6.4. In the simplified case of a THIN-TARGET BREMSSTRAHLUNG process in which each incident electron of a parallel, initially monoenergetic, beam is assumed to impinge upon a target so thin that the individual electrons lose no appreciable energy in atomic ionization, experience no significant elastic scattering deflections, and have a vanishingly small likelihood of making successive radiative encounters, so that at most *one* photon is emitted per incident electron, the DIFFERENTIAL BREMSSTRAHLUNG CROSS SECTION (in square centimeters per nucleus) follows from a calculation of the transition probability W_{fi}, since in a large box of volume L^3 the familiar quantum-mechanical treatment of Section 3.6.1 furnishes the relation

$$d\sigma_B = \frac{L^3}{v_i} W_{fi} \tag{4-157}$$

where v_i is the initial electron velocity and, as in Section 3.4.2,

$$W_{fi} = \frac{2\pi}{\hbar} \rho(E_f)|H'_{fi}|^2 \tag{4-158}$$

The density of final states $\rho(E_f)$ is simply the product of the density functions for the electron and the photon in the final system, namely

$$\rho(E_f) = \rho_e \rho_\gamma \, d(h\nu) = \left[\left(\frac{L}{2\pi}\right)^3 \frac{E_f p_f c}{(\hbar c)^3} \, d\Omega_e\right]\left[\left(\frac{L}{2\pi}\right)^3 \frac{(h\nu)^2}{(\hbar c)^3} \, d\Omega_\gamma\right] d(h\nu) \tag{4-159}$$

$$= \left(\frac{L}{2\pi\hbar c}\right)^6 E_f p_f c(h\nu)^2 \, d(h\nu) \, d\Omega_e \, d\Omega_\gamma \tag{4-160}$$

with E_f the total electron energy $(E_f = E_{kin(f)} + m_e c^2)$ and p_f the electron momentum in the final state. Substitution of (4-158) and (4-160) in (4-157), with $v_i = p_i c^2 / E_i$, yields

$$d\sigma_B = \frac{L^9 E_i E_f}{(2\pi)^5 (\hbar c)^7} \frac{p_f}{p_i} |H'_{fi}|^2 \, d(h\nu) \, d\Omega_e \, d\Omega_\gamma \qquad (4\text{-}161)$$

An exact evaluation of $|H'_{fi}|^2$, the squared transition matrix element pro-
portional to L^{-9}, cannot be performed for relativistic electrons (because the
Dirac wave equation for an electron in a Coulomb field cannot be solved in
closed form to obtain an "exact" wave function), but various approximations,
such as the Born approximation used by Bethe and Heitler [Be 34a, Be 34b]
(see also [He 54, pp. 242 ff.]), have been employed to advantage in deriving
solutions which agree moderately well with experimental findings. The rather
involved resulting expression for $d\sigma_B$ (which had also independently been
derived by Racah and by Sauter) indicates that at very low electron energies
the radiation intensity is maximal in a direction perpendicular to the incident
beam, but that as the electron energy is increased, the maximum intensity
appears at increasingly forward angles. In the limit of very high electron
energies, the emission of bremsstrahlung essentially occurs as a narrow pencil
in the forward direction, since irrespective of the photon energy, the average
angle of emission is

$$\bar{\theta}_\gamma \cong \frac{m_e c^2}{E_i} \qquad (4\text{-}162)$$

An interesting feature of the bremsstrahlung distribution at high incident
energies (E_i, $E_f \gg m_e c^2$) is that, according to calculations in which the screening
of the nuclear field by the atomic electrons was neglected by Sommerfeld
[So 39], the small component of the radiation which is emitted at *large* angles
has a distribution $\sim (1 - \cos \theta_\gamma)^{-2} \sim (\sin \frac{1}{2}\theta_\gamma)^{-4}$, which resembles that for
Rutherford scattering (although a more recent result by Hough [Ho 48]
which furnishes a large-angle distribution, essentially $\sim \cos^2 (\frac{1}{2}\theta_\gamma) \sin^{-4} (\frac{1}{2}\theta_\gamma)$,
sheds possible doubt upon this in the absence of conclusive experimental
verification).

Here, and in the following considerations, it is necessary to distinguish
carefully between various types of bremsstrahlung cross sections, expressions
for which have been collected in a detailed survey article by Koch and Motz
[Ko 59]. The cross section $d\sigma_B$ is, namely, *triply* differential in that it contains
the product $d(h\nu) \, d\Omega_\gamma \, d\Omega_e$. When integrated over the electron-emission
angles, it becomes *doubly* differential, viz. in the photon energy and angle,
and thus furnishes information on the energy and angular distribution of the
bremsstrahlung. A further integration over the γ-emission angles yields a
singly differential cross section representing the probability for a photon of
energy $h\nu$ to $h\nu + d(h\nu)$ to be emitted, without reference to its direction of
emergence. The ENERGY DISTRIBUTION of the bremsstrahlung is, accordingly,

expressed by this cross section, which we term the DIFFERENTIAL RADIATIVE CROSS SECTION, $d\sigma_{\text{rad}}$:

$$d\sigma_{\text{rad}} \equiv \int_{d\Omega_\gamma,\, d\Omega_e} d\sigma_{\text{B}} \qquad (4\text{-}163)$$

As evaluated in Born approximation by Bethe and Heitler [Be 34b], the

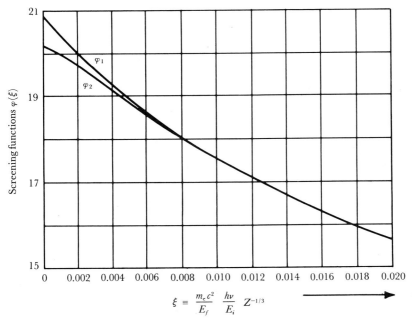

Fig. 4-79. Values of the screening functions φ_1, φ_2 in the Bethe-Heitler bremsstrahlung theory [Be 34b], expressed in function of the parameter

$$\xi = (m_e c^2)\,(h\nu)\,E_f^{-1}\,E_i^{-1}\,Z^{-1/3}$$

(from [Be 34b]).

explicit form of this when screening of the nuclear Coulomb field is taken into account is

$$d\sigma_{\text{rad}} = \alpha Z^2 r_e^2 \frac{d(h\nu)}{h\nu} \left\{ \left[1 + \frac{E_f^2}{E_i^2}\right]\left[\varphi_1(\xi) - 4\ln\left(Z^{-1/3}\right)\right] - \frac{2}{3}\frac{E_f}{E_i}\left[\varphi_2(\xi) - 4\ln\left(Z^{-1/3}\right)\right] \right\}$$

$$(4\text{-}164)$$

with $\alpha \equiv e^2/\hbar c = 1/137$ the fine-structure constant, $r_e \equiv e^2/m_e c^2$ the classical electron radius, and the screening functions φ_1, φ_2 which depend upon the parameter $\xi \equiv (m_e c^2/E_f)\,(h\nu/E_i)\,Z^{-1/3}$ and are depicted graphically in Fig. 4-79. The condition $\xi \gg 1$ corresponds to the absence of screening, and $\xi = 0$ (i.e., $E_i \approx \infty$) to "complete" screening. In the latter limit, the screening functions are just

$$\varphi_1(0) = 4 \ln 183 \quad \text{and} \quad \varphi_2(0) = 4 \ln 183 - \tfrac{2}{3} \qquad (4\text{-}165)$$

which in turn yields a simple expression for the radiative cross section at high energies ($E_i \gg m_e c^2$):

$$d\sigma_{\text{rad}} = 4\alpha Z^2 r_e^2 \frac{d(h\nu)}{h\nu}\left[\left(1 - \frac{2}{3}\frac{E_f}{E_i} + \frac{E_f^2}{E_i^2}\right)\ln 183 Z^{-1/3} + \frac{1}{9}\frac{E_f}{E_i}\right] \qquad (4\text{-}166)$$

which may be compared with the high-energy radiative cross section in the *absence* of screening:

$$d\sigma_{\text{rad}} = 4\alpha Z^2 r_e^2 \frac{d(h\nu)}{h\nu}\left[1 - \frac{2}{3}\frac{E_f}{E_i} + \frac{E_f^2}{E_i^2}\right]\left[\ln\left(\frac{2E_f}{m_e c^2}\frac{E_i}{h\nu}\right) - \frac{1}{2}\right] \qquad (4\text{-}167)$$

Instead of defining the TOTAL RADIATIVE CROSS SECTION as the integral of $d\sigma_{\text{rad}}$, we follow Heitler and, in accordance with current usage, define it as

$$\sigma_{\text{rad}} = \int_{h\nu=0}^{(h\nu)_{\text{max}} = E_{\text{kin}(i)}} \frac{h\nu}{E_i} d\sigma_{\text{rad}} \qquad (4\text{-}168)$$

thereby obtaining the following expressions for different incident energy regions:

(a) Nonrelativistic region ($E_{\text{kin}(i)} \ll m_e c^2$),

$$\sigma_{\text{rad}} = \frac{16}{3}\alpha Z^2 r_e^2 \qquad \text{cm}^2/\text{nucleus} \qquad (4\text{-}169)$$

(b) Relativistic region ($E_{\text{kin}(i)} \approx m_e c^2$), complicated power series formula;
(c) High relativistic region ($m_e c^2 \ll E_{\text{kin}(i)} \ll 137 m_e c^2 Z^{-1/3}$) with no screening correction,

$$\sigma_{\text{rad}} = 8\alpha Z^2 r_e^2\left[\ln\left(\frac{E_i}{m_e c^2}\right) - \frac{1}{6}\right] \qquad \text{cm}^2/\text{nucleus} \qquad (4\text{-}170)$$

(d) Extreme relativistic region ($E_{\text{kin}(i)} \gg 137 m_e c^2 Z^{-1/3}$) with screening correction,

$$\sigma_{\text{rad}} = 4\alpha Z^2 r_e^2[\ln(183 Z^{-1/3}) + \tfrac{1}{18}] \qquad \text{cm}^2/\text{nucleus} \qquad (4\text{-}171)$$

The variation of σ_{rad} (normalized to $\alpha Z^2 r_e^2$) with the initial kinetic energy $E_{\text{kin}(i)}$ of the electrons is shown in Fig. 4-80 which, for comparison, also depicts the corresponding probabilities for energy loss by *ionization*. The rise in the radiative cross section with increasing incident energy causes the energy loss by bremsstrahlung to predominate over other types of energy dissipation. Experimental studies have confirmed this feature and shown that not only in the case of light elements is there fair quantitative agreement with the theoretical predictions, but even for heavy elements when the validity of the

Born approximation is open to question the discrepancies between theory and experiment are not large. For example, Fig. 4-81 shows such a comparison for the elements aluminum and gold.

Of interest, too, in this connection is a comparison of the AVERAGE ENERGY LOSSES PER UNIT PATH LENGTH. For bremsstrahlung, the RADIATIVE LOSS

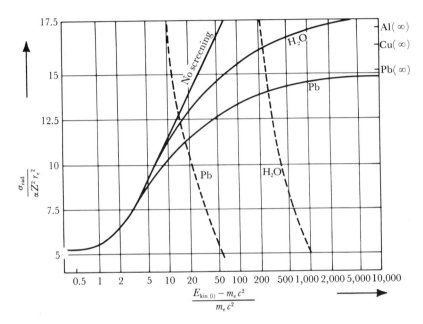

Fig. 4-80. Variation of the total radiative cross section σ_{rad} (normalized to $\alpha Z^2 r_e^2$) with normalized initial electron kinetic energy. Asymptotic values for Al, Cu, and Pb are shown on the right-hand ordinate. The straight line at high energies represents the radiative energy loss per cm path neglecting the effect of screening (i.e., is universally valid) while the dotted curves show for comparison the energy loss by inelastic collisions for electrons passing through water and lead (from [He 54]). (Used by permission of The Clarendon Press, Oxford.)

$(-dE_i/dx)$ is given essentially by integrating over the intensity:

$$-\frac{dE_i}{dx} = n^\square \int_0^{(h\nu)_{max}} h\nu \, d\sigma_{rad} = n^\square E_i \sigma_{rad} \tag{4-172}$$

where $n^\square$ is the number of atoms per cubic centimeter in the target. It may be noticed that the energy loss due to bremsstrahlung (divided by the primary energy E_i) is roughly proportional to E_i and to Z^2, so that its behavior in fact shows strong similarity to that of the γ-ray absorption coefficient κ for pair creation. Although ionization losses dominate over bremsstrahlung losses at low incident electron energies, as may be seen from Fig. 4-82(a), their relative importance changes drastically above a critical energy, which is approximately $20\ m_e c^2 \approx 10$ MeV for lead, and $200\ m_e c^2 \approx 100$ MeV for water or air, as is evident from Fig. 4-82(b).

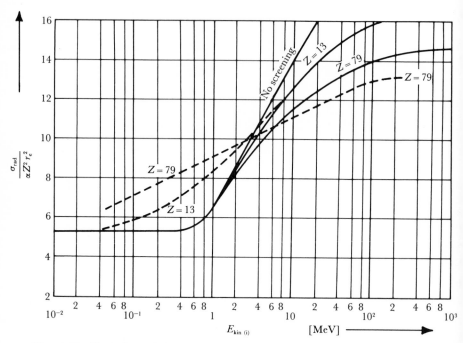

Fig. 4-81. Variation of the normalized total radiation cross section with initial electron kinetic energy. The solid curves depict the theoretical trend calculated by Koch and Motz [Ko 59] for (i) no screening, (ii) aluminum ($Z = 13$), and (iii) gold ($Z = 79$), while the dashed curves represent the most probable actual trend, derived from a synthesis of experimental data up to 1 MeV and values furnished by exact theory in the extreme-relativistic region above 10 MeV (adapted from [Ko 59]).

As pointed out by Landau and Rumer [La 38], the emission of bremsstrahlung can also occur in the Coulomb field of an atomic *electron*, as well as in that of the nucleus. As might be expected, the screening corrections become considerably more involved; in the limit of complete screening, Wheeler and Lamb [Wh 39] obtained the following expression for the differential radiative cross section:

$$d\sigma_{\mathrm{rad}(e)} = 4\alpha Z r_e^2 \frac{d(h\nu)}{h\nu}\left[\left(1 - \frac{2}{3}\frac{E_f}{E_i} + \frac{E_f^2}{E_i^2}\right)\ln\left(\frac{1440}{Z^{2/3}}\right) + \frac{1}{9}\frac{E_f}{E_i}\right] \quad (4\text{-}173)$$

in which the replacement of Z^2 by Z is especially to be noted. Along this line of reasoning, one often finds that for targets with high atomic numbers the total (nuclear + electron) bremsstrahlung is written in terms of the Born-approximation formulae with Z^2 replaced by $Z(Z+1)$. A review of the rather complicated evaluations of electron-electron bremsstrahlung has been

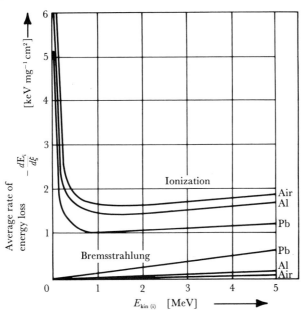

Fig. 4-82(a). Relative dominance of ionization over bremsstrahlung energy losses along the actual path traversed by the electrons in function of the kinetic energy E_{kin}: mass-absorption losses for electrons in air, Al, and Pb, due to ionization compared with the lower values due to bremsstrahlung (from [Ev 55]). (Used by permission of McGraw-Hill Book Co.)

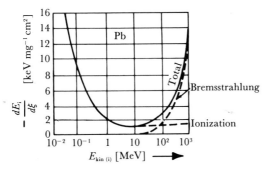

Fig. 4-82(b). The same, but at higher energies, showing how radiation losses begin to dominate above about 10 MeV for electrons in Pb (adapted from [He 44]). (Used by permission of The Clarendon Press, Oxford.)

given by Joseph and Rohrlich [Jo 58] and additional discussion is presented by Koch and Motz [Ko 59, pp. 947–949].

Throughout the preceding, the tacit assumption has been made that all processes responsible for a diminution in the energy of electrons as they traverse a target (e.g., scattering) have no appreciable influence upon

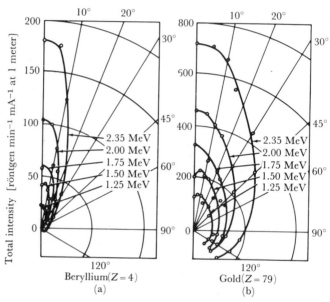

Fig. 4-83. Angular dependence of the thick-target bremsstrahlung intensity integrated over photon energy, shown in polar representation for electrons having initial kinetic energies between 1.25 and 2.35 MeV incident upon beryllium ($Z = 4$) and gold ($Z = 79$). The theoretical curves were derived by Koch and Motz [Ko 59] and the experimental data by Buechner *et al.* [Bue 48] (from [Ko 59]).

bremsstrahlung production. Though this is valid for thin targets it ceases to apply in the practical situation of THICK-TARGET BREMSSTRAHLUNG, and may be taken as a criterion to distinguish between thin and thick targets in the above sense. The transition from the idealized to the practical case introduces complications into the treatment which in many cases devolves upon the introduction of empirical numerical constants that take multiple-collision effects into account. Thus the *angular distribution* cannot be expressed analytic- ally, and is to a considerable extent affected by the target geometry. Some results for the thick-target bremsstrahlung due to the passage of electrons with incident kinetic energies between 1.25 and 2.35 MeV through beryllium and gold targets are presented in Fig. 4-83, in which the radiation intensity has been integrated over photon energy. The theoretical estimates of the distribution for *extreme-relativistic* electron energies agree fairly well with experimental data. The results of theoretical predictions for tungsten are shown in Fig. 4-84.

For the SPECTRAL DISTRIBUTIONS in the *nonrelativistic and intermediate* region no general analytical expressions exist which accurately indicate the variation

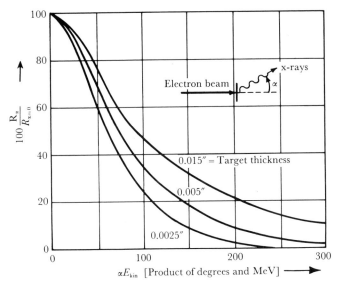

Fig. 4-84. Theoretically predicted angular distribution of thick-target bremsstrahlung for electrons passing through tungsten ($Z = 74$) at relativistic energies. The abscissa scale is a product αE_{kin} (in deg MeV) and the ordinate represents percentage radiated intensity (R_α is the fraction of the total incident electron kinetic energy that is radiated at a mean angle α per steradian) (from [Ko 59]).

of photon energy with angle, but in specific cases good agreement between theory and experiment has been attained, as may be seen from Fig. 4-85. For thick targets, one finds that the number of photons in the high-energy region *increases* as the angle of emission becomes smaller—an opposite trend to that observed in thin-target spectra. When integrated over the photon emission angles, the spectrum of energy radiated in the interval $h\nu$ to $h\nu + d(h\nu)$ is, according to a nonrelativistic semiclassical calculation by Kramers [Kr 23] in which thick-target bremsstrahlung is considered as a superposition of thin-target contributions,

$$dS_{\mathrm{B}} = kZ(h\nu_{\mathrm{max}} - h\nu)\,d(h\nu) \tag{4-174}$$

with k a constant of proportionality which can be adjusted to yield satisfactory order-of-magnitude agreement with experimental data. The total brems-strahlung energy (in MeV per incident electron, of kinetic energy $E_{\text{kin}(i)}$ in MeV), as obtained by integrating the above expression, is then

$$S_{\text{B}} = \int_{h\nu=0}^{(h\nu)_{\max}=E_{\text{kin}}} dS_{\text{B}} = KZE_{\text{kin}(i)}^2 \qquad (4\text{-}175)$$

where K is a new constant whose numerical value is roughly $K \approx 7 \times 10^{-4}$

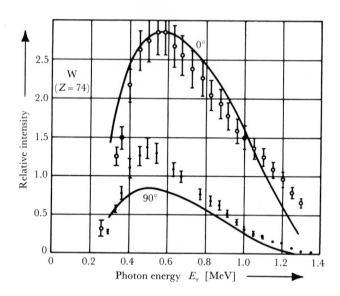

Fig. 4-85. Relative spectral intensities at 0° and 90° for 1.4-MeV electrons incident on a thick tungsten target. The figure is taken from Koch and Motz [Ko 59], who analyzed the results of Miller *et al.* [Mi 54] and presented the data in normalized form (to derive absolute intensities, the theoretical curves must be multiplied by 10^{-3} and the experimental points by 2.1×10^{-3}).

MeV^{-1} for electron energies up to 2.5 MeV. Thus the fractional energy loss $S_{\text{B}}/E_{\text{kin}(i)}$ of 1-MeV electrons traversing a thick target of lead ($Z = 82$) is about 6 percent.

Formulae giving the thick-target bremsstrahlung energy spectrum at relativistic incident energies have been derived by Penfold [Ko 59, ref. 53] and Hisdal [Hi 57a, Hi 57b]. These yield results very similar to those obtained from thin-target calculations: an example of an intensity spectrum in the forward direction for 11.3-MeV electrons on a thick tungsten target ($Z = 74$) is compared with measured values in Fig. 4-86.

Finally, we draw attention to the fact that although we have so far concentrated exclusively on EXTERNAL BREMSSTRAHLUNG, emitted as a continuous spectrum of x radiation when electrons from an external source experience changes of motion as they traverse a target, it is also possible to observe INTERNAL BREMSSTRAHLUNG in atoms whose nuclei experience a transformation by β decay or by capture of electrons from inner orbital shells. An appropriate theory has been developed whose predictions are in fair agreement with

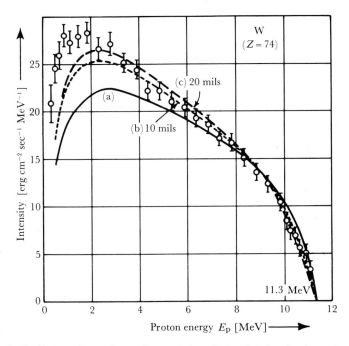

Fig. 4-86. Comparison of experimental data [Mo 53e] for the intensity of bremsstrahlung emitted in the forward direction when 11-MeV electrons impinge upon a thick tungsten target with theoretical predictions: (a) theoretical thin-target spectrum in Born approximation; (b) theoretical spectrum for a 10-mil thick target; (c) theoretical spectrum for a 20-mil thick target (from [Ko 59]).

experimental data; the emission probability of internal bremsstrahlung, expressed in photons per β particle is of the order of 1/137. The intensity spectrum is actually a continuous one, but it is overlaid by the presence of INTERNAL CHARACTERISTIC X RAYS having a discrete energy in consequence of a competing process of internal ionization of inner atomic electron shells.

The emission of radiation as charged particles pass through matter can also be manifested in form of the Čerenkov effect, which we discuss next.

4.9.4. ČERENKOV RADIATION

The passage of fast charged particles in or near a medium with a speed v which exceeds c_n, the phase velocity of light in that medium, viz.

$$v > c_n \equiv c/n \tag{4-176}$$

where n is the index of refraction, gives rise to characteristic remarkable phenomena, such as the emission of electromagnetic ČERENKOV RADIATION, as

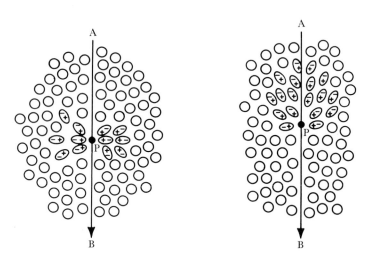

Fig. 4-87. Diagram to illustrate the production of Čerenkov radiation: (a) symmetrical polarization of a dielectric upon passage of a *slow* electron, producing no net dipole field beyond the electron's immediate vicinity; (b) axially asymmetrical polarization upon passage of a *fast* electron, giving rise to a resultant dipole field and the emission of Čerenkov radiation.

discovered by Čerenkov [Če 34, Če 37a, Če 37b] at the instigation of Vavilov [Va 34] and explained theoretically by Frank and Tamm [Fr 37] (see also [Mo 53c, Je 58]). Its counterpart in acoustics is the sonic boom emitted when bodies move with supersonic velocities.

To visualize this, consider a *slow* electron moving through a transparent dielectric along a path such as that shown in Fig. 4-87(a). The atoms of the dielectric may be taken to be roughly spherical except in the immediate vicinity of the electron, where they are distorted by the Coulomb field of the latter into an elongated shape with a nonhomogeneous charge distribution. Thus, in the *polarized* region the atoms behave as electric dipoles, and as the electron moves along its path, polarized regions are formed in its vicinity in step with the motion. In regions remote from the path, however, no resultant dipole field is set up since the polarization field around the electron is symmetrical in the axial and azimuthal directions, and hence no electromagnetic radiation is emitted. When, by contrast, we consider a *fast*-moving electron, we find that the azimuthal symmetry is retained but the axial symmetry is lost, with the consequence that a resultant dipole field acts even at large distances from the electron's track (Fig. 4-87(b)), since there is no chance for a polarization field to be built immediately in front of the fast electron, as the electron's rapid motion outstrips, as it were, the electric field propagated with

Fig. 4-88. Geometric Huygens-type wavelet construction to indicate the formation of a coherent wavefront BC which determines the Čerenkov angle θ that constitutes the semiangle of a cone within which the entire Čerenkov intensity is confined.

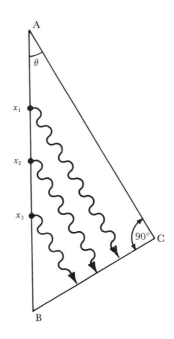

the phase velocity c_n. This dipole field along the direction of motion initiates the emission of brief electromagnetic pulses whose frequencies cover the frequency spectrum of the resultant Čerenkov radiation.

In more sophisticated terms there is, in general, *destructive* interference between the wavelets propagated from adjacent track segments, causing cancellation of remote contributions to the field intensity, but when the particle's speed exceeds the phase velocity of light, there can be a direction θ in which *constructive* interference takes place, all wavelets from all the track segments being in phase. According to the elementary Huygens consideration in wave optics, the wavelets originating from segments x_1, x_2, ... of the track AB shown in Fig. 4-88 present a *coherent* wavefront BC when the time t required by the particle to cover the distance AB is equal to that for light to cover the distance AC. In terms of the reduced relativistic particle velocity $\beta \equiv v/c$, where v is the particle speed, and c is the velocity of light *in vacuo*, this time interval is

$$t = \frac{AB}{\beta c} = \frac{AC}{c_n} \equiv \frac{AC}{c/n} \qquad (4\text{-}177)$$

Since in the right-angled triangle ABC,

$$\frac{AC}{AB} = \cos \theta \qquad (4\text{-}178)$$

it follows that Čerenkov radiation is observable at an angle θ (sometimes called the ČERENKOV ANGLE) given by

$$\cos \theta = \frac{1}{\beta_n} \qquad (4\text{-}179)$$

Several features of Čerenkov radiation may be deduced from this relation, namely:

(i) In a medium of given constant refractive index n, there is a THRESHOLD VELOCITY $\beta_{\min} = 1/n$ below which no Čerenkov radiation can be emitted, and at which the emission direction θ coincides with the particle's direction of motion.

(ii) There is a MAXIMUM ANGLE OF EMISSION

$$\theta_{\max} = \text{arc cos } (1/n) \qquad (4\text{-}180)$$

at the extreme-relativistic limit $\beta = 1$ of the particle velocity.

(iii) Čerenkov radiation frequencies in the main correspond to the high-frequency (blue) region of the visible and near-visible electromagnetic spectrum, but do not extend to the x-ray region since $n < 1$ there. A characteristic manifestation comprises the bluish-white glow surrounding the fission core of a swimming-pool nuclear reactor, due to Compton electrons which propagate through the water at velocities $v \approx c > c/n$.

(iv) The *mass* of the particle in motion has no influence upon the radiation emission angle.

(v) The emission angle θ depends on the frequency (i.e., the wavelength) of the Čerenkov radiation since n is wavelength-dependent.

(vi) The emission angle *increases* with increasing velocity β, which renders the Čerenkov radiation distribution altogether unlike that due to the entirely unrelated process of bremsstrahlung.

If the wavelets are to be in coherent superposition, two conditions have to be satisfied in addition to those mentioned above, namely:

(a) The particle's track length must be large compared with the wavelength

λ of the Čerenkov radiation in the medium (as otherwise diffraction effects may occur), and

(b) The particle's velocity in the medium must remain constant, $v = \beta c = \text{const.}$

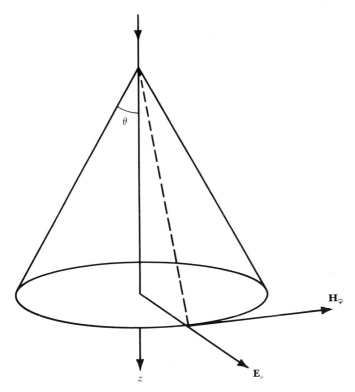

Fig. 4-89. Emission of Čerenkov radiation within a cone of semiangle θ, showing the radial electric vector $\mathbf{E}_\rho$ and the tangental magnetic vector $\mathbf{H}_\varphi$ perpendicular to one another as the primary particle moves in the z direction.

Because of the retention of azimuthal symmetry, the emission of Čerenkov radiation takes place in a conic element of semiangle θ, with the radiation's electric vector $\mathbf{E}$ everywhere perpendicular to the surface of the cone, and the magnetic vector $\mathbf{H}$ tangential, as shown in Fig. 4-89.

A slight modification to the classical results ensues in a quantum-mechanical treatment of the process [Gi 40a, Gi 40b] which takes into account the reaction of the emitted radiation on the motion of the particle. This can be illustrated by considering momentum and energy conservation in the case of a particle having a rest mass m_0

which, while moving at a velocity $v_i = \beta_i c$, emits a photon of energy $h\nu$ at an angle θ_γ. Its energy is thereby reduced, which corresponds to a reduction in speed to a new value $v_f = \beta_f c$. Its motion is at the same time deflected along the new direction θ_p, as shown in Fig. 4-90.

Writing $\gamma_i \equiv (1 - \beta_i^2)^{-1/2}$ and $\gamma_f \equiv (1 - \beta_f^2)^{-1/2}$, the following relations ensue from the conservation of momentum and energy:

$$\gamma_f m_0 v_f \cos \theta_p + (h/\lambda) \cos \theta_\gamma = \gamma_i m_0 v_i \tag{4-181}$$

$$\gamma_f m_0 v_f \sin \theta_p - (h/\lambda) \sin \theta_\gamma = 0 \tag{4-182}$$

$$\gamma_f m_0 c^2 \qquad\quad + h\nu \qquad\qquad = \gamma_i m_0 c^2 \tag{4-183}$$

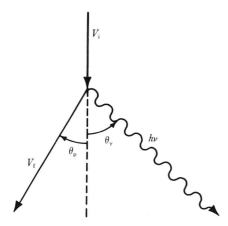

Fig. 4-90. Emission of a photon of energy $h\nu$ at an angle θ_γ as a particle p originally moving with a relativistic velocity V_i is deflected through an angle θ_p and continues with a diminished relativistic velocity V_f.

Eliminating θ_p and writing $\nu = c/(n\lambda)$, since λ is the wavelength in the medium, we find

$$\cos \theta_\gamma = \frac{c}{n v_i} + \frac{h(n^2 - 1)}{2 \gamma_i m_0 v_i n^2 \lambda} \tag{4-184}$$

In terms of the de Broglie wavelength of the particle,

$$\lambda_B = \frac{h}{\gamma_i m_0 v_i} = \frac{1}{\gamma_i \beta_i} \lambda_C \tag{4-185}$$

where λ_C is the Compton wavelength (for electrons, $\lambda_C = h/(m_e c) = 2.4621 \times 10^{-10}$ cm), Eq. (4-184) can be written as

$$\cos \theta_\gamma = \frac{1}{\beta_i n} + \left(\frac{\lambda_B}{\lambda}\right)\left(\frac{n^2 - 1}{2n^2}\right) \tag{4-186}$$

The second term represents a small "reaction correction." Thus for 0.26-MeV electrons ($\beta_i = 0.75$) at the Čerenkov threshold passing through water ($n = 1.3$) the value of λ_B/λ is about 5×10^{-6} and the magnitude of the correction term is thus around 10^{-6}.

Classical theory and quantum theory give the same expression for the ENERGY LOSS experienced by the particle in a short element ds of its path as a result of the small transfers of energy to distant atoms:

$$\frac{dE}{ds} = \frac{4\pi^2 z^2 e^2}{c^2} \int \left(1 - \frac{1}{\beta^2 n^2}\right) \nu \, d\nu \qquad (4\text{-}187)$$

where ze is the charge on the moving particle giving rise to emission of Čerenkov radiation of frequency ν, and the integral extends over all frequencies for which $\beta n > 1$. This kinetic energy loss, equal in magnitude to the total energy radiated per unit path length, is quite small (e.g., about 10^3 eV/cm for singly-charged swift particles moving through glass or lucite), but can be detected because comparatively many low-energy quanta of light are emitted. The average number of such quanta can be derived by carrying out the integration of Eq. (4-187) over a frequency interval ν_1 to ν_2 (corresponding to an interval λ_1 to λ_2 of wavelength *in vacuo*, e.g., $\lambda = c/\nu$) about a mean value $\bar{\nu} = \frac{1}{2}(\nu_1 + \nu_2)$:

$$N \equiv \frac{1}{h\bar{\nu}} \left(\frac{dE}{ds}\right) = \frac{4\pi^2 z^2 e^2}{hc^2} \left(1 - \frac{1}{\beta^2 n^2}\right)(\nu_2 - \nu_1)$$

$$= \frac{2\pi z^2}{137} \left(1 - \frac{1}{\beta^2 n^2}\right)\left(\frac{1}{\lambda_2} - \frac{1}{\lambda_1}\right) \qquad \text{quanta/cm} \qquad (4\text{-}188)$$

where n is the *average* refractive index of the medium over the frequency interval ν_1 to ν_2. In terms of the Čerenkov angle $\theta = \theta_\gamma = \arccos\left(1/\beta_n\right)$, the above can be written

$$N = \frac{2\pi z^2}{137c} \Delta\nu \sin^2 \theta \qquad \text{quanta/cm} \qquad (4\text{-}189)$$

which indicates that for singly charged particles the emission over the visible spectrum $(\Delta\nu \equiv \nu_2 - \nu_1 \simeq 3 \times 10^{14} \ \text{sec}^{-1})$ corresponds to about $450 \sin^2 \theta$ photons/cm. Thus for extreme-relativistic electrons passing through glass $(z = -1, \ \beta \simeq 1, \ n \simeq 1.5)$ the Čerenkov angle is about $\theta \simeq 48°$, and per centimeter of path some 200 Čerenkov photons are emitted within the visible spectrum. It will be noted that particles moving just above the Čerenkov threshold velocity bring about very little emission of Čerenkov radiation, since $\sin^2 \theta$ is then very small. The number of quanta per *wavelength* interval is, according to Eq. (4-187), proportional to $1/\lambda^2$, which explains why the *short* wavelengths at the blue end of the spectrum are predominantly present in Čerenkov radiation. The above formulae also show that the emitted intensity is independent of the rest mass of the moving particle, but depends only upon the velocity and charge state of the latter. This property has been extensively utilized in ČERENKOV DETECTORS which can be used not only to *detect* swiftly

moving charged particles but also to determine their velocity from a measurement of the Čerenkov angle.

A further advantage which these detectors possess is that of fast time discrimination, down to intervals of about 10^{-10} sec in practice (ideally, in a

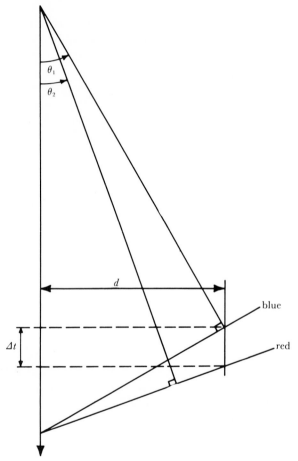

Fig. 4-91. Dispersion of Čerenkov wave trains resulting in slightly different
Čerenkov angles θ_1 and θ_2 for the extreme frequencies ν_1 (blue) and ν_2 (red).

nondispersive medium, the duration of the light pulse at any point within
the Čerenkov cone is vanishingly short). In a *dispersive* medium the Čerenkov
angle varies slightly with wavelength, and this causes the wave trains to spread
out, as shown in Fig. 4-91. The duration Δt of the light pulse for the extreme
frequencies ν_1 and ν_2 (having corresponding Čerenkov angles θ_1 and θ_2)

observed parallel to the particle track at a perpendicular separation d is

$$\Delta t = \frac{d}{\beta c}\left[(\beta^2 n^2 \nu_1 - 1)^{1/2} - (\beta^2 n^2 \nu_2 - 1)^{1/2}\right] = \frac{d}{\beta c}\left[\tan\theta_2 - \tan\theta_1\right] \quad (4\text{-}190)$$

Thus the light flash 10 cm to the side of the track of an extreme-relativistic electron ($\beta \cong 1$) passing through water (mean $\bar{n} = 1.33$) is, for the typical wavelength limits $\lambda = 4000\text{–}6000$ Å, of theoretical duration $\Delta t = 5 \times 10^{-12}$ sec. This, conjoined with the high efficiency of the Čerenkov detector, provides for high counting rates and renders the arrangement particularly suitable for coincidence measurements. In addition to their capability of resolving

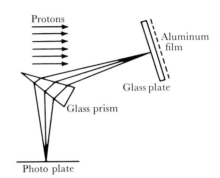

Fig. 4-92. Schematic diagram of an early type of Čerenkov detector, used by Mather [Ma 51a] to determine the energies of fast protons.

energies, Čerenkov detectors can make use of THRESHOLD DISCRIMINATION to distinguish between particles having different mass, but the same momentum or the same range in an absorber. The energy thresholds for particles rise as the rest mass increases; e.g., for glass ($n = 1.50$) the thresholds range from 0.175 MeV for electrons through 36 MeV for muons and 48 MeV for pions to 322 MeV for protons.

To illustrate the application of Čerenkov counters, an early design as used by Mather [Ma 51a] to determine the energy of protons in the external beam of the Berkeley synchrocyclotron is shown schematically in Fig. 4-92. The protons were brought to pass through an optically flat sheet of dense flint glass ($n = 1.88$) only 0.67 mm thick in which the Čerenkov radiation originated. The sheet, whose rear surface was aluminized, was set at an angle to the incident beam so that a part of the Čerenkov cone would, after reflection at the rear face, emerge normal to the front face and thus experience no first-order refraction effects. By ingenious introduction of a small prism whose opening angle was selected to provide an exact compensation of the first-order dispersion of the Čerenkov radiation, accurate registration of the Čerenkov images upon the film of a 35-mm camera arranged near the glass plate but clear of the proton beam could be achieved. The results, in form of a calibrated microphotometric trace of the film image, are depicted in Fig. 4-93. They show the spread of proton velocities about $\beta = 0.680$, corresponding to a kinetic energy

$$E_{\text{kin}} = m_p c^2 (1 - \beta^2)^{-1/2} - m_p c^2 = 341 \text{ MeV} \qquad (4\text{-}191)$$

the overall resolution being below 1 percent, an accuracy which has not been significantly surpassed even in more recent instruments.

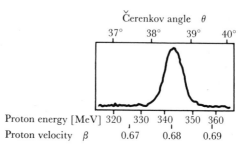

Fig. 4-93. Results obtained by Mather [Ma 51a] for the energy distribution of protons in the external beam of the Berkeley synchrocyclotron, clustering around $\beta \equiv v/c = 0.68$. The resolution of the microphotometric trace shown in the figure is below 1 percent.

Other types of Čerenkov counter are described in detail by Jelley [Je 58]. A historically interesting application to velocity selection featured in experiments by Chamberlain *et al.* [Ch 55] which led to the detection of the antiproton. By arranging for a proton beam accelerated to 6.3 GeV in the Berkeley synchrotron to impinge upon an internal copper target which served as a source of target protons, it was expected that, in addition to mesons, some antiprotons would be produced by the reaction $p + p \rightarrow p + p + (p + \bar{p})$.

An ordinary Čerenkov unit was employed as a threshold discriminator between antiprotons and simultaneously present pions, and this was followed by a special Čerenkov unit, illustrated in Fig. 4-94, which responded only to particles whose velocities lay within the narrow velocity range $0.75c < v < 0.78c$. The Čerenkov pulses generated in a cylinder of fused quartz ($n = 1.46$) by particles moving parallel to the axis were refracted on emerging from the flat rear face of the cylinder into a wide-angled cone. This impinged upon the inside of a cylindrical mirror which contained a blackened baffle arranged to intercept any radiation originating from particles whose velocities lay outside the desired range. By reflection, the Čerenkov radiation was reconcentrated into a small area. In practice instead of focusing upon a point on the axis of the system, which would have had the disadvantage of requiring that a detecting photomultiplier be located within the high-energy particle beam, an

arrangement of three plane mirrors was interposed in order to split the light
into three components and direct these to three separate photomultipliers set
away from the axis. The overall counting efficiency for particles of the correct
velocity was 97 percent, but dropped to 3 percent for a 3-percent deviation
of velocity from the desired value. Thus although about 62,000 mesons were
generated by the p-p reaction for every single antiproton, the rigorous
selection of particles having the desired velocity at a magnetically selected
momentum of 1.19 GeV/c enabled some 60 antiprotons to be registered, and
their mass to be established as $m_{\bar{p}} = (1824 \pm 51)m_e = 932.1 \pm 26.1$ MeV,
which tallies with the expected value of 938 MeV.

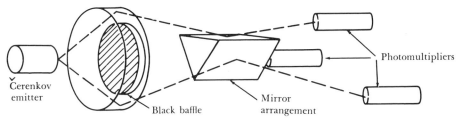

Fig. 4-94. Arrangement used by Chamberlain *et al.* [Ch 55] in the Čerenkov unit
employed as a threshold discriminator between sought-for antiprotons and contaminant π mesons.

4.10. Energy Loss of Heavy Ions

Heavy nuclear projectiles, with $Z > 2$, which can carry a multiple charge
have come to play an increasingly prominent role in nuclear studies upon the
advent of heavy-ion cyclotrons and heavy-ion linear accelerators. By the
term "heavy ion," one designates all atoms beyond helium which carry an
abnormal complement of electrons (in general, heavy ions therefore carry a
positive charge, but negatively charged ions have also been produced and
accelerated following charge-exchange processes). Their reaction properties
will be discussed in detail in Volume II (Section 12.7); the present section is
confined exclusively to pointing out certain characteristics which complicate
heavy-ion behavior upon passage through matter.

The Bethe-Bloch theory would also have been applicable to heavy ions
were it not for the fact that their *charge states do not remain constant* as they traverse
matter. Even under rarefied conditions, e.g., down to about 10^{-5} mm Hg,
heavy ions are continually subject to charge-exchange processes, and the
conditions under which some ultimate effective equilibrium charge is attained
differ from situation to situation (but on an average entail passage through a
layer 100 atomic diameters thick). This effect, which vitiates the theory, is

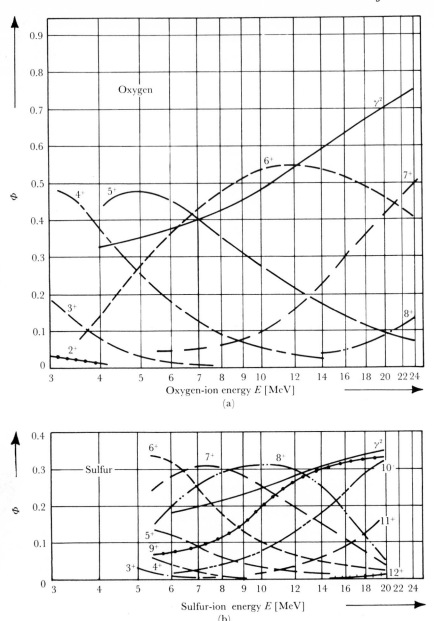

Fig. 4-95. Experimental charge distributions for multiply charged heavy ions: (a) oxygen and (b) sulfur shown in plots of Φ *vs* energy E, where Φ is the fraction of the net time that an ion spends in a particular charge state $(\gamma \, z)$. The solid curves depict γ^2 as a function of E (from [Be 66b]).

similar to optical electronic transitions in atoms, and is a function of the path traversed by the ion. The charge-exchange process accordingly depends in a complicated manner upon the density of the material through which the ion passes, and no theory has, up to the present, been constructed which is able to take this effect quantitatively into account. There is no comprehensive treatment of charge exchange which might furnish absolute values for the

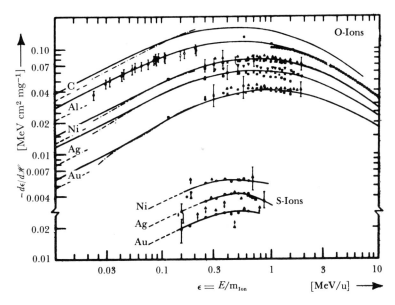

Fig. 4-96. Comparison between semiempirical estimates and experimental results for specific energy losses of O ions and S ions in various elements. The symbols are defined in the text (from [Be 66b]).

empirically introduced parameter γ which, weighting z in a modified Bethe formula [cf. (4-101)],

$$-\frac{dE}{dx} = \frac{4\pi e^4}{m_e v^2} (\gamma z)^2 n^\square \, Z \left[\ln \left(\frac{2m_e v^2}{I} \right) \right] \qquad (4\text{-}192)$$

represents the ratio of the instantaneous charge of an ion to its nuclear charge. An experimental way of determining (γz) is to compare the measured energy loss of the heavy ion with that of protons having the same velocity and passing through the same absorber, since

$$\left(\frac{dE}{dx} \right)_{\text{ion}} = (\gamma z)^2 \left(\frac{dE}{dx} \right)_{\text{p}} \qquad (4\text{-}193)$$

Writing the fraction of the time that an ion spends in the charge state (γz) as Φ, one can calculate the effective charge from the charge components according to the relation

$$(\gamma z)^2 = \sum_{n=1}^{z} \Phi \, n^2 \qquad (4\text{-}194)$$

Experimentally determined charge distributions [Be 66b] for multiply

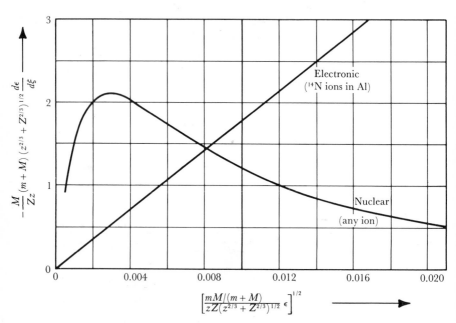

Fig. 4-97. Universal nuclear stopping-power curve for heavy ions at low velocities represented in terms of dimensionless parameters [Li 63] and compared with a typical electronic stopping-power graph (for ^{14}N ions in Al). The mass and charge, m and z, refer to the heavy ion, whose kinetic energy as expressed in units of its rest mass is $\epsilon = E_{kin}/m$ and whose range in an absorber of mass M and charge Z is ξ[mg cm^{-2}] (from [No 63]).

charged oxygen and sulfur ions up to about 20 MeV are shown, together with the corresponding γ^2 curves, in Fig. 4-95(a, b). By using these values of γ in conjunction with tabulated proton stopping powers [Ba 64a] in the relation (4-193) one can carry out a comparison between the semiempirical numerical estimate and the experimental data. Figure 4-96 shows such a representation in which results for O and S ions in several elements derived by several groups of investigators have been combined. The specific energy losses, expressed in terms of $\epsilon \equiv E/m_{ion}$ and $\mathscr{H} = (xz)/(\rho_{absorber}m_{ion})$ where x is the integrated

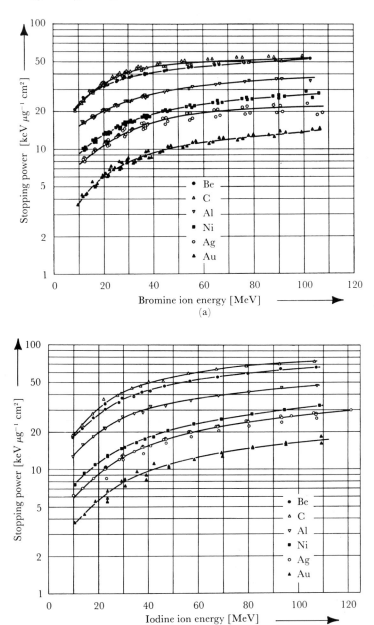

Fig. 4-98. Measured stopping powers of (a) bromine ions, (b) iodine ions in various elements (from [Mo 66a]).

path length in cm, agree well with the semiempirical Northcliffe estimate [No 60].

For heavy ions having a low velocity, the loss of energy due to collisions with *electrons* has been expressed [Li 61c] through the formula

$$-\frac{dE}{dx} = \kappa_e 8\pi \frac{\hbar^2 v}{m_e e^2} n^\Box z Z (z^{2/3} + Z^{2/3})^{-1/2} \qquad \text{erg cm}^{-1} \qquad (4\text{-}195)$$

where κ_e is a numerical factor of the order of $z^{1/6}$. The loss due to collisions with target *nuclei* can be appreciable at low velocities (when the average net ion charge approaches zero), and can exceed the *electronic* stopping power, as may be seen from a comparison of the universal curve [Li 63] depicted in Fig. 4-97 with a typical energy-loss curve (e.g., for ^{14}N ions in Al).

At higher energies, the nuclear collision effect becomes negligible, while the electronic stopping power is large. Measured stopping powers for bromine and iodine ions in several elements [Mo 66a] are shown by way of example in Fig. 4-98(a, b).

Reviews of heavy-ion penetration in matter have been presented by Whaling [Wha 58] and Northcliffe [No 63, No 64], which include a discussion of energy-range curves, straggling, and stopping power as functions of energy.

EXERCISES

4-1 In an experiment using lead as a characteristic absorber of density $\rho = 11.35$ g cm^{-3}, one seeks to discriminate between two γ-ray lines of energy 85 keV and 90 keV through preferential absorption above the K edge. What thickness of lead is needed if one is to attain an intensity reduction ratio of 10, and what is the respective attenuation of each line?

4-2 In which of the following three situations does a γ quantum suffer the largest energy loss, and in which case the least: (a) one single Compton scattering through 180 degrees, (b) two scattering events through 90 degrees each, (c) three scattering events through 60 degrees each?

4-3 In an experiment designed to measure a 10-keV γ-ray line with a detector which has as window a pure beryllium foil whose thickness is $d = 0.05$ cm, and density is $\rho = 1.85$ g cm^{-3} (mass attenuation $\mu/\rho = 0.554$ cm^2 g^{-1}), one observes an 18 percent attenuation when a second foil of identical thickness and presumably identical material is interposed between the source and the detector. Show that this finding is incompatible with the assumption that the beryllium foil is pure. What percentage impurity by weight is present if the contaminant is taken to be iron of density $\rho' = 7.86$ g cm^{-3} and mass attenuation $\mu'/\rho' = 169$ cm^2 g^{-1}?

4-4 The Klein-Nishina formula is valid only in such cases as entail a momentum transfer which is *small* compared with the momentum of the orbital electron. For which scattering angle in the case of Compton scattering of 1-MeV unpolarized γ quanta on lead is the momentum transfer exactly equal to the momentum of one of the K electrons? (Neglect the change in γ-ray energy upon scattering.)

4-5 Using the invariance of $(E_{tot}^2 - p^2c^2)$, verify that the threshold energy for pair production in the field of an electron lies at $4\ m_ec^2$.

4-6 If two-quantum annihilation occurs of a positron having a kinetic energy $E_{(kin)}$ with a stationary electron, what is the energy of the annihilation radiation observed in the incident positron direction? Specialize the result to the extreme-relativistic case $E_{(kin)} \gg m_ec^2$.

4-7 The two-quantum annihilation cross section for nonrelativistic positrons has the form

$$\sigma = \pi r_e^2 \left(\frac{c}{v}\right) \qquad \text{cm}^2/\text{electron}$$

where r_e is the classical electron radius. Calculate the mean lifetime of positrons in $^{27}_{13}\text{Al}$ of density $\rho = 2.7$ g cm^{-3}.

4-8 Prove that for sufficiently high γ-ray energies the separation between the two β detectors of a pair spectrometer having a homogeneous magnetic field is independent of the way the net energy is distributed between the two β particles.

4-9 Deduce analytical expressions as a function of energy for the stopping power and range of deuterons and α particles in a material for which these quantities have been determined in the case of passage of *protons*.

4-10 An approximate formula for the range of protons in air [Wi 47b] is given by Eq. (4-112):

$$R_p^{[m]} = \left(\frac{E_p^{[\text{MeV}]}}{9.3}\right)^{1.8}$$

Deduce an equivalent expression for the range R_α of α particles in air.

4-11 As a means of degrading the energy of a proton beam from 6 MeV to half this value, the beam was passed through two adjacent metal foils having the same thickness but different composition. The requisite slowing down was obtained when the foils were so arranged that the beam first passed through copper and then through gold. Would it have made any difference if the foils had been interchanged, and if so, what would have been the mean energy of the emergent protons? How large is the energy straggling and angle straggling in each case? [Base your calculation upon Al as standard absorber and use the relation $(-dE/d\xi)_{Al} = a/E$.]

4-12 By way of a rough determination of the energy of a monochromatic proton beam, a pencil of rays is allowed to pass through an aluminum foil window of thickness $d = 60 \ \mu\text{m}$ and density $\rho = 2.7 \ \text{g cm}^{-3}$ into air at STP and the length of the luminous emergent conical beam is measured. If this is $L = 140$ cm, what are the magnitudes of (a) the energy of the incident beam, (b) the mean energy of the beam as it emerges from the window, (c) the energy straggling upon emergence?

4-13 A thin nickel foil ($\rho = 8.6 \ \text{g cm}^{-3}$) is arranged to slow down a mono-energetic 10-MeV beam of deuterons.

(a) What is the maximal thickness of the foil if the energy straggling α amounts to at most 1 percent of the mean energy $\bar{E}$ of the emergent beam?

(b) Calculate the values of $\bar{E}$ and α absolutely in MeV.

(c) Calculate the mean angle of divergence $\bar{\theta}$ according to the formula (4-121). [See note to Exercise 4-11.]

4-14 In Section 4.6.5 we derived the bremsstrahlung formula (4-85) for the energy radiated per second by an electron which is subjected to a uniform acceleration $\ddot{\mathbf{x}}$. Applying this expression to the case of an atomic electron moving in a circular orbit whose radius is $R = 10^{-8}$ cm initially, deduce the classical expectation for the "lifetime" of an atom.

4-15 By way of an idealized model for the annihilation of a positron with one of the orbital electrons in an atom, one may contend that the annihilation probability is maximal when both particles have the same velocity.

(a) On the basis of this assumption, evaluate the (discrete) energy distribution and angular distribution of the annihilation quanta when the positrons in a parallel beam interact with electrons in the K, L, or M shell of $^{239}_{94}\text{Pu}$.

(b) If in place of this model purely kinematic considerations are applied to the annihilation with K electrons in a $^{239}_{94}\text{Pu}$ atom, what (continuous) energy distribution ensues?

4-16 Consider a γ-ray source in the shape of a homogenous sphere of radius R throughout which there is a homogeneous temperature distribution. Calculate the factor by which the intensity at large distances from the sphere is reduced through self-absorption. Express this as a function of (a) source radius, (b) source temperature.

What is the magnitude of these reductions in the case of an ^{198}Au source ($\mu = 2.9 \ \text{cm}^{-1}$) of radius $R = 0.15$ cm when the temperature is changed by $\Delta T = 1000°C$ (the linear coefficient of expansion being $\alpha = 17 \times 10^{-6}$ per $°C$)?

4-17 In an experiment using a Čerenkov counter, one measures the kinetic energy of a given particle species as $E_{(kin)} = 420$ MeV and observes that the Čerenkov angle in flint glass (refractive index $n = 1.88$) is $\theta = \text{arc cos} (0.55)$. What particles are being detected? (Calculate their mass in m_e units.)

Chapter 5

NUCLEI AND PARTICLES AS QUANTUM-MECHANICAL SYSTEMS

Quantum Properties of Nuclei and Particles

5.1. The Need to Treat Nuclei and Particles Quantum-Mechanically

What basis exists for the tacit assumption that the description of nuclei does indeed entail a quantum-mechanical, rather than a classical, approach? Their small size provides a first indication, since a nucleon within a nucleus whose radius is at most $R \approx 10^{-12}$ cm has, according to the uncertainty relation

$$\Delta p \, \Delta q \gtreqless \hbar \qquad (5\text{-}1)$$

a velocity indeterminacy

$$\Delta v \gtreqless \frac{\hbar}{m_{\mathcal{N}} \Delta o} \simeq \frac{\hbar}{m_{\mathcal{N}} R} = \frac{1.055 \times 10^{-27}}{1.673 \times 10^{-24} \times 10^{-12}} \approx 10^9 \text{ cm/sec} \qquad (5\text{-}2)$$

which is far too large to be compatible with the classical viewpoint. (Incidentally, the same argument holds for an electron in an atom, since with $R \approx 10^{-8}$ cm and $m_e = 9.109 \times 10^{-28}$ g it follows that $\Delta v \gtrsim 10^8$ cm/sec.)

Similar, more rigorous considerations have been applied by Blochinzew [Bl 62, § 34] to the scattering of a wave packet in the field of force of an atom. These show that classical methods suffice if only Coulomb forces are involved (and hence vindicate Rutherford's treatment) but become inadequate when nuclear forces are taken into account (as in "anomalous" α scattering).

The need to treat the bound state of nucleons inside a nucleus by quantum rather than classical theory is also evident from an examination of the mathematical reduction of quantum wave mechanics to classical Hamilton-Jacobi theory. This limiting reduction bears a close affinity to that between wave optics and geometrical optics in electrodynamic theory. When the electromagnetic wave system is described by a function of the form

$$u = ae^{i\phi} \tag{5-3}$$

with a the amplitude and ϕ the phase of the wave (both real quantities), the limiting case of geometrical optics is that represented by small wavelengths, or, equivalently, by the requirement that the eikonal ϕ vary by a large amount over small regions of space, whence it follows that ϕ must be large in absolute value.

By analogy, if one assumes that in a quantum-mechanical representation the nucleons are described by a wave function

$$\psi = ae^{i\phi} \tag{5-4}$$

the classical situation is denoted by the limit of large ϕ. Furthermore, just as Fermat's principle in optics requires that the difference in phases be an extremum at the ends of the path, so does Hamilton's principle of least action in mechanics suggest that the dimensionless quantity ϕ be represented by a dimensionless function of action whose numerical magnitude provides an index of the classicality of the system. Since $\phi \to \infty$ as $\lambda \to 0$, ϕ might be represented by the ratio $S/\hbar$ so that $\phi \to \infty$ as $\hbar \to 0$ in the classical limit, where S is the mechanical action of the system.

The *Ansatz*

$$\psi = ae^{iS/\hbar} \tag{5-5}$$

as due to Wentzel [We 26] and Brillouin [Br 26] can thus be used as a criterion of classicality, namely, only for large ratios $S/\hbar \gtrsim 100$ is classical theory appropriate. For a nucleon bound within a nucleus with a mean energy $E_B \approx 8$ MeV, we can write

$$\frac{S}{\hbar} = \frac{\oint p \, dq}{\hbar} \approx \frac{pR}{\hbar} = \frac{p}{m_{\mathcal{N}}c} \frac{R}{\lambda_C} = \left(\frac{2E_B}{m_{\mathcal{N}}c^2}\right)^{1/2} \frac{R}{\lambda_C} \tag{5-6}$$

where $\lambda_C \equiv \hbar/m_{\mathcal{N}}c = 2.1 \times 10^{-14}$ cm is the rationalized Compton wavelength of a nucleon of mass $m_{\mathcal{N}}$. Numerical substitution of $m_{\mathcal{N}}c^2 = 938$ MeV and $R \approx 10$ fm indicates that

$$\frac{S}{\hbar} \approx 6.2 \tag{5-7}$$

and points to the need to apply quantum-mechanical considerations to the nuclear system.

This approach may, incidentally, also be used to show that electrons cannot be contained within nuclear dimensions, since on substituting $p/m_ec = 1$, $R \approx 10$ fm and $\lambda_C = 3.86 \times 10^{-11}$ cm, one obtains the physically unacceptable result

$$\frac{S}{\hbar} \simeq 0.026 \ll 1 \tag{5-8}$$

Further indications are provided, of course, by quantum effects in nuclear phenomena (such as the passage of particles through nuclear potential barriers, as in α decay— a form of "tunneling" which does not lend itself to a classical explanation but can readily be accounted for quantum-mechanically) or, indeed, in atomic phenomena. Thus the observation by Stern and Gerlach [St 22] that a neutral beam of silver atoms is split into two discrete components as it passes through an inhomogeneous magnetic field implies the quantization of the magnetic moment associated with a quantized spin of the single valence electron.

5.2. Quantization of Angular Momentum

Not only electrons, but also protons and neutrons have a quantized intrinsic angular momentum, i.e., spin, whose magnitude (along a direction parallel or antiparallel to a distinguished direction of quantization such as the direction of an external applied magnetic field) is characterized by the SPIN QUANTUM NUMBER $s = \frac{1}{2}$, corresponding quantitatively to an angular momentum $[s(s+1)]^{1/2}\hbar = \hbar\sqrt{3}/2$. As their behavior thus falls within the realm of Fermi-Dirac statistics they, and other particles having a half-integer spin $(s = \frac{1}{2}, \frac{3}{2}, \ldots)$ are called FERMIONS, whereas those having an integer spin $(s = 0, 1, \ldots)$ and describable by Bose-Einstein statistics are termed BOSONS.

Arranging particles in classes according to their spin quantum number, we obtain the following scheme:

BARYONS (strongly interacting fermions, viz. hyperons and nucleons)

$$s = \tfrac{3}{2}: \quad \Omega$$

$$s = \tfrac{1}{2}: \quad \Xi, \Sigma, \Lambda, \text{n, p}$$

MESONS (strongly interacting bosons)

$$s = 0: \quad \eta, \text{K}, \pi$$

LEPTONS (weakly interacting fermions)

$$s = \tfrac{1}{2}: \quad \mu, \text{e}, \nu_\mu, \nu_e$$

MASSLESS BOSONS (electromagnetic interaction)

$$s = 1: \quad \gamma$$

(gravitational interaction)

$$s = 2: \quad \text{graviton (?)}$$

After Uhlenbeck and Goudsmit [Uh 25, Uh 26] had introduced the idea of electron spin, the connection between spin and statistics was elucidated quantum-theoretically by Pauli [Pa 27]. At that time, following a proposal by Hund [Hu 27], Dennison [De 27] showed that protons are also spin-$\frac{1}{2}$ particles. (A resumé of these developments has been presented by Goudsmit [Go 61, Go 65b]). Stemming from Pauli's conception [Pa 24] of hyperfine structure as due essentially to a magnetic coupling between the nucleus and orbital electrons, which provided a theoretical basis to explain the Zeeman and Paschen-Back effects, a classical investigation by Back and Goudsmit [Ba 27] provided a unique determination of the angular momentum quantum number of a bismuth nucleus. Thus, when a nucleus is regarded as a many-fermion system, the total spin is simply the vector sum of the individual particle momenta. It follows that even-A nuclei have integer spin, and odd-A nuclei have half-integer spin, as is indeed observed. (An account of some direct experimental methods to determine nuclear spins and moments has been given by Kopfermann [Ko 56] and Ramsey [Ra 53, Ra 56]).

One may note in passing that the presence of electrons within a nucleus would give rise to disturbing inconsistencies. Thus, for example, the even-A nuclei ^2H, ^6Li, ^{14}N would be expected on the basis of an electron-proton model of the nucleus to have the respective constitution $2p + 1e$, $6p + 3e$, $14p + 7e$, and therefore to have a *half*-integer spin due to the *odd* number of fermions in each case, yet experimentally they are all found to have *unit* spin in their normal state.

To avoid confusion, one should also note that the *total* angular momentum of a nucleus is quite legitimately termed the NUCLEAR SPIN, even though it is a vectorial synthesis of the orbital and the intrinsic (spin) angular momenta of the constituent particles, for the total momentum of a stationary nucleus is indeed just the intrinsic momentum. The ORBITAL ANGULAR MOMENTUM of a body is meaningful only if based upon a point of reference: Thus the nucleons have quantized orbital momenta with respect to the center of mass of the nucleus. We have already considered the absolute cross section associated with particles of different orbital momentum impinging upon a massive target (Section 3.5.5).

When a nucleus is in an excited state, its spin in general differs from that of the ground state, for the increase in internal energy is associated with a change in the particle configuration and hence with a change (not necessarily an increase) in the spin. The fact that the ground-state spins of all even-A nuclei (except for the four stable o-o nuclides with $Z = 1, 3, 5, 7$, viz. ^2H, ^6Li, ^{10}B, ^{14}N) are invariably zero suggests that there is a tendency for the spin settings and orbital momenta of the constituent nucleons to be such as to cancel each other out. This is further borne out by the finding that the remaining nuclei all have fairly low ground-state spins ($\lesssim 11/2$), whereas in the absence of such cancellation the nuclear spin could well assume large values.

5.3. Quantum Numbers of Individual Particles

Quantum-mechanically, the state of a body as determined by such physically observable quantities as energy, momentum, and angular momentum is in principle uniquely defined by its wave function. This, a solution of the quantal wave equation for the system (e.g., the Schrödinger equation), is characterized by specific values of parameters termed QUANTUM NUMBERS, following a nomenclature which has been taken over from atomic physics. The most commonly used scheme is as follows.

5.3.1. ORBITAL QUANTUM NUMBER l

This specifies the value of orbital angular momentum with respect to a center of reference, e.g., the orbital momentum of a particle with respect to the nucleus upon which it impinges. Historically, its introduction goes back to Sommerfeld's extension of Bohr's atomic model, in which the "azimuthal" quantum number n_ϕ has been later replaced by the orbital quantum number l. As is proved in texts on quantum theory, the numerical value of momentum characterized by an orbital quantum number l is $\hbar[l(l+1)]^{1/2}$ and l runs over the positive integer values 0, 1, 2, 3, 4, ... (corresponding to S, P, D, F, G, ... waves: see Section 3.5.5).

5.3.2. MAGNETIC ORBITAL QUANTUM NUMBER m_l

This establishes the component of l in a specified direction (e.g., that of an applied external magnetic field), termed the "quantization direction" and usually taken as the z axis of a right-handed orthogonal system. This quantum number can take on any of the $(2l+1)$ possible integer values $-l, -l+1, ...,$ $-1, 0, +1, ..., l-1, l$. On the basis of the momentum VECTOR MODEL (Fig. 5-1), the range of values of m_l is represented by the quantal projections of l along the z axis.

5.3.3. SPIN QUANTUM NUMBER s

The intrinsic angular momentum of a particle or a body is termed the SPIN and represented by a quantum number s associated with an absolute momentum $\hbar[s(s+1)]^{1/2}$. The quantum number s takes on half-integer values in the case of particles subject to Fermi-Dirac statistics and to the Pauli exclusion principle (FERMIONS), but integer values for particles described by Bose-Einstein statistics (BOSONS) to which the Pauli principle does not apply.

Although mechanistically one would associate spin with a rotation of an extended body, this viewpoint is not applicable to a particle or nucleus. (If it were, classical theory would indicate that a point on the surface of a spinning electron would move faster than light!)

5.3.4. MAGNETIC SPIN QUANTUM NUMBER m_s

This is the component of s in an arbitrary quantization direction taken as z axis; when $s = 0$ the only component is $m_s = 0$, whereas for $s = \frac{1}{2}$ the possible values of m_s are $+\frac{1}{2}$ and $-\frac{1}{2}$, corresponding to parallel and antiparallel spin alignment (spin "up" or "down") with respect to the positive z axis.

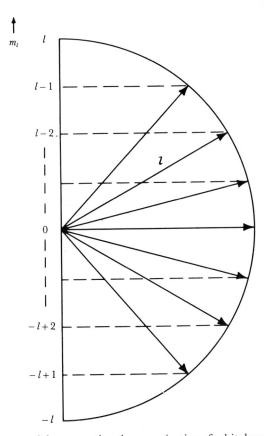

Fig. 5-1. Vector model representing the quantization of orbital momentum l (which has an absolute magnitude $\hbar[l(l+1)]^{1/2}$) into discrete observable components $m_l = -l, -l+1, \ldots, +l-1, +l$ in a distinguished direction, such as that of an applied magnetic field. The components have a magnitude $m_l \hbar$.

5.3.5. TOTAL ANGULAR MOMENTUM QUANTUM NUMBER j

For a single particle, this is given by the vector sum of orbital and spin momenta,

$$\mathbf{j} = \mathbf{l} + \mathbf{s} \qquad (5\text{-}9)$$

and characterizes an absolute total angular momentum of magnitude $\hbar[j(j+1)]^{1/2}$. The quantum number j is always positive and, with l integer and s half-integer ($\frac{1}{2}$), can take on the values $\frac{1}{2}$, $\frac{3}{2}$, $\frac{5}{2}$, ...

5.3.6. Magnetic Total Angular Momentum Quantum Number m_j

Often written simply as m, this is the component of **j** in the quantization direction according to the vector model. It can take on negative or positive values in integral steps. Thus for spin-$\frac{1}{2}$ particles it can have any of the $(2j+1)$ possible values $-j, -j+1, \ldots, -\frac{1}{2}, +\frac{1}{2}, \ldots, j-1, j$.

5.3.7. Radial (n_r) and Principal Quantum Number n

In the atomic theory due to Bohr [Bo 13a], the energy levels of an atom consisting of Z electrons revolving with angular velocity in a circular orbit of radius r around a nucleus under the influence of a Coulomb force and subjected to the quantum requirement that their angular momentum $m_e \omega r^2$ must be an integer (n) times $\hbar \equiv h/2\pi$ are given by the formula

$$E = E_{\text{kin}} + E_{\text{pot}} = -\frac{1}{2}\frac{Ze^2}{r} = -\frac{1}{2}Ze^2\left(\frac{Ze^2 m_e}{n^2 \hbar^2}\right) = -\frac{1}{n^2}\frac{Z^2 e^4 m_e}{2\hbar^2} \qquad (5\text{-}10)$$

wherein the minus sign betokens *bound* states. The principal quantum number n takes on the values 1, 2, 3, ... Sommerfeld [So 16] modified this model to take account of fine structure in spectral lines by generalizing to *elliptical* electron orbits governed by the quantum conditions $\oint p_\phi \, d\phi = n_\phi h$ and $\oint p_r \, dr = n_r h$ for the angular ($p_\phi = m_e r \omega^2$) and radial ($p_r = m_e \, dr/dt$) components of electron momentum in terms of polar coordinates r, ϕ, which caused Bohr's quantum number n to be replaced by the sum of the two numbers

$$n_\phi + n_r = n \qquad (5\text{-}11)$$

In modern practice, the old connotations have been slightly modified: In place of the original azimuthal quantum number n_ϕ ($= 1, 2, 3, \ldots, n$, the value $n_\phi = 0$ being excluded, since this would correspond to a radial oscillation of the electron through the nucleus) we use the orbital quantum number l ($= 0, 1, 2, 3, \ldots, n-1$), and accordingly redefine the range of the radial quantum number n_r to be 1, 2, 3, ..., n (in place of 0, 1, 2, 3, ..., $n-1$) or alternatively write the latter as n' and use the relation $n = n' + l + 1$ with $n' = 0, 1, 2, 3, \ldots, n-1$. The radial quantum number essentially determines the number of nodes in the radial part of the wave function describing motion in terms of a radius vector **r** in a centrally symmetric (Coulomb) field of force.

For convenience, we collate the quantum numbers considered thus far, together with their properties, in Table 5-1, before going on to deal with rather more specialized types of quantum number which have more recently assumed prominence in connection with reaction phenomena evinced by the strong interactions.

Table 5-1. QUANTUM NUMBERS FOR INDIVIDUAL PARTICLES

Name	Symbol	Absolute magnitude	Relation	Value
Orbital	l	$\hbar[l(l+1)]^{1/2}$		$0, 1, 2, \ldots, (n-1)$
Spin	s	$\hbar[s(s+1)]^{1/2}$		$\begin{cases} \frac{1}{2}, \frac{3}{2} \text{ (fermions)} \\ 0, 1, (2) \text{ (bosons)} \end{cases}$
Total angular momentum	j	$\hbar[j(j+1)]^{1/2}$	$\mathbf{j} = \mathbf{l} + \mathbf{s}$	$\begin{cases} \frac{1}{2}, \frac{3}{2}, \frac{5}{2}, \ldots \text{ (fermions)} \\ 0, 1, 2, \ldots \text{ (bosons)} \end{cases}$
Magnetic orbital	m_l	$\hbar m_l$	$-l \leqslant m_l \leqslant +l$	$-l, -l+1, \ldots, -1, 0, 1,$ $\ldots, l-1, l$
Magnetic spin	m_s	$\hbar m_s$	$-s \leqslant m_s \leqslant +s$	$\begin{cases} \left.\begin{array}{c} -\frac{3}{2}, -\frac{1}{2}, \frac{1}{2}, \frac{3}{2} \\ -\frac{1}{2}, \frac{1}{2} \end{array}\right\} \text{ (fermions)} \\ \left.\begin{array}{c} 0 \\ -1, 0, 1 \end{array}\right\} \text{ (bosons)} \end{cases}$
Magnetic (total)	m or m_j	$\hbar m_j$	$-j \leqslant m_j \leqslant +j$ $m_j \equiv m = m_l + m_s$	$-j, -j+1, \ldots, j-1, j$
Radial	n_r (n')			$1, 2, 3, \ldots, n$ $(0, 1, 2, 3, \ldots, n-1)$
Principal	n		$n = n_r + l$ $(n = n' + l + 1)$	$1, 2, 3, \ldots$

5.3.8. ISOSPIN QUANTUM NUMBERS T AND T_z

The classification of fundamental particles already shows that they fall into distinct groups when arranged according to mass.

Thus among the baryons there is, in ascending order, a *doublet* of nucleons (p,n), whereas the lambda hyperon forms a *singlet* (Λ^0), the sigma hyperons a *triplet* ($\Sigma^+, \Sigma^0, \Sigma^-$), the xi hyperons a *doublet* (Ξ^0, Ξ^-) and the omega hyperon a *singlet* (Ω^-). Similarly, the mesons form multiplets, e.g., pions a *triplet* (π^+, π^0, π^-), kaons a *doublet* (K^+, K^0), and the eta meson a *singlet* (η^0). Within a multiplet, the masses differ but very slightly compared with the mass difference between multiplets; thus the Σ^+, Σ^0, and Σ^- have respective rest energies of 1189, 1193, and 1197 MeV, compared with the rest energy of the neighboring Λ^0 singlet (1115 MeV) and the Ξ^0, Ξ^- doublet (1315 and 1321 MeV).

The labeling of these multiplets and a consistent mathematical formalism to account for reaction properties can be achieved through the introduction of an ISOSPIN QUANTUM NUMBER. Historically, the basic similarity between a proton and a neutron led Heisenberg [He 32] to conceive of these particles as just different charge manifestations of the same inherent particle, the nucleon, and to describe their quantum state by an appropriate quantum number which, because of its overt similarity to the spin quantum numbers s and m_s, while applied to ensembles analogous to the isotopic groupings of elements, was termed the ISOTOPIC SPIN quantum number by Wigner [Wi 37]. However, when the concept was subsequently found to have a very direct relevance to systems of nucleons in isobaric nuclei, it was renamed ISOBARIC SPIN [Aj 52, In 53] and, still more recently, simply ISOSPIN.

As we have seen, there is a multiplicity $(2s + 1)$ of states which have a spin s: thus $s = \frac{1}{2}$ yields a doublet $(m_s = -\frac{1}{2}, +\frac{1}{2})$ and $s = 1$ a triplet $(m_s = -1, 0, +1)$, etc., with a difference of 1 unit between possible orientations. On this analogy, if we go from "spin space" to a hypothetical "isospin space," we could conceive of $T = 0$ singlets, $T = \frac{1}{2}$ doublets, $T = 1$ triplets, etc., and regard these quantum numbers as vectors whose quantized projections T_z represent components which classify the particle state in a multiplet along essentially similar lines to the magnetic quantum numbers. Thus the nucleon doublet is characterized by $T = \frac{1}{2}$, while by present convention $T_z = +\frac{1}{2}$ for the proton state and $T_z = -\frac{1}{2}$ for the neutron state of the fundamental nucleon. On the basis of this current arbitrary convention, one can build up an isospin classification scheme as in Table 5-2, assigning a consistent set of T_z values to the particles and the corresponding T_z's of opposite sign to the antiparticles.

5.3.9. STRANGENESS S

The concept of strangeness has found wide application in particle physics, but hitherto has played no role in nuclear physics. It has been invoked in order to explain certain characteristics of particle reactions, and is closely related to the HYPERCHARGE Y. It actually originated through several independent proposals [Pa 52, Ge 53, Na 53, Ge 55], in which an *ad hoc* extension of the isospin formalism was invoked in order to account for the "strange" discrepancy between mean hyperon production and decay times together with the phenomenon of "associated production." Whereas it needs but approximately 10^{-23} sec (i.e., about the time taken by a relativistic particle to traverse the diameter of a nucleus) in order to produce a hyperon on an average, the latter, once formed, lives for about 10^{-10} sec before decaying into lighter particles and π mesons. The sole exception to this remarkable factor-of-10^{13} difference between production and decay rate (a ratio equivalent to 1 second compared to 1 million years) is to be found in the case of the very short-lived Σ^0 hyperon, distinguished by the fact that its sole decay mode $\Sigma^0 \rightarrow \Lambda^0 + \gamma$ is of the electromagnetic type, whereas all other *hyperon decays* involve a *weak interaction*. Their *production*, on the other hand, proceeds by a *strong* interaction, and is accordingly a fast process. Another otherwise inexplicable feature was the phenomenon of ASSOCIATED PRODUCTION, namely the observation that hyperons and K mesons were never produced separately, but always in

Table 5-2. Isospin Classification Scheme

Class	Particles	Antiparticles	Spin	T	T_z	Mass (rest energy) [MeV]
Mesons	π^+					$+1$	140
	π^0		0	1	0	135
		π^-			-1	140
	K^+				$+\frac{1}{2}$	494
	K^0				$-\frac{1}{2}$	498
			0	$\frac{1}{2}$		
		$\overline{K^0}$			$+\frac{1}{2}$	498
		$\overline{K^-}$			$-\frac{1}{2}$	494
		η^0	0	0	0	549
Baryons	p^+				$+\frac{1}{2}$	938
	n^0				$-\frac{1}{2}$	939
			$\frac{1}{2}$	$\frac{1}{2}$		
		$\overline{n^0}$			$+\frac{1}{2}$	939
		$\bar{p}^-$			$-\frac{1}{2}$	938
	Λ^0				0	
			$\frac{1}{2}$	0		1115
		$\overline{\Lambda}^0$			0	
	Σ^+				$+1$	1189
	Σ^0				0	1193
		Σ^-			-1	1197
			$\frac{1}{2}$	1		
		$\overline{\Sigma}^+$			$+1$	
		$\overline{\Sigma}^0$			0	
		$\overline{\Sigma}^-$			-1	
	Ξ^0				$+\frac{1}{2}$	1315
		Ξ^-			$-\frac{1}{2}$	1321
			$\frac{1}{2}$	$\frac{1}{2}$		
		$\overline{\Xi}^+$			$+\frac{1}{2}$	
		$\overline{\Xi}^0$			$-\frac{1}{2}$	
	Ω^-				0	
			$\frac{3}{2}$	0		1672
		$\overline{\Omega}^+$			0	

association with each other, as illustrated by the occurrence of the typical high-energy collision reaction

$$\pi^- + p^+ \to \Lambda^0 + K^0 \qquad (5\text{-}12)$$

(and the *absence* of the otherwise compatible process $\pi^- + p^+ \nrightarrow \Lambda^0 + \pi^0$).

The arrangement of the BARYONS in Table 5-2 reveals a systematic variation of the *mean charge* of the particles which comprise a multiplet: For the nucleon multiplet, it is $\bar{q} = +\frac{1}{2}$, the Λ^0 singlet has $\bar{q} = 0$, as has also the Σ triplet, whereas the Ξ doublet has $\bar{q} = -\frac{1}{2}$, and the Ω^- singlet has $\bar{q} = -1$. If we take the charge center of the nucleon doublet arbitrarily to be a reference origin, that of the Λ^0 is shifted by $\Delta\bar{q} = \bar{q}_\Lambda - \bar{q}_{\mathcal{N}} = -\frac{1}{2}$. Similarly the shift for the triplet of Σ hyperons is also $\Delta\bar{q} = -\frac{1}{2}$, but that for the Ξ doublet is $\Delta\bar{q} = -1$ and for the Ω^- particle is $\Delta\bar{q} = -\frac{3}{2}$. The corresponding shifts for the *antiparticle* multiplets are simply of reversed sign. By defining a STRANGENESS QUANTUM NUMBER as

$$S = 2\Delta\bar{q} \qquad (5\text{-}13)$$

one obtains integer values which are zero for the nucleons and nonzero for the hyperons, e.g., the "strange particles." Thus the Λ^0 and the Σ's have $S = -1$, the Ξ's have $S = -2$, and the Ω^- has $S = -3$.

The analogy to the isospin formalism comes about through the introduction of a BARYON NUMBER B, which is set to $B = +1$ for each baryon, $B = -1$ for each antibaryon, and $B = 0$ for any other particle. Essentially, the introduction of isospin brings about an increase in the number of distinct quantum states within the provisions of the Pauli exclusion principle, and the T_z component is simply a representation in isospin space of the ordinary electric charge q (in units of the elementary charge e), since, e.g., for the members of the nucleon "charge multiplet," one can write

$$q = \tfrac{1}{2}B + T_z \qquad (5\text{-}14)$$

or, in still more general form,

$$q = \tfrac{1}{2}Y + T_z \equiv \tfrac{1}{2}(B + S) + T_z \qquad (5\text{-}15)$$

where the HYPERCHARGE Y is defined as the sum of the baryon number B and the strangeness S, and is in fact just twice the average charge of a multiplet, e.g., for baryons,

$$Y = 2\bar{q} = 2\Delta\bar{q} + 1 = S + 1 \equiv S + B \qquad (5\text{-}16)$$

Although in the original formulation, this quantity was called ISOPARITY U, the more recent term "hypercharge" is preferable for a quantum number which, like all the others considered so far, is *additive* rather than multiplicative (an example of the latter is parity) and which is intimately concerned with charge states.

The formalism has also been extended to embrace MESONS in order to cover the whole range of strongly interacting particles, but this necessitates a modification of the strangeness assignment. With the strangeness of the Λ^0 hyperon set by convention to $S = -1$ and on the assumption that strangeness is a conserved quantity (like the charge) in a reaction such as

$$p^+ + p^+ \rightarrow p^+ + \Lambda^0 \quad + K^+$$
$$S: \qquad 0 \; + 0 \; = 0 \; + (-1) + S_{K^+}$$

(5-17)

a strong interaction whose threshold lies at 1.57 GeV, it follows that the strangeness $S_{K^+} = +1$ must be assigned to the K^+ meson. Similar arguments establish the strangeness of the neutral kaon K^0 also to be $S = +1$ and that of the antiparticles $\bar{K}^0$ and $\bar{K}^-$ to be $S = -1$, which is consistent with all the experimentally observed data. Then from the associated production reaction

$$\pi^- + p^+ \rightarrow \Lambda^0 + K^0$$
$$S: \qquad S_{\pi^-} + 0 \; = -1 \; + 1$$

(5-18)

it follows that $S_{\pi^-} = 0$, an assignment which also applies to π^+ and π^- mesons and to the η^0 meson. Since for mesons the baryon number B is zero, it follows that the strangeness S is equal to the hypercharge Y, and we obtain the assignments collated in Table 5-3.

Table 5-3. STRANGENESS AND HYPERCHARGE OF STRONGLY INTERACTING PARTICLES ("HADRONS")

Class	Mesons ($B = 0$)				Baryons ($B = +1$)				
Particle	$\pi^\pm, \pi^0$	K^+, K^0	$\bar{K}^0, \bar{K}^-$	η^0	p^+, n^0	Λ^0	$\Sigma^\pm, \Sigma^0$	Ξ^0, Ξ^-	Ω^-
Strangeness S	0	+1	−1	0	0	−1	−1	−2	−3
Hypercharge $Y = S + B$	0	+1	−1	0	+1	0	0	−1	−2

On the assumption of STRANGENESS CONSERVATION in *fast* (*strong*) interactions, the phenomenon of associated production can readily be explained by the fact that hyperons, having negative strangeness, can be produced only in association with K mesons of positive strangeness. Once these particles have been produced and are no longer under the influence of the strong creation interaction, they cannot decay individually by a strong interaction into products which have zero strangeness, since this would violate strangeness conservation. They accordingly have no option but to wait until decay by a

weak interaction can take place, as the concept of strangeness and its conservation cannot meaningfully be employed in the case of weak or electromagnetic interactions on the one hand, or for leptons and massless bosons on
the other. In applying the concept exclusively to the strongly interacting
HADRONS ($\ddot{\alpha}\delta\rho o\nu$, strong) and postulating that for fast reactions, strangeness
must be conserved, it follows as a general rule that the *decays of kaons and
hyperons are very slow because they involve a breakdown of strangeness conservation.*
Accordingly, the decays $\varXi^- \to n + \pi^-$ and $\varXi^0 \to p^+ + \pi^-$ which both involve
an unusually large change in strangeness, $|\varDelta S| = 2$, would be expected to be
exceptionally slow, and the decays $\varXi^- \to \varLambda^0 + \pi^-$ and $\varXi^0 \to \varLambda^0 + \pi^0$ with
$|\varDelta S| = 1$ to be slow. The latter are the epitome of weak decays and, since
charge and baryon number are *always* conserved, it follows that one may set
up a fairly general rule for *weak decay* processes:

$$|\varDelta S| = 1 = 2|\varDelta T_z|$$

i.e.,

$$|\varDelta T_z| = \tfrac{1}{2} \tag{5-19}$$

By some authors [Ge 57], this $|\varDelta T_z| = \tfrac{1}{2}$ rule has been interpreted in the case
of weak decays in which leptons are *not* produced as a manifestation of a more
general rule for *weak nonleptonic* decay, viz.

$$|\varDelta T| = \tfrac{1}{2} \tag{5-20}$$

where T is the *total* isospin of the system. On the basis of this latter rule one
would, for instance, expect the branching ratios for the two modes of $\varLambda^0$ decay,

$$\varLambda^0 \begin{array}{c} \nearrow\ p^+ + \pi^- \\ \searrow\ n + \pi^0 \end{array} \tag{5-21}$$

to be in the ratio $2:1$, and this has been vindicated experimentally [Cr 59].

5.3.10. PARITY π

When a particle or quantum-mechanical system of particles is described
by a wave function ψ which defines its quantum state in terms of position
coordinates collectively specified by $\mathbf{r}$ and spin orientation represented by
the axial spin vector $\mathbf{s}$ (whose sign is left unchanged under space inversion
$\mathbf{r} \to -\mathbf{r}$), certain of its properties, such as decay or interaction probability,
are affected by the character of ψ with respect to space inversion. It is appropriate to distinguish between *spatially symmetric* functions

$$\psi(\mathbf{r},\mathbf{s}) = +\psi(-\mathbf{r},\mathbf{s})$$

and *spatially antisymmetric* functions

$$\psi(\mathbf{r},\mathbf{s}) = -\psi(-\mathbf{r},\mathbf{s})$$

and to describe these as having positive (or even) and negative (or odd) PARITY respectively under inversion of the polar coordinate vector $\mathbf{r} \rightarrow -\mathbf{r}$. This transformation property of the wave function affects the phase of transition or interaction matrix elements even though it ceases to be evident in the *square* of the wave function, which represents the quantum-mechanical probability of locating the system at a point $\mathbf{r}$ with spin $\mathbf{s}$.

For a system in (nonrelativistic) motion, the parity of its wave function may be factorized (since it is a *multiplicative* entity) into the product of the particles' INTRINSIC PARITY, uniquely assigned by convention to each species of particle, and the PARITY associated with orbital momentum with respect to a reference center. The parity π of a state comprising N particles having altogether $(N-1)$ definite orbital momenta l_i is decomposable into the product

$$\pi = \left[\prod_{i=1}^{N} \pi_i \right] \left[(-1)^{\sum_{i=1}^{N-1} l_i} \right] \tag{5-22}$$

the first factor representing the product of individual intrinsic parities and the second factor the parity associated with orbital motion.

If the production of a particle takes place through an interaction which conserves parity (e.g., a strong interaction) and in which no other particles are created or annihilated, a knowledge of the relevant orbital momentum suffices to establish the intrinsic parity of the nascent particle uniquely within the convention. If, however, any other particles are created or destroyed in the reaction, only the *relative* parity of the two or more particles can be determined, i.e., the product of the intrinsic parities. Since strange particles are basically subject to *associated production*, it is not possible to determine their intrinsic parities individually, and parity assignments degenerate into arbitrary allocation according to a secondary convention.

INTRINSIC PARITY ASSIGNMENTS of particles are based upon the fundamental (but arbitrary) convention that *nucleons* (as also electrons) have *positive* (*even*) *intrinsic parity*. This then ordains that *pions* have *negative* (*odd*) *parity* and therefore in technical parlance constitute *pseudoscalar particles*. In point of fact, the experimental findings indicate only that the parity product for the (n-π^--p) triad is negative, but on recourse to the convention that the neutron and the proton both have positive parity, one assigns negative parity to the π^-, as also by similar stages of reasoning to the π^0 and π^+ mesons. A fairly complicated process of reasoning leads to the allocation of negative parity to the K mesons and to the η^0 meson. Hyperon parities cannot be related uniquely to the nucleon/pion convention since no strong interaction has as yet been found which links these particles, and associated production precludes individual parity determinations. A secondary convention has therefore been invoked, according to which the Λ parity is taken as positive, and since evidence exists from reactions such as $\Sigma^0 \rightarrow \Lambda^0 + e^+ + e^-$ indicating the (Λ, Σ) relative parity to be positive, the assignment of *positive* parity to the Σ hyperons ensues. The Ξ hyperons are also assumed to have positive parity (in principle, one should be able to deduce relative parities from such reactions as $K^- + p^+ \rightarrow \Xi^- + K^+$): this permits their incorporation into an octet of baryons in their ground states ($N\Lambda\Sigma\Xi$) according to a group-theoretical classification scheme

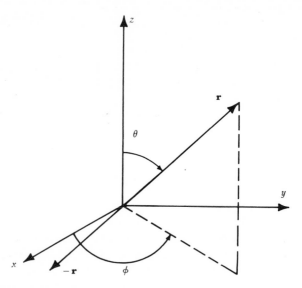

Fig. 5-2. Diagram illustrating the polar coordinates (r, θ, ϕ) corresponding to the tip of a vector **r**, together with the operation of spatial inversion ($\mathbf{r} \rightarrow -\mathbf{r}$).

(discussed in Volume II), wherein the above particles all have spin $\frac{1}{2}$ and positive parity. The parity of the Ω hyperon has not yet been established, but there are grounds for taking it to be *positive*. Current means of establishing the intrinsic parity of particles have been surveyed, e.g., by Lichtenberg [Li 65] and Tripp [Tr 65].

The PARITY associated with *orbital motion* can be elucidated as follows: If the spatial coordinates **r** in the wave function describing the state are expressed in polar representation, $\mathbf{r} = (r, \theta, \phi)$, the spatial inversion $\mathbf{r} \rightarrow -\mathbf{r}$ is represented by the parity transformation

$$\mathbf{r} \rightarrow -\mathbf{r}: \qquad r \rightarrow r, \qquad \theta \rightarrow \pi - \theta, \qquad \phi \rightarrow \pi + \phi \qquad (5\text{-}23)$$

as is evident from Fig. 5-2. Decomposing the wave function $\psi(\mathbf{r}, \mathbf{s})$ into spatial and spin components,

$$\psi(\mathbf{r}, \mathbf{s}) = \Phi(\mathbf{r})\chi(\mathbf{s}) \qquad (5\text{-}24)$$

we see that the spin-dependent part is fundamentally of even parity, and, furthermore, that the parity character of the remaining part $\Phi(\mathbf{r})$ (which, on the assumption of a central interaction may be separated further, e.g.,

$$\Phi(\mathbf{r}) = f(r)\, Y(\theta,\phi) \tag{5-25}$$

where $f(r)$ is the parity-symmetric radial component and $Y(\theta,\phi)$ the angular constituent) reposes entirely in the angle-dependent SPHERICAL HARMONIC $Y(\theta,\phi)$, which is also a function of the orbital momentum l and magnetic quantum number m of the form

$$Y_{lm}(\theta,\phi) = (-1)^m \left[\frac{2l+1}{4\pi} \frac{(l-|m|)!}{(l+|m|)!} \right]^{1/2} P_l^m(\cos\theta)\, e^{im\phi} \tag{5-26}$$

where $P_l^m(\cos\theta)$ is an associated Legendre function.† Under the transformation $\theta \to \pi - \theta$ this becomes

$$P_l^m[\cos(\pi-\theta)] = P_l^m(-\cos\theta) = (-1)^{l-m} P_l^m(\cos\theta) \tag{5-27}$$

whereas the transformation $\phi \to \pi + \phi$ causes the exponential to become

$$e^{im(\pi+\phi)} = e^{im\pi}\, e^{im\phi} = (-1)^m\, e^{im\phi} \tag{5-28}$$

while the other factors remain unchanged. The net effect of the parity transformation is, therefore, to multiply $Y_{lm}(\theta,\phi)$ by $(-1)^l$. Accordingly, $\Phi(\mathbf{r})$ and hence $\psi(\mathbf{r},\mathbf{s})$ have the parity character

$$\psi(\mathbf{r},\mathbf{s}) = (-1)^l \psi(-\mathbf{r},\mathbf{s}) \tag{5-29}$$

whence it follows that the parity of a state does not depend on m at all, but is positive (even) if l is even, and negative (odd) if l is odd.

Extension of this argument to a system of N particles shows that the (orbital) parity of the ensemble is

$$(-1)^{\sum\limits_{i=1}^{N-1} l_i} \tag{5-30}$$

It must be stressed that the simple result as derived above holds only for nonrelativistic motion: If the particles are relativistic, the parity of the state is no longer determined by so simple a dependence upon l.

---

† In the usual phase convention, as used, e.g., by Condon and Shortley [Co 51] the phase factor $(-1)^m$ is omitted when m is *negative*. This is but one manifestation of the practical problems with which one can be confronted when dealing with negative orders $m < 0$, for even though $Y_{l-m}(\theta,\phi) = (-1)^m Y_{lm}^*(\theta,\phi)$ one encounters the difficulty that the associated Legendre functions $P_l^m(\cos\theta)$ are essentially defined only for $m \geqslant 0$. We treat these functions in more detail in Appendix E.5.

5.4. Quantum Properties of Nuclear States

5.4.1. Nuclear Energy Levels

Just as the electrons in an atom can occupy only certain states of discrete energy, so also can the nucleons comprising a nucleus make up a system whose possible energy states are restricted to certain fixed values (within the precision ordained by Heisenberg's principle of indeterminacy). These constitute

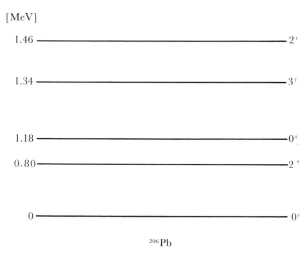

Fig. 5-3. Simplified level scheme of the stable even-even nucleus ^{206}Pb, showing the ground state and first four excited states, together with the excitation energy and spin-parity assignments.

the eigenvalues E_n of a nuclear Hamiltonian operator H_n acting on the energy eigenfunction ψ_n of the nucleus:

$$H_n\psi_n = E_n\psi_n \tag{5-31}$$

The Schrodinger wave function here depends upon spatial coordinates and also spin and isospin quantum numbers appropriate to the overall system. The GROUND STATE of a nucleus constitutes the lowest of the entire set of energy levels, depicted as NIVEAUX in a level scheme such as that shown in Fig. 5-3. Even though the ground state possesses "latent" energy, in the form of potential energy, it is normalized arbitrarily to zero in a level representation, and the EXCITATION ENERGIES of the respective higher levels (BOUND STATES and SCATTERING STATES or CONTINUUM STATES) are referred to this ground-state (zero) energy. Thus to activate the second excited state of ^{206}Pb an energy of 1.18 MeV in the center-of-mass system has to be supplied. The next level cannot be activated until an energy at of least 1.34 MeV can be applied externally. It follows that on deexcitation of a nucleus back to its ground state, energy emission evinces discrete values only. The description of a nuclear energy level accordingly requires specification of (a) excitation energy E^*, (b) nuclear spin (total angular momentum) J, (c) parity π, (d) lifetime τ, and (e) isospin T.

5.4.2. Nuclear Angular Momentum (Spin and Coupling Schemes)

The total angular momentum of a given nuclear state is the resultant of the individual momenta of the constituent nucleons. As described by the NUCLEAR SPIN QUANTUM NUMBER J (or, in an alternative notation, I) its absolute magnitude is $\hbar[J(J+1)]^{1/2}$; the values of J range over a set of integer values for even-A nuclei and half-integer values for odd-A nuclei. The vector summation of the individual constituent momenta to a resultant $\mathbf{J}$ can involve direct summation,

$$\mathbf{J} = \sum_i \mathbf{j}_i \tag{5-32}$$

or alternatively may be built from the vector sum of resultant orbital and intrinsic angular momenta,

$$\mathbf{J} = \mathbf{L} + \mathbf{S} \tag{5-33}$$

with

$$\mathbf{L} = \sum_i \mathbf{l}_i \quad \text{and} \quad \mathbf{S} = \sum_i \mathbf{s}_i \tag{5-34}$$

wherein the summation extends over the ensemble of component nucleons i.

Although the final result must of necessity be the same in either case, the choice as to which of the above two alternative COUPLING SCHEMES is adopted in the evaluation of a particular problem may in certain instances crucially affect the degree of complexity of the ensuing calculations. SPIN-ORBIT (j-j) COUPLING is more especially suited to consideration of problems in a nuclear interaction having "noncollective" character, namely one in which the orbital and spin momenta of each individual nucleon i are strongly coupled to a resultant angular momentum $\mathbf{j}_i = \mathbf{l}_i + \mathbf{s}_i$, while the respective $\mathbf{j}_i$'s can be regarded as essentially separate quantities whose vector sum builds $\mathbf{J}$:

$$\mathbf{J} = \sum_i \mathbf{j}_i, \quad \text{with} \quad j_i = l_i \pm \tfrac{1}{2} \tag{5-35}$$

It is therefore assumed that under these conditions the separate orbital momenta l_i do *not* compound to a resultant $\mathbf{L}$ for the given nuclear state, nor do the spins compound directly.

When, on the other hand, there *is* collective interaction of orbital and intrinsic momenta one may, within the framework of the RUSSELL-SAUNDERS (L-S) COUPLING SCHEME, treat the spin-orbit interaction as negligibly weak and couple the orbital momenta separately from the spin angular momenta:

$$\mathbf{L} = \sum_i \mathbf{l}_i \quad \text{and} \quad \mathbf{S} = \sum_i \mathbf{s}_i \qquad (5\text{-}36)$$

before building the vector sum

$$\mathbf{J} = \mathbf{L} + \mathbf{S} \qquad (5\text{-}37)$$

therefrom. Thus each nuclear state has a particular value of L and S associated with it, and can be identified by the spectroscopic notation, of which an example would be $^2F_{5/2}$ for a level having $L = 3$ (an "F level"), a multiplicity $(2S + 1) = 2$, whence $S = \frac{1}{2}$, and a total angular momentum $J = \frac{5}{2}$ $(= 3 - \frac{1}{2})$. In this scheme, one conceives of states having different L values as differing in energy too. Furthermore, energy subdistinctions arise for different S values within the spin multiplets of multiplicity $(2S + 1)$ associated with each L value.

The above two coupling schemes by no means exhaust the theoretical possibilities; they merely represent simple extreme forms. One can, of course, employ INTERMEDIATE COUPLING SCHEMES of varying degrees of complexity. Their description falls outside the scope of the present treatment, however.

Classification of nuclear ground-state spins reveals a definite pattern, as is evident from Table 5-4.

Table 5-4. NUCLEAR GROUND-STATE SPINS[a]

Nuclear species	Number observed [stable + longlived $(T_{1/2} > 10^9\mathrm{y})$]	Ground-state spin
Odd-odd	$6 + 4 = 10$	Integer
Odd-even	$51 + 3 = 54$	Half-integer
Even-odd	$55 + 3 = 58$	Half-integer
Even-even	$166 + 11 = 177$	Integer (zero)
	$278 + 21 = 299$	

[a] Data read off the 1965 GEC "Chart of the Nuclides" [Go 65a].

The remarkable finding that *all* e-e nuclei have spin $J = 0$ in their ground states prompts the supposition that the momenta compensate each other and cancel out. Thus protons or neutrons pair off as twins with mutually opposite spin. Accordingly, odd-A nuclei might be regarded as comprising an e-e core (of spin 0) plus one extra proton or neutron whose angular momenta determine the value of the half-integer nuclear spin. Further, o-o nuclei are analogous to an e-e core plus a single additional proton and neutron, the combined momenta of which furnish the (integer) nuclear spin. By treating the nucleus as an agglomerate of nucleon pairs and unpaired single particles, many properties of nuclear ground and excited states can be accounted for in terms of this SINGLE (or INDIVIDUAL) PARTICLE MODEL.

Quantization of nuclear spin J also implies the existence of a NUCLEAR MAGNETIC QUANTUM NUMBER m_J which represents the projection of $\mathbf{J}$ along a given quantization axis, e.g., the direction of an external magnetic field. The absolute value of this observable quantized momentum component is $m_J \hbar$, with m_J ranging over the values $-J, -J+1, \ldots, J-1, J$. Thus the maximum observable nuclear spin is $J\hbar$, and there are $(2J+1)$ distinct MAGNETIC SUBSTATES associated with each spin J.

5.4.3. NUCLEAR PARITY

Since, essentially to a first approximation, the wave function Ψ for each nuclear state can be written as a product of individual particle wave functions ψ_i, each of which has a definite parity,

$$\Psi = \prod_i \psi_i \qquad (5\text{-}38)$$

or as a linear combination of these, the NUCLEAR PARITY π is uniquely determined by the product of individual parities. The wave function of a state having positive parity is thus symmetric with respect to inversion of nucleon spatial coordinates, whereas negative parity implies an antisymmetric character. On the basis of the single-particle model, one would expect e-e nuclei in their ground states to have positive parity ($J^\pi = 0^+$), and this is borne out by experiment. The parity of odd-A nuclear ground states is given by $(-1)^{l_i}$, where l_i is the orbital momentum of the unpaired nucleon. Odd-odd nuclei in their ground states have a spin J which lies within the range

$$|j_n - j_p| \leqslant J \leqslant j_n + j_p \qquad (5\text{-}39)$$

where the j_n and j_p are odd-neutron and odd-proton total angular momenta, and a parity which is the product of odd-neutron and odd-proton parities, i.e.,

$$\pi = (-1)^{l_n + l_p} \qquad (5\text{-}40)$$

The few exceptions to this rule all admit of a satisfactory explanation.

The parity character of nuclear ground and excited states plays an important role in restrictions upon otherwise energetically possible transitions, as a later consideration of selection rules for parity (and isospin) will show.

5.4.4. ISOSPIN

It is also possible to associate an isospin with each nuclear state, though this does not lend itself to visualization in pictorial terms. The isospin embodies as it were the relative insensitiveness in the energy levels of isobaric nuclei toward interchange of a proton and a neutron. Thus the energy states of the mirror nuclei $^{7}_{3}\text{Li}$ and $^{7}_{4}\text{Be}$ bear a close similarity to one another with respect to energy, spin, and parity as can be seen from Fig. 5-4(a), and the same holds for the isobars $^{13}_{6}\text{C}$, $^{13}_{7}\text{N}$ depicted in Fig. 5-4(b).

Recently, there has been increasing recognition of the importance of ISOBARIC ANALOGUE (ISOANALOGUE) STATES. For a given nucleus, the value of

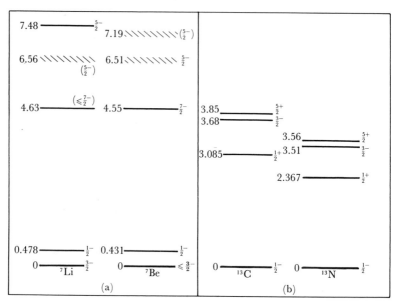

Fig. 5-4. Similarity in the level schemes of the isobaric mirror nuclei (a) $^{7}_{3}\text{Li}_4$ and $^{7}_{4}\text{Be}_3$ and (b) $^{13}_{6}\text{C}_7$ and $^{13}_{7}\text{N}_6$, indicating the occurrence of isobaric analogue states.

T_z is the same for all levels and is just minus one-half of the neutron excess,

$$T_z = \tfrac{1}{2}(Z - N) \equiv -\tfrac{1}{2}(N - Z) \qquad (5\text{-}41)$$

the net isospin being simply that of the unpaired neutrons. In a set of isobars of given A, some member X will have an isospin $T_{z\,\text{max}}$ which is largest among that set, and accordingly corresponds with the maximum component of a system for which $T = T_{z\,\text{max}} = \tfrac{1}{2}(Z_{\text{X}} - N_{\text{X}})$, comprising $(2T + 1)$ states which form an ISOBARIC MULTIPLET. An example of this is provided by the adjacent isobars $^{14}_{6}\text{C}$, $^{14}_{7}\text{N}$, $^{14}_{8}\text{O}$ for which T_z is -1, 0 and $+1$ respectively, as illustrated in Fig. 5-5. Certain of their energy levels can be observed to display similarities in such characteristics as spin and parity, as also in energy when Coulomb contributions are subtracted. Such matching levels can accordingly be regarded as members of isobaric multiplets, e.g., the ground states of ^{14}C and ^{14}O together with the first level of ^{14}N all have spin 0 and form an isobaric triplet ($T = 1$), while the ground state of ^{14}N is a singlet with $T = 0$. The wave functions for the nucleon systems in an isobaric multiplet are closely akin to one another except for the substitution of a neutron for a proton: This exception is, for instance, borne out in the case of the above multiplet by the fact that ^{14}O much more readily decays by β^+ emission to the $T = 1$ *first* level of ^{14}N than to the $T = 0$ ground state (99.4 percent as against 0.6 percent).

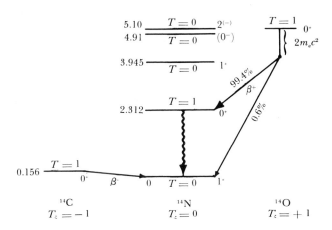

Fig. 5-5. An isobaric triplet comprised of states in ^{14}C, ^{14}N and ^{14}O, characterized by the isospin quantum number $T = 1$ (spin-parity $= 0^+$), shown together with additional levels in ^{14}N. The low branching ratio of 0.6 percent for β^+ decay to the *ground* state of ^{14}N as compared with that (99.4 percent) to the isoanalogue state is noteworthy. When allowance is made for differences in Coulomb energy, the $T = 1$ levels line up almost perfectly. (Taking the ground state of ^{14}N as energy origin, the $T = 1$ levels lie at 2.36 MeV [^{14}C], 2.312 MeV [^{14}N], and 2.44 MeV [^{14}O].)

Clearly, such assignments of T can be made only through detailed comparison of nuclear properties of corresponding levels in a group of isobars, taking into consideration the known T_z value and theoretical predictions based upon definite nuclear models. By treating isospin as a well-defined quantum number in strong interactions, it is possible to classify energy states of nuclei. Even elementary energy considerations can sometimes provide conclusive information: Thus the ^{4}He nucleus ($\alpha = 2p + 2n$) has $T_z = 0$ and could in principle have a ground-state isospin $T = 0$, 1, or 2. The $T = 0$ assignment is evidently to be favored over the other possibilities by the fact that if it were otherwise the existence of a stable nucleus ^{4}H with practically the same binding energy as ^{4}He could not be ruled out†—a situation at variance with observation (see Section 12.6). Similarly, the mirror nuclei ^{3}H and ^{3}He must have $T = \frac{1}{2}$ in their ground state (and not $\frac{3}{2}$), since otherwise one would

† *Note added in proof:* Evidence in support of the transient existence of (unstable) excited states of ^{4}H has recently been derived [Zi 68, Mi X] from the ^{7}Li (π^-, t) ^{4}H reaction.

expect stable nuclei composed of three protons or three neutrons to exist also. Furthermore, the absence of a stable di-neutron state precludes any other assignment than $T = 0$ to the ground state of the deuteron. It is found in practice that the ground states of all light nuclei ($A \leqslant 20$) have the smallest isospin consistent with their T_z value, viz. $T = T_z$. Isospin assignments to *excited* nuclear levels can be established through scattering or reaction studies.

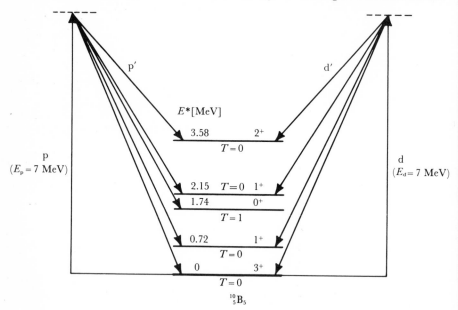

Fig. 5-6. Elastic and inelastic proton and deuteron scattering at 7 MeV to various levels in ^{10}B. The absence of (d, d′) scattering to the second excited level is significant in that it illustrates the operation of an isospin selection rule forbidding d′ transitions to $T = 1$ states while permitting them to $T = 0$ states.

We illustrate this by considering the even-A nucleus ^{10}B, whose ground-state isospin is $T = T_z = 0$. By combining inelastic scattering data for $T = 0$ particles (d or α) with that for $T = \frac{1}{2}$ particles (p or n), one can distinguish the $T = 0$ levels in ^{10}B from those with $T = 1$: Isospin conservation dictates that scattering of $T = 0$ particles can proceed only to levels having the same isospin as the initial state, whereas the scattering of $T = \frac{1}{2}$ particles can lead to levels whose isospin differs by 0 or 1 unit from that of the original state. By so distinguishing processes with $\Delta T = 0$ from those with $\Delta T = 1$, a T-assignment can be made as indicated in Fig. 5-6. Whereas (p, p′) scattering occurs to all the levels shown, there is a conspicuous absence of (d,d′) scattering to one of these levels, namely, the 0^+ level at 1.74 MeV excitation energy (see Fig. 5-7, which shows the spectra of particles incident at 7 MeV and inelastically scattered at 90 degrees). Hence this constitutes a $T = 1$ level while all the others are $T = 0$ (further supporting evidence is provided by data from the ^{6}Li(α,γ)^{10}B reaction). It is this state which is the isospin analogue state to

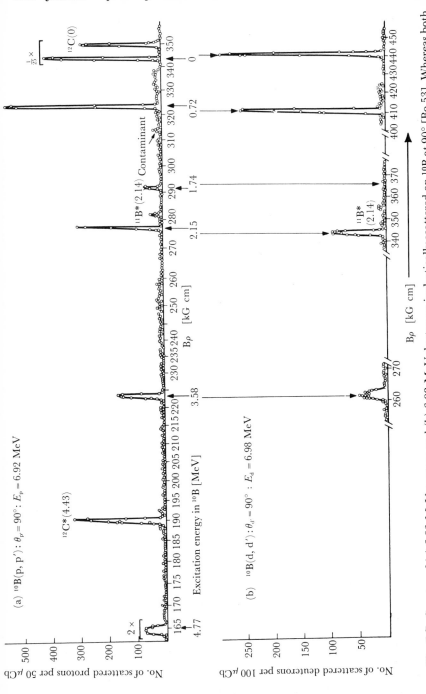

Fig. 5-7. Spectra of (a) 6.92-MeV protons and (b) 6.98-MeV deuterons inelastically scattered on ^{10}B at 90° [Bo 53]. Whereas both spectra display peaks which correspond to transitions to final states at 0.72, 2.15, and 3.58 MeV in ^{10}B, there is conspicuously no deuteron peak corresponding to scattering to the 1.74-MeV level in ^{10}B, as discussed in the text.

the ^{10}B and ^{10}C ground states, and thus the third member of an isospin triplet with $T = 1$ and spin 0^+ (for ^{10}Be, $T_z = -1$, while for ^{10}B, $T_z = 0$ and for ^{10}C, $T_z = +1$). Energy considerations confirm this conclusion: The level diagrams for ^{10}Be and ^{10}C can be brought into isospin juxtaposition with that for ^{10}B by taking account of Coulomb energy differences† and the (atomic) mass difference $\delta M \equiv m_n - M_H = 0.783$ MeV which arises when a neutron is replaced by a proton. Since the Coulomb contributions $a_c Z(Z-1)A^{-1/3}$ to the binding energies of ^{10}Be, ^{10}B and ^{10}C are respectively 3.314, 5.524, and 8.285 MeV, it follows that their differences (δE_C) with respect to ^{10}B are 2.210, 0 and 2.761 MeV. The appropriate shift is then

$$\varDelta = \delta E_\beta \pm (\delta E_C - \delta M) \qquad (5\text{-}42)$$

in terms of the difference in atomic masses (expressed in energy units as δE_β and numerically equal to the ground-state β-transition energy), which is $+0.555$ MeV for ^{10}Be and $+3.606$ MeV for ^{10}C. Hence the shift for ^{10}Be is

$$\varDelta = 0.555 + 2.210 - 0.783 = 1.982 \text{ MeV}$$

and that for ^{10}C is‡

$$\varDelta = 3.606 - (2.761 - 0.783) = 1.628 \text{ MeV}$$

The resulting scheme is shown, somewhat simplified, in Fig. 5-8, wherein the dashed lines connect levels which form likely isospin multiplets. The assignment $T = 1$ to the lower levels of ^{10}Be is compatible with reaction data: The problem is to match these niveaux with the corresponding states of commensurate energy in ^{10}B. Thus the 5.167-MeV level of spin 2^+ in ^{10}B is indicated to be the analogue to the first level (2^+) in ^{10}Be and probably the first level (spin unknown) in ^{10}C by (a) energy considerations, (b) the observation of ^{10}B (p,p′) scattering to this level, (c) data from the ^{9}Be(d,n)^{10}B reaction, conjoined with (d) the absence of ^{10}B(d,d′), and (e) the absence of ^{12}C(d,α)^{10}B processes. Similar, albeit weaker, evidence serves to link the other presumed multiplets of higher energy.

This detailed example should by no means, however, give the impression that analogue states are to be found only in light nuclei: Indeed, the pheno-

---

† From semiempirical expressions proposed by Sengupta [Se 60] and Jaenecke [Jae 60], Anderson *et al.* [An 65] derived a formula for the Coulomb displacement energy in the medium-mass region, $\delta E_C = E_1 \bar{Z}A^{-1/3} + E_2$, with $\bar{Z}$ the mean atomic number and the energy coefficients equal to $E_1 = 1.444 \pm 0.005$ MeV, $E_2 = -1.13 \pm 0.04$ MeV. This provides an excellent fit to (p, n) isobaric reaction and β-decay data.

‡ In contrast to other current representations in which relative atomic mass differences are implicitly taken into account through a vertical displacement δE_β of the respective nuclear ground states in a level scheme, our representation has been chosen to line up the isobaric multiplet ground states horizontally, which is equivalent to taking
$$\varDelta = \pm \delta E_\beta \pm (\delta E_C - \delta M) = Q_{(p,n)} + \delta E_C = Q_{(n,p)} - \delta E_C$$
rather than simply $\varDelta = \pm(\delta E_C - \delta M)$.

menon of ISOBARIC RESONANCES was revealed [Fo 64] through a study of the
(p,p) and (p,n) reactions initiated by proton bombardment of the heavy nuclei
^{88}Sr and ^{89}Y. Thus, resonances in the (p,p) excitation functions (shown
in Fig. 5-9, viz. p-yields in function of incident energy at 90 degrees and
125 degrees, point to levels in the intermediate nuclear systems ^{88}Sr + p = ^{89}Y*
and ^{89}Y + p = ^{90}Zr* which are isoanalogues of corresponding low-lying
states in ^{89}Sr and ^{90}Y, respectively. The resonances were closely related to
spectra observed in (d,p) reactions with the same target nuclei, and were
energetically in excellent agreement with predicted Coulomb displacements.
Whereas they refer to isoanalogues in *intermediate* nuclear systems, an earlier
investigation [An 62b] of (p,n) data on medium-heavy nuclei had revealed
isoanalogues in the *residual* nuclei corresponding to the ground states of the
target nuclei.

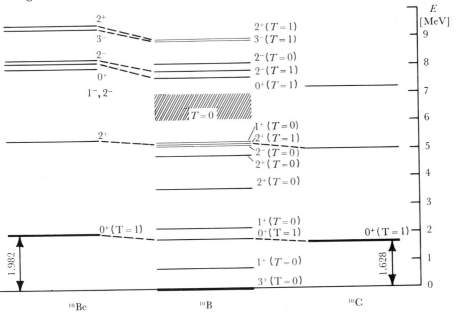

Fig. 5-8. Juxtaposition of isoanalogue states in ^{10}Be, ^{10}B, and ^{10}C, as discernible
in a somewhat simplified energy-level diagram (cf. [Lau 66, pp. 110, 111 and 129]).

At this stage we may point to some interesting features encountered in the
study of analogue states; more detailed descriptions are to be found in several
excellent reviews of this subject, e.g., [Bu 55a, Da 59, Schi 60, Fo 66, Ro 66;
He 66, especially the invited papers by J. D. Fox, D. Robson, and D. A.
Bromley]. The ISOBARIC MASS FORMULA† (see D. H. Wilkinson's paper in Fox

† Through isospin considerations, Garvey and Kelson [Ga 66] (see also [Ke 66])
have derived more general relationships between nuclidic masses.

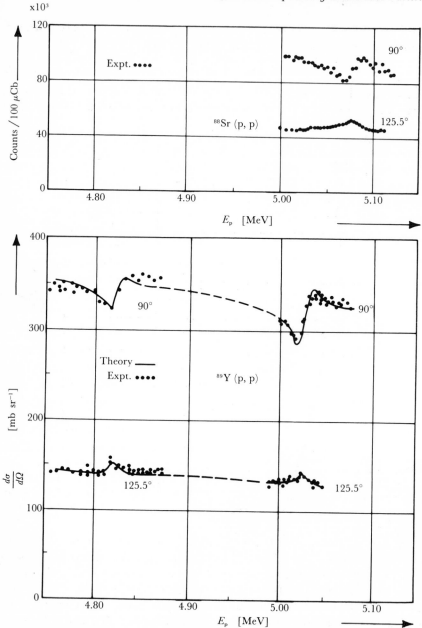

Fig. 5-9. Excitation functions at 90° and 125.5° for the reactions ^{88}Sr(p, p) and ^{89}Y(p, p) showing characteristic elastic-scattering resonances which indicate the existence of levels in the intermediate nuclei ^{89}Y* and ^{90}Zr* that have corresponding isoanalogues in the low-lying states of ^{89}Sr and ^{90}Y (from [Fo 64]).

and Robson [Fo 66]) states that the nuclear masses of members of an iso-multiplet are given by the "parabolic" expression

$$M = a + b T_z + c T_z^2 \qquad (5\text{-}43)$$

without any even-Z, odd-Z alternations due to pairing, where a, b, and c are numerical parameters, whose values have not yet been determined with

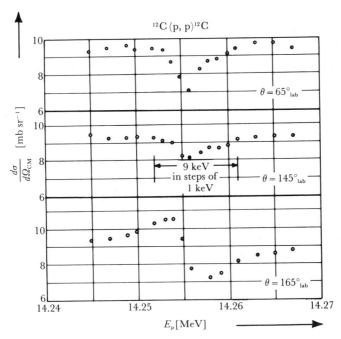

Fig. 5-10. Corresponding resonances at several scattering angles in the ^{12}C(p, p) reaction observed in high-resolution measurements around $E_p = 14.255$ MeV, pointing to the presence of an excited state at 15.047 ± 0.018 MeV in the intermediate nucleus ^{13}N* whose isospin $T = \frac{3}{2}$ suggests that some isospin mixing may be occurring (from [Fo 66]).

conclusive accuracy. To check the quadratic formula it is necessary to insert data for multiplets with $T > 1$. So far $T = \frac{3}{2}$ quartets have been studied ([Ce 66] and earlier references therein) for nuclei of mass $A = 9$, 13, and 37, while some preliminary work has been done on $T = 2$ quintets in nuclei with $A = 16$. The results indicate strikingly good agreement: Thus for $A = 9$, the predicted mass excess for ^{9}C calculated from ^{9}Li, ^{9}Be, and ^{9}B data is 29.00 ± 0.05 MeV, whereas the measured value is 28.99 ± 0.07 MeV, and for $A = 13$ the mass excess of ^{13}O as calculated from ^{13}N, ^{13}C and ^{13}B is 23.10 ± 0.05 MeV, in excellent agreement with the measured value, 23.11 ± 0.07 MeV.

The location of the $T = \frac{3}{2}$ levels in the $A = 13$ multiplet has followed along similar lines and yielded comparable results [Ad 65, Ce 66]. An interesting countercheck of the $T = \frac{3}{2}$ level in ^{13}N, isoanalogue to the ground state of ^{13}B, was provided by the observation of a resonance anomaly in the excitation function for elastic and inelastic proton scattering on ^{12}C—processes which

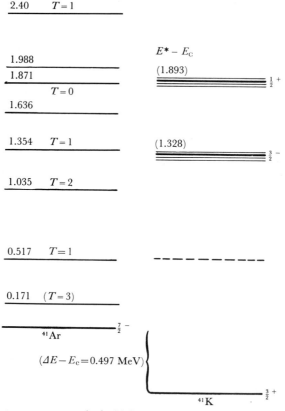

Fig. 5-11. States at comparatively high excitation in ^{41}K which are isoanalogues of the fourth and sixth levels (and possibly the second level, as shown dashed) in ^{41}Ar, observed in studies of proton-induced reactions on ^{40}Ar (adapted from [Fo 66]).

could not take place via such a level if the isospin of the latter were pure and strictly conserved. The fact that a spin-$\frac{3}{2}^-$ level in ^{13}N *was* indeed observed [Fo 66, pp. 472 ff.] at the appropriate energy of 15.047 ± 0.018 MeV (which compares well with the value 15.068 ± 0.008 MeV deduced from $^{11}B(^{3}He,n)$ ^{13}N data), as is evident from Fig. 5-10, indicates that some mixing of the ^{12}C and ^{13}N isospins must have occurred, presumably through the Coulomb interaction, involving both $T = 1$ admixture in ^{12}C and $T = \frac{1}{2}$ admixture in ^{13}N.

Still more stringent tests of the formula will become feasible when members of higher-T multiplets become identified and measured: Thus, for example, (p,n) reaction studies have enabled a $T = 6$ level in $^{90}_{40}Zr_{50}$ ($T_z = -5$) to be located and investigations are now under way to spot the analogue states to this.

On the other hand, data which elucidate the structure of analogue resonances at fairly high excitation (10–11 MeV) have now become available from a study of proton-induced reactions on ^{40}Ar. Thus, elastic proton scattering goes by way of an intermediate ^{41}K system whose levels correspond to

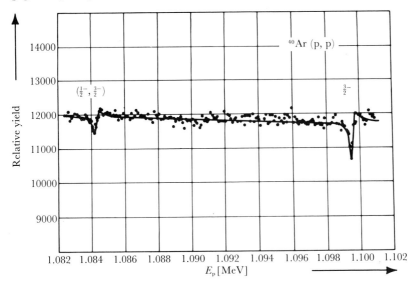

Fig. 5-12. Elastic proton scattering resonances indicating the existence of two very narrow levels in ^{41}K whose energies, conjoined with the probable spin-parity assignments shown, render them possible candidates for the isoanalogue states corresponding to the second level of ^{41}Ar (from [Fo 66]).

the fourth and sixth levels, at 1.354 and 1.871 MeV, respectively, of ^{41}Ar, as may be seen from Fig. 5-11, in which the isoanalogues appear in juxta-position when allowance has been made for the Coulomb displacement energy. Also shown therein is the position of a possible analogue to the second level of ^{41}Ar: Recent (p,p) data [Wi 66] have indicated the presence of two levels in an appropriate energy region having reasonable spin values, as shown in Fig. 5-12. Of particular interest, however, is the (p,p) data obtained at very high resolution in small energy steps by Bilpuch [Fo 66, pp. 235 ff.] which reveal the fine structure of the analogue resonances, at $E_p = 1.87$ and 2.45 MeV (lab), respectively, corresponding to the fourth and sixth ^{41}Ar levels. Whereas earlier work [Ba 61b] at medium resolution indicated these resonances as anomalies in the excitation function [Fig. 5-13(a)], the fact that they were not single isolated states but rather a cluster of levels was revealed by the fine structure evinced by the high-resolution measurements shown on an extended scale in Fig. 5-13(b) and in still more detail in Fig.

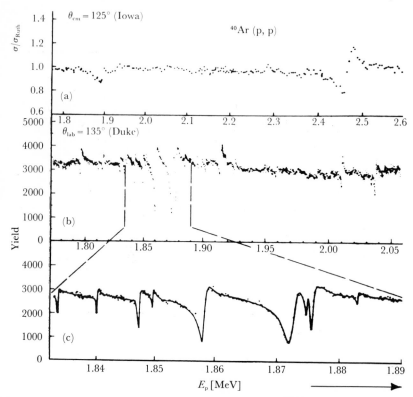

Fig. 5-13. Elastic proton yields from the ^{40}Ar(p, p) reaction:
(a) medium-resolution data obtained by the Iowa group [Ba 61b] with a 6-KeV target, indicating resonances which are shown in (b) and (c) to have multi-resonance structure;
(b) high-resolution data from repetition of the measurements by the Duke group (Bilpuch *et al.*, in Fox and Robson [Fo 66, pp. 235 ff.]) showing fine structure at a comparable scattering angle in the low-energy region from 1.8 to 2.5 MeV;
(c) expansion of the data around the 1.87-MeV analogue resonance (from [Fo 66]).

5-13(c) for the analogue at $E_p = 1.87$ MeV. Theoretical analysis of a small portion of this resonance data gave the fit shown in Fig. 5-14 when spin $\frac{3}{2}^-$ was assigned to the levels, whose mean separation was determined to be 5 keV. Similarly, for the other analogue resonance at $E_p = 2.45$ MeV, the data [Fig. 5-15(a, b)] showed individual levels, with clustering around the analogue resonance energy. In expanded form this is depicted in Fig. 5-15(c). The solid curve represents the theoretical fit to the data in the analogue region, commensurate with a spin assignment of $\frac{1}{2}^+$ for the main levels whose mean spacing is 10 keV (the ratio of spacings for the two sets of analogue data is thus as $2J + 1$). We see that what appeared to be resonances associated with

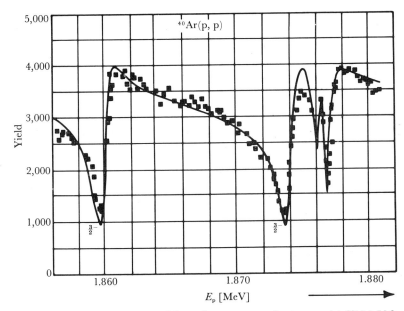

Fig. 5-14. Theoretical multilevel fit to the resonance data around 1.87 MeV for the ^{40}Ar(p, p) reaction studied at high resolution (from [Fo 66]).

two individual analogue levels are indeed two clusters of many closely spaced levels acting as a pair of single levels, an observation which is not at variance with theoretical expectations.

A final example will serve to illustrate the clarity and unambiguity of the information which can be derived from isoanalogue resonance reactions. This concerns measurement of the polarization of protons after elastic scattering on ^{90}Zr in the vicinity of the 6.8-MeV resonance [Te 66; Fo 66, pp. 333 ff.]. Figure 5-16 shows the experimental data over a range of angles which, when compared with the theoretical predictions for an analogue state of spin $\frac{3}{2}$ or $\frac{5}{2}$, very evidently supports a $\frac{3}{2}$ assignment and indicates maximal polarization to occur at an angle of 140 degrees. When the polarization was measured at this particular angle over a range of energies about the 6.8-MeV resonance, the results again tallied with the theoretical prediction for spin $\frac{3}{2}$, as can be seen from Fig. 5-17. Measurements of this sort [Ad 66], when carried out at an appropriate angle and energy can therefore in a very simple way give a clear indication of the spin of the analogue resonance state in the intermediate nucleus.

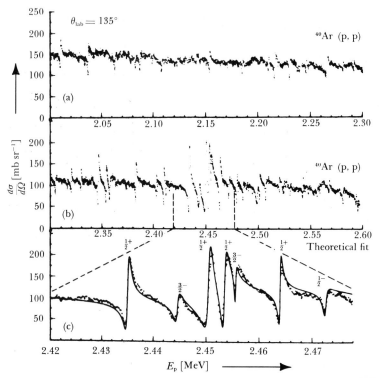

Fig. 5-15. High-resolution data showing fine detail in what appeared to be a single resonance at 2.45 MeV in Fig. 5-13(a) for the ^{40}Ar(p, p) reaction:
  (a) the energy region from 2.0 to 2.3 MeV;
  (b) the energy region from 2.3 to 2.6 MeV, showing the strong band of levels around 2.45 MeV which is the analogue region corresponding to the sixth level of ^{41}Ar;
  (c) theoretical fit to the expanded data around 2.45 MeV,
presented by the Duke group (Bilpuch *et al.*, in Fox and Robson [Fo 66, pp. 235 ff.]).

5.4.5. MAGNETIC AND ELECTRIC MOMENTS OF PARTICLES AND NUCLEI

Though classical ideas provide a *qualitative* basis for the occurrence of magnetic and electric multipole moments in particles and nuclei, which give rise to an interaction with externally applied electromagnetic fields causing precession and alignment or polarization, classical physics cannot even

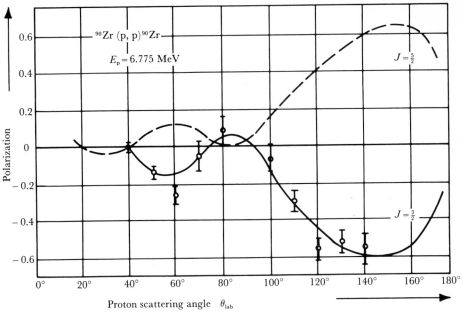

Fig. 5-16. The angular distribution of polarized protons from elastic scattering on ^{90}Zr at 6.775 MeV. The comparison between experimental data and theoretical curves evaluated on the assumption of an analogue state of spin $\frac{3}{2}$ or $\frac{5}{2}$ indicates the $\frac{3}{2}$ spin assignment to be valid and excludes the possibility of a spin-$\frac{5}{2}$ level (from [Te 66]).

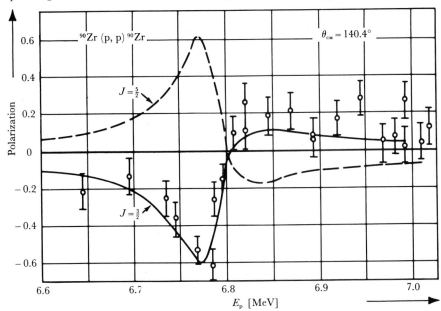

Fig. 5-17. A plot of proton polarization in function of incident energy for the reaction ^{90}Zr(p, p) at a fixed scattering angle $\theta_{\mathrm{CM}} = 140.4°$ corresponding to maximum polarization at 6.78 MeV again indicates an assignment of $J = \frac{3}{2}$ to furnish a good fit to experimental data, whereas $J = \frac{5}{2}$ fails to account for the observed behavior (from [Ad 66]). (Used by permission of North-Holland Publishing Co.)

roughly account *quantitatively* for the observed phenomena. Indeed, it is one of the triumphs of quantum-electrodynamic theory that it furnishes values in excellent agreement with those derived by high-precision experimental techniques.

We note first that symmetry (parity) considerations indicate that spherical or ellipsoidal systems with axial symmetry should have an *alternating* sequence of multipole moments, namely electric monopole ($E0$), equivalent simply to an electric charge, magnetic dipole ($M1$), electric quadrupole ($E2$), magnetic octupole ($M3$), electric hexadecapole ($E4$), ... , etc., but *not* the conjugate moments $M0$, $E1$, $M2$, $E3$, $M4$, ... The presence of an electric monopole in charged particles, corresponding as it does to a point elementary charge, needs no further consideration. The complementary entity, the MAGNETIC MONOPOLE has been the subject of theoretical speculations by Dirac [Di 31, Di 48] and experimental searches [Pu 63, Am 63, Go 63, Fo 63] but has so far eluded detection (though current investigations with deep-sea sediment appear to offer some tentative evidence [Ko 67, 68]). So also has, for example, the ELECTRIC DIPOLE MOMENT OF FREE NEUTRONS, whose value Smith *et al.* ([Sm 57], quoted by Segrè [Se 64, pp. 221–222]; earlier details by Ramsey [Ra 56, pp. 201–202]) experimentally showed to be compatible with zero. By observing the neutron precession frequency in a weak magnetic field and looking for a possible change in frequency when an external electric field is applied, the authors were led to conclude that the neutron's $E1$ moment referred to the elementary charge e has a value $(-0.1 \pm 2.4) \times 10^{-20}$ cm/e. Meanwhile, more precise measurements have furnished still smaller limits; e.g., Miller *et al.* [Mi 67] obtained the value $(-2 \pm 3) \times 10^{-22}$ cm/e, while Shull and Nathan [Shu 67] deduced the value to be $+(2.4 \pm 3.9) \times 10^{-22}$ cm/e. Nelson *et al.* [Ne 59] have experimentally derived an upper limit for the ELECTRIC DIPOLE MOMENT OF THE ELECTRON, and cite limits for other particles.

A MAGNETIC DIPOLE MOMENT is associated with a point charge moving along a closed path, which constitutes a current loop, or with the rotation (spin) of an extended electric charge. Thus all fundamental particles carrying a charge and having some angular momentum would be expected to evince an $M1$ moment. In particular that of the *electron* was demonstrated and measured by Stern and Gerlach [St 22] using an atomic-beam technique involving the passage of neutral silver atoms through a strongly inhomogeneous magnetic field. Later experiments [Ku 47, La 47a, Ku 48] showed that the magnetic dipole moment of an electron due to its spin is slightly larger than predicted by Dirac's relativistic quantum theory of the electron, which led Schwinger [Schw 48, Schw 49] to set about a reformulation of quantum electrodynamics that constitutes one of the monumental achievements of modern theoretical physics. The correction factor supplied by his theory, $[1 + (\alpha/2\pi) = 1.001\ 16]$, accounted exactly for the observed discrepancy. A simple historical review has been given by Kusch [Ku 66].

The MAGNETIC DIPOLE MOMENT OF FREE PROTONS due to their spin can also be determined directly by techniques of the Stern-Gerlach type (reviewed by Estermann *et al.* [Es 37, Es 46]) or less directly but with greater precision by more modern atomic and molecular-beam techniques, viz.

(a) molecular beam magnetic resonance methods, due to Rabi *et al.* [Ra 38, Ra 39, Ku 40], reviewed in [Ha 41, Ke 46, Ra 56];

(b) nuclear paramagnetic-resonance absorption methods, due to Purcell *et al.* [Pu 46], reviewed in [Pa 50, Po 52, Ra 56]; and

(c) nuclear resonance induction methods, due to Bloch *et al.* [Bl 46a, Bl 46b].

The proton spin magnetic dipole moment is used as a standard to which other dipole moments are referred. Two precision methods which gave essentially the same magnitude for this quantity, though based on different principles, involved (a) a comparison of the proton LARMOR PRECESSION FREQUENCY $\nu_L^{(p)}$ with the CYCLOTRON FREQUENCY $\nu_C^{(p)}$ [Hi 50], and (b) a determination of the ratio of proton Larmor frequency $\nu_L^{(p)}$ to *electron* cyclotron frequency $\nu_C^{(e)}$ in a magnetic field of the same strength [Ga 51], followed subsequently by high-precision corrections; the data have been summarized by Walchli [Wa 53] and Fuller and Cohen [Fu 65, pp. 19 ff.]. When a magnetic field **B** is applied to a body possessing an absolute magnetic dipole moment **μ**, a torque

$$\mathbf{T} = [\boldsymbol{\mu} \times \mathbf{B}] \tag{5-44}$$

acts on the system, tending to align the dipole in the field direction. Newton's second law of motion indicates that this must be equal to the rate of change of angular momentum **L**,

$$\mathbf{T} = \frac{d\mathbf{L}}{dt} = [\boldsymbol{\mu} \times \mathbf{B}] \tag{5-45}$$

Since the change $d\mathbf{L}$ is always perpendicular to **L**, with **L** describing the surface of a cone having **B** as axis (Fig. 5-18), this gives a measure of the LARMOR PRECESSION FREQUENCY

$$\nu_L \equiv \frac{\omega_L}{2\pi} = \frac{1}{2\pi} \frac{d\varphi}{dt} = \frac{1}{2\pi} \frac{dL}{L \sin\theta \, dt} \tag{5-46}$$

Since, from (5-45), $dL/dt = \mu B \sin\theta$ it follows that

$$\nu_L = \frac{1}{2\pi} \frac{\mu B}{L} \tag{5-47}$$

Classically, the angular momentum of a body of mass m and charge ze moving with speed v in a circular orbit of radius r is

$$L = mvr \qquad (5\text{-}48)$$

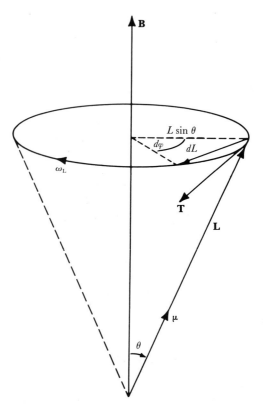

Fig. 5-18. Vector momentum diagram to illustrate Larmor precession with a frequency ω_L about an axis in the direction of an applied magnetic field **B**. The other symbols are explained in the text.

and the corresponding circular Ampère current is

$$I = \frac{zev}{2\pi r} \qquad (5\text{-}49)$$

Inserting a factor $1/c$ to relate e in electrostatic units to μ in electromagnetic units, we obtain

$$\mu = \frac{1}{c}\,(\text{current}) \times (\text{area}) = \frac{ze}{c}\,\frac{v}{2\pi r}\,\pi r^2 = \frac{ze}{2mc}\,(mvr) \qquad (5\text{-}50)$$

or vectorially,

$$\boldsymbol{\mu} = \pm \frac{ze}{2mc} \mathbf{L} \tag{5-51}$$

according as $\boldsymbol{\mu}$ is parallel or antiparallel to the angular momentum vector $\mathbf{L}$. The magnitude of μ is directly proportional to L. The motion is analogous to that of a precessing gyroscope.

The CYCLOTRON FREQUENCY ν_C in a magnetic field of the same strength B is the number of revolutions per second executed by a free charged particle describing a circular path of radius ρ. Equating the centrifugal force with the magnetic centripedal force, we find

$$\left| \frac{d\mathbf{p}}{dt} \right| = \frac{mv_C^2}{\rho} = (ze)|[\mathbf{v}_C \times \mathbf{B}]| = ze\,v_C B \tag{5-52}$$

whence

$$\nu_C = \frac{v_C}{2\pi\rho} = \frac{1}{2\pi} \frac{zeB}{m} \tag{5-53}$$

Thus the ratio of Larmor to cyclotron frequency is

$$\frac{\nu_L}{\nu_C} = \frac{\mu m}{zeL} \tag{5-54}$$

in which m is respectively either the proton or the electron rest mass.

The existence of a NEUTRON MAGNETIC DIPOLE MOMENT indicates that although the neutron as a whole carries no net electric charge, it nevertheless has an extended electric charge distribution in its internal structure. The value of the neutron's dipole moment was determined by Alvarez and Bloch [Al 40] and Arnold and Roberts [Ar 46, Ar 47], who used the MAGNETIC RESONANCE technique to relate its value to that of the proton as standard. A beam of slow neutrons from a reactor was passed through a magnetized block of iron to obtain a partially polarized neutron beam by differential absorption (BLOCH EFFECT). This was then sent through a radio-frequency coil situated in a constant magnetic field, and the radio frequency necessary to depolarize it completely was compared with the proton resonance frequency under similar conditions. This measurement was followed by later determinations by Bloch *et al.* [Bl 48] and Cohen *et al.* [Co 56a].

The accuracy with which the magnetic dipole moments of particles and nuclei can be determined by the above methods is not matched by that of the corresponding electric quadrupole measurements, based on hyperfine-structure investigations in atoms, molecules, or crystalline solids, even though high-resolution radio-frequency or microwave spectroscopy techniques are currently employed. The methods are in many cases basically an extension of the Bloch-Purcell techniques and have been described by Ramsey [Ra 53,

Ra 56], but their description and discussion lies outside the scope of the present text. We concentrate rather on outlining the quantum-theoretical approach to nuclear moments and the explanation of the anomalous observed values for nucleons.

The transition from the classical viewpoint considered so far to that employed in quantum mechanics lies in replacing the classical angular momentum **L** by its quantum-mechanical substitute, viz.

$$\mathbf{L} \to \mathbf{l}\hbar \text{ for orbital momentum} \tag{5-55}$$

$$\mathbf{L} \to \mathbf{s}\hbar \text{ for spin momentum} \tag{5-56}$$

and introducing an appropriate dimensionless correction factor g which normally incorporates the charge number z and is called the "g FACTOR" or "LANDÉ FACTOR."

Thus, with (5-51) and (5-55) we arrive at the ORBITAL MAGNETIC DIPOLE MOMENT

$$\boldsymbol{\mu}_l = g_l \frac{e\hbar}{2mc} \mathbf{l} \tag{5-57}$$

and with (5-51) and (5-56) at the SPIN MAGNETIC DIPOLE MOMENT

$$\boldsymbol{\mu}_s = g_s \frac{e\hbar}{2mc} \mathbf{s} \tag{5-58}$$

Each of these is directly proportional to the respective angular momentum, the constant of proportionality being termed the GYROMAGNETIC RATIO

$$\gamma_l \equiv \frac{\mu_l}{l} = g_l \frac{e\hbar}{2mc} \quad \text{and} \quad \gamma_s \equiv \frac{\mu_s}{s} = g_s \frac{e\hbar}{2mc} \tag{5-59}$$

For convenience, the MAGNETON UNITS have been introduced, having the absolute value

$$\mu_B \equiv \frac{e\hbar}{2m_e c} = 9.273\ 2 \times 10^{-21} \text{ erg/Gauss} \tag{5-60}$$

for the BOHR MAGNETON and

$$\mu_N \equiv \frac{e\hbar}{2m_p c} = \frac{1}{1836} \mu_B = 5.050\ 5 \times 10^{-24} \text{ erg/Gauss} \tag{5-61}$$

for the NUCLEAR MAGNETON, and in terms of these, magnetic moments are expressed as dimensionless quantities. We make the important observation at this stage that *by convention all magnetic moments are quoted as maximum values in the external field direction*: They correspond to the *maximum* value of the projected quantum number along the quantization direction, viz. $m_l = l$.

and $m_s = s$ (though there is, respectively, a $(2l+1)$ or $(2s+1)$ multiplicity of values) and are taken as *positive* if the $\boldsymbol{\mu}$ direction referred to that of its determining angular momentum vector corresponds to that of a rotating *positive* electric charge.

The ORBITAL ELECTRON MOMENT $\mu_l^{(e)}$ was found to be 1 Bohr magneton, so that $g_l^{(e)} = 1$. With the recognition of electron intrinsic spin $s = \frac{1}{2}$, the expectation for the SPIN ELECTRON MOMENT $\mu_s^{(e)}$ was the value 2 Bohr magnetons, so that $g_s^{(e)} = 2$, but as already discussed, subsequent theoretical and experimental developments in fact yielded the value

$$g_s^{(e)} = 2\left(1 + \frac{\alpha}{2\pi}\right) = 2 \times 1.001\ 16 = 2.002\ 32 \qquad (5\text{-}62)$$

5.4.6. ANOMALOUS NUCLEON SPIN MAGNETIC DIPOLE MOMENTS

NUCLEON MAGNETIC MOMENTS proved still more difficult to explain theoretically. The orbital moments were found to have reasonable absolute values, commensurate with absolute momenta $\mathbf{l} \simeq [l(l+1)]^{1/2}\ \hbar$, but the magnitudes of the SPIN MAGNETIC MOMENTS were *anomalous*: Instead of the values $\mu_s^{(p)} = 1\mu_N$ and $g_s^{(p)} = 2$, while $\mu_s^{(n)} = g_s^{(n)} = 0$ as required by Dirac's relativistic theory, one found that (citing the latest accepted values)

$$\mu_p \equiv \mu_s^{(p)} = 2.792\ 76\ \mu_N \qquad (5\text{-}63)$$

and

$$\mu_n \equiv \mu_s^{(n)} = -1.913\ 148\ \mu_N \qquad (5\text{-}64)$$

The discrepancy between the expected and the observed values can to a large extent be explained by picturing the internal structure of nucleons along the lines of the MESON-NUCLEOR MODEL [Sa 52].

In a rather over-simplified, rough approximation we can view the proton as made up of a core (the NUCLEOR) which carries a positive charge of magnitude $(1-\alpha)e$ and which is surrounded by a π-mesic cloud having a positive charge αe. The neutron can then be regarded as composed of a positive nucleor of charge $+\alpha e$ surrounded by a π-meson cloud which carries the compensating negative charge $-\alpha e$. The arbitrary parameter α (whose magnitude according to charge-symmetric meson theory is $g^2/\hbar c \approx \frac{1}{10}$), expresses the distribution of the charge which, for spinning $s = \frac{1}{2}$ particles, gives rise to a contribution to the net spin moment of $(1-\alpha)\mu_N$ from the proton core, $\alpha\mu_N$ from the neutron core, and $\alpha\mu_\pi \equiv \alpha(m_p/m_\pi)\mu_N$ from the mesic cloud. Hence the proton magnetic moment becomes

$$\mu_p = \left(1 - \alpha + \alpha\frac{m_p}{m_\pi}\right)\mu_N \qquad (5\text{-}65)$$

and the neutron dipole moment takes on the nonzero negative value

$$\mu_n = -\left(\alpha + \alpha\frac{m_p}{m_\pi}\right)\mu_N \qquad (5\text{-}66)$$

On setting these expressions equal to the experimentally observed "anomalous" values, one obtains agreement when the parameters are set to $\alpha = 0.06$, (which tallies with $g^2/\hbar c \approx \frac{1}{10}$) and $m_{\mathrm{p}}/m_\pi = 31$ (as against the actual ratio of 6.5). It is evident that the model cannot altogether be reconciled with experience—its basis would seem to be acceptable, but its detailed features incorrect. The actual model, as based upon charge-symmetric meson theory, is slightly more complicated than that outlined above, in that it has *two* parameters α and β in addition to the mass ratio m_{p}/m_π. It views the neutron as if for a fraction α of the time it were a proton nucleor surrounded by a π^--meson cloud and for a fraction β it were a neutron nucleor surrounded by a π^0 cloud, while for the remaining portion $(1 - \alpha - \beta)$ it is a "bare" neutron nucleor. The proton is regarded as being a neutron nucleor with a π^+ cloud for a time α, and a proton nucleor with a π^0 cloud for a time β, and a "bare" proton for the remaining time $(1 - \alpha - \beta)$. The meson clouds are taken to have momentum $l = 1$ and it is found that the coupling scheme requires one to set $\alpha = 2\beta$. Agreement with the measured moments ensues when one takes $\alpha = 0.28$, $\beta = 0.14$, and $m_{\mathrm{p}}/m_\pi = 6.8$, which does not involve the above shortcomings. Yet the theory can also go far toward accounting for the neutron-proton mass difference [Ho 54] and can readily be elaborated to take spin orientations into more detailed consideration or to take cognisance of the nonvanishing (albeit small) probability of nucleon dissociation into two or more pions, e.g.,

$$p \to \begin{array}{l} \nearrow \; \mathrm{n} + \pi^+ \\ \mathrm{n} + \pi^+ + \pi^0 \\ \searrow \; \text{-------} \end{array} \tag{5-67}$$

The charge distribution is affected by not only the meson cloud itself but also by the core motion, e.g., recoil following emission or absorption of mesons and inherent jitter (*Zitterbewegung*—a relativistic effect for free particles). Furthermore, the effects of virtual creation of nucleon-antinucleon pairs also has to be taken into consideration. When these influences are treated as higher-order perturbations, the agreement between theory and observation is vastly improved, but nonetheless small, disturbing discrepancies remain unaccounted for.

The TOTAL MAGNETIC DIPOLE MOMENT is the vector sum of the orbital and the spin moments,

$$\boldsymbol{\mu} = \boldsymbol{\mu}_l + \boldsymbol{\mu}_s \tag{5-68}$$

For nucleons, the *orbital* moments take on nonanomalous values $\boldsymbol{\mu}_l^{(\mathrm{p})} = \mathbf{1}$ and $\boldsymbol{\mu}_l^{(\mathrm{n})} = 0$ (since the *orbital* motion of a neutron represents zero current), while the *spin* moments have the anomalous values $\boldsymbol{\mu}_s^{(\mathrm{p})} = \mu_{\mathrm{p}} = 2.79$ and $\boldsymbol{\mu}_s^{(\mathrm{n})} = \mu_{\mathrm{n}} = -1.91$. The respective *orbital* g factors are thus $g_l^{(\mathrm{p})} = 1$, $g_l^{(\mathrm{n})} = 0$ while the *spin* g factors are $g_s^{(\mathrm{p})} = 2 \times 2.79 = 5.58$ and $g_s^{(\mathrm{n})} = 2 \times -1.91 = -3.82$. Because of the latter anomaly, even though $\boldsymbol{\mu}_l$ and $\boldsymbol{\mu}_s$ lie along $\mathbf{1}$ and $\mathbf{s}$ respectively, in the case of nucleons the resultant total vector dipole moment $\boldsymbol{\mu}$ does not lie along $\mathbf{j} = \mathbf{1} + \mathbf{s}$, but precesses about the $\mathbf{j}$ axis in such a way that on building a time average, only the component of $\boldsymbol{\mu}$ in the $\mathbf{j}$ direction remains constant. This conserved quantity, shown in Fig. 5-19, is what is actually measured; it can be written as

$$\boldsymbol{\mu}_j = g_j \, \mathbf{j} \, \mu_{\mathrm{N}} \tag{5-69}$$

and evaluated as follows:

$$\mu_j = \mu \cos(\mathbf{\mu}, \mathbf{j}) = \mu_l \cos(\mathbf{l}, \mathbf{j}) + \mu_s \cos(\mathbf{s}, \mathbf{j}) \qquad (5\text{-}70)$$

whence

$$g_j \mathbf{j} = g_l \mathbf{l} \frac{\mathbf{j}^2 + \mathbf{l}^2 - \mathbf{s}^2}{2(\mathbf{j} \cdot \mathbf{l})} + g_s \mathbf{s} \frac{\mathbf{j}^2 + \mathbf{s}^2 - \mathbf{l}^2}{2(\mathbf{j} \cdot \mathbf{s})} \qquad (5\text{-}71)$$

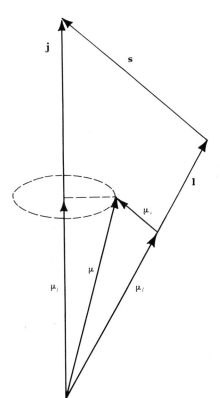

Fig. 5-19. Vector diagram depicting the precession of the dipole moment vector $\mathbf{\mu} = \mathbf{\mu}_l + \mathbf{\mu}_s$ about the axis of $\mathbf{j} = \mathbf{l} + \mathbf{s}$ with a constant projection μ_j on to the $\mathbf{j}$ axis.

using the law of cosines. Thus, employing Kramer's rule for quantum-mechanical squares,

$$\mathbf{j}^2 = j(j+1), \qquad \mathbf{l}^2 = l(l+1), \qquad \mathbf{s}^2 = s(s+1) \qquad (5\text{-}72)$$

and considering first the situation with *parallel* spin setting $(j = l + s)$, one arrives at the result

$$g_j^{(+)} = g_l \frac{l}{l+s} + g_s \frac{s}{l+s} \qquad (5\text{-}73)$$

whence

$$\mu_j^{(+)} = \mu_l + \mu_s \qquad (5\text{-}74)$$

whereas for an *antiparallel* spin setting one has $j = l - s$ and thus

$$g_j^{(-)} = g_l \frac{l+1}{l-s+1} - g_s \frac{s}{l-s+1} \tag{5-75}$$

whence

$$\mu_j^{(-)} = \frac{j(l+1)}{l(j+1)} \mu_l - \frac{j}{j+1} \mu_s \tag{5-76}$$

a result less intuitively obvious than that for parallel spin orientation. An equivalent, more convenient, form which embraces *both* spin orientations $j = l \pm s$ when $s = \frac{1}{2}$ is

$$\mu_j^{(\pm)} = j \left[g_l \pm \frac{(g_s - g_l)}{2l+1} \right] \mu_N \tag{5-77}$$

5.4.7. Magnetic Dipole Moments of Nuclei

This result can be used to derive the MAGNETIC MOMENTS OF NUCLEI when the latter are treated from the viewpoint of the single particle model, ascribing nuclear properties to those of an unpaired particle, so that $\mu_J = \mu_j$. In *even-even nuclei* the pairing of protons and neutrons causes the nuclear spin J and hence also the magnetic moment μ_J to be zero. In *odd-even* and *even-odd nuclei*, the value of μ_J is determined by that of the unpaired proton or neutron respectively, whence, inserting the values $g_l^{(p)} = 1, g_s^{(p)} = 5.58$ or $g_l^{(n)} = 0, g_s^{(n)} = -3.82$ in (5-77), one obtains the SCHMIDT FORMULAE [Schm 37] for the magnetic dipole moment of odd-A nuclei in their ground states:

odd-even nuclei (proton contribution only)

$$\mu_J^{(o\text{-}e)} = \begin{cases} J \left(1 + \dfrac{2.29}{J}\right) \mu_N & J = l + \frac{1}{2} \tag{5-78} \\[2ex] J \left(1 - \dfrac{2.29}{J+1}\right) \mu_N & J = l - \frac{1}{2} \tag{5-79} \end{cases}$$

even-odd nuclei (neutron contribution only)

$$\mu_J^{(e\text{-}o)} = \begin{cases} J \left(-\dfrac{1.91}{J}\right) \mu_N & J = l + \frac{1}{2} \tag{5-80} \\[2ex] J \left(\dfrac{1.91}{J+1}\right) \mu_N & J = l - \frac{1}{2} \tag{5-81} \end{cases}$$

Plotting these expressions as a function of spin quantum number J yields the "SCHÜLER-SCHMIDT LINES," along which one would expect all measured magnetic moments to lie. As is evident from Fig. 5-20, they are found not to do so, but rather to lie between them and, apart from a very small number of exceptions, (^{75}As, ^{127}I, ^{165}Ho, ^{3}He) in fact lie between the Schüler-Schmidt limits and the so-called DIRAC LIMITS which would ensure from straightforward application of relativistic Dirac theory to nucleons instead of electrons, thereby failing to account for the anomaly in measured values of μ_p and μ_n and using the g factors $g_l^{(p)} = 1$, $g_s^{(p)} = 2.0023$ or $g_l^{(n)} = g_s^{(n)} = 0$.

It will be noted that in most instances a measurement of μ_J can be used to discriminate between $J = l + \frac{1}{2}$ and $J = l - \frac{1}{2}$ orientations, and hence from a known value of J to establish the appropriate l. This in turn enables the level parity $\pi = (-1)^l$ to be determined in a very direct manner.

The above formulae for the magnetic moments of nuclei in their ground states differ from the corresponding expressions derived from a consideration of the whole set of constituent nucleons or of collective motion of nucleons in deformed nuclei; the appropriate treatment has been relegated to a later chapter concerned with the collective model.

Some evidence points to the existence of HIGHER-ORDER MAGNETIC MOMENTS for certain nuclei: Thus, the hyperfine splitting observed in ^{79}Br, ^{81}Br, ^{115}In, and ^{127}I does not admit of explanation in terms of magnetic dipole and electric quadrupole moments alone but would seem to necessitate the introduction of a nuclear magnetic octupole moment (see Fuller and Cohen [Fu 65], especially pp. 55 ff.]).

5.4.8. ELECTRIC MOMENTS

We have already mentioned the preclusion of an ELECTRIC DIPOLE MOMENT, and the experimental ratification in the case of neutrons (Section 5.4.5). Though this property can be shown to follow automatically from the requirement of time-reversal invariance, we shall derive the result in this section on the basis of the definite parity of particle or nuclear states and the existence of a plane of symmetry passing through the center of mass of the quantum system. In the case of non-axial "pear-shaped" nuclei, as currently under study by Davydov [Da 58b], these considerations require some modification.

In terms of a charge density $\rho(x,y,z)$ whose volume integral $\int \rho(x,y,z)\, d\Omega$ gives the total nuclear electric charge Ze, the z component of the nuclear electric dipole moment is defined as

$$D_z = \int z\, \rho(x,y,z)\, d\Omega \qquad (5\text{-}82)$$

with respect to the center of the charge distribution, which coincides with the center of mass. The density ρ at the point (x,y,z) in the nucleus can in turn be expressed through the spatial part of the nuclear wave function

$$\Psi(r_1^{(n)}, r_2^{(n)}, \ldots, r_N^{(n)}, r_{A-N}^{(p)}, \ldots, r_A^{(p)})$$

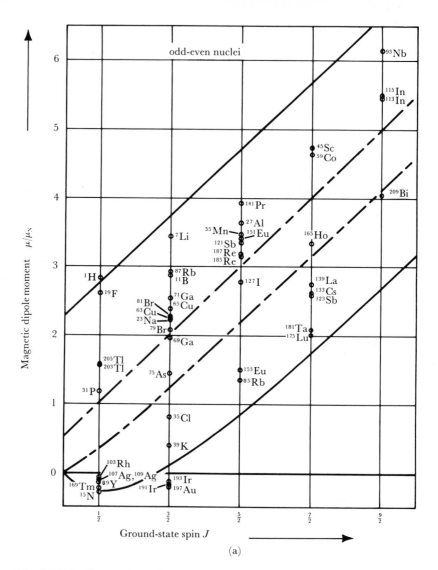

Fig. 5-20(a). Comparison of measured nuclear magnetic dipole moments (expressed in units of the nuclear magneton $\mu_N \equiv e\hbar/2m_p c = 5.05 \times 10^{-24}$ erg G^{-1}) in function of nuclear spin J with the Schüler-Schmidt lines (solid) and the Dirac limits (dashed) for odd-Z, even-N nuclei (from [von B 64]). (Used by permission of Akademische Verlagsgesellschaft, Frankfurt.)

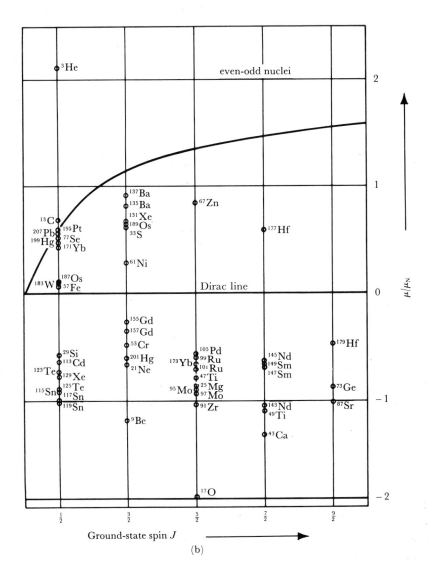

Ground-state spin J

(b)

Fig. 5-20(b). Comparison of measured nuclear magnetic dipole moments (expressed in units of the nuclear magneton $\mu_N \equiv e\hbar/2m_pc = 5.05 \times 10^{-24}$ erg G^{-1}) in function of nuclear spin J with the Schüler-Schmidt lines and the Dirac line (abscissa) for even-Z, odd-N nuclei (from [von B 64]). (Used by permission of Akademische Verlagsgesellschaft, Frankfurt.)

wherein the first N variables denote *neutron* position vectors and the remaining $(A - N)$ variables *proton* vectors referred to the center of mass of the nucleus composed of A nucleons:

$$D_z = e \sum_{k=A-N}^{A} \int z_k |\Psi(\dots, \mathbf{r}_k, \dots)|^2 \, d\Omega \qquad (5\text{-}83)$$

The integral extends over all space (i.e., all coordinates) while the sum is confined to the $(A - N)$ *proton* coordinates $\mathbf{r}_k^{(p)}$ only. Since Ψ has definite parity $(\pm)$, $|\Psi|^2$ must have *even* parity $(+)$, whereas by definition the z_k have *odd* parity $(-)$. Consequently, each integrand $z_k |\Psi|^2$ has odd parity and therefore each integral $\int z_k |\Psi|^2 \, d\Omega$ must vanish identically, since the integration extends over a region which has a well-defined plane of symmetry. Hence the electric dipole moment D_z of symmetric nuclei in the quantization direction must also be zero.

With regard to the ELECTRIC QUADRUPOLE MOMENT, it may be mentioned at the outset that there is a dearth of information as to its existence in electrons or nucleons. The detection of a small but definite quadrupole moment of the deuteron [Ke 39, Ke 40, Ko 52, Au 61] ($Q_d = 2.82$ mb) in addition to its magnetic dipole moment [Wi 53] ($\mu_d = 0.857\ 393\ \mu_N$) had far-reaching consequences in that it could not be explained through central forces, but necessitated the introduction of a *tensor component* into the internucleon force [Ra 41]. This is discussed in detail in Section 12.4.9. The quadrupole moment of nuclei gives information on the nuclear shape. It takes on a zero value only for *spherically-symmetric* nuclei, and for those which have spin $J = 0$ or $\frac{1}{2}$ (see Fermi [Fe 50, pp. 15 ff.]). It is of major importance as regards the nuclear *collective model*, which deals with nonspherical deformed nuclei.

The Coulomb expression for the potential V_C at a point distant $\mathbf{r}$ from the center of a *spherical* nucleus carrying a charge Ze is

$$V_C = \frac{Ze}{r} \qquad (5\text{-}84)$$

This can be extended by analogy to give the potential at a point S_1 having the polar coordinates (r_1, θ_1, ϕ_1) with respect to the central origin of a *nonspherical* (deformed) nucleus whose axis of symmetry is taken as z axis (thereby suppressing the ϕ_1-dependence),

$$V(r_1, \theta_1) = \int \frac{\rho(r_2)}{d} \, d\Omega \qquad (5\text{-}85)$$

by building the volume integral of the charge density ρ at some point $S_2 = (r_2, \theta_2, \phi_2)$ on the surface of the nucleus, distant d from S_1. This generalized potential can also be expressed in terms of a Legendre-polynomial sum (of odd and even order L),

$$V(r_1, \theta_1) = \frac{1}{r_1} \sum_{L=0}^{\infty} a_L \frac{1}{(r_1)^L} P_L(\cos \theta_1) \tag{5-86}$$

Then if the location of S_1 is specialized to lie on the z axis specified by the coordinates $S_z = (z_1, 0, 0)$, the potential at S_z is, with $P_L(\cos \theta_1) = P_L(1) = 1$,

$$V_z = \frac{1}{z_1} \sum_{L=0}^{\infty} a_L \frac{1}{(z_1)^L} \tag{5-87}$$

An identity involving Legendre polynomials whose argument Θ is the angle subtended by $\mathbf{d}$ between $\mathbf{r}_1$ and $\mathbf{r}_2$,

$$\frac{1}{d} \equiv \frac{1}{(r_1^2 + r_2^2 - 2r_1 r_2 \cos \Theta)^{1/2}} \equiv \frac{1}{r_1} \sum_{L=0}^{\infty} \left(\frac{r_2}{r_1}\right)^L P_L(\cos \Theta) \tag{5-88}$$

can now be invoked for the situation depicted in Fig. 5-21, substituting $r_1 = z_1$, $r_2 = r$, and $\Theta = \theta_2 = \theta$,

$$\frac{1}{d} = \frac{1}{z_1} \sum_{L=0}^{\infty} \left(\frac{r}{z_1}\right)^L P_L(\cos \theta) \tag{5-89}$$

so that (5-85) becomes

$$V_z = \int \sum_{L=0}^{\infty} \rho(r) \frac{1}{z_1} \left(\frac{r}{z_1}\right)^L P_L(\cos \theta) \, d\Omega \tag{5-90}$$

Equating this with (5-87), we find that

$$a_L = \int \rho(r) \, r^L P_L(\cos \theta) \, d\Omega \tag{5-91}$$

Setting $r \cos \theta = z$ and noting that

$$P_0(\cos \theta) = 1, \qquad P_1(\cos \theta) = \cos \theta, \qquad P_2(\cos \theta) = \tfrac{1}{2}(3 \cos^2 \theta - 1), \text{ etc.}, \tag{5-92}$$

we obtain the coefficients,

$$L = 0: \quad a_0 = \int \rho \, d\Omega = Ze = \text{charge (electric monopole)} \tag{5-93}$$

$$L = 1: \quad a_1 = \int \rho z \, d\Omega = \text{electric dipole moment} = 0 \tag{5-94}$$

$$L = 2: \quad a_2 = \int \tfrac{1}{2}\rho(3z^2 - r^2) \, d\Omega = \tfrac{1}{2}eQ \tag{5-95}$$

where Q is the QUADRUPOLE MOMENT, expressed as an area in *barn* units
($1\ b = 10^{-24}\ cm^2$). In terms of the nuclear wave function $\Psi(x,y,z)$, normalized
for the charge distribution due to the protons, Q can be written

$$Q = \int (3z^2 - r^2)|\Psi(x,y,z)|^2\ d\Omega \qquad (5\text{-}96)$$

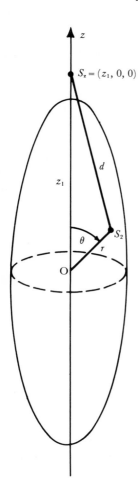

Fig. 5-21. Illustration of the geometrical relationships considered in the
evaluation of the potential V_z at the
point S_z on the symmetry axis of an
ellipsoidal deformed nucleus which has
an electric quadrupole moment Q.

For a *spherically* symmetric distribution of charge the expectation values are
equal, $\langle x^2 \rangle = \langle y^2 \rangle = \langle z^2 \rangle = \frac{1}{3}r^2$, and hence the integral in (5-96) vanishes.
Thus *spherical nuclei have zero quadrupole moment.*

On the other hand, nuclei in the shape of a *prolate* spheroid (cigar-shaped
along the z axis, so that $3\langle z^2 \rangle > \langle r^2 \rangle$) have *positive* moments Q, while *oblate-*

spheroid nuclei ($3\langle z^2 \rangle < \langle r^2 \rangle$, i.e., discus-shaped) have *negative* moments, viz.

prolate-spheroid deformed nuclei:	$Q > 0$	(5-97)
spherical undeformed nuclei:	$Q = 0$	(5-98)
oblate-spheroid deformed nuclei:	$Q < 0$	(5-99)

A compilation of numerical values has been made by Fuller and Cohen [Fu 65].

5.5. Symmetries, Invariances, and Conservation Laws

The quantum numbers discussed in the earlier sections represent conserved quantities in appropriate physical interactions. The conservation laws which govern such interactions can be derived from invariance principles and are directly associated with symmetries inherent in the formulation of the physical process. From the viewpoint of group theory, the *classical* conservation properties derive from *continuous* symmetry groups and yield *additive quantum numbers*, whereas the *quantum theoretical* conservation properties ensue from *discrete* symmetry groups and give rise to *multiplicative quantum numbers*. Thus:

Invariance with respect to *time translation* betokens *energy conservation*;
Invariance with respect to *spatial translation* betokens *momentum conservation*;
Invariance with respect to *spatial rotation* betokens *angular momentum conservation*;
Invariance with respect to *spatial inversion* betokens *parity conservation*.

We give a more detailed list of related quantities in Table 5-5, and at this point illustrate the connections in just a few special cases.

A system can be described by a wave function Ψ which may contain space and time variables,

$$\Psi = \Psi(\mathbf{r},t) = \Psi(x,y,z,t) \tag{5-100}$$

and its behavior will then be described by the general time-dependent Schrödinger equation

$$-\frac{\hbar}{i}\frac{\partial \Psi}{\partial t} = \underset{\sim}{H}\Psi \tag{5-101}$$

in terms of the Hamilton operator $\underset{\sim}{H}$ (the subtilde denotes an operator). Any invariances in the behavior of the system with respect to a given transformation are then evinced as invariances in the Hamiltonian. This yields conditions which have a direct bearing upon conservation laws. A *conserved quantity* is one which does not change with time. This is formally expressed by the statement that the expectation value derived from the appropriate quantum mechanical operator is independent of time, i.e., its time derivative is zero. The expectation value $\bar{A}$ of a quantity A furnished by the operator $\underset{\sim}{A}$ is the matrix element (volume integral)

$$\bar{A} = \int \Psi^* \underset{\sim}{A}\Psi \, d\Omega \tag{5-102}$$

Table 5-5. Invariances, Symmetries, and Conserved Quantities

Category	Invariant transformation	Symmetry group	Conserved quantity	Conserved in S	Conserved in EM	Conserved in W
C1[a]	Time translation	$t \to t + \Delta t$	T_1 — Energy E	✓	✓	✓
C2	Space translation	$\mathbf{r} \to \mathbf{r} + \Delta\mathbf{r}$	T_3 — Linear momentum $\mathbf{p}$	✓	✓	✓
C3	Space rotation	$\phi \to \phi + \Delta\phi$	$SO(3) = SU(2)_s$ — Angular momentum $\mathbf{j}, \mathbf{s}$	✓	✓	✓
C4	Isospin rotation	$\phi_T \to \phi_T + \Delta\phi_T$	$SU(2)_T$ — Isospin $\mathbf{T}, T_z$	✓	✗	✗
G1	Hypercharge gauge	$\psi \to e^{i\alpha}\psi$	$U(1)_Y$ — Hypercharge Y	✓	✓	✗
G2	Baryon gauge	$\psi \to e^{i\alpha}\psi$	$U(1)_B$ — Baryon "charge" number B	✓	✓	✓
G3	Electric charge gauge (implied by C4 and G1)	$\psi \to e^{i\alpha}\psi$	$U(1)_q$ — Electric charge q	✓	✓	✓
G4	Strangeness gauge (implied by G1 and G2)	$\psi \to e^{i\alpha}\psi$	$U(1)_s$ — Strangeness $S = Y - B$	✓	✓	✗
G5	Lepton gauge	$\psi \to e^{i\alpha}\psi$	Lepton number	✓	✓	✓
			Symmetry operation			
D1	Space inversion	$\mathbf{r} \to -\mathbf{r}$	Parity $\underline{P}$ — Parity π	✓	✓	✗
D2	Particle (charge) conjugation	$\mathbf{X} \to \overline{\mathbf{X}}$	Charge conjugation $\underline{C}$ — Charge parity	✓	✓	✗
D3	Time reversal	$t \to -t$	Time-reversal $\underline{T}$ — (detailed balance)[b] (microscopic reversibility)	✓	✓	✓
	Lorentz transformation of a local Hermitian field		PCT — Lagrangian PCT-invariant	✓	✓	✓

[a] Category C embraces continuous transformations (additive quantum numbers); category G embraces gauge transformations (additive quantum numbers), and category D embraces discrete transformations (multiplicative quantum numbers). Thus C and G are of a quasi-classical nature, whereas D, leading to multiplicative conserved entities, is quantum-theoretical in nature.

[b] Because the time-reversal transformation is *antilinear* (i.e., it involves complex conjugation: $\underline{T}$ is therefore termed an ANTIUNITARY OPERATOR, although it is actually unitary), the symmetry operation has no corresponding conserved quantity.

and therefore its time derivative is

$$\frac{d\bar{A}}{dt} = \int \frac{\partial \Psi^*}{\partial t} \underline{A}\Psi \, d\Omega + \int \Psi^* \frac{\partial \underline{A}}{\partial t} \Psi \, d\Omega + \int \Psi^* \underline{A} \frac{\partial \Psi}{\partial t} \, d\Omega \qquad (5\text{-}103)$$

This vanishes if two conditions are fulfilled, viz.

(a)
$$\frac{\partial \underline{A}}{\partial t} = 0$$

e.g., the physical quantity expressed by $\underline{A}$ must not *explicitly* depend upon the time t, and

(b)
$$\int \frac{\partial \Psi^*}{\partial t} \underline{A}\Psi \, d\Omega + \int \Psi^* \underline{A} \frac{\partial \Psi}{\partial t} \, d\Omega = 0 \qquad (5\text{-}104)$$

With (5-101) and its complex conjugate, (5-104) can be rewritten as

$$\frac{i}{\hbar} \int \underline{H}^* \Psi^* \underline{A}\Psi \, d\Omega - \frac{i}{\hbar} \int \Psi^* \underline{A}\underline{H}\Psi \, d\Omega = 0 \qquad (5\text{-}105)$$

which, since the Hamilton operator is Hermitian $[(\underline{H}\Psi)^* = \Psi^*\underline{H}]$ becomes

$$\frac{i}{\hbar} \int \Psi^*(\underline{H}\underline{A} - \underline{A}\underline{H})\Psi \, d\Omega \equiv \frac{i}{\hbar} \overline{[\underline{H},\underline{A}]} = 0 \qquad (5\text{-}106)$$

where $\overline{[\underline{H}\underline{A}]}$ is the expectation value of the commutator $[\underline{H},\underline{A}] \equiv (\underline{H}\underline{A} - \underline{A}\underline{H})$. Thus the subsidiary condition (b) is equivalent to requiring the operator $\underline{A}$ to commute with the Hamilton operator $\underline{H}$.

Hence the quantity A is *conserved* when:

(a) Its operator $\underline{A}$ does not explicitly depend on time: $\partial \underline{A}/\partial t = 0$, and
(b) Its operator $\underline{A}$ commutes with the Hamiltonian: $[\underline{H},\underline{A}] = 0$.

Applying these criteria first to ENERGY CONSERVATION, we note that the appropriate operator giving the total energy E of the system as a whole is the Hamiltonian itself $(\underline{H}\Psi = E\Psi)$, and this clearly satisfies both (a) and (b), unless of course the system is subjected to an external field which *varies* with time.

MOMENTUM CONSERVATION in the absence of external forces follows along similar lines from invariance with respect to space displacement $(x_i \to x_i + \Delta x_i$, where i labels the Cartesian coordinates $x, y, z)$. For a single free particle, the momentum operator is

$$\underline{p}_i = \frac{\hbar}{i} \frac{\partial}{\partial x_i} = \frac{\hbar}{i} \nabla_i \qquad (5\text{-}107)$$

which clearly has no explicit time-dependence. The Hamiltonian operator in absence of external forces $(V = 0)$ takes the form

$$\underline{H} = \frac{1}{2m} \underline{p}^2 = -\frac{\hbar^2}{2m} \nabla^2 \qquad (5\text{-}108)$$

which commutes with $\underline{p}_i$, since

$$[\underline{H},\underline{p}_i] = -\frac{\hbar^3}{2im} (\nabla^2 \nabla_i - \nabla_i \nabla^2) = 0 \qquad (5\text{-}109)$$

Extending these considerations to *two* particles 1 and 2 whose interaction is described by a potential $V = V(|x_i^{(1)} - x_i^{(2)}|)$, which depends only on the spatial *separation* and hence is unchanged when both particles experience the same displacement, it is clear that

$$\left(\frac{\partial}{\partial x_i^{(1)}} + \frac{\partial}{\partial x_i^{(2)}}\right) H - H \left(\frac{\partial}{\partial x_i^{(1)}} + \frac{\partial}{\partial x_i^{(2)}}\right) = 0 \tag{5-110}$$

and that

$$\frac{\partial(p_i^{(1)} + p_i^{(2)})}{\partial t} = 0 \tag{5-111}$$

so that *total linear momentum is conserved* again.

ANGULAR MOMENTUM CONSERVATION under invariant rotations ($\phi \to \phi + \Delta\phi$) can also be demonstrated in a similar way on noting that the angular momentum operator

$$L = \frac{\hbar}{i} [\mathbf{r} \times \nabla] \tag{5-112}$$

can be split into components; the z component when expressed in terms of polar coordinates takes the form

$$L_z = \frac{\hbar}{i} \frac{\partial}{\partial\phi} \tag{5-113}$$

The vanishing of the commutator,

$$\left(\frac{\partial}{\partial\phi} H - H \frac{\partial}{\partial\phi}\right) = 0 \tag{5-114}$$

and of the partial time-derivative

$$\frac{\partial L_z}{\partial t} = 0 \tag{5-115}$$

in consequence of the absence of an *explicit t dependence* despite a rotational transformation $\phi \to \phi + \Delta\phi$ indicates that the laws of motion do not depend upon the orientation of the system in (isotropic, homogeneous) space. Thereby, angular momentum remains, together with energy and linear momentum, a rigorously conserved quantity in the absence of external influences upon the overall system.

Other quantities, however, are not universally conserved. Thus parity and charge conjugation are violated in weak interactions—indeed, the violation has been found to be maximal in this case—while isospin ceases to be a meaningful quantity outside the context of *strong*-interaction physics. Table 5-5 indicates whether or not such fundamental quantities are conserved in the various interaction modes, and shows that seemingly no violation occurs in the strong interactions. (More specifically, very general analyses of the characteristics of strong-interaction forces make provision for an admixture of "impure," nonconserving components and in many cases set vanishingly small upper limits upon their relative contribution.)

EXERCISES

5-1 What is the parity of a P electron? Is it the same as that of an α particle with total angular momentum $j = 4$?

5-2 Specify the spins and parities of nuclear energy levels to which a D-wave neutron transition can take place from the following states:

(a) $J^{\pi} = 0^+$ (e) $\frac{5}{2}^+$
(b) $\frac{1}{2}^-$ (f) $\frac{5}{2}^-$
(c) 1^+ (g) $\frac{11}{2}^-$
(d) 1^-

What l values are involved in

(h) a proton transition between states $0^+ \rightarrow \frac{5}{2}^+$,
(i) ,,   ,,   ,,   ,,   ,, $0^+ \rightarrow \frac{5}{2}^-$,
(j) ,,   ,,   ,,   ,,   ,, $1^- \rightarrow \frac{7}{2}^+$,
(k) an α-particle ,,   ,,   ,, $1^- \rightarrow 4^+$?

Can a $1^- \rightarrow 4^-$ α transition occur? If not, why not?

5-3 The ground-state spin of $^{41}_{20}$Ca is $J = \frac{7}{2}$. If the nucleus is regarded as a rigid sphere with a radius given by $R = 1.4\, A^{1/3}$ fm, what would be the speed of a point on the surface of this nucleus?

5-4 Determine the parity of *ortho-* and *para*-positronium. Demonstrate that parity conservation in electromagnetic interactions rules out two-quantum decay of *ortho*positronium.

5-5 The energy of a proton beam can be determined by passing it between the poles of a magnet whose field strength H_0 is sufficient to deflect the beam through 90 degrees, the radius ρ of the trajectory being fixed by a system of slits. The field H_0 is measured using the method of proton resonance, which operates as follows:

The probe, containing water as a source of quasi-free protons, is inserted into the magnetic field while subjected to an alternating field $H_1 \sin \omega t$ which acts perpendicularly to H_0. When ω tallies exactly with the Larmor precession frequency of protons, ω_L, the reversal of the proton spins can be observed.

Calculate the relationship between this Larmor resonance frequency $\omega = \omega_L$ and the proton energy E, using relativistic formulae [e.g., (1-10)].

5-6 Calculate the gyromagnetic ratio for a spherical nucleus of mass number A and atomic number Z which rotates about its diameter (assuming that the nuclear mass and charge are homogeneously distributed throughout the nucleus).

5-7   Compare the Earth's gyromagnetic ratio to that of the electron.
(The Earth has a magnetic dipole moment $\mu = 8.1 \times 10^{25}$ erg/Gauss.
For simplicity, assume a homogeneous distribution of mass.)

5-8   Calculate the electrostatic interaction energy between a homogeneously
charged nucleus which carries a net charge $\int \rho d\Omega = Ze$ and a point
charge e, assuming the separation r between the center of mass of the
nucleus and the point charge to be much larger than the nuclear
diameter. Express this energy as a multipole expansion (e.g., in terms
of dipole moment, quadrupole moment, etc.). How large is the quadrupole moment of a homogeneously charged ellipsoid having a major
semiaxis a and a minor semiaxis b?

5-9   Transitions between the Zeeman hyperfine levels of the hydrogen
nucleus can be induced by subjecting protons to an oscillating magnetic
field whose frequency corresponds exactly to the energy difference
between levels.

  Calculate the magnitude of the proton's magnetic moment if this
resonance occurs in a 5-kG magnetic field at a frequency of 21.28
megacycles/sec. Further, specify the ratio of the occupation numbers
of these Zeeman levels at $T = 300°$K, assuming Boltzmann statistics
to apply. Discuss what happens to the relative occupation of these levels
when subjected to the resonance field.

5-10   The ground state of the deuteron ($J^\pi = 1^+$) may theoretically comprise four possible configurations, for which the respective magnetic
moments are:

$^1P_1$: $\mu_d = \frac{1}{2}$
$^3S_1$: $\mu_d = \mu_p + \mu_n$
$^3P_1$: $\mu_d = \frac{1}{2}(\mu_p + \mu_n + \frac{1}{2})$
$^3D_1$: $\mu_d = \frac{1}{2}(-\mu_p - \mu_n + \frac{3}{2})$.

Experimentally one finds $\mu_d = 0.857,393$ in units of the nuclear
magneton. Show that this moment can be explained through only *one*
of the possible combinations and determine the mixing ratio.

  (Spectroscopic notation: The upper index indicates the multiplicity
($2S + 1$) of the state, viz. 1 = singlet and 3 = triplet; the lower index
(1) indicates the total angular momentum J, and the letters S, P, D
correspond to the relative net orbital momentum $L = 0, 1, 2$.)

Chapter 6

RADIOACTIVITY

Radioactive Decay

6.1. Mean Lifetime toward Radioactive Decay

An isolated system unstable toward particle or photon emission will on an average remain in its state of relatively elevated energy for one MEAN LIFETIME τ before proceeding to decay by one or more decay modes. The spontaneous or induced transmutation of a "parent" nucleus into its "daughter" product can take place only when the discrete TRANSITION ENERGY difference is positive. The characteristics of the transition then determine the decay probability and hence the mean lifetime. Thus ^{24}Na evinces a mean lifetime of $\tau = 21.6$ h before undergoing decay to an excited state of ^{24}Mg by β^- emission. The β^- decay of a neutron,

$$n \rightarrow p^+ + e^- + \bar{\nu}_e \tag{6-1}$$

takes place on an average in roughly 16.9 min. A muon has a mean lifetime $\tau_\mu = 2.2 \times 10^{-6}$ sec toward the weak-interaction decay process

$$\left. \begin{aligned} \mu^+ &\rightarrow e^+ + \nu_e + \bar{\nu}_\mu \\ \mu^- &\rightarrow e^- + \bar{\nu}_e + \nu_\mu \end{aligned} \right\} \tag{6-2}$$

The range of lifetimes for the different types of radioactive transitions depicted in Fig. 6-1 extend from about 10^{-8} sec to 10^{11} y. The processes may therefore be viewed as comparatively slow on the nuclear scale. The law which represents the time dependence of such decay processes is by no means confined solely to radioactive transitions. For example, the emission of light by free excited atoms or the kinematic rate of a chemical reaction essentially follows a similar law (unless perturbed by collective effects, as in the stimulated emission of radiation in the case of a laser, for instance).

The development of new experimental techniques has rendered a new range of lifetimes accessible to measurement. Whereas hitherto only comparatively slow decay processes could be studied directly, much effort has of

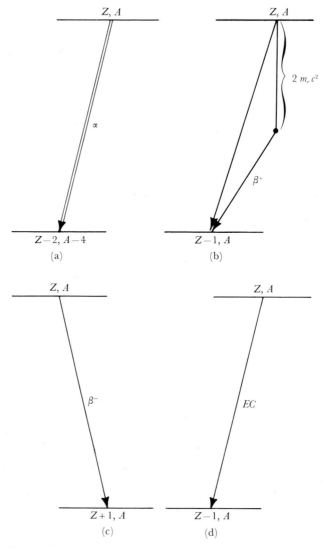

Fig. 6-1. Schematic representation of various types of radioactive decay transitions from a parent nucleus of atomic number Z and mass number A: (a) α decay to a daughter nucleus $(Z-2, A-4)$; (b) β^+ decay to a daughter nucleus $(Z-1, A)$; (c) β^- decay to a daughter nucleus $(Z+1, A)$; (d) electron capture by the parent nucleus leading to the formation of a product nucleus $(Z-1, A)$.

late been devoted to extending the investigations to embrace very short and very long lifetimes. Thus the use of fast coincidence techniques or methods based upon Doppler-shift attenuation has made it possible to measure lifetimes down to 10^{-12} sec or even lower, while at the other end of the scale low-level counting techniques have made it feasible to determine lifetimes as long as 10^{16} y. The various methods applicable to the measurement of life-times [De 60c, Da 63b] are indicated in Fig. 6-2. We shall concern ourselves principally with the characteristics of the comparatively slow α- and β-decay processes, which fall within the following range of mean lifetimes:

$$\tau_\alpha \approx 10^{-6} \text{ sec to } 10^{10} \text{ y} \tag{6-3}$$

$$\tau_\beta \approx 10^{-2} \text{ sec to } 10^{10} \text{ y} \qquad (\text{cf. } \tau_\beta(^{115}\text{In}) \simeq 7 \times 10^{14} \text{ y}) \tag{6-4}$$

6.1.1. LEVEL WIDTH AND DECAY PROBABILITY

Unless a given nuclear state has an infinite lifetime, it will, according to the Heisenberg uncertainty principle, be characterized by a corresponding nonzero ENERGY LEVEL WIDTH Γ, which betokens an energy indeterminacy,

$$\Delta E \equiv \Gamma = \frac{\hbar}{\tau} = \frac{1.0545 \times 10^{-27}}{\tau^{[\text{sec}]}} \text{ erg} = \frac{0.658 \times 10^{-15}}{\tau^{[\text{sec}]}} \text{ eV} \tag{6-5}$$

Hence for lifetimes ranging from 10^{-7} sec to 10^{11} y, the level widths will lie in the range 10^{-8} to 10^{-34} eV, and accordingly represent rather narrow levels (We defer the consideration of multiple decay modes of any given level until somewhat later, and depict the level width of a particular nuclear state as in Fig. 6-3 by a Lorentzian curve of width Γ at half-height.) It is sometimes pos-sible to deduce the mean lifetime of very short-lived levels indirectly from a measurement of their level width. Thus lifetimes ranging from about 10^{-13} to 10^{-15} sec ($\Gamma \simeq 0.007$ to 0.7 eV) in the case of γ decay of excited levels have been established [Boo 64] from a determination of level widths via a measure-ment of the γ-ray self-absorption cross section, which is proportional to Γ. However, this method would not readily lend itself to a determination of α- or β-decay lifetimes, for which activity measurements prove to be more suitable. The activity of a radioactive sample constitutes a measure of its mean rate of decay and hence is directly related to the nuclear lifetime.

Because a given nucleus remains unchanged up to the moment of decay, i.e., of α, β, or γ emission, the probability at any particular instant for it to decay within an *immediately following interval of time* is a constant. *Microscopically,* no sharp prediction of when an individual unstable nucleus will succumb to

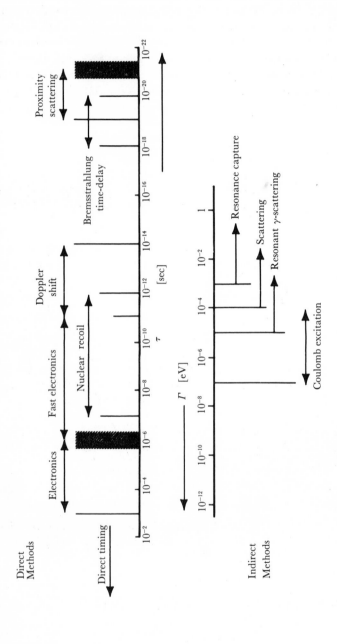

Fig. 6-2. Schematic indication of the various methods employed in the determination of nuclear lifetimes τ by direct methods, and by indirect methods via a determination of level widths Γ.

decay can be made. Only a *macroscopic* description of the likelihood of decay among a large ensemble of identical radioactive nuclei is valid. In other words, the decay probability has a meaning only as a statistical average over a large collection of similar unstable nuclei, and although it can be reduced to a *mean decay probability per unit time per nucleus* it is merely a measure of the decay rate of a many-body sample and *not* of any particular nucleus within that sample. It is expressed by the DECAY CONSTANT λ, which is simply the reciprocal of the mean lifetime,

$$\lambda \equiv \frac{1}{\tau} \qquad (6\text{-}6)$$

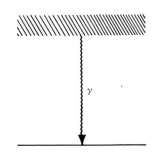

Fig. 6-3. The width Γ of an exaggeratedly broad level for ground-state γ decay. The value Γ represents the full width at half-maximum (FWHM) of the Lorentzian peak.

Consequently, in an ensemble composed of N identical radioactive nuclei or unstable particles the total number of decays per unit time will on an average be λN. For radioactive sources, this quantity is termed the ACTIVITY of the particular sample.

Unless replenished artificially, the nuclei in a sample decay at a rate given by $-dN/dt$ (the minus sign denotes *decay* in which N diminishes as t increases), and this is numerically equal to the activity:

$$-\frac{dN}{dt} = \lambda N \qquad (6\text{-}7)$$

The integration of this differential equation between the limits $N = N_0$ at $t = 0$ and $N = N$ at a later time t gives, upon separation of the variables and the assumption that λ is constant with time,

$$N = N_0\, e^{-\lambda t} \equiv N_0\, e^{-t/\tau} \qquad (6\text{-}8)$$

or, expressed in terms of the width of the decaying level,

$$N = N_0 \exp\left(-\frac{\Gamma t}{\hbar}\right) \qquad (6\text{-}9)$$

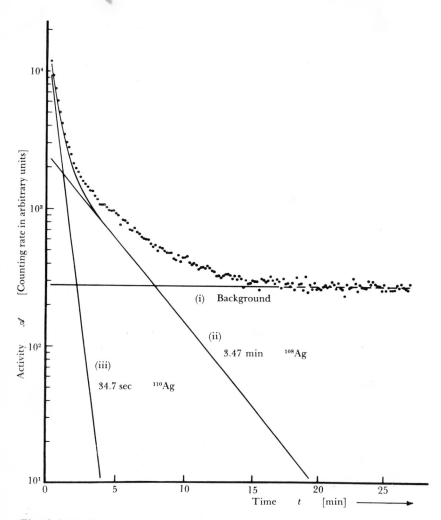

Fig. 6-4. Radioactive decay curve of activity in function of time. The diminution in activity measured after a silver sample (of natural isotopic abundance 51.8 percent ^{107}Ag and 41.2 percent ^{189}Ag) had been exposed to a weak source of slow neutrons for 2 min which produced the β^--decaying isotopes ^{108}Ag ($\tau = 3.47$ min) and ^{110}Ag ($\tau = 34.7$ sec) through (n, γ) reactions. The net decay curve, displaying statistical counting-rate fluctuations, is shown decomposed into its three components: (i) a background contribution, constant in time; (ii) an activity due to the more long-lived isotope ^{108}Ag ($\tau = 3.47$ min) formed from the radiative capture process ^{107}Ag(n, γ)^{108}Ag; (iii) a steeper activity due to the short-lived isotope ^{110}Ag ($\tau = 34.7$ sec) produced by slow-neutron capture in ^{109}Ag.

This EXPONENTIAL DECAY LAW (cf. Fig. 6-4) indicates that in principle a determination of the decay constant λ might be made from a plot of $\log N$ vs t. In practice, however, it is much easier to measure the relative ACTIVITY $\mathscr{A}_t$ of a sample as a function of time t, expressed by the relation

$$\mathscr{A}_t \equiv \left(\frac{dN}{dt}\right)_t = \left(\frac{dN}{dt}\right)_{t=0} e^{-\lambda t} \equiv \mathscr{A}_0 \, e^{-\lambda t} \tag{6-10}$$

A plot of $\log \mathscr{A}$ vs t yields a straight line (Fig. 6-5) whose slope furnishes the decay constant λ and hence the lifetime τ for the given pure sample. For

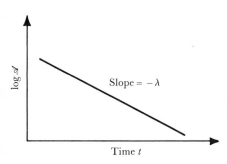

Fig. 6-5. Plot of activity $\mathscr{A}$ on a *logarithmic* scale *vs* time t yields a linear graph when only a single (pure) activity is present. From the slope, the mean lifetime $\tau \equiv 1/\lambda$ can be determined.

different radioactive nuclei, λ ranges from roughly $10^7 \ \text{sec}^{-1}$ to $10^{-9} \ \text{sec}^{-1}$. For stable nuclei λ is, of course, zero. In connection with this method, it should be pointed out that it breaks down for lifetimes *large* compared to the time of observation. In such cases λ has to be deduced from an absolute measurement of the decay rate dN/dt of a sample containing a known number of radioactive nuclei, using radiochemical techniques, for instance.

6.1.2. HALF-LIFE AND SPECIFIC ACTIVITY

The MEAN LIFETIME τ can be interpreted as the time needed on an average for the activity of a sample to decrease to $(1/e)$th of its initial value. It is more convenient to define a related quantity $T_{1/2}$ which corresponds to the mean time required for the activity to diminish to *one half* of its value. This is called the HALF-LIFE, $T_{1/2}$, of a radioactive nuclide: after a time $T_{1/2}$ the number of active nuclei will on an average have been halved by the particular mode of decay. Substituting $N = \frac{1}{2}N_0$ at a time $t = T_{1/2}$ in the radioactive decay relation (6-8) shows that

$$T_{1/2} = \frac{\ln 2}{\lambda} = \frac{0.693}{\lambda} \equiv 0.693 \ \tau \tag{6-11}$$

Originally, the unit of activity was defined with reference to the amount of radium emanation in equilibrium with 1 g of pure radium, and termed the

Curie (C). The modern definition breaks away from the reference to a radium standard while retaining approximately the same magnitude by respecifying the *Curie* (Ci) to be *that activity of a pure radioactive sample which has a mean decay rate of* 3.70×10^{10} *decays per sec*. It might be mentioned in passing that the activity of a source is not necessarily related directly to the number of counts registered in a detection system of known efficiency. Some sources can in a single decay transition emit more than one particle or photon, and for this reason an exact knowledge of the *decay scheme* of the nuclide under consideration is necessary before one embarks upon an absolute activity determination.

Sources used in nuclear experimentation are usually of fairly low activity. For instance, sources used in the calibration of scintillation counters or solid-state spectrometers normally have activities of no more than a few micro-curies (μCi) or even appreciably less. On the other hand, γ sources having a strength of several kCi are employed in radiotherapy, while very strong sources, whose activities lie in the MCi range, are used in irradiation units.

For some applications, as in β spectrometry, high SPECIFIC ACTIVITIES are desirable, e.g., high values of the *activity per gram of source material*, given by the relation

$$\mathscr{A}' \equiv \frac{\mathscr{A}}{M} = \frac{\lambda N}{M} = \lambda \frac{N_{\mathrm{A}}}{A} \tag{6-12}$$

where M is the mass in grams, $N_{\mathrm{A}} = 6.023 \times 10^{23}$ is the Avogadro constant, and A is the atomic weight. The requirement of high λ and low A is met by short-lived light nuclides: Thus, the specific activity of ^{17}F, whose half-life toward β^+ emission is $T_{1/2} = 66$ sec, is

$$\mathscr{A}' = \frac{0.693 N_{\mathrm{A}}}{A T_{1/2}} = \frac{4.174 \times 10^{23}}{17 \times 66} \simeq 3.72 \times 10^{20} \text{ decays sec}^{-1} \text{ g}^{-1} \tag{6-13}$$

$$\simeq 10^{10} \text{ Ci g}^{-1} \tag{6-14}$$

It is clear that the mass of a carrier-free source (that is, a source containing only radioactive atoms of one species) having an activity of 1 Ci depends on the atomic weight and the half-life:

$$M = 8.864 \times 10^{-14} A \, T_{1/2} \tag{6-15}$$

Thus a 1-Ci source of ^{238}U ($T_{1/2} = 4.51 \times 10^9$ y) has a mass of some 3000 kg, whereas a 1-Ci source of ^{13}N ($T_{1/2} = 9.96$ m) has a mass of only 6.89×10^{-10} g.

6.2. Branching Ratios (Partial Widths)

In many instances, the decay of a nucleus or particle can proceed in more than just one mode. For example, ^{64}Cu can decay either by β^- emission to ^{64}Zn or by β^+ emission to the ground state of ^{64}Ni, a situation examined in

Section 2.4.4. The same applies to particle decay modes, e.g., the usual decay process of Ξ^- hyperons leads to the products $(\Lambda^0 + \pi^-)$, but very occasionally the $(\Lambda^0 e^- \bar{\nu}_e)$ and still more seldom the $(n\pi^-)$ modes are found to occur with the following relative likelihood:

$$\Xi^- \begin{array}{l} \xrightarrow{\;\approx 100\%\;} \Lambda^0 + \pi^- \\ \xrightarrow{\;0.3\%\;} \Lambda^0 + e^- + \bar{\nu}_e \\ \searrow^{\;0.5\%} \\ \qquad\quad n + \pi^- \end{array} \qquad (6\text{-}16)$$

while Λ^0 decay can follow either of two principal branches:

$$\Lambda^0 \begin{array}{l} \nearrow^{\;68\%} p^+ + \pi^- \\ \searrow_{\;32\%} \\ \qquad n + \pi^0 \end{array} \qquad (6\text{-}17)$$

The BRANCHING RATIO in nuclear and particle physics is usually expressed as a percentage of the total decays. The term can also be applied to the description of γ-decay modes in the de-excitation of an excited nuclear state. The γ decay may, subject to selection rules, proceed as a cascade made up of stepwise transitions between adjacent levels or may take the form of a cross-over transition to a nonadjacent level of lower energy. This is illustrated in Fig. 6-6 for the γ decay of excited levels in ^{56}Fe. (Selection rules exclude γ transitions other than those shown.)

Denoting by $\lambda_1, \lambda_2, \lambda_3, \ldots$ the PARTIAL DECAY CONSTANTS of each specific decay mode, we see that the net overall decay constant λ for the given nuclear state is just the sum of the partial constants,

$$\lambda = \lambda_{\text{total}} = \lambda_1 + \lambda_2 + \lambda_3 + \cdots \qquad (6\text{-}18)$$

i.e.,

$$\frac{1}{\tau} = \frac{1}{\tau_{\text{total}}} = \frac{1}{\tau_1} + \frac{1}{\tau_2} + \frac{1}{\tau_3} + \cdots \qquad (6\text{-}19)$$

where the τ_i denote mean PARTIAL LIFETIMES. It is not possible to measure the λ_i or τ_i directly, but their values can be ascertained indirectly from a knowledge of the net constants λ or τ and the respective branching ratios. The fastest partial decay modes thereby provide the main contribution to the decay of the parent state and represent the strongest constituent in the branching.

By analogy, it is possible to express a level width Γ in terms of PARTIAL WIDTHS Γ_i such that the sum of these (hypothetical) quantities, one for each decay channel, builds the net width,

$$\Gamma = \Gamma_{\text{total}} = \Gamma_1 + \Gamma_2 + \Gamma_3 + \cdots \qquad (6\text{-}20)$$

and

$$N = N_0 \exp\left[-\frac{(\Gamma_1 + \Gamma_2 + \Gamma_3 + \cdots)t}{\hbar}\right] \qquad (6\text{-}21)$$

This decomposition proves particularly useful in the consideration of nuclear reactions, when with each level of a nucleus one can associate partial widths

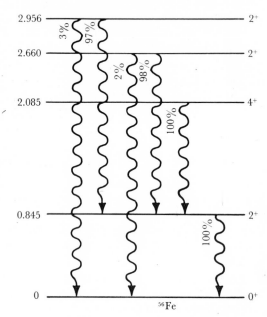

Fig. 6-6. Branching ratios of various γ-decay transitions between the lower levels of ^{56}Fe. Note in particular the absence of a direct (crossover) ground-state transition from the 4^+ level due to the large spin difference ($\Delta J = 4$): the γ decay proceeds as a two-step cascade by way of the first 2^+ level.

corresponding to *competing processes* such as, e.g., α emission, γ emission, p emission, n emission, etc., and conceive the total width to be the sum of these, viz.

$$\Gamma = \Gamma_\alpha + \Gamma_\gamma + \Gamma_p + \Gamma_n + \cdots \qquad (6\text{-}22)$$

6.3. Radioactive Decay: Daughter Activity

If a β-active nucleus has Z differing by more than 1 unit from Z_{stable}, the atomic number of the most stable isobar, the considerations of Section 2.4.4 indicate that successive β decays will take place until stability is attained

($Z \rightarrow Z_{\text{stable}}$). Moreover, the α instability of heavy nuclei causes these to decay by a sequence of successive mixed α and β transitions, giving rise to the decay chains depicted in Fig. 6-7. The decay sequence terminates with the formation of a stable nuclide.

The characteristics of such decay stages can be derived from a consideration of the simple example of a two-step process in which a radioactive parent nuclide decays to an unstable daughter product and thence to a grand-daughter, which is taken to be stable:

$$P \xrightarrow{\lambda_P} D \xrightarrow{\lambda_D} \underline{\underline{G}} \tag{6-23}$$

For convenience, we represent the number of parent, daughter, and grand-daughter nuclei present at some arbitrary instant of time t by P, D, and G, and assume that $P = P_0$ and $D = D_0 = G = G_0 = 0$ at the time origin $t = 0$. If the decay constants of the parent nuclide and the daughter nuclide are λ_P and λ_D respectively (they are not equal, of course), it follows that after a time t the (diminishing) parent activity will be $\lambda_P P$ and the (increasing) daughter activity will be $\lambda_D D$. Hence the rate of change of D is

$$\frac{dD}{dt} = \lambda_P P - \lambda_D D \tag{6-24}$$

i.e.,

$$\frac{dD}{dt} = P_0 \lambda_P \, e^{-\gamma_P t} - \lambda_D D \tag{6-25}$$

the first term on the right giving the growth rate of D due to the decay of P, and the second term the dissipation rate due to decay of D.

The solution of this differential equation with the boundary conditions $P = P_0$ and $D = 0$ at time $t = 0$ is

$$D = \frac{\lambda_P}{\lambda_D - \lambda_P} \, P_0 (e^{-\lambda_P t} - e^{-\lambda_D t}) \tag{6-26}$$

Thus the activity of the daughter nuclide is, by definition, $\lambda_D D$ (and *not* dD/dt, since this includes a parent activity term), i.e.,

$$\mathscr{A}_D \equiv \lambda_D D = \frac{\lambda_D \lambda_P}{\lambda_D - \lambda_P} \, P_0 (e^{-\lambda_P t} - e^{-\lambda_D t}) \tag{6-27}$$

which, since the parent activity is

$$\mathscr{A}_P = \lambda_P P = \lambda_P P_0 \, e^{-\lambda_P t} \tag{6-28}$$

may be written

$$\mathscr{A}_D = \mathscr{A}_P \frac{\lambda_D}{\lambda_D - \lambda_P} \, (1 - e^{-(\lambda_D - \lambda_P)t}) \tag{6-29}$$

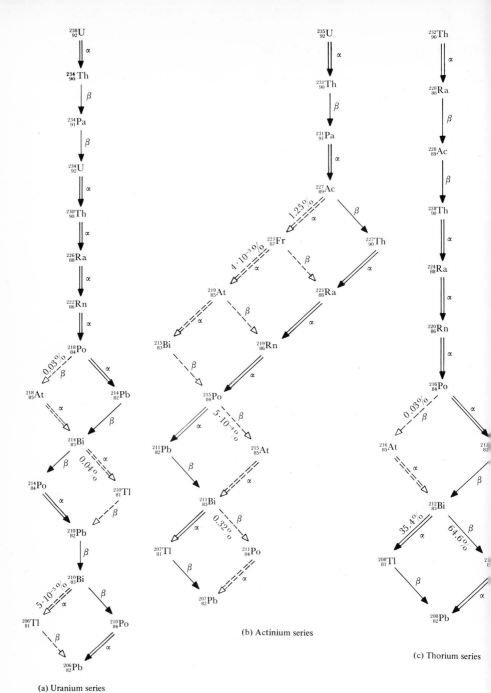

Fig. 6-7. Decay chains for the (a) uranium, (b) actinium, and (c) thorium series, terminating in the three stable isotopes of lead. A fourth chain, the neptunium series, involving artificially produced nuclides (e.g., $^{237}_{93}$Np) and terminating in stable $^{209}_{83}$Bi after a sequence of α and β transitions is not shown in the figure, which features the decay processes of *natural* unstable nuclides.

To illustrate the various possibilities, which have been discussed exhaustively in other textbooks (e.g., [Ev 55, pp. 477 ff.]), some examples of special parent-daughter activities are presented in Fig. 6-8 (a–d).

A straightforward extension of this treatment can be undertaken for decay processes in which great-granddaughters, etc., are formed. For instance, one finds for the decay sequence

$$P \xrightarrow{\lambda_P} D \xrightarrow{\lambda_D} G \xrightarrow{\lambda_G} \cdots \qquad (6\text{-}30)$$

that the number of nuclei G after a time t is

$$G = \lambda_P \lambda_D P_0 \left[\frac{e^{-\lambda_P t}}{(\lambda_D - \lambda_P)(\lambda_G - \lambda_P)} + \frac{e^{-\lambda_D t}}{(\lambda_G - \lambda_D)(\lambda_P - \lambda_D)} \right.$$

$$\left. + \frac{e^{-\lambda_G t}}{(\lambda_P - \lambda_G)(\lambda_D - \lambda_G)} \right] \qquad (6\text{-}31)$$

and the granddaughter activity is given by

$$\mathscr{A}_G = \lambda_G G \qquad (6\text{-}32)$$

6.3.1. Daughter Activity in Special Cases

Confining ourselves to a consideration of daughter formation and decay, we note that the MAXIMUM ACTIVITY $\mathscr{A}_{D\max}$ is evinced at the instant of time $t_{\mathscr{A}_{D\max}}$ when $dD/dt = 0$, for then D itself is a maximum, as in Fig. 6-9. From the time differential of (6-26) we derive the relation

$$\lambda_P e^{-\lambda_P t}\big|_{t=t_{\mathscr{A}_{D\max}}} = \lambda_D e^{-\lambda_D t}\big|_{t=t_{\mathscr{A}_{D\max}}} \qquad (6\text{-}33)$$

whence

$$t_{\mathscr{A}_{D\max}} = \frac{1}{\lambda_D - \lambda_P} \ln\left(\frac{\lambda_D}{\lambda_P}\right) \equiv \tau_D\left[\frac{\tau_P}{\tau_P - \tau_D} \ln\left(\frac{\tau_P}{\tau_D}\right)\right] \qquad (6\text{-}34)$$

This rather cumbersome expression can be simplified to an approximate relationship when $\tau_P \approx \tau_D$, for one can then set

$$\tau_P = \tau_D(1 + \epsilon) \qquad \text{with} \quad \epsilon \ll 1 \qquad (6\text{-}35)$$

If in (6-34) we insert the logarithmic expansion

$$\ln\left(\frac{\tau_P}{\tau_D}\right) = \ln(1 + \epsilon) = \epsilon - \frac{1}{2}\epsilon^2 + \frac{1}{3}\epsilon^3 - \cdots \qquad (6\text{-}36)$$

we see that

$$t_{\mathscr{A}_{D\max}} = \tau_D(1 + \epsilon)(1 - \tfrac{1}{2}\epsilon + \cdots) \qquad (6\text{-}37)$$

$$\cong \tau_D(1 + \tfrac{1}{2}\epsilon) \qquad (6\text{-}38)$$

i.e.,

$$t_{\mathscr{A}_{D\max}} \cong (\tau_P \tau_D)^{1/2} \qquad (6\text{-}39)$$

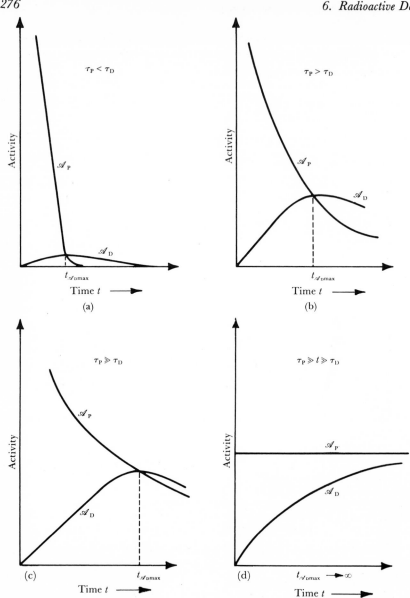

Fig. 6-8. Activity curves for various parent-daughter relationships: (a) short-lived parent $(\tau_P < \tau_D)$; (b) long-lived parent $(\tau_P > \tau_D)$; (c) very long-lived parent $(\tau_P \gg \tau_D)$; (d) almost stable, or constantly replenished, parent $(\tau_P \gg t \gg \tau_D)$.

The instant of ideal equilibrium, when the daughter activity reaches a maximum and is numerically equal to the parent activity is in each case denoted by $t_{\mathscr{A}_{D\,max}}$, which shifts to progressively longer times t as one deals with progressively longer-lived parent nuclides.

Since at the time of maximum daughter activity the time derivative dD/dt vanishes, it follows from (6-24) that at the instant of IDEAL EQUILIBRIUM parent and daughter activities are equal, viz.

$$\mathscr{A}_P = \mathscr{A}_D = \lambda_P P_0 \left(\frac{\lambda_P}{\lambda_D}\right)^{\lambda_P/(\lambda_D - \lambda_P)} \qquad \text{when } t = t_{\mathscr{A}_D \max} \qquad (6\text{-}40)$$

While at earlier instants $\mathscr{A}_P$ transcends $\mathscr{A}_D$, the situation reverses thereafter

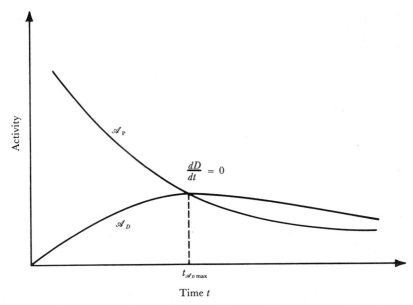

Fig. 6-9. Illustration of the conditions at ideal equilibrium, as would apply to the case of a fairly long-lived parent nuclide.

and the ratio $\mathscr{A}_P/\mathscr{A}_D$ diminishes from unity, tending to zero as $t \to \infty$. Clearly, a momentary state of ideal equilibrium always occurs except for the limiting case of extremely long-lived parent nuclides ($\lambda_P \to 0$) or for a situation in which the parent is constantly being replenished in some way.

In the case of a very *short-lived parent*, the system soon passes beyond the stage of ideal equilibrium, and thereafter behaves essentially as though the initial stock P_0 of the parent species had been rapidly converted to an equal amount P_0 of the daughter nuclide, since when t is sufficiently large and $\lambda_P \gg \lambda_D$, Eq. (6-26) indicates that

$$\mathscr{A}_D = \lambda_D D \simeq P_0 \lambda_D e^{-\lambda_D t} \qquad (6\text{-}41)$$

As illustrated in Fig. 6-10, the α decay of ^{211}Bi with a mean lifetime $\tau_P = 3.10$ m

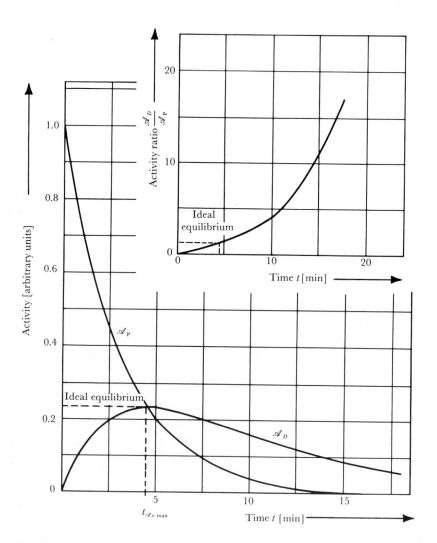

Fig. 6-10. Activity curves for the decay of a short-lived parent nuclide, as exemplified by the sequence $^{211}\text{Bi} \xrightarrow{\alpha} {}^{207}\text{Tl} \xrightarrow{\beta^-} {}^{207}\text{Pb}$ which forms the end of the actinium chain. The ordinate has been normalized to *unit parent activity* at the time $t = 0$.

to $^{207}\mathrm{Tl}$, whose β^- decay ($\tau_D = 6.90$ m) to stable $^{207}\mathrm{Pb}$ terminates the actinium series, reaches ideal equilibrium after a time $t_{\mathscr{A}_D\max} = 4.5$ m and the ratio $\mathscr{A}_D/\mathscr{A}_P$ *increases continuously with time.*

However, for the contrasting case of a *long-lived parent,* the activity ratio $\mathscr{A}_D/\mathscr{A}_P$ displays an altogether different trend, for after a momentary initial increase, it levels off to a *constant value* (> 1) in due course, since from (6-27) one has

$$\frac{\mathscr{A}_D}{\mathscr{A}_P} = \frac{\lambda_D D}{\lambda_P P} = \frac{\tau_P}{\tau_P - \tau_D}\left(1 - e^{-(\lambda_D - \lambda_P)t}\right) \to \frac{\tau_P}{\tau_P - \tau_D} = \text{const} \qquad (6\text{-}42)$$

The neglect of the exponential term is valid only when the time of observation exceeds τ_P and τ_D, and the system then attains a state of TRANSIENT EQUILIBRIUM. An example of this situation is provided by the end of the thorium series,

$$^{212}\mathrm{Bi} \xrightarrow[\tau_P = 87.4\,\mathrm{m}]{\alpha} {}^{208}\mathrm{Tl} \xrightarrow[\tau_D = 4.47\,\mathrm{m}]{\beta^-} {}^{208}\mathrm{Pb} \qquad (6\text{-}43)$$

as illustrated by the activity curves in Fig. 6-11. Although *ideal* equilibrium occurs at time $t_{\mathscr{A}_D\max} = 14$ m, *transient* equilibrium, characterized by the onset of a *constant* activity ratio

$$\frac{\mathscr{A}_D}{\mathscr{A}_P} = \frac{\tau_P}{\tau_P - \tau_D} = \frac{87.4}{87.4 - 4.5} = 1.054 > 1 \qquad (6\text{-}44)$$

is attained only after some 25 m.

Finally, a rather special situation obtains when the parent nuclide is *extremely long-lived or constantly replenished.* When $\lambda_P \ll \lambda_D$ one can reduce (6-26) in first approximation to

$$D \cong (P_0\, e^{-\lambda_P t})\frac{\lambda_P}{\lambda_D} \qquad (6\text{-}45)$$

which yields the interesting (approximate) activity relation

$$\mathscr{A}_D = \lambda_D D \cong \lambda_P P_0 e^{-\lambda_P t} = \lambda_P P = \mathscr{A}_P \qquad (6\text{-}46)$$

It is as though the daughter's lifetime toward decay roughly approximates that of the parent. From the more accurate expression for the activity ratio as furnished by (6-27),

$$\frac{\mathscr{A}_D}{\mathscr{A}_P} = \frac{\lambda_D}{\lambda_D - \lambda_P}\left(1 - e^{-\lambda_D t}\right) \qquad (6\text{-}47)$$

it is evident that the activities in fact become equal only at large times t (see Fig. 6-8c) and may, indeed, for sufficiently small λ_P tend to this limit only asymptotically. A permanent, rather than momentary, state of ideal equilibrium is then attained, to which the name SECULAR EQUILIBRIUM has been

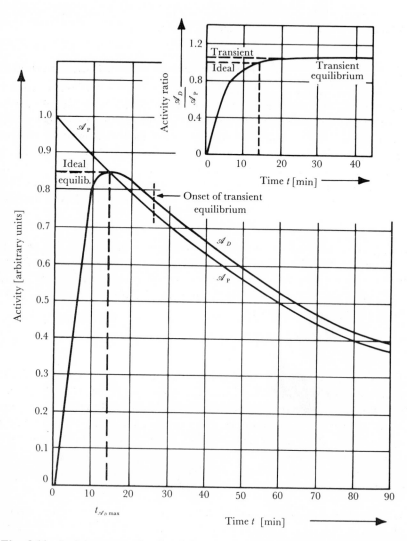

Fig. 6-11. Activity curves for the decay of a long-lived parent nuclide, illustrated by the sequence $^{212}\text{Bi} \xrightarrow{\alpha} {}^{208}\text{Tl} \xrightarrow{\beta^-} {}^{208}\text{Pb}$ which occurs at the end of the thorium series. Some time after the instant of ideal equilibrium, a state of transient equilibrium sets in characterized by a constant ratio of parent-daughter activities.

given. It occurs only when the parent's lifetime greatly exceeds the available duration of observation,

$$\tau_P \gg t \gg \tau_D \qquad (6\text{-}48)$$

so that essentially the parent activity remains constant throughout—a situation realized for parents which are so long-lived as to be well-nigh stable, or which are artificially replenished at a constant rate.

6.3.2. PRODUCTION OF RADIOACTIVE SOURCES (INDUCED RADIOACTIVITY)

The artificial production of an unstable radioelement which serves as a source of induced radioactivity can be accomplished by exposing a stable or comparatively long-lived target nuclide to a particle or γ-ray beam. The latter may stem from natural sources, as in the early transmutation experiments [Cu 34, Fe 34a] or may be obtained from accelerators or reactors. Under the assumption that the irradiating beam intensity does not vary with time—a condition not always realized in practice—the number of radioactive nuclei produced artificially in a short time interval dt is (cf. Section 3.5.1)

$$dR = \sigma T B \, dt \qquad (6\text{-}49)$$

when B particles per second bombard a target composed of T nuclei per square centimeter. The total reaction cross section represents an effective value integrated over the entire range of the incident particles. The overall rate of change of R is given by combining the above production rate (σTB) with the decay rate $(\lambda_R R)$, viz.

$$\left(\frac{dR}{dt}\right)_{net} = \left(\frac{dR}{dt}\right)_{prod} - \lambda_R R = \sigma T B - \lambda_R R \qquad (6\text{-}50)$$

The starting rate of build-up of radionuclei, at time $t \to 0$ when hardly any have already been formed (and hence $\lambda_R R \to 0$), is just equal to the production rate σTB, but thereafter the build-up rate diminishes until in the limit it drops to zero because the decay rate exactly compensates the production rate, so that no further increase or diminution in R is possible beyond the secular equilibrium value

$$R = \frac{1}{\lambda_R} \left(\frac{dR}{dt}\right)_{prod} \qquad (6\text{-}51)$$

We can treat this as akin to the "decay" of a very large number of parent nuclei ($T \triangleq P \approx P_0 \to \infty$) which are extremely long-lived ($\lambda_P \to 0$), so that the ensuing parent activity $\mathscr{A}_P \equiv \lambda_P P$ is finite and essentially constant. The formation of the radionuclide daughter ($R \triangleq D$) can again be derived from (6-26) with the condition $\lambda_P \ll \lambda_D$, but in place of the approximation (6-45), which ceases to be valid when $\lambda_P \approx 0$, we can instead reduce (6-26) in the limit to

$$D \cong \frac{\lambda_P}{\lambda_D} P_0 (1 - e^{-\lambda_D t}) \tag{6-52}$$

and thence deduce the growth of the source activity to be

$$\mathscr{A}_D = \lambda_D D = \mathscr{A}_P (1 - e^{-\lambda_D t}) \tag{6-53}$$

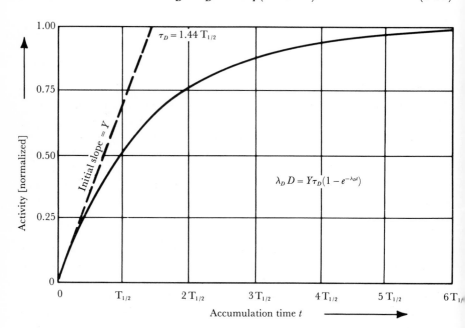

Fig. 6-12. Build-up of daughter activity in a target subjected to constant irradiation which produces a radioactive daughter nuclide whose activity increases at a constant rate expressed by the yield Y. The latter is equal to the initial slope of the growth curve at time $t = 0$. The extrapolation of this tangent intersects the asymptotic-activity line (which in our normalization lies at $\mathscr{A}_D = 1.00$) at a time $t = \tau_D = 1.44\, T_{1/2}$, where $T_{1/2}$ is the half-life of the daughter species (from [Ev 55]). (Used by permission of McGraw-Hill Book Co.)

the total net activity being

$$\mathscr{A}_{\text{total}} = \mathscr{A}_P + \mathscr{A}_D \approx \mathscr{A}_P (2 - e^{-\lambda_D t}) \tag{6-54}$$

The build-up of daughter activity is depicted in Fig. 6-12. The rate of formation of new radionuclide *activity* is termed the YIELD (which therefore has the dimensions of *activity per unit time*, and which depends upon the irradiation conditions). In the present special case of a first daughter product we are dealing with a constant yield equal to the initial slope of the daughter activity growth curve,

$$Y = \frac{d\mathscr{A}_D}{dt}\bigg|_{t=0} \tag{6-55}$$

We can build the time derivative of (6-27) to get the following simple expression for the yield:

$$Y = \lambda_P \lambda_D P_0 = \lambda_D \mathscr{A}_P \tag{6-56}$$

whence

$$\mathscr{A}_P = Y\tau_D \tag{6-57}$$

and

$$\mathscr{A}_D = Y\tau_D(1 - e^{-\lambda_D t}) \tag{6-58}$$

which shows the daughter activity to attain a maximal value equal to $Y\tau_D = \mathscr{A}_P$. In one daughter half-period exactly one half of this ultimate maximal activity can be accumulated. Figure 6-12 shows that after about two half-periods it is no longer worthwhile to endeavour to accumulate activity further, and that after about six half-periods there is virtually no further accretion, since when $t = 6\,T_{1/2} = 4.16\tau$ the exponential factor almost vanishes $(e^{-\lambda t} = 0.016)$. To attain 90 percent activity, the irradiation time must be $t = 2.3\tau = 3.3\,T_{1/2}$.

6.3.3. MIXTURE OF ACTIVITIES

The determination of the decay constant or the half-life of a pure source composed of but a single radioactive species presents no great problems in general, and may be effected along the lines indicated in Section 6.1.1. However, when two or more activities due to dissimilar radioelements are simultaneously present in an "impure" sample, a complication arises in that the measured activity data yield a decay curve which comprises the sum of exponential functions. Moreover, the detection apparatus inherently introduces a background counting rate, made up of a combination of the radioactivity in the counter itself, the radioactivity of the environment, and the ever-present cosmic radiation. Normally, the background radiation is fairly constant and can accordingly either be subtracted out or treated as an activity of virtually infinite half-life.

In the idealized situation when the net activity $\mathscr{A}$ due to a mixture of two radioactive components can be entirely separated from the background contribution, the plot of $\log \mathscr{A}$ vs t no longer remains linear over its entire range. Thus Fig. 6-13 shows its appearance when one constituent is considerably shorter-lived than the other. The nonlinear portion of the plot extends up to a time after which the short-lived component has to all intents vanished, and thereafter the straight-line portion has a slope equal to the decay constant of the longer-lived nuclide. By extrapolating the ultimate linear portion back to earlier times, the admixture due to the longer-lived component can be determined. Then the net activity left after subtraction of the latter contribution represents that due to the shorter-lived species only, whose decay constant can hence be determined also.

This procedure can also in principle be adopted when more than two radionuclides are present. On extrapolating back from the longest-lived

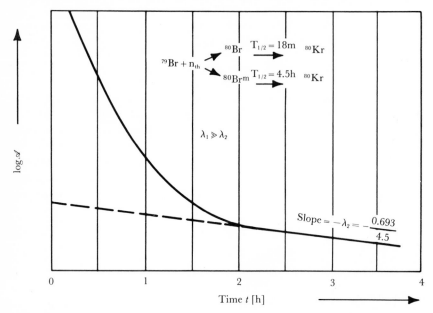

Fig. 6-13. Semilogarithmic plot of activity *vs* time showing the deviation from linearity when more than one activity is simultaneously present, as illustrated by the β^- decay of the ground state and an isomeric state in ^{80}Br whose half-lives are 18 min and 4.5 h respectively. After some 2 hours, the activity of the former shorter-lived component is no longer discernible, and the subsequent decay is essentially that of the metastable isomeric state, which yields a linear decay graph whose slope corresponds to the longer-lived activity (from [Sm 65]). (Reprinted with permission of Pergamon Press, Ltd.)

activity, the penultimate activity can be established and thence by successive re-extrapolation each component resolved in turn. However, in practice such a decomposition of a decay curve into linear constituents presents some difficulty and ambiguity, especially when two or three simultaneously present activities stem from radionuclides having rather similar half-lives. The attendant inaccuracies are considerable. The use of computer and least-squares-fit methods may effect some improvement in accuracy, but much depends upon the way the half-lives in a particular sample are related to one another. An example of such a decomposition is presented in Fig. 6-14.

Whenever possible, the method of counting should be suitably chosen so as to help in separating the individual contributions to the overall activity. This can sometimes be achieved through energy discrimination or chemical separation. A judicious choice of the irradiation time may also serve to enhance one of the activities and thereby render its identification more straightforward.

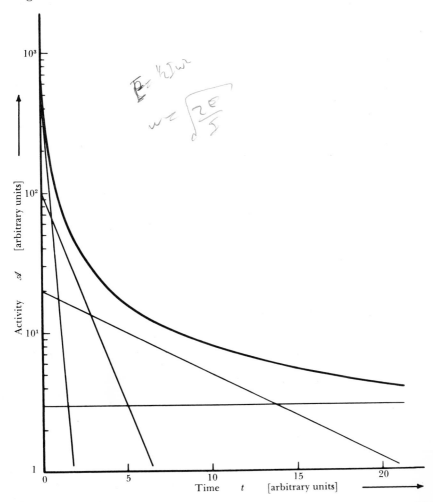

Fig. 6-14. Illustrative decay curve for a hypothetical radioactive source comprising *three* activities $\mathscr{A}_1$, $\mathscr{A}_2$, $\mathscr{A}_3$ whose initial strengths are in the ratio $500:100:20$ and whose relative half-lives are $2:10:50$. The semilogarithmic activity plot is seen to be distinctly nonlinear, and its precise decomposition into separate constituents can well be problematic or inaccurate, even when as few as three activities are involved.

6.4. Decay Schemes of
Widely Used Radioactive Sources

In Figs. 6-15 to 6-21 we have assembled somewhat simplified decay schemes of some widely used radioactive sources, listing on the left the principal type of radiation emitted by the radionuclide and on the right the energies and branching ratios (i.e., relative intensities) of the competing radiative transitions. Since β decay yields a continuous but finite β-energy spectrum, the

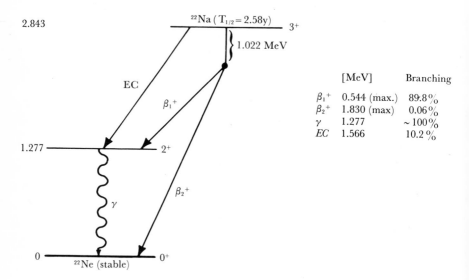

Fig. 6-15. Simplified decay scheme of the β^+ source ^{22}Na, showing the levels of the daughter nuclide ^{22}Ne and listing the energies (in MeV) and branching ratios of the principal radiations. The abbreviation (EC) refers to electron capture. The vertical part of the β^+-transition arrows corresponds to an energy $2m_e c^2 = 1.022$ MeV which is required before positron decay is energetically permitted.

quoted β-transition energy represents the maximum kinetic energy of emitted β radiation (Section 8.1.3). In electron capture (EC) transitions the available energy is shared between the excitation and recoil energy of the product nucleus and the kinetic energy of the emitted neutrino (Sections 8.1.2 and 8.5.1). A compilation of radiations emitted from frequently used radioisotopes has been prepared by Slack and Way [Sl 59] and a tabulation of γ rays from radionuclides in increasing order of energy has been assembled by Hawkings et al. [Ha 61] and Slater [Sl 62]. Such tabulations indicate that

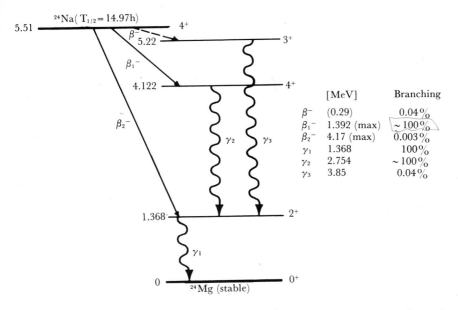

Fig. 6-16. Decay scheme of the β^- source ^{24}Na, listing transition energies and branching ratios.

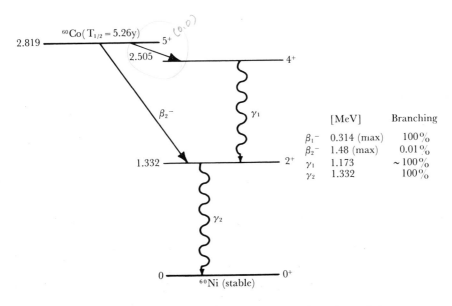

Fig. 6-17. Decay scheme of ^{60}Co, giving rise to β^- and γ emission.

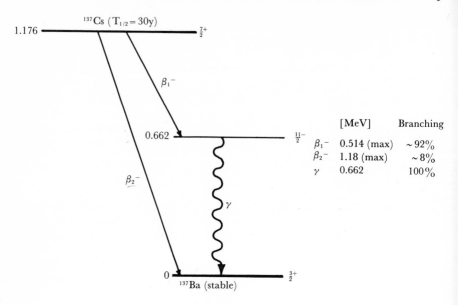

Fig. 6-18. Decay scheme for ^{137}Cs.

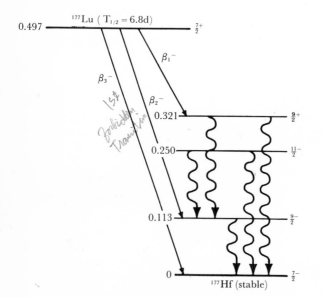

Fig. 6-19. Decay scheme of the β^- emitter ^{177}Lu.

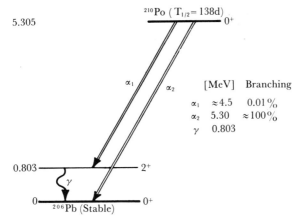

Fig. 6-20. Decay scheme of the α source ^{210}Po.

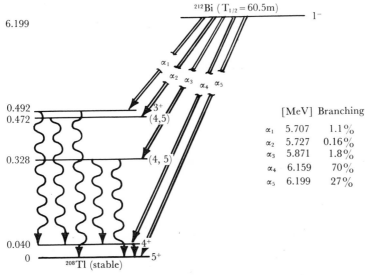

Fig. 6-21. Decay scheme of the α emitter ^{212}Bi.

most γ-ray transitions have energies below 1 MeV and that hardly any exceed 3 MeV (cf. Exercise 6-3).†

† Because of energy and yield restrictions, *radioactive* γ-ray sources are being complemented by artificial sources: e.g., bremsstrahlung sources, which can provide a continuous spectrum of γ radiation up to a specific maximal energy, or reaction sources, as in the technique developed by Jarczyk *et al.* [Ja 61] in which essentially monoenergetic γ radiation of discrete energy within the range typically of 3-11 MeV is derived from (*n*,γ) reactions induced by reactor neutrons which impinge upon suitable target elements.

6.5. Parent-Daughter Relationships in Radioactive Dating

Dating has posed age-old problems in parent-daughter relationships, and its nuclear aspects are no exception. Although radioactivity has provided the means to establish a dating scale which has shed considerable light on the past history of our planet and the solar system, it is paradoxical that it vitiated some of the early estimates of the age of the Earth because ignorance of the heating effect of terrestrial radioelements very appreciably upset the result inferred from estimates of the rate of cooling. For instance, Kelvin's figure of 40 million years for the effective age of the Earth proved to be badly at variance with the geological estimate of 400 million years derived from data for the salt accumulation rate or sediment deposition. The modern methods of age determination from relative parent-daughter abundances in long-lived radioactive decay processes involving α and/or β emission have been refined to yield results which agree well with those derived from other dating methods.

The basis of the various methods is provided by the fact that the decay relation

$$N = N_0 e^{-t/\tau} \tag{6-59}$$

yields a formula for a time period

$$t = \tau \ln\left(\frac{N_0}{N}\right) \tag{6-60}$$

which can be expressed as

$$t = \tau_P \ln\left(\frac{P+D}{P}\right) \tag{6-61}$$

in terms of the lifetime τ_P of the parent species and the abundances P and D of the unstable parent and stable daughter elements respectively. These can be measured by mass-spectroscopic or microchemical-analysis methods to a comparatively high degree of precision. If, then, one makes the assumption that at the time origin $(t = 0)$ only the parent species existed and that its decay over an interval of time t produced D daughter nuclei and left P parent nuclei in a given sample, the original parent abundance can be set to $P + D$, and the sample's age deduced from the formula (6-61). Alternatively, an age determination is also possible in cases for which independent evidence enables one to evaluate the initial amount D_0 of daughter nuclei present as "contaminant" from the very beginning, for then the original concentration of the decaying parent species is $P + D - D_0$.

An assumption inherent in the calculation is the constancy of the decay constant λ_P over long periods of time, but this is vindicated by a wealth of evidence ranging back over 10^9 y, such as the pleochroic halos of uranium and thorium in mica. The information deducible from such halos can be of appreciable cosmological significance [Ko 61] and may also provide indications of now-extinct radioactivity [Ge 67].

The radioactive dating methods, which involve the use of delicate precision techniques for trace analysis of specimens such as artifacts, minerals, or meteorites, depend upon a knowledge of whether or not any loss of daughter nuclei has occurred since their formation, and if so, exactly how much.

The principal dating methods are:

(a) The radiocarbon technique developed by Libby [Li 55a] for the dating of organic matter between 500 and 30,000 y old. The specific activity of organic carbon is due to its ^{14}C content which depreciates by β^- decay ($\tau = 8,000$ y) to stable ^{14}N unless replenished by the continuous exchange with atmospheric carbon, which contains a natural equilibrium abundance of 1 part of ^{14}C (formed by nuclear reactions in the atmosphere initiated by cosmic rays) in about 10^{12} parts of stable ^{12}C (see [Mü 63]). In a living specimen the equilibrium ^{14}C specific activity is $\mathscr{A}' = 15$ decays g^{-1} sec^{-1}, but from the moment of death, the ^{14}C concentration progressively diminishes, which results in a proportionate diminution in the specific activity, enabling the latter to be used as an index of age subsequent to death upon calibration against archaeological specimens of known antiquity (Fig. 6-22).

(b) Determination of the U/Pb content of uranium-bearing minerals (based upon the uranium decay chain depicted in Fig. 6-7) or the corresponding thorium-containing rocks.

(c) Measurement of the helium content of minerals or meteorites, since in α decay ^4He particles are evolved (eight, for example in the decay of ^{238}U to ^{206}Pb) and, unless they escape from the sample, are manifested as an accumulation of helium gas in the specimen.

(d) Determination of parent-daughter abundances in mineral specimens (e.g., K/Ar, K/Ca, Rb/Sr).

(e) Photometric examination of pleochroic halos in mica due to α radiation.

(f) In certain applications, notably in glaciology studies, involving comparatively young specimens, measurement of tritium activity

$$^3H \xrightarrow[T_{1/2}\,=\,12.26\ y]{\beta^-} {}^3He$$

These methods have, except for a few as yet unresolved discrepancies, been demonstrated to tally among themselves as well as against other established dating techniques (e.g., magnetic, thermoluminescence, or paleological methods). They indicate the ages of the oldest terrestrial minerals in pre-Cambrian rocks to approach 3×10^9 y and of the oldest meteorites to be about 4.5×10^9 y. Some of the results, as discussed in various review articles

[Ah 56, Al 58a, Su 58, An 62a, Bu 62] are tabulated in Table 6-1. The table incorporates a revision of the age of the universe from 5×10^9 y (disconcertingly similar to the age of our planet) to the value 13×10^9 y derived [Sa 65, Ta 66] from cosmological models (e.g., Einstein, de Sitter) on substituting the emended value of the Hubble constant as deduced from observations on the Cepheid variables.

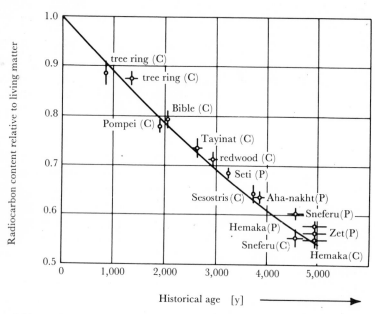

Fig. 6-22. Radiocarbon calibration curve for dating objects with the ^{14}C method ($T = 5568 \pm 30$ y). The samples range from 600 to nearly 5000 y and are labeled according to dates as determined by the Chicago group (C) and the Pennsylvania group (P). This figure was presented by W. F. Libby in his Nobel Lecture in 1960 on the occasion of the conferring of the Chemistry Prize (from [Se 64]).

The evidence from radioactivity measurements constitutes a valuable adjunct to studies in optical, neutrino, and radio astronomy which strive to establish the age of the universe with enhanced precision as a means toward distinguishing between such alternative current cosmological hypotheses as the "primeval atom" theory of Lemaître *et al.*; the "big bang" theory of Gamow *et al.*; the currently discarded "steady state" theory (see [Ho 66c, Da 66a]) of Bondi, Gold, Hoyle, Narlikar, Sciama, etc.; and such "pulsating" theories of the universe as those of Dicke *et al.* (see Fig. 6-23). A critical comparison between theory and observation has recently been made [Da 66a] (see also [Ta 66]).

Table 6-1. AGE DETERMINATION BY RADIOACTIVITY ANALYSIS[a]

Source	Determination method	Age (10^9 y)	Reference
Oldest terrestrial minerals			
Rhodesia-Manitoba	Pb (Sr) {in monazites and uranitites}	2.68 (2.50–3.35)	[Lo 55]
Klerksdorp, Transvaal	Pb in uranitite	2.93	[Ah 55]
Rhodesia	Sr in lepidolite	2.57–2.82	[Schr 55]
Earth ores and rocks			
Galena	Pb isotopes	4.20	[Ah 56]
Granite	Pb	3.5–4.0	[Ah 56]
Earth	Estimated maximum	5.3–5.5	
Stony meteorites	Pb in chondrites	4.4–4.6	[Pa 55]
	Ar in chondrites	4.6	[Wa 55]
Universe	Estimated	≈ 13	[Sa 65]

[a] Table adapted from Ahrens [Ah 56].

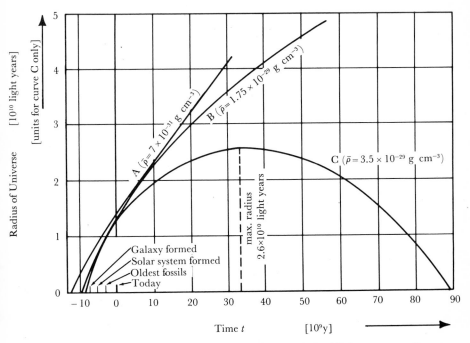

Fig. 6-23. Possible alternatives for the evolution of the Universe according to different cosmological models in dependence of the present average mass density $\bar{\rho}$. Curves A and B represent a continuously expanding Universe whereas curve C corresponds to a pulsating Universe (from [Di 66]).

6.6. Nuclear Stability Limits
according to the Liquid Drop Model

The semiempirical formulae furnished by the liquid drop model give quantitative information on nuclear binding energies as functions of A and Z. They therefore enable one to calculate the value of A beyond which β-stable nuclei become α active, and, indeed, to estimate the limits beyond which an onset of any particular type of disintegration is to be expected. The results

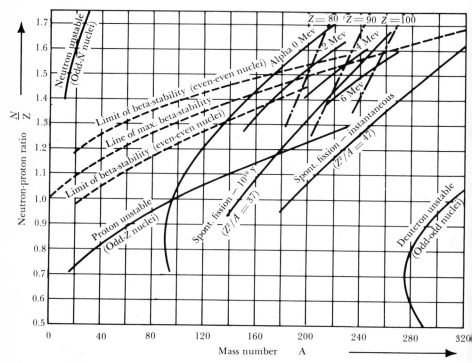

Fig. 6-24. Nuclear stability limits predicted by the liquid-drop model for β, p, n, d, and α emission, and for spontaneous fission (from [Ha 59]).

obtained by Hanna [Ha 59] in 1959, with nuclear mass values of that time, are depicted in Fig. 6-24 in terms of the mass number A. These show stability limits for β, p, n, d, and α emission, as also for spontaneous fission, and indicate that α instability sets in above about $A = 150$. Because of considerations which we present in the next chapter, spontaneous α decay is, with but few exceptions, observed only above $A = 200$, whereas β emission constitutes a prevalent decay mode for much lighter nuclei, and the γ decay of excited nuclear states is found throughout the entire range of nuclides.

EXERCISES

6-1 Calculate the volume at STP of a 1000-Ci source of tritium gas $(T_{1/2} = 12.26$ y$)$.

6-2 A 1-liter container that has been filled with pure tritium at STP is hermetically sealed and left to stand at $0°C$ for a time $t = T_{1/2} = 12.26$ y. What, if any, is the change in (a) activity, (b) specific activity and (c) pressure of the contents over this period? (Noting that gaseous *molecular* tritium changes to *atomic* ^{3}He upon β^- decay, assume for simplicity that nuclei decay pairwise so as to maintain the *molecular* constitution.)

6-3 Although there are many levels in nuclei with an excitation energy exceeding 12 MeV, one practically never observes a 12-MeV γ transition. Why?

6-4 Calculate the specific activity of carrier-free ^{90}Sr $(T_{1/2} = 28$ y$)$.

6-5 The electric power needed by a satellite is 20 W. If this is provided by converting nuclear decay energy (with a conversion efficiency of 5 percent) originating from the α decay $(E_\alpha = 5.5$ MeV$)$ of a ^{238}Pu fuel element $(T_{1/2} = 89$ y$)$, what activity must the latter have? What mass of ^{238}Pu is required?

6-6 (a) What is the mass of a 1-Ci carrier-free source of ^{60}Co, given that its half-life is $T_{1/2} = 5.26$ y? (b) What is the volume in cubic centimeters at STP of the helium gas produced by 1 g of pure ^{226}Ra in 1 year (leaving aside α decay by other radionuclides and taking the half-life of ^{226}Ra to be $T_{1/2} = 1622$ y)?

6-7 In order to produce the radioisotope ^{24}Na $(T_{1/2} = 15$ h$)$, stable ^{23}Na is exposed to a flux of thermal neutrons. For how long must the irradiation be maintained if one desires to produce (a) one half or (b) three quarters of the maximum attainable activity?

6-8 A radioactive source consisting of ^{24}Na with a half-life $T_{1/2} = 15$ h is to be prepared by bombarding 500 g of NaF with thermal neutrons. The production rate of ^{24}Na amounts to 50 Ci/h. What is the maximum attainable activity and what fraction of the overall number of sodium atoms is then radioactive? What value does the activity attain after irradiation for a time $t = n T_{1/2}$?

6-9 How long an irradiation period in a nuclear reactor is required if a 10-mg sample of ^{176}Lu is to yield a ^{177}Lu source $(T_{1/2} = 6.8$ d$)$ for laboratory experiments necessitating an activity of 2 mCi? (On very long irradiation, one can produce a specific activity $\mathscr{A}' = 1$ Ci g^{-1} in ^{177}Lu. Assume that the transport of the source to the laboratory requires 2 d).

6-10   Apart from α decay, with a partial half-life $T_{1/2}^{(\alpha)} = 8.91 \times 10^8$ y, ^{235}U nuclei can undergo spontaneous fission (mean lifetime $\tau_f \cong 3 \times 10^{17}$ y). Estimate how many nuclei in 1 g of pure ^{235}U decay spontaneously in the course of 1 h. How many of these disintegrations are due to fission and how many to α decay?

6-11   A silver foil weighing 1 g is exposed to a thermal neutron flux $\Phi = 10^6$ cm^{-2} sec^{-1} for a time $t = 6$ min. The natural abundance of ^{107}Ag is 51.9 percent, and its neutron capture cross section is 45 b. If the half-life of ^{108}Ag is $T_{1/2} = 2.3$ min, calculate the ^{108}Ag activity upon cessation of neutron irradiation.

6-12   In order to prepare a sample of radioactive ^{140}La, which constitutes an intermediate stage in the decay chain

$$^{140}\text{Ba} \xrightarrow[T_{1/2}=300\,\text{h}]{\beta^-} {}^{140}\text{La} \xrightarrow[T_{1/2}=40.2\,\text{h}]{\beta^-} {}^{140}\text{Ce} \;\;\text{(stable)}$$

a "milking" procedure is adopted in which every time the ^{140}La activity reaches a maximum, the ^{140}La is separated from its parent ^{140}Ba by fast chemical extraction. This procedure is maintained until the ^{140}La samples so derived fall below 1 mCi activity. (a) How many samples can be taken if the original activity of the parent specimen is 5 mCi? (b) What activity does the last ^{140}La sample have at the moment of its chemical separation, and how large is the net activity of all the samples at that instant?

6-13   A ^{124}Sb source ($T_{1/2} = 60.2$ d) can be prepared by thermal neutron irradiation in a reactor. The capture cross section for ^{123}Sb is $\sigma_{(n,\gamma)} = 2.5$ b. How long must a 500-mg sample of natural antimony be irradiated in a thermal neutron flux $\Phi = 2 \times 10^{14}$ cm^{-2} sec^{-1} in order to obtain (a) an activity $\mathscr{A}_1 = 1$ Ci in ^{122}Sb *alone*, (b) an activity $\mathscr{A}_2 = 1$ Ci in ^{124}Sb *alone*?

(Natural isotopic abundances: 57.25 percent ^{121}Sb; 42.75 percent ^{123}Sb. Thermal-neutron capture cross section for ^{121}Sb is $\sigma = 6.8$ b.)

6-14   A researcher planning an experiment with radioactive ^{61}Cu ($T_{1/2} = 3.3$ h) notes that a lecturing commitment will oblige him to have the measurements complete by 4 P.M. the following day. The investigation demands that the ^{61}Cu activity must at the least be $\mathscr{A}_{\min} = 5$ mCi, whereas health regulations forbid him from exceeding an activity $\mathscr{A}_{\max} = 10$ mCi. To derive the optimum results from his investigation he aims to start with as strong a source as permitted and to extend his measurements over as long a period as possible. The source is to be prepared by bombarding stable ^{61}Ni with deuterons, the maximal production rate of ^{61}Cu being $P = 5 \times 10^8$ nuclei sec^{-1}. At what time of the morning will he have to commence the irradiation?

6-15 In living matter the specific activity of ^{14}C combined within organic molecules is $\mathscr{A}' = 15$ disintegrations/sec per gram of carbon. The half-life of ^{14}C is $T_{1/2} = 5570$ y. What is the percentage of ^{14}C atoms in living biochemical material?

6-16 What proportion of ^{235}U was present in a mineral at the time of formation (5×10^9 y ago), given that the present ratio by weight of ^{235}U to ^{238}U is $1/140$? [$T_{1/2}(^{235}$U$) = 7.13 \times 10^8$ y, $T_{1/2}(^{238}$U$) = 4.51 \times 10^9$ y.]

6-17 If a rock contains 100 g of natural uranium together with its decay products, how much ^{235}U and ^{238}U did it contain when it was formed (a time $t = 5 \times 10^9$ y ago)? How much would it have when its age is double the present value? [Natural isotopic abundances: 99.27 percent ^{238}U, 0.72 percent ^{235}U.]

6-18 The ^{206}Pb content of a mineral which contains uranium provides a means of dating the specimen on the assumption that at the time origin $t = 0$ no ^{206}Pb existed. Its formation may be attributed to a decay chain from ^{238}U to ^{206}Pb in which eight α particles are liberated:

$$^{238}\text{U} \rightarrow {}^{206}\text{Pb} + 8\alpha$$

A rough calculation in which the admixture of ^{235}U is neglected and the ^{238}U content is taken to remain constant then yields the age formula

$$t \cong \frac{M(^{206}\text{Pb})}{M(^{238}\text{U})} \times 7.5 \times 10^9 \text{ y}$$

where $M(^{206}$Pb$)$ is the Pb mass measured at time t and $M(^{238}$U$)$ is the mass of U at that time.

Verify this formula and (a) calculate the age of a mineral sample for which $M(\text{Pb})/M(\text{U}) = 0.01$, (b) repeat for $M(\text{Pb})/M(\text{U}) = 0.1$, (c) show that if the assumption of constancy in ^{238}U content is abandoned, the formula becomes

$$t = \left[\frac{M(\text{Pb})}{M(\text{U}) + 1.15 \, M(\text{Pb})} \right] \times 7.5 \times 10^9 \text{ y}$$

(d) determine the percentage correction to the age t that this gives in cases (a) and (b) above, (e) determine how many grams of helium are formed per year if the lead content in case (a) is 200 g.

Chapter 7

ALPHA DECAY

Alpha Decay

7.1. Introduction

This mode of radioactive decay was not only historically the first to be detected and investigated, but it has led to most valuable information on nuclear structure and the characteristics of nuclear reactions. The fact that α particles ($= {}^4\text{He}$ nuclei $= 2n + 2p$) have spin 0 and positive intrinsic parity very considerably simplifies the analysis of reaction and scattering processes. For example, a basically simple test of PARITY CONSERVATION IN STRONG INTERACTIONS was carried out by Tanner [Ta 57] who studied the products of the reaction

$$ {}^{19}\text{F} + \text{p} \rightarrow {}^{20}\text{Ne*} \rightarrow {}^{16}\text{O} + \alpha \tag{7-1} $$

proceeding by way of a known 1^+ state in the intermediate nucleus ${}^{20}\text{Ne*}$. In the absence of an intrinsic spin, α particles must have a *total* angular momentum equal to their *orbital* momentum l, and a parity given by $(-1)^l$, viz., 0^+, 1^-, 2^+, 3^-.... Since the product nucleus ${}^{16}\text{O}$ has a ground-state spin-parity 0^+, it follows that if parity is a quantity conserved during the transition from the intermediate to the final state, α decay to the ground state of ${}^{16}\text{O}$ could occur only for 0^+, 1^-, 2^+, 3^-...levels in ${}^{20}\text{Ne}$ and *not* for a state established to be 1^+. A careful search for any such "parity-violating" α decay particles failed to detect any, and thereby within the accuracy of the experiment (1 part in 10^7) verified parity conservation in strong nuclear interactions (cf. Exercise 7-5).

7.2. Semiempirical Mass Formula Applied to α Decay

The systematics of α decay, and in particular the specification of the range of nuclides energetically able to emit α particles, can be derived from a consideration of the semiempirical mass formula.

The total energy liberated in α decay, namely, the Q value of the α-decay process, is

$$Q_\alpha = [M(Z, A) - M(Z - 2, A - 4) - M(^4\text{He})] c^2 \qquad (7\text{-}2)$$

in terms of atomic masses M. Neither the total number of neutrons nor of protons changes thereby, and hence the equation reduces to the binding-energy relation

$$Q_\alpha = B(^4\text{He}) + B(Z - 2, A - 4) - B(Z, A) \qquad (7\text{-}3)$$

This can be simplified further, since the difference between the pairing-energy terms δ in the Weizsäcker expression is negligible, and since one can to a good degree of approximation write

$$\Delta B \equiv B(Z - 2, A - 4) - B(Z, A) = \frac{\partial B}{\partial Z} \Delta Z + \frac{\partial B}{\partial A} \Delta A = -2 \frac{\partial B}{\partial Z} - 4 \frac{\partial B}{\partial A} \qquad (7\text{-}4)$$

Inserting the experimental value $B(^4\text{He}) = 28.3$ MeV, one arrives at the result

$$\begin{aligned} Q_\alpha &= B(^4\text{He}) + \Delta B \\ &\cong 28.3 - 4a_v + \frac{8}{3} \frac{a_s}{A^{1/3}} + 3 \frac{a_c Z}{A^{1/3}} \left(1 - \frac{Z}{3A}\right) - 4a_a \left(1 - \frac{2Z}{A}\right)^2 \end{aligned} \qquad (7\text{-}5)$$

where Z and A now represent mean values between the parent and daughter nuclei and the Weizsäcker coefficients a carry their usual significance.

Numerical substitution indicates that Q_α takes on *positive* values for heavy nuclei only, e.g., those with $A \gtrsim 150$ (beyond Sm), which are therefore α unstable. It is also apparent that there is an approximately linear rise in Q_α with increasing Z among a family of isobars.

The fact that only comparatively few nuclei between $A = 150$ (Sm) and $A = 210$ (Pb) have actually been found to be α active may be attributed to the smallness of the available decay energies, which so greatly reduces the probability of an α particle escaping from the parent nucleus that α-decay lifetimes rise beyond 10^{16} y and therefore render detection unfeasible. Above $A \approx 210$, decay energies tend to be higher, as can be seen from Fig. 7-1, and hence the heavy nuclei decay predominantly by α emission. All α emitters are found to have very long half-lives (10^{-6} sec to 10^{17} sec) compared with the typical period of 10^{-21} sec needed for a fast particle to traverse the nucleus. To illustrate this enormous half-life range we might cite the case of ^{213}Po, whose half-life for the emission of 8.336-MeV α particles is $T_{1/2} = 4.2 \times 10^{-6}$ sec $= 1.33 \times 10^{-13}$ y, as against ^{232}Th, whose half-life of 1.39×10^{10} y for the emission of 3.98-MeV α radiation is a factor 10^{23} greater.

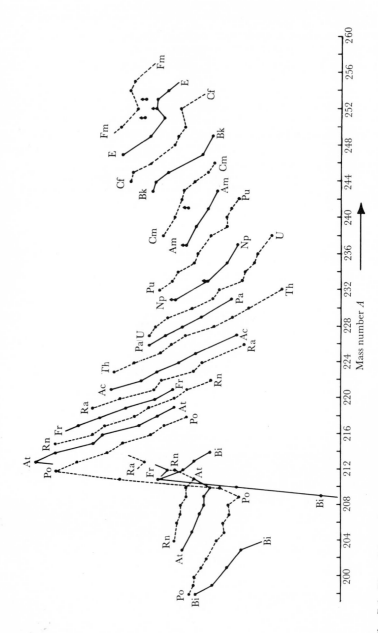

Fig. 7-1. Plot of the α-decay energy E_α vs mass number A for α-unstable nuclides showing systematic trends except for the large discontinuity at the major shell of 120 neutrons (magic-number nuclei) and a small discontinuity at a subshell of 152 neutrons (from [Ra 65a]). (Used by permission of North-Holland Publishing Co.)

7.3. Relation between α Energy and Decay Half-Life

From the examples given in Table 7-1, it will be seen that the decay half-life depends markedly upon the α energy, as one might expect from transition probability considerations.

Table 7-1. α-DECAY ENERGIES AND HALF-LIVES OF SOME TYPICAL NATURAL α EMITTERS

α Emitter	E [MeV]	$T_{1/2}$
^{212}Po (ThC′)	8.8	3×10^{-7} sec
^{214}Po (RaC′)	7.7	1.6×10^{-4} sec
^{210}Po	5.3	1.38×10^2 d
^{226}Ra	4.7	1.62×10^3 y
^{238}U	4.1	4.5×10^9 y

The pioneer investigations of Geiger and Nuttall [Gei 11, Gei 12], who measured radioactive decay constants $\lambda(= 0.693/T_{1/2})$ and α ranges R_α in air, led them to propose a simple linear relationship between $\log \lambda$ and $\log R_\alpha$ of the form

$$\log \lambda = a + b \log R_\alpha \qquad (7\text{-}6)$$

where a and b are empirical constants which can be derived from a log-log plot (Fig. 7-2). The linear range R_α is a direct measure of the α energy E. The plot shows distinct differences between the three decay series. Geiger was further able to show that R_α bears a simple approximate relationship to the initial velocity v of the α particles, which can be expressed by the formula

$$R_\alpha \cong cv^3 \qquad (7\text{-}7)$$

where c is a constant. This is now known as GEIGER'S RULE. Hence one expects a direct relation between λ and v. An empirical relation of this type was found by Swinne [Sw 12, Sw 13] who expressed it in the form

$$\log \lambda = d + e v^f \qquad (7\text{-}8)$$

where d and e are constants and f lies between 1 and 2.

Only with the advent of wave mechanics did it become possible to provide a satisfactory explanation for the strong relationship between λ and E or v. In terms of penetration of α particles through nuclear potential barriers, a sound theoretical basis was provided for the interpretation of α decay which gave results in excellent numerical agreement with experimental data. The theory, as applied by Gamow [Ga 28] and independently by Condon and Gurney [Co 28b, Co 29], yields a relation of the form

$$\log \lambda = g - h\left[\left(1 + \frac{4}{A}\right)^{-1/2} ZE^{-1/2}\right] \qquad (7\text{-}9)$$

where the first term g is almost constant and the coefficient of the second term is $h \equiv (2.303) \, (4\pi e^2) \, (m_\alpha/2\hbar^2)^{1/2}$. On substituting for this numerically and

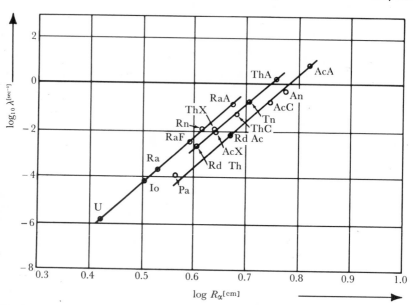

Fig. 7-2. Geiger-Nuttall type [Gei 11, 22] log-log plot of decay constant λ *vs* range R_α for α particles in air, as emitted by natural radioactive α sources. The original nuclide nomenclature (e.g., Io = Ionium) has been retained. (*Key:* U = UI, UII $\equiv$ ^{238}U, ^{234}U; Io = ^{230}Th; Ra = ^{226}Ra; RaF = ^{210}Po; Rn = ^{220}Rn; RaA = ^{218}Po; RdTh = ^{228}Th; AcX = ^{223}Ra; ThX = ^{224}Ra; ThC = ^{212}Bi; Tn = ^{220}Rn; ThA = ^{216}Po; Pa = ^{231}Pa; RdAc = ^{227}Th; AcC = ^{211}Bi; An = ^{219}Rn; AcA = ^{215}Po.)

inserting the typical $A = 220$, one obtains an expression of the Geiger-Nuttall form:

$$\log_{10} \lambda = g - 1.7037(ZE^{-1/2}) \qquad (7\text{-}10)$$

with the kinetic energy E in MeV. On adjusting g to a suitable value, the expression is found to agree remarkably well with α-decay data for even-even nuclei, as can be seen from Fig. 7-3.

A convenient expression derived semiempirically by Taagepera and Nurmia [Ta 61] to fit data for even-even nuclei beyond $N = 126$ can be written as

$$\log_{10} \lambda = 36.24 - 1.61(ZE^{-1/2} - Z^{2/3}) \qquad (7\text{-}11)$$

in terms of the daughter atomic number Z and the α-particle kinetic energy E.

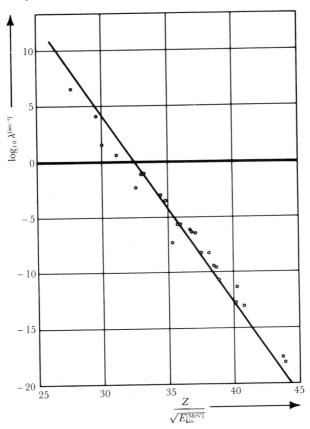

Fig. 7-3. Semiempirical plot of a relation of the Geiger-Nuttall type connecting the α-decay constant with the α energy for even-even nuclei, according to Eq. (7-10), using $g = 55.5$ (from [Gr 55]). (Used by permission of McGraw-Hill Book Co.)

7.4. Penetration of Potential Barriers

The field of force exerted on a particle by a nucleus can be represented by a NUCLEAR POTENTIAL which has a characteristic shape in function of the radial distance r from the center of the nucleus of radius R. Thus a positively charged particle may be subjected to a combination of short-range strong nuclear attractive forces (represented by a square-well potential) and long-range electromagnetic repulsive forces (represented by a Coulomb potential). This situation is schematized in Fig. 7-4, which for simplicity depicts the shape of the spherically symmetric potential in an orthogonal representation against r. The actual potential is generated by a rotation about the V axis.

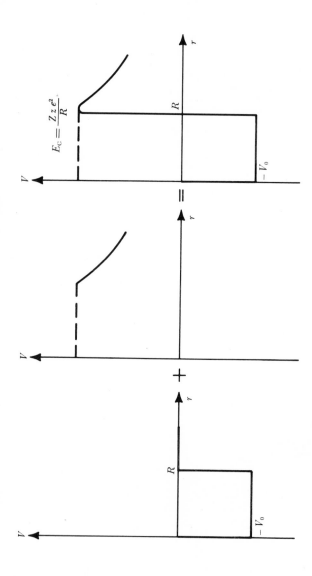

Fig. 7-4. Schematic superposition of a repulsive Coulomb barrier on an attractive deep rectangular-well nuclear potential whose depth V_0, though reduced in scale for graphical convenience in the diagram, is actually much larger in magnitude than the height E_c of the Coulomb barrier. The ensuing radial dependence of the net nuclear potential is indicated on the right.

One would be drawn to conclude that at radial separations less than R the α particle would be trapped within the nucleus by the strong potential barrier. There would be no chance for it to burrow through the potential wall which confronts it, and hence no possibility of nuclear α decay at energies less than those corresponding to the peak of the potential barrier. Classically, this viewpoint would be valid, and α emission at energies less than the Coulomb energy $E_C = Zze^2/R$ would be impossible. However, according to quantum theory there is a certain *nonzero* probability of TUNNELING through the barrier at lower α energies, as may be demonstrated qualitatively with the aid of the uncertainty relation

$$\Delta p \, \Delta x \approx \hbar \tag{7-12}$$

Since the uncertainty in the particle's momentum Δp cannot exceed its actual momentum p, it follows that

$$\Delta p \leqslant p \tag{7-13}$$

and hence

$$\Delta x \geqslant \frac{\hbar}{p} = \frac{\hbar}{mv} = \lambda \tag{7-14}$$

where Δx is the localization uncertainty of a particle which has a reduced mass m, nonrelativistic velocity v, and rationalized de Broglie wavelength λ. Then if the velocity v (i.e., the energy E) of the particle is such as to cause $\Delta x(\geqslant \lambda)$ to equal or exceed the barrier width at that energy, there will be a certain finite probability that the particle wave may find itself on either side of the barrier, and thereby have penetrated the barrier by "tunneling."

Regarded from an alternative viewpoint, there is an "unsharpness" in the energy ΔE given by the Heisenberg uncertainty relation

$$\Delta E \, \Delta t \approx \hbar \tag{7-15}$$

which, manifested over a time interval Δt, confers sufficient energy temporarily to the particle to enable it to surmount the barrier. The equivalence of these two approaches may be demonstrated as follows:

If the particle momentum ranges from p to $p + \Delta p$, the corresponding range of kinetic energy will be E to $E + \Delta E$ where, assuming $\Delta p \ll p$ and $\Delta E \ll E$,

$$E + \Delta E = \frac{(p + \Delta p)^2}{2m} \cong \frac{p^2}{2m} + \frac{p \, \Delta p}{m} = E + v \, \Delta p \tag{7-16}$$

with $v = p/m$ the particle velocity corresponding to a momentum p. From (7-16) we see that

$$\Delta E \approx v \, \Delta p \tag{7-17}$$

and from our earlier considerations we can relate the momentum uncertainty to a width Δx which can be traversed with velocity v in a time Δt:

$$\Delta p \approx \frac{\hbar}{\Delta x} = \frac{\hbar}{v \, \Delta t} \tag{7-18}$$

Substituting this in (7-17) we arrive back at the energy-time uncertainty relation

$$\Delta E \, \Delta t \approx \hbar \tag{7-19}$$

and thus vindicate our viewpoint.

7.4.1. Rectangular Barrier

It is instructive first to consider the penetration of a RECTANGULAR BARRIER before proceeding to more complicated situations. This constitutes a standard exercise in wave mechanics, and its solution provides evidence for the validity of the quantum-mechanical approach.

In the case of the one-dimensional potential wall, depicted in Fig. 7-5, we distinguish between three regions:

$$\left. \begin{array}{llll} \text{Region} & \text{I:} & V = 0 & -\infty < x < 0 \\ \text{Region} & \text{II:} & V = U & 0 \leqslant x \leqslant b \\ \text{Region} & \text{III:} & V = 0 & b < x < \infty \end{array} \right\} \tag{7-20}$$

For a particle having a kinetic energy E which is less than U this barrier would classically be impenetrable. Quantum-mechanically, however, it will have a certain nonzero TRANSPARENCY, which we proceed to calculate.

When the particle is treated as a de Broglie wave of wavelength $\lambda = \hbar/mv = \hbar/(2mE)^{1/2}$, it can be conceived to be in part transmitted through the wall at the point $x = 0$ and in part reflected back. Thus if, as in Fig. 7-6, a plane wave of unit amplitude impinges upon the barrier at $x = 0$, a part having an amplitude $A^{(-)}$ will be reflected back while the remainder of amplitude $B^{(+)}$ will proceed in the positive x direction into Region II. This in its turn will be partly reflected back (amplitude $B^{(-)}$) at the interface at $x = b$ and partly transmitted (amplitude $C^{(+)}$) into Region III, so that of the original incident unit amplitude, a net amplitude $C^{(+)}$ emerges from beyond the barrier. In terms of intensity, viz. (amplitude)^2, the ratio of transmitted to incident intensity is $|C^{(+)}|^2$ and hence the barrier PENETRATION FACTOR is

$$P = \frac{\text{probability current density in the transmitted wave}}{\text{probability current density in the incident wave}} \tag{7-21}$$

$$= \frac{\text{emergent intensity}}{\text{incident intensity}} = \frac{|C^{(+)}|^2}{1} \tag{7-22}$$

The barrier TRANSPARENCY, or the particle TRANSMISSION COEFFICIENT of the barrier is closely related to this, being the ratio

$$T = \frac{\text{emergent flux}}{\text{incident flux}} = \frac{|C^{(+)}|^2}{1} \frac{k_{\mathrm{III}}}{k_{\mathrm{I}}} \qquad (7\text{-}23)$$

where the k's are the wave numbers $(1/\lambda)$ in Regions III and I, whose ratio is

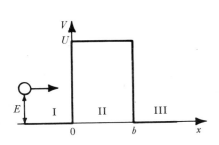

Fig. 7-5. The different potential regions encountered by a particle of kinetic energy E as it tunnels through a repulsive rectangular potential barrier of height U and breadth b.

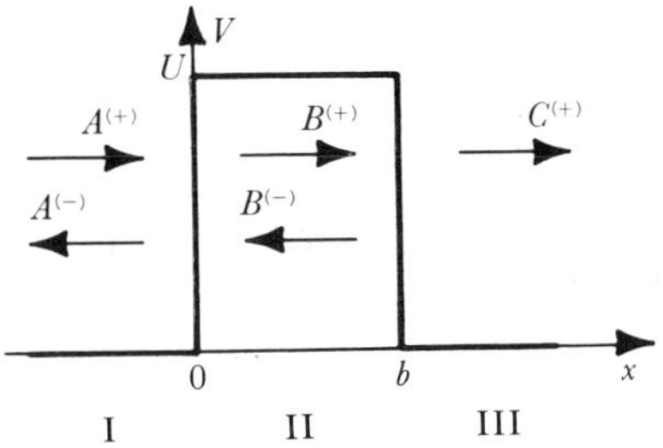

Fig. 7-6. Transmitted and reflected wave amplitudes in the one-dimensional treatment of rectangular barrier penetration.

thus the ratio of particle velocities, and in the present situation this is unity:

$$k_{\mathrm{I}} = k_{\mathrm{III}} \qquad \text{and} \qquad T = P = |C^{(+)}|^2 \qquad (7\text{-}24)$$

The plane waves in each of the three regions are described by wave functions ψ_{I}, ψ_{II}, ψ_{III} which are solutions of the one-dimensional time-independent Schrödinger equation for each region,

$$\frac{\partial^2 \psi}{\partial x^2} + \frac{2m}{\hbar^2}(E - V)\psi = 0 \qquad (7\text{-}25)$$

In Regions I and II, the net wave is a linear superposition of components moving in the positive and the negative x direction,

$$\psi_{\mathrm{I}} = e^{ik_{\mathrm{I}} x} + A^{(-)} e^{-ik_{\mathrm{I}} x} \qquad \text{with } k_{\mathrm{I}} = \frac{(2mE)^{1/2}}{\hbar} \qquad (7\text{-}26)$$

$$\psi_{\mathrm{II}} = B^{(+)} e^{ik_{\mathrm{II}} x} + B^{(-)} e^{-ik_{\mathrm{II}} x} \qquad \text{with } k_{\mathrm{II}} = \frac{[2m(E - V)]^{1/2}}{\hbar} \qquad (7\text{-}27)$$

In the present problem the energy deficiency $E - V = E - U$ is a *negative* quantity equal to the (positive) binding energy in magnitude, but of opposite sign. It therefore leads to an *imaginary* wave number k_{II}, which for mathematical convenience we replace by the real quantity

$$K_{II} \equiv ik_{II} = \frac{[2m(U - E)]^{1/2}}{\hbar} \tag{7-28}$$

and write

$$\psi_{II} = B^{(+)} e^{K_{II} x} + B^{(-)} e^{-K_{II} x} \tag{7-29}$$

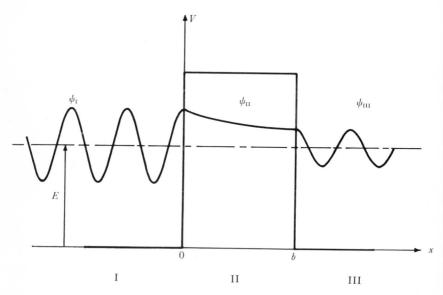

Fig. 7-7. Graphical representation of the wave function at some arbitrary instant of time as a particle tunnels through a rectangular potential barrier. The wave function and its first derivative are continuous at $x = 0$ and $x = b$ (adapted from [Ev 55]). (Used by permission of McGraw-Hill Book Co.)

In Region III, only the transmitted wave remains, travelling in the positive x direction,

$$\psi_{III} = C^{(+)} e^{ik_{III} x} \qquad \text{with } k_{III} = k_I = \frac{(2mE)^{1/2}}{\hbar} \tag{7-30}$$

Further constraints upon the values of the coefficients $A^{(-)}$, $B^{(\pm)}$, $C^{(+)}$ are imposed by the physical requirement of continuity in the wave functions and their first derivatives with respect to x at the boundaries $x = 0$ and $x = b$ (see Fig. 7-7). The ensuing four conditions can be written as

$$\left. \begin{array}{ll} \psi_I(0) = \psi_{II}(0) & \psi_{II}(b) = \psi_{III}(b) \\ \psi_I'(0) = \psi_{II}'(0) & \psi_{II}'(b) = \psi_{III}'(b) \end{array} \right\} \tag{7-31}$$

where the prime denotes differentiation with respect to x and the argument inside the brackets denotes the coordinate. Still more simply, they can be expressed by the requirement that the logarithmic derivatives ψ'/ψ must be continuous at $x = 0$ and $x = b$ for a wave function ψ which vanishes at $x = \pm\infty$. These serve to determine $A^{(+)}$, $B^{(\pm)}$, and $C^{(+)}$, yielding the result

$$T = |C^{(+)}|^2 = \frac{4k_{\mathrm{I}}^2 K_{\mathrm{II}}^2}{(k_{\mathrm{I}}^2 + K_{\mathrm{II}}^2)^2 \sinh^2 bK_{\mathrm{II}} + 4k_{\mathrm{I}}^2 K_{\mathrm{II}}^2} \qquad (7\text{-}32)$$

The right-hand side is identical to the expression

$$\frac{4k_{\mathrm{I}}^2 k_{\mathrm{II}}^2}{(k_{\mathrm{I}}^2 - k_{\mathrm{II}}^2)^2 \sin^2 bk_{\mathrm{II}} + 4k_{\mathrm{I}}^2 k_{\mathrm{II}}^2} \qquad (7\text{-}33)$$

but has the merit of featuring *real* quantities throughout. The REFLECTIVITY of the barrier, i.e., the probability of reflection, is simply $|A^{(-)}|^2$ and, as is indeed to be expected, when the calculated value is added to the above expression for the transparency T, the sum is unity. It is interesting to note that the reflectivity depends also on the barrier thickness. The entire depth of the barrier contributes quantum-mechanically toward reflecting the incident wave.

The exact formula (7-32) for the particle transmission coefficient of the barrier can be written in terms of energy as

$$T = \left\{ 1 + \frac{U^2}{4E(U-E)} \left[\frac{1}{4} (e^{2bK_{\mathrm{II}}} + e^{-2bK_{\mathrm{II}}}) - \frac{1}{2} \right] \right\}^{-1} \qquad (7\text{-}34)$$

and simplified in the limit of THICK RECTANGULAR BARRIERS, for which $bK_{\mathrm{II}} \gg 1$ and the term in square brackets reduces in first approximation to $\frac{1}{4} e^{2bK_{\mathrm{II}}}$:

$$T \to \frac{16E(U-E)}{U^2} e^{-2bK_{\mathrm{II}}} \approx e^{-2b\,[2m(U-E)]^{1/2}/\hbar} \qquad (7\text{-}35)$$

which is GAMOW'S FORMULA.

It is evident that for particles of small mass m the transparency is large (thus electrons encounter little difficulty in penetrating potential barriers) and that in the classical limit, i.e., as $\hbar \to 0$, the transparency tends to zero as expected. It is interesting to note, however, that even an *attractive* potential, such as a rectangular well, causes some of the incident wave to be reflected and therefore its transparency is less than unity except in the limit of high incident energies (the value can be deduced from Eq. (7-33); cf. Exercise 7-7).

7.4.2. BARRIER OF ARBITRARY SHAPE

For a barrier of arbitrary shape, such as that depicted in Fig. 7-8, the transparency can in rough approximation be obtained by a simple generalization of Gamow's formula in which the barrier is effectively replaced by an "equivalent" rectangular barrier having an appropriate effective height and width:

$$T \cong \exp\left\{ -\frac{2}{\hbar} \int_{a}^{b} [2m(V(x) - E)]^{1/2} \, dx \right\} \tag{7-36}$$

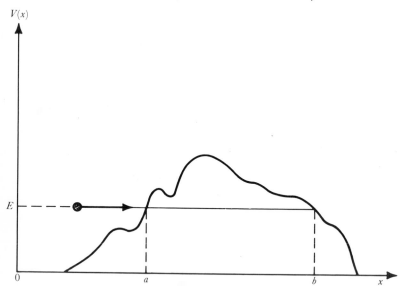

Fig. 7-8. Representation of an arbitrarily shaped one-dimensional barrier extending from $x = a$ to $x = b$ at the energy E at which an incident particle impinges upon it.

7.4.3. Nuclear Potential Barrier

The nuclear potential felt by a charged particle is represented by a THREE-DIMENSIONAL SPHERICALLY SYMMETRIC BARRIER. Since it is radially symmetric, it can most easily be considered by expressing the Schrödinger equation for the system in polar coordinates (r, θ, ϕ), with the wave function split into a product of functions of each variable,

$$\psi = f(r)\, \Theta(\theta)\, \Phi(\phi) \tag{7-37}$$

The Schrödinger equation can then be subdivided into three separate differential equations (as in the quantum-mechanical treatment of the hydrogen atom). Thus the overall equation

$$\left[\frac{1}{r^2} \frac{\partial}{\partial r}\left(r^2 \frac{\partial}{\partial r}\right) + \frac{1}{r^2 \sin\theta} \frac{\partial}{\partial \theta}\left(\sin\theta \frac{\partial}{\partial \theta}\right) + \frac{1}{r^2 \sin^2\theta} \frac{\partial^2}{\partial \phi^2}\right]\psi$$

$$+ \frac{2m}{\hbar^2}(E_{\text{tot}} - V)\psi = 0 \tag{7-38}$$

can be split into a part depending only on the radial variable r, and a part which depends only on the angular variables θ, ϕ. Each part may accordingly be set equal to a common constant, which may be taken to be of the form $l(l+1)$, with l a positive integer or zero (cf. Eqs. (5-24) ff. in Section 5.3.10 or pp. 70 ff. in [Schi 55]). This gives us a RADIAL WAVE EQUATION

$$\frac{1}{r^2}\frac{d}{dr}\left(r^2\frac{df(r)}{dr}\right)+\frac{2m}{\hbar^2}\left[(E_{\text{tot}}-V)-\frac{l(l+1)\,\hbar^2}{2mr^2}\right]f(r)=0 \qquad (7\text{-}39)$$

and two ANGULAR WAVE EQUATIONS

$$\frac{1}{\sin\theta}\frac{d}{d\theta}\left(\sin\theta\frac{d\Theta}{d\theta}\right)+\left[l(l+1)-\frac{m_l^2}{\sin^2\theta}\right]\Theta=0 \qquad (7\text{-}40)$$

$$\frac{d^2\Phi}{d\phi^2}+m_l^2\,\Phi=0 \qquad (7\text{-}41)$$

of which the latter two stem from a combined equation involving the spherical harmonic $Y_{lm_l}(\theta,\phi)\equiv\Theta(\theta)\,\Phi(\phi)$, viz.

$$\frac{1}{\sin\theta}\frac{\partial}{\partial\theta}\left(\sin\theta\frac{\partial Y}{\partial\theta}\right)+\frac{1}{\sin^2\theta}\frac{\partial^2 Y}{\partial\phi^2}+l(l+1)\,Y=0 \qquad (7\text{-}42)$$

and are of no further direct concern at this stage.

We concentrate upon the *radial* wave equation, and note that its derivation introduces the orbital quantum number l (which also features in the angular wave equations, together with its projection, the magnetic orbital quantum number m_l). It also contains the "physics" of the problem inasmuch as it includes the total energy E_{tot} and the potential $V=V(r)$ of the system. We can rewrite the radial equation on introducing a MODIFIED RADIAL WAVE FUNCTION $u\equiv rf(r)$ as a MODIFIED RADIAL WAVE EQUATION

$$\frac{d^2 u}{dr^2}+\frac{2m}{\hbar^2}\left[(E_{\text{tot}}-V(r))-\frac{l(l+1)\,\hbar^2}{2mr^2}\right]u=0 \qquad (7\text{-}43)$$

and compare this with the Schrödinger equation for the one-dimensional case, namely,

$$\frac{d^2\Psi}{dx^2}+\frac{2m}{\hbar^2}(E_{\text{tot}}-V)\,\Psi=0 \qquad (7\text{-}44)$$

Clearly, the one-dimensional potential V has in the three-dimensional case been replaced by

$$V(r)+\frac{l(l+1)\,\hbar^2}{2mr^2} \qquad (7\text{-}45)$$

The new term $l(l+1)\,\hbar^2/2mr^2$ has the dimensions of energy, its denominator having the appearance of a moment of inertia and its numerator suggesting the square of an angular momentum.† It can therefore be regarded as a form of rotational energy associated with the motion of colliding particles about their common center of mass, that is to say, a CENTRIFUGAL POTENTIAL whose effect is to *augment* the potential barrier of a nucleus whenever there is mutual orbital momentum present. The CENTRIFUGAL BARRIER therefore enters into all problems concerned with disintegration or collision of a noncentral nature

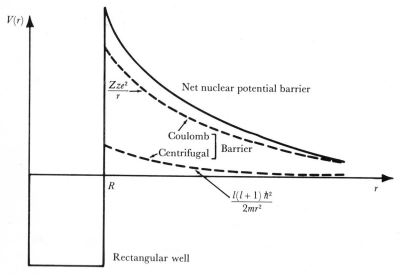

Fig. 7-9. Schematic representation of the potential barrier of a nucleus. In the extranuclear region it comprises a superposition of a Coulomb potential barrier (for all particles which carry a positive charge ze) and a centrifugal potential barrier (for all particles, charged or uncharged, which have an orbital momentum $l > 0$ with respect to the nucleus of radius R, charge Ze, and mass m).

† The classical expression for the centrifugal force,

$$F = mr\omega^2 = \frac{L^2}{mr^3}$$

where $L = mr^2\omega$ represents the orbital momentum of a particle of mass m moving about a radius r with orbital velocity ω, gives rise to a classical CENTRIFUGAL POTENTIAL

$$V_{\mathrm{c}}(r) = \frac{L^2}{2mr^2}$$

since

$$F = -\frac{\partial V_{\mathrm{c}}}{\partial r} = \frac{L^2}{mr^3}$$

This classical expression for a centrifugal potential corresponds exactly to the quantum-mechanical form when one makes the plausible substitution $L^2 \to l(l + 1)\hbar^2$ for the magnitude of the square of the orbital angular momentum.

($l \neq 0$, e.g., P, D, F, ... waves), even in the case of *uncharged* particles such as neutrons penetrating the nuclear barrier. It acts in such a way as to diminish the probability of penetration (see Fig. 7-9), and especially in the case of interactions involving high orbital momenta, its effect may be quite appreciable, though for α decay it can be shown to have but slight influence upon the decay rate. (For example, for α's with $l = 5$, the effect is to increase the lifetime

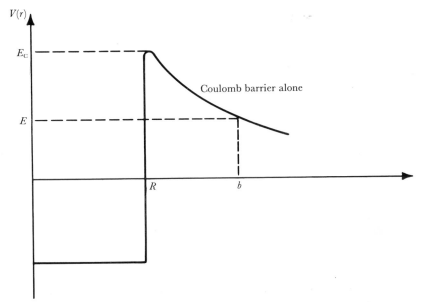

Fig. 7-10. Representation of the Coulomb wall which confronts a positively charged particle of energy E seeking to enter or emerge from a nucleus of radius R.

by only a factor of 10 beyond that which a pure Coulomb barrier would entail; cf. p. 577 of [Bl 52b].) The evaluation of the transparency of a centrifugal barrier is not as simple an analytic procedure as in the case of a Coulomb barrier alone, and in general is performed numerically by electronic computation.

For a purely electrostatic COULOMB POTENTIAL BARRIER such as is encountered by a positively charged particle approaching a nucleus centrally, i.e., with $l = 0$, the potential has the form shown in Fig. 7-10, and the top of the barrier is equal in height to the COULOMB ENERGY

$$E_{\mathrm{C}} = \frac{Zze^2}{R} \tag{7-46}$$

where R is the radius of the nucleus of charge Ze upon which a particle of charge ze impinges. The Coulomb barrier height therefore varies with z, the charge of the incident particle, as does E_C. For example the Coulomb energy for a proton incident on a ^{238}U nucleus is $E_C^{(p)} = 13.02$ MeV, whereas for an α particle carrying double the charge and impinging on (or escaping from) the same nucleus it is nearly twice as large, $E_C^{(\alpha)} = 23.97$ MeV. The Coulomb

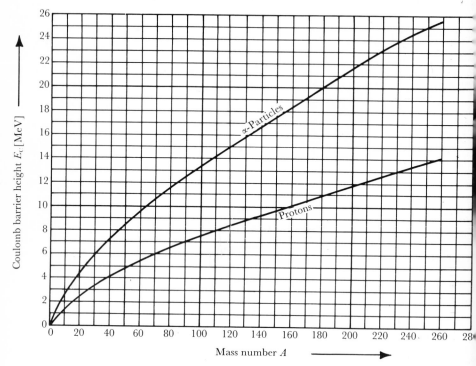

Fig. 7-11. Plots of the nuclear Coulomb barrier heights E_c *vs* mass number A for protons and α particles, making allowance for the particle and target radii (using numerical values tabulated by Martin [Ma 54b]).

barrier heights E_C for various incident particles including p's and α's have been tabulated by Martin [Ma 54b] and are plotted in Fig. 7-11. The transparency of the barrier depends on the *width* rather than the height of the potential at the energy in question, and therefore it is determined by b and R. The explicit form of this relationship will be derived below. We first presuppose a knowledge of the transparency T and relate this to the decay constant λ.

If we assume that at time $t = 0$ a wave of α particles having an initial intensity I_0 impinges on the potential barrier, the latter will transmit a certain proportion T out and reflect the remainder $(1 - T)I_0$ back into the nuclear well. After n such "attempts" to traverse the barrier, the wave will have an intensity within the nucleus given by

$$I_n = (1 - T)^n I_0 = I_0 e^{n \ln(1-T)} \tag{7-47}$$

If, as we have reason to expect from the long α-decay lifetimes, the barrier transparency is small, we can write to a first approximation

$$\ln(1 - T) \cong -T \tag{7-48}$$

and therefore

$$I_n = I_0 e^{-nT} \tag{7-49}$$

The average time for traversal of a nucleus is equal to the time interval between successive attempts to escape, and is

$$t_0 = \frac{2R}{v} \tag{7-50}$$

where v is the α velocity. Then, if there are n such attempts in a time interval t, it follows that

$$n = \frac{t}{t_0} = \frac{v}{2R} t \tag{7-51}$$

and hence

$$I_n = I_0 e^{-(v/2R) T t} \tag{7-52}$$

Since the escape repetition rate is very high, viz.

$$\frac{dn}{dt} = \frac{1}{t_0} = \frac{v}{2R} \approx 10^{21} \text{ sec}^{-1} \tag{7-53}$$

we are justified in treating it as continuous rather than discrete. We can therefore write

$$I_n \rightarrow I(t) = I_0 e^{-\lambda t} \tag{7-54}$$

where the DECAY CONSTANT λ is given by comparison with (7-52) as

$$\lambda = \frac{v}{2R} T = \frac{dn}{dt} T \tag{7-55}$$

Hence the decay probability per unit is equal to the product of the repetition rate and the barrier transparency. We can introduce the designation

$$\lambda_0 \equiv \frac{dn}{dt} = \frac{v}{2R} \approx 10^{21} \text{ sec}^{-1} \tag{7-56}$$

and write

$$\lambda = \lambda_0 T \tag{7-57}$$

so that

$$I(t) = I_0 e^{-\lambda_0 T t} \tag{7-58}$$

We see, for example, that within the mean lifetime of the ^{238}U nucleus, $\tau \cong 6.5 \times 10^9 y$, the number of attempts to penetrate the potential barrier is $n \approx 10^{38}$.

Applying the foregoing considerations to α tunneling through a Coulomb barrier, we note that in the general approximate expression for transparency given in Section 7.4.2 we must replace the variable x by the radial distance r

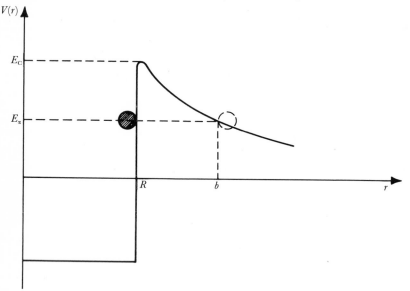

Fig. 7-12. Schematic representation of the tunnelling of an α particle of energy E_α through a nuclear Coulomb potential barrier.

and in place of the barrier height V write the Coulomb energy $E_C = Zze^2/R$. Thus, since

$$\lambda = \lambda_0 T \tag{7-59}$$

and

$$T = \exp\left\{-\frac{2}{\hbar} \int [2m(E_C - E)]^{1/2} dr\right\} \tag{7-60}$$

we can define the integration limits according to Fig. 7-12 as $r = R$ to $r = b$ and write

$$\lambda = \lambda_0 e^{-G} \tag{7-61}$$

where

$$G \equiv \frac{2}{\hbar} \int_R^b [2m(E_C - E)]^{1/2} dr = \left(\frac{8m}{\hbar^2}\right)^{1/2} \int_R^b \left(\frac{Zze^2}{r} - E\right)^{1/2} dr \tag{7-62}$$

$$= \left(\frac{8Zze^2 m}{\hbar^2}\right)^{1/2} \int_R^b \left(\frac{1}{r} - \frac{E}{Zze^2}\right)^{1/2} dr \tag{7-63}$$

i.e.,

$$G = \left(\frac{8Zze^2\,mb}{\hbar^2}\right)^{1/2}\left[\arccos\left(\frac{R}{b}\right)^{1/2} - \left(\frac{R}{b} - \frac{R^2}{b^2}\right)^{1/2}\right] \tag{7-64}$$

with

$$b \equiv \frac{Zze^2}{E} \tag{7-65}$$

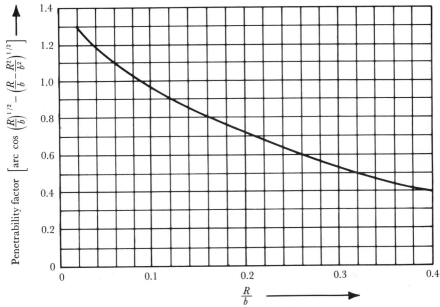

Fig. 7-13. Plot of the penetrability factor in Eq. (7-64) *vs* the ratio R/b of nuclear radius R to Coulomb barrier width b for α particles ($b = Zze^2/E_\alpha$), from data tabulated by Martin [Ma 54b].

It has become customary to term the power G in (7-61), as given by (7-64), the GAMOW FACTOR, although strictly speaking this designation should be confined to the consideration of *thick* barriers, i.e., to the limiting case $R \to 0$. Numerical values of the expression in square brackets in (7-64) have been tabulated by Martin [Ma 54b] over a wide range of values of R/b and are plotted in Fig. 7-13.

Nevertheless, (7-64) remains a somewhat cumbersome expression; it can be simplified when $E \ll E_C$, for then $b \gg R$, as can be seen from Fig. 7-10, and the square bracket in (7-64) reduces in first approximation to

$$\left[\arccos\left(\frac{R}{b}\right)^{1/2} - \left(\frac{R}{b} - \frac{R^2}{b^2}\right)^{1/2}\right] \to \frac{\pi}{2} - \left(\frac{R}{b}\right)^{1/2} \tag{7-66}$$

On the one hand, the Gamow factor G can then be written in the form

$$G \cong A - BR^{1/2} \tag{7-67}$$

where the terms A and B are independent of the nuclear radius R and are, explicitly,

$$A \equiv \left(\frac{2\pi^2 \, Zze^2 \, mb}{\hbar^2}\right)^{1/2}, \qquad B \equiv \left(\frac{8Zze^2 \, m}{\hbar^2}\right)^{1/2} \tag{7-68}$$

On the other hand, one can ignore the radius dependence altogether and simply set the term in square brackets in (7-64) to unity, as deducible from (7-66), to obtain a greatly simplified expression for G, correct to within ± 30 percent,

$$G \approx \left(\frac{8Zze^2 \, mb}{\hbar^2}\right)^{1/2} \qquad \text{with } b \equiv \frac{Zze^2}{E} \tag{7-69}$$

Numerically, for the α decay of heavy nuclei to daughter nuclides of atomic number Z (with b in cm),

$$G \approx 4.7 \times 10^6 (Zb)^{1/2} \tag{7-70}$$

It will be noted that the decay constant

$$\lambda = \lambda_0 \, e^{-G} \approx 10^{21} \exp\left[-\left(\frac{8Zze^2 \, mb}{\hbar^2}\right)^{1/2}\right] \tag{7-71}$$

is thereby related to b and thus to the kinetic energy E. The very approximate expression (7-71) can be transformed into an equation of the Geiger-Nuttall type,

$$\log_{10}\lambda = \left(21 - \frac{(8m)^{1/2} ze^2}{2.303 \, \hbar}\right) ZE^{-1/2} = (21 - 1.0941)\, ZE^{-1/2} \tag{7-72}$$

which may be compared with the formulae presented in Section 7.3. Thus the quantum-mechanical theory of barrier penetration is able to account well for α decay. More exact treatments have been presented by Winslow and Simpson [Wi 52], while Preston [Pr 47] has given a rigorous solution avoiding the use of WKB approximation methods—though his solution of the wave equation essentially corresponds with the WKB solution. A survey of all such treatments and of α-decay systematics in general has been given by Perlman and Rasmussen [Pe 57]. Other good review articles in this field are those by Hanna [Ha 59], Mang [Ma 64], and Rasmussen [Ra 65a].

Many of the features of α decay can be explained satisfactorily by treating the nucleus on the basis of the liquid-drop model, but in several instances, shell-model effects point to the necessity of using an independent-particle approach. There has recently also been some progress in the application of

Bardeen–Cooper–Schrieffer superconductivity theory to the nuclear properties involved in α decay [Ma 64], but these considerations lie beyond the
scope of the present account.

One should note, however, that the simple theory as outlined here presupposes that α particles are actually formed within the nucleus and are
subsequently (i.e., after some 10^{38} abortive attempts) emitted by tunneling

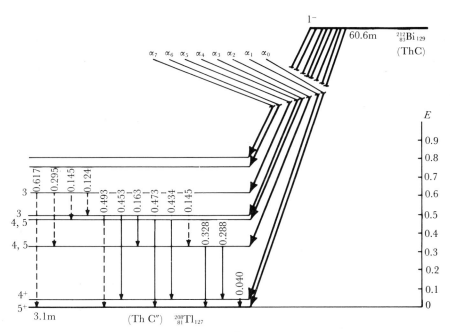

Fig. 7-14. Decay scheme of ^{212}Bi ($T_{1/2} = 60.6$ m) to ^{208}Tl levels showing the α
transitions listed in Table 7-2 and the ensuing α transitions. The ground-state energy
difference is 6.09 MeV. The spin-parity assignments are somewhat tentative.

through the potential barrier. The extent to which α particles are pre-formed
[Pe 50] inside a nucleus is still a matter for dispute. Basically, in α-particle
pre-formation, it is required that the wave functions of two protons and two
neutrons in a nucleus overlap simultaneously. In view of the comparatively
high nucleon density, this is quite likely to occur frequently. In different
theories the probability for a nucleus under various circumstances to be
composed of nucleon clusters which resemble α particles ranges from 10 to
100 percent. Hence the contention of 100 percent pre-formation constitutes
a weak point in Gamow's theory. Nevertheless, it *does* have some backing from
empirical evidence, and it does provide an explanation—admittedly, somewhat pragmatic—for the observation that there are occasional exceptions to

the rule that the relative intensity I_E of α particles emitted with an energy E rises with increasing E. We can examine this in the case of the FINE STRUCTURE of α-ray spectra. Apart from α transitions from the ground state of the parent nucleus to the *ground* state of the daughter nuclide, transitions to *excited* states can also take place, as illustrated in Fig. 7-14 for the well-known case of ^{212}Bi (ThC) decay. Because the daughters lie close together, the spectrum of the α group shows several discrete lines corresponding to but slightly differing α energies E up to the maximum decay energy $E_0 = 6.090$ MeV. One would expect the intensities to obey a relation of the form

$$I_E = I_{E_0} \exp\left[-170\, E_0^{-3/2}(E_0 - E)\right] \tag{7-73}$$

and to diminish as E decreases, but this is not observed. The intensities fluctuate in the manner shown in Table 7-2.

Table 7-2. ENERGIES AND RELATIVE INTENSITIES
OF THE FINE-STRUCTURE α GROUP FROM ^{212}Bi (ThC)

Radiation	α Energy E [MeV]	Relative intensity I_E (percent)
α_0	6.090	27.2
α_1	6.050	70.0
α_2	5.764	1.7
α_3	5.621	0.1
α_4	5.601	1.0
α_5	5.480	0.01
α_6	5.341	0.001
α_7	5.298	0.0001

7.5. Short- and Long-Range α Radiation

The designation SHORT-RANGE α RAYS refers to those α groups which make a transition to an *excited* state of the daughter nucleus and therefore, because their energy is less than that of the decay radiation proceeding to the daughter ground state, have a lower range than their companions.

On the other hand, LONG-RANGE α RAYS are those emitted from excited states of the decaying nucleus and therefore of higher energy and range than those which stem from ground-state decay. Thus, for example, it is possible for *excited* levels of ThC' ($^{212}_{84}$Po) formed by β decay of ThC($^{212}_{83}$Bi) to decay not only by γ emission but also occasionally by an α transition to the ground state of ThD ($^{208}_{82}$Pb), which is stable.

7.6. Application of the Gamow Formula to α Decay

7.6.1. α Energy and Intensity

The Gamow formula (7-35) shows that the α emission probability increases rapidly with increasing energy. One would therefore expect that in the α decay of any particular nuclide, those α particles would emerge most readily,

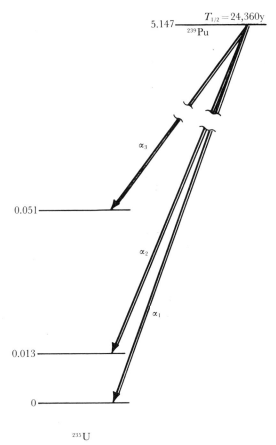

Fig. 7-15. The decay of ^{239}Pu ($T_{1/2} = 24{,}360$ y) to stable ^{235}U, in which three α-particle groups participate, as listed in Table 7-3.

i.e., with the highest intensity, that have the highest energy, since for them the barrier is commensurately narrow. As foreshadowed in the previous paragraph, this is in general also the case. Thus, in the α decay of ^{239}Pu to ^{235}U one can observe three main groups of α particles, as shown in Fig. 7-15, whose properties are listed in Table 7-3.

Thus, even though the difference in energy between α_1 and α_2 is only 0.25 percent, it suffices to occasion an approximately 4-fold difference in relative intensity.

Table 7-3. ENERGIES AND RELATIVE INTENSITIES OF THE MAIN α GROUPS IN THE DECAY OF ^{239}Pu TO ^{235}U

Transition	α Energy E [MeV]	Relative intensity [percent]
α_1	5.147	72.5
α_2	5.134	16.8
α_3	5.096	10.7

7.6.2. NUCLEAR RADIUS CONSTANT

From the form of the Gamow relation which applies at low energies, when $E \ll E_C$,

$$\lambda = \lambda_0 e^{-G} = \lambda_0 \exp\left[-(A - BR^{1/2})\right] \qquad (7\text{-}74)$$

where A and B are radius-independent parameters given by (7-68), it is evident that if the decay constant λ is known from α-decay lifetime measurements, the nuclear radius R can be determined for each α-active nuclide, and thence the radius constant $r_0 = RA^{-1/3}$ evaluated. This was in fact the way r_0 was first determined. The value so derived tends to be a little high. To illustrate this we collect in Table 7-4 some of the values of r_0 obtained by Kaplan [Ka 51] when he applied a somewhat more rigorous theoretical approach than that above to some even-even α emitters. Kaplan's data yield the result

$$r_0 = 1.57 \pm 0.015 \, \text{fm} \qquad (7\text{-}75)$$

Table 7-4. RADIUS CONSTANT r_0 AS DETERMINED FROM THE DECAY CONSTANT λ AND α ENERGY OF HEAVY EVEN-EVEN α-EMITTING NUCLEI

Parent nucleus	Decay energy E [MeV]	$\log_{10} \lambda$	r_0 [fm]
^{208}Po	5.24	-9.867	1.43
^{214}Po	7.83	3.666	1.56
^{222}Ra	6.62	-1.745	1.59
^{226}Th	6.41	-3.426	1.58
^{236}Pu	5.85	-8.090	1.56
^{238}U	4.25	-17.312	1.59
^{240}Pu	5.20	-11.437	1.60

7.6.3. SPONTANEOUS NUCLEAR DISINTEGRATION

Spontaneous disintegration of a nucleus, as by α decay, is repressed by a high Gamow factor G. The large values Z and b for heavy nuclei tend to make G so large that α decay is practically prohibited. For example, we can examine the likelihood of α decay in the region of nuclei with $A = 200$. Specifically, let us examine the hypothetical decay

$$^{200}_{82}\text{Pb} \xrightarrow{?} {}^{196}_{80}\text{Hg} + {}^{4}_{2}\text{He} + 3.3 \text{ MeV} \tag{7-76}$$

If we set the kinetic energy of the α particle equal to the Q-value and insert values for the daughter nuclide, we obtain the result

$$b = \frac{Zze^2}{Q} = 69.81 \text{ fm} \tag{7-77}$$

The thickness of the barrier can be appreciated on comparing this value with the nuclear radius

$$R = r_0 A^{1/3} = 1.4 \times (196)^{1/3} \text{ fm} = 8.13 \text{ fm} \tag{7-78}$$

We thus expect a very large Gamow factor, corresponding to an extremely low transparency. On numerical substitution in (7-70) we find

$$G \approx 111 \tag{7-79}$$

so that

$$\lambda \simeq 10^{21} e^{-111} \tag{7-80}$$

giving a mean lifetime

$$\tau \equiv 1/\lambda \simeq 10^{27} \text{ sec} \approx 3 \times 10^{19} \text{ y} \tag{7-81}$$

Even with the more correct expression (7-64), one finds (using Martin's table [Ma 54b]) that $G \simeq 99$ and $\tau \simeq 10^{22} \text{ sec} \approx 3 \times 10^{14}$ y. Hence ^{200}Pb is very stable against α decay. It can, however, decay to $^{200}_{81}$Ti by electron capture, for which process it has a mean lifetime of about 30 h.

Of course, an increase in Q would have led to a marked decrease in b and thence to an appreciable diminution in the Gamow factor G. The Q-values (α-disintegration energies) for α decay of the next isobars $^{200}_{83}$Bi, $^{200}_{84}$Po, $^{200}_{85}$At are, respectively, 5.190, 5.980, and 6.547 MeV. When *these* higher values are inserted in the calculation, one finds that ^{200}Bi is only just α stable, whereas the next members are α unstable (the respective values of b are 45, 39, and 36 fm, and of G in the above rough approximation are 90, 84, and 81).

Furthermore, with increasing Z, as one comes into the region of the trans-uranic elements Pu, Am, Cm, Bk, Cf, Es, Fm, Md, Lo, Lw, and Kh, the α-decay instability is complemented by the still greater likelihood of spontaneous fission. There is therefore only faint hope of finding any very heavy *stable* nuclides in the future (the lifetimes of those as yet undetected are expected generally to be of the order of 1 sec or less), but at the same time this expectation should not be interpreted too dogmatically. There are, indeed, some indications that on going beyond khurchatovium ($Z = 104$) one may encounter fairly stable nuclides at $Z = 114$ and $Z = 126$. These might be produced by bombarding very heavy nuclei with high-energy heavy ions, e.g., via reactions [Wo 67] such as ^{231}Pa $(^{81}$Br, 2n$)^{310}$ 126 or ^{251}Cf $(^{62}$Ni, 3n$)^{310}$ 126, or by bringing about fusion reactions (with a lower Q-value) between medium-heavy nuclei. Such projects are now underway. The production of element $Z = 105$ through bombardment of ^{242}Am by ^{22}Ne has already been reported.

EXERCISES

7-1   Estimate the width of the α line of ^{222}Rn at room temperature, given that $T_{1/2} = 3.823$ d and $E = 5.49$ MeV.

7-2   What is the probability for P-wave thermal neutrons ($E_n = 0.025$ eV) to penetrate into a lead nucleus? [Standard integral:

$$\int \frac{(a^2 - x^2)^{1/2}}{x} \, dx = (a^2 - x^2)^{1/2} - a \ln\left(\frac{a + (a^2 - x^2)^{1/2}}{x}\right)]$$

7-3   The nucleus ^{212}Po in its ground state has a half-life $T_{1/2} = 3 \times 10^{-7}$ sec toward emission of 8.78-MeV α particles. What is the probability for such α particles to penetrate into ^{208}Pb upon making a central collision?

7-4   The majority of even-even nuclei have an α-decay half-life which can be represented by the relation (cf. Eq. (7-72))

$$\log_{10} T_{1/2}^{[y]} = 1.61\left(\frac{Z}{(E^{[\text{MeV}]})^{1/2}} - Z^{2/3}\right) - 28.9$$

Substantiate this formula with the aid of Gamow Coulomb-barrier penetrability theory, using a suitable approximation for the nuclear radius. What numerical values of the coefficients would be expected on the basis of this simple theory?

7-5   Parity conservation in nuclear interactions can be tested experimentally by searching for the occurrence of parity-forbidden α-decay transitions. (a) In the first such investigation, Tanner [Ta 57] studied the ^{19}F$(p,\alpha)^{16}$O reaction at the resonance energy $E_p = 0.340$ MeV (lab). Use energy, momentum, and parity considerations applied to the data below to deduce which of the states in ^{16}O are accessible to α decay of the intermediate 1^+ (13.19 MeV) state in ^{20}Ne*. (b) A variant of this experiment [Al 61, Se 61c, Se 61d] consists of studying the α decay of ^{16}O following the β^- decay of the ground state of ^{16}N. At which (lab) energy in the α-decay spectrum would the presence of a peak provide evidence of parity violation?

(c) What interpretation can be placed upon the observation (cf. Fig. 3 of [Se 61d]) of a pronounced peak at 1.73 MeV (lab) in the measured α-particle spectrum, flanked by essentially zero counts between 0.9 and 2.3 MeV?

(d) Look up the original references (together with [Wi 58]) for information and deductions which well-deserve attention. [Ground-state spin-particles: $J^{\pi}(^{16}\text{N}) = 2^-$, $J^{\pi}(^{19}\text{F}) = \frac{1}{2}^+$. Q-value of the ^{19}F(p,α) reaction: $Q = 8.118$ MeV. Level scheme of ^{16}O:

$E^*[\text{MeV}] = 0,\ 6.05,\ 6.13,\ 6.92,\ 7.12,\ 8.88,\ 9.59,\ 9.85,\ 10.36$

$J^{\pi} = 0^+,\ 0^+,\ 3^-,\quad 2^+,\quad 1^-,\quad 2^-,\quad 1^-,\quad 2^+,\quad 4^+$

On taking the ^{16}O ground state as energy origin, the ^{16}N ground state lies at 10.40 MeV and (^{12}C + α) at 7.15 MeV.]

7-6 Perform an exact calculation of the transmission coefficient for a particle of energy $E = \frac{1}{2}U_0$ to tunnel through a rectangular barrier of height U_0 and breadth b, and compare the result with the expression derived by using the approximate formula

$$T = \exp\left\{-2\int_a^b \left[\frac{2M}{\hbar^2}(V(r) - E)\right]^{1/2} dx\right\}$$

7-7 Because of reflection effects, one finds that even for *attractive* potential wells the transmission coefficient for a wave of incident low-energy particles is *less* than unity. Confirm this through a comparison of the transmission coefficients for particles of mass m and energy $0 \leqslant E < U_0$ impinging upon (a) a repulsive, (b) an attractive rectangular potential of height U_0 and width b.

7-8 What is the orbital momentum l (and how large is the absolute momentum) of a neutron emitted in decay of an excited ^{58}Ni nucleus whose potential barrier has exactly the same height as that for an S-wave proton?

7-9 The decay chain of $^{226}_{88}$Ra passes through three α-decay stages and terminates in the α stable nuclide $^{214}_{82}$Pb:

$$^{226}_{88}\text{Ra} \xrightarrow[\alpha_1]{4.78 \text{ MeV}} {}^{222}_{86}\text{Rn} \xrightarrow[\alpha_2]{5.49 \text{ MeV}} {}^{218}_{84}\text{Po} \xrightarrow[\alpha_3]{6.00 \text{ MeV}} {}^{214}_{82}\text{Pb}$$

Using these data and the atomic masses cited below, demonstrate how it comes about that the direct transition

$$^{226}_{88}\text{Ra} \to {}^{214}_{82}\text{Pb} + {}^{12}_{6}\text{C}$$

is energetically permitted but is nevertheless unobservable.

[Nuclear radius: $R = r_0 A^{1/3} = 1.4\, A^{1/3}$ fm.

Atomic masses: $M(^{12}\text{C}) = 12.000\ 000$ u;

$M(^{226}\text{Ra}) = 226.023\ 528$ u;

$M(^{214}\text{Pb}) = 213.999\ 766$ u.]

7-10 The ground state of ^{212}Po decays by α emission ($T_{1/2} = 3 \times 10^{-7}$ sec; $E = 8.78$ MeV). Furthermore, this nucleus has an isomeric excited state at 2.93 MeV which has a half-life $T'_{1/2} = 46$ sec and emits 11.65-MeV α particles. The explanation of the fact that the isomeric state has a much longer decay lifetime than the ground state despite its larger α-decay energy lies in the higher orbital momentum with which the α particles are emitted. Estimate how large this orbital momentum actually is.

[In calculating the dependence of the decay probability on the orbital momentum, assume that the energies of the α particles and also the magnitude of the centrifugal barrier are small compared with the Coulomb barrier height. For the nuclear radius, use the expression $R = r_0 A^{1/3} = 1.4\, A^{1/3}$ fm. It may be useful to employ the following standard integral:

$$\int \frac{dx}{x(bx + cx^2)^{1/2}} = \frac{-2(bx + cx^2)^{1/2}}{bx}]$$

Chapter 8

BETA DECAY
The Weak Beta-Decay Interaction

8.1. Introduction

The investigations of β decay and weak interactions have ushered in some of the most significant changes through which physics has passed in this century. The very foundations of certain physical ideas and concepts have been shaken and, in some cases, radically modified. Even now the subject of weak interactions may well represent the Enigma Variations in the realm of physics; many of its features have hitherto resisted elucidation or, at best, lent themselves only to contrived explanation.

Thus at the outset of β-decay studies, it was the principle of energy conservation, the very cornerstone of physical theory, that appeared to be in jeopardy. Rather than abandon this fundamental support as some of the leading physicists were reluctantly reconciled to doing (e.g., Bohr in his Faraday Lecture in 1930 [Bo 32]), Pauli was impelled to seek a way out of the difficulty by the bold measure of postulating the existence of a particle, the (anti)neutrino, possessed of remarkable properties and not susceptible to detection by the methods of that time. Indeed, although the introduction of this provisionally hypothetical particle resolved the energy dilemma and enabled Fermi to formulate a theory of β decay which proved to be outstandingly successful, it needed a quarter century of intensive effort before the neutrino's existence was actually verified and a satisfactory explanation of the interaction mode derived.

Just as it began to seem that we truly understood the β-decay process and other weak interactions, the totally unexpected phenomenon of parity violation once more upset the conceptual basis. So profound was its influence that it rendered much of our "pre-parity" thinking invalid and demanded a re-examination of many other tenets that had hitherto been treated as virtually self-evident. The tremendous importance now accorded to symmetry principles, conservation rules, and applications of group theory to nuclear and particle physics is a direct consequence of the new and novel way of thought in physics which stemmed from attempts to cope with the problems of β decay, weak interactions, and electron-muon kinship.

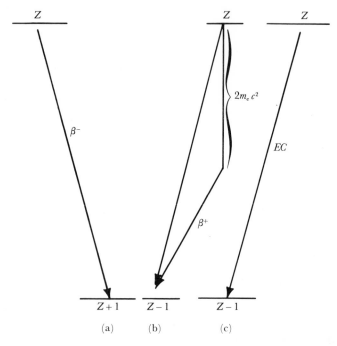

Fig. 8-1. Schematic representation of decay schemes in which the mass number A is left unchanged: (a) a β^- transition; (b) a β^+ transition, indicated in two alternative representations (cf. Fig. 8-3); (c) an electron capture transition.

Beta decay has formed the subject of many extensive review articles, including several of recent origin [Ko 62, Si 65, Scho 66, Wu 66] which provide and amplify much of the information featured in the present chapter.

8.1.1. DECAY MODES

The term β decay embraces all modes of nuclear decay in which the atomic number Z (i.e., the nuclear charge) changes by one unit while the mass number A remains constant. It therefore covers not only β^- and β^+ decay, but also electron capture, as depicted in Fig. 8-1(a–c).

In β^- DECAY, a negative electron is emitted from the nucleus, the nuclear charge changing from Z to $Z + 1$ in units of e. Hence the β-active element is displaced one position to the right in the periodic system. The *nuclear* emission of negative electrons is, of course, not to be confused with the emission of electrons from *atomic* orbits as in the conversion or Auger effect.

In β^+ DECAY, a positive electron, viz. a positron, is emitted, and the nuclear charge decreases by one unit from Z to $Z - 1$. The element is displaced one position to the left in the periodic system. To preserve atomic charge neutrality, a negative electron is also ejected from an outer *atomic* orbit.

In ELECTRON CAPTURE, the nucleus absorbs one of the negative electrons from an inner *atomic* shell and thereby alters its charge from Z to $Z - 1$. The atom as a whole remains neutral, but is left in an excited state because a vacancy has been created in one of the inner shells. Electron capture from the innermost atomic shell, the K shell (K-ELECTRON CAPTURE), dominates over that from L, M, etc. shells, since there is a higher expectation probability for K electrons to approach within the nuclear volume than for those from outer orbits.

In many cases, β decay does not lead directly to the ground state of the daughter nuclide, but rather to one of the excited states which then de-excites by emission of γ quanta and/or conversion electrons as discussed in the next chapter.

The vacancies which arise in inner atomic electron shells through an electron capture or conversion process engender the emission of characteristic x radiation (or of Auger electrons) by the daughter element, viz. a line spectrum of discrete energy. Also, the sudden change in nuclear charge affects the atomic electron cloud. Hence the actual process of nuclear β decay can be associated with the emission of atomic electrons (Auger electrons, conversion electrons), and characteristic electromagnetic radiation. All these extranuclear radiations are emitted with discrete energies, in contrast to the actual nuclear β-decay radiation whose energy ranges over a continuous spectrum which terminates at a definite maximal energy equal to the transition energy.

8.1.2. MASS-ENERGY BALANCE

Beta decay can take place only if the binding energy of the daughter nuclide exceeds that of the parent, as discussed in Section 2.4.4 in connection with isobaric mass parabolas. The calculations presented there require slight modification in that whereas they treated the mass differences between neutral *atoms*, the present considerations necessitate examination of mass differences between *nuclei*, and these may differ from the above in certain modes of β decay.

Thus in the case of β^- DECAY, the mass of the parent nucleus $M(Z, A) - Zm_e$ decomposes into that of the daughter nucleus $M(Z + 1, A) - (Z + 1)m_e$ and that of the β^- particle m_e. Hence the transition energy is

$$\Delta E^{(\beta^-)} = \{[M(Z, A) - Zm_e] - [M(Z + 1, A) - (Z + 1)\, m_e + m_e]\}\, c^2 \quad (8\text{-}1)$$

i.e.,

$$\Delta E^{(\beta^-)} = [M(Z, A) - M(Z + 1, A)]\, c^2 \qquad\qquad (8\text{-}2)$$

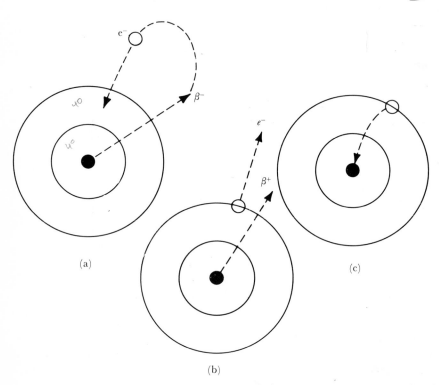

(a)

(b)

(c)

Fig. 8-2. Symbolic representation of the occurrence of (a) β^- emission, (b) β^+ emission, (c) electron capture by the nucleus.

the difference of the nuclear masses being the same as the difference of the atomic masses.

In β^+ DECAY, however, this simple relation does not apply, for as the nuclear charge decreases by 1 unit, an electron must also be emitted from the atomic cloud to preserve charge neutrality. The mass of the parent nucleus $M(Z, A) - Zm_e$ decomposes into that of the daughter, $M(Z - 1, A) - (Z - 1)\, m_e$ and that of the positron, m_e. The transition energy is, accordingly,

$$\Delta E^{(\beta^+)} = \{[M(Z, A) - Zm_e] - [M(Z - 1, A) - (Z - 1)\, m_e + m_e]\}\, c^2 \quad (8\text{-}3)$$

i.e.,

$$\Delta E^{(\beta^+)} = [M(Z, A) - M(Z - 1, A) - 2m_e] c^2 \qquad (8\text{-}4)$$

the atomic mass difference being diminished by two electron masses.

In ELECTRON CAPTURE, the combined mass of the parent nucleus $M(Z, A) - Zm_e$ and of the captured electron m_e goes over into the mass of the daughter nucleus, $M(Z - 1, A) - (Z - 1) m_e$. The transition energy is therefore

$$\Delta E^{(\text{EC})} = \{[M(Z, A) - Zm_e + m_e] - [M(Z - 1, A) - (Z - 1) m_e]\} c^2 \qquad (8\text{-}5)$$

i.e.,

$$\Delta E^{(\text{EC})} = [M(Z, A) - M(Z - 1, A)] c^2 \qquad (8\text{-}6)$$

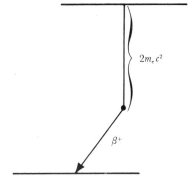

Fig. 8-3. Conventional representation of a β^+ transition.

In a purely symbolic representation these three processes and their mass characteristics can be depicted by the atomic schemes shown in Fig. 8-2(a–c).

These processes can occur only if ΔE is *positive*. This condition is fulfilled in the case of β^- decay and electron capture whenever the mass of the parent atom exceeds that of the daughter, whereas β^+ decay requires the atomic mass difference to be at least $2m_e \triangleq 1.022$ MeV. For this reason, the latter process is by convention frequently depicted in the way shown in Fig. 8-3. As β^+ decay and electron capture both lead to the same daughter nuclide $(Z - 1, A)$, electron capture inevitably accompanies β^+ decay as a competing process. However, when the atomic mass difference lies between zero and $2m_e$ only electron capture can take place, since β^+ decay is energetically forbidden.

8.1.3. BETA-ENERGY SPECTRUM

As depicted in Fig. 8-4, the energy spectrum of β particles forms a smooth continuum on which lines due to Auger electrons and conversion electrons are superimposed. The high-energy end of the continuum terminates abruptly at a specific energy which equals the transition energy, as expressed by Eqs. (8-2), (8-4), or (8-6).

The experimental finding that β particles, instead of being emitted with a single discrete energy equal to the transition energy, are in fact emitted over

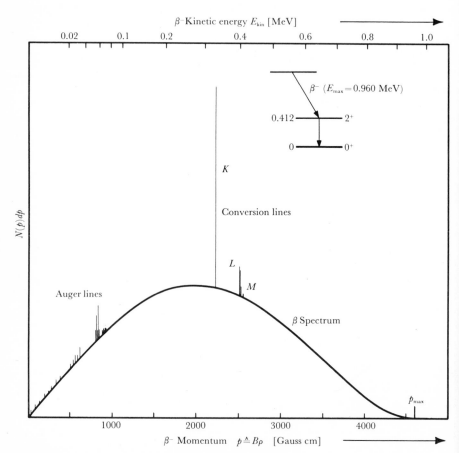

Fig. 8-4. The energy spectrum (in momentum representation) of β^- particles from a specimen nucleus in the vicinity of $Z = 80$ (the data actually refer to the decay of $^{198}_{79}$Au to $^{198}_{80}$Hg). Superposed on the β-spectrum are lines (depicted on a greatly reduced scale) due to the Auger effect (below $B\rho \simeq 1000$ Gauss cm) and to K-, L-, and M-shell internal conversion of the 0.412-MeV transition to the ground state of ^{198}Hg. The line spectra had to be drawn on a reduced scale in order to accommodate them, since the area under the ensemble of peaks is commensurate with that under the continuous spectrum.

an entire continuous energy range up to a maximal energy $E_{max} = E_{transition}$ presented great difficulties in the interpretation and comprehension of the decay process. It even led eminent authorities to consider abandoning the fundamental principle of energy conservation. What proved so perplexing was the fact that the other decay modes, namely, α and γ decay, showed no such inexplicable features. Emission occurred at discrete energies equal to the transition energy (diminished by an appropriate recoil correction) between initial and final states of definite energy. Why should β particles display such an evident energy spread, totally at variance with two-body kinematics?

Yet at the same time, the spectral cutoff at a maximal energy equal to the sharp transition energy between discrete nuclear states constituted evidence showing that energy must be conserved in some manner. Arguments in support of energy conservation could also be adduced from measurements for such processes as the following:

$$
\begin{array}{ccc}
 & {}^{212}_{84}\mathrm{Po} & \\
\overset{\beta^-}{\nearrow} & & \overset{\alpha}{\searrow} \\
{}^{212}_{83}\mathrm{Bi} & & {}^{208}_{82}\mathrm{Pb} \\
\underset{\alpha}{\searrow} & & \nearrow \\
 & {}^{208}_{81}\mathrm{Tl} \;\; \overset{\beta^- + \gamma}{} &
\end{array}
\tag{8-7}
$$

The sum of the decay energies for the upper sequence ${}^{212}\mathrm{Bi} \overset{\beta^-}{\longrightarrow} {}^{212}\mathrm{Po} \overset{\alpha}{\longrightarrow}$ ${}^{208}\mathrm{Pb}$ proved empirically to be exactly equal to that for the lower sequence ${}^{212}\mathrm{Bi} \overset{\alpha}{\longrightarrow} {}^{208}\mathrm{Tl} \overset{\beta^-}{\longrightarrow} {}^{208}\mathrm{Pb}$ if for the respective β transitions the *maximal*, and not the mean, β energy is inserted.

The possibility that β particles may indeed have been emitted with a discrete energy, but that some of the latter was lost and "smeared out" by absorption in the source was disproved by calorimetric measurements [El 27, Me 30, Zl 35, Zl 41, Zu 48, Je 50]. The β source was arranged inside a lead sheath which completely absorbed the entire kinetic energy of the emitted β particles (see Fig. 8-5). The heat developed in the source and sheath corresponded to the *mean* β energy, of course. If, therefore, *all* β particles were emitted with one and the same energy E_{max}, the quantity of heat generated would have been expected to equal the area bounded by the rectangle in Fig. 8-6, and not that under the shaded curve.

A way out of these difficulties was suggested by Pauli first in a communication to the radioactivity group at Tübingen in December 1930 (see [Wu 66, p. 385]), then at a Pasadena conference in 1931, and later at the Solvay Congress in 1933 [Pa 33]. As *two*-body kinematics could not be reconciled with the experimental data, Pauli postulated that a third particle features in the outgoing products of β decay, a particle of necessity electrically neutral and of a mass and magnetic moment which must either be zero or vanishingly small. It was accordingly termed the NEUTRINO, and was supposed to carry away the "missing" energy difference $E_{max} - E_\beta$. *Three*-body kinematics would be perfectly compatible with the observed energy distribution; apart from the small recoil energy of the daughter nucleus, the available transition energy would be shared between the β particle and the neutrino in such a way as to yield a continuum for the β energy alone (and a complementary continuum for the neutrino energy alone). On the basis of this hypothesis, the properties to be assigned to the neutrino could then be established, and a search for the evascent particle commenced.

Absorber (Pb)   Sample

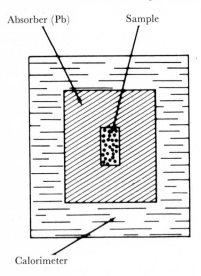

Calorimeter

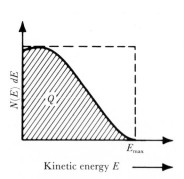

Kinetic energy E ⟶

Fig. 8-5. Experimental arrangement used to measure (mean) β-decay energies by total absorption in a lead sheath surrounded by a calorimeter.

Fig. 8-6. Diagram to indicate discrepancy between a calorimetric measurement of the β-decay energy and the energy magnitude to be expected if all β particles were emitted with an energy E_{max}.

8.2. The Neutrino

8.2.1. Neutrino Properties

If the neutrino emitted simultaneously with the β particle in $\beta^{\pm}$ decay has a certain energy but zero rest mass ($m_\nu = 0$), its velocity v must equal that of light (c). This follows directly from the relativistic expression for the total energy,

$$E = m_\nu c^2 \left(1 - \frac{v^2}{c^2}\right)^{-1/2} \tag{8-8}$$

Experiments show that the neutrino rest mass is at most very small indeed, $m_\nu < 250$ eV (i.e., less than 0.05 percent of the electron's rest mass), and therefore it seems reasonable to set it to zero:

$$m_\nu \cong 0 \qquad (8\text{-}9)$$

Since the calorimetric experiments and other investigations failed to provide any indication whatsoever that the β's were accompanied by other energetic particles, one may conclude that such particles, if they exist, can have only a *weak interaction* with matter. The weakness of this interaction was, in fact, responsible for the neutrino's having eluded detection then and for a long time afterward.

On the assumption of angular momentum conservation the intrinsic spin of the neutrino can be established to be $s = \frac{1}{2}$. In β decay, the mass number A does not change. Therefore the spin of the nucleus cannot change from an integer to a half-integer value, or *vice versa*. Since the β particle has spin $\frac{1}{2}$, momentum conservation demands that the neutrino's spin quantum number must also be half integral. An examination of β-decay reactions shows that only the lowest half-integer value is permissible, viz. $s = \frac{1}{2}$. An example of such a process is radiocarbon decay,

$$^{14}\text{C} \rightarrow {}^{14}\text{N} + \beta^- + \bar{\nu} \qquad (8\text{-}10)$$

Spin: $\qquad 0 \qquad\quad 1 \quad\; \frac{1}{2} \quad\; \frac{1}{2}$

As regards electric charge, magnetic moment, rest mass, and relativistic velocity, the neutrino and the photon can be regarded as equal. They differ only in their spin and therefore in the type of statistics which describes their behavior. Neutrinos, being spin-$\frac{1}{2}$ particles, are fermions, whereas photons, whose spin is 1, are bosons. Neutrinos are therefore subject to the Pauli exclusion principle, while photons, being bosons, are not.

Subsequent studies have shown that there are in fact *two* kinds of neutrino, each with its respective antiparticle, namely,

the BETA-NEUTRINO, ν_e, emitted in β^+ decay: $\qquad p \rightarrow n + \beta^+ + \nu_e$

the BETA ANTINEUTRINO, $\bar{\nu}_e$, emitted in β^- decay: $\qquad n \rightarrow p + \beta^- + \bar{\nu}_e$

the MU NEUTRINO, ν_μ, emitted in $\pi^+ \rightarrow \mu^+$-decay: $\qquad \pi^+ \rightarrow \mu^+ + \nu_\mu$

the MU ANTINEUTRINO, $\bar{\nu}_\mu$, emitted in $\pi^- \rightarrow \mu^-$ decay: $\pi^- \rightarrow \mu^- + \bar{\nu}_\mu$

Each of these neutrinos is a LEPTON. The β-decay process from which they ensue can be regarded as essentially the conversion of a proton to a neutron in the nucleus, or vice versa. It should be noted that the lifetime of *bound* neutrons within a nucleus is entirely unrelated to that of *free* neutrons ($\tau = 1.01 \times 10^3$ sec); according to the Heisenberg-Yukawa theory, bound neutrons spend a fraction of their time as protons and are therefore no longer in the same way susceptible to decay. Hence no valid direct connection can be envisaged between β-decay lifetimes and the free neutron lifetime.

8.2.2. Neutrino Hunting

Even indirect methods designed to detect neutrinos are inherently complicated and difficult to carry out. The detection of a massless, momentless, uncharged relativistic particle that enters into reaction only through a very *weak* interaction posed considerable problems and involved the development of special delicate, high-efficiency techniques. The earliest successful methods, which provided only an indirect verification of the neutrino's existence, studied the recoil of a nucleus upon electron capture or the momentum distribution among the products of a β^--decay process.

The RECOIL ENERGY of a nucleus as it emits a neutrino on electron capture is given by the nonrelativistic relation

$$E_R = \frac{p_\nu^2}{2Am_\mathcal{N}} = \frac{E_0^2}{2Am_\mathcal{N}c^2} \qquad (8\text{-}11)$$

where $p_\nu \equiv E_0/c$ is the relativistic linear momentum of the (anti) neutrino and E_0 is the transition energy. Since a low value of A is desirable in order that the small quantity E_R be as large as possible, the electron-capture reaction was studied with light nuclei. Thus Davis [Da 52a] investigated K capture in the reaction

$$_4^7\text{Be} + e_K^- \rightarrow {}_3^7\text{Li} + \bar{\nu}_e + 0.86 \text{ MeV} \qquad (E_R = 56 \text{ eV}) \qquad (8\text{-}12)$$

and obtained results compatible with antineutrino emission. However, the varying degree of freshness and purity of the beryllium surface influenced his measurements; also, the interpretation of his data was affected by the complication that when a molecule breaks up, it is difficult to establish which molecular fragment acquires the electric charge by which its recoil energy is experimentally determined. On the other hand, the advantages of studying a process in which the momentum is distributed between only *two* emergent particles led to attempts to overcome the molecular and solid-state drawbacks by using monatomic specimens. Thus, more recent recoil studies [Lau 58, Ba 58, La 59, Bu 59] have used ^{8}Li and ^{24}Na. Also, since the rare gases are monatomic, Rodeback and Allen [Ro 52a] elected to study the K-capture process with ^{37}Ar,

$$_{18}^{37}\text{Ar} + e_K^- \xrightarrow[\tau = 50.6 \text{ d}]{} {}_{17}^{37}\text{Cl} + \bar{\nu}_e + 0.814 \text{ MeV} \qquad (8\text{-}13)$$

in which the difficulty of measuring ions that have a very small recoil energy ($E_R = 9.6$ eV) is offset by the above advantage and by the convenience of using a gaseous source whose daughter product is formed in the *ground state* without any interfering β^+ decay or γ decay. The ion velocity was determined from the flight time as deduced by a delayed coincidence technique. A sharp peak in excellent accord with that expected on the basis of neutrino co-emission was obtained, as shown in Fig. 8-7. The maximum time of flight so determined over the measured distance, $t = 8.9 \pm 0.9$ μsec, corresponds to a neutrino momentum of 0.8 ± 0.1 MeV/c in good agreement with the ^{37}Ar-^{37}Cl mass difference of 0.814 MeV.

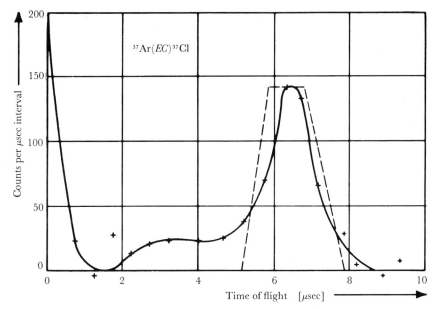

Fig. 8-7. Distribution of ^{37}Cl recoil velocities, measured by a time-of-flight technique over a 6-cm path by Rodeback and Allen [Ro 52a]. The dashed spectrum indicates the theoretical expectation for co-emission of a neutrino in the electron capture process ^{37}Ar $\xrightarrow{EC}$ ^{37}Cl. The measurements, with 20-channel timing circuits triggered by emission of an Auger electron, indicate the recoil velocity to be 0.71 ± 0.06 cm μsec^{-1}, which corresponds to an energy of 9.7 ± 0.8 eV for which the dashed spectrum has been calculated. The counts below 5 μsec are specious coincidences due in the main to scattered Auger electrons. The presence of negative counts can be attributed to the subtraction of a background contribution from the measured counting rate.

A detailed survey of these and other recoil experiments, such as those in which the recoil momentum is shared among several particles in a β^--decay process, has been given by Kofoed-Hansen [Ko 62, Ko 65]. All these studies furnish evidence in support of neutrino existence, but do not constitute unequivocal proof in that they do not observe or identify neutrinos as such directly. Neutrinos in their free state long eluded DIRECT DETECTION, despite many attempts, extending over a quarter century from the time of Pauli's postulate, to observe neutrino interactions directly via such effects as a possible magnetic-moment ionization interaction in elastic collisions with atomic electrons [Ch 34, Ba 50, Hou 54].

The painstaking investigations by Reines and Cowan, with co-workers, originally also proved inconclusive (cf. initial proposal [Re 53a] and report [Re 53b]), but then provided the first conclusive results [Co 56b, Re 57], which were subsequently refined [Re 59, Ca 59, Re 60, Ne 66] in accuracy as improved techniques became available. The experiments studied the INVERSE β-DECAY process, which can be represented as

$$p^+ + \bar{\nu}_e \rightarrow n + \beta^+ \tag{8-14}$$

and which constitutes a reaction that can be initiated directly *only* by a free antineutrino, though with a vanishingly small cross section, around 10^{-43} cm^2, at the energies under consideration. This necessitated elaborate detection techniques with very high incident antineutrino fluxes (around 10^{13} $\bar{\nu}_e$ cm^{-2} sec^{-1}) and elaborate shielding in order to derive unambiguous verification of the direct $p(\bar{\nu}_e, \beta^+)n$ process. The first definitive experiments [Re 59, Re 60] were identical in principle, but the later version [Re 60] featured improved details which we describe. A search was undertaken (using a high-intensity antineutrino beam from the fission products in a high-flux reactor at Savannah River impinging upon the protons of two large water targets, yielding about 3×10^{28} target protons, in which some CdCl$_2$ had been dissolved) for the following events: (i) the creation of a β^+ particle which is slowed down from within a given initial energy interval until it ·has been brought to rest and combines annihilatively with an atomic electron; (ii) the emission of the two simultaneous annihilation γ quanta, each of energy 0.511 MeV, which traverse the target in opposite directions and are registered· in ˙coincidence by two large scintillation detectors arranged on opposite sides of the target (because two target tanks were used, three scintillation detectors were needed); (iii) the production of a neutron which moves comparatively slowly without causing appreciable ionization and which after being slowed down further by multiple collisions is finally captured by a cadmium nucleus within about 10 μsec of its production. The occurrence of neutron capture in cadmium causes excitation of the nucleus to an energy which is dissipated by γ decay. Several cadmium-capture gammas whose energies sum to 9.1 MeV were detected in prompt coincidence in one or other of the two large scintillators some 10 μsec after the preceding β^+-annihilation gamma pulse.

The setup is shown schematically in Fig. 8-8. With this arrangement, about 1 antineutrino in 10^{20} interacts with a target proton. Signals from the scintillation detectors A and C arranged on either side of the 200-l water target B were observed by photomultipliers (110 to each scintillator tank) and fed by way of an elaborate electronic circuit to a twin-beam recording oscilloscope. Because of the high sensitivity of detection, special provision had to be made to screen the arrangement and to reject undesired events with the aid of anticoincidence suppression techniques. Only those combinations of coincidence signals which corresponded to two simultaneous 0.511-MeV annihilation pulses in A and C followed 10 μsec later by capture-gamma pulses totaling 9.1 MeV in both detectors were retained as evidence that a $\bar{\nu}_e$-capture inverse β-decay process had occurred. With this arrangement,

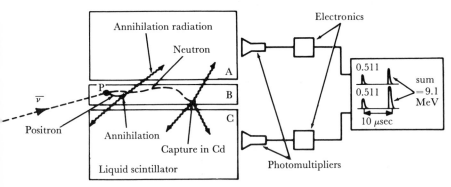

Fig. 8-8. Schematic representation of the equipment and operating principle utilized by Reines and Cowan [Co 56b, Re 57, 60, 59] in the experiments designed to detect free antineutrinos from the fission process in a high-flux reactor via the inverse β-decay process $p^+ + \bar{\nu}_e \rightarrow n + \beta^+$ taking place (e.g., at the point P) in a target chamber B containing an aqueous solution of CdCl$_2$ that served both as a source of hydrogen target nuclei and of neutron-absorbing Cd nuclei. To illustrate the basic principle, the figure has been simplified to show only one target chamber sandwiched between scintillation detectors, although in later investigations, *two* 200-1. target chambers, each containing $N = 1.1 \times 10^{28}$ target protons, were used, flanked by three large detector tanks to register annihilation γ quanta and neutron-capture γ radiation.

using two water targets sandwiched between three scintillator tanks of capacity 5400 l, an average counting rate per triad equal to $R = 1.5 \pm 0.1$ per hour was attained.

With the values $\epsilon_n = 0.17 \pm 0.06$ and $\epsilon_\beta = 0.15 \pm 0.02$ for the neutron- and β^+-detection efficiency as determined experimentally, and using calculated values of the number of target protons in each tank ($N = 1.1 \times 10^{28}$) in conjunction with an average antineutrino flux $\Phi = 1.2 \times 10^{13}$ $\bar{\nu}$ cm^{-2} sec^{-1} at the target, 15 m away from the reactor face, one obtains an experimental $\bar{\nu}_e$-absorption cross section,

$$\sigma_{\text{exp}} = (R/3600)/N\Phi\epsilon_n \epsilon_\beta = (1.2^{+0.7}_{-0.4}) \times 10^{-43} \text{ cm}^2 \qquad (8\text{-}15)$$

A later refinement of this experiment [Ne 66] attained far better energy resolution through the use of NaI(Tl) detectors to measure the annihilation gammas and a gadolinium-loaded hydrogeneous liquid scintillator target in which neutron capture was signaled by the emission of four gammas on an average, each with a mean energy around 2 MeV. With thick lead shielding and special circuit precautions, an appreciable reduction in background was obtained, and a further improvement in accuracy resulted from improved determinations of detection efficiency. With a net count rate of $R = 0.187 \pm 0.021$ h^{-1} and an enhanced antineutrino flux, $\Phi = 7.2 \times 10^{13}$ $\bar{\nu}$ cm^{-2} sec^{-1} incident on $N = 2.56 \times 10^{26}$ protons, the cross section for absorption of a fission $\bar{\nu}_e$ was determined as

$$\sigma_{\text{exp}} = (0.94 \pm 0.13) \times 10^{-43} \text{ cm}^2 \tag{8-16}$$

in fine agreement with the theoretically expected value

$$\sigma_{\text{th}} = (1.07 \pm 0.07) \times 10^{-43} \text{ cm}^2 \tag{8-17}$$

derived from the semiempirical formula [Le 57, He 64] for proton capture of essentially monoenergetic $\bar{\nu}_e$'s of energy $E^{[\text{MeV}]}$:

$$\sigma(E) = (0.223 \pm 0.005) \times 10^{-43} \times \left\{ \frac{E - 1.29}{0.26} [(E - 1.29)^2 - 0.26]^{1/2} \right\} \tag{8-18}$$

This is based on the assumption that 6.1 $\bar{\nu}_e$ are produced per fission and that the antineutrino can exist only with *one* of the two possible spin orientations in respect of the direction of its linear momentum (viz. either parallel *or* antiparallel spin, but not both): if antineutrinos could have *both* spin orientations, the theoretical cross section would be only one half of the above value [Le 57]. We know now that the *antineutrino* spin vector has the *same* direction as the linear momentum vector. The antineutrino has *right*-handed, or *positive* HELICITY, while the *neutrino* has *antiparallel* spin, i.e., *negative* helicity (Fig. 8-9). Neutrinos ν_e, $\bar{\nu}_e$ or ν_μ, $\bar{\nu}_\mu$ are therefore described by a TWO-COMPONENT THEORY, rather than a *four*-component theory which would correspond to ν's with parallel and antiparallel spins and $\bar{\nu}$'s with a similar dichotomy.

Detection of the ν_e is accordingly a *different* undertaking than an identification of an $\bar{\nu}_e$. As prototype of a ν_e-capture process, we may take the INVERSE β DECAY leading to β^- emission,

$$n + \nu_e \rightarrow p^+ + \beta^- \tag{8-19}$$

If ν_e and $\bar{\nu}_e$ were one and the same particle, one would expect *both* to initiate a reaction such as

$$^{35,37}\text{Cl} + \binom{\bar{\nu}_e}{\nu_e} \rightarrow {}^{35,37}\text{Ar} + \beta^- \tag{8-20}$$

An early attempt by Crane [Cr 39] to investigate the possibility of occurrence of the $^{35}\text{Cl}(\nu_e, \beta^-)\,^{35}\text{Ar}$ reaction with a laboratory ν source set an upper limit on the cross section ($\sigma < 10^{-30}$ cm^2), but its null result called for reinvestigation with stronger ν_e fluxes, and these were not available. In the absence of terrestrial ν_e sources or an adequate flux of solar ν_e's, Davis [Da 55, Da 56, Da 57b] used $\bar{\nu}_e$'s from a reactor at Brookhaven to search for the $^{37}Cl(\bar{\nu}_e, \beta^-)^{37}Ar$ reaction. A negative outcome, indicating that at most the cross section was 0.9×10^{-45} cm^2, i.e., well below $\approx 10^{-43}$ cm^2, demonstrated the *non*identity of neutrino and antineutrino,

$$\nu_e \quad \neq \quad \bar{\nu}_e \qquad (8\text{-}21)$$

Helicity: $-$ $+$

Spin: antiparallel   parallel

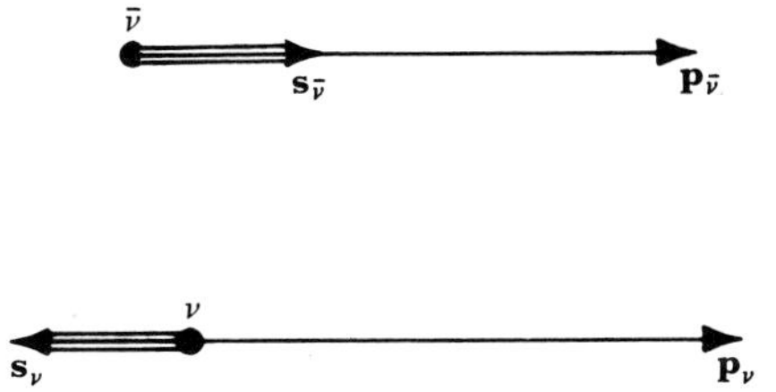

Fig. 8-9. Schematic diagram depicting the *positive helicity* $\mathcal{H} = \mathbf{p} \cdot \boldsymbol{\sigma}/|\mathbf{p}||\boldsymbol{\sigma}|$ of the antineutrino and the *negative* helicity of the neutrino.

The complementary experiment $^{37}Cl(\nu_e, \beta^-)\,^{37}Ar$ is currently being performed by Davis [Da 64]. The solar ν_e flux (mainly from β^+ decay of ^8B) on the Earth's surface $\Phi \approx (2^{+2}_{-1}) \times 10^7$ ν_e cm^{-2} sec^{-1}, when incident on a target comprising about 3.8×10^5 l of CCl$_4$ may create sufficient radioactive ^{37}Ar ($\tau = 50.6$ d) to be detected radiochemically. On sweeping this out with the aid of a stream of helium gas into a small but very sensitive low-level counter, a counting rate of roughly 5 counts per day was anticipated, but preliminary measurements taken at a depth of one mile beneath the Earth's surface in a South Dakota mine to obviate cosmic-ray background, etc. have up to the present shown no significant increase over the normal background noise of about 1 event per 3 days.† While a negative result would point to a solar-level activity smaller by a factor of 10 below that expected, with consequent serious repercussions upon solar models, a positive outcome would require sophisticated analysis to verify its solar origin, such as a measurement of a 7-percent inverse-square variation of the signal which has been predicted on the basis of the Earth's slightly elliptical orbit around the Sun.

Whereas the radiochemical experiments of Davis does not aim to yield information on the energy of the captured neutrinos, other direct-counting

† *Note added in proof:* Current results [Ba 68, Da 68] indicate that the solar neutrino flux appears to be smaller by a factor of 7 than that expected theoretically; e.g., the flux of neutrinos from ^8B decay is $\Phi \leqslant 2 \times 10^6$ ν_e cm^{-2} sec^{-1} at the Earth's surface.

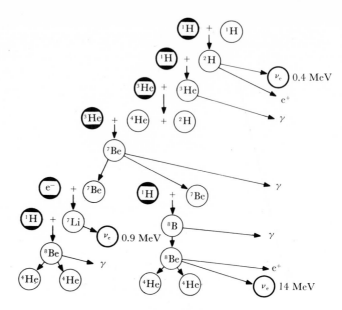

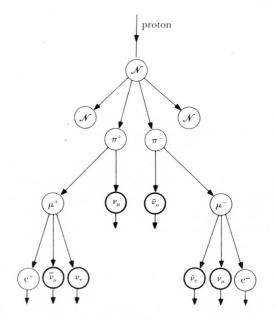

Fig. 8-10a.

Fig. 8-10b.

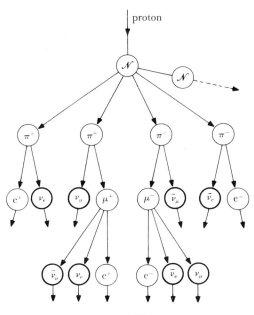

Fig. 8-10c.

Fig. 8-10. Pictorial indication of the production of neutrinos and antineutrinos ν_e, $\bar{\nu}_e$, ν_μ, $\bar{\nu}_\mu$ in (a) solar and stellar nuclear reaction processes, involving thermonuclear fusion of light nuclei and emission of low-energy ν_e's in β^+-decay and electron-capture processes; (b) cosmic-ray collisions in space, giving rise to a low flux of all four neutrino species with energies ranging up to very high values; (c) atmospheric cosmic-ray collisions, again producing a low flux of neutrinos whose energies extend over a wide range.

experiments have been designed to measure the energy of each neutrino-induced event as it occurs. One such approach as proposed by Reines and Woods [Re 65b], and independently by Bahcall, utilizes the reaction $^7\mathrm{Li}(\nu_e, \beta^-)^7\mathrm{Be}$ (an alternative possibility, namely, $^{11}\mathrm{B}(\nu_e, \beta^-)^{11}\mathrm{C}$ having been set aside for the time being). A detector made up of $\frac{3}{4}$ ton of lithium arranged as slabs within a 7600-l liquid scintillation counter which will serve to register the β^-'s is being set up in a salt mine at a depth of about 600 m surrounded by a large anticoincidence shield. It is hoped on an average to detect about 1 β^- per week. The β^- energy will be measured and thereby the energy of the solar ν_e's within the range 6 to 15 MeV deduced. An alternative suggestion, being pursued by Reines and Kropp [Re 64], is a reversion to the study of elastic neutrino scattering on orbital electrons. From the experimental results of Nezrick and Reines [Ne 66] for fission *antineutrino* capture by protons, one can deduce a cross section of $(8.4 \pm 2.8) \times 10^{-46}$ cm^2 for the production of recoil electrons whose energies lie above 1.5 MeV by the elastic antineutrino-scattering reaction $\bar{\nu}_e + e^-$. Since the cross section for the process $\nu_e + e^- \rightarrow \nu_e + e^-$ is expected to be of the same order of magnitude, the ν_e-scattering experiment is anticipated to yield some 20 counts per year for solar neutrinos having energies between 6 and 15 MeV.

These investigations are all concerned with a study of ν_e interactions. However, the other species, namely, $\bar{\nu}_e$, ν_μ, and $\bar{\nu}_\mu$ also impinge with presumably random angular distribution upon the Earth and have energies which extend over a very considerable range.

Table 8-1. NEUTRINO SOURCES AND ESTIMATED FLUXES PASSING THROUGH THE EARTH

Origin	Neutrino species	Energy range	Flux Φ [cm^{-2} sec^{-1}]
Cosmic-ray collisions in intergalactic space	$\nu_e, \bar{\nu}_e, \nu_\mu, \bar{\nu}_\mu$	Extensive	10^{-5}–10^{-7}
Solar and stellar nuclear reactions	ν_e	Low	10^7–10^{11}
Cosmic-ray interactions in the Earth's atmosphere	$\nu_e, \bar{\nu}_e, \nu_\mu, \bar{\nu}_\mu$	Extensive	3×10^{-3}
Terrestrial radioactive β^- decay	$\bar{\nu}_e$	Low	10^5–10^7
Powerful nuclear reactor	$\bar{\nu}_e$	Low	10^{14}

Solar and stellar nuclear reactions are considered to produce *only low-energy* ν_e's ($E \lesssim 15$ MeV) by β^+-decay processes such as the comparatively rare, but experimentally important, reaction $^8B \rightarrow \alpha + \alpha + \beta^+ + \nu_e$, which forms part of the fusion cycle depicted in Fig. 8-10(a).

The URCA process may also be responsible for an appreciable production of neutrinos. It consists of a cycle in which high-energy electrons are first captured by nuclei and subsequently re-emitted at lower energy, accompanied by the ejection of a neutrino and an antineutrino:

$$e^- + (Z, A) \rightarrow (Z-1, A) + \nu_e$$
$$\downarrow$$
$$(Z, A) + e^- + \bar{\nu}_e \qquad (8\text{-}22)$$

The process essentially creates ν_e's and $\bar{\nu}_e$'s at the cost of electron energy. (The loss of electron energy with nothing material to show for its expenditure prompted an analogy with the dwindling of one's means in the Casino de URCA in Rio de Janeiro [Ga 41, Chi 61], hence the designation.)

In addition to neutrinos so produced, the Earth is irradiated by the entire gamut of neutrino species, ν_e, $\bar{\nu}_e$, ν_μ, $\bar{\nu}_\mu$, proceeding from cosmic-ray collision processes in intergalactic space and in the Earth's atmosphere, as indicated in Fig. 8-10(b, c), and furthermore by a considerable $\bar{\nu}_e$ flux originating from the β^- decay of natural radioactive sources within the Earth. Solar neutrino fluxes are shown in Fig. 8-11. Estimates of these and other relative fluxes have been listed in Table 8-1. The values are based in part upon recent experimental findings and in part upon tentative theoretical considerations.

Evidence showing the distinction between neutrinos and antineutrinos has also been sought through studies of DOUBLE β DECAY, a second-order

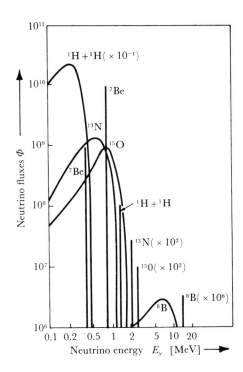

Fig. 8-11. Pictorial indication of the production of neutrinos and antineutrinos ν_e, $\bar{\nu}_e$, ν_μ, $\bar{\nu}_\mu$ in log–log plot of the solar neutrino spectrum, expressing the estimated flux at the Earth's surface Φ (in units ν_e cm^{-2} sec^{-1} MeV^{-1} for continuum sources which experience β^+ decay and in units ν_e cm^{-2} sec^{-1} for line sources in which electron capture takes place; data calculated by Bahcall [Ba 64c, Ba 66]).

process which has different characteristics depending on whether or not ν_e and $\bar{\nu}_e$ are identical particles. If identical, the two stages of double β decay from a nucleus (A, Z) to $(A, Z \mp 2)$ would combine in such a way that emission of a (virtual anti-) neutrino is compensated by absorption of the neutrino, so that no neutrinos are emitted in the overall process

$$n \rightarrow p^+ + e^- + (\bar{\nu}_e \equiv \nu_e)$$
$$\underline{\nu_e + n \rightarrow p^+ + e^-}$$
$$2n \rightarrow 2p^+ + 2e^- \qquad (8\text{-}23)$$

When ν_e and $\bar{\nu}_e$ are regarded as different (conjugate) particles, however, the two-step process results in the emission of two electrons *accompanied by two* antineutrinos

$$n \rightarrow p^+ + e^- + \bar{\nu}_e$$
$$n \rightarrow p^+ + e^- + \bar{\nu}_e$$

$$\overline{2n \rightarrow 2p^+ + 2e^- + 2\bar{\nu}_e} \qquad (8\text{-}24)$$

and the available transition energy is now shared statistically between the two e^-'s and the two $\bar{\nu}_e$'s, so that the sum of electron energies is no longer unique but shows a broad distribution, as illustrated in Fig. 8-12 for the

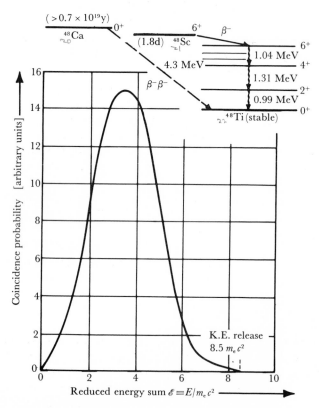

Fig. 8-12. Calculated coincidence spectrum for the two-neutrino double β decay of ^{48}Ca plotted in terms of the reduced net kinetic energy of both emitted electrons. Because of the large mutual spin difference, no direct β^- transition takes place between the ground states of ^{48}Ca and ^{48}Sc (hence the long half-life of ^{48}Ca) (from [Wu 66]).

much-studied case [McCa 55, Aw 56, Do 59, Gr 60b, Sh 65a, Sh 65b] of ^{48}Ca.

As might be expected, the half-lives for both alternatives are long, the approximate predicted values being [Ko 55, Pr 59, corrigenda Pr 61], for neutrinoless double β decay,

$$T_{1/2}^{(\nu_e = \bar{\nu}_e)} \simeq 10^{15\pm2} \left(\frac{8m_e c^2}{E_0 - 2m_e c^2}\right)^6 \left(\frac{60}{Z}\right)^2 \left(\frac{A}{150}\right)^{2/3} \quad \text{years} \quad (8\text{-}25)$$

and for two-neutrino double β decay,

$$T_{1/2}^{(\nu_e \neq \bar{\nu}_e)} \simeq 6 \times 10^{17\pm2} \left(\frac{8m_e c^2}{E_0}\right)^{10} \left(\frac{60}{Z}\right)^2 [1 - \exp(\pm 2\pi\alpha Z)]$$

$$\times \left(\Delta E + \frac{E_0 + 2m_e c^2}{2m_e c^2}\right)^2 \quad \text{years} \quad (8\text{-}26)$$

where the $(\pm)$ sign refers to $\beta^\pm \beta^\pm$ decay and ΔE represents the average energy difference between the virtual intermediate and the initial nuclear state. Under comparable conditions, the latter process is the slower by a factor 10^5–10^6. The error limits in the exponents reflect uncertainties in the estimates of the nuclear matrix element. The calculations have been reviewed in survey articles [De 60a, Ro 65b] and seem to be reasonable in predicting half-lives of 10^{15}–10^{16} and 10^{20}–10^{22} y, respectively. (Formulae for double K capture have also been given. These appear to give even longer half-lives.) Despite many experimental investigations (some 48 nuclei would be expected to be unstable toward double β decay) no clear-cut evidence has so far come to light, though the data on the lower limit of the lifetime definitely point to two-neutrino rather than neutrinoless double β-decay (see [Scho 66, pp. 155 ff.] or [Wu 66, pp. 200 ff.]).

This concludes the consideration of the difference between ν_e's and $\bar{\nu}_e$'s, together with experiments aimed at the detection of these particles, and leads us to examine the DISTINCTION BETWEEN ELECTRON-TYPE AND MUON-TYPE NEUTRINOS, viz.

$$\nu_e \neq \nu_\mu \quad \text{and} \quad \bar{\nu}_e \neq \bar{\nu}_\mu \quad (8\text{-}27)$$

A distinctness between the two species was already strongly indicated by the absence of the much-sought-for decay mode $\mu^\pm \to e^\pm + \gamma$. Whereas one readily observed the normal μ-decay process $\mu^\pm \to e^\pm + \nu + \bar{\nu}$, in which the total number of leptons is conserved, all attempts to detect the equally feasible reaction $\mu \to e + \gamma$ in which neutrino and antineutrino annihilate in the field of the electron to yield a photon failed. (A recent upper limit for the branching ratio was established experimentally as $\lesssim 2 \times 10^{-8}$ by Parker *et al.* [Pa 64].) Since $\nu\bar{\nu}$ pair annihilation *should* occur if the above two particles are mutually conjugate, one is driven to conclude that there is a dissimilarity and that, just as electron and muon are related but nonconjugate particles, so also are the neutrinos in β and μ decay. One species is associated with the electron and the other with the muon. In μ decay the neutrino emission is of mixed character,

$$\mu^- \to e^- + \bar{\nu}_e + \nu_\mu \tag{8-28}$$

$$\mu^+ \to e^+ + \nu_e + \bar{\nu}_\mu \tag{8-29}$$

The distinction between ν_e and ν_μ, which prevents mutual annihilation, was first demonstrated in 1962 by a Columbia-Brookhaven group [Da 62b] with the aid of the Brookhaven proton synchrotron. On inserting a beryllium target into the 15-GeV proton beam, a high-energy beam of pions and kaons was produced and extracted which, as it decayed in flight, in its turn gave rise to a neutrino/antineutrino "beam"

$$p + Be \to \begin{cases} \pi^+ \to \mu^+ + \nu_\mu \\ \pi^- \to \mu^- + \bar{\nu}_\mu \\ K^+ \to \mu^+ + \nu_\mu \\ K^- \to \mu^- + \bar{\nu}_\mu \end{cases} \tag{8-30}$$

in traveling a path of 21 m. The resulting beam of assorted particles then impinged upon a 13.5-m thick iron shield wall which served to screen off a neutrino detector in the shape of a 10-ton aluminum spark chamber. The shield gave 10^{24}-fold attenuation for strongly interacting particles (and thus essentially allowed only neutrinos to pass through) whose probable energy spectrum is shown in Fig. 8-13. (A correction for the cosmic-ray background could be made in dummy runs.) The spark chamber responded to nucleon-neutrino interactions and, with its associated circuitry, could discriminate between electrons and muons so produced. The experiment then lay in determining the relative numbers of e's and μ's produced, since if $\nu_e = \nu_\mu$ these should be produced in equal amounts. On the other hand, if $\nu_e \neq \nu_\mu$, then *no* electrons should be created, since the experimental beam contained only ν_μ's and $\bar{\nu}_\mu$'s (but no ν_e's and $\bar{\nu}_e$'s), and these are not able to initiate the creation of e's but only of μ's, viz.

$$
\nu_\mu + n \begin{array}{c} \nearrow p^+ + \mu^- \\ \not\searrow p^+ + e^- \end{array}
\qquad
\bar{\nu}_\mu + p^+ \begin{array}{c} \nearrow n + \mu^+ \\ \not\searrow n + e^+ \end{array}
\tag{8-31}
$$

Of the 34 observed muon events, 5 were considered to be cosmic-ray background, which left 29 definitive μ-creation events. Instead of 29 e-creation events which would be expected to ensue if $\nu_e = \nu_\mu$, only 6 possible candidates were observed, and these could all be explained away (e.g., through neutron-background considerations, the admixture of such decays as $K^+ \to e^+ + \pi^0 + \nu_e$, $K^0 \to e^\pm + \pi^\mp + \nu_e$, etc.). Thus $\nu_e \neq \nu_\mu$ and $\bar{\nu}_e \neq \bar{\nu}_\mu$, whence it is evident that experiments to search for "natural" ν_μ's and $\bar{\nu}_\mu$'s from sources of the type listed in Table 8-1 become very desirable as a complement to those undertaken to detect solar ν_e's.

The DETECTION OF "NATURAL" MUON-TYPE NEUTRINOS (also called "NEUTRETTOS"), ν_μ and $\bar{\nu}_\mu$, through the creation of μ's in underground experiments has been the subject of investigation by two groups since 1964, one of which (the Case-Witwatersrand group) under Reines and Sellschop

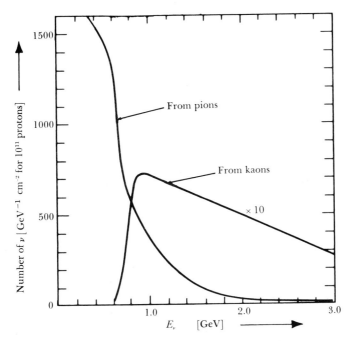

Fig. 8-13. The neutrino energy spectrum expected in the Columbia-Brookhaven experiment [Da 62b] when a flux of 10^{11} protons having an energy of 15 GeV impinged upon a beryllium target.

has a huge liquid-scintillator array arranged under 3200 m of rock in the East Rand Proprietary Mines near Johannesburg in South Africa, and the other (the Durham-Tata-Osaka group) operates at a depth of 2300 m in a Kolar goldmine near Bangalore in India.

The Case-Witwatersrand experiment [Re 65a, Re 66a, Re 66b] contains a double bank of interconnected scintillators, of area 110 m² and in process of being lengthened from 33 m to 67 m, which register incident μ's. Statistically, those μ's which ensue from ν_μ and $\bar{\nu}_\mu$ interactions in the surrounding rock can be distinguished from those in the cosmic-ray background by the fact that the former are expected to arrive isotropically (with slight peaking in the horizontal direction) whereas the latter descend preponderantly from a vertical direction. (Thus of the estimated 150 "background" muons which are expected to traverse the detector in the course of a year's operation, less than 1 would be expected to make an angle of more than 45 degrees with the vertical.) A muon found to be traveling vertically *upwards* would clearly have been created by a ν_μ or $\bar{\nu}_\mu$ interaction in the rock, since extraneous muons could not be expected to penetrate the entire thickness of the Earth. In their preliminary report [Re 65a], the group lists 7 likely neutrino-induced events and 7 miscellaneous events; in the meantime the number of the former has [Re 66b] been brought up to 10 (the number of "extraneous" muons detected so far being 80). There are some indications that in the high-energy range here being examined, the interaction cross section is somewhat larger than would be predicted from extrapolation of results at laboratory energies. An ancillary point at issue is the question as to whether or not cosmic-ray neutrinos have sufficient energy to create the long-sought-for INTER-MEDIATE BOSON. As will be recalled from Section 3.2.3, this particle has been postulated in order to account for the weak-interaction force as an "exchange quantum," analogous to the role of the photon as a mediator of the electro-magnetic force. With a lifetime of less than 10^{-17} sec (and a mass exceeding 800 MeV, and probably $\gtrsim 1500$ MeV) it can be deemed to decay in either of the modes

$$W^+ \rightarrow e^+ + \nu_e \quad \text{or} \quad W^+ \rightarrow \mu^+ + \nu_\mu \qquad (8\text{-}32)$$

The latter mode constitutes a crucial diagnostic criterion to establish the possibility of W creation by cosmic-ray ν_μ's as they interact with nucleons in solid matter, for if the intermediate vector boson W is *not* formed in the interaction, the neutrino capture process gives rise to *one* muon per reaction,

$$\nu_\mu + \mathcal{N} \rightarrow \mathcal{N} + p^+ + \mu^- \qquad (8\text{-}33)$$

whereas if it *is* formed, *two* muons are released in the overall process, according to the scheme

$$\nu_\mu + \mathcal{N} \rightarrow \mathcal{N} + W^+ + \mu^-$$
$$\downarrow \approx 10^{-17} \text{ sec}$$
$$\nu_\mu + \mu^+ \qquad (8\text{-}34)$$

Hence the "signature" for intermediate-boson creation in the experiment is the occurrence of a pair of nearly parallel muon tracks starting from a common origin and penetrating the scintillators at two places on each side of the detector array. Only one possible occurrence of this type has been observed to date, the interpretation of which, since it was not possible to specify the common point of origin with any precision, remains ambiguous. The Durham-

Tata-Osaka group have also recorded one pair-event having suitable characteristics, but it is possible that this was a neutrino-induced pion-muon combination, and again no unique identification of an intermediate boson could be made.

This completes the current information on neutrino detection, though more details can be found in survey articles [Ko 62, Fr 63, Fr 65a, Ko 65, Ru 65, Le 66a (especially [Schw 66 and Le 66b]), Fr 66, Fa X] and relevant papers [Ok 65, Ku 65a, Ku 65b, Ed 66, Ba 66]. A bibliographic compilation of publications dealing with the neutrino has been prepared by Kuchowicz [Ku X]. The firm establishment of neutrino existence not only vindicates Pauli's original contention and preserves energy conservation in all interactions, but provides strong support for the validity of the theory of β decay, which we discuss next.

8.3. Beta-Decay Theory

8.3.1. FORMULATION

Pauli's neutrino hypothesis paved the way to a quantitative theory of β decay as formulated in 1933/4 by Fermi [Fe 33, Fe 34b]. The theory provided a dynamical description of β decay which accounted for the observed form of the β-momentum (or energy) spectrum and the lifetimes of β-active nuclei as well as providing a classification of β-decay transitions and their corresponding selection rules.

Fermi drew upon an analogy with the emission of light or γ radiation from the atomic electron cloud, which had been described through perturbation theory in terms of a transition probability per unit time for an initial system i, consisting of an excited atom, to change to a final system f, consisting of the same atom in a less excited state together with a photon. The photon could be represented as an electromagnetic radiation field, and the transition was effected through a field interaction process expressed through a Hamilton operator which carried the initial state over into the final state, creating a photon thereby. Along similar lines, the β-decay process

$$\left.\begin{array}{l} n \rightarrow p^+ + e^- + \bar{\nu} \\ p^+ \rightarrow n + e^+ + \nu \end{array}\right\} \text{ viz. } \mathcal{N} \rightarrow \mathcal{N}' + e + \nu \qquad (8\text{-}35)$$

was represented through the creation of leptons within the decaying nucleus and expressed in terms of an electron-neutrino field. The Hamiltonian H' was thereby an agent (in the form of an interaction energy operator) transforming a nuclear system from an initial state i into one of several possible final states f through lepton creation by means of a *weak* interaction. The strength of the latter is expressed by the matrix element H'_{fi}, equivalent to a volume integral

$$H'_{fi} \equiv \int \Psi_f^* \, H' \, \Psi_i \, d\Omega \qquad (8\text{-}36)$$

In terms of this interaction, when the density of accessible states in the final system within a small energy interval dE_0 about a mean final energy E_f is

$$\rho(E_f) \equiv \frac{dn}{dE_0} \qquad (8\text{-}37)$$

the transition probability per unit time is derived from Fermi's "Golden Rule No. 2" as set up in Section 3.4.2,

$$W_{fi} = \frac{2\pi}{\hbar} \rho(E_f) \, |H'_{fi}|^2 \qquad (8\text{-}38)$$

8.3.2. PROBABILITY FUNCTION AND THE BETA-MOMENTUM SPECTRUM

If the linear momenta of the emitted electrons corresponding to the energy interval dE_0 in the vicinity of E_f range from p to $p + dp$, it follows that the above transition probability per unit time simply gives the mean number of β particles emitted in unit time with momenta within this range. Hence it can be set equal to a PROBABILITY FUNCTION $N(p) \, dp$ which determines the momentum distribution and therefore the shape of the β-MOMENTUM SPECTRUM. To arrive at a more explicit expression for the form of the spectrum, we examine the terms in the probability formula

$$N(p) \, dp \equiv W_{fi} = \frac{2\pi}{\hbar} \frac{dn}{dE_0} \, |H'_{fi}|^2 \qquad (8\text{-}39)$$

individually, basing our considerations upon the presupposition of neutrino existence and strict conservation of total energy, momentum, and angular momentum. Our only modification of Fermi's basic approach is to extend the formalism to make provision for nonconservation of parity.

Physically, we envisage the process as a transition from the parent-nucleus state i through the action of a weak perturbation to a final quantum-mechanical system f consisting of a daughter nucleus in one particular state out of very many possible states, together with two leptons treated as an electron-neutrino field. While the daughter nucleus is left with a recoil energy plus possibly a sharp excitation energy, the remaining available energy can be shared between the emergent leptons in many different ways. The apportionment is not unique and, indeed, net leptonic energy is not sharp, but is defined only to within an uncertainty dE_0 which is numerically equal to the energy uncertainty of the original system. The latter is $\Delta E_i = \hbar/\tau_i$, where τ_i is the mean decay lifetime of the parent nucleus. Since we assume the daughter to be in a state of *sharp* energy, this spread must be evinced by the leptonic energy if energy is to be a strictly conserved quantity. The density of accessible daughter states over this interval dE_0 represents the STATISTICAL FACTOR dn/dE_0, and this we examine next.

8.3.3. STATISTICAL FACTOR (FINAL-STATE DENSITY)

By statistical factor we specifically mean the density of possible final states in the phase space of the electron-neutrino field. This can be derived with the aid of the Heisenberg uncertainty relation (in the form used by Fermi [Fe 34b]), which tells us that the state of a particle such as an electron cannot be confined to an area in phase space (see Fig. 8-14) smaller than $\Delta x \, \Delta p_x \approx h$,

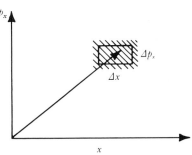

Fig. 8-14. Sketch of a cell in phase space as represented in the x dimension only, in which we assume the state of the electron to be confined.

and similarly for the y and z components. In the 6-dimensional phase space defined by (x, y, z, p_x, p_y, p_z), the state of the electron accordingly cannot be specified to a greater precision than that betokened by the volume element

$$\Delta x \, \Delta y \, \Delta z \, \Delta p_x \, \Delta p_y \, \Delta p_z \approx h^3 \qquad (8\text{-}40)$$

This indicates the restriction on the number of distinct states that can occupy a given phase volume. Conversely, it enables the phase volume of an electron in a given state to be calculated. We suppose the electron to be localized within a spatial volume V and a momentum interval $p \rightarrow p + dp$. Since in phase space this momentum range is represented as a spherical shell of volume $4\pi p^2 \, dp$, the net phase volume of the electron is $V(4\pi p^2 \, dp)$. Noting that a *single* cell (viz. a given state) in phase space occupies a volume h^3, we derive the number of electron states in phase space as

$$dn_e = V \frac{4\pi p^2 \, dp}{h^3} \qquad (8\text{-}41)$$

This is just the probability of encountering an electron within a spatial volume $V = dx \, dy \, dz$ with a linear momentum between p and $p + dp$. Equivalently, the number of neutrino states in phase space is

$$dn_\nu = V \frac{4\pi p_\nu^2 \, dp_\nu}{h^3} \qquad (8\text{-}42)$$

for neutrino momenta p_ν to $p_\nu + dp_\nu$. The total probability of encountering *both* the simultaneously emitted leptons is then the product of the individual probabilities:

$$dn = dn_e \, dn_\nu \tag{8-43}$$

with dn denoting the number of accessible states in the phase space of the electron-neutrino field. The state density referred to an energy interval dE_0 is

$$\frac{dn}{dE_0} = \frac{16\pi^2 \, V^2}{h^6} p^2 \, p_\nu^2 \frac{dp_\nu}{dE_0} \, dp \tag{8-44}$$

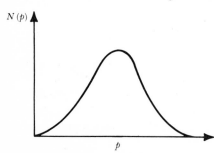

$N(p)$

p

Fig. 8-15. Basic shape of a β-momentum spectrum, terminating at a sharp cut-off momentum whose value is determined by the β-transition energy.

To eliminate all terms containing p_ν, we note that

$$p_\nu = \frac{E_\nu}{c} \tag{8-45}$$

and that, neglecting the small recoil energy of the massive daughter nucleus, the overall decay energy E_0 is shared between the electron and the neutrino:

$$E_0 = E + E_\nu \tag{8-46}$$

Hence

$$p_\nu = \frac{1}{c}(E_0 - E) \tag{8-47}$$

and

$$p_\nu^2 \left(\frac{dp_\nu}{dE_0}\right) = \frac{1}{c^3}(E_0 - E)^2 \tag{8-48}$$

so that

$$\frac{dn}{dE_0} = \frac{16\pi^2 \, V^2}{c^3 \, h^6} p^2 (E_0 - E)^2 \, dp \tag{8-49}$$

The relativistic relation between momentum p and *kinetic* energy E,

$$p = \frac{1}{c}[E(E + 2m_e c^2)]^{1/2} \tag{8-50}$$

as derived in Section 1.4 applies here to the electron's motion.

When the interaction $|H'_{fi}|$ is independent of E or p (namely, when $Z = 0$ in the case of allowed transitions), the form of the β spectrum is given solely by the value of the statistical factor. The shape, as shown in Fig. 8-15, is given by the expression

$$N(p)\, dp = Cp^2(E_0 - E)^2\, dp \tag{8-51}$$

However, $|H'_{fi}|$ is in many cases not independent of E or p and therefore requires more detailed investigation.

8.3.4. Interaction Matrix Element

The matrix element

$$H'_{fi} \equiv \int \Psi_f^* H' \Psi_i\, d\Omega \tag{8-52}$$

is expressed in terms of initial and final-state normalized energy eigenfunctions Ψ_i and Ψ_f which can in turn be decomposed further. If we write the wave function of the nucleons in the parent nucleus as ψ_i and in the daughter nucleus as ψ_f, representing the electron-neutrino field by a product of wave functions $\varphi_e(\mathbf{r})\, \varphi_\nu(\mathbf{r})$, we can write the total wave functions for the process

$$P \rightarrow D + e + \nu \tag{8-53}$$

as

$$\Psi_i = \psi_i, \qquad \Psi_f = \psi_f \varphi_e(\mathbf{r})\, \varphi_\nu(\mathbf{r}) \tag{8-54}$$

and use the definition (8-52) to express the interaction matrix element as

$$H'_{fi} = g_F \int [\psi_f^* \varphi_e^*(\mathbf{r})\, \varphi_\nu^*(\mathbf{r})]\, M\psi_i\, d\Omega \tag{8-55}$$

with g_F an empirical fundamental constant and M a dimensionless Hamilton operator. The Fermi coupling constant [He 61]

$$g_F = 1.41 \times 10^{-49} \text{ erg cm}^3 = 0.9 \times 10^{-4} \text{ MeV fm}^3$$

expresses the strength of the β interaction. Its character therefore resembles that of the gravitational constant or the fine-structure constant when expressed in dimensionless terms, as discussed in Section 3.2, particularly in 3.2.3. In the analogous treatment of light emission the electron charge e stands in its place.

We now set about simplifying the matrix element H'_{fi}. Because the interaction between the nucleus and the leptons is so weak, the lepton waves comprising the field are undistorted by a nuclear potential and may be taken as plane waves. (For the moment, we neglect the distortion of the electron wave through electromagnetic interaction with the nuclear charge. We shall return to this point in the next section.) Thus

$$\varphi_e(\mathbf{r}) = N_e e^{i(\mathbf{k}_e \cdot \mathbf{r})}, \qquad \varphi_\nu(\mathbf{r}) = N_\nu e^{i(\mathbf{k}_\nu \cdot \mathbf{r})} \tag{8-56}$$

where $\mathbf{k}_e$ and $\mathbf{k}_\nu$ represent lepton wave vectors, $\mathbf{k} = \mathbf{p}/\hbar = \lambdabar^{-1}$. Normalizing the wave functions within a volume V which is equal to that considered in the previous section, namely, the spatial volume element $V = dx\,dy\,dz$, so that

$$\int_V \varphi_e^* \, \varphi_e \, d\Omega = 1 \quad \text{and} \quad \int_V \varphi_\nu^* \, \varphi_\nu \, d\Omega = 1 \tag{8-57}$$

we get the amplitude constants,

$$N_e = N_\nu = V^{-1/2} \tag{8-58}$$

The nuclear wave functions ψ_i and ψ_f are nonzero only within the dimensions of the respective nuclei. For this reason, the integral in (8-55) extends only over V. However, since the extension of the nuclei is small compared with the volume within which the leptons can be localized, the wave functions $\varphi_e(\mathbf{r})$ and $\varphi_\nu(\mathbf{r})$ can be expanded as a power series about the origin $\mathbf{r} = 0$:

$$\varphi_e(\mathbf{r}) = V^{-1/2}[1 + i\,(\mathbf{k}_e \cdot \mathbf{r}) + \cdots] \tag{8-59}$$

$$\varphi_\nu(\mathbf{r}) = V^{-1/2}[1 + i\,(\mathbf{k}_\nu \cdot \mathbf{r}) + \cdots] \tag{8-60}$$

In general it suffices to consider only the first term in these series (e.g., in the case of allowed transitions and $Z = 0$) and set

$$\varphi_e(0) = \varphi_\nu(0) = V^{-1/2} \tag{8-61}$$

since the next term in the expansion is roughly 50 times smaller for medium-heavy nuclei, so that the expression

$$H_{fi}' = g\varphi_e^*(0)\,\varphi_\nu^*(0) \int \psi_f^* \, M\psi_i \, d\Omega \equiv g\varphi_e^*(0)\,\varphi_\nu^*(0)\, M_{fi} \tag{8-62}$$

reduces to

$$H_{fi}' = \frac{g}{V} M_{fi} \tag{8-63}$$

with the rather imprecisely known quantities ψ_f, M, and ψ_i all gathered into the one matrix element M_{fi}. This matrix element is the overlap integral of the final and initial wave functions of the nucleons comprising the daughter and parent nucleus. In the case of allowed transitions it is independent of the electron energy E and its magnitude is of order 1, whereas for so-called forbidden transitions it tends to zero in this approximation. Its value can be computed when the structure of parent and daughter nuclei is known; thus for neutron decay, it is 1.

Substituting in (8-39) for the statistical factor from (8-49) and for $|H_{fi}'|^2$ from (8-63) we finally arrive at the spectral formula

$$N(p)\,dp = \frac{g^2}{2\pi^3 c^3 \hbar^7} |M_{fi}|^2 \, (E_0 - E)^2 \, p^2 \, dp \tag{8-64}$$

In conclusion, we modify this by incorporating a correction factor to take account of the Coulomb distortion of the electron wave.

8.3.5. COULOMB CORRECTION FACTOR

Through Coulomb interaction with the nuclear charge, the velocity, and therefore the energy, of emitted β^- particles is decreased and that of β^+ increased, which affects the form of the β spectrum and the decay probability. In comparing the situation for β^- with that for β^+ decay at the same transition energy E_0 it may at first sight appear paradoxical that this Coulomb effect serves to *stimulate* the emission of β^-'s and *hinder* that of β^+'s, as is evident from Fig. 8-16, but this can readily be explained as follows: The decay probability $N(p)\,dp$ is proportional to $p^2\,dp$. The quantity p in the spectral formula betokens the asymptotic value of electron momentum, as measured far from the emitting nucleus. The term $p^2\,dp$, however, stems from the statistical factor which is concerned with a state density determined by the electron state immediately after creation, e.g., while in the immediate vicinity of the nucleus. In the case of β^+ decay, the momentum of the positron is *smaller* in the vicinity of the nucleus than when distant, and therefore the statistical factor which determines the decay probability is also smaller than it would be if Coulomb repulsion did not take place, when we bear in mind that the momentum spectrum is measured far from the nucleus. For β^- decay, the converse holds.

Classically, no positrons would be expected to emerge below an energy of about Ze^2/R, where R is the nuclear radius. Nevertheless, the experimental data reveal that positrons *are* emitted with lower energies, albeit not in large numbers. The β^+ spectrum goes exponentially to zero as p tends to zero. This should occasion no surprise, however, in the light of our discussion of barrier tunneling in Section 7.4, in which it was pointed out that a nuclear potential presents hardly any hindrance to a particle appreciably lighter than a nucleon.

On the other hand it should be noted that the upper bound E_0 of the β spectrum is not displaced by the action of a Coulomb effect, since the Coulomb energy has already been taken into account in the value of nuclear binding energy. By definition, the binding energy refers to the dissociation of a nucleus into *free* particles, e.g., into leptons at great separation from the nuclear origin.

The change of the momentum in the Coulomb field is enhanced when the β particle's motion is slow. In consequence, the influence is greatest at the lower end of the spectrum, so that the actual shape of the spectrum is deformed.

We can regard the perturbation as causing the electron wave to become distorted. This necessitates a modification in the general matrix element $|H'_{\mathrm{fi}}|^2$ to one which contains an electron wave function $|\varphi_e(0)|_Z$ at the nuclear origin modified by the nuclear charge Ze, i.e.,

$$|H'_{\mathrm{fi}}|^2 = g^2|\varphi_e(0)|^2\,|\varphi_\nu(0)|^2\,|M_{\mathrm{fi}}|^2 \to g^2|\varphi_e(0)|^2_Z\,|\varphi_\nu(0)|^2\,|M_{\mathrm{fi}}|^2 \quad (8\text{-}65)$$

The appropriate adjustment to φ_e can be made by introducing the FERMI

FUNCTION, which is simply the ratio of the electron density at the daughter nucleus to the density at infinity:

$$F(E, Z) \equiv \frac{|\varphi_{\mathrm{e}}(0)|^2_Z}{|\varphi_{\mathrm{e}}(0)|^2} \tag{8-66}$$

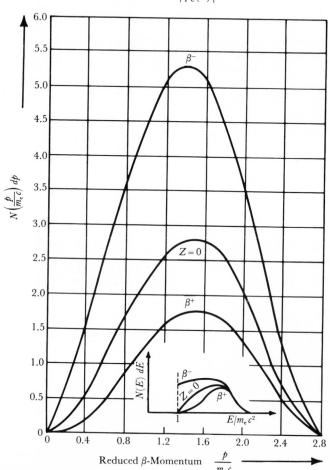

Fig. 8-16. Nuclear Coulomb effect on the β-momentum spectrum (represented in relativistic units) of allowed shape (Section 8.4.1). The $Z = 0$ curve depicts the unmodified spectrum, as determined by the value of the statistical factor, while the other curves indicate the altered shape as evaluated for constant $|M_{\mathrm{fi}}|^2$ from data [NBS 52] for β^+ decay of calcium ($Z = 20$) with an end-point momentum $(p/m_e c)_{\max} = 2.8$ which corresponds to an energy $E_{\max} = 1$ MeV. The augmentation or suppression effect is even more perceptible in *energy* spectra (shown inset) in consequence of the additional velocity factor $dE_{\mathrm{kin}}/dp = v$ (valid relativistically as well as non-relativistically; adapted from [Ev 55]). (Used by permission of McGraw-Hill Book Co.)

so that the energy-dependent function F takes on the following values:

$$\left.\begin{array}{lll} F \geqslant 1 & \text{for} & \beta^- \text{ emission} \\ F = 1 & \text{for} & Z = 0 \\ F \leqslant 1 & \text{for} & \beta^+ \text{ emission} \end{array}\right\} \tag{8-67}$$

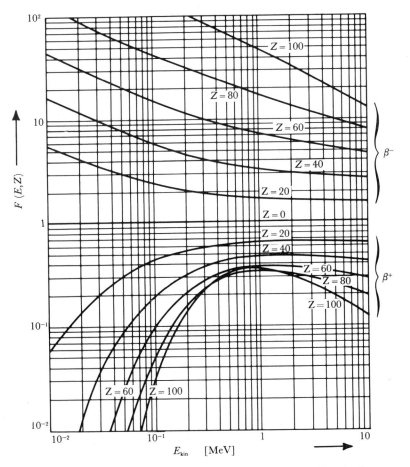

Fig. 8-17. Log-log plots of the Fermi function $F(E, Z)$ vs kinetic β energy for electrons and positrons and for $Z = 0$–100, compiled from data in [NBS 52].

The explicit expression for F is fairly complicated and involves a ratio of Γ functions. Suitable expressions are given by Siegbahn [Si 65, pp. 1338 and 1339]. (The earlier edition of Siegbahn's text [Si 55] quotes the formulae on pp. 280–282 and also tabulates numerical values of a modified Fermi function $G \equiv (p/E)F$ as Appendix II on pp. 875–883, as taken over from

Rose *et al.* [Ro 52b, Ro 53].) A graphical representation of the values tabulated in [NBS 52] is given in Fig. 8-17, which may be compared with somewhat similar curves by Yuasa and Laberrigue-Frolow [Yu 53] reproduced by Segrè [Se 59b, p. 251]. An extensive set of plots has also been prepared by Schopper [Scho 66, Appendix E].

The relativistic expressions for F simplify considerably in a nonrelativistic approximation and reduce to

$$F(E, Z) = x(1 - e^{-x})^{-1} \tag{8-68}$$

where

$$x \equiv \pm \frac{2\pi Z c}{137 v} \qquad \text{for } \beta^{\pm} \text{ emission with velocity } v \tag{8-69}$$

This elucidates the effect of the Coulomb correction especially at the lower end of the β-momentum spectrum, since when $x \gg 1$ one can write

$$F \approx x \sim \frac{1}{p} \qquad \text{for } \beta^- \text{ decay} \tag{8-70}$$

and

$$F \approx |x| e^{-|x|} \sim \frac{1}{p} e^{-(K/p)} \qquad \text{for } \beta^+ \text{ decay} \tag{8-71}$$

where K is a constant. Whereas the lower end of the momentum spectrum has a parabolic shape for $Z = 0$ [since $N(p) \sim p^2$], it rises linearly for β^- decay [since $N(p) \sim p$] and is exponentially depressed for β^+ decay [since $N(p) \sim p e^{-(K/p)}$], as shown in Fig. 8-16.

An effect not taken into account in the Fermi function F (which is also called the COULOMB FACTOR) is the screening of the nuclear charge by the electrons in the atomic cloud. This additional correction is of opposite sign to F and seldom exceeds 2 or 3 percent except for the heaviest elements.

We thus arrive at the ultimate formula for the β-momentum spectrum on replacing

$$|\varphi_e(0)|^2 \qquad \text{by} \qquad F(E, Z)|\varphi_e(0)|^2 \equiv |\varphi_e(0)|_Z^2 \tag{8-72}$$

viz.

$$N(p)\, dp = \frac{g^2}{2\pi^3 c^3 \hbar^7} |M_{\mathrm{fi}}|^2 \, F(E, Z) \, (E_0 - E)^2 p^2 \, dp \tag{8-73}$$

8.3.6. KURIE PLOT

The standard procedure for comparing experimental β spectra with theory is to employ a linear representation of the data, which is termed a KURIE PLOT [Ku 36], FERMI PLOT, or F-K PLOT. The spectral formula (8-73) can be cast into the form

$$\left[\frac{N(p)}{p^2 F(E, Z)} \right]^{1/2} = C(E_0 - E) \tag{8-74}$$

where in the case of allowed transitions the term C contains the generally unknown but energy-independent matrix element $|M_{\text{fi}}|$. Hence a plot of the expression on the left against the kinetic energy E of the emitted β particle should yield a straight line with a clearly defined intercept E_0 and a slope which can yield information as to the value of M_{fi}. When the experiments are conducted with sufficiently thin radioactive sources, linearity is indeed observed, as can be seen from the examples of Kurie plots shown in Fig. 8-18(a–d). These give a quantitative indication of the extent of agreement with theory and show up any deviations more clearly than a conventional spectral plot. The discrepancies at the lowest energies can usually be attributed to instrumental errors.

When more than one transition is concurrently included in a measured spectrum, as is the case for a branched decay process such as that depicted in Fig. 8-19, the Kurie plot is kinked in the way shown in Fig. 8-20 for the β^- decay of ^{59}Fe to ^{59}Co [Me 52] and permits the individual contributions to be distinguished, whereas in the conventional spectral plot there is hardly perceptible deviation from a Fermi shape. The END-POINTS E_0 are clearly established in the Kurie representation whereas the parabolic form of the upper end of the conventional spectrum permits no such precise determination.

8.3.7. NEUTRINO REST MASS

Determination of the neutrino rest mass m_ν from kinematics presents formidable experimental difficulties because of its vanishingly small magnitude. In the advantageous case of ORBITAL ELECTRON CAPTURE, the momenta of the recoil nucleus and of the emitted neutrino can be equated and thence an upper limit for m_ν deduced. Thus the recoil experiments by Snell and Pleasonton [Sn 55] on the process ^{37}Ar $+ e_{\text{K}}^- \rightarrow {}^{37}$Cl yield a value $m_\nu < 6 \,\text{keV}/c^2$. However, a still more precise determination is afforded by a Kurie-plot method.

The *shape* of the KURIE PLOT in the vicinity of the end point E_0 provides information on the magnitude of m_ν. Before considering this in more detail, we note that it is fallacious to imagine that m_ν can be established from the Kurie-plot *intercept*, since it can readily be shown that an extrapolation from measured points which do not lie close to the end point gives an abscissal intercept equal to the Q value (namely, the total energy release $Q = E + E_\nu$) rather than to E_0, whence agreement between the value of the intercept and the Q value as determined from nuclear reactions conveys no information as to the magnitude of m_ν.

We have thus far calculated the form of the β spectrum on the assumption that $m_\nu c^2 \ll E_0$, which is equivalent to setting m_ν to zero. If m_ν is nonzero, we would accordingly expect the spectrum to be affected most where $E_\nu \simeq m_\nu c^2$, namely, at *small* values of E_ν, and therefore at *large* values of E. The possible small end-point shift $(m_\nu \ll m_{\text{e}})$ would be unidentifiable in a

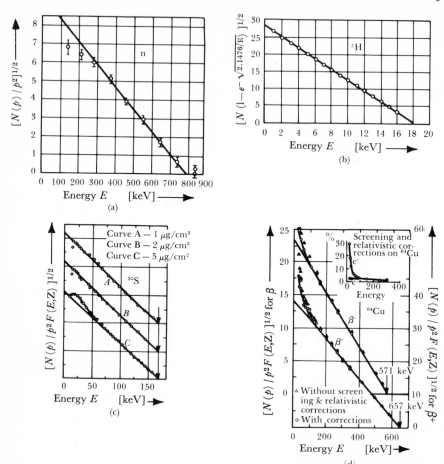

Fig. 8-18. Kurie plots for several β^- and β^+ emitters: (a) the β^- decay of free neutrons, n → p^+ + β^- + $\bar{\nu}_e$, for which the Fermi function F is unity (from [Ro 51a]); (b) β^- decay of tritium, ^{3}H → ^{3}He + β^- + $\bar{\nu}_e$, which provides an instance of a low-energy transition, with E_{max} = 18.1 keV (from [Cu 49]); (c) β^- decay of ^{35}S ($T_{1/2}$ = 86.7 d) to stable ^{35}Cl, yielding Kurie plots for three source thicknesses which show progressively less deviation from linearity at the low-energy end as the source thickness is reduced. The remaining deviation can be ascribed to instrumental errors (from [Al 48]); (d) β^- and β^+ decay of ^{64}Cu, corrected for screening and relativistic effects which particularly influence the β^+ spectrum, as can be seen from the inset. This example is of particular interest in that it simultaneously features β^- and β^+ spectra of approximately equal intensity and maximal energy; the figure is taken from the emended second paper of Wu and Albert [Wu 49], and displays departure from linearity only below 130–140 keV as a result of employing a specially thin, uniform source.

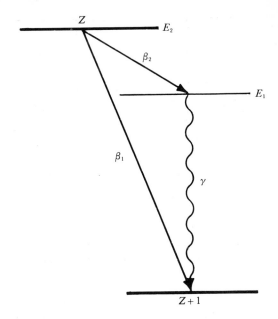

Fig. 8-19. Mixed decay scheme featuring two β^- transitions.

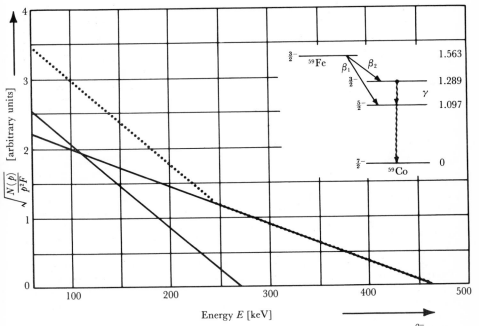

Fig. 8-20. Kurie plot and its decomposition for the mixed decay scheme $^{59}\text{Fe} \xrightarrow{\beta^-} {}^{59}\text{Co}$ which features two β^- transitions (from [Me 52]).

conventional spectral plot, but would be expected to show up in a Kurie plot as a deviation from linearity in the upper energy region where $E \approx E_0 - m_\nu c^2$. The transition energy E_0 now has to include the neutrino rest energy,

$$E_0 = E + E_\nu + m_\nu c^2 \tag{8-75}$$

so that the β spectrum terminates at an energy

$E + E_\nu = E_0 - m_\nu c^2$

$$E_{\max} = E_0 - m_\nu c^2 < E_0 \tag{8-76}$$

$E_{\max}$ when $E_\nu = 0$

which betokens an end-point shift.

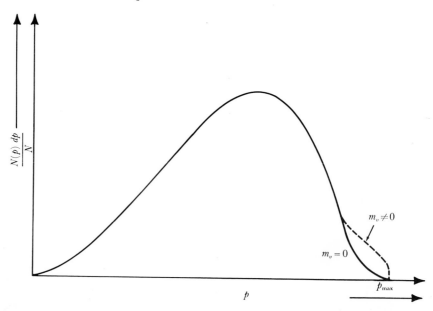

Fig. 8-21. Deformation of a β-decay momentum spectrum to be expected if the rest mass of the (anti) neutrino were nonzero. The end-point would remain unchanged, but the curve would become parabolic at its upper end, and would intersect the abscissa perpendicularly.

Near the end point, $E_0 - E \ll m_\nu c^2$, while $E\ (\approx E_0)$, $E_\nu\ (\approx m_\nu c^2)$, and p_ν do not change appreciably, so that the spectral shape is essentially determined by $p_\nu = (1/c)\,[(Q - E)^2 - (m_\nu c^2)^2]^{1/2}$. Hence, on substituting $Q = E_0 + m_\nu c^2$ we see that

$$p_\nu = \frac{1}{c}\,[(E_0 - E)^2 + 2m_\nu c^2(E_0 - E)]^{1/2} \cong \frac{1}{c}\,[2m_\nu c^2(E_0 - E)]^{1/2}$$

$$= [2m_\nu(E_0 - E)]^{1/2} \tag{8-77}$$

i.e., p_ν takes on approximately its nonrelativistic value. Consequently the β-momentum spectrum becomes deformed, as shown in Fig. 8-21, and acquires the form of a parabola intersecting the abscissa at $E = E_0$ with a vertical tangent. On the other hand, the ordinary extrapolated Kurie plot $(m_\nu = 0)$ intersects at $E = Q$, whence a determination of the distance between these two points furnishes a value for m_ν. A rigorous discussion also has to take account of a possible change in the β-interaction matrix element, which

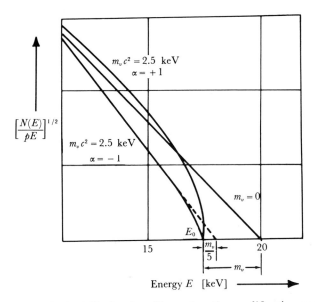

Fig. 8-22. Hypothetical Kurie plots illustrating the modification ensuing from a finite neutrino mass, (e.g., $m_\nu c^2 = 2.5$ keV) depicted for extreme values of a so-called relativistic spinor term α which indicates the limits of deviation from linearity at the high-energy end. The curves with $\alpha = \pm1$ respectively refer to data obtained from calculations with *odd*-parity and with *even*-parity wave functions on the assumption of parity *conservation*. The corresponding curve assuming parity violation would be between these two extremes (from [Wu 65]). (Used by permission of North-Holland Publishing Co.)

imposes still more stringent restrictions upon the possible shape of the Kurie plot when maximum violation of parity is assumed [Sa 58, Scho 60]. Under these conditions, the Kurie plot would be expected to have a shape of the type shown in Fig. 8-22.

An equivalent way to arrive at the same result is to start out from the relativistic energy-momentum relation

$$p_\nu = \frac{1}{c} [E_\nu(E_\nu + 2m_\nu c^2)]^{1/2} \qquad (8\text{-}78)$$

and thence obtain

$$p_\nu^2 \frac{dp_\nu}{dE_0} = \frac{1}{c^3}(E_\nu + m_\nu c^2)\left[E_\nu(E_\nu + 2m_\nu c^2)\right]^{1/2} \tag{8-79}$$

which, on replacing E_ν by $E_0 - E - m_\nu c^2$ reduces to

$$p_\nu^2 \frac{dp_\nu}{dE_0} = \frac{1}{c^3}(E_0 - E)^2\left[1 - \left(\frac{m_\nu c^2}{E_0 - E}\right)^2\right]^{1/2} \tag{8-80}$$

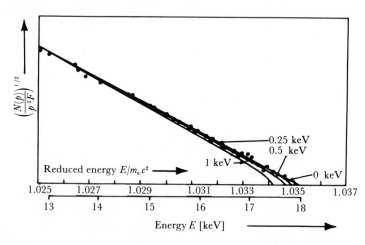

Fig. 8-23. Precision measurements at the high-energy end of the ³H spectrum (cf. Fig. 8-18(b)) taken with a high resolution magnetic spectrometer and compared with Kurie plots commensurate with an antineutrino rest energy of 0, 0.25, 0.5 and 1.0 keV. The data are compatible with a rest mass of zero and set an upper limit of 0.25 keV upon the possible rest energy (from [La 52]).

The term in square brackets represents the modification for $m_\nu \neq 0$ (the above reduces to the expression (8-48) for $m_\nu = 0$) which causes the Kurie plot to be deformed. The latter is now given by

$$\left[\frac{N(p)}{p^2 F(E, Z)}\right]^{1/2} = C(E_0 - E)\left[1 - \left(\frac{m_\nu c^2}{E_0 - E}\right)^2\right]^{1/4} \tag{8-81}$$

The correction factor in square brackets has an influence only near the end point. It drops from unity when $E \ll E_0$ to zero at $E = E_0 - m_\nu c^2$ and its derivative at this latter point becomes infinite, i.e., the tangent is normal to the abscissa.

A determination of m_ν thus necessitates extremely precise measurement of the upper end of the spectrum. The smallest perceptible neutrino mass corresponds roughly to the momentum resolution Δp of the β spectrometer.

Hence to minimize Δp it is advantageous to select decays in which p is *small* near the endpoint, e.g., cases with small E_0. Furthermore, it is of advantage to choose a case in which the deformation of the electron wave φ_e is small, namely one in which $F \simeq 1$, as in low-Z nuclei. These criteria indicate the β^- decay of ^{3}H to ^{3}He, with its low end point of 18.1 keV, to be particularly suitable. The earlier experiments have been summarized by Allen [Al 58b]; these have been superseded by the more precise measurements taken by Langer and Moffat [La 52] and Hamilton *et al.* [Ha 53]. Figure 8-23 shows the Kurie plot obtained by Langer and Moffat, which can be interpreted (on the assumption of maximum parity violation in β decay) to dictate an upper limit to the neutrino mass as

$$m_\nu < 0.25 \text{ keV}/c^2 \simeq 5 \times 10^{-4} \, m_e \tag{8-82}$$

which is compatible with $m_\nu = 0$.

The upper limit of the ν_μ rest mass is less accurately known. The experiments by Barkas *et al.* [Ba 56] on energy-momentum balance in pion decay set a limit:

$$m_{\nu_\mu} < 3.5 \text{ MeV}/c^2 \simeq 6.8 \, m_e \tag{8-83}$$

8.3.8. NEUTRINOS AND COSMOLOGY

Since neutrinos interact but weakly with matter, it is possible that quite an appreciable amount of cosmic energy is reposed in galactic neutrinos. This may at least in part account for a hitherto unresolved discrepancy between the estimated density of visible matter in the universe ($\rho \simeq 3 \times 10^{-31}$ g cm^{-3}) and the value furnished by current cosmological models ($\rho \approx 10^{-29}$ g cm^{-3}) [Bo 64]. A significant difference between neutrinos and photons is to be found not only in their respective interaction strengths but in the fact that whereas photons are bosons, neutrinos are subject to *Fermi-Dirac statistics* and can accordingly be treated as a DEGENERATE FERMI GAS along similar lines to those used for nucleons in the Fermi-gas model of the nucleus, to be discussed later. From this viewpoint, Weinberg [Wei 62] has estimated the Fermi energy E_F up to which the available energy levels are filled according to various cosmological models. A measurement of the Fermi energy would provide information on the expansion of the Universe. This could, in principle, be derived from the Kurie plot for allowed β^- decay determined with a detector which has no energy threshold. The upper end of the plot corresponds to maximal β^- energies and hence to minimal $\bar{\nu}_e$ energies. But all $\bar{\nu}_e$'s emitted in β^- decay whose energies are smaller than E_F would find the available levels already occupied, and hence their emission would be considerably inhibited, with the result that the Kurie plot would terminate at an appropriately diminished β^- energy $E_e = E_0 - E_F$, as shown in Fig. 8-24(a). But if the degeneracy applies to the *neutrinos*, rather than to the antineutrinos, an enhancement effect should be observed, as in Fig. 8-24(b), since the decay could take place via *absorption* of a neutrino from the Fermi sea rather than by way of $\bar{\nu}_e$ emission. This would be evinced as an *augmented* β-endpoint energy $E_e = E_0 + E_F$.

An upper limit for the Fermi energy E_F can be derived on the assumption of zero neutrino rest mass from a Kurie plot of the type shown in Fig. 8-23 as used to derive information on m_ν from the $β^-$ spectrum of ^{3}H, since the effect is similar when one supposes the $\bar{\nu}_e$ gas to be degenerate. This indicates that $E_F \lesssim 250$ eV, a figure in good agreement with that $(E_F \lesssim 500$ eV) obtained

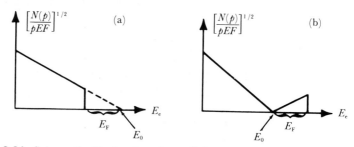

Fig. 8-24. Schematic Kurie plots for a β-decay process featuring a degenerate Fermi gas of (a) antineutrinos, (b) neutrinos, indicating the modification to the end-point energy by an amount E_F, the Fermi energy of the degenerate gas (from [Scho 66]). (Used by permission of North-Holland Publishing Co.)

for neutrinos by Drever, as reported by Weinberg [Wei 62], and not incompatible with the estimate $(E_F \lesssim 2$ eV) for ν_μ's [Ma 63c, Co 64]. An inference which can, for example, be drawn from this data is that if the Universe is pulsating (cf. Section 6.5) the radii at the two extremes must stand in the ratio of at least 1 : 10,000. The theoretical situation has been surveyed by Chiu [Chi 65, Chi 66], to whom the interested reader is referred for additional details.

8.3.9. Beta-Decay Lifetime

The decay constant λ, equal to the reciprocal of the mean decay lifetime τ, is just the probability per unit time for emission of a β particle with *arbitrary* energy between 0 and E_0:

$$\lambda = \int_{p=0}^{p \triangleq E=E_0} N(p)\,dp = \frac{g^2}{2\pi^3 c^3 \hbar^7} \int_0^{p \triangleq E_0} |M_{fi}|^2\, F(E,\,Z)\,(E_0 - E)^2 p^2\, dp \quad (8\text{-}84)$$

It is convenient to rewrite this in terms of the *total* energy referred to the electron rest energy $m_e c^2$:

$$W \equiv \frac{E + m_e c^2}{m_e c^2} \xrightarrow{E=E_0} W_0 \equiv \frac{E_0 + m_e c^2}{m_e c^2} \quad (8\text{-}85)$$

viz.

$$\lambda = \left(\frac{g^2}{2\pi^3}\frac{m_e^5 c^4}{\hbar^7}\right) |M_{fi}|^2 \int_1^{W_0} F(W, Z)\,(W^2 - 1)^{1/2}(W_0 - W)^2\,W\,dW \quad (8\text{-}86)$$

so that

$$\lambda \equiv \frac{|M_{fi}|^2}{\tau_0} f(W_0, Z) \quad (8\text{-}87)$$

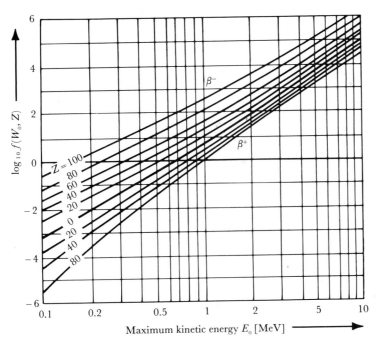

Fig. 8-25. Log-log plot of the Fermi integral function $f(W_0, Z)$ for β^- and β^+ decay in terms of the maximal kinetic energy $E_0 = W_0 - m_e c^2$, as calculated for emission from a point-charge nucleus (from [Fee 50], redrawn by Evans [Ev 55]). (Used by permission of McGraw-Hill Book Co.)

This expresses λ through the matrix element $|M_{fi}|^2$ multiplied by a calculable function f and divided by a constant termed the UNIVERSAL β-DECAY TIME CONSTANT,

$$\tau_0 \equiv \frac{2\pi^3}{g^2}\frac{\hbar^7}{m_e^5 c^4} \approx 7000 \text{ sec} \quad (8\text{-}88)$$

which, containing as it does the coupling constant g, provides a measure of the β-interaction strength. The FERMI INTEGRAL FUNCTION $f(W_0, Z)$ is determined by the statistical factor and Coulomb effects. It can be expressed in analytic form only when $Z = 0$, because then $F(W, Z = 0) = 1$, but has been evaluated numerically and tabulated over a wide range of W_0 and Z values [Fee 50, Fei 50, Ro 52b, NBS 52, Dz 56, Wa 59].

Since the values of $f(W_0, Z)$ extend over a large range, it is customary to plot them logarithmically as in Fig. 8-25. At the highest energies $f(W_0, Z)$ is roughly proportional to $(W_0)^5$, while at lower energies it increases even

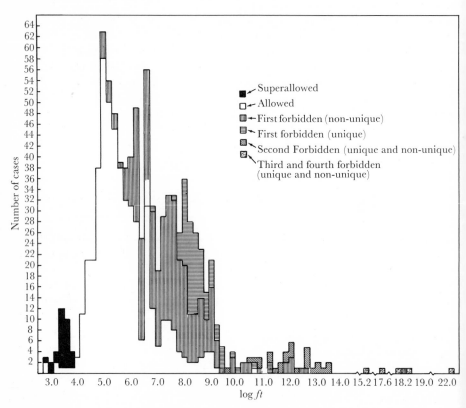

Fig. 8-26. Frequency histogram for $\log ft$ values, compiled by Gleit *et al.* [Gl 63].

more rapidly with W_0. This strong dependence of f on W_0 betokens that λ is also very dependent upon the end-point energy E_0.

The HALF-LIFE $T_{1/2}$ follows from λ in the usual way, being given by the expression

$$T_{1/2} = \frac{\ln 2}{\lambda} = \frac{\tau_0 \ln 2}{|M_{\mathrm{fi}}|^2} \frac{1}{f(W_0, Z)} \tag{8-89}$$

As essentially determined by f, the half-life $T_{1/2}$ is thus also strongly energy dependent, a feature which can be circumvented by introducing in its place the COMPARATIVE HALF-LIFE, defined as the product $f \times T_{1/2}$ and customarily called the ft VALUE:

$$ft \equiv f(W_0, Z) \, T_{1/2} = \frac{\tau_0 \ln 2}{|M_{\mathrm{fi}}|^2} \simeq \frac{5000}{|M_{\mathrm{fi}}|^2} \tag{8-90}$$

The ft value accordingly provides an inverse measure of the transition matrix element. It is expressed in sec and covers a very wide range of values, viz. 10^3 to 10^{23} sec, most of which cluster around 10^4–10^9 sec. Various tabulations exist [Fe 51, Ro 55, Ro 60, Zy 61, Gl 63], which may be compared with values expected theoretically, as listed by Blatt and Weisskopf [Bl 52b] or derived from suitable nomograms [Mo 51b]. The representation of the data by Gleit *et al.* [Gl 63], as in Fig. 8-26, indicates that there is no clear-cut grouping of ft values into well-defined categories. Nevertheless a classification scheme can be drawn up into which transitions are collected into groups on the basis of their ft values supplemented by further spectral information and model predictions.

8.4. Classification of Beta Transitions

At first there was reason to think that β-decay lifetimes fell into two categories and to classify transitions into those which were ALLOWED and those which were FORBIDDEN, the latter being roughly a factor of 100 more sluggish than the former. The evidence from a plot of the decay constant λ against the maximum β energy E_0, called a SARGENT DIAGRAM [Sa 33] and closely akin to the Geiger-Nuttall plots for α decay, appeared to indicate a separation into two distinct groups, as depicted in Fig. 8-27. Gamow [Ga 34a, Ga 34b] thereupon proposed, on the basis of the newly propounded Fermi theory, that this could be explained through a consideration of angular momenta and parity changes. The ALLOWED transitions proceeded readily because they involved *no change in nuclear spin or parity* $(\Delta J = 0, \Delta \pi = \mathrm{no})$, whereas the FORBIDDEN transitions were hindered by their involving a *change in nuclear spin and parity* $(\Delta J = \pm 1, \Delta \pi = \mathrm{yes})$. However, these conclusions, based on the sparse and inadequate data available at that time solely from the natural radioactive uranium, thorium, and actinium series, were found to be too naive to represent the actual situation as further data were amassed from artificial β emitters, and had to be replaced by a more elaborate classification scheme based upon a more complicated evaluation of various diagnostic criteria.

Chief among these criteria, but nevertheless *not* the sole arbiter, is the *comparative half-life* $ft \equiv f T_{1/2}$, which represents a more suitable quantity to describe the DEGREE OF FORBIDDENNESS than f or $T_{1/2}$ alone, but which has

Table 8-2. ROUGH CLASSIFICATION OF β TRANSITIONS
ACCORDING TO *ft* VALUE

Category	Range of log *ft*
Superallowed	2.9–3.7
Allowed	4.4–6.0
First forbidden (nonunique)	6–9
First forbidden (unique)	8–10
Second forbidden	10–13
Third forbidden	>15

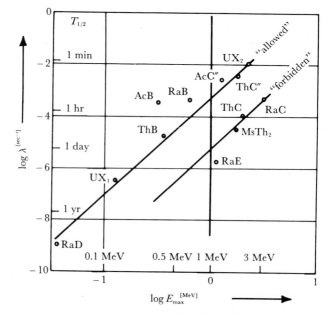

Fig. 8-27. Adapted form of the Sargent diagram [Sa 33] for some "allowed" and "forbidden" natural β emitters in which the logarithm of the partial decay constant for the principal β transition is plotted against the logarithm of the end-point energy E_{max}. The ordinate also indicates the corresponding half-lives $T_{1/2}$.

to be supplemented by other spectral and theoretical considerations before a given transition can be assigned to a definite category.

The *ft* histogram (Fig. 8-26) displays a cluster of values around log *ft* = 3.5. Such transitions proceed even more readily than the majority of the allowed decays and are accordingly termed SUPERALLOWED. The bulk of the ALLOWED TRANSITIONS are to be found for log *ft* = 5–6 and thereafter the FORBIDDEN TRANSITIONS predominate. Thus allowed transitions have *ft* values ranging from 3000 to 100,000 sec, while for (first) forbidden transitions ft is about one thousand times bigger. An indication of a rough classification into categories according to ft value is given in Table 8-2.

The naming of the categories is derived from a consideration of the interaction matrix element. In Section 8.3.4 we considered only the first term in a series expansion in powers of $(\mathbf{k}_e \cdot \mathbf{r})$ and $(\mathbf{k}_\nu \cdot \mathbf{r})$ for the lepton wave functions $\varphi_e(\mathbf{r})$ and $\varphi_\nu(\mathbf{r})$. The integration to derive H'_{fi} could be confined to the nuclear volume, since the *nuclear* wave functions ψ_f and ψ_i vanish outside this region, and it accordingly sufficed to know the values of $\varphi_e(\mathbf{r})$ and $\varphi_\nu(\mathbf{r})$ for the range $0 \leqslant r \leqslant R$ where R, the nuclear radius, is about 10^{-12} cm.

8.4.1. DEGREES OF FORBIDDENNESS

The plane-wave product $\varphi_e^* \varphi_\nu^*$ is proportional to $\exp\{-i[(\mathbf{k}_e + \mathbf{k}_\nu) \cdot \mathbf{r}]\}$. Taking the lepton wave vectors to be roughly equal and to correspond to an energy of 1 MeV we set

$$k_e \approx k_\nu \cong 10^{11} \text{ cm}^{-1} \tag{8-91}$$

whence we find

$$(k_e + k_\nu)\, r \cong (k_e + k_\nu)\, R \cong 10^{11} \times 10^{-12} \lesssim \tfrac{1}{10} \tag{8-92}$$

Since the transition probability is proportional to $|H'_{fi}|^2 \sim |\varphi_e \varphi_\nu|^2$, it follows that consecutive terms in the wave-function expansion differ by almost *two* orders of magnitude. They therefore diminish rapidly but do not vanish. It will be recalled that the first term of the expansion generates the ALLOWED TRANSITION matrix element and that the successive terms refer to FORBIDDEN TRANSITIONS of increasing order. The latter thus have *diminishing* but *nonvanishing* probability, and might more felicitously have been termed "2nd-order," "3rd-order," etc., transitions. An nth-order transition would require the application of the first $(n + 1)$ terms of the power expansion in kR in order to render $|H'_{fi}|$ nonzero, resulting in a commensurately diminished transition probability (increased ft value). The situation is similar to that in optical spectroscopy, in which an expansion in higher multipoles is adopted, but the fact that in β decay $kR = R/\lambda \approx 1/10$ whereas for optical transitions the corresponding quantity is much smaller, viz. $R/\lambda \approx 10^{-8}$ cm$/10^{-5}$ cm $\approx 1/1000$, causes one to encounter transitions with higher degrees of forbiddenness much more readily in β spectroscopy than in optical spectroscopy.

An additional influence ensues from the intrinsic motion of the nucleons with a mean velocity v_N within the nucleus. This gives rise to an extra relativistic correction term to the matrix element, of magnitude v_N/c smaller than the original uncorrected element. Since v_N/c and kR are of roughly equivalent magnitude, the additional relativistic term can exert an appreciable influence. In addition to these two "parameters of smallness," a third parameter $\alpha Z = Z/137$ enters when the Coulomb effect of the nucleus is taken into consideration. Systematic investigation of these corrections indicates that the largest terms surviving (viz. $|H'_{fi}| \neq 0$) for n-times forbidden (unique) transitions, i.e., for nth-order transitions, give rates of order R^{2n}. The name UNIQUE is associated with those forbidden transitions governed by a *single* β moment, as against *parity*-forbidden transitions which are governed by a combination of *several* β moments.

To summarize, ALLOWED decays are characterized by the absence of inhibiting factors and by the absence of a parity change in the transition ($\Delta J = 0$, $\Delta\pi = $ no). FIRST FORBIDDEN transitions are associated with a change in parity ($\Delta J = 0, \pm 1, \pm 2$, $\Delta\pi = $ yes) and contain *one* of the two factors kR or v_N/c. In the case of the SECOND FORBIDDEN matrix elements, one has to distinguish between a group of order 0 or 1 which represents *corrections to allowed transitions* since they obey the same selection rules as the allowed transitions ($\Delta J = 0, \pm 1, \Delta\pi = $ no) and are invariably accompanied by the latter although much smaller in magnitude since they are modified by the factors kR and/or v_N/c, and a group with $\Delta J = \pm 2, \pm 3$, $\Delta\pi = $ no which constitutes *true second forbidden transition matrix elements*, untainted by interference from allowed transitions. Similarly, an admixture of THIRD FORBIDDEN decays to first forbidden can occur or not, depending upon the spin change ΔJ. Transitions with $\Delta J^\pi = 2^-, 3^+, 4^-, 5^+, \ldots$ are termed UNIQUE FORBIDDEN, while those with $\Delta J^\pi = 0^-$, and $1^-, 2^+, 3^-, 4^+, 5^-, \ldots$ are NONUNIQUE FORBIDDEN, or PARITY FORBIDDEN. An n-TIMES FORBIDDEN TRANSITION has $\pi_i \pi_f = (-1)^n$ and $\Delta J = n$ or $n + 1$.

8.4.2. SUPERALLOWED TRANSITIONS

The large value of the overlap integral M_{fi} indicates that the initial- and final-state nuclear wave functions ψ_i and ψ_f are very similar. The perfect overlap which would yield the largest possible value of M_{fi} is never realized in practice, but an extensive overlap is readily attainable, in the case of mirror nuclei, for example. To assist in appreciating this point, we consider a highly simplified idealization of the connection between the energy levels of mirror nuclei. According to the nuclear shell model, the proton states and neutron states are filled up progressively, starting at the lowest vacant level. Each fermion state is restricted by the Pauli exclusion principle to containing no more than two particles of each kind whose spins are antiparallel. This yields the situation depicted in Fig. 8-28. Assuming that nuclear forces are of a universal strength, irrespective of whether they act on a proton or a neutron, we conclude that the wave functions and the energies of the protons and the neutrons in the same quantum states are equal, apart from a small Coulomb perturbation. Since a β transition between mirror nuclei betokens an interchange in the character of two nucleons in the same quantum state, it can be expected to be associated with a large overlap integral M_{fi}.

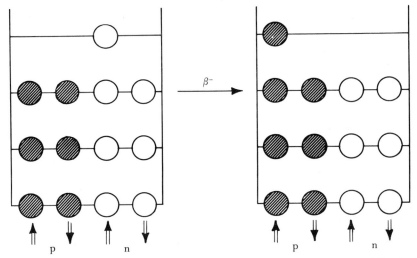

Fig. 8-28. Schematic illustration of the change in level occupation following a β^- transition between mirror nuclei. Protons are indicated by darkened circles, neutrons by open circles. The arrows indicate spin orientation.

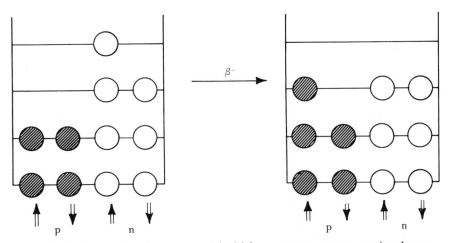

Fig. 8-29. A β^- transition between nuclei which are *not* mirror partners involves a more radical change in the nuclear wave functions than one between mirror nuclei (Fig. 8-28). Hence the overlap integral, and therefore the transition matrix element M_{fi}, is smaller by about one order of magnitude.

8.4.3. Allowed Transitions

In the case of β transitions between nuclei which are not mirror partners, the transformation of a neutron into a proton carries the system over into a new quantum state whose wave function is appreciably different from that for the initial state (Fig. 8-29). Hence there is far less overlap in the wave functions ψ_i and ψ_f, which results in the matrix element M_{fi} being reduced by an order of magnitude. This constitutes the normal situation, whose transition rate as characterized by the ft value is roughly a factor of 10^2 smaller than that for superallowed transitions.

8.4.4. FORBIDDEN TRANSITIONS

Through the need to consider higher-order terms proportional to kR and/or v_N/c in order to render $|H'_{fi}|$ nonzero, the magnitude of the matrix

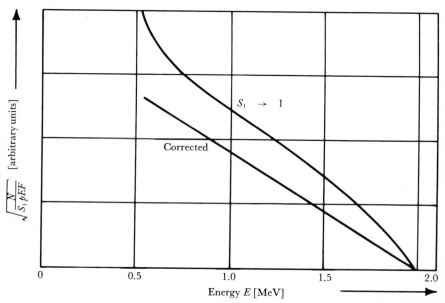

Fig. 8-30. Correction of the Kurie plot for ^{91}Y effected with the aid of the unique first-forbidden shape correction factor S, defined in Eq. (8-93). The curves follow experimental data points obtained from the measurements of Langer and Price [La 49a, La 49b].

element becomes about 10-fold smaller (or even less) than that for normal allowed transitions, which indicates that there is hardly any overlap in the nuclear wave functions. Accordingly the quantum numbers which characterize the states differ very markedly. This is evinced through the large spin differences ΔJ which apply to forbidden transitions. The occurrence of the factor kR indicates that the corresponding matrix element depends upon the kinetic energy of the electron E, because kR depends directly upon the electron momentum p through its wave-number factor $k \equiv p/\hbar$. The shape of the spectrum, and therefore of the Kurie plot, differs for forbidden transitions from that for allowed transitions (though in some cases it is hardly possible to discriminate between forbidden and allowed shapes [La 49b].) It is possible to calculate analytic SHAPE CORRECTION FACTORS S_n by which an allowed distribution, as given by the spectral formula (8-73), must be multiplied in order to obtain the shape of the forbidden spectrum [Ko 41, Ko 43, Sm 51]. These are discussed by Konopinski and Rose [Si 65, pp. 1351 ff.]. By applying an appropriate correction to the β-momentum spectrum of ^{91}Y, Langer and Price [La 49a, La 49b] were able to derive a linear Kurie plot in place of the deformed plot shown in Fig. 8-30. The value $ft = 5 \times 10^8$ sec would indicate this to be a second forbidden transition, but shell-model considerations indicate that $\Delta J = 1$ and $\Delta \pi = $ yes, which places it in the category of unique first forbidden transitions, for which the shape correction factor is

$$S_1 = (W^2 - 1) + (W_0 - W)^2 \tag{8-93}$$

This effects the requisite linearization.

8.5. Electron Capture

The process of electron capture, first suggested by Yukawa and Sakata [Yu 35b] (see also [Al 37, Al 38, Ab 39]), invariably accompanies β^+ decay, which it closely resembles.

8.5.1. DISCRETE NEUTRINO ENERGY SPECTRUM

Although the process wherein an atomic electron in venturing inside the nuclear volume is captured by a proton can be viewed as similar to an inverse β^+ decay,

$$p^+ + e^- \rightarrow n + \nu \tag{8-94}$$

the energies of the *two* product particles do *not* have a continuous spectral distribution, but are *discrete*. The monoenergetic neutrinos are emitted with a kinetic energy

$$E_\nu = E_0 + m_e c^2 - E_B \tag{8-95}$$

where E_0 is the available transition energy corresponding to the mass difference between the parent and daughter *nuclei* (not atoms—cf. the energy balance equation for $E_0 = \Delta E$ in Section 8.1.2, where *atomic* masses were used), and E_B is the binding energy of the electron in its respective atomic shell before capture (we have neglected the small recoil energy of the daughter nucleus). This gives rise to a LINE SPECTRUM for the neutrino energy. If capture takes place simultaneously from more than one shell, several discrete lines are featured in the neutrino energy spectrum at appropriate energies, as indicated in Fig. 8-31, which also shows the continuous neutrino energy spectrum corresponding to a concurrent competing β^+-decay process. Although the preceding β-decay theory cannot therefore meaningfully be applied to a consideration of the spectrum, it does have relevance to the evaluation of capture lifetime and ft value.

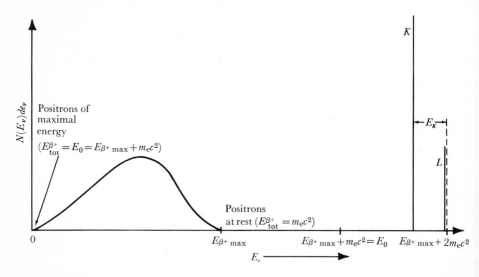

Fig. 8-31. Continuous neutrino energy spectrum from competing β^+ decay, shown together with the line spectrum due to K and L electron capture, where E_0 is the transition energy and $E_{\beta+\max}$ is the maximum energy of the β^+ spectrum (adapted from [Se 59b]).

8.5.2. CAPTURE LIFETIME AND ft VALUE

A "decay constant" λ and mean lifetime τ for electron capture can be derived along analogous lines to those adopted hitherto, viz.

$$\lambda \equiv \frac{1}{\tau} = \frac{2\pi}{\hbar} |H_{\text{fi}}'|^2 \frac{dn}{dE_\nu} \tag{8-96}$$

Here, the statistical factor is determined solely by the neutrino energy, since the electron is in a definite quantum state before capture, i.e.,

$$dn \to dn_\nu = V \frac{4\pi p_\nu^2 \, dp_\nu}{h^3} \tag{8-97}$$

whence, with $p_\nu = E_\nu/c$,

$$\frac{dn}{dE_\nu} = V \frac{E_\nu^2}{2\pi^2 c^3 \hbar^3} \tag{8-98}$$

The matrix element is the same as before. Hence for allowed transitions it takes the form

$$|H_{fi}'|^2 = g^2 |\varphi_e(0)|^2 |\varphi_\nu(0)|^2 |M_{fi}|^2 \tag{8-99}$$

As before, the neutrino field can be expressed as a plane wave, and

$$|\varphi_\nu(0)|^2 = \frac{1}{V} \tag{8-100}$$

However, the wave function of the electron before capture has to be set to that appropriate to the given shell. Thus for K electrons it can be written (see, e.g., [Le 59, p. 183]) in terms of the Bohr radius $a_0 \equiv \hbar^2/m_e e^2$ as

$$\varphi_e(\mathbf{r})_K = \frac{1}{\pi^{1/2}} \left(\frac{Z}{a_0}\right)^{3/2} \exp\left(-\frac{Z}{a_0} r\right) \tag{8-101}$$

so that

$$|\varphi_e(0)_K|^2 = \frac{1}{\pi}\left(\frac{Z}{a_0}\right)^3 \equiv \frac{1}{\pi}\left(\frac{Zm_e e^2}{\hbar^2}\right)^3 \tag{8-102}$$

Substituting (8-100) and (8-102) in (8-99), we get

$$|H_{fi}'|_K^2 = g^2 \frac{1}{V}\frac{1}{\pi}\left(\frac{Zm_e e^2}{\hbar^2}\right)^3 |M_{fi}|^2 \tag{8-103}$$

and thence the decay constant for K capture, on inserting a factor of 2 to take account of the presence of *two* electrons in the K shell,

$$\lambda_K = \left[\frac{2g^2}{\pi^2 \hbar^{10}}\left(\frac{Zm_e e^2}{c}\right)^3\right]|M_{fi}|^2 E_\nu^2 \tag{8-104}$$

On expressing the neutrino energy in relativistic units,

$$W_\nu \equiv \frac{E_\nu}{m_e c^2} = \frac{E_0}{m_e c^2} + 1 - \frac{E_B}{m_e c^2} \equiv W_0 - \epsilon_B \tag{8-105}$$

where

$$W_0 \equiv \frac{E_0 + m_e c^2}{m_e c^2} \quad \text{and} \quad \epsilon_B \equiv \frac{E_B}{m_e c^2} \tag{8-106}$$

one obtains the decay constant for K capture in terms of the universal time constant $\tau_0 \equiv 2\pi^3 \hbar^7/g^2 m_e^5 c^4$ as

$$\lambda_K = \frac{|M_{fi}|^2}{\tau_0} 4\pi \left(\frac{Ze^2}{\hbar c}\right)^3 (W_0 - \epsilon_B)^2 \tag{8-107}$$

We observe in passing that electron capture takes place more readily in heavy nuclei than in those with low Z since two factors contribute to augmenting the overlap of the electron wave function with that of the parent nucleus: namely, the radii of the orbits decrease as Z increases (and λ_K is proportional to Z^3), while the nuclear radius increases as $A^{1/3}$.

If we introduce a Fermi integral function

$$f_K^{(EC)} \equiv 4\pi \left(\frac{Ze^2}{\hbar c}\right)^3 (W_0 - \epsilon_B)^2 \tag{8-108}$$

we can write (8-107) in the familiar form

$$\lambda_K = \frac{|M_{fi}|^2}{\tau_0} f_K^{(EC)} \tag{8-109}$$

This form of the Fermi integral function can be rewritten in terms of the fine-structure constant $\alpha \equiv e^2/\hbar c$, since the K-shell binding energy ϵ_K in

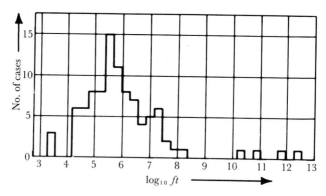

Fig. 8-32. Frequency histogram for $\log ft$ values in the case of electron capture, as given in the tables of Major and Biedenharn [Ma 54a].

relativistic units is approximately

$$\epsilon_K \simeq \frac{1}{2}\left(\frac{Ze^2}{\hbar c}\right)^2 = \frac{1}{2}(\alpha Z)^2 \tag{8-110}$$

as

$$f_K^{(EC)} = 4\pi(\alpha Z)^3 [W_0 - \tfrac{1}{2}(\alpha Z)^2]^2 \tag{8-111}$$

However, it has to be modified in practice by the inclusion of two corrections which play an important role in the case of heavy nuclei. It is necessary to correct for relativistic effects and for screening of the nuclear charge by atomic electrons. The effect of screening on the Coulomb-distorted electron wave can in the case of K electrons be taken into account by substituting an EFFECTIVE NUCLEAR CHARGE, $Z_K' = Z - 0.35$ for the charge Z of the parent nucleus (cf. for L electrons, $Z_L' = Z - 4.15$) to obtain an EFFECTIVE FERMI INTEGRAL FUNCTION

$$f_K' = 4\pi(\alpha Z')^3 [W_0 - \tfrac{1}{2}(\alpha Z')^2]^2 \tag{8-112}$$

The *ft* values built from the product of *f'* with $T_{1/2}$,

$$ft \equiv f' \, T_{1/2} = \frac{\tau_0 \ln 2}{|M_{\mathrm{fi}}|^2} \qquad (8\text{-}113)$$

agree fairly well with those for the corresponding β^+ transitions. In particular, the good agreement between the *ft* value for electron capture and that for β decay in the case of the superallowed transition $^7\mathrm{Be} \rightarrow \, ^7\mathrm{Li}$ ($ft = 2000$ sec) indicates that the strength of the interaction must be approximately the same in both processes.

The various DEGREES OF FORBIDDENNESS also apply to electron-capture transitions according to the magnitude of the interaction matrix element, but in practice almost exclusively only superallowed and allowed electron-capture transitions have been observed with $\log ft \approx 4.5\text{--}6.5$. The histogram of $\log ft$ values for K capture [Ma 54a], as shown in Fig. 8-32, is very similar in appearance to that for β decay (Fig. 8-26).

8.5.3. ELECTRON CAPTURE RATIOS

In practice, a complication arises in that several shells may contribute to a capture process, and for each shell (K, L_{I}, L_{II}, L_{III}, M_{I}, ..., M_{V}, ...) the appropriate wave functions φ_e must be inserted into the calculations together with the requisite values of p_ν and the screening factor has to be adjusted. Some indication of relative magnitudes can be gained by considering the form of the nonrelativistic radial wave function for S electrons ($l = 0$) in a hydrogenlike atom in terms of the principal quantum number n,

$$\varphi_e^{(n)} \sim \left(\frac{\alpha Z}{n} \right)^{3/2} \qquad (8\text{-}114)$$

The prediction of this simplified model, which cannot differ radically from that furnished by more rigorous relativistic considerations, is that the decay constant λ falls off as n^{-3}, viz. the contribution from the L_{I} shell ($n = 2$) to K-shell capture ($n = 1$) is about $\frac{1}{8}(\cong 12$ percent) provided the available transition energy is large compared with the K-shell binding energy E_K. This appreciable contribution can become even more significant when K capture is precluded by energy restrictions ($W_0 < |\epsilon_K|$) which permit capture from the L or higher shells only.

Calculations of ATOMIC SHELL CAPTURE RATIOS [Ro 49b, Br 55, Br 58, Wa 59] indicate the L_{II}-to-L_{I} branching ratio to be small. Its magnitude increases from about 0.002 for $Z = 20$ to 0.1 for $Z = 100$. The smallness of the ratio of decay constants $\lambda_{L_{\mathrm{II}}}/\lambda_{L_{\mathrm{I}}}$ in its turn causes the ratio $\lambda_{L_{\mathrm{II}}}/\lambda_K$ to be very small. The ratio $\lambda_{L_{\mathrm{I}}}/\lambda_K$ is appreciably larger—indeed, it becomes infinite at the threshold for K capture, as can be seen from Fig. 8-33. Some calculations of higher-shell ratios $\lambda_{M+\cdots}/\lambda_K$ have also been performed in the above references, and a review of the experimental results for these ratios has been presented by Robinson and Fink [Ro 60]. A correction by Bahcall [Ba 62a, Ba 63], in which exchange among the bound electrons participating in a capture process is taken into account, increases the theoretical value of λ_{L_I}/λ_K to bring it into agreement with measured values, as can be seen in Table 8-3.

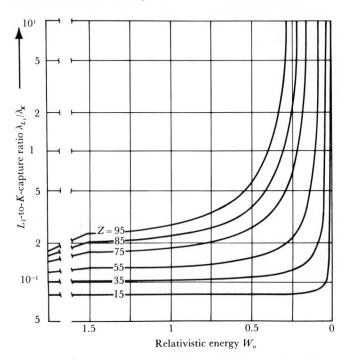

Fig. 8-33. Semilogarithmic plots of L_I-to-K electron-capture branching ratios as functions of W_0, defined by Eq. (8-106), for various values of Z, using values of the partial "decay constants" λ as calculated by Brysk and Rose [Br 55] with relativistic wave functions including the effect of nuclear screening.

An interesting, if small, effect associated with electron capture from the L shell is the influence of an external factor upon the NUCLEAR LIFETIME [De Be 66]. As was independently noted by Segrè [Se 47] and Daudel [Da 47], the L electrons in the $_4^7$Be atom are the *valence* electrons and therefore the nuclear lifetime for L capture, which depends on the electron density at the nucleus, should be slightly sensitive to chemical binding effects. A small difference in the ^{7}Be lifetime should therefore be evinced when the beryllium is chemically combined in different compounds. Experimental studies of this effect [Bo 49, Se 49b, Le 49, Le 51, Se 51, Kr 53, Bo 56] showed that the

Table 8-3. COMPARISON OF SOME THEORETICAL AND EXPERIMENTAL L/K ELECTRON CAPTURE RATIOS[a]

Parent nucleus	$(\lambda_L/\lambda_K)_{th}$[b]	$(\lambda_L/\lambda_K)_{th}$[c]	$(\lambda_L/\lambda_K)_{exp}$	Reference
^{36}Cl	0.0808	0.0994	0.112 $\pm$ 0.008	[Do 62]
^{37}Ar	0.0820	0.100	0.103 $\pm$ 0.003	[Po 49, La 54, Ki 59, Sa 60b, Ma 61, Do 62]
^{51}Cr	0.0885	0.1034	0.1026 $\pm$ 0.0004	[Fa 62]
^{54}Mn	0.0899	0.1043	0.098 $\pm$ 0.006	[Mo 63]
^{55}Fe	0.0937	0.1078	0.106 $\pm$ 0.003	[Sco 59, Ma 61, Mo 63]
^{57}Co	0.0916	0.1044	0.099 $\pm$ 0.011	[Mo 63]
^{58}Co	0.0908	0.1035	0.107 $\pm$ 0.004	[Mo 63]
^{65}Zn	0.0965	0.1090	0.119 $\pm$ 0.007	[Sa 62]
^{71}Ge	0.1032	0.1156	0.1187 $\pm$ 0.0008	[Dr 60, Ma 61]
^{79}Kr	0.102	0.111	0.108 $\pm$ 0.005	[Ra 52b, La 54, Dr 60]

[a] From [Scho 66].
[b] From [Br 55].
[c] From [Ba 63].

Table 8-4. SOME THEORETICAL AND EXPERIMENTAL K/β^+ RATIOS[a]

Parent nucleus	$(\lambda_K/\lambda_{\beta+})_{th}$	$(\lambda_K/\lambda_{\beta+})_{exp}$
^{18}F	0.029	0.030 $\pm$ 0.002
^{48}V	0.066	0.068 $\pm$ 0.02
^{52}Mn	1.77	1.81 $\pm$ 0.07
^{107}Cd	310	320 $\pm$ 30
^{111}Sn	1.5	2.5 $\pm$ 0.25

[a] From Bouchez and de Pommier [Bo 60].

lifetime of ^{7}Be in BeF$_2$ is about 0.07 percent less than that in beryllium metal, whereas in BeO it is only about 0.01 percent less than in beryllium metal.

An altogether different type of branching ratio is the K/β^+ RATIO, since β^+ decay can in many instances compete with K capture. In terms of initial and final *nuclear* masses m_i and m_f, K capture can occur when $m_i + m_e > m_f$, whereas β^+ emission requires that the inverse need not always apply. The K/β^+ RATIO $\lambda_K/\lambda_{\beta+}$ is independent of the matrix element M_{fi} and has the value (cf. Sections 8.5.2 and 8.3.9)

$$\frac{\lambda_K}{\lambda_{\beta+}} = \frac{4\pi}{f(W_0, Z)} \left(\frac{Ze^2}{\hbar c}\right)^3 (W_0 - \epsilon_K)^2 = \frac{4\pi}{f(W_0, Z)} (\alpha Z)^3 (W_0 - \epsilon_K)^2 \qquad (8\text{-}115)$$

This has been plotted logarithmically as a function of the transition energy E_0 for various values of Z in Fig. 8-34. It is infinite when only electron capture is energetically permitted. More sophisticated theoretical calculations have been performed for allowed [Fe 50, Zw 57] and forbidden transitions [Br 55, Be 63] including corrections for screening and finite nuclear size effects [Pe 58, De 60b] and the data have been critically reviewed by Bouchez

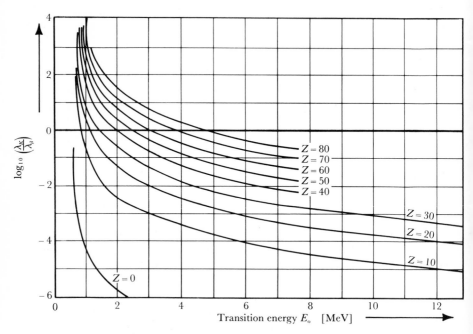

Fig. 8-34. Branching ratios for K capture to positron emission, for various values of the atomic number Z of the product nucleus, plotted semilogarithmically as functions of the transition energy (adapted from [Fee 50]).

[Bo 52, Bo 60] and Berenyi [Be 63]. Some of the results for allowed transitions as discussed in [Bo 60] are summarized in Table 8-4.

The acquisition of experimental data on electron capture presents greater difficulties than in the case of normal $\beta^\pm$ decay since the emitted neutrinos are practically unobservable and the simultaneous presence of β^+ decay complicates the interpretation of measurements. Also, the ionization of the daughter atoms may be unspecific. In practice, the detection of electron capture entails either the study of the γ decay of the excited daughter *nucleus* or the observation of atomic processes in the daughter *atom* triggered by the creation of a vacancy in an inner electron shell. As electrons cascade from outer orbits to fill the vacancy, characteristic x-ray photons or Auger electrons are emitted, and from measurements of their emission rate the electron capture rate (λ_K) can be derived when the K-fluorescence yield is known. On comparing this with the measured rate of positron emission by the same source the K/β^+ RATIO $\lambda_K/\lambda_{\beta^+}$ can be determined and used to test the validity of capture theory. This method is best suited to cases in which the ratio is small. When it is large, it is often more straightforward to determine the total number of disintegrations in a certain time, e.g., by measuring the daughter (or, in some cases, the parent) activity, and to compare this with the number of positrons emitted over the same time interval. The latter can be determined with a β^+ counter (magnetic spectrometer) or from a measurement of the number of annihilation quanta. Together with other experimental techniques, these methods and the results they furnish have been reviewed by Robinson and Fink [Ro 60]. The data, while not highly accurate, nevertheless agree reasonably well with the theory, as can be seen from the selection in Table 8-4. The above review also includes a survey of the rather sparse experimental data on the L_I/K RATIO. An instance of an early, but essentially reliable, investigation of this [Po 49] which was repeated later with higher precision [La 55b] is provided by studies of the ground-state transition

$$^{37}\text{Ar} + e^-_{K,L} \rightarrow {}^{37}\text{Cl} \qquad (8\text{-}116)$$

for which the transition energy $E_0 = 0.816$ MeV far exceeds the binding energy of the electrons. The gaseous radioactive ^{37}Ar was introduced into a large proportional counter at a pressure sufficient to ensure that the daughter's x rays were practically completely absorbed within the counter. Pulses from K and L_I capture were recorded as peaks in a pulse-height (energy) spectrum. The ratio of the peak areas directly gave the experimental L/K ratio, $(\lambda_{L_I}/\lambda_K)_{\exp} = 0.092 \pm 0.010$, in good agreement with the theoretical prediction, $(\lambda_{L_I}/\lambda_K)_{\text{th uncorr}} = 0.082$ or $(\lambda_{L_I}/\lambda_K)_{\text{th corr}} = 0.100$ (Table 8-3 lists the data average).

The determination of the TRANSITION ENERGY E_0, though fairly simple in the case of $\beta^\pm$ decay, becomes appreciably more difficult for electron capture, but constitutes an essential adjunct to the comparison with theory. Its determination from nuclear recoil measurements is inexact, and its evaluation from the threshold energy of the inverse (p, n) reaction is indirect. When the K-capture transition energy is of roughly the same magnitude as the K-electron binding energy, the value of E_0 can be deduced from the L/K ratio. A less restrictive method to arrive at the value of E_0 is to measure the INTERNAL (or "INNER") BREMSSTRAHLUNG spectrum and to derive E_0 by extrapolation in a quasi-"Kurie plot." Internal bremsstrahlung [Pe 65] is produced in normal β decay as well as in electron capture, since the emission of electromagnetic radiation over a wide spectral range accompanies the sudden ejection or capture of a charged electron by a charged nucleus. The theoretical spectral formula for the bremsstrahlung energy E_γ (momenta p_γ to $p_\gamma + dp_\gamma$) in the case of allowed K capture can be written in somewhat simplified form valid at all but the lowest energies [Mo 40, Ja 51, Cu 54] when Coulomb effects are neglected as

$$N(p_\gamma)\, dp_\gamma = C(E_\gamma)\ \frac{\alpha}{\pi}\frac{E_\gamma}{(m_e c^2)^2}\left(1 - \frac{E_\gamma}{E_0}\right)^2 dE_\gamma \qquad (8\text{-}117)$$

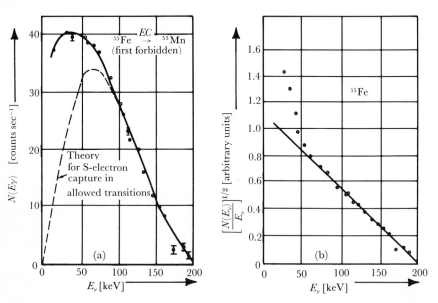

Fig. 8-35. Energy spectrum of internal bremsstrahlung for ^{55}Fe depicted as a function of γ energy E_γ in conventional representation in (a) and as a Jauch plot in (b), giving an end-point energy $E_0 = 206 \pm 20$ keV in good agreement with that $(217 \pm 10$ keV) derived from reaction Q values. The discrepancies at the low-energy end of the spectrum can be attributed to impurities and poor resolution of the NaI(Tl) scintillation spectrometer (from [Mae 51, Be 62]). (Used by permission of North-Holland Publishing Co.)

The function $C(E_\gamma)$ varies only slightly with energy except at very low energies, so that when regarded as a constant and combined with the fine-structure constant α divided by $\pi(m_e c^2)^2 (E_0)^2$ one would expect a quasi-Kurie plot (JAUCH PLOT) of $[N(p_\gamma)/E_\gamma]^{1/2}$ vs E_γ to yield a straight line intersecting the E_γ axis at E_0. In practice, it is more usual to employ $N(E_\gamma)$, the number of observed γ counts per unit energy interval, in a plot of $[N(E_\gamma)/E_\gamma]^{1/2}$ vs E_γ and derive E_0 from the abscissal intercept. By way of illustration, Fig. 8-35(a) shows the internal bremsstrahlung spectrum for electron capture in ^{55}Fe, the Jauch plot from which is depicted in Fig. 8-35(b). The theoretical curve in Fig. 8-35(a) was derived from Morrison-Schiff theory [Mo 40]. The method is particularly useful for the determination of the decay energy in cases of pure RADIATIVE ELECTRON CAPTURE in which positron emission is energetically excluded. In the above approximation the ratio of the total

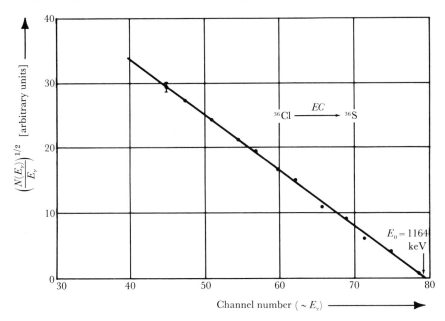

Fig. 8-36. Linear Jauch plot derived by Lipnik *et al.* [Li 64] for the internal bremsstrahlung spectrum resulting from electron capture by ^{36}Cl (a second-forbidden transition), as registered with a 5.1×5.1 cm NaI(Tl) detector, giving an end-point energy $E_0 = 1164 \pm 15$ keV. With a 3.8×3.8 cm NaI(Tl) detector, a linear Jauch plot again ensued, yielding the value $E_0 = 1193 \pm 20$ keV. The mean of these two values, viz. $E_0 = 1178 \pm 15$ keV has been cited in the text. An alternative decay mode of ^{36}Cl is β^- decay to ^{36}Ar (branching ratio 98.3 percent). A β^+-decay mode of very low intensity was also detected. (Used by permission of North-Holland Publishing Co.)

number of photons emitted with energies up to E_0 to the total number of K captures is

$$\frac{N_\gamma}{N_K} = \frac{\lambda_\gamma}{\lambda_K} = \frac{\alpha}{12\pi} \left(\frac{E_0}{m_e c^2} \right)^2 \qquad (8\text{-}118)$$

The theory has been extended to include a Coulomb interaction in a fully relativistic treatment [Gl 54, Gl 56, Ma 58a], which greatly improves the agreement between predicted and observed values, and has even been tested [Mi 56, Bi 62, Zy 63] for the capture of S and P electrons, separately. A study [Li 64] of the decay of ^{36}Cl, on the one hand by a second-forbidden β^-

transition to ^{36}Ar and on the other by radiative electron capture to ^{36}S furnished data which gave a linear Jauch plot (Fig. 8-36) having an inter-cept at $E_0 = 1178 \pm 15$ keV, in good agreement with a previous deter-mination [Be 62]. Whereas in electron capture only *internal* bremsstrahlung is produced, normal $\beta^{\pm}$ decay also gives rise to EXTERNAL BREMSSTRAHLUNG produced when the charged β particles after emission by the parent nucleus interact with the Coulomb field of *another* nucleus. It can be minimized by using very thin samples, and its contribution to the total bremsstrahlung can

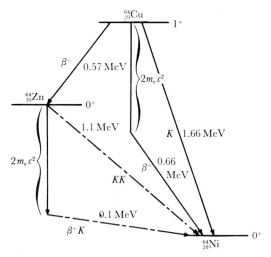

Fig. 8-37. Energy-level scheme for double K capture, designated as a KK transition, from $^{64}_{30}$Zn to $^{64}_{28}$Ni (from [Pr 59]). (Used by permission of The Institute of Physics and The Physical Society, London.)

be estimated by using samples of various thicknesses and extrapolating to zero thickness.

It remains but to mention that DOUBLE K CAPTURE, though theoretically a possible process of the type depicted in Fig. 8-37, has so long a lifetime [Pr 59] as to place it beyond the likelihood of experimental investigation with present techniques.

Before we recommence discussion of further important experiments which provide information on the β-decay and electron-capture interaction, we are obliged to outline the more detailed formulation of β-interaction theory, which forms the subject of the next section.

8.6. Forms of Beta Interaction

The outline of the theory up to this point represents an over-simplification of the basic approach developed by Fermi and others. The formulation has to be explored and exploited further if one is to appreciate and, indeed, to savor, the fascinating interplay of theory with experiment, so characteristic of β decay.

The actual physics of the problem reposes in the interaction Hamiltonian H' and so far we have examined this but cursorily. It can be developed to describe the interaction process in a more comprehensive and general way as follows.

The first step is to combine the separate processes

$$n \rightarrow p^+ + e^- + \bar{\nu} \quad \text{and} \quad p^+ \rightarrow n + e^+ + \nu \qquad (8\text{-}119)$$

into a generalized symmetric form

$$n + \nu \rightleftharpoons p^+ + e^- \qquad (8\text{-}120)$$

which suggests their being conceived as a simultaneous transformation of one nucleon into another (e.g., n → p) and one lepton into another (e.g., $\nu \rightarrow$ e). Mathematically this transformation can be effected through the agency of an interaction operator $\mathcal{O}_i$, which carries the label i to denote which of several *a priori* equally acceptable forms it takes. The same subscript then has to be applied to the corresponding coupling constant g_i, since one has no reason to assume that all the available types of interaction are of the same strength. The subscript also labels the Hamiltonian density $\mathcal{H}_i$ which stands in a simple relationship to the Hamilton operator used hitherto,

$$\mathcal{H}_i \equiv \psi_f^* H' \psi_i \qquad (8\text{-}121)$$

The Hamiltonian density for β^- decay,

$$\mathcal{H}_i^{(\beta^-)} = g_i(\psi_p^* \mathcal{O}_i \psi_n)(\psi_e^* \mathcal{O}_i \psi_\nu) \qquad (8\text{-}122)$$

can be combined with that for β^+ decay

$$\mathcal{H}_i^{(\beta^+)} = g_i(\psi_n^* \mathcal{O}_i \psi_p)(\psi_\nu^* \mathcal{O}_i \psi_e) = \text{Hermitian conjugate of } \mathcal{H}_i^{(\beta^-)} \qquad (8\text{-}123)$$

into the GENERALIZED HAMILTONIAN DENSITY FOR $\beta^\pm$ DECAY,

$$\mathcal{H}_i = g_i(\psi_p^* \mathcal{O}_i \psi_n)(\psi_e^* \mathcal{O}_i \psi_\nu) + \text{Hermitian conjugate (hc)} \qquad (8\text{-}124)$$

The operator $\mathcal{O}_i$ is taken to act in the first bracket upon the nucleon variables only, and in the second bracket on the lepton parameters only. The types of interaction labeled by i are prescribed by Dirac's relativistic wave mechanics for spin-$\frac{1}{2}$ particles, which makes provision for five different classes. These can be described through products of the Dirac γ matrices γ_μ, γ_ν, with μ, $\nu = 0$, 1, 2, 3 and $\gamma_5 \equiv \gamma_0 \gamma_1 \gamma_2 \gamma_3$, as defined in Appendix D.5.1, and classified according to their transformation properties as in Table 8-5 (see Appendix D.5.2).

The second of the five classes, namely, V, stands in direct analogy to the interaction density in electromagnetic theory (the quantum-mechanical transition current density having the form $j_\mu = \psi_f^* \gamma_\mu \psi_i$), while the other categories have more complicated electromagnetic counterparts, corresponding to the emission of *two* quanta [Se 55]. In the absence of theoretical exclusion criteria, a linear combination of all five classes must be considered [Ko 53]:

$$\mathscr{H} = g \sum_i C_i \mathscr{H}_i \quad \text{with} \quad gC_i \equiv g_i \quad \text{and} \quad \sum_i |C_i|^2 = 1 \qquad (8\text{-}125)$$

Table 8-5. Types of β Interaction

Operator $\mathcal{O}_i$	Number of independent matrices	Transformation property of $\psi_b^* \mathcal{O}_i \psi_a$	Label i
$1 \equiv \gamma_\mu \gamma_\mu$	1	Scalar	S
γ_μ	4	Vector	V
$i\gamma_\mu \gamma_\nu$	6	Tensor (of 2nd rank, antisymmetric)	T
$i\gamma_5 \gamma_\mu$	4	Axial vector	A
γ_5	1	Pseudoscalar	P

in terms of respective coupling constants C_i which define the relative strength of each interaction species. Moreover, in addition to *this* set of so-called *even* coupling forms $(\mathscr{H}_i \equiv \mathscr{H}_i^{\text{even}})$, there exists an alternative, equally justified, set of five *odd* coupling forms,

$$\mathscr{H} = g \sum_i C_i' \mathscr{H}_i^{\text{odd}}$$

where $\qquad gC_i' \equiv g_i' \quad \text{and} \quad \mathscr{H}_i^{\text{odd}} = g_i'(\psi_p^* \mathcal{O}_i \psi_n)(\psi_e^* \mathcal{O}_i \gamma_5 \psi_\nu) + \text{hc} \qquad (8\text{-}126)$

The introduction of γ_5 into the lepton bracket induces an interchange of couplings, viz.

$$S \leftrightarrow P \qquad V \leftrightarrow A \qquad \text{but } T \leftrightarrow T \qquad (8\text{-}127)$$

$$1 \leftrightarrow \gamma_5 \qquad \gamma_\mu \leftrightarrow \gamma_\mu \gamma_5 \qquad \gamma_\mu \gamma_\nu \leftrightarrow \gamma_\mu \gamma_\nu \gamma_5 \equiv \gamma_\mu \gamma_\nu \qquad (8\text{-}128)$$

Whereas prior to 1956 the two forms were regarded as mutually exclusive, since they could give rise to mixed terms $\mathscr{H}_i^{\text{even}} \mathscr{H}_i^{\text{odd}}$ of *negative* parity, their coexistence had to be allowed as soon as it was realized that parity was *not* conserved in β decay. Thus a more general Hamiltonian density which makes provision for parity violation takes the form

$$\mathcal{H} = g \sum_i (C_i \mathcal{H}_i^{\text{even}} + C_i' \mathcal{H}_i^{\text{odd}}) + \text{hc} \qquad (8\text{-}129)$$

with

$$\sum_i |C_i|^2 + |C_i'|^2 = 1 \qquad (8\text{-}130)$$

8.6.1. RESTRICTIONS UPON THE COUPLING STRENGTHS

The determination of the coupling strengths C_i, C_i' through theoretical and experimental investigations constitutes one of the central problems in the elucidation of β decay. Certain of the restrictive conditions can be perceived immediately. Thus parity conservation would require that for all i either $C_i = 0$ or $C_i' = 0$, whereas partial parity nonconservation would be characterized by $C_i \neq C_i' \neq 0$ and the actual physical situation of *maximal parity violation* by

$$C_i = \pm C_i' \qquad (8\text{-}131)$$

Similar limitations follow from other symmetry considerations. Invariance with respect to *charge conjugation*, e.g. the equal feasibility of a commensurate process with antiparticles, such as $\bar{n} \rightarrow \bar{p}^- + e^+ + \nu$, imposes the restriction that *all C_i are real and all C_i' are imaginary quantities, (or vice versa)*.

In fact, time-reversal invariance inherently suffices to restrict the C_i and C_i' to *real* quantities (since the complex conjugate quantities betoken the inverse process, the interchange requires that $C_i = C_i^*$ and $C_i' = C_i'^*$).

Thus although the formulation involves 10 different complex coupling strengths, their values are restricted to some extent by basic considerations. On the other hand, when lepton conservation is taken into account the theory becomes more complicated and 20 coupling strengths C_i, C_i', D_i, D_i' play a decisive role [Pa 57, Scho 66, Wu 66]. Their discussion would take us beyond the scope of this text, however. It suffices to indicate the manner in which some of the main numerical results and conclusions have been derived by combining theoretical criteria with experimental data.

8.6.2. FERMI AND GAMOW-TELLER TRANSITIONS

ALLOWED (especially SUPERALLOWED) TRANSITIONS offer the means to derive quantitative information on the coupling strengths and the Fermi coupling constant g_F. It is customary to divide the allowed transitions, for which the net orbital angular momentum of the electron-neutrino pair is zero,

$$\mathbf{l}_{e\nu} = 0 \qquad (8\text{-}132)$$

into two categories whose transition rates are approximately (but not exactly) the same:

(a) FERMI TRANSITIONS, in which the electron and neutrino are emitted with mutually *antiparallel* spins ($\uparrow\downarrow$) in a relative *singlet* state (net lepton spin $\mathbf{s}_{e\nu} = 0$), so that $\mathbf{j}_{e\nu} = \mathbf{l}_{e\nu} + \mathbf{s}_{e\nu} = 0$; and

(b) GAMOW-TELLER TRANSITIONS, in which the electron and neutrino are emitted with mutually *parallel* spins ($\uparrow\uparrow$) in a relative *triplet* state (net lepton spin $\mathbf{s}_{e\nu} = 1$), so that $\mathbf{j}_{e\nu} = 1$.

The total angular momentum $\mathbf{j}_{e\nu}$ carried off by the lepton pair is related to the nuclear spins $\mathbf{J}_i$, $\mathbf{J}_f$ between which the transition proceeds: momentum conservation implies

$$\mathbf{J}_i = \mathbf{J}_f + \mathbf{j}_{e\nu} \qquad (8\text{-}133)$$

whence

$$|\Delta J| \equiv |\mathbf{J}_i - \mathbf{J}_f| = |\mathbf{j}_{e\nu}| \leqslant \mathbf{j}_{e\nu} = \begin{cases} 0 \text{ for Fermi transitions} \\ 0, \pm1 \text{ for Gamow-Teller transitions} \end{cases} \qquad (8\text{-}134)$$

At the same time, the PARITY SELECTION RULE prescribes in both cases that

$$\pi_{e\nu} = (-1)^{l_{e\nu}} = (-1)^0 = +, \quad \text{whence } \pi_i \pi_f = + \quad \text{and} \quad \Delta\pi = \text{no} \quad (8\text{-}135)$$

In terms of the interaction forms $i = S, V, T, A, P$, it turns out that the possibilities S and/or V are associated with Fermi transitions, while T and/or A appertain to Gamow-Teller transitions. The pseudoscalar interaction P contributes in lowest order only to first-forbidden transitions, such as between $0^- \to 0^+$ levels, and therefore has no role to play in the present considerations.

The nuclear matrix element M_{fi}, takes on slightly different values for Fermi and Gamow-Teller transitions, M_F and M_{GT}, respectively. Its absolute square determines the interaction probability; however, mixed terms of the type $C_S C_V$ in the case of Fermi transitions, or $C_T C_A$ in the case of Gamow-Teller transitions, which could give rise to a spectral INTERFERENCE TERM [Fie 37] and (except in the case of the so-called 2-component theory) are theoretically permissible, have been ruled out by experimental evidence [Ma 52a, Da 53, Sh 54, Dr 56, Da 58a, Ge 58, Ko 58b, Le 61, Ra61].

Clearly, an allowed transition with $\Delta J = 1$, $\Delta\pi = \text{no}$ is *pure Gamow-Teller*, whereas one with $\Delta J = 0$, $\Delta\pi = \text{no}$ can be either, except in the case of a $0^\pm \to 0^\pm$ transition, which is *pure Fermi*, since $\mathbf{j}_{e\nu} = \mathbf{J}_i - \mathbf{J}_f = 0$. The latter situation is exemplified by the β^+ decay of ^{10}C and ^{14}O, or the β^- decay of ^{36}Cl, etc. Such transitions are particularly well suited to the derivation of numerical parameters whose interpretation is relatively clear-cut.

8.6.3. ft Values and Nuclear Matrix Elements

When only one class of allowed transition prevails, as is the case for the $0^+ \to 0^+$ pure Fermi β^+ transitions listed in Table 8-6, the ft value bears a simple relationship, through Eq. (8-90), to the overlap integral M_{fi},

$$\frac{1}{ft} = \frac{|M_{fi}|^2}{\tau_0 \ln 2} = \frac{g^2}{2\pi^3} \frac{m_e^5 c^4}{\hbar^7 \ln 2} |M_{fi}|^2 \qquad (8\text{-}136)$$

Hence a knowledge of ft and the nuclear matrix element M_{fi} suffices to establish the value of g. Thus a selection of fairly accurate, corrected ft values

Table 8-6. Selected ft Values of Pure Fermi $0^+ \to 0^+$ β^+ Transitions

Transition	Half-life $T_{1/2}$ [sec]	Maximum β^+ energy [MeV]	ft [sec]	Reference
$^{14}O \to {}^{14}N$	71.36 ± 0.09	1.8126 ± 0.0014	3111 ± 15	[Bü 65a]
$^{26}Al \to {}^{26}Mg$	6.374 ± 0.016	3.2080 ± 0.0023	3086 ± 12	[Wu 64]
$^{34}Cl \to {}^{34}S$	1.565 ± 0.007	4.4600 ± 0.0045	3140 ± 20	[Wu 64]
$^{50}Mn \to {}^{50}Cr$	0.286 ± 0.001	6.6090 ± 0.0026	3125 ± 9	[Fr 65b]
$^{54}Co \to {}^{54}Fe$	0.194 ± 0.001	7.2290 ± 0.0050	3134 ± 18	[Wu 64]

as presented in Table 8-6 shows very little scatter [Wu 64, Scho 66]; the most reliable value pertains to the $^{14}O \to {}^{14}N$ allowed β^+ transition between isobaric analogue states and yields the result

$$g_F = (1.4061 \pm 0.0034) \times 10^{-49} \text{ erg cm}^3 \qquad (8\text{-}137)$$

In principle, a similar method could be used to derive the Gamow-Teller coupling constant g_{GT}, but in the case of pure Gamow-Teller transitions with $\Delta J = 1$, $\Delta \pi = $ no, the matrix elements are not known sufficiently well. A more reliable value can be derived from mixed transitions ($\Delta J = 0$, $\Delta \pi = $ no) in special cases. For these, the resultant decay probability is the sum of the individual probabilities, so that one has

$$\frac{1}{ft} = \frac{1}{(ft)_F} + \frac{1}{(ft)_{GT}} = \frac{g^2}{2\pi^3} \frac{m_e^5 c^4}{\hbar^7 \ln 2} (C_F^2 |M_F|^2 + C_{GT}^2 |M_{GT}|^2) \qquad (8\text{-}138)$$

i.e.,

$$B = ft[(1-x)|M_F|^2 + x|M_{GT}|^2] \qquad (8\text{-}139)$$

with

$$B \equiv \frac{2\pi^3 \hbar^7 \ln 2}{g^2} \frac{1}{m_e^5 c^4} \frac{1}{C_F^2 + C_{GT}^2} \quad \text{and} \quad x \equiv \frac{C_{GT}^2}{C_F^2 + C_{GT}^2} \qquad (8\text{-}140)$$

The relation (8-139) should be simultaneously satisfied for all transitions and should yield a linear *B-x* plot in each case. When ft and M are known, such plots can be constructed and from their common intersection the values of g and C_{GT}/C_F determined. Unfortunately, the strongly model-dependent matrix elements M_F and M_{GT} are not sufficiently accurately known except in the case of neutron decay and for certain mirror transitions in light nuclei (e.g., ^{3}He, ^{15}O, ^{17}F, ^{39}Ca, ^{41}Sc, composed of closed nucleon shells plus or

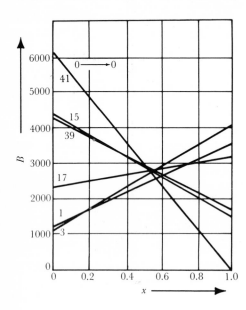

Fig. 8-38. *B-x* diagram for neutron decay, $0 \to 0$ transitions and certain β transitions between light mirror nuclei (^{3}He, ^{15}O, ^{17}F, ^{39}Ca, ^{41}Sc). The quantities B and x are defined in Eq. (8-140). The transitions are designated according to mass number A (from [Ko 62]). (Used by permission of Springer-Verlag, Berlin.)

minus a single "valence" nucleon which undergoes β decay). For these special mirror transitions, the theoretical matrix elements are

$$|M_F|^2 = 1 \tag{8-141}$$

$$|M_{GT}|^2 = \begin{cases} \sqrt{\dfrac{J+1}{J}} & \text{when } J = l + \tfrac{1}{2} \tag{8-142} \\[2ex] \dfrac{J}{J+1} & \text{when } J = l - \tfrac{1}{2} \tag{8-143} \end{cases}$$

whence neutron decay, with $J = j = l + \tfrac{1}{2} = \tfrac{1}{2}$, has

$$|M_F|^2 = 1, \qquad |M_{GT}|^2 = 3 \tag{8-144}$$

The *B-x* diagram for the above special mirror transitions, shown in Fig. 8-38, displays a deceptively sharp intersection region, as becomes evident when compared with the corresponding *B-x* plots for the remaining mirror transitions, as depicted in Fig. 8-39. (We note in passing that a similar situation prevails in the case of nuclear magnetic moments. Those for the special single-valence-particle nuclei coincide fairly well with the Schüler-Schmidt

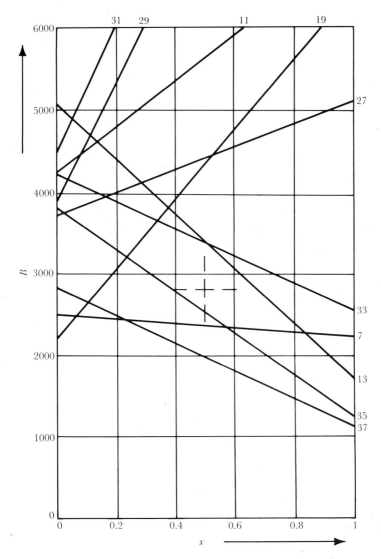

Fig. 8-39. *B-x* diagram for the remaining mirror transitions between light nuclei, showing no evidence of a common point of intersection. The lines are labeled according to mass number *A* (from [Se 59b]).

predictions, whereas discrepancies arise for the remaining nuclei. "Improved" values of the relatively less well-known *Gamow-Teller* matrix element can be derived from a semiempirical relation between μ and $|M_{\mathrm{GT}}|^2$ which, when employed for B-x plots yield considerably better consistency).

Combining the neutron data with that for $^{14}\mathrm{O}$ (which has $|M_{\mathrm{GT}}|^2 = 0$ and $|M_{\mathrm{F}}|^2 = 2$, since the two protons involved in the β^+ transition are both in the same quantum state), we derive the ratio of coupling strengths, the most recent value being

$$\left|\frac{C_{\mathrm{GT}}}{C_{\mathrm{F}}}\right|^2 = 1.38 \pm 0.08 \tag{8-145}$$

and, with $ft = 1211 \pm 37$ sec for neutron decay [Bu 65a],

$$\left|\frac{g_{\mathrm{GT}}}{g_{\mathrm{F}}}\right|^2 = 1.38 \pm 0.04 \tag{8-146}$$

The studies of neutron β decay also provide information on the relative sign of C_{GT} to C_{F} which, as we shall see later, shows C_{V} to be of opposite sign to C_{A} (i.e., a *V minus A*, (written $V - A$) interaction is responsible for β decay).

8.6.4. Electron-Neutrino Angular Correlation

The probability that when an electron is emitted with an energy E a neutrino will be simultaneously emitted at an angle θ to it with the residual available energy can be expressed [Ja 57a, Al 57] in the form

$$N(E, \theta)\, dE = N(E)\,[1 + a(\mathbf{p}_e \cdot \mathbf{p}_\nu)]\, dE = N(E)\left[1 + a\frac{v}{c}\cos\theta\right] dE \tag{8-147}$$

The anisotropy coefficient a in terms of the F/GT mixing ratio,

$$a = (a_{\mathrm{F}} - a_{\mathrm{GT}})y + a_{\mathrm{GT}}, \quad \text{with } y \equiv \frac{|M_{\mathrm{F}}|^2}{|M_{\mathrm{F}}|^2 + |M_{\mathrm{GT}}|^2} \tag{8-148}$$

is strongly dependent on the relative coupling strengths C_i. The coefficient for a pure Fermi transition is

$$a_{\mathrm{F}} = \frac{C_{\mathrm{V}}^2 - C_{\mathrm{S}}^2}{C_{\mathrm{V}}^2 + C_{\mathrm{S}}^2} \tag{8-149}$$

and for a pure Gamow-Teller transition is

$$a_{\mathrm{GT}} = \frac{1}{3}\frac{C_{\mathrm{T}}^2 - C_{\mathrm{A}}^2}{C_{\mathrm{T}}^2 + C_{\mathrm{A}}^2} \tag{8-150}$$

where the factor $\frac{1}{3}$ arises because of the three possible triplet-state orientations $0, \pm1$. Substituting (8-149) and (8-150) in (8-148), one obtains the anisotropy coefficient for the individual couplings,

$$a = \begin{cases} -1 \text{ for } S \text{ coupling} \\ +1 \text{ for } V \text{ coupling} \\ +\frac{1}{3} \text{ for } T \text{ coupling} \\ -\frac{1}{3} \text{ for } A \text{ coupling} \\ (-1 \text{ for } P \text{ coupling: forbidden transitions only}) \end{cases} \qquad (8\text{-}151)$$

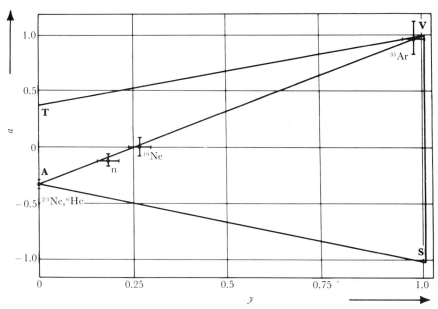

Fig. 8-40. Plots of the anisotropy parameter a *vs* mixing ratio y for ^{1}n, ^{6}He, ^{12}Ne, ^{23}Ne, and ^{35}Ar indicating the β interaction to be of the type $\mathbf{V} - \mathbf{A}$ (from [Scho 66]).

Since it is not feasible to measure the e-ν correlation directly because the observation of the neutrino is not practicable, the experimental determination of a involves simultaneous detection of the directions and energies of the electron and recoil nucleus. The experimental results [Al 59, Jo 63, Ca 63] are shown in the form of a plot of a vs y in Fig. 8-40. They clearly rule out an S-T interaction and point to a *pure V-A interaction*.

8.7. Parity Nonconservation in Beta Decay

Despite the many experimental investigations of β decay prior to 1956, none provided an indication of parity nonconservation because:

(i) an ambiguity arising from the neutrino's vanishing rest mass renders information on parity conservation from β spectra, etc., inconclusive [Ya 50];

and

(ii) it does *not* suffice to study only the parities of the nuclear levels and the radiation involved in a β transition. (Note that transition assignments include *parity* selection rules.)

In their critical analysis, Lee and Yang [Le 56] pointed to the need to study the parity conservation of the decay process *as a whole*, by designing experiments which are sensitive to right-left spatial asymmetries in the overall decay

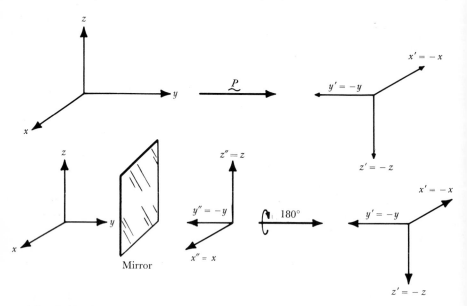

Fig. 8-41. Comparison of the parity-inversion operation with that of mirroring followed by a rotation through 180°.

situation. Although tests of this kind had been performed for strong-interaction and electromagnetic processes, none of the previous β-decay experiments had been sensitive to *pseudo*scalar quantities such as might be detected and measured in experiments which investigate (a) β-emission asymmetry from polarized nuclei, or (b) the longitudinal polarization of β particles, or (c) β-γ circular polarization.

It is interesting to note that as long ago as 1928 Cox *et al.* [Co 28a] obtained a nonzero value for the longitudinal polarization of electrons emitted in β decay, but this finding was passed over.

In principle a decisive test requires that measurements from an experimental arrangement used to observe β decay be compared directly with similar measurements performed upon a system which, except that its geometry is a mirror image of the original arrangement, corresponds exactly with it. It can readily be shown, as in Fig. 8-41, that the parity inversion operation

$$\underline{P} : \mathbf{r} \to -\mathbf{r}, \quad \text{viz.} \quad (x, y, z) \to (-x, -y, -z) \qquad (8\text{-}152)$$

is equivalent to a mirror inversion $(x, y, z) \to (x, -y, z)$, since the latter differs from the system $(-x, -y, -z)$ only through the (physically invariant) rotation of ± 180 degrees about the $y' = -y$ axis. A right-handed coordinate system is changed by the inversion operation (of any one or all three of its spatial coordinates) into a left-handed system, and an ensuing right-left asymmetry betokens the nonconservation of parity.

8.7.1. TEST OF PARITY VIOLATION

The famous experiment undertaken by Wu *et al.* [Wu 57] used an arrangement which enabled a direct comparison to be made of experimental behavior in a system with the behavior in the corresponding mirror system. Although by visualizing the situation in this way, the experimental basis can be presented and understood most easily, a word of warning is called for, in that simplicity is thereby gained at the cost of rigor (the mathematical formalism gives a different picture because of the 180-degree rotation difference).

Wu *et al.* studied the emission probability of electrons from polarized ^{60}Co nuclei which decayed by an allowed Gamow-Teller transition to ^{60}Ni ($\Delta J = 1$, $\Delta \pi = $ no, electron spin parallel to antineutrino spin). For the system shown symbolically in Fig. 8-42 together with its mirror image, they found emission to occur preferentially in the direction opposite to that of the nuclear spin vector—the pseudoscalar $\langle \mathbf{v} \cdot \mathbf{J} \rangle$ built from the scalar product of the electron velocity $\mathbf{v}$ and the nuclear spin $\mathbf{J}$ was established to *differ* from zero.

In Fig. 8-42, mirroring reverses the spin sense and hence inverts the spin vector but leaves the emission direction unchanged. Parity conservation would require the decay probabilities to be the same in both cases, and hence the β intensity *antiparallel* to the nuclear spin direction should be the same as that *along* the nuclear spin direction. The measurements, however, showed a clear asymmetry, incompatible with parity conservation. The experimental arrangement is schematically shown in Fig. 8-43, the external magnetic field serving to line up the radioactive ^{60}Co nuclei in the field direction along which a β counter was arranged. The alignment was produced by the Rose-Gorter technique [Ro 48, Ro 49a, Go 48], employed at a temperature of $0.01°$K (achieved by adiabatic demagnetization of a cerium-magnesium-nitrate crystal, $2Ce(NO_3)_3 \cdot 3Mg(NO_3)_2 \cdot 24H_2O$, on the surface of which a very thin layer of ^{60}Co had been deposited). As a paramagnetic ion, cobalt has a very high (atomic) magnetic moment and an extremely strong internal magnetic field at the nucleus. Thus a moderately powerful external magnetic field, produced by passage of current through a solenoid, suffices to induce alignment of the atomic electrons and this in turn lines up the nuclei for such time as the temperature can be held low enough to prevent depolarization through thermal agitation. An index of the degree of alignment was given

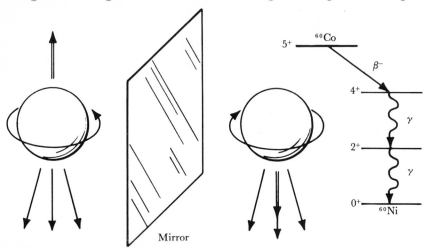

Fig. 8-42. Schematic representation of a ^{60}Co nucleus undergoing β^- decay and its mirror image, wherein the reversal of the spin sense entails inversion of the nuclear spin vector, which in the mirror system appears parallel to the predominant direction of β emission, as against the original antiparallel orientation.

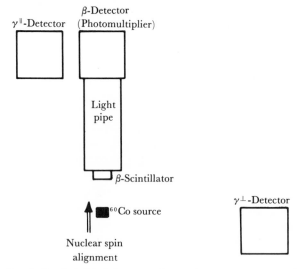

Fig. 8-43. Symbolic sketch of the experimental arrangement used by Wu *et al.* [Wu 57] which first indicated parity to be violated in β decay.

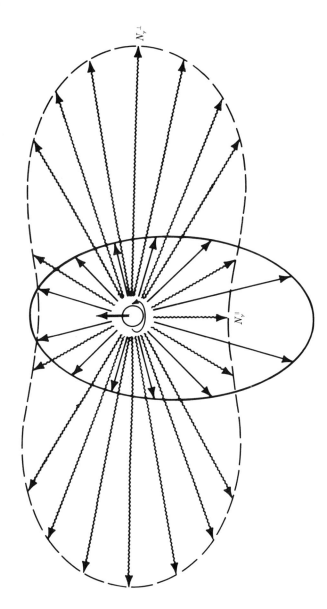

Fig. 8-44. Comparison of β-emission asymmetry with γ-emission anisotropy in the interpretation of the Wu parity-violation experiment.

from simultaneous measurements of anisotropy in emitted γ radiation, taken with an axial and a radial scintillation counter,

$$\xi = \frac{N_\gamma^\perp - N_\gamma^\|}{N_\gamma^\perp} \tag{8-153}$$

Such an asymmetry does *not* define a screw sense, since symmetry is still preserved about the *equatorial* plane, and hence does *not* indicate a breakdown of parity conservation. The test of parity violation lay in establishing an asymmetry in the emission rate of β particles along the axial direction and antiparallel to it (Fig. 8-44). Experimentally it sufficed to set up *one* β counter inside the cryostat to measure just the *upward* β-emission rate and to compare readings for two nuclear spin settings, namely, "up" or "down" according to the direction of the solenoid current which establishes the direction of the polarizing field. Reversing the current was therein equivalent to mirroring the system.

The arrangement did not permit continuous maintenance of the source at maximal polarization at $0.01°$ K, for the source warmed up slightly after 6–8 min of operation and lost its polarization through thermal agitation. Readings of the two γ counters accordingly showed an anisotropy which fell from its initial maximal value, $\xi \simeq 0.4$, corresponding to a maximal nuclear polarization $P = 65$ percent, to zero within about 7 min, as depicted in Fig. 8-45. The corresponding β asymmetry, initially approximately ± 24 percent, exactly followed this trend, as is evident from Fig. 8-45. The outcome that the normalized asymmetry *exceeded* unity for H (i.e., nuclear spin) *down*, and was *less than unity* for H *up* not only confirmed parity violation, but also showed that *preferential emission of β^-'s takes place in a direction anti-parallel to the nuclear spin orientation.*

The numerical results moreover furnished a value for the ANISOTROPY PARAMETER A in the electron angular distribution referred to a β-emission angle θ_e with respect to the spin (i.e., H) direction,

$$W(\theta_e) = 1 + \left[AP\left(\frac{v}{c}\right) \cos \theta_e \right] \tag{8-154}$$

where the electron velocity v was such as to make $v/c \simeq 0.6$. On introducing a correction for electron back-scattering from the specimen, which was measured to be $k \approx \frac{2}{3}$ and noting from Fig. 8-45 that the β anisotropy was about 24 percent, one sees on substitution that

$$\frac{W(0) - W(\pi)}{W(0) + W(\pi)} = -0.24 = kAP\left(\frac{v}{c}\right) = \frac{2}{3} \times A \times 0.65 \times 0.6 \tag{8-155}$$

whence

$$A \simeq -1 \tag{8-156}$$

This value is significant. For a pure Gamow-Teller transition, associated with an *axial-vector* interaction, the anisotropy parameter A is related to the coupling constants C_A, C_A' via the expression

$$A = -2 \frac{\text{Re}(C_A^* C_A')}{|C_A|^2 + |C_A'|^2} \qquad (8\text{-}157)$$

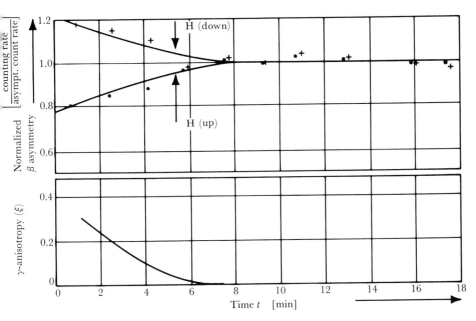

Fig. 8-45. Experimental findings which show the β-emission asymmetry to be directly correlated with the γ anisotropy which provides an index of the nuclear polarization. The results indicate that β emission takes place preferentially in a direction antiparallel to the nuclear spin orientation and point to the breakdown of parity conservation in weak-interaction processes (from [Wu 57]).

and therefore an asymmetry $A = -1$ implies that

$$C_A \approx C_A' \qquad (8\text{-}158)$$

which betokens:

(i) that there is *maximum violation of parity*;

(ii) that the emitted *antineutrino has positive helicity*, i.e., it is *right-handed* (the sign of A is that of the neutrino helicity and hence is opposite to that of the antineutrino) and can be described by a *two-component massless-neutrino theory* [Le 57, La 57, Sa 57];

(iii) that invariance with respect to *charge conjugation is not preserved* ($C_A C_A'$ is not pure imaginary, cf. Section 8.6.1).

The same conclusions would have held in the case of a pure Gamow-Teller *tensor* interaction, with $C_T \approx -C_T'$, but this latter possibility can be dismissed in the light of subsequent deductions outlined in Section 8.6.4.

8.7.2. Neutrino Helicity

The Wu experiment already indicated that antineutrinos have definite (positive) helicity, and this finding was corroborated by later investigations

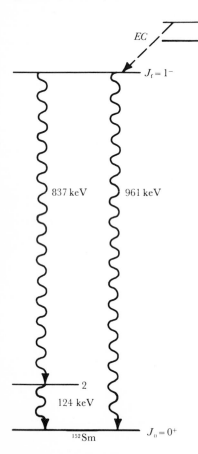

Fig. 8-46. Decay scheme of ^{152}Eu by K-electron capture to ^{152}Sm followed by γ transitions in ^{152}Sm. The measurement of the circular polarization of the 961-keV γ radiation in a $0 \to 1 \to 0$ spin sequence provides information on neutrino helicity, as described in the text.

of parity nonconservation through study of longitudinal electron polarization and β-γ circular polarization correlation. (A review of this research and its implications, together with an exhaustive list of references, can be found in Siegbahn [Si 65, pp. 1451 ff.] and Schopper [Scho 66, pp. 88 ff.].) An elegant experiment to determine the helicity of the neutrino directly was performed by Goldhaber *et al.* [Go 58], who studied the polarization of the neutrinos emitted on electron-capture decay of an isomeric excited state of ^{152}Eu. The decay of this to ^{152}Sm, depicted in Fig. 8-46, is accompanied by de-excitation γ radiation whose circular polarization is related to the neutrino helicity in a particularly simple way for the special $0^- \to 1^- \to 0^+$ transition sequence. Angular momentum conservation in the electron-capture process with $l_e = 0$, viz.

$$^{152}\text{Eu} + e_K^- \to {}^{152}\text{Sm} + \nu_e \to {}^{152}\text{Sm} + \nu_e + \gamma \qquad (8\text{-}159)$$

spin: $\qquad\qquad 0 + \tfrac{1}{2} \to \qquad 1 - \tfrac{1}{2} \to \qquad 0 - \tfrac{1}{2} + 1$

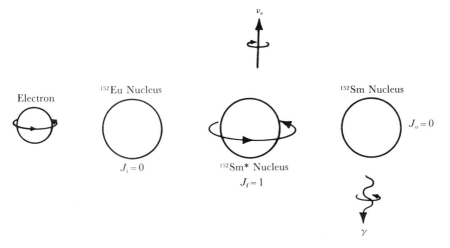

Fig. 8-47. Symbolic representation of the manner in which a measurement of the circular polarization of the de-excitation γ rays in ^{152}Sm establishes the helicity of the neutrino emitted following electron capture in an isomeric state of ^{152}Eu.

requires the spins of the daughter nucleus and of the neutrino to be mutually *anti*parallel, and the ν_e emission to be in the opposite direction to the recoiling nucleus. Hence the daughter nucleus must have the same helicity as the neutrino. Similarly, the momentum balance for de-excitation photons sent out *parallel* to the recoil direction requires their helicity to be the same as that of the daughter nucleus and of the neutrino (Fig. 8-47).

The γ helicity was determined by measuring the circular polarization of just those photons which were emitted in the *same* direction as the recoil nucleus ^{152}Sm and hence in the *opposite* direction to the neutrinos emitted in the preceding transition. The measurement made use of the phenomenon that the absorption of a γ-ray beam on passage through a magnetized block of iron differs slightly according as it is right-circular polarized or left-circular polarized. The selection of the appropriate γ direction was achieved by observing only those photons which were scattered *resonantly* from a $^{152}Sm_2O_3$ scatterer (Fig. 8-48). Such scattering evinces a *resonance* only when there is some form of exact compensation for nuclear kinetic energy changes which are brought about by emission or absorption of a γ quantum and which represent

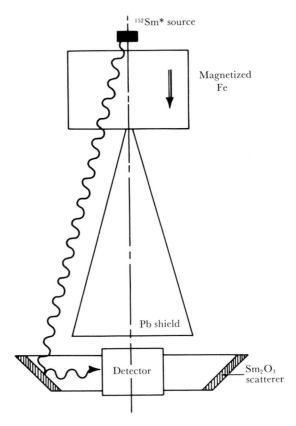

Fig. 8-48. Experimental arrangement used by Goldhaber *et al.* [Go 58] to determine the helicity of the neutrino.

a recoil effect (in momentum and energy) that renders the γ energy slightly (but significantly) smaller than the respective γ-transition energy. In the helicity experiment, the requisite compensation obtained precisely when the emission of a γ quantum immediately followed the emission of a neutrino, but in the opposite direction, with $E_\gamma = E_\nu$, for then the motion of the ^{152}Sm nucleus in its intermediate 1^- short-lived ($\tau = 7 \pm 2 \times 10^{-14}$ sec) state, due to its recoil in a direction opposite to ν emission, exactly compensates the γ-energy shift.

From the measured change in the counting rate of resonantly scattered γ quanta (superimposed upon the normal Compton-scattered γ radiation) on reversal of the saturation magnetic field applied to the iron absorber (i.e., upon reversal of the transmission of right-circular polarized γ radiation),

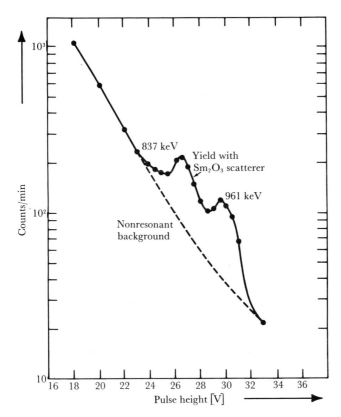

Fig. 8-49. Pulse-height spectrum of γ rays scattered into the scintillation spectrometer used as detector by Goldhaber *et al.* [Go 58].

as shown in Fig. 8-49, the degree of circular polarization was calculated to be

$$P_\gamma = 0.67 \pm 0.15 \tag{8-160}$$

in fair agreement with the theoretical expectation of 0.84, and the helicity of the γ quanta was identified as negative. The above reasoning, conjoined with the fact that lepton conservation requires *neutrinos*, and not antineutrinos, to be emitted in the electron-capture process, establishes the neutrino helicity as *negative* (i.e., left-handed) also. In the pure Gamow-Teller transition, the dominant interaction mode is therefore axial-vector (A), and not tensor (T), in agreement with the $V-A$ interaction conclusions from β-ν angular correlation studies discussed in Section 8.6.4.

8.8. Beta Decay
Coupling Strengths and Interaction Characteristics

Viewing the evidence as a whole, we see that the possible interactions featuring in the Hamilton density through the coupling strengths C_i, C_i' reduce simply to V, A and that the breakdown of parity and charge conjugation further demands that $C_A = \pm C_A'$, $C_V = \pm C_V'$, with all quantities *real*. Indeed, the various parity experiments indicate that

$$C_A = +C_A' \quad \text{and} \quad C_V = +C_V' \tag{8-161}$$

with

$$C_S = C_S' = C_T = C_T' = C_P = C_P' = 0 \tag{8-162}$$

(and

$$D_i = D_i' = 0 \quad \text{for } i = S, V, T, A, P) \tag{8-163}$$

so that the interaction energy density reduces from the original general form derived in Section 8.6 to the much simpler expression

$$\mathcal{H} = \frac{g_F}{2^{1/2}} \left[\psi_p^* \gamma_\mu \left(1 - \frac{C_A}{C_V} \gamma_5 \right) \psi_n \right] \left[\psi_e^* \gamma_\mu (1 + \gamma_5) \psi_\nu \right] \tag{8-164}$$

which implies negative helicity for all particles and positive helicity for all antiparticles.

The ratio of the coupling strengths has been derived in Section 8.6.3 as

$$\left| \frac{C_{GT}}{C_F} \right|^2 = \left| \frac{C_A}{C_V} \right|^2 = 1.38 \pm 0.08 \tag{8-165}$$

i.e.,

$$|C_A| = (1.18 \pm 0.02)|C_V| \tag{8-166}$$

the relative sign of C_A to C_V being expressible in terms of a phase angle ϕ,

$$C_A = 1.18 C_V \, e^{i\phi} \tag{8-167}$$

For an interaction recognized to be of pure V, A form, the requirements of time-reversal invariance dictate that $\phi = 0°$ for a $(V + A)$ interaction, or $\phi = 180°$ for a $(V - A)$ interaction. The experiments on neutron β decay carried out by Burgy *et al.* [Bu 58a] (see also [Scho 59, Bu 60]) in which asymmetries in the electron angular distribution from decay of a polarized neutron beam were measured, not only vindicated time-reversal invariance, but provided the value

$$\phi = 180° \pm 8° \tag{8-168}$$

the error limits being due to statistical uncertainties. Confirmation of this was provided by β-γ angular correlation results [Vl 59, Ya 59]. We therefore conclude that β decay is mediated by a *V minus A interaction*, with

$$C_A = -(1.18 \pm 0.02)\,C_V \qquad (8\text{-}169)$$

and

$$g_F = (1.4061 \pm 0.0034) \times 10^{-49}\ \text{erg cm}^3 \qquad (8\text{-}170)$$

The fact that the ratio of C_A to C_V differs slightly from unity has been attributed to strong-interaction perturbative influence on the axial-vector weak inter-action associated with the coupling between nucleons and mesons [Fe 58]. Various further general developments arising from the similarity between

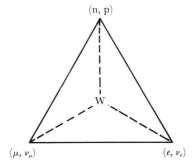

Fig. 8-50. Puppi triangle, connecting participants in weak-interaction processes arranged pairwise at each apex of a triangle in the center of which is inserted the intermediate vector boson W pre-sumed to act as agent of a weak-interac-tion exchange force.

β decay and other weak-interaction processes (such as the universal Fermi interaction, the conserved vector current hypothesis and time-reversal invariance) are described in later chapters dealing with meson decay and weak-interaction phenomena.

A useful symbolic scheme to represent the linkage between certain pairs of fermions through a universal $V-A$ weak interaction of the type we have been considering is the PUPPI TRIANGLE [Pu 62], shown in Fig. 8-50. The fermion pairs are arranged at the vertices of a triangle whose sides denote the linkage of the participants in a weak Fermi interaction: thus the coupling

$(n, p)\ (e, \nu_e)$ depicts beta decay, e.g., $n \rightarrow p^+ + e^- + \bar{\nu}_e$

$(n, p)\ (\mu, \nu_\mu)$ represents muon capture by a nucleon, e.g., $p^+ + \mu^- \rightarrow n + \bar{\nu}_\mu$

$(\mu, \nu_\mu)\ (e, \nu_e)$ corresponds to muon decay, e.g., $\mu^- \rightarrow \nu_\mu + e^- + \bar{\nu}_e$

Provision can also be made for the possible existence of the intermediate vector boson W postulated to act as agent of the weak interaction by including it at the center of the Puppi triangle in the way shown in Fig. 8-50. Current evidence indicates that if such charged vector bosons (also termed "schizons") exist at all, they must have a rest energy in excess of 1.3 GeV and a lifetime below about 10^{-17} sec.

8-1   There is a simple numerical relationship between the average β energy $\bar{E}$ and the cut-off energy E_0 in a β spectrum corresponding to a single β transition. Deduce this ratio on taking the spectrum to be of the form (cf. (8-51))

$$N(p)\, dp \sim (E_0 - E)^2 p^2 \, dp$$

and using the nonrelativistic approximation for the momentum p,

$$p = (2m_e E)^{1/2}$$

(On using the exact relativistic energy-momentum relation, a similar numerical ratio is obtained, but the integration over the β spectrum appropriate to finding the energy average is much more complicated. On the other hand, the extreme-relativistic relation $p = E/c$ makes for easy calculation, but furnishes a different numerical ratio. The reader may care to examine these alternatives.)

8-2   Use the result for $\bar{E}/E_0$ derived in the previous problem to calculate the average energy carried away by antineutrinos in the β-decay process

$$^3\text{H} \rightarrow {}^3\text{He} + \beta^- + \bar{\nu}_e$$

which has an end-point energy $E_0 = 18.1$ keV.

8-3   The decay of ^{210}Bi leads to the formation of stable ^{206}Pb, the sequence being

$$^{210}_{83}\text{Bi} \xrightarrow[\substack{E_0 = 1.17\text{ MeV} \\ T_{1/2} = 5.02\text{ d}}]{\beta^-} {}^{210}_{84}\text{Po} \xrightarrow[\substack{E = 5.297\text{ MeV} \\ T_{1/2} = 138\text{ d}}]{\alpha} {}^{206}_{82}\text{Pb}$$

Use the following data to calculate the mean energy of the β spectrum of ^{210}Bi from first principles, and thence show that its value is in accord with the relation deduced in Exercise 8-1:

A sample of pure ^{210}Bi is placed at time $t = 0$ within a lead container which completely absorbs the α and β radiation from the radioactive source. The rate of heat production $\dot{Q}$ in the lead container is measured calorimetrically over a period of days. The values at various times are:

t [days]:	0	3.20	7.20	17.20
$\dot{Q}$ [arbitrary units]:	25	20.8	17.8	14.2

8-4 The following β spectrum was measured for the β decay of ^{114}In:

E_β [MeV]:	0.4	0.8	1.2	1.6
Count rate [arbitrary units]:	386	425	323	116

Draw a Kurie plot and from this determine the end-point energy E_0. Also calculate the mean energy of the electrons, for simplicity taking the Fermi function F to be constant and using the extreme-relativistic relation $p = E/c$ (cf. Exercise 8-1).

8-5 In the electron-capture transition ^{7}Be $\xrightarrow{\text{EC}}$ ^{7}Li, the mass difference between the two atoms is equivalent to 0.864 ± 0.003 MeV. Using an electrostatic spectrometer, Davis [Da 52a] measured the energy of the recoil nuclei to be $E_R = 55.9 \pm 1.0$ eV. The source in this experiment comprised a monatomic layer of ^{7}Be on a tungsten substrate. What conclusions can be drawn from this measurement as regards the neutrino rest mass? With what accuracy would it be necessary to measure the recoil energy in order to be in a position to establish whether or not the neutrino mass exceeds one-tenth of the electron mass?

8-6 Calculate the momentum distribution of electrons on the hypothesis that in each β decay event *two* massless neutrinos, rather than one, are emitted. (Only the statistical factor is to be taken into consideration; Coulomb effects are to be neglected and the calculation is to be carried out for an *allowed* transition.)

8-7 Having worked late to take measurements of β spectra for nuclei in the region around $Z = 50$, you discover afterwards that in your exhaustion you forgot to specify the nucleus from which you derived the following set of β data:

$B\rho$ [Gauss cm]:	3000	4000	5000	6000
$N(p)$ [arbitrary units]:	135	135	97	60

Identify the requisite nucleus and calculate (in non-relativistic approximation—see Exercise 8-1) the mean β energy and mean antineutrino energy. (Assume that you had samples of the following β emitters around $Z = 50$ (the end-points in MeV are given in parentheses):

^{105}Ru (1.15)	^{107}Ru (3.2)	^{109}Pd (1.03)
^{108}Ag (1.65)	^{115}Cd (1.63)	^{114}In (1.98)
^{115}In (0.84)	^{121}Sn (0.38)	^{124}Sb (0.61)
^{129}Te (1.45)	^{133}I (1.4)	^{137}Xe (4.2)
^{138}Cs (3.40)	^{140}Ba (1.02)	^{140}La (1.34).)

8-8 Classify the following β transitions according to degree of forbiddenness:

Parent nucleus	Decay	E_0 [MeV]	$T_{1/2}$	Spin-Parity Initial	Final
$^{1}_{0}\text{n}$	β^-	0.78	12 m	$\frac{1}{2}+$	$\frac{1}{2}+$
$^{17}_{9}\text{F}$	β^+	1.74	66 sec	$\frac{5}{2}+$	$\frac{5}{2}+$
$^{35}_{16}\text{S}$	β^-	0.168	86.7 d	$\frac{3}{2}+$	$\frac{3}{2}+$
$^{75}_{32}\text{Ge}$	β^-	1.18	82 m	$\frac{1}{2}-$	$\frac{3}{2}-$
$^{87}_{37}\text{Rb}$	β^-	0.27	4.7×10^{10} y	$\frac{3}{2}-$	$\frac{9}{2}+$
$^{91}_{39}\text{Y}$	β^-	1.55	59 d	$\frac{1}{2}-$	$\frac{5}{2}+$

8-9 For the following *allowed* β transitions deduce whether pure Fermi,
pure Gamow-Teller, or mixed coupling is operative:

Transition	Spin-Parity Initial	Final	$\log ft$
$^{1}\text{n} \xrightarrow{\beta^-} {}^{1}\text{H}$	$\frac{1}{2}+$	$\frac{1}{2}+$	3.07
$^{3}\text{H} \xrightarrow{\beta^-} {}^{3}\text{He}$	$\frac{1}{2}+$	$\frac{1}{2}+$	3.00
$^{6}\text{He} \xrightarrow{\beta^-} {}^{6}\text{Li}$	$0+$	$1+$	2.92
$^{14}\text{O} \xrightarrow{\beta^+} {}^{14}\text{N}$	$0+$	$0+$	3.50
$^{15}\text{O} \xrightarrow{\beta^+} {}^{15}\text{N}$	$\frac{1}{2}-$	$\frac{1}{2}-$	3.64

Using the known values for matrix elements, viz.

$$^{14}\text{O}: |M_F|^2 = 2 \quad \text{and} \quad |M_{GT}|^2 = 0$$
$$^{1}\text{n}: |M_F|^2 = 1 \quad \text{and} \quad |M_{GT}|^2 = 3$$

check the calculation of the Fermi coupling constant g_F and the ratio
of coupling strengths $|C_{GT}|^2/|C_F|^2$ in Section 8.8 and determine the
nuclear matrix elements for the remaining β-decay transitions.

8-10 In what energy interval is K capture precluded but L capture possible?

8-11 Down to what temperature would one need to cool a ^{60}Co source
having an activity of 1 mCi in order that a 20-kG magnetic field would
produce alignment in 60 percent of the nuclei?
Assuming the source to be clad in a spherical mantle of natural Co

whose radius is just sufficient for all the electrons in the main β transition ($E_0 = 0.314$ MeV) to be stopped, what is the rate of increase of temperature in °C sec^{-1} if the outside of the mantle is perfectly insulated?

(Use the following data:

> density of Co, $\rho = 8.9$ g cm^{-3};
> specific heat of Co, $c = 0.05$ cal g^{-1} °C^{-1};
> Landé factor for ^{60}Co, $g = 0.762$;
> ground-state spin-parity of ^{60}Co, $J^\pi = 5^+$.

Assume that there is no magnetic interaction between nuclei or between nuclei and electrons. The "brute force" method is accordingly to be used to produce nuclear alignment.)

8-12 Bethe has suggested [Be 39] the following reaction cycle to account for energy production in the Sun:

$$
\begin{aligned}
p + {}^{12}C &\rightarrow {}^{13}N + \gamma \\
{}^{13}N &\rightarrow {}^{13}C + \beta^+ + \nu \\
p + {}^{13}C &\rightarrow {}^{14}N + \gamma \\
p + {}^{14}N &\rightarrow {}^{15}O + \gamma \\
{}^{15}O &\rightarrow {}^{15}N + \beta^+ + \nu + \gamma \\
p + {}^{15}N &\rightarrow {}^{12}C + {}^4He
\end{aligned}
$$

Calculate the neutrino flux $\Phi[\nu$ cm^{-2} sec$^{-1}]$ expected at the surface of the Earth on the basis of this model if the mean energy of the neutrinos is in each case two-thirds of the threshold energy Q of the respective β^+-decay process.

(For ^{13}N decay, $Q = 1.2$ MeV; for ^{15}O decay, $Q = 1.68$ MeV; solar constant $K = 2.00$ cal min^{-1} on 1 cm^2 of the Earth's surface.)

8-13 The positrons which originate from β^+ decay of a radioactive nucleus are slowed down on passage in a metal to thermal energies. Annihilation takes place with free conduction electrons when energy equilibrium is attained. The two annihilation quanta are therefore emitted at almost exactly 180 degrees to one another. However, precise measurements of the angular correlation [DeBe 50] between the two annihilation γ's established the mean deviation from 180° to be $\delta\alpha = 0.7°$.

What value does this indicate for the mean energy at which positron-electron annihilation occurs? (Apply momentum- and energy-balance considerations to the situation in which positron and electron move in the same direction with the same velocity at the moment of annihilation and assume that the two γ quanta are emitted with the same energy.)

Chapter 9

RADIATIVE TRANSITIONS IN NUCLEI

Gamma Decay

9.1. Multipole Character of Gamma Radiation

The classical representation of an electromagnetic radiation source in terms of an oscillating distribution of electric or magnetic charges which constitutes a MULTIPOLE has been taken over into the quantum-mechanical formalism [He 54] used to describe nuclear moments and to classify radiative transitions in nuclei. The radiation modes are quantized (the transition energy $E_0 = \hbar\omega$, where ω is the multipole angular frequency) and are represented in terms of spherical harmonics $Y_{LM}(\theta, \varphi)$ of rank $L = 0, 1, 2, 3, 4, \ldots$. The MULTIPOLE ORDER is expressed through the rank L, e.g., *radiation represented by the rank L has* MULTIPOLARITY 2^L. The classification of some multipolarities in accordance with this scheme is given in Table 9-1.

Table 9-1. ELECTROMAGNETIC RADIATION MULTIPOLARITY

Rank L	Multi-polarity 2^L	Designation	Illustrative transition $J_i \to J_f$
0	1	Monopole	$0 \leftrightarrow 0$ Excluded
1	2	Dipole	$1 \to 0$
2	4	Quadrupole	$2 \to 0$
3	8	Octupole	$3 \to 0$
4	16	Hexadecapole	$4 \to 0$
⋮	⋮	⋮	⋮

The transverse (divergence-free) nature of electromagnetic radiation absolutely rules out monopole γ radiation and precludes single γ rays from effecting $0 \to 0$ transitions. (The only way in which γ rays can participate in $0 \to 0$ transitions is through second-order processes, whose probability is low, such as double-γ emission or the concurrent emission of a γ ray and a conversion electron; this is discussed by Goldhaber and Sunyar [Si 65, pp. 947 ff.].)

Consequently, in a γ transition the multipolarity L can assume nonzero values only (whereas in internal conversion, or double conversion, $L = 0$ is also allowed). As Heitler [He 36] has shown, L represents the *total angular momentum* (of absolute magnitude $\hbar[L(L + 1)]^{1/2}$) carried by 2^L-pole γ radiation with respect to the source of the radiation field. Since angular momentum is a conserved quantity in electromagnetic interactions, it follows that in the nuclear transition $\mathbf{J}_i \xrightarrow[L]{\gamma} \mathbf{J}_f$ a vector triangle relation applies to the triad $(\mathbf{J}_i, \mathbf{J}_f, \mathbf{L})$, so that

$$\mathbf{J}_i - \mathbf{J}_f = \mathbf{L} \tag{9-1}$$

Written in nonvectorial form,

$$\Delta J \equiv |J_i - J_f| \leqslant L \leqslant \Sigma J \equiv J_i + J_f \tag{9-2}$$

This constitutes a MOMENTUM SELECTION RULE which curbs the permitted range of multipolarities L.

Convergence of the expansion in multipoles of γ radiation is in any case assured by the fact that the rationalized γ wavelength λ is invariably much larger than the dimensions of the emitter nucleus, so that $R/\lambda \ll 1$ (see, e.g., [Schi 55, pp. 252 ff.]): for example, $\lambda \cong 200$ fm for 1-MeV γ rays. Since the emission or absorption probability decreases rapidly, as we shall see, with increasing order L (roughly as $(R/\lambda)^{2L}$) there is an abrupt effective cut-off to even those higher-order multipoles admissible by the momentum selection rule. Thus in practice one encounters multipolarities $L = |\Delta J|$, or sometimes in appropriate cases $L = |\Delta J|$ admixed with $L = |\Delta J| + 1$, where $\Delta J \equiv J_i - J_f$.

The quantity M, as z component of $\mathbf{L}$, plays the role of *total magnetic quantum number* of the radiation, and is related to the corresponding quantum numbers m_i and m_f of the nuclear states between which the transition takes place. As the absolute magnitudes in a common z direction are, respectively, $M\hbar$, $m_i\hbar$ and $m_f\hbar$, it follows that

$$M = m_f - m_i \equiv -\Delta m \tag{9-3}$$

A distinction is, moreover, made between ELECTRIC and MAGNETIC MULTI-POLES according to the parity associated with the electromagnetic radiation. This in turn is directly determined by whether or not there is a difference in the parities of the nuclear states between which the transition occurs. Electric multipole radiation of order L has opposite parity to that of magnetic radiation

of the same multipolarity L in accordance with the PARITY RULE:
electric multipole radiation of order L has parity

$$\pi_E = (-1)^L \tag{9-4}$$

magnetic multipole radiation of order L has parity

$$\pi_M = -(-1)^L = (-1)^{L+1} \tag{9-5}$$

so that radiation EL or ML of *even* parity is

M1, E2, M3, E4, ...

radiation EL or ML of *odd* parity is

E1, M2, E3, M4, ...

One notes that electric multipoles EL of even order L have even parity, and that the alternation of parity corresponds to that already considered in Section 5.4.5 in connection with the multipole moments of nuclei. A qualitative demonstration of this characteristic can be derived from the electric multipole expansion coefficients discussed in Section 5.4.8, viz.

$$a_L = \int \rho(r)\, r^L P_L(\cos\theta)\, d\Omega \tag{9-6}$$

where

$$a_0 = \int \rho\, d\Omega = \text{electric monopole (E0)} \tag{9-7}$$

$$a_1 = \int \rho z\, d\Omega = \text{electric dipole (E1)} \tag{9-8}$$

$$a_2 = \int \tfrac{1}{2}\rho(3z^2 - r^2)\, d\Omega \sim \text{electric quadrupole (E2)} \tag{9-9}$$

with $\rho(r)$ representing the charge density at a point defined by the radius vector $\mathbf{r}$ from the origin of the electric field system at an angle θ to the quantization axis z. The spatial inversion betokened by the parity operation here corresponds to the transformation $z \to -z$. It is evident that since $a_0, a_2, \ldots$ corresponding to E0, E2, ... multipoles are *even* in z their parity character is even, whereas the *odd* a_L's, e.g., a_1, which corresponds to an E1 multipole, are *odd* in z, and therefore of *odd* parity. Another way of deriving the same result is to note that the expression (9-6) is odd or even with respect to the parity transformation $\mathbf{r} \to -\mathbf{r}$ according as L is odd or even. The same conclusion applies to the more basic multipole function $\int \rho(r)\, r^L Y_{LM}(\theta, \varphi)\, d\Omega$ for multipolarity EL, irrespective of the value of M.

Similar arguments apply to the parities of magnetic multipoles ML of order L, arising from electric currents due to motion of charges within the nucleus. In the simplest case of a MAGNETIC DIPOLE (M1) associated with the motion of a charged particle which we assume to be spinless, the z component

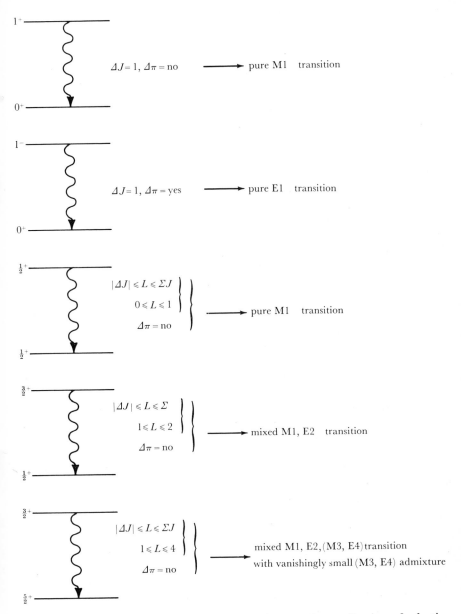

Fig. 9-1. Examples of γ transitions which illustrate the application of selection rules to permit progressively more complicated multipolarities.

of the magnetic moment produced by the electric current corresponding to the transition is described by the angular-momentum operator $\underline{L}_z = (\underline{x}\underline{p}_y - \underline{y}\underline{p}_x)$. As the $\underline{x}_i$ and the $\underline{p}_i$ *both* change sign on parity inversion, the operator $\underline{L}_z$ is left invariant by this operation and therefore has *even* parity for an M1 transition. The z component of the M1-transition matrix element, proportional to

$$\int \Psi_f^*(\underline{x}\underline{p}_y - \underline{y}\underline{p}_x)\,\Psi_i\,d\Omega \tag{9-10}$$

hence vanishes unless the states i and f have the same parity, according to the PARITY RULE

$$\Delta\pi = \text{no for odd-order M}L \text{ or even-order E}L \text{ radiation} \tag{9-11}$$

$$\Delta\pi = \text{yes for even-order M}L \text{ or odd-order E}L \text{ radiation} \tag{9-12}$$

Some examples which serve to illustrate the operation of the MOMENTUM and PARITY SELECTION RULES are depicted in Fig. 9-1.

Three further ISOSPIN SELECTION RULES obtain when isospin T is a good quantum number [Aj 52, Ra 52a, Mo 58a, Wa 58b], viz.:

(i) all γ transitions are forbidden when $|\Delta T| > 1$;

(ii) E1 γ transitions are forbidden between two states in a self-conjugate nucleus (i.e., one with $N = Z$) when $\Delta T = 0$;

(iii) there may be an inhibitory effect on other multipoles (e.g., on M2) when $\Delta T = 0$.

The MULTIPOLE CHARACTER of γ radiation can be determined only indirectly through comparing experimental data with theoretical predictions of angular distributions, polarizations, conversion coefficients, or level lifetimes. Accounts of the various methods can be found in Korsunski [Ko 57], Ajzenberg-Selove [Aj 60], and Siegbahn [Si 65].

9.2. Multipole Transition Probability

As already mentioned, the probability of emission or absorption of radiation with multipolarity L decreases with increasing L roughly as $(R/\lambda)^{2L}$ and hence, since $R \ll \lambda$, becomes vanishingly small for higher-order multipoles. To date the highest orders observed are M4 and E5. Furthermore, the relative probability of emission of EL radiation is appreciably higher than that of the corresponding ML radiation. A rough indication of this can be gleaned from the uncertainty relation

$$\Delta x\,\Delta p_x \approx \hbar \tag{9-13}$$

applied to the simple case of a charge e confined to a spherical volume element of radius R. If the particle carrying the charge has mass m and velocity v, it can be shown to give rise to an electric-dipole field proportional to $eR/r\lambda^2$ at a point distant r from the origin. Thus, setting

$$R\,mv \approx \hbar \qquad (9\text{-}14)$$

i.e.,

$$R \approx \frac{\hbar}{mv} \qquad (9\text{-}15)$$

it follows that the electromagnetic field $\mathscr{E}^{(\mathrm{E})}$ due to the electric dipole is

$$\mathscr{E}^{(\mathrm{E})} \sim \frac{e}{r\lambda^2}\frac{\hbar}{mv} \qquad (9\text{-}16)$$

On the other hand, the same system generates a magnetic moment of order $e\hbar/mc$ and therefore gives rise to an electromagnetic field $\mathscr{E}^{(\mathrm{M})}$ due to a magnetic dipole, where

$$\mathscr{E}^{(\mathrm{M})} \sim \frac{e\hbar}{mc}\frac{1}{r\lambda^2} \qquad (9\text{-}17)$$

Hence

$$\frac{\mathscr{E}^{(\mathrm{M})}}{\mathscr{E}^{(\mathrm{E})}} = \frac{v}{c} \qquad (9\text{-}18)$$

which shows that the ML transition probability is reduced by a factor v/c compared with that for EL radiation, a result which holds even for higher values of L, though here demonstrated for $L = 1$ only. This finding is, as we shall see, equivalent to approximately equating the transition probability for ML radiation to that for $\mathrm{E}(L + 1)$ radiation. Hence multipole mixtures of the type (M1, E2) or (M2, E3) are frequently encountered, whereas mixtures of the type (E1, M2) are not found, being inhibited by the multipole-order factor $(R/\lambda)^{2L}$ and further by the M/E factor (v/c).

Schematically, the decay half-lives for γ multipoles which effect transitions of ≈ 1 MeV energy are compared in Table 9-2.

The theoretical treatment of multipole transition probability is rather complicated and, even at best, still rather approximate, so that the results one derives constitute only rough estimates. The detailed calculation as presented, e.g., by Weisskopf [Wei 51] and Blatt and Weisskopf [Bl 52b, pp. 627 ff.], have been based upon a nuclear model which assumes such weak coupling between the constituent nucleons that in a γ transition only a single nucleon (a proton) experiences a change in its quantum state, the specific change assumed in the calculation being from $J_i = j_i = l + \frac{1}{2}$ to $J_f = j_f = \frac{1}{2}$. The transition probability for emission or absorption of electric multipole radiation having an energy $E_\gamma = \hbar\omega$ can be expressed as a reciprocal mean lifetime which very approximately (i.e., only to within a factor of 10 or 100) is

$$\frac{1}{\tau_E} = \alpha \omega \left(\frac{R}{\lambda}\right)^{2L} S \tag{9-19}$$

where $\alpha = 1/137$ is the fine-structure constant and S is a statistical factor, of value

$$S = \frac{18[1 + (1/L)]}{\{[(2L + 1) \times (2L - 1) \times \cdots \times 5 \times 3 \times 1][1 + (L/3)]\}^2} \tag{9-20}$$

Numerical substitution shows S to vary from 0.25 for $L = 1$ to 3.1×10^{-9} for $L = 5$ and to decrease by roughly a factor 10^2 for unit increase in L. The

Table 9-2. COMPARISON OF DECAY HALF-LIVES FOR γ MULTIPOLE TRANSITIONS OF ENERGY ≈ 1 MeV

$T_{1/2}$ for $E_\gamma \approx 1$ MeV:	10^{-12} sec		$10^{-7.5}$ sec	10^{-3} sec
EL radiation:	Dipole	Quadrupole	Octupole	Hexadecapole
$\left.\begin{array}{l} \|\Delta J\| \equiv \|J_i - J_t\| \leqslant \\ \sum J \equiv J_i + J_t \geqslant \end{array}\right\} L$	1	2	3	4
$\Delta\pi$	Yes	No	Yes	No
ML radiation		Dipole	Quadrupole	Octupole
$\left.\begin{array}{l} \|\Delta J\| \equiv \|J_i - J_t\| \leqslant \\ \sum J \equiv J_i + J_t \geqslant \end{array}\right\} L$		1	2	3
$\Delta\pi$		No	Yes	No

value of τ_E so calculated should not be viewed as more than an *estimate of the minimum mean lifetime* of the nuclear state J_i against a γ transition of multipolarity EL. Numerical evaluation of τ_E is facilitated by rearranging Eq. (9-19) to

$$\tau_E = \left(\frac{137}{E_\gamma/m_e c^2}\right)^{2L+1} \left(\frac{e^2/m_e c^2}{r_0 A^{1/3}}\right)^{2L} \frac{\hbar}{m_e c^2} \frac{1}{S} \tag{9-21}$$

$$\cong \left(\frac{70}{E_\gamma \text{ [MeV]}}\right)^{2L+1} \left(\frac{2}{A^{1/3}}\right)^{2L} \frac{1.29 \times 10^{-21}}{S} \text{ sec}$$

$$\cong \frac{0.645 \times 10^{-21}}{S} \left(\frac{140}{E_\gamma \text{ [MeV]}}\right)^{2L+1} A^{-(2/3)L} \text{ sec} \tag{9-22}$$

wherein the classical electron radius $r_e \equiv e^2/m_e c^2 = 2.82$ fm has for convenience been set to $2r_0$ and we have substituted $\hbar/m_e c^2 = 1.29 \times 10^{-21}$ sec. The mean lifetime τ_E increases with L rapidly, due to the very sharp decrease of $(R/\lambda)^{2L}$ and of S. A typical value might be of the order 10^{-11} sec for de-excitation of a

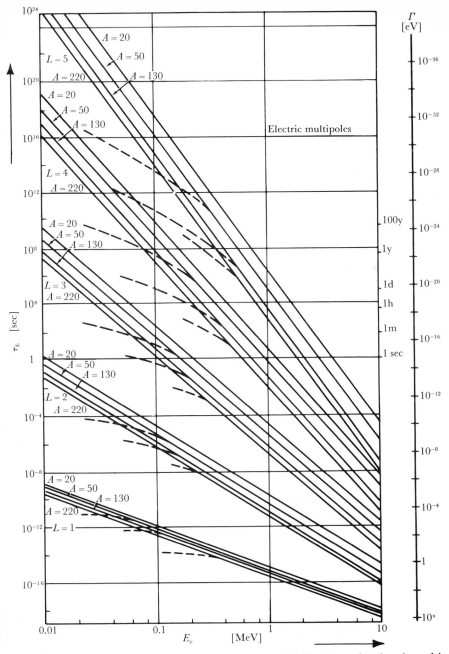

Fig. 9-2. Log-log plots of the calculated mean partial lifetime τ_E for electric multipole γ transitions *vs* the transition energy E_γ, according to the single-particle formula (9-21) with $R = 1.4\,A^{1/3}$ fm. Also shown are the corresponding level widths $\Gamma \equiv \hbar/\tau_E$ and, as dashed curves, the reduction of the partial lifetime when internal conversion is also taken into account (adapted from [Co 58]). (Used by permission of McGraw-Hill Book Co.)

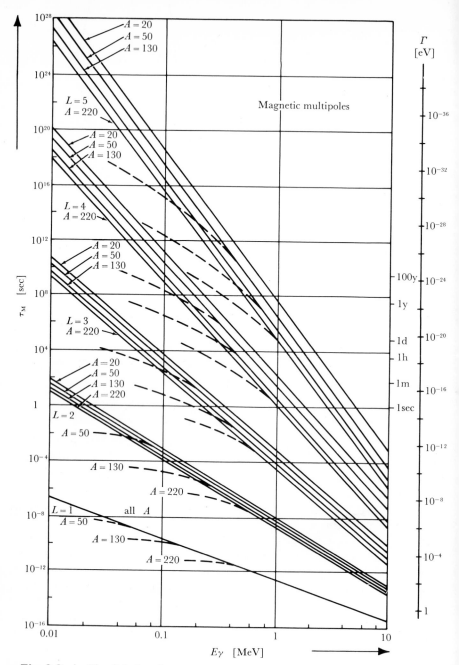

Fig. 9-3. As Fig. 9-2, but for magnetic multipoles, which by comparison can be observed to differ in lifetime from the faster electric transitions of commensurate multipolarity by roughly a factor 1000 (adapted from [Co 58]). (Used by permission of McGraw-Hill Book Co.)

level by 1-MeV E2 γ radiation, such as corresponds roughly to the first excited
state in medium-heavy e-e nuclei. The factor $A^{-(2/3)L}$ indicates that light
nuclei might be expected to display rather longer lifetimes toward γ decay than
those higher up the mass scale. Plots of the mean lifetime τ_E calculated from
the single-particle Weisskopf formula (9-22) for nuclei of mass $A = 20$, 50,
130, and 220 emitting γ radiation in the energy range 10 keV to 10 MeV
of multipolarity E1 to E5 are presented in Fig. 9-2.

A similar calculation for MAGNETIC MULTIPOLE TRANSITIONS of order L
yields the Weisskopf estimate

$$\tau_M \cong \frac{2.90 \times 10^{-21}}{S} \left(\frac{140}{E_\gamma \text{ [MeV]}} \right)^{2L+1} A^{-(2/3)(L-1)} \text{ sec} \qquad (9\text{-}23)$$

whence for the same multipole order L the ratio of partial mean lifetimes is

$$\frac{\tau_M}{\tau_E} \cong 4.5 A^{2/3} \qquad (9\text{-}24)$$

Estimates of this type were also carried out in slightly more detailed calculations by Moszkowski [Mo 53b] (see also Siegbahn [Si 65, pp. 863 ff.]) based
upon a similar approach. They yielded essentially the same formula for τ_E
and an expression for τ_M which differs numerically from Weisskopf's estimate
by less than one order of magnitude, viz.

$$\tau_M \cong \frac{29.3 \times 10^{-21}}{S} \left(\frac{140}{E_\gamma \text{ [MeV]}} \right)^{2L+1} A^{-(2/3)(L-1)} \left(\frac{\mu_p L}{2} - \frac{L}{L+1} \right)^{-2} \text{ sec} \qquad (9\text{-}25)$$

where $\mu_p = 2.79$ is the proton magnetic moment in nuclear magnetons and
the statistical factor S has $[1 + (L/3)]$ replaced by $[(2/3) + (L/3)]$ in the
denominator. Figure 9-3 shows values of τ_M calculated from this formula in
function of the γ energy for magnetic multipoles M1 to M5 and for nuclei
with $A = 20$, 50, 130, 220. Also included in Figs. 9-2 and 9-3 as dashed curves
are the respective contributions furnished by internal conversion.

The ordinate scale on the right of Figs. 9-2 and 9-3 shows the PARTIAL
WIDTH for the corresponding decay of the state J_i, as calculated from the
formula $\Gamma = \hbar/\tau$. Experimentally determined level widths are often related
to these "standard" values and expressed in "WEISSKOPF UNITS" as an aid to
their theoretical interpretation in terms of model predictions.

An indication of the measure of agreement between experimental and
theoretical γ-transition probabilities is provided by Fig. 9-4(a, b) in which
some of the experimental data for E2, E3, and M4 transitions [Go 51, Go 52]
(corrected for internal conversion contributions) are compared with
theoretical predictions in the logarithmic representation used by Hayward
[Co 58, p. 9–106 ff.]. In this, the mass dependence of the partial lifetime is
eliminated by plotting $\tau_E A^{(2/3)L}$ and $\tau_M A^{(2/3)(L-1)}$ vs E_γ. The extent of

agreement improves with increasing multipole order, but the experimental lifetimes tend to be higher, on the whole, than the values predicted by Weisskopf single-particle theory. The exceptions, marked as crosses in the E2 data, have been identified with enhanced transitions between rotational

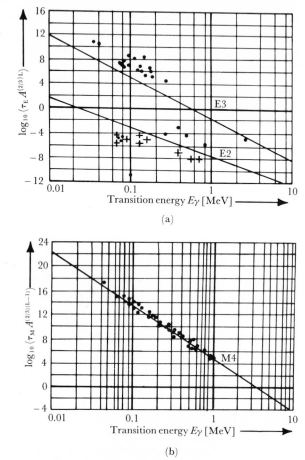

(a)

(b)

Fig. 9-4. Comparison of experimental and theoretical partial γ transition lifetimes, in a representation which eliminates the mass dependence, as functions of the transition energy E_γ for (a) E2 and E3 multipolarity and (b) M4 multipolarity.

The crosses in the E2 data indicate enhanced γ transitions between rotational states. The experimental points stem from measurements by Goldhaber *et al.* [Go 51, Go 52] and have been corrected for internal conversion. In the case of M4 radiation, a statistical correction has also been incorporated in the experimental data. The solid lines represent the Weisskopf single-particle estimates, which predict too low a value for the lifetime (adapted from [Co 58]). (Used by permission of McGraw-Hill Book Co.)

collective nuclear states. Special types of surface motion in nuclei also enhance the E2-decay probability. A systematic survey of these and other multipole transition properties for deformed odd-mass nuclei has recently been presented [Lö 66b].

The situation with E1 transitions is less certain because of the relative paucity of experimental data occasioned by the very short lifetimes conjoined with the fact that there are rather few low-lying nuclear levels which could give rise to E1 radiation (as one would expect from the single-particle or collective model). Viewed from the opposite extreme, namely, according to the liquid-drop model, the electric center of charge cannot move under the action of internal forces and hence the E1 moment is zero. Thus E1 radiation would be absolutely forbidden—in practice, it is indeed much weaker in many low-energy transitions than expected.

At high energies, around 20 MeV, however, the internal motion seems to favor oscillation of the proton ensemble relative to the neutron ensemble, which gives rise to a large E1 moment and hence to large cross sections for photo-decay reactions (γ, n) and (γ, p). In this energy range, the cross sections show a broad resonance termed the GIANT DIPOLE RESONANCE.

9.2.1. REDUCED TRANSITION PROBABILITY

Another significant quantity used in comparing experiment with theory is the REDUCED TRANSITION PROBABILITY $B(EL)$ or $B(ML)$, which is essentially the square of a multipole transition matrix element. Writing the transition probability as

$$W_E(LM, J_i \rightarrow J_f) \equiv \frac{1}{\tau_E} = \frac{4\pi}{\hbar}\left(\frac{E_\gamma}{\hbar c}\right)^{2L+1} S|Q_{LM}|^2 \qquad (9\text{-}26)$$

$$W_M(LM, J_i \rightarrow J_f) \equiv \frac{1}{\tau_M} = \frac{4\pi}{\hbar}\left(\frac{E_\gamma}{\hbar c}\right)^{2L+1} S|M_{LM}|^2 \qquad (9\text{-}27)$$

the electric multipole factor Q_{LM} is a matrix element summed over all constituent protons labeled by k:

$$Q_{LM}(i \rightarrow f) = e \sum_{k=1}^{Z} \int (r_k)^L Y_{LM}^*(\theta_k, \varphi_k) \, \Psi_f^* \Psi_i \, d\Omega \qquad (9\text{-}28)$$

and the commensurate magnetic multipole factor is

$$M_{LM}(i \rightarrow f) = -\frac{1}{L+1}\frac{e\hbar}{m_p} \sum_{k=1}^{Z} \int (r_k)^L Y_{LM}(\theta_k, \varphi_k) \operatorname{div}\left\{\Psi_f^*\left[\frac{1}{i}\,\mathbf{r}_k \times \nabla_k\right]\Psi_i\right\} d\Omega$$

$$(9\text{-}29)$$

These factors might in principle be calculated for all possible transitions

$J_i m_i \rightarrow J_f m_f$ with $|M| \leqslant L$, though in practice this is prevented by inadequate knowledge of the nuclear wave functions in all but the simplest cases. As experimental measurements are in general insensitive to magnetic substates defined by m_i and m_f, the above quantities are for convenience of comparison

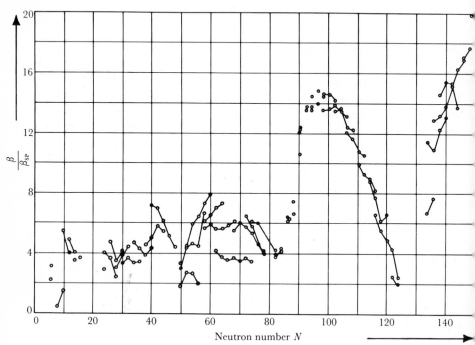

Fig. 9-5. Variation of the ratio of the quadrupole deformation parameter β to the single-particle prediction $\beta_{\rm SP}$ with the neutron number N for nuclei ranging from $_4$Be to $_{96}$Cm (from [St 65]).

with experimental findings averaged over all m_i, m_f in order to give the REDUCED TRANSITION RATES

$$B(EL, J_i \rightarrow J_f) = (2J_i + 1)^{-1} \sum_{m_i} \sum_{m_f} |Q_{LM}(i \rightarrow f)|^2 \qquad (9\text{-}30)$$

$$B(ML, J_i \rightarrow J_f) = (2J_i + 1)^{-1} \sum_{m_i} \sum_{m_f} |M_{LM}(i \rightarrow f)|^2 \qquad (9\text{-}31)$$

in terms of which the transition probabilities are

$$W_{\rm E}(L, J_i \rightarrow J_f) \equiv \frac{1}{\tau_{\rm E}} = \frac{4\pi}{\hbar} \left(\frac{E_\gamma}{\hbar c}\right)^{2L+1} S B(EL, J_i \rightarrow J_f) \qquad (9\text{-}32)$$

$$W_{\rm M}(L, J_i \rightarrow J_f) \equiv \frac{1}{\tau_{\rm M}} = \frac{4\pi}{\hbar} \left(\frac{E_\gamma}{\hbar c}\right)^{2L+1} S B(ML, J_i \rightarrow J_f) \qquad (9\text{-}33)$$

One most frequently encounters the reduced transition rate for E2 radiation, $B(\text{E2})$, normally expressed in units $e^2 \times 10^{-48}$ cm^4, in considerations of collective nucleon motion within deformed nuclei, since from it one can deduce the deformation parameter β according to the expression

$$\beta = \frac{4\pi[B(\text{E2})]^{1/2}}{3Z(r_0)^2} \qquad (9\text{-}34)$$

When $B(\text{E2})$ has been determined experimentally—through Coulomb excitation studies, for example—it can be inserted in the analysis of related data, such as that from internal-conversion or γ-decay investigations. Stelson and Grodzins [St 65] (see also [Ha 65b]) have tabulated the most recent data on $B(\text{E2})$ (in units of e^2) for e-e nuclei together with the corresponding quadrupole deformation parameter β and present a curve, reproduced as Fig. 9-5, comparing derived β's with the single-particle assignments β_{SP} to show the measure of agreement between experiment and single-particle model theory. This will be discussed in more detail when we come to consider the collective model in Volume II.

9.2.2. "Forbidden" Transitions

It is interesting to compare the situation in nuclear physics, involving γ spectroscopy, with that in atomic physics, which involves optical spectroscopy. The allowed transitions in the latter case are confined to E1 multipolarity ($L = 1$, $\pi = -$) by the selection rules for electronic transitions in atoms ($\Delta J = 0, \pm 1$, but not $0 \rightarrow 0$; $\Delta M = 0, \pm 1$; $\Delta \pi = $ yes). All other multipolarities correspond to forbidden transitions and are seldom observed. However, nuclei cannot be de-excited by such nonradiative processes as thermal agitation collisions which serve to dissipate the excitation energy of atoms, and therefore they much more readily include what would otherwise be "forbidden" transitions—especially E2 multipolarity—in their decay schemes.

9.2.3. Nuclear Isomerism

When γ emission is so greatly inhibited by selection rules that the mean lifetime of a particular excited level exceeds the arbitrary limit of 0.1 sec, the long-lived nuclear state is termed an ISOMERIC STATE and is assumed to be in metastable equilibrium. Thus of the γ-active states whose lifetimes range from 10^{-16} to 10^8 sec, those with $\tau > 10^{-1}$ sec are associated with strongly inhibited γ decay of high multipole order. One therefore expects such isomeric states to evince *large spin differences* and *small energy differences* from lower-lying levels, since a combination of high multipolarity $L \cong \Delta J$ and low energy E_γ yields a long lifetime toward γ decay. Situations of this type, involving E3, M3, E4, M4, ... transitions, which take place mainly by internal conversion, are encountered only in medium-mass and heavy nuclei ($A \gtreqqless 39$), and then preferentially among certain clusters of nuclei having particular single-particle properties. The single-particle model can therefore provide an explanation for the occurrence of "islands of isomerism" as depicted in Fig. 9-6.

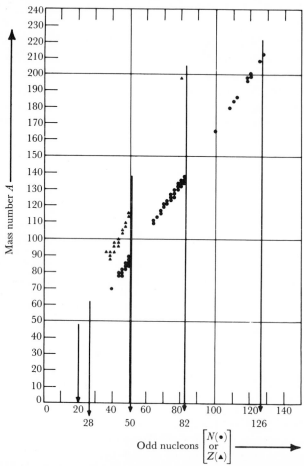

Fig. 9-6. Representation of "islands of isomerism" in a distribution plot of long-lived odd-A isomers. The dots denote odd-N (e-o) nuclei, while the squares depict odd-Z (o-e) nuclei. Attention is drawn to the locations of closed-shell nuclei having N or Z equal to the "magic numbers" 20, 28, 50, 82, or 126 (from [Go 52]).

9.2.4. Multipole Mixing

Spin and parity selection rules prescribe the multipole character of γ transitions and frequently permit multipole mixing in the γ radiation. Thus, unless J_i or J_f is zero, the rule $|\Delta J| \leqslant L \leqslant \Sigma J$ in principle permits the occurrence of several multipolarities L. In practice, however, the marked differences in partial lifetimes of different orders L, and of E or M character, reduce the possibilities to mixtures of at most *two* components in general, viz.

$$ML + E(L+1), \qquad \text{e.g., } M1 + E2 \quad \text{or} \quad M2 + E3,$$

where $L = |\Delta J|$ or (if $\Delta J = 0$) $L = \Delta J + 1 = 1$. One does not encounter such mixtures as E1 + M2 outside exceptional circumstances because the relative emission probability of M2 is so much smaller than that of E1 unless E1 is strongly inhibited.

Such inhibition, arising, for instance, from the operation of isospin selection rules when $\Delta T = 0$ in self-conjugate nuclei, as mentioned in Section 9.1, can in principle lead to abnormal multipole mixtures which have been the subject of recent investigations. Thus there is reason to expect that the de-excitation γ transition from the 5.10-MeV $(2^-, T = 0)$ state of $^{14}_{7}N_7$ to the 1^+ $(T = 0)$ ground state might have no E1 component and a greatly inhibited M2 component [Wa 58b], whereas enhancement of E3 multipolarity may occur through the collective motion of the nucleons to such an extent that it could represent a significant contribution [Wa 60]. The multipolarity of the γ radiation was therefore studied experimentally by Blake *et al.* [Bl 65] and theoretically by Bishop [Bi 65], who found evidence to support the occurrence of an (E1 + M2 + E3) mixture. (The presence of an E1 component, though precluded by the isospin selection rule, can be justified through the assumption that the rule no longer holds strictly when current contributions, which are normally neglected, are incorporated in the transition matrix element.)

Another suggestion along similar lines, purporting to adduce evidence [Gü 65] for an irregular (M1 + M3) mixture devoid of an E2 component to occur in the 1.095-MeV γ transition in ^{172}Yb has since been refuted by Ramaswamy [Go 66, pp. 112 and 113] and Kleinheinz *et al.* [Kl 67], who claim that the data is compatible with an (M1 + E2) mixture.

A very stimulating conjecture along somewhat different lines is currently under investigation. If the strong nuclear binding force were to be very slightly "tainted" by a parity-violating weak-force admixture, γ transitions might in special circumstances evince improper multipolarity, such as M1 + E1 when γ decay took place between states of *mixed* parity. An E1 "contaminant" might consequently appear, and be detectable, e.g., through circular polarization measurements. Following an analysis by Michel [Mi 64] which set upper limits on the admixture of a parity non-conserving force [Ge 57, Ge 59, Bl 60a, Bl 60b, Fe 64], experiments were undertaken to detect and measure possible circular polarization of the observed γ radiation in the $\frac{5}{2}^+ \rightarrow \frac{7}{2}^+$ de-excitation transition of ^{181}Ta following β^- decay of ^{181}Hf [Boe 64, Boe 65, Bo 65a, Lo 66a] (see also [Ab 64]). These, augmented by theoretical considerations [Wa 65] (see also [Szy 66, Ma 67b]), point to the existence of a very slight effect, compatible with a parity-violating "contaminant" of amplitude $F \cong -10^{-7}$, viz. intensity $F^2 \approx 10^{-14}$, but the magnitude of the observed polarization is so small that, taken in conjunction with error limits, it still needs to be regarded with some caution (cf. $P_{\text{th [Ma 67b]}} \approx -0.7 \times 10^{-4}$, while $P_{\text{exp [Boe 65]}} = -(2.0 \pm 0.4) \times 10^{-4}$, $P_{\text{exp [Bo 65a]}} = +(0.3 \pm 2.1) \times 10^{-4}$, $P_{\text{exp [Lo 66a]}} < 0.2 \times 10^{-4}$). The results of still more recent measurements of this type [Boe 68] on ^{175}Lu, ^{181}Ta and ^{203}Tl are throughout compatible with zero circular polarization.

The admixture of $L' = L + 1$ electric multipolarity in a mixed $ML + EL'$ transition is expressed by the MIXING RATIO whose square gives the ratio of intensities $I(L')/I(L)$ defined by the ratio of (reduced) matrix elements averaged over magnetic substates in the bra-ket notation for matrix elements:

$$\delta \equiv \frac{\langle J_f \| L' \| J_i \rangle}{\langle J_f \| L \| J_i \rangle} \Rightarrow \delta^2 = \frac{|\langle J_f \| L' \| J_i \rangle|^2}{|\langle J_f \| L \| J_i \rangle|^2} \equiv \frac{I(L')}{I(L)} \qquad (9\text{-}35)$$

Hence a value of δ below unity betokens that L is predominantly present,

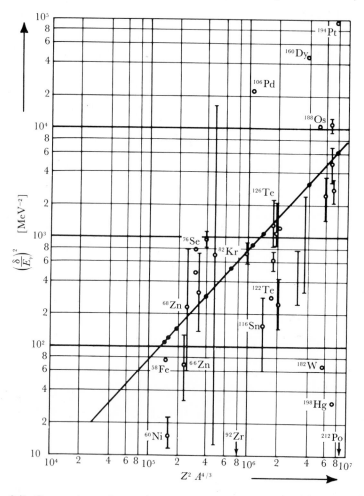

Fig. 9-7. Comparison of experimental data with the theoretical Davydov prediction [Da 58b] of reduced intensity mixing ratio *vs* $Z^2 A^{4/3}$ according to Eq. (9-36), as undertaken by Singhal and Trehan [Si 63].

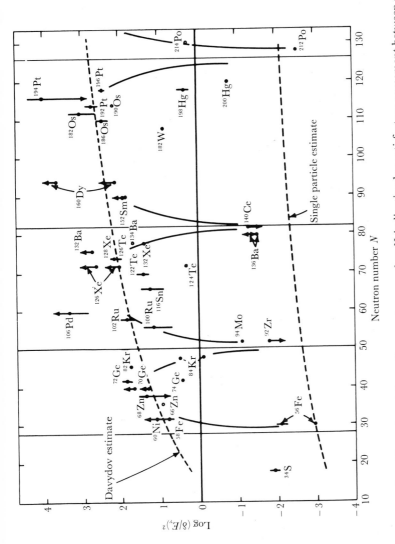

Fig. 9-8. Plot of the reduced intensity mixing ratio *vs* neutron number *N*, indicating less satisfactory agreement between experiment and Davydov theory. The largest deviations can be observed to occur at closed neutron shells, in the vicinity of which the single-particle estimate might be expected to be more valid. The closed circles represent data from γ-γ angular correlations, while the open circles represent data derived from reported percentage admixtures of E2 radiation (from [Ma 59a]). (Used by permission of North-Holland Publishing Co.)

whereas when δ exceeds unity, there is an excess of L'. The value $\delta = 0$ characterizes pure L radiation, and $\delta = \infty$ pure L' radiation. The percentage admixture of ML intensity is, accordingly $(1 + \delta^2)^{-1}$, and of EL' is $\delta^2(1 + \delta^2)^{-1}$. While δ^2 is always positive, the sign of δ, expressing as it does the relative phase of the matrix elements, can assume considerable theoretical importance; many different conventions are used, since no absolute sign can be allocated to the reduced matrix elements, but that given above is most widely employed and is consistent (see [Sa 64, Ro 67]).

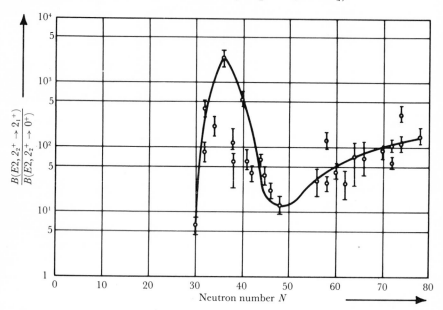

Fig. 9-9. Comparison between theory and experiment for the ratio on $B(E2)$ transition probabilities as a function of neutron number N (from [Si 63]).

The values of δ derived from the reduced transition probabilities $B(EL)$ and $B(ML)$ calculated on the basis of Weisskopf theory do not tally well with experimental data, as might be expected; more reliable values can be obtained by building the ratios of *empirical* $B(EL)$'s and $B(ML)$'s. The bulk of the presently available data on intensity mixing ratios $\delta^2(E2/M1)$ stems from $2^+ \rightarrow 2^+$ transitions in medium-mass deformed nuclei whose properties are best described by the collective model. The Davydov theory [Da 58b] of *asymmetric* nuclei predicts a value

$$\delta^2(E2/M1) = 8.1 \times 10^{-5} \, Z^2 \, A^{4/3} \, E_\gamma^2 \tag{9-36}$$

where E_γ is the transition energy in MeV. As can be seen from the comparison with experimental data undertaken by Singhal and Trehan [Si 63] and presented in Fig. 9-7, the agreement is reasonable in many cases. When plotted directly against the neutron number N, however, the comparison appears less satisfactory [Ma 59a, Po 63], as can be seen from Fig. 9-8, which contrasts the Davydov and the Weisskopf estimates with the experimental data and shows up shell effects by the occurrence of systematic dips.

In this connection, it is also interesting to compare the ratios of $B(E2)$ for the E2 admixture in the mixed M1 + E2 γ transition from the second 2^+ to the first level (2^+) to the $B(E2)$ for the pure E2 crossover transition from the second 2^+ state to the ground state (0^+) in the case of e-e nuclei over the range $N = 30-80$. When this ratio $B(E2, 2_2^+ \to 2_1^+)/B(E2, 2_2^+ \to 0^+)$ is plotted against N as in Fig. 9-9, it compares favorably with the theoretical trend.

9.3. Nuclear Level Scheme Compilation

By way of illustrating how a nuclear scheme can be inferred from γ-decay data, we give two instances which may serve to show up some of the features that beset the interpretation. It will be noted that a level scheme is deduced by coordinating evidence from a very much wider field than that provided by γ transitions alone, but that γ-ray energy and distribution measurements provide valuable corroborative information on the locations and spins of nuclear levels.

Table 9-3. DIFFERENCES OF γ ENERGIES IN CASCADE TRANSITIONS BETWEEN ^{175}Lu STATES

γ Energies	0.114'[a]	0.138"	0.145'''	0.251 ''''	0.283 '''''	0.396 ''''''
0.114	—					
0.138	0.024					
0.145	0.031	0.007	—			
0.251	0.137"	0.113'	0.106	—		
0.283	0.169	0.145'''	0.138"	0.032	—	
0.396	0.282''''	0.258	0.251''''	0.145'''	0.113'	—

[a] The primes serve to match energy differences (i.e., transition energies) with measured γ energies.

The first example concerns ^{175}Lu, which can be formed in an excited state by β^- decay of ^{175}Yb. As it de-excites, the following γ radiations are observed spectroscopically (the multipolarities are also listed as adduced from other measurements): E_γ: 0.11381 MeV (80 percent M1 + 20 percent E2), 0.13765 MeV (M1 + E2), 0.14485 (E1), 0.2513 (E2), 0.28257 (98 percent E1 + 2 percent M2), 0.3961 (80 percent E1 + 20 percent M2).

These can for convenience be arranged, together with differences between pairs of values, as a partial matrix of the form of Table 9-3.

If we then compare the energy *differences* with the set of measured energies, we find that certain values correspond. We indicate such matching by an appropriate number of primes. Thus, for example, while 0.396-MeV γ radiation could be identified with a ground-state transition from the top observed level, the next-highest energy of 0.283 MeV could be interpreted as corresponding to a transition from the top level to a level at 0.114 MeV, and similarly other differences could betoken specific transition energies. From such correspondences we could set up a plausible self-consistent energy-level scheme, such as that shown in Fig. 9-10(b), from among the entire set of possibilities depicted in Fig. 9-10(a). Thus we could associate the 0.283-MeV γ''''

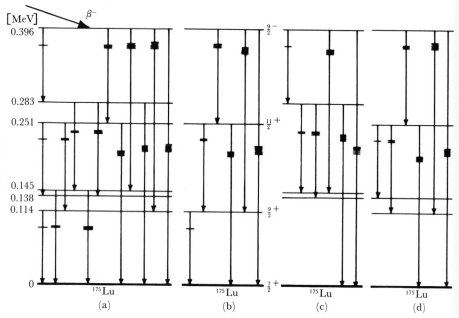

Fig. 9-10. Energy level schemes for ^{175}Lu drawn up as alternative versions compatible with the measured energies of γ lines in cascade.

with a 0.396 $\rightarrow$ 0.114 MeV transition, and thereby account simultaneously for the 0.114-MeV γ' ray. If, then, the 0.145-MeV γ''' radiation involves a transition from the top level to one at 0.251 MeV, the remaining γ'' and γ'''' radiations are also accounted for. Having arrived at this scheme, however, we must recognize that *it is not unique*. Other schemes, such as those indicated in Fig. 9-10(c, d), might equally well apply in principle, though they appear a little more contrived because they involve more intermediate levels. Some of the additional possibilities may be excluded once we begin to make tentative spin-parity assignments to the levels, combining the information on multipolarity with that derived from additional considerations. For instance, a $\frac{7}{2}^+$ assignment is indicated for the ground state by measurements of the atomic hyperfine structure and determination of the magnetic moment. The (E1 + M2) multipole character of the 0.396-MeV γ transition then points to a $\frac{9}{2}^-$ assignment to the top level, which is compatible with β-transition data from ^{175}Yb to this niveau. Thereafter the argument for scheme (b) spin assignments might take the form: γ''' is pure E1 and hence the 0.251-MeV level might be $\frac{7}{2}^+$, $\frac{9}{2}^+$, $\frac{11}{2}^+$, but of these all but the last is ruled out by the pure E2 γ'''' transition, whereas the (M1 + E2) character

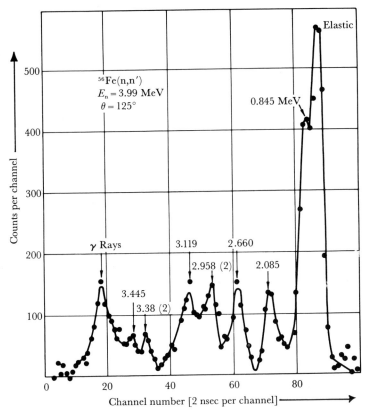

Fig. 9-11. Time-of-flight spectrum of neutrons inelastically scattered by ^{56}Fe at 125°, for an incident energy $E_n = 3.99$ MeV. The flight path used by Gilboy and Towle [Gi 65] was 171.2 cm. The corresponding level energies are indicated over each peak; the groups at 2.958 MeV and 3.38 MeV being doublets. (Used by permission of North-Holland Publishing Co.)

of γ' and γ'' rules out all assignments other than $\frac{9}{2}^+$ for the 0.114-MeV level. For the scheme (c) a rather questionable but not incompatible set of spin assignments could be made, as becomes evident upon trial, but β-γ coincidence measurements indicate a β transition to a first excited state of ^{175}Lu at 0.114 MeV, which is incompatible with this scheme. For (d) internal discrepancies arise which destroy this possibility, e.g., γ''' and γ'''' taken in conjunction rule out all but $\frac{11}{2}^+$ for the 0.251-MeV level, while γ'' and γ'''' in conjunction suggest $\frac{9}{2}^+$ for the 0.114-MeV niveau. However, this is

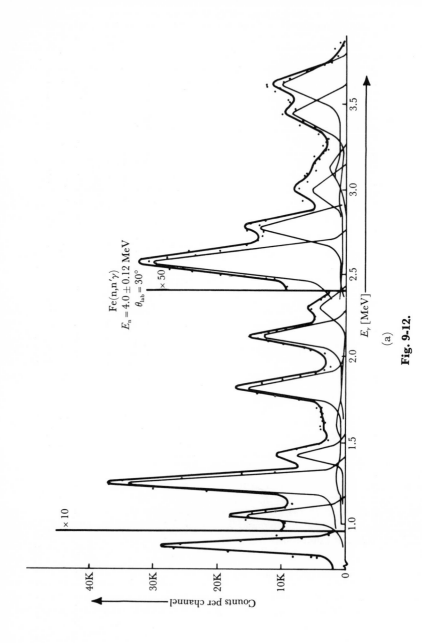

Fig. 9-12.

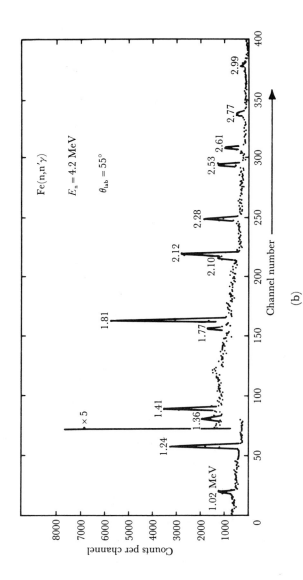

Fig. 9-12. Pulse-height γ-ray spectra from the Fe(n, n' γ) reaction around 4 MeV: (a) using an enriched ⁵⁶Fe target and NaI(Tl) detectors in a γ-ray spectrometer of the Raboy-Trail type set up to measure γ radiation at 30° to the 4.0-MeV incident neutron beam (from [Be 66a]); (b) using a natural iron target bombarded with 4.2-MeV neutrons. The γ spectrum as taken with Ge (Li) detectors after a flight path of 105 cm at 55° shows appreciably higher resolution. (Courtesy G. H. Williams, Texas Nuclear Corporation.)

inconsistent with the absence of an (M1 + E2) transition from the latter to the ground state. In point of fact, the scheme (b) is the one which best fits the data, and it is this which has been adopted.

In an alternative example, we proceed along different lines. We illustrate the compilation of a level scheme for the lowest excited states of ^{56}Fe, and again remark that extraneous information can shed invaluable light upon the level sequence and spin assignments. For instance, the level structure can be revealed through energy measurements (using a time-of-flight technique)

Table 9-4. INTEGRATED EXPERIMENTAL $(n, n'\gamma)$ CROSS SECTIONS FOR NATURAL Fe (in mb ± 20 percent)[a]

E_γ [MeV]	E_n [MeV]					
	0.95	1.05	1.5	3.1	3.5	4.0
0.845	261	400	557	1198	1141	1399
1.03					25	49
1.24				159	154	240
1.40				36	39	50
1.81				170	163	180
2.12				92	113	172
2.30					55	78
2.60					41	106
2.76						36
2.99						<25
3.12						<12
3.45						49
3.60						36

[a] From [Be 66a].

upon neutrons inelastically scattered from a ^{56}Fe target. A spectrum of this type [Gi 65] is depicted in Fig. 9-11; from this, level energies can readily be deduced. Measurements of the angular distributions of low-energy inelastically scattered neutrons can also furnish information on the final nuclear levels. On the other hand, analysis of γ energies and angular distributions can also serve to elucidate the level structure, and in this connection it is instructive to note how successive γ lines start to appear as the incident neutron energy is raised, since an increase in the incident energy permits progressively higher levels to be populated by n' transitions. Table 9-4 lists a set of such results as determined [Be 66a] for several incident neutron energies over the range 0.95 to 4.0 MeV$_{\text{lab}}$.

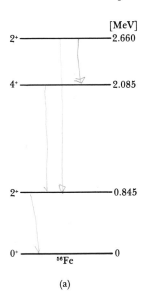

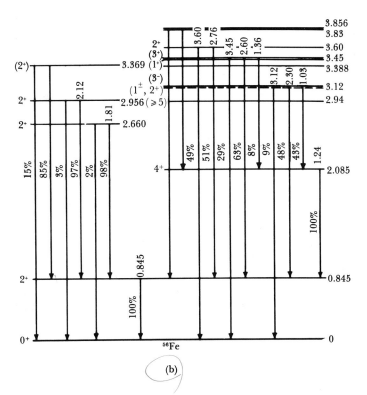

Fig. 9-13. Proposed level scheme for ^{56}Fe in which the energies and branching ratios of cascade γ rays are indicated, and tentative spin-parity assignments made (from [Be 66a]). (Used by permission of North-Holland Publishing Co.)

To give some indication of the way in which these γ lines, plus their energies and relative strengths, can be deduced from the observed γ spectrum, an example of a measured spectrum from which background counts have been subtracted is given in Fig. 9-12(a) for the highest incident neutron energy. Also shown (more faintly) is the decomposition of the overall spectrum into individual peaks whose energies are indicated in each case. The data were obtained with NaI(Tl) detectors using a Raboy-Trail total absorption γ-ray spectrometer, which gives good resolution but which nevertheless does not furnish as much fine detail as still more recent techniques using Ge(Li) detectors. (For comparison, Fig. 9-12(b) shows commensurate data so derived.)

Evidently, the 0.845-MeV γ transition which is observed at all neutron energies (even, quite strongly, down to $E_n = 0.95$ MeV) corresponds to de-excitation of the first level. The γ line of next higher energy, $E_\gamma = 1.03$ MeV, is absent until the neutron energy is raised to 3.5 MeV and even then shows up rather weakly; it is therefore interpreted as due to a transition between appreciably higher levels. On the other hand, the 1.24-, 1.81-, and 2.12-MeV γ radiations which one first observes quite strongly at $E_n = 3.1$ MeV are most likely to correspond with transitions of low-lying states to the first level. Such an interpretation, conjoined with the recognition that the weaker 1.40-MeV γ ray is but an intruder from ^{54}Fe decay, yields the scheme shown in Fig. 9-13(a), which compares well with the currently adopted overall level scheme depicted in Fig. 9-13(b). The energy and spin assignments are fully compatible with those derived from (n, n') data and from an extensive set of extraneous considerations (see, e.g., [Ma 67a]).

9.4. Angular Distributions and Correlations

Measurements of the angular distribution of γ radiation emitted from an excited nucleus, or of the angular correlation between two radiations observed in coincidence—i.e., emitted within so brief a time interval of one another that the intermediate nuclear state is not disturbed by relaxation effects—can yield valuable information on multipole mixing ratios or nuclear level spins, which in turn can be related to model predictions in order to derive insight into the behavior, structure, shape, and configuration of the nucleus.

Whereas purely kinematical considerations when applied to the decay process in the center-of-mass system might induce one to expect an isotropic angular distribution of the emitted γ radiation, one, in fact, finds the γ emission to be *symmetric* about a plane normal to the quantization axis, but markedly *anisotropic*. This feature has already been mentioned in Section 8.7.1 where the Wu experiment was discussed, and is in no way at variance with parity conservation. Indeed, the above *symmetry property* derives from the conservation of parity in an electromagnetic interaction process observed under such conditions that measurements make no distinction between left- and right-handed coordinate systems. (As discussed by Beltrametti [Be 58], measurements of fermion polarization or γ-circular polarization *are* sensitive to this distinction; hence the use of the latter to detect possible parity impurities in excited states of nuclei, as described in Section 9.2.3.) When one writes the differential cross section of a process in terms of the transition amplitude, i.e.,

$$\frac{d\sigma}{d\Omega} \sim \sum |a(\theta, \varphi)|^2 \tag{9-37}$$

and, in the mirror system,

$$\frac{d\sigma'}{d\Omega} \sim \sum |a(\pi - \theta, \pi + \varphi)|^2 \tag{9-38}$$

the summations run over the spins involved in the process and the transition amplitudes are of either even or odd relative parity when the nuclear parity is definite and pure, viz.

$$a(\theta, \varphi) = \pm a(\pi - \theta, \pi + \varphi) \tag{9-39}$$

Since the experimental arrangement is such as to effectively average out over the spins, the azimuthal dependence on φ over an entire plane is removed and consequently one derives a plane of symmetry, with

$$a(\theta) = \pm a(\pi - \theta) \tag{9-40}$$

i.e.,

$$\frac{d\sigma}{d\Omega}(\theta) = \frac{d\sigma'}{d\Omega}(\pi - \theta) \tag{9-41}$$

so that the differential cross section is symmetrical about $\theta = 90°$.

Anisotropy, however, ensues through restrictions upon γ transitions at the magnetic substate level. More explicitly, although selection rules might permit given γ transitions between nuclear states of spin J_1 and J_2, restrictions curb their taking place between certain of their respective magnetic substates characterized by sets of quantum numbers m_1 $(= -J_1, \ldots, +J_1)$ and m_2 $(= -J_2, \ldots, +J_2)$. Thus, for instance, orientation of nuclei through polarizing techniques destroys the otherwise uniform population of the magnetic substates—some are preferentially occupied with respect to others and consequently only *these* can participate in transitions to the substates of another niveau. On the other hand, a γ transition in nonoriented nuclei can serve to "pick out" a particular magnetic substate, so that a subsequently ensuing γ transition can proceed only from that special sublevel rather than from the entire set of $(2J_1 + 1)$ magnetic sublevels, with the result that its angular distribution is anisotropic. To render this evident, we consider a particularly simple, albeit hypothetical, transition sequence, namely,

$$J_0 = 0 \rightarrow J_1 = 1 \rightarrow J_2 = 0$$

involving a pure dipole γ_1-γ_2 cascade. Then only the intermediate level $J_1 = 1$ is degenerate, its three magnetic substates having the quantum numbers $m_1 = -1, 0, +1$. Under normal conditions, they are uniformly populated. (Their relative population can be expressed through the Boltzmann factor $\sim \exp[(-\mu H/kT)(m_1/J_1)]$, which is essentially unity for all m_1 except at the high fields and low temperatures needed to

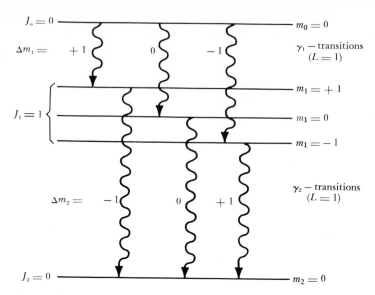

Fig. 9-14. Diagram depicting the magnetic substates corresponding to levels $J_0 = 0$, $J_1 = 1$, $J_2 = 0$ between which a $\gamma_1 - \gamma_2$ cascade of pure dipole character proceeds, as split into partial modes (adapted from [Ev 55]). (Used by permission of McGraw-Hill Book Co.).

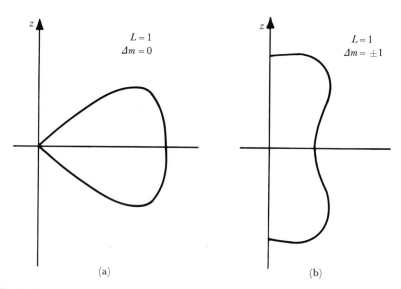

Fig. 9-15. Polar representation of the dipole distribution for $\Delta m = 0$ and $\Delta m = \pm 1$ modes (from [Sh 64a]).

polarize nuclei.) Hence the transition "modes" $\Delta m_1 = +1$, 0, and -1, as also correspondingly $\Delta m_2 = -1$, 0, and $+1$, depicted in Fig. 9-14 all occur with equal abundance. The overall distribution of either γ_1 or γ_2 is the synthesis of the "partial" distributions for each Δm, and is accordingly constant (i.e., isotropic), since the sum over each radiation magnetic mode $M = \Delta m$ can be shown to be constant as follows: The general expression for the emission probability of electromagnetic radiation having multipolarity L and magnetic quantum number m in a direction specified by polar angles (θ, φ) is (see [Bl 52b, p. 594])

$$W_M(\theta)\, d\Omega = \frac{1}{2}\left[1 - \frac{M(M+1)}{L(L+1)}\right]|Y_{L,\,M+1}|^2 + \frac{1}{2}\left[1 - \frac{M(M-1)}{L(L+1)}\right]|Y_{L,\,M-1}|^2$$

$$+ \frac{M^2}{L(L+1)}|Y_{L,\,M}|^2 \tag{9-42}$$

Hence, substituting $L = 1$ and simplifying the spherical harmonics Y, we find the "partial" probabilities to take the form (see Fig. 9-15):

$$\text{for } M = \Delta m = +1: \quad W_{+1}(\theta)\, d\Omega = \frac{3}{16\pi}(1 + \cos^2\theta)\, d\Omega \tag{9-43}$$

$$\text{for } M = \Delta m = 0: \quad W_0(\theta)\, d\Omega = \frac{3}{8\pi}\sin^2\theta\, d\Omega \tag{9-44}$$

$$\text{for } M = \Delta m = -1: \quad W_{-1}(\theta)\, d\Omega = \frac{3}{16\pi}(1 + \cos^2\theta)\, d\Omega \tag{9-45}$$

$$\text{whence for } \sum_M = \sum_{\Delta m}: \quad W\, d\Omega = \left[\frac{3}{8\pi} + \frac{3}{8\pi}(\sin^2\theta + \cos^2\theta)\right]d\Omega = \frac{3}{4\pi}\, d\Omega \tag{9-46}$$

The probability for emission into unit solid angle therefore takes the constant value $W = 3/(4\pi)$ for *all* directions θ. (The numerator is 3 because there are 3 "partial transition modes," and the denominator is 4π because the probability refers to *unit* solid angle and hence has to be reduced by a factor 4π from the unit value it assumes per mode over all space.) Thus individually the radiations γ_1 or γ_2 have an isotropic angular distribution. However, this is no longer true when γ_1 and γ_2 are taken in conjunction; for if we take the direction of γ_1 to define our quantization axis, the mode $M = 0$ is automatically excluded since $W_0 = 0$ for $\theta = 0$, and therefore the substate $m_1 = 0$ cannot be "fed" by γ_1 decay. This preferential selection of the modes $\Delta m_1 = \pm 1$, $\Delta m_2 = \mp 1$ causes γ_2 to have an anisotropic $(1 + \cos^2\theta)$ distribution with respect to γ_1 of the form shown in polar representation in Fig. 9-16(a) and as a distribution function in Fig. 9-16(b). Similar findings ensue for other, more complicated spin sequences, as described in [Sh 64a].

Thus when an angular distribution is expressed as a Legendre polynomial expansion over all permitted orders ν,

$$\frac{d\sigma}{d\Omega} = \sum_\nu A_\nu P_\nu(\cos\theta) \tag{9-47}$$

it follows that:

(i) the ANISOTROPY is established by the occurrence of orders ν higher than zero;

(ii) the SYMMETRY is established by the occurrence of only *even* orders $\nu = 0, 2, 4, \ldots$ whose range is bounded by angular momentum considerations;

(iii) the MAGNITUDE is essentially determined by the zeroth-order coefficient A_0, since $P_0 = 1$ and the higher-order contributions progressively diminish.

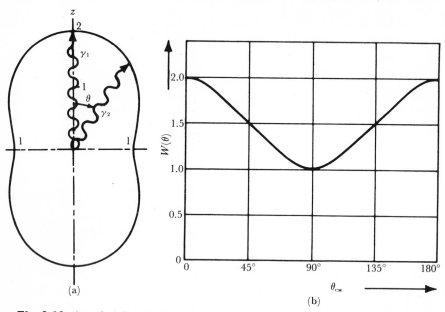

(a)

(b)

Fig. 9-16. Angular distribution of γ_2 radiation with respect to the direction of a preceding γ_1-ray in a dipole-dipole cascade transition: (a) in polar representation, in which the length of the γ_2 vector is a measure of the probability per unit solid angle Ω that the γ_2 radiation will be emitted at a mean angle θ with respect to the γ_1-direction; (b) plotted as a distribution function normalized to unity at $\theta = 90°$. In common with all such distributions for unpolarized γ radiation, the function is perfectly symmetric about $\theta = 90°$, as discussed in the text.

The expansion coefficients A_ν involve rather complicated products of Racah functions (see Appendix E) whose arguments are the nuclear spins and other angular momenta featured in the transitions [Ha 52, Bl 52a, Bie 53, Sa 54, Sa 56, De 57, Bie 60, Sh 66]. Hence when the numerical values of the A_ν, as determined, e.g., through fast computer programs [Sh 67] for various surmised spin sequences are compared with those evaluated from a least-squares fit to experimental data, the actual spin sequence can in many cases be identified uniquely, and thereby a definitive or probable spin assignment to particular nuclear levels made. When the nuclear spins are known, an angular distribution measurement of mixed-multipole γ radiation enables the mixing ratio δ to be established from a comparison of the normalized expansion coefficients A_2/A_0 and A_4/A_0 determined experimentally with those evaluated theoretically over the entire δ range. In the example shown in Fig. 9-17 the γ distribution following inelastic scattering of 5.92-MeV

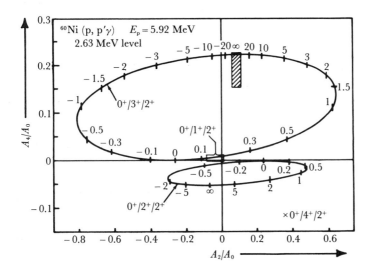

Fig. 9-17. Theoretical δ-multipole ellipses for the γ radiation from a ^{60}Ni(p, p' γ) reaction at 5.92 MeV effecting the transition between excited states at 2.63 and 2.16 MeV in ^{60}Ni. The cross-hatched area indicates the average result of three angular distribution measurements [Van P 64], which is evidently compatible with 3+ and 2+ spin assignments respectively for these levels and indicates the γ radiation to be of essentially pure E2 multipolarity (from [Sh 66]).

protons on ^{60}Ni is indicated to stem from a $3^+ \rightarrow 2^+$ transition of essentially pure E2 multipolarity.

ANGULAR CORRELATION studies involve the occurrence and observation of *three* transitions, as in a γ_1-γ_2-γ_3 cascade. Since the directions of γ_1 and γ_2 define a plane whose normal gives the orientation of the nucleus in the excited state from which γ_3 emission proceeds, this latter transition is qualitatively equivalent to one in an aligned nucleus, for basically the observation of the direction of γ_2 with respect to γ_1 "picks out" from among the entire ensemble of randomly oriented nuclei just those which happen to be aligned in a particular direction. When γ_2 and γ_3 are measured in coincidence over various angles θ_{γ_2} and θ_{γ_3} with respect to the γ_1 direction, the correlation is essentially akin to a γ_3 distribution from polarized nuclei, and is no longer therefore exactly symmetric about $\theta_{\gamma_a} = 90°$. Angular correlation measurements, especially when expressed in the form of an absolute double-differential cross section, furnish a very useful and sensitive means of elucidating not only spins and parities of nuclear levels, but also the mechanism of nuclear reactions [Li 61a, Br 63, Sh 63, Ha 65c], despite their greater theoretical and experimental complexity, particularly when extended to include particle transitions.

9.5. Recoil-Free Gamma Spectroscopy

Whereas fluorescence radiation is readily observed when *atoms* undergo electromagnetic transitions from resonantly excited niveaux, the corresponding phenomenon with *nuclei* long escaped detection. Atomic resonance fluorescence, predicted by Rayleigh at the turn of the century and discovered by Wood in 1904, was subjected to extensive investigation (see, e.g., the review by J. G. Winans and E. J. Seldin in Condon and Odishaw [Co 58]) and formally explained through the quantum theory of radiation [Wei 30, Wei 31, He 54]. It has recently awakened renewed interest in connection with the orientation phenomenon of "optical pumping." The search for *nuclear resonance fluorescence* was commenced by Kuhn [Ku 29] in 1929, but eluded observation despite many attempts up to 1950 [Mo 50, Mo 51a, Mo 53a] because of difficulties posed by the altogether different nature of nuclear transitions.

9.5.1. NUCLEAR RESONANCE ABSORPTION AND FLUORESCENCE

NUCLEAR RESONANCE ABSORPTION is in principle expected to occur when γ radiation emitted in a ground-state transition is reabsorbed by another nucleus of the same kind, according to the scheme depicted in Fig. 9-18.

Fig. 9-18. Schematic representation of γ emission and resonant re-absorption occurring between an excited level at an energy E_0 and the ground state of two nuclei of the same species.

In practice the process has so far been observed to take place only between the first level and the ground state; its cross section can be unusually high, but only within an extremely narrow energy interval about the resonance energy.

If the lifetime τ of the first niveau toward γ decay is large, the level width Γ_0 is small, since $\Gamma_0 = \hbar/\tau$ and therefore the "natural line width" of the γ radiation is correspondingly small (e.g., $\Gamma_0 \simeq 6.55 \times 10^{-9}$ eV for $\tau = 10^{-7}$ sec). Consequently, any slight perturbation is able to displace the delicate overlap of emission and absorption lines, and thereby destroy resonance absorption. Two perturbative influences do in fact change the position and shape of the emission line and the absorption line, viz.:

(i) A NUCLEAR RECOIL EFFECT serves to *shift* the lines away from the resonance (transition) energy E_0; and

(ii) NUCLEAR THERMAL MOTION serves to *broaden* the lines; i.e., DOPPLER BROADENING increases the width of each line from Γ_0 to a larger value Γ.

The action of the first of these effects upon the emission line can be inferred from considerations of momentum and energy conservation, since when a nucleus of mass m emits a photon of energy E_γ, its mass diminishes to $[m - (E_\gamma/c^2)]$ and it recoils with a velocity v_R (energy E_R). Therefore

$$\left(m - \frac{E_\gamma}{c^2}\right) v_R = \frac{E_\gamma}{c} \tag{9-48}$$

and

$$E_\gamma = E_0 - E_R \tag{9-49}$$

where the displacement of the actual emission line from the resonance energy E_0 shown in Fig. 9-19(a) to the lower energy E_γ depicted in Fig. 9-19(b) is

$$E_R = \frac{1}{2} m v_R^2 \simeq \frac{1}{2} \frac{E_0^2}{mc^2} \tag{9-50}$$

The Doppler broadening through thermal agitation of the emitter nuclei can very approximately be derived as follows. In the first place, a Maxwell distribution of thermal velocities already indicates the line width at $(1/e)$ height corresponding to a kinetic energy spread

$$\Delta E_{\text{kin}} = 2E_0 \left(\frac{2kT}{mc^2}\right)^{1/2} \tag{9-51}$$

to be very appreciably larger than Γ (at half height), except at the very lowest temperatures. An alternative approach is via an extension of the formalism which considers not only the final (recoil) energy of the nucleus

$$E_R \simeq \frac{1}{2m}\left(\frac{E_0}{c}\right)^2 \equiv \frac{p_0^2}{2m} \tag{9-52}$$

but also the initial energy E_i which the nucleus has through thermal motion of momentum $\mathbf{p}_i$,

$$E_i = \frac{\mathbf{p}_i^2}{2m} \tag{9-53}$$

The net change in energy can be written as

$$\Delta E \simeq \frac{\mathbf{p}_i^2}{2m} - \frac{(\mathbf{p}_i - \mathbf{p}_0)^2}{2m} = \frac{(\mathbf{p}_i \cdot \mathbf{p}_0)}{m} + \frac{\mathbf{p}_0^2}{2m} \tag{9-54}$$

wherein the first term depicts the Doppler broadening and the second term the recoil displacement. Some idea of the magnitude of the first term can be derived by writing it as

$$\frac{(\mathbf{p}_i \cdot \mathbf{p}_0)}{m} = \frac{p_i p_0 \cos \alpha}{m} \equiv E_D \cos \alpha \tag{9-55}$$

with $E_D \equiv p_i p_0 / m$, and the angle α between $\mathbf{p}_i$ and $\mathbf{p}_0$ assumed to vary from 0 to 2π. Then

$$E_\gamma = E_0 - \Delta E = E_0 - E_R \mp E_D \tag{9-56}$$

where

$$E_D = \frac{1}{m} P_i p_0 \simeq \frac{1}{m} (2mE_i)^{1/2} (2mE_R)^{1/2} = 2(E_i E_R)^{1/2} \tag{9-57}$$

is a measure of the energy spread of the emission line shown in Fig. 9-19(c).

Similar influences affect the absorption process, as a result of which the absorption peak is shifted to a higher energy $E_0 + E_R$ and Doppler broadened, so that the actual situation is represented by Fig. 9-19(d), and resonance absorption takes place only when there is an overlap, indicated by the shaded region about E_0. This overlap is at best small in the nuclear case (whereas the large *atomic* energy widths cause it to be large in the case of atomic resonance fluorescence), and much experimental effort has been expended to increase it artificially, either by broadening or displacing the emission line. Broadening can be achieved by heating the emitter and/or absorber. A way of displacing the emission line with the aid of a *Doppler shift* was exploited by Moon, who plated a radioactive ^{198}Au source (which decayed to the 0.412-MeV level of ^{198}Hg) on to the tips of a steel rotor and spun the rotor at maximum revolutions to attain tip speeds up to 800 m/sec. Tangentially emitted γ rays thereby experienced a Doppler shift which to a large extent compensated the recoil displacement and permitted resonance absorption to be observed. The intensity of scattered γ quanta attained a maximum at peripheral velocities around 700 m/sec.

We might at this point warn against drawing the fallacious conclusion that recoil-free γ emission occurs when the peripheral rotor velocity $\mathbf{v}$ is equal but opposite to the recoil velocity $\mathbf{v}_R$, for basically we need to provide *energy* compensation and not momentum compensation. Equating *total* energies before and after emission of a photon at the resonance energy E_0,

$$mc^2 + \frac{1}{2} mv^2 = \frac{1}{2} \left(m - \frac{E_\gamma}{c^2} \right) (v - v_R)^2 + E_0 + \left(m - \frac{E_\gamma}{c^2} \right) c^2 \tag{9-58}$$

we see that recoil-free emission occurs when, approximately,

$$v \cong \tfrac{1}{2}v_{\mathrm{R}} \tag{9-59}$$

Hence for resonance absorption, emitter and absorber must move toward one another with an overall relative velocity

$$u = 2v \cong v_{\mathrm{R}} \cong \frac{E_0}{mc} \tag{9-60}$$

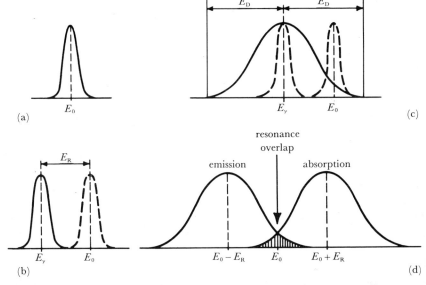

(a)

(b)

(c)

(d)

Fig. 9-19. Schematic representation of various situations pertaining to resonance absorption of γ radiation: (a) perfect resonance corresponds to total overlap of "natural" unperturbed emission and absorption lines at an energy E_0; (b) recoil causes the emission line to be displaced from the resonance energy E_0 by an amount E_{R} given by Eq. (9-50); (c) moreover, thermal motion of emitter nuclei produces a Doppler-effect broadening of the emission line which may cause it to extend in part beyond the energy E_0 even though still centered at $E_\gamma = E_0 - E_{\mathrm{R}}$; (d) the overlap of broadened and displaced emission and absorption lines corresponds to resonance absorption.

The rotor method of providing energy compensation through a Doppler shift then entails a peripheral velocity u which, for a transition energy $E_0 = 0.412$ MeV in ^{198}Hg, is

$$u \cong \frac{E_0}{mc} = 0.67 \times 10^5 \text{ cm/sec} = 670 \text{ m/sec} \tag{9-61}$$

in good accord with Moon's observation of a resonance around $u \approx 700$ m/sec. The thermal and rotor methods of attaining resonance overlap have since been developed considerably and found to give consistent results for level-width determinations. Another technique to attain energy compensation has consisted of studying the γ radiation emitted by a nucleus which, immediately prior to γ decay, had been set in motion by a preceding transition or nuclear reaction in such a way as to compensate the recoil. Also, fluorescence studies have been initiated [Se 61b], which use a bent-crystal diffraction spectrometer technique to obtain monochromatic γ radiation of continuously variable energy (e.g., from 0.03 to 0.25 MeV), which is particularly well suited to the investigation of the widths of low-lying nuclear levels.

The various methods and underlying theory have been described in a number of survey articles [Me 59, De 60c, Ma 65c]. However, with the advent of the Mössbauer technique to attain recoilless γ radiation of extremely sharp energy, these methods have been relegated to a subordinate position, and the majority of the more recent investigations have used Mössbauer transitions to study not only level widths but also many otherwise inaccessible effects.

9.5.2. Mössbauer Effect

Mössbauer's discovery of a method whereby emission and absorption of recoil-free γ radiation at precisely the resonance energy E_0 within the extremely narrow natural line-width Γ_0 might be achieved came about almost by accident. While investigating the nuclear resonance scattering of the 0.129-MeV γ radiation from ^{191}Ir, for which the recoil energy of the free nucleus would be $E_R = 0.047$ eV and the Doppler broadening at room temperature about 0.1 eV, Mössbauer set about trying to reduce the resonance scattering (whose cross section is determined by the overlap of the γ emission and absorption spectrum) by cooling the emitter and absorber with liquid air; but instead of decreasing the resonance scattering, the cooling brought about an increase. His experimental results [Mö 58c] are shown in Fig. 9-20 and compare well with the theoretical curve calculated by Vissher [Vi 60]. The explanation lay in the strong binding of emitter nuclei within a crystal lattice, which prevented rigidly bound source and absorber nuclei from individually taking up an energy E_R and thereby left the energy and width of the natural γ line unperturbed. By applying to his results an adapted form of calculations which Lamb had carried out earlier [La 39] for the effect of lattice binding on slow-neutron capture cross sections, Mössbauer was able to vindicate his observations and furnish a quantitative explanation for the occurrence in a certain percentage of γ decays of recoilless resonance γ lines having a natural width Γ_0. A crude way of visualizing this phenomenon might be to conceive of the recoil upon γ emission as being taken up by the whole lattice ensemble of 10^5–10^9 nuclei, rather than by an individual "free" nucleus, and thus as becoming negligibly small since the "shock absorber" is relatively so much more massive. More accurately, the imparting of energy to a crystal lattice is subject to quantum selection rules: If the recoil energy is less than the fairly large (quantized) energy associated with lattice vibrations ("single or multiple phonon transition"), no energy at all can be transferred either to or from the crystal lattice as a whole, and consequently the emission and absorption spectra display in common a recoilless component having the natural line width (Fig. 9-21). Concomitant with the narrow, high "Möss-BAUER LINE" is a wide Doppler continuum as background, which, although

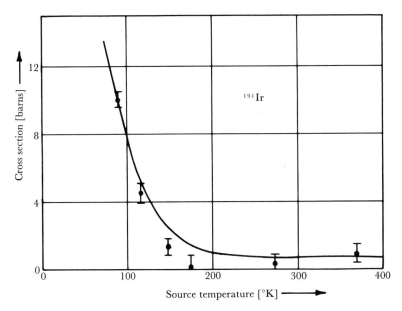

Fig. 9-20. The effective absorption cross section per ^{191}Ir nucleus corresponding to an absorber maintained at 88 °K while the temperature of the emitter is varied over the range shown. The experimental points [Mö 58c] are well fitted by the theoretical curve [Vi 60].

it can be appreciably reduced by cooling the source and absorber, cannot ever be fully eliminated (even at $\approx 0°$K, because of the zero-point energy). The relative areas under the peaks establish whether or not the Mössbauer effect predominates over the background with sufficient strength to warrant investigation.

With the arrangement shown in Fig. 9-22, Mössbauer found that a considerable fraction ($f \approx 40$ percent) of the observed 0.129-MeV γ decays from ^{191}Ir at 88°K fell within that part of the emission and absorption spectra which constituted the resonance line (Fig. 9-21). The effect was not only measurable, but proved to be discernible over a small range of artificially introduced relative velocities between source and absorber, with a distinct

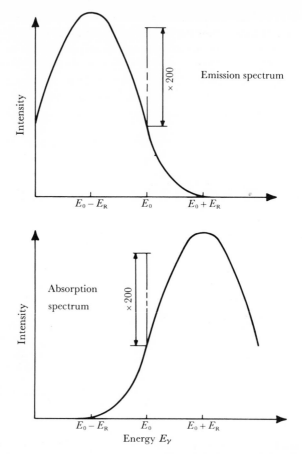

Fig. 9-21. Shape of the emission and absorption spectra in which a high, narrow resonance line is accompanied by broad Doppler continua which slightly overlap (from [Mö 59a]).

maximum at exactly zero velocity, as can be seen from Fig. 9-23. From the profile of the resonance line so determined, the width at half height could be read off as about 2 cm/sec in terms of relative velocity, or $9.2 \pm 1.2\ \mu eV$ in terms of energy, as calculated from the first-order Doppler-shift relation

$$\Delta E = \frac{\Delta v}{c} E_0 \qquad (9\text{-}62)$$

for a relative velocity Δv. Interpreting this value as *twice* the natural width $\Gamma_0 = 4.6 \pm 0.6\ \mu eV$ (because the analysis involved folding an emission spectrum into an absorption spectrum), Mössbauer obtained the value

$\tau = (1.4^{+0.2}_{-0.1}) \times 10^{-10}$ sec for the lifetime of the 0.129-MeV niveau in ^{191}Ir. The theoretical analysis, following Lamb's treatment, used Debye continuum theory to describe lattice vibration behavior in terms of a model based on a system of independent linear oscillators having a continuous frequency

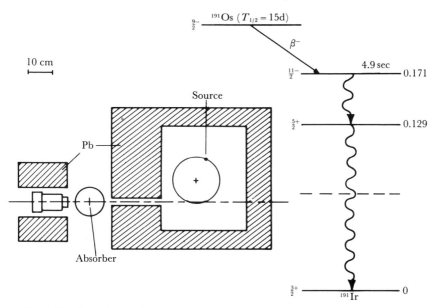

Fig. 9-22. Experimental arrangement used by Mössbauer [Mö 58c, Mö 59a] in a study of nuclear resonance scattering in which an en*hancement* of the recoil-less absorption was observed, contrary to expectation, on cooling the source to 88 °K. The "Mössbauer γ line" used was that from the 0.129-MeV ground-state transition in ^{191}Ir shown in the level scheme on the right side of the figure. (Used by permission of Springer-Verlag, Berlin.)

distribution up to a maximum ω_D and proportional to ω^2. In terms of the characteristic quantity termed the Debye temperature Θ, as given by the relation

$$\hbar\omega_D = k\Theta \qquad (9\text{-}63)$$

where k is the Boltzmann constant, the fraction of recoilless γ transitions is

$$f = f(T) \cong \exp\left\{-\frac{3}{2}\frac{E_R}{k\Theta}\left[1 + \frac{2}{3}\left(\frac{\pi T}{\Theta}\right)^2\right]\right\} \qquad (9\text{-}64)$$

and therefore depends markedly upon the absolute temperature T of the emitter, increasing to a maximum as T tends to zero. Setting $\Theta = 316°$K, one gets good agreement with Mössbauer's measurements at 88°K, which gave a self-absorption cross section of 10.0 ± 0.5 barns. The decisive quantity *f*, which is sometimes called the LAMB-MÖSSBAUER FACTOR, corresponds to the DEBYE-WALLER FACTOR used in atomic physics [Bl 33a] to express the temperature dependence of x-ray intensity after scattering on crystals without loss of energy to the lattice.

After some delay in receiving recognition (see Table 1-1 in Frauenfelder [Fr 62, p. 13]) the Mössbauer technique schematized in Fig. 9-24 won universal acclaim when recoilless emission of an extremely narrow 14.4-keV

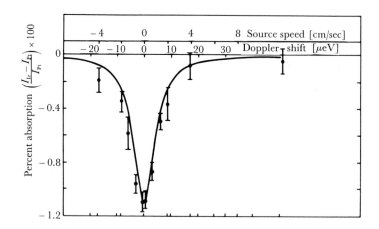

Fig. 9-23. Experimental findings of the resonance absorption in [191]Ir in terms of the relative velocity v between source and absorber. The corresponding Doppler-energy shift is also shown. The ordinate represents a relative intensity ratio $(I_{Ir} - I_{Pt})/I_{Pt}$ in which the γ intensity from β-γ decay of a 65-mCi [191]Os source as transmitted through an Ir absorber is compared with that which penetrates a representative Pt absorber of comparable thickness. The "Mössbauer line" evinces a sharp maximum at $v = 0$ and has a width which corresponds to the natural width of the 0.129-MeV level (it is actually 2Γ, as discussed in the text). The curve is a theoretical fit [Vi 60] to Mössbauer's later experimental data [Mö 59a].

γ line from the first level of [57]Fe was discerned independently at several research centers (Harvard, Harwell, Illinois, and Argonne) and gave the impetus to hitherto unfeasible investigations, as in the relativity experiment described in Section 3.2.4. These studies demanded such extreme energy precision (e.g., to 1 part in 10^{14}) that only the ultrasharp Mössbauer effect in [57]Fe (for which $\Gamma_0/E_0 = 3 \times 10^{-13}$ and $f \simeq 63$ percent at room temperature) and, as subsequently discovered [Cr 60a, Ak 61], in [67]Zn (for which the 93-keV γ line gives an even smaller ratio, $\Gamma_0/E_0 = 5 \times 10^{-16}$) provided the means for them to be carried through successfully. In order to utilize the Mössbauer effect, one has to seek transitions in which:

(a) the proportion f of recoilless transitions is adequately large (this corresponds to a high Debye temperature Θ conjoined with not too high a transition energy E_0 in order to attain low $E_R = E_0^2/mc^2$ for a fairly large mass m (see Fig. 9-25));

(b) nevertheless, E_0 must be fairly large in order to furnish good precision (low Γ_0/E_0);

(c) there must be few competing γ transitions and small likelihood of internal conversion;

(d) if possible, fine-structure splitting should not occur.

Even when these requirements are satisfied, the sensitiveness of the effect calls for precautions against undesired disturbances arising from, e.g.,

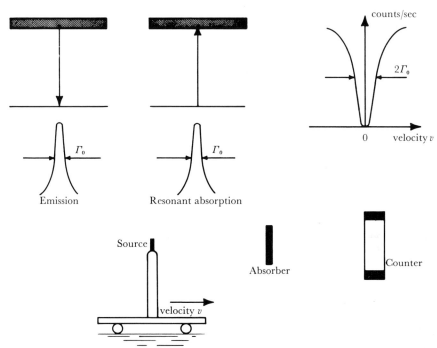

Fig. 9-24. Principle of the Mössbauer method (from [Si 65]). (Used by permission of North-Holland Publishing Co.)

(i) changes in nuclear mass on emission and absorption of γ rays, which cause the γ energy to change by an amount [Jo 60]

$$\Delta E_\gamma = -\frac{E_\gamma}{mc^2}\langle E_{\text{kin}}\rangle \tag{9-65}$$

where $\langle E_{\text{kin}}\rangle$ is the expectation value of the kinetic energy per atom of the lattice;

(ii) differences in the average isotopic mass number A (e.g., in the case of ^{67}Zn, a 1 percent difference brings about an energy shift of magnitude Γ_0);

(iii) differences in chemical constitution of presence or lattice defects;

(iv) differences in the Debye temperature Θ between source and absorber (for ^{67}Zn, $\Delta E_\gamma = \Gamma_0$ when $\Delta\Theta = 1.3°$K);

(v) differences in the absolute temperature T between source and absorber [Po 60, Jo 60]. (The mass reduction on emission, $\Delta m = -E_\gamma/c^2$, as considered

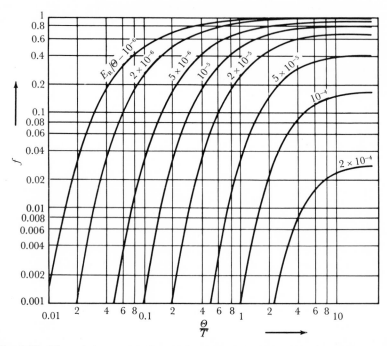

Fig. 9-25. Plot of the proportion f of recoilless transitions *vs* the ratio of Debye temperature Θ to source-absorber temperature T for various values of E_R/Θ (in units eV $°$K^{-1}). For iron, the Debye temperature is conventionally taken to be $\Theta_{Fe} = 420\ °$K, but in Mössbauer studies one takes (*cf.* [Fr 62]) $\Theta_{Fe} = 490\ °$K (data by A. H. Muir, Jr., Atomics International Report AI-6699, reproduced by Frauenfelder [Fr 62]).

in (i) causes the kinetic energy of the atom to *increase* at constant thermal momentum by an amount

$$\Delta E_{\text{kin}} = \frac{dE_{\text{kin}}}{dm}\,\Delta m = \left(-\frac{p^2}{2m^2}\right)\left(-\frac{E_\gamma}{c^2}\right) = \frac{1}{2}\frac{v^2}{c^2}E_\gamma \qquad (9\text{-}66)$$

as can also be deduced from considerations of a *second*-order Doppler effect upon the distribution of thermal velocities. This is matched by a corresponding *decrease* in the energy of the emitted γ ray which is balanced out in the absorption process except when there is a mutual temperature difference. In that case the temperature sensitivity for ^{57}Fe at room temperature is $\Delta E_{\gamma}/E_{\gamma} = -2.21 \times 10^{-15}/°C$ difference, and therefore perceptible);

(vi) relative velocity between emitter and absorber (the relative velocity v_{Γ_0} causing a shift of one line-width Γ_0 is given by the formula for the *first-order* Doppler effect,

$$\frac{v_{\Gamma_0}}{c} = \frac{\Delta E}{E_0} = \frac{\Gamma_0}{E_0} \Rightarrow \begin{cases} v_{\Gamma_0} = 3 \text{ cm/sec for } ^{191}\text{Ir} \\ v_{\Gamma_0} = 0.02 \text{ cm/sec for } ^{57}\text{Fe} \\ v_{\Gamma_0} = 3 \times 10^{-5} \text{ cm/sec for } ^{67}\text{Zn} \end{cases} \qquad (9\text{-}67)$$

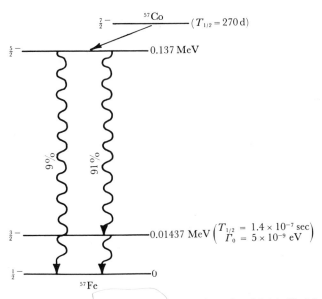

Fig. 9-26. Decay scheme of ^{57}Co $\rightarrow$ ^{57}Fe featuring the 14.4-keV Mössbauer γ transition in ^{57}Fe.

Apart from applications to the study of relativistic, chemical, and solid-state effects, the process of recoilless γ emission and absorption has been found to constitute a useful tool in several diverse fields of nuclear physics, e.g., the investigation of

(1) level widths, lifetime, and internal conversion;
(2) nuclear moments;
(3) chemical shifts ("isomer shifts"); and
(4) parity conservation in strong interactions.

Some examples of such studies, in the main based upon the 14.4-keV Mössbauer transition in ^{57}Fe at room temperature (Fig. 9-26), are presented and discussed briefly below:

(1) The Mössbauer technique lends itself best to the measurement of fairly short lifetimes, since when the lifetime exceeds about 10^{-10} sec, extra-nuclear fields can cause line broadening and hence lead to a spuriously low

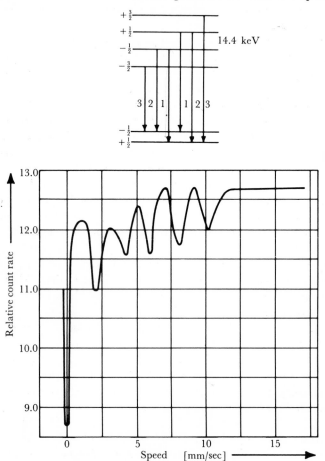

Fig. 9-27. Nuclear hyperfine structure in ^{57}Fe as measured with the Mössbauer technique by Hanna *et al.* [Ha 60b]; the corresponding transition modes are indicated in the upper part of the figure.

value of τ. Thus although Mössbauer obtained valid results with ^{191}Ir ($\tau = 1.4 \times 10^{-10}$ sec [Mö 58c, emended in Mö 59a]) and ^{187}Re ($\tau = 1.5 \times 10^{-11}$ sec [Mö 60b]), the appreciably longer lifetime of the first level in ^{57}Fe ($\tau = 1.4 \times 10^{-7}$ sec) rendered the latter difficult to measure correctly. Basically, the range 10^{-13} sec $\lesssim \tau \lesssim 10^{-10}$ sec seems best suited to such determinations.

(2) The spin, g factor, and magnetic moment of the first niveau in a "Mössbauer nucleus" can be determined from NUCLEAR ZEEMAN SPLITTING resulting from the application of an external magnetic field H to either the source or the absorber. To illustrate this, we take a particularly simple example, namely, a nucleus having a ground state of spin 0 and a first excited state of spin J, Landé factor g, and magnetic moment μ such that no electric or magnetic field gradient perturbs the state. On applying an external field H, the $(2J + 1)$-fold degeneracy is removed and the formerly unsplit γ line is resolved into $(2J + 1)$ components having a uniform separation $\Delta E = g\mu_N H$ which, providing it exceeds the width $2\Gamma_0$ of the overlap line (Fig. 9-24), can be separately distinguished. Then from a count of the number of components, J can be established, and from a determination of the separation ΔE at known field strength H, the Landé factor g can be evaluated. Thence the magnetic moment $\mu = gJ\mu_0$ can be calculated. The nuclear hyperfine lines for ^{57}Fe as measured by Hanna *et al.* [Ha 60b] are shown in Fig. 9-27; they were derived with the aid of the strong internal magnetic field at the ^{57}Fe nucleus, whose effective value was thereby established to be $(3.33 \pm 0.10) \times 10^5$ Oersted. The technique has allowed not only magnetic dipole moments, but also electric quadrupole moments of excited nuclear states to be investigated [Ki 60, DeBe 61].

(3) An energy shift occasioned by minute effects of chemical binding on the nuclear level structure when source and absorber are situated in different chemical environments has been shown up through Mössbauer techniques [Ki 60, DeBe 61, Wa 61, Sh 61, Bo 66, Ca 68]. From it information can be gleaned on differences in the nuclear radius appropriate to the ground and first excited state, and also on changes in the wave function of S electrons in various substances. Thus, for instance, there is a specific change $\Delta r/r = 1.8 \times 10^{-3}$ in ^{57}Fe, wherein the effective ground-state radius is the larger [Wa 61].

(4) The admixture of a possible parity-violating component in a strong interaction can be elucidated from the intensity and polarization of the separate hyperfine components obtained when an external magnetic field is applied to the source parallel to the direction of observation. Parity non-conservation would evince itself through a net difference in emission rates for the $\Delta m = \pm 1$ components parallel and antiparallel to the direction of nuclear polarization. A comparison of this type [Gr 61] indicated that any possible parity nonconserving admixture in the nuclear wave function of ^{57}Fe cannot exceed 10^{-5}. Tests of time-reversal invariance in strong interactions utilizing the Mössbauer effect have recently been surveyed by Hannon and Trammel [Ha 68].

For further information on these and allied effects, the reader is referred to review articles [Fr 62, Bo 62, Mö 65] and Mössbauer conference reports [Fr 60, Co 62]. An index to Mössbauer-effect data has been compiled by Muir *et al.* [Mui 66].

EXERCISES

9-1 The 3^+ first excited level at $E^* = 2.184$ MeV in ^{6}Li can decay either by a γ transition to the 1^+ ground state or by emission of a deuteron to ^{4}He. Relative to the ground state of ^{6}Li, the energy for the combination (^{4}He + d) is $E = 1.471$ MeV. If the total width of the first level in ^{6}Li is $\Gamma = 21$ keV, what is the ratio of probabilities for the two competing transitions?

9-2 Determine the excitation energies and probable spin-parity assignments of levels in ^{161}Tb (which has a $\frac{3}{2}^+$ ground state), given the following observed γ-transition energies and multipolarities:

0.057 MeV	(M1)	0.077 MeV	(M1)
0.103	(E1)	0.105	(?)
0.134	(E2?)	0.166	(E1)
0.181	(E2)	0.258	(M1)
0.273	(?)	0.284	(E1)
0.315	(M1)	0.361	(E1)
0.481	(?)	0.529	(?)

9-3 Following the β decay of ^{181}Hf to ^{181}Ta, the following γ lines were measured with a crystal spectrometer (the corresponding multipolarities are given in parentheses):

3.90 ± 0.10 keV	(M1)	133.02 ± 0.02 keV	(E2)
136.02 ± 0.02	(M1)	136.86 ± 0.04	(M1)
345.85 ± 0.20	(E2)	482.0 ± 0.2	(M1)
615.5 ± 0.5	(M3)		

Use these results to construct the level scheme of ^{181}Ta (which has a $\frac{7}{2}^+$ ground state), making appropriate spin and parity assignments, given that the spin-parity of the ^{181}Hf ground state is $J^\pi = \frac{1}{2}^-$.

9-4 On decay of the β-active radionuclide ^{115}Cd, whose half-life is $T_{1/2} = 2.3$ d, the following β and γ transitions are observed:

$E_{\beta\,\text{max}}$ [keV]	Intensity [percent]	E_γ [keV]
1110 ± 10	60	528
860	4	492
630	10	262
590	26	230
		35

The decay proceeds via an isomeric level for which $E_\gamma = 335$ keV and $T_{1/2} = 4.5$ h; the K-conversion coefficient is $\alpha_K = 0.84$. Every one of the β spectra with the exception of that for an end-point energy of 860 keV yields a straight Kurie plot. The angular correlations of the 35–492 keV and 230–262 keV γ cascades are *not* isotropic. The ground-state spin parities of $^{115}_{48}$Cd and $^{115}_{49}$In are respectively $\frac{1}{2}^+$ and $\frac{9}{2}^+$.

Compile the decay scheme and, so far as possible, assign spin-parity values to the levels in ^{115}In using the above data in conjunction with *ft* values which you should calculate.

What prediction does the Weisskopf formula make for the lifetime of the isomeric state?

[N.B.: (a) First-forbidden β transitions, with the exception of the case $\Delta I = 2$, frequently evince "allowed" shape. (b) If the level to which β-decay occurs and which de-excites by a γ-γ cascade has a spin 0 or $\frac{1}{2}$, or if the penultimate level in the γ-γ decay has either of these spin values, the γ-γ angular correlation is automatically isotropic (see Appendix E.8).]

9-5 A beam of excited ^{20}Ne ions is slowed down in aluminum. The excitation energy is $E^* = 1.63$ MeV and the velocity of the particles as they enter the aluminum attenuator is $v_i = 4 \times 10^8$ cm sec^{-1}. In this velocity region the stopping power can to a good degree of approximation be taken to be directly proportional to the velocity,

$$-\frac{dE}{d\xi} = Kv$$

with $K = 8 \times 10^{-9}$ keV cm sec μg^{-1} for ^{20}Ne in aluminum.

(a) Calculate the energy $E(\theta, t)$ of the γ quanta emitted at time t and at an angle θ to the incident beam direction.

(b) Also calculate the observable mean energy distribution $E(\theta)$ by building an average of $E(\theta, t)$ over the time t, assuming t to be the flight time of an ion from the moment of entry into the attenuator to the instant of γ emission.

(c) If one experimentally observes a Doppler shift of 0.65 percent between $\theta = 0°$ and $\theta = 90°$, viz.

$$\frac{\Delta E}{E} = \frac{E(0°) - E(90°)}{E_0} = 0.65 \text{ percent}$$

what is the mean lifetime τ of the excited level in ^{20}Ne?

9-6 Consider the emission of γ rays in flight by nuclei in an excited state, the excitation energy being E_0 and the kinetic energy being E_1.

(a) Calculate the specific energy shift $\Delta E/E$ of the emitted γ quanta as a function of the emission angle θ referred to the direction of flight of the nucleus. (The reference energy E is that of γ radiation emitted by a rigidly fixed nucleus at rest.)

(b) At what emission angle θ_0 does the energy shift become zero?

(c) Compare the result for forward emission with an estimate for the longitudinal first-order Doppler effect (cf. Appendix A.1.5).

(d) Examine numerically the energy shift in the measured γ distribution from the 0.48-MeV level in ^{7}Li excited by way of inelastic scattering of 2.4-MeV neutrons. The incident neutron makes a central collision with a ^{7}Li nucleus which is at rest up to the moment of impact; the inelastically scattered neutron emerges in the backward direction. What is the energy shift for γ rays emitted in the forward direction? In what direction is the energy shift zero?

$\checkmark$ **9-7** The 14.4-keV γ line of ^{57}Fe is to be used in a resonance fluorescence experiment in which the iron atoms in a stationary absorber are *free*, and cannot therefore participate in Mössbauer absorption. By what amount will the energy of the absorbed γ quanta deviate from the resonance energy E_0, and what speed would have to be imparted to the source (which is *not* inherently recoil-free in this experiment) in order to achieve maximal absorption? What speed would be necessary when the source (but not the absorber) may be taken to be recoil-free?

When the same γ transition is used in a Mössbauer resonance experiment, a line shift of Γ (full width at half height) is observed to occur when the source moves with a speed $v = 10^{-2}$ cm sec^{-1}. What is the value of the width Γ and of the mean lifetime τ of the excited state in ^{57}Fe?

9-8 Upon γ emission by a nucleus there is a change in the mass of the latter and a corresponding change in the energy of the lattice vibrations, which occasions a small shift in the energy of the emitted γ quanta. In resonance absorption this energy diminution compensates out when emitter and absorber are at exactly the same temperature. However, a small temperature difference suffices to yield a perceptible energy difference between the emission and absorption lines.

Show that a relative temperature difference $\Delta T \equiv T_{\text{emitter}} - T_{\text{absorber}} = 1°$K can produce roughly the same line displacement as that due to a gravitational shift in the Harwell experiment [Cr 60b, Cr 64; cf. Section 3.2.4], in which the 14.4-keV γ rays from ^{57}Fe were sent through a height difference of $h = 12.5$ m. What source velocity would yield an equivalent shift?

(Assume that
(a) all lattice atoms have the same mass;
(b) the lattice energy is uniformly distributed over the entire lattice;
(c) the lattice coupling is harmonic in character, namely, the vibrational kinetic energy is half the magnitude of the total internal lattice energy;
(d) the specific heat of iron is $C_{Fe} = 0.115$ cal g^{-1} °K^{-1}.)

Chapter 10

INTERNAL CONVERSION
Internal Conversion and Internal Pair Formation

As a competing process to γ decay, INTERNAL CONVERSION and, at higher excitation energies $(E^* > 2m_e c^2 \cong 1 \text{ MeV})$, INTERNAL PAIR FORMATION or PAIR INTERNAL CONVERSION serves to dissipate the excitation energy of a nucleus which is prevented through energy or selection-rule prohibitions from decaying by particle emission. By interacting *directly* with one or more of the orbital electrons that venture within the nuclear region, the nucleus can impart its excess energy to the conversion electron in a one-step process and cause it to be ejected from the atom with a comparatively high energy equal to the transition energy minus the electronic binding energy in its respective atomic shell. The vacancy thereby created in an atomic shell is filled through an electron cascade process from outer orbits, accompanied by the emission of x radiation and possibly Auger electrons from the *atom*. It is important to distinguish a single-stage conversion process in which the nucleus transfers energy directly to electrons in K, L, M, ... shells, viz. "K CONVERSION," "L CONVERSION," "M CONVERSION," etc., from *second*-order processes in which photon emission from the nucleus causes subsequent electron ejection from the atom. The same selection rules apply to conversion as to the competing γ-decay process, except that $0 \to 0$ transitions between nuclear levels are *permitted in conversion*, although forbidden in γ decay.

In several respects modern developments are rendering it as easy, if not easier, to acquire high-precision internal-conversion data as to obtain data for the accompanying γ radiation and at the same time to derive unambiguous information therefrom on the spins *and parities* of nuclear levels, on the validity of various models of nuclear structure, and on the distribution of electrons within an atom. The measure of agreement between theory and experiment has of late been vastly improved, and the introduction of high-precision magnetic spectrometers [Si 65] has led to marked progress in experimental techniques—a particularly valuable improvement in view of the complexity of the respective electron spectra. Since the higher electron shells have sub-shells for each of which the electron binding energy E_B differs slightly, the energy spectrum of conversion electrons, given by the relation

$$E_e = E_\gamma - E_B \qquad (10\text{-}1)$$

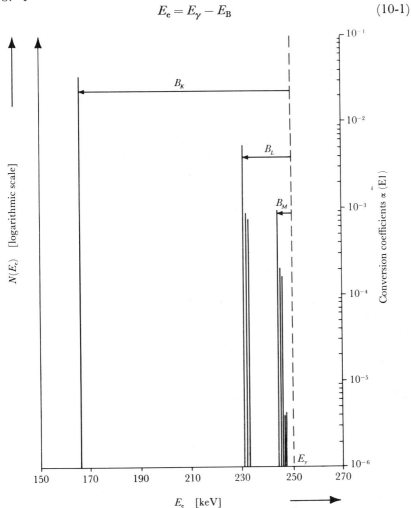

Fig. 10-1. Schematic representation (on a logarithmic scale) of the line energy spectrum for K-, L-, and M-shell conversion electrons. The K-singlet, L-triplet and M-quintuplet groups are displaced from the transition energy E_γ by an amount corresponding to their respective binding energies. The heights of the lines have been scaled to be proportional to the respective conversion coefficients (defined in Section 10.1.1) for a hypothetical electric dipole transition of energy $E_\gamma = 250$ keV in a nucleus for which $Z = 80$. The conversion coefficients α were taken from the tabulation compiled by Pauli [Pa 67].

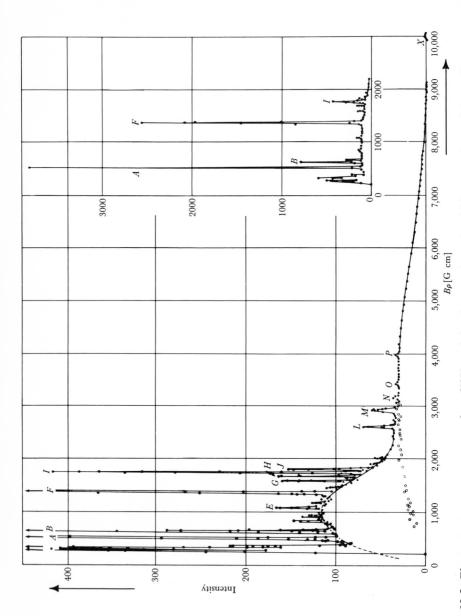

Fig. 10-2. Electron momentum spectrum from ^{212}Pb and daughter products, in which conversion and Auger lines appear super-imposed upon the β continuum (from [Fl 39]). (Used by permission of Springer-Verlag, Berlin.)

displays 1 component for the K shell, 3 for the L shell, 5 for the M shell, etc., as schematized in Fig. 10-1. The actual measured spectrum is often super-imposed upon a normal β spectrum when, as most frequently happens, conversion de-excites a state in a residual nucleus formed by a preceding β-decay transition. The resulting spectrum can therefore be quite com-plicated, as in Fig. 10-2, which shows the momentum spectrum of electrons observed from the β^- decay of ^{212}Pb (ThB) in equilibrium with its daughter products, which include two other β^- emitters and two α emitters, together with internal-conversion transitions. The analysis of spectra to derive infor-mation on the relative conversion probability is discussed in Section 10.1.2. In general, the probability of K conversion surpasses that from higher shells, and normally only K, L, and M conversion lend themselves to observation unless specially sensitive high-resolution techniques are adopted.

The probability of INTERNAL PAIR FORMATION is appreciably lower: thus, if it occurs in competition with γ emission, its rate is of the order 10^{-4} times the γ rate. Although internal pair formation can occur anywhere within the Coulomb field of the nucleus, it is most likely to take place [Jae 35] at a distance from the nuclear center of about $(\alpha Z)^2 \times$ (radius of atomic K shell). The absolute probability of internal pair formation is greatest when that for normal conversion is least (viz. large transition energy E_0, low atomic number Z, low multipolarity L).

10.1 Conversion Coefficients

As conversion and γ decay compete, the branching ratio provides a measure of the relative probability in a given transition, and one accordingly defines the CONVERSION COEFFICIENT as

$$\alpha = \frac{\text{conversion probability}}{\gamma\text{-emission probability}} \equiv \frac{\lambda_e}{\lambda_\gamma} = \frac{N_e}{N_\gamma} \tag{10-2}$$

in terms of decay constants λ, where N_e and N_γ are the number of conversion electrons and photons, respectively, observed per unit time.

10.1.1. PARTIAL AND TOTAL CONVERSION COEFFICIENTS

One can distinguish between PARTIAL INTERNAL CONVERSION COEFFICIENTS according to the shell from which the electron is ejected, viz.

$$\alpha_K = N_K/N_\gamma, \quad \alpha_L = N_L/N_\gamma, \quad \alpha_M = N_M/N_\gamma \tag{10-3}$$

Their sum then gives the TOTAL CONVERSION COEFFICIENT as defined above,

$$\alpha_K + \alpha_L + \alpha_M + \ldots = \alpha = N_e/N_\gamma \tag{10-4}$$

Inasmuch as in the past one was not as a rule easily able to resolve the in-dividual subshell contributions, the practice has developed of regarding α_K

as the partial conversion coefficient for *both K* electrons, α_L as that for *all L* shell electrons, etc. Nevertheless, precision measurements enable the partial subshell conversion coefficients to be determined to a fair degree of accuracy, viz.

$$\left.\begin{array}{llllll} \alpha_K & & & & \\ \alpha_{L_I} & \alpha_{L_{II}} & \alpha_{L_{III}} & & \\ \alpha_{M_I} & \alpha_{M_{II}} & \alpha_{M_{III}} & \alpha_{M_{IV}} & \alpha_{M_V} \\ \text{etc.} & & & & \end{array}\right\} \qquad (10\text{-}5)$$

where

$$\alpha_L = \sum_{i=I}^{III} \alpha_{L_i}, \qquad \alpha_M = \sum_{i=I}^{V} \alpha_{M_i}, \qquad \text{etc.} \qquad (10\text{-}6)$$

and

$$\alpha = \sum_{k=L,M,N,\ldots} \alpha_k \qquad (10\text{-}7)$$

It should be observed that the α_k's are basically defined with respect to N_γ and *not* $(N_\gamma + N_e)$, so that α can assume positive values greater or less than unity. In the case of strong conversion, N_γ is small and therefore the lifetime toward γ decay alone is commensurately large; the converse holds for weak conversion. The TOTAL DECAY CONSTANT λ for a given transition is, then, the sum of the individual partial decay constants,

$$\lambda = \lambda_\gamma + \lambda_e = \lambda_\gamma(1 + \alpha) \qquad (10\text{-}8)$$

10.1.2. EXPERIMENTAL STUDY OF CONVERSION

In the case of a simple γ transition, the total conversion coefficient $\alpha = N_e/N_\gamma$ can be determined from a knowledge of N_e derived from the intensity of the conversion line or lines in the electron energy spectrum, conjoined with a knowledge of N_γ derived either directly, e.g., with two NaI(Tl) counters, or indirectly from the intensity of photoelectrons ejected by the emitted γ quanta through a photoeffect process on a radiator. However, conversion electrons and photoelectrons from the *same* element cannot be distinguished directly from one another, since they have the same energy if they stem from the same (sub) shell. Consequently, it is preferable to arrange that the photoelectrons be produced in a thin foil of some other element than that whose decay is under investigation. The arrangement is sketched in Fig. 10-3: the converter foil usually consists of a high-Z element, such as lead or uranium, since the cross section for the photoeffect varies as Z^4 or Z^5 (see Section 4.4.1). Provision has to be made for screening out primary conversion electrons emanating from the source by encapsulating the latter in low-Z material of appropriate thickness. The high-Z radiator, in the form of a thin foil, is arranged as close as possible to the source. From the photoelectron intensity, after provision has been made for anisotropy in the angular emission

probability, N_γ can be determined if the counter has previously been calibrated with γ sources of known intensity and different energy. Such a technique combines a direct measurement of N_e with an indirect determination of
N_γ in order to derive the conversion coefficient α.

A complementary procedure may also be employed, in which N_e is determined indirectly and N_γ directly. This makes use of a comparison in a measured γ spectrum of the relative intensities of the γ line and the characteristic
x-ray line corresponding to the filling of the vacancy created by the ejection of
a conversion electron. The method, involving as it does a knowledge of the
fluorescence yield (defined in Section 4.5) to establish the number of emitted
photons per vacancy, is only of rather limited accuracy and applicability,

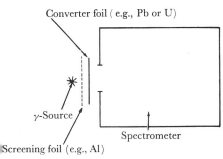

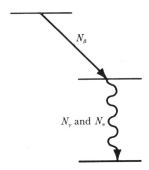

Fig. 10-3. Sketch of the type of arrangement used in the investigation of external conversion electrons, a screening foil being interposed between source and converter in order to screen out primary conversion electrons.

Fig. 10-4. Sequential β-γ transition in which N_β, the number of β particles emitted per unit time, is equal to the sum of the emission rates of γ quanta and conversion electrons, $N_\gamma + N_e$.

since even for the K shell the fluorescence yields are not known to a high
degree of precision, and for the remaining shells are even less definite.

Considerably simpler to investigate experimentally are those instances
where γ decay of a given level is preceded or followed by β decay (or α decay)
to that level, as exemplified in Fig. 10-4. Since the number of β particles
emitted per unit time is equal to the sum $(N_\gamma + N_e)$ for the associated transition, the area F_β under the β-continuum spectrum, as registered with a
magnetic spectrometer, provides a measure of the value of $(N_\gamma + N_e)$. At
the same time, the spectrum is overlaid with conversion peaks (Fig. 10-5),
whose net area F_e can readily be determined, whence

$$\frac{F_e}{F_\beta} = \frac{N_e}{N_\gamma + N_e} = \frac{1}{(N_\gamma/N_e) + 1} \tag{10-9}$$

and therefore

$$\alpha \equiv \frac{N_e}{N_\gamma} = \frac{1}{(F_\beta/F_e) - 1} = \frac{F_e}{F_\beta - F_e} \qquad (10\text{-}10)$$

This method offers the advantage that no absolute calibration is needed and that at sufficiently good resolution the individual partial conversion coefficients

$$\alpha_K, \ \alpha_L, \ \alpha_M, \ \ldots \text{ or even } \alpha_{L_I}, \ \alpha_{L_{II}}, \ \alpha_{L_{III}}, \ \ldots$$

can be determined in a direct manner.

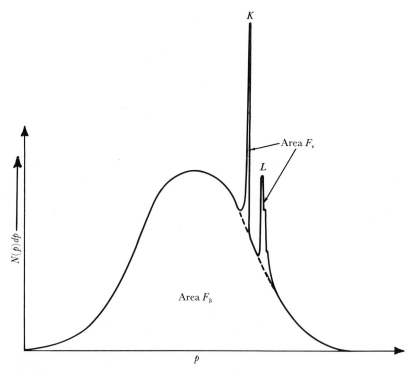

Fig. 10-5. Momentum spectrum for a single β transition overlaid by K- and L-conversion peaks. By comparing the area under these peaks, F_e, with the area under the β continuum, F_β, the conversion coefficient can be derived.

With the advent of high-resolution β spectroscopy and its utilization in internal conversion studies [Ew 65] some hitherto unenvisaged facets have become evident. This is illustrated graphically by the example in Fig. 10-6, which contrasts the conversion electron spectrum of ^{239}Np as measured with a thick-lens spectrometer [Ew 57] with the considerably more detailed results obtained on using an iron-free double-focusing $\pi\sqrt{2}$ spectrometer [Ew 59,

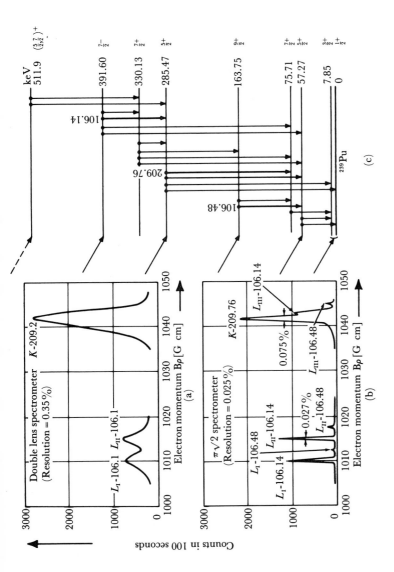

Fig. 10-6. Conversion electron spectra corresponding to transitions in ^{239}Pu fed by β decay of ^{239}Np: (a) momentum spectrum taken by Ewan *et al.* [Ew 57] with a thick lens β spectrometer; (b) the much more detailed β-momentum spectrum obtained at higher resolution by Ewan *et al.* [Ew 59] with a double-focusing spectrometer; (c) the level scheme of ^{239}Pu as fed by β decay of ^{239}Np in which the observed conversion transitions are indicated by heavy lines [Ew 59]). (Used by permission of North-Holland Publishing Co.)

Gr 60a], which confers a more than 10-fold improvement in resolution. The simple spectrum, Fig. 10-6(a), is revealed in Fig. 10-6(b) to consist of several additional components corresponding to *L*-conversion transitions between higher levels, and the lines are displayed in their natural width. The *K* 209.76 line from the 285.47 → 75.71 keV transition (see Fig. 10-6(c)) is shown on the one hand to have an unexpectedly large width and to be complex in that it overlaps with the L_{III} 106.14 line from the 391.60 → 285.47 keV transition which had not shown up in earlier investigations, even though the corresponding L_I and L_{II} lines had been found. Also the three small peaks for the first time revealed the L_I, L_{II}, and L_{III} lines corresponding to the weak 106.48-keV E2 transition to the second excited state at 57.27 keV from a level at 163.75 keV which can be identified with a $\frac{9}{2}^+$ highest member of a group of (rotational) levels having a monotonic spin sequence down to

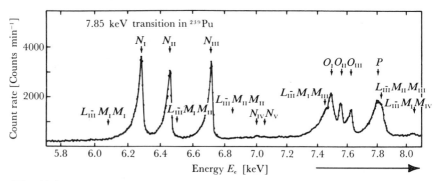

Fig. 10-7. Conversion electron lines of the 7.85-keV ground-state transition from the first level in ^{239}Pu, admixed with *L-MM* Auger electron lines (from [Ew 65]). (Used by permission of North-Holland Publishing Co.)

the $\frac{1}{2}^+$ ground state. The excellent resolution of the lines provides for an accurate determination of conversion coefficients and identification of the multipole character of the respective transitions. Thus additional evidence could be adduced for a hitherto sparsely supported conjecture [As 60] that certain *strongly hindered* E1 *transitions have anomalously high conversion coefficients for the K, L_I, and L_{II} subshells, but evince normal values for L_{III} conversion.* (A related phenomenon is the occurrence of *anomalous K-conversion coefficients for hindered* M1 *transitions* [Ge 61, Ge 65].) The 106.14-keV transition is of E1 multipolarity and highly hindered. The experimental L_I and L_{II} conversion coefficients prove to exceed the theoretical values by factors of 1.5 and 2.7, respectively, but the L_{III} coefficient is evidently unaffected, for α_{exp}/α_{th} = 1.1 ± 0.2. The anomalous natural width of the *K* 209.76 line is attributable [Ew 65] to a natural energy spread in the excitation of comparatively wide higher *atomic* states. (The *nuclear* states have lifetimes $\tau \gtrsim 10^{-12}$ sec and therefore widths $\Gamma \lesssim 0.08$ eV.) The quality of the high-resolution measurements on ^{239}Np conversion is also manifest in the ability to discriminate the N and O subshell conversion lines of the 7.85-keV ground-state transition, as depicted in Fig. 10-7 (together with "contaminant" Auger L-MM lines.) This dramatically exemplifies the power of conversion measurements to provide otherwise inaccessible experimental data to assist the theoretical interpretation of details of nuclear and atomic structure.

10.1.3. EVALUATION OF INTERNAL CONVERSION COEFFICIENTS

Of the many theoretical studies of internal conversion [Se 49a] which include a nonrelativistic early treatment by Rasetti ([Ra 36], discussed in

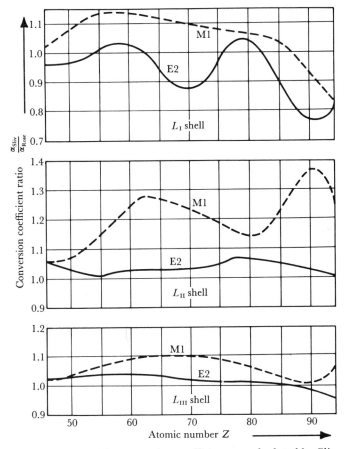

Fig. 10-8. Comparison of the conversion coefficients as calculated by Sliv and Band [since revised in Sl 61] with those computed by Rose [Ro 51b, revised in Ro 58] for transitions having an energy $E = 51$ keV and for various multipolarities (from [Ja 65]).

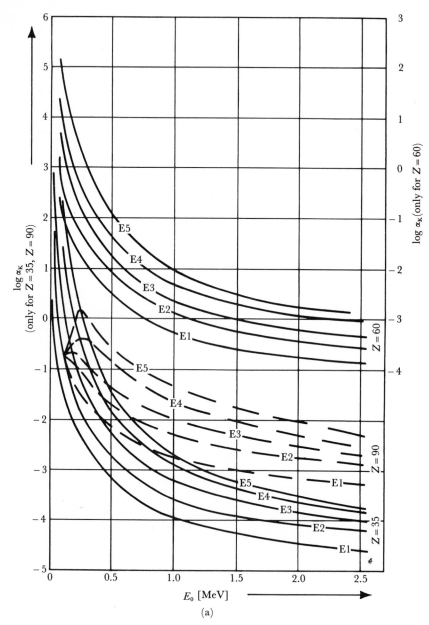

Fig. 10-9. Semilogarithmic plots of *K*-conversion coefficients for (a) electric and (b) magnetic multipole transitions in nuclei with atomic number $Z = 35$, 60, and 90, using data tabulated by Sliv and Band [Si 65], in terms of the transition energy E_0.

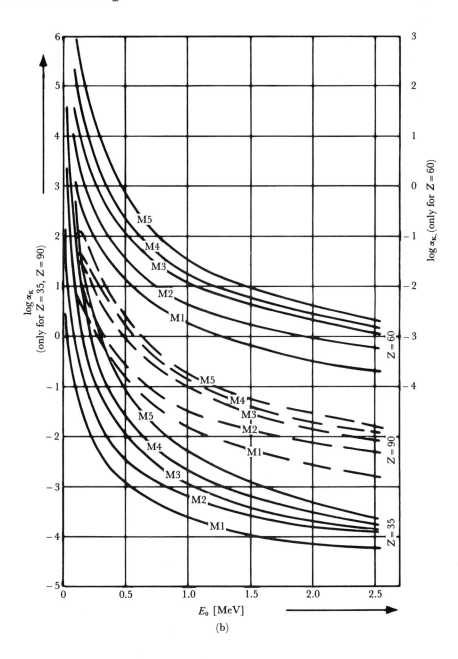

(b)

[Fe 50, pp. 101 ff.]) and a relativistic approximation by Dancoff and Morrison [Da 39], we confine ourselves to indicating some of the main results and refer the reader to more detailed recent surveys of this field [Ja 65, Ha 66, Ho 66b] and to the comprehensive review by Rose [Ro 65a]. None of the approaches to date has succeeded in deriving universal expressions for conversion coefficients in closed analytic form and the most recent and reliable numerical calculations depend upon fast electronic computers for the numerical evaluation of integrals, etc. Although in the past, rather disturbing discrepancies existed between the coefficients as evaluated by different groups (see Fig. 10-8), the most recent tabulations by Rose [Ro 58], which improve upon the earlier calculations involving a point nucleus [Ro 51b] in that they are based upon the consideration of nuclei having a finite extension $R = 1.2A^{1/3}$ fm, agree tolerably well with revised values published by Sliv and Band [Sl 61, Sl 65] which were derived from a similar, but not identical, screened finite-nucleus treatment.

Discrepancies between the two sets of values can be attributed to the authors' differing approach to the consideration of *dynamic* nuclear effects [Chu 56b, Chu 60] which enter into play as soon as the nucleus is regarded as no longer merely a point charge. Both in the Rose and in the Sliv-Band formulations, *static* effects, which are associated with a smearing out of the charge distribution within the nucleus and with a static deformation of the electron wave functions that can appreciably influence the magnitude of the conversion coefficients (i.e., by some 40–50 percent for large Z) are explicitly taken into account. These static effects involve no assumption as to details of the nuclear structure and hence are model-independent, whereas the *dynamic effects are model-dependent.* This feature led Rose to bypass such considerations in his "universal" approach, and to relegate their explicit evaluation to subsidiary calculations based upon appropriate nuclear models [Gr 58], whereas in the Sliv-Band treatment an attempt to take them into account was made in first approximation. The conversion interaction in the main takes place within the nucleus and its immediate vicinity. However, it has a different form in the interior to that in the extranuclear region. The former gives rise to "penetration" matrix elements which account for the dynamic intranuclear processes and are thereby strongly model-dependent. Sliv and Band introduce surface currents as a means of making provision for penetration effects, and thereby derive values which might be deemed to have somewhat greater validity for transitions that are not expected to evince an abnormal character, i.e., are not strongly hindered. (Appreciable penetration effects are, on the basis of current data, to be expected only in the case of E0 and retarded M1 and E1 transitions.) In the absence of independent evaluations of matrix elements [Ba 65a] and conversion coefficients, or of more detailed comparisons with measured values no clear-cut criterion of reliability can be assigned to the Sliv-Band values as against those of Rose, but a preliminary indication is furnished by some calculations which are still in progress, based upon a new reformulation of conversion theory [Im 66]. In this, nuclear structure influences are confined to but a few parameters which remain the same for all conversion processes in a given transition. (In a crude approximation, a single parameter suffices.) Initial calculations of conversion matrix elements have been carried out for the K and L shells of heavy nuclei ($Z = 75$–95) at low energy ($\lesssim 0.5$ MeV) for the first four multipole orders, and a markedly better measure of agreement has been found with the results of Sliv *et al.* than with the Rose values.

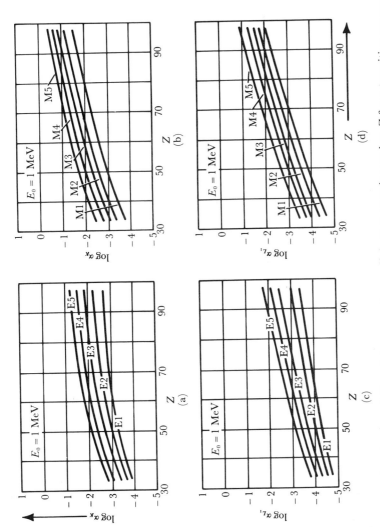

Fig. 10-10. Semilogarithmic plots of K- and L_I-conversion coefficients *vs* atomic number Z for a transition energy $E_0 = 1$ MeV using the data tabulated by Sliv and Band [Si 65]: (a) α_K for electric multipoles; (b) α_K for magnetic multipoles; (c) α_{L_I} for electric multipoles; (d) α_{L_I} for magnetic multipoles.

The Rose tabulation [Ro 58] lists partial conversion coefficients from the "no-penetration" model for K and L subshells at transition energies from 0.1 to 4 MeV in nuclei with $Z \geqslant 25$ (electric and magnetic multipolarities $\leqslant 5$). Subsidiary tables list M-subshell coefficients for point nuclei ($40 \leqslant Z \leqslant 95$) without screening corrections (and hence may be too large by as much as 40 percent), and K-shell radial matrix elements, again for unscreened point nuclei. The Sliv-Band tables [Sl 61, Sl 65] contain K- and L-subshell coefficients for electric and magnetic multipole transitions ($\leqslant 5$) in nuclei with $Z = 33$–98 for energies 0.1–10 MeV (K and L_{I}) or 0.1–4 MeV (L_{II} and L_{III}).

Graphical plots of the Rose conversion coefficients have been given in several sources (e.g., [Pr 62, Ro 65a]). These may be compared with Figs. 10-9 and 10-10, which respectively show the energy- and Z-dependence of typical Sliv-Band conversion coefficients. A noteworthy feature is the distinct difference in absolute value between coefficients for electric and magnetic transitions of the same multipolarity L, which in many instances enables an unambiguous assignment of parity to be made to a level involved in a conversion transition, when no such inference is possible from a study of the corresponding γ radiation. This useful property can also be exploited by using the K/L ratios α_K/α_L or, still better, L-subshell ratios $\alpha_{L_{\mathrm{I}}} : \alpha_{L_{\mathrm{II}}} : \alpha_{L_{\mathrm{III}}}$ for straightforward identification of multipolarity and parity.†

10.1.4. Approximate Analytic Expressions for Conversion Coefficients and Mean Lifetime

While, admittedly, the earlier approximate treatments have since been superseded by nonanalytic computer evaluations, the simple formulas they yield, though of insufficient accuracy to be quantitatively reliable, are nevertheless able to indicate the rough qualitative trend. Thus, Blatt and Weisskopf [Bl 52b, pp. 614 ff.] outline a somewhat over-simplified derivation of K-conversion coefficients for transitions of multipolarity L and energy E_0, taken to be small with respect to $m_e c^2 = 0.511$ MeV. (At higher energies, it would have been necessary to use *relativistic* electron wave functions, as in the Dancoff-Morrison treatment. The K-shell electron binding energy was neglected in comparison with E_0.) For K conversion in a (parity-favored) transition of *electric multipolarity* L, the partial conversion coefficient takes the form

$$\alpha_K^{(\mathrm{EL})} \approx \frac{L}{L+1} Z^3 \left(\frac{e^2}{\hbar c}\right)^4 \left(\frac{2 m_e c^2}{E_0}\right)^{L+(5/2)} \qquad \text{for} \qquad B_K \ll E_0 \ll m_e c^2 \quad (10\text{-}11)$$

† A very recent tabulation of (static) internal conversion coefficients for the K, L and M subshells of heavy nuclei ($Z = 60$–96), together with M-shell conversion matrix elements and phases for $Z = 72$–96 over the energy range 0.01–0.5 MeV has been prepared by Pauli [Pa 67] from computations which took the screening of atomic electrons and the finite nuclear size into account. A further tabulation of dynamic corrections and particle parameters is contemplated.

and for *magnetic multipolarity L,*

$$\alpha_K^{(ML)} \approx Z^3 \left(\frac{e^2}{\hbar c}\right)^4 \left(\frac{2m_e c^2}{E_0}\right)^{L+(3/2)} \qquad \text{for} \qquad B_K \ll E_0 \ll m_e c^2 \quad (10\text{-}12)$$

The trend is fairly similar: α_K increases with increasing Z and L, but decreases as E_0 increases. Higher-subshell coefficients involve a more complicated treatment: a good description of the formal theory has been given by Tralli and Goertzel [Tr 51], and references to the more rigorous treatments can be found in Blatt and Weisskopf [Bl 52b], and Rose [Ro 65a].

In the special case of $0 \to 0$ monopole transitions, Blatt and Weisskopf also give a derivation of the transition probability per unit time, which yields the approximate expression

$$W_{fi}^{(0 \to 0)} = \frac{32\sqrt{2}}{9} Z^3 \left(\frac{m_e c e^2}{\hbar^2}\right) \left(\frac{\overline{R^2}}{a_0^2}\right)^2 \left(\frac{E_0 - B_K}{m_e c^2}\right)^{1/2} \quad (10\text{-}13)$$

where $\overline{R^2}$ corresponds to the mean square nuclear radius (it is actually a matrix element which vanishes when the levels differ in parity) and a_0 is the Bohr radius. Substituting $\overline{R^2} = \frac{3}{5}R^2 = \frac{3}{5}(r_0 A^{1/3})^2$ and building the reciprocal, we derive the *K* CONVERSION MEAN LIFETIME for $0 \to 0$ transitions between levels of the same parity as

$$\tau_K^{(0 \to 0)} \approx \frac{0.37}{Z^3 A^{4/3}} \left(\frac{m_e c^2}{E_0 - B_K}\right)^{1/2} \text{sec} \quad (10\text{-}14)$$

Substitution of typical values, e.g., $Z = 60$, $A = 150$, $E_0 = 1$ MeV, yields a lifetime $\tau_K \approx 2 \times 10^{-9}$ sec. Clearly, conversion constitutes another instance of a process in which the *nuclear* lifetime is slightly, but perceptibly, influenced by *atomic* electron-configuration effects (cf. Section 8.5.3, which deals with electron capture). A change in the chemical environment can therefore produce a slight change in the lifetime. A recent such observation [Ho 66b] has been the finding by Cooper *et al.* that by purely chemical means, the lifetime of the metastable nuclear isomer ^{90m}Nb ($\tau = 34.5$ sec) can be altered by as much as 3.5 percent. This may be compared with an earlier measurement [Ba 53] in which a difference of about 0.3 percent was observed between the lifetime of ^{99}Tc^m in the form KTcO$_4$ as against Tc$_2$S$_7$. At low energy ($E_0 \approx 2$ keV) an appreciable fraction of the conversion occurs in the *N* shell, which carries the valence electrons. Since the two compounds have different ionization energy, the density of available *N* electrons is changed sufficiently to give rise to an observable effect, which corresponds to a slight *diminution in the decay lifetime* as a consequence of conversion. The dependence of internal conversion in ^{169}Tm on the chemical environment has lately been studied by Carlson *et al.* [Ca 68], who found that the relative intensities of the conversion electrons from all but the outermost P_I shell of ^{169}Tm were unaffected by chemical variants in the composition of the host medium; however, the P_I conversion was 38 percent lower for a Tm_2O_3 host and 46 percent lower for a WO_3 host than for a pure tungsten metal host.

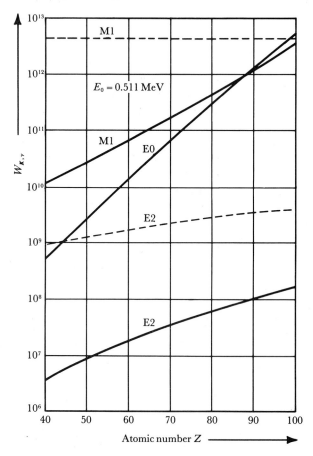

Fig. 10-11. Comparison of calculated probabilities for K conversion (dashed curves) with those for γ emission (solid curves) in terms of atomic number Z. The values have been calculated for M1 and E2 transitions having an energy $E_0 = m_e c^2 = 0.511$ MeV. The plot for E0 conversion is also shown (from [Chu 56a]).

It is also interesting to make a rough comparison between transition probabilities for, say, K conversion and those for γ emission. To this end, we depict in Fig. 10-11 some calculated values [Chu 56a], which show that although conversion probabilities tend to be lower than the "Weisskopf single-particle estimates" of the γ-emission probability for a transition of the same multipolarity and energy (the authors took $E_0 = m_e c^2 = 0.511$ MeV), they can readily compete, particularly in the case of heavy nuclei. It will also be noted that E0 conversion competes favorably with M1 and E2 components, especially at low energies, since it has a weaker energy-dependence than that for commensurate γ emission.

10.2. Selection Rules

Though in essence the momentum and parity selection rules for internal conversion are the same as those for γ transitions, some special features deserve attention. A distinction which has already been pointed out is the *occurrence of $0 \rightarrow 0$ conversion transitions*, strictly forbidden for γ radiation, which cannot have an $L = 0$ component. As has long been recognized [Fo 30b, Yu 35c] it is, however, perfectly feasible for electrons whose wave function penetrates within the nuclear volume to take up energy from the nucleus in a direct conversion process. If this occurs, however, the transition must be of E0, and *not* M0, character, for no M0 multipole interaction field can exist. ($\Delta \pi =$ yes is ruled out for conversion transitions.) To appreciate this point, we note that a spin-0 nucleus can be regarded as a uniformly charged sphere with no other electric moments than the E0 monopole (while the lowest *magnetic* moment is M1). For a transition to another spin-0 state, the only classical equivalent is a periodic spherical pulsation. Viewed from an outside point, the charge acts as though concentrated at the sphere's center, so that even if it pulsates, it cannot set up an alternating potential *outside* the sphere, and hence cannot give rise to γ radiation. *Inside* the sphere, however, potential fluctuations *can* occur and direct transfer of energy from nucleus to electrons *can* take place. (This argument also precludes the occurrence of such two-step processes as e.g., an internal photoeffect.)

Thus we can summarize the selection rules as:

(a) MOMENTUM SELECTION RULE FOR MULTIPOLARITY

$$\Delta J \equiv |J_i - J_f| \leqslant L \leqslant J_i + J_f \equiv \Sigma J \tag{10-15}$$

(b) PARITY SELECTION RULE FOR MULTIPOLE CHARACTER

$$\Delta \pi = \text{no: betokens E} \qquad \Delta \pi = \text{yes: betokens M}$$

(c) $0 \rightarrow 0$ TRANSITIONS

E0 allowed   M0 forbidden

(d) MULTIPOLE MIXING

In practice, restricted to 2 orders, e.g., M1 + E2; however, an E0 contribution can add incoherently to the (M1 + E2) conversion rate.

10.2.1. Mixed Multipolarity

When a transition involves a mixture of multipoles, e.g., $ML + EL'$, the net conversion coefficient is simply

$$\alpha = I_{ML}\,\alpha(ML) + I_{EL'}\,\alpha(EL') \tag{10-16}$$

where I_{ML} and $I_{EL'}$ are the relative intensities of the γ rays (or conversion electrons), normalized to

$$I_{ML} + I_{EL'} = 1 \tag{10-17}$$

and the $\alpha(ML)$, $\alpha(EL')$ are conversion coefficients for the appropriate shell, multipolarity, and transition energy.

A measurement of L-subshell intensities, for example, can yield direct information on multipole mixing ratios, and this technique is currently being widely exploited [Ho 66b]. Some caution has to be exercised in that, at present, theory lags somewhat behind experiment. Thus in the case of pure E2-multipole transitions there is some disagreement between the experimental data on L-subshell ratios and the values predicted by Rose or Sliv-Band theory. Recent studies by Mladjenović and others [Ha 66] of pure E2 transitions ($2^+ \rightarrow 0^+$), e.g., in the mass region $A = 152$–188, indicate that experimentally determined L_I/L_{II} and L_I/L_{III} ratios exceed theory by as much as 25 percent while L_{II}/L_{III} ratios appear to agree fairly well with calculated values.

10.3. Conversion Distributions and Correlations (Particle Parameters)

As a very valuable adjunct to the information which can be derived from direct comparison of experimental and theoretical values of the conversion coefficients, the determination of angular distribution and correlation parameters has recently received increasing attention. It is important to note a modification in the underlying theory. In Section 9.4 we remarked that the angular distribution of γ radiation in a sequence of γ transitions can be expressed as a Legendre polynomial expansion. The weighting coefficients comprise essentially a product of "transition parameters," one for each step in the decay sequence. When a particular step involves particle emission rather than γ emission, the corresponding transition parameter simply needs to be modified by multiplication with an appropriate "particle parameter." This formulation [Bie 53] accordingly permits one to write down expressions for the weighting coefficients which essentially determine the angular distribution or correlation for particles, as well as for γ rays. In particular, for the distribution of electrons from K, L_I, and L_{II} conversion, the requisite particle parameters have been established [Bie 53], and have since been tabulated in more extended form, corrected for screening and finite nuclear size [Ba 65b]. This tabulation takes cognizance of a basic error [Bie 53] in the sign of the interference term between mixed multipoles which has since been brought to light [Chu 64, Bie 64]. A still more recent tabulation in [Ha 66] lists particle parameters for L_I, L_{II}, *and* L_{III} conversion, corrected for screening and finite nuclear size for $Z = 49$–98, as derived from partially independent sets of calculations. Work is now in progress to extend these to higher shells and to test the values against more wide-ranging experimental data. Preliminary analyses indicate good agreement in the case of K conversion and fair agreement for L subshell conversion. In some cases, ratios of K- to L-shell particle parameters have been found to tally with the theoretical predictions even though the individual subshell parameters appear to be somewhat at variance with absolute calculated values [Ho 66b]. A critical comparison between theory and experiment [Ho 67] indicates that the K- and L-shell parameters in E1, E2, E3, and M4 transitions are largely independent of details of nuclear structure and that a good approximation is furnished by the point-nucleus model in most cases.

EXERCISES

10-1 In magnetic β-spectrometry measurements employing a ^{60}Co source and using a homogeneous magnetic field of strength 265.5 or 293.5 Gauss for a radius of curvature $\rho = 20$ cm maximum yields of K electrons stemming from K conversion in ^{60}Ni were registered. What is the value of the two corresponding γ-transition energies, if the binding energy of K electrons in Ni is $E_B = 8.5$ keV?

10-2 The electron capture process on $^{75}_{34}$Se leads to the formation of excited $^{75}_{33}$As nuclei. Decay to the ground state of ^{75}As occurs through γ emission and internal conversion; the following conversion lines are observed:

23.2 keV	24.4 keV
54.3	64.6
68.9	85.0
95.3	96.4
109.4	124.3
134.7	136.0
186.9	197.2
253.3	263.6
268.2	278.5
293.4	303.4
390.0	400.5

The 68.9- and 109.4-keV lines stem from K conversion.

Determine the energies of the emitted γ quanta and set up a tentative level scheme for ^{75}As. (The electron binding energy in the K, L, and M shells of arsenic is, respectively, approximately 11.9, 1.5, and 0.2 keV.)

10-3 The following γ and conversion-electron transitions have been observed in the stable nuclide ^{131}Xe, which has a $\frac{3}{2}^{+}$ ground state.

ΔE [keV]	Observation
80	$\alpha_K = 0.3$, α_K/α_L ratio $= 7.5$ no coincident β radiation observed
164	$\alpha_K = 30$, $\alpha_K/\alpha_L \approx 1.8$
273	
284	E2
359	
365	in coincidence with other γ's and with β radiation having $\log ft = 6.6$
638	$\alpha_K = 0.004$, $\alpha_K/\alpha_L = 7.5$ coincident β radiation with $\log ft = 6.8$
644	
724	$\alpha_K \approx 10^{-3}$, $\alpha_K/\alpha_L \approx 7.5$ coincident β radiation with $\log ft = 6.9$

On the basis of these data, compile a possible level scheme for ^{131}Xe and estimate the lifetime of the isomeric state.

Chapter 11

FUNDAMENTAL CHARACTERISTICS
OF NUCLEAR REACTIONS
Nuclear Reaction Characteristics

A comprehensive discussion of nuclear reaction behavior entails two distinct approaches, namely the general and the particular. While one aspect relates essentially to fundamental principles, the other falls more aptly within the purview of specific models and mechanisms—a dichotomy similar to the division of this book into Volume I, which stresses basic general principles, and Volume II, in which the main emphasis is placed upon structural and interaction models. For this reason the present chapter is devoted solely to general considerations concerning nuclear reaction characteristics, e.g., dynamic and probability aspects, which serve as a link with the model-dependent treatment that falls within the context of Volume II.

11.1. Reaction Energetics

11.1.1. Energy and Momentum Conservation in Nuclear Reactions

The fundamental principle of energy and momentum conservation has a decisive influence upon the characteristics of all physical processes including nuclear reactions, whose energetics we consider in the present section. The related kinematic considerations have been presented in Appendix C. (Elastic scattering is treated in detail in Appendix B.)

Most nuclear reactions involve two-body interactions of the type depicted in Fig. 11-1. In this situation a projectile of mass m_1 impinges with a velocity $\mathbf{v}_1$ upon a stationary target ($\mathbf{v}_2 = 0$) of mass m_2 and gives rise to an interaction which releases two reaction products, one of mass m_3 (ejected with a velocity $\mathbf{v}_3$ at an angle θ) and the other of mass m_4 (emitted with a velocity $\mathbf{v}_4$ at an angle φ). The normal usage is to associate m_3 with the lighter outgoing particle and m_4 with a residual nucleus. As the *total* energies of initial and final systems must be equal, it follows that

$$E_{\text{tot}}^{(1)} + E_{\text{tot}}^{(2)} = E_{\text{tot}}^{(3)} + E_{\text{tot}}^{(4)} \tag{11-1}$$

i.e.,

$$(E_{\text{kin}}^{(1)} + E_0^{(1)}) + (0 + E_0^{(2)}) = (E_{\text{kin}}^{(3)} + E_0^{(3)}) + (E_{\text{kin}}^{(4)} + E_0^{(4)}) \tag{11-2}$$

in terms of the kinetic and rest energies of the particles involved (noting that for the stationary target $E_{\text{kin}}^{(2)} = 0$). Separating the variables, we can recast

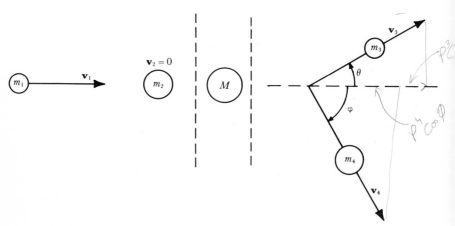

Fig. 11-1. Diagram depicting the initial, intermediate, and final systems involved in the kinematics of a nuclear reaction.

this expression into a form which contains the Q value, as introduced in Section 2.1 to take account of the energy balance,

$$Q \equiv E_0^{\text{(initial)}} - E_0^{\text{(final)}} = (E_0^{(1)} + E_0^{(2)}) - (E_0^{(3)} + E_0^{(4)}) \tag{11-3}$$

$$Q = [(m_1 + m_2) - (m_3 + m_4)]\,c^2 \tag{11-4}$$

whence

$$E_{\text{kin}}^{(1)} + Q = E_{\text{kin}}^{(3)} + E_{\text{kin}}^{(4)} \tag{11-5}$$

Also, momentum conservation is expressed through the vector relation

$$\mathbf{p}^{(1)} = \mathbf{p}^{(3)} + \mathbf{p}^{(4)} \tag{11-6}$$

which can be resolved into components along the incident direction and perpendicular thereto,

$$p^{(1)} = p^{(3)} \cos\theta + p^{(4)} \cos\varphi \tag{11-7}$$

$$0 = p^{(3)} \sin\theta - p^{(4)} \sin\varphi \tag{11-8}$$

11.1.2. NONRELATIVISTIC Q EQUATION

When, as is generally the case in low- and medium-energy nuclear physics, the particle velocities lie in the nonrelativistic region (see Appendix A.3) for which $E_{\text{kin}}^{(i)} \ll E_0^{(i)}$ the classical energy-momentum relation

$$p^{(i)} = (2m_i E_{\text{kin}}^{(i)})^{1/2} \tag{11-9}$$

can be used in the momentum expressions (11-7) and (11-8) which, in conjunction with (11-5), can be used to eliminate the energy $E_{\text{kin}}^{(4)}$ and angle of emergence φ of the residual nucleus m_4 in order to derive the NONRELA-TIVISTIC Q EQUATION which plays a decisive role in particle kinematics:

$$Q = E_{\text{kin}}^{(3)}\left(1 + \frac{m_3}{m_4}\right) - E_{\text{kin}}^{(1)}\left(1 - \frac{m_1}{m_4}\right) - \frac{2(m_1 E_{\text{kin}}^{(1)} m_3 E_{\text{kin}}^{(3)})^{1/2}}{m_4}\cos\theta \tag{11-10}$$

with $E_{\text{kin}}^{(1)}$, $E_{\text{kin}}^{(3)}$, and θ all measured in the *laboratory* system. Clearly, when $\theta = \pi/2$ the final term vanishes. This mechanism- and model-independent formula can be applied to all types of two-body nonrelativistic reaction processes. In particular, it can be utilized for:

(a) *elastic scattering*, in which $m_1 = m_3 = m$ and $m_2 \doteq m_4 = m_T$

$$Q = 0 \tag{11-11}$$

(N.B.: $E_{\text{kin}}^{(1)} \neq E_{\text{kin}}^{(3)}$. There is a change in kinetic energy associated with the deflection of the projectile through an angle θ by momentum transfer which imparts a recoil energy $E_{\text{kin}}^{(4)}$ to the target nucleus);

(b) *inelastic scattering*, in which $m_1 = m_3 = m$ and $m_2 = m_4 = m_T$, but

$$Q = -E^* \tag{11-12}$$

where E^* is the excitation energy imparted to the target nucleus m_T.

Of course, *endothermic* reactions have *negative* Q's, while *exothermic* processes have *positive* Q's. In parametric form, the general solution of the Q equation which gives the outgoing kinetic energy in terms of the incident energy for any two-body process can be written as

$$(E_{\text{kin}}^{(3)})^{1/2} = s \pm (s^2 + t)^{1/2} \tag{11-13}$$

where

$$s \equiv \frac{(m_1 m_3)^{1/2}}{m_3 + m_4}(E_{\text{kin}}^{(1)})^{1/2}\cos\theta \tag{11-14}$$

and

$$t \equiv \frac{m_4 Q + E_{\text{kin}}^{(1)}(m_4 - m_1)}{m_3 + m_4} \tag{11-15}$$

Consequently, the energy $E_{\text{kin}}^{(4)}$ of the residual nucleus must lie between the two extreme values given by (11-13), and energetically possible processes are characterized by *real, positive* values of $(E_{\text{kin}}^{(3)})^{1/2}$.

11.1.3. Relativistic Q Equation

When particle motion has to be treated relativistically, the calculation becomes more complicated. A shortcut to the final formulae is provided by the useful "prescription" that all nonrelativistic equations are rendered relativistically valid when the expression $[m_0^{(i)} + (E_{kin}^{(i)}/2c^2)]$ is throughout substituted for the rest mass $m_0^{(i)}$ of the particle i.

With the aid of this, we can derive the relativistic Q equation as follows: Squaring (11-6) and using the relativistic energy-momentum relation [cf. Eqs. (A-66) and (A-67)] we get

$$(p^{(i)})^2 = \frac{(E_{kin}^{(i)})^2}{c^2} + 2m_i E_{kin}^{(i)} \tag{11-16}$$

Expressing (11-5), which serves to define Q, as

$$Q = \sum_{\substack{\text{final} \\ \text{states}}} E_{kin}^{(i)} - \sum_{\substack{\text{initial} \\ \text{states}}} E_{kin}^{(i)} \tag{11-17}$$

and eliminating $E_{kin}^{(4)}$, we obtain the RELATIVISTIC Q EQUATION (in which the m_i are rest masses),

$$Q = E_{kin}^{(3)} - E_{kin}^{(1)} + m_4 c^2 \left\{ \left[1 + W^{(1)} d^{(1)} \left(\frac{m_1}{m_4} \right)^2 + W^{(3)} d^{(3)} \left(\frac{m_3}{m_4} \right)^2 \right. \right.$$
$$\left. \left. - \frac{2m_1 m_3}{m_4^2} (W^{(1)} d^{(1)} W^{(3)} d^{(3)})^{1/2} \cos\theta \right]^{1/2} - 1 \right\} \tag{11-18}$$

with

$$W^{(i)} \equiv \frac{E_{kin}^{(i)}}{m_i c^2} \tag{11-19}$$

and

$$d^{(i)} \equiv 2 + W^{(i)} \tag{11-20}$$

This reduces to the nonrelativistic form in the limit $E_{kin}^{(i)} \ll m_i c^2$ (so that $W^{(i)} \to 0$).

11.1.4. Threshold Energetics

The minimal incident energy required to bring about a (necessarily endothermic) reaction is the THRESHOLD ENERGY E_{thresh}, expressed in the laboratory system. The threshold energy just suffices to initiate the process in which the reaction products are formed with *zero mutual velocity in the center-of-mass (CM) system*, whence it follows that at threshold the *final net kinetic energy in the CM system is zero* (but $E_{kin}^{(3)}$ and $E_{kin}^{(4)}$ are *not* zero in the laboratory system, because the velocity of the CM is itself not zero in the laboratory system).

We now make a distinction between corresponding quantities in the laboratory and CM systems by affixing *primes* upon the latter, while noting that Q, being essentially a mass difference, is independent of the reference system. Thus the total kinetic energy is, in the laboratory system,

$$E_{\text{kin}}^{(1)} \tag{11-21}$$

and, in the CM system,

$$E_{\text{CM}} \equiv E_{\text{kin}}^{(1)'} + E_{\text{kin}}^{(2)'} = E_{\text{kin}}^{(1)} \frac{m_2}{m_1 + m_2} \tag{11-22}$$

The net *final* kinetic energy in the CM system is obtained by simply adding Q to this, which yields the THRESHOLD CONDITION

$$\left[E_{\text{kin}}^{(1)} \left(\frac{m_2}{m_1 + m_2} \right) + Q \right]_{\text{thresh}} = 0 \tag{11-23}$$

Hence the positive threshold energy in the laboratory system is

$$[E_{\text{kin}}^{(1)}]_{\text{thresh}} = -Q \left(\frac{m_1 + m_2}{m_2} \right) \tag{11-24}$$

with $Q < 0$, and at threshold the emergent particles are emitted in the incident direction ($\theta = 0$). Strictly, the above formula contains a slight approximation in that by expressing the initial energy in the CM system by the formula (11-22) we have ignored the influence of Q upon the mass ratio. The exact formula is

$$[E_{\text{kin}}^{(1)}]_{\text{thresh}} = -Q \left(\frac{m_3 + m_4}{m_3 + m_4 - m_1} \right) \tag{11-25}$$

which reduces to (11-24) when $m_2 c^2 \gg Q$. In the limit $m_1, m_3 \ll m_2, m_4$, it follows that

$$[E_{\text{kin}}^{(1)}]_{\text{thresh}} \to |Q| \tag{11-26}$$

a result quoted in Section 2.1.

Furthermore, the Q equation can be applied to reactions with *vanishingly small incident energy*, e.g., thermal neutron capture, with $E_{\text{n}} = \frac{1}{40}$ eV—a consideration restricted to exothermic reactions, since the available energy in endothermic reactions does not suffice to make up the deficiency in the rest energy expressed by the negative Q value, i.e.,

$$(E_{\text{kin}}^{(3)})^{1/2} \text{ is imaginary when } Q < 0 \text{ and } E_{\text{kin}}^{(1)} \to 0 \tag{11-27}$$

From (11-13) with $Q > 0$ and $E_{\text{kin}}^{(1)} \to 0$ it follows that

$$s \to 0 \quad \text{and} \quad t \to \frac{m_4}{m_3 + m_4} Q \tag{11-28}$$

whence

$$E_{\text{kin}}^{(3)} \cong \left(\frac{m_4}{m_3 + m_4}\right) Q \tag{11-29}$$

independently of the emission angle θ. In this case,

$$E_{\text{kin}}^{(3)} + E_{\text{kin}}^{(4)} = Q \qquad \text{and} \qquad \theta + \varphi = 180° \tag{11-30}$$

11.1.5. ENERGY-CORRELATION ANALYSIS

   Momentum analysis involving the deflection of charged particles in a magnetic field as a means of sorting out reaction products into groups having the same specific charge Ze/m constitutes one of the principal techniques concerned with the kinematics of particle motion. Energy analysis can also serve a most useful role in the evaluation of measured data, particularly when used in conjunction with coincidence techniques. Thus, for example, it provides a means of distinguishing between particles of the same species which originate from different alternative competing reactions. For a given incident energy $E_{\text{kin}}^{(1)}$, the energy-balance relation (11-5), namely,

$$E_{\text{kin}}^{(1)} + Q = E_{\text{kin}}^{(3)} + E_{\text{kin}}^{(4)} \tag{11-31}$$

indicates that for a two-body nuclear reaction which can follow different outgoing channels a measurement of the emergent particles in coincidence enables them to be assigned to different groups according to the Q value in a straight-line plot of $E_{\text{kin}}^{(3)}$ vs $E_{\text{kin}}^{(4)}$, as shown schematically in Fig. 11-2. A clear distinction can thereby be made between different products of closely similar mass which might otherwise prove difficult to distinguish. Particularly in the case of reactions induced by heavy ions, such procedures prove invaluable both in planning and in interpreting an experimental investigation. Even more potentially informative is an examination of emission energies and angles. To illustrate this we show a typical full kinematic plot in Fig. 11-3. This features the overall kinematic relationships in the laboratory system for the reactions

$$^{16}\text{O} + {}^{12}\text{C} \begin{cases} \longrightarrow {}^{24}\text{Mg} + {}^{4}\text{He} & (Q = 6.769 \text{ MeV}) \\ \longrightarrow {}^{25}\text{Mg} + {}^{3}\text{He} & (Q = 6.478 \text{ MeV}) \end{cases} \tag{11-32}$$

and considers outgoing channels to certain states in ^{24}Mg (at $E^* = 0$, 4.24, 8.50, and 12.342 MeV) and ^{25}Mg (at $E^* = 0$, 0.98, and 2.80 MeV). The structure of the corresponding curves for other competing reactions is similar. By undertaking an analysis of the density distribution of measured data points which refer to appropriate discrete lines, direct information can be derived concerning the relative probabilities of competing processes.

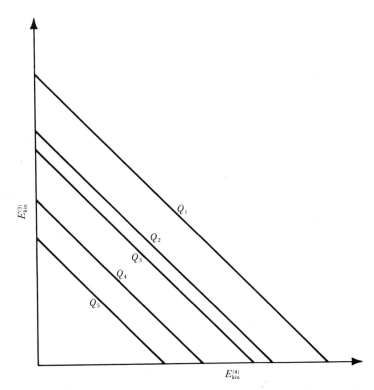

Fig. 11-2. Illustration of the linear $E_{\text{kin}}^{(3)}$ *vs* $E_{\text{kin}}^{(4)}$ plots having a slope equal to -1 and separated according to Q value when the reaction product particles are detected in coincidence.

A still more sophisticated application of this basic principle can be utilized for a study of SEQUENTIAL PROCESSES leading to *three-body* [You 64, Mo 64a, Mo 64b] or *four-body* [Wa 64, Et 64] final states. Thus, for example, it has been used [You 64] to discriminate between α particles emanating from different sequential reactions in the scheme

$$^{6}\text{Li} + {}^{3}\text{He} \left\langle \begin{array}{l} {}^{8}\text{Be} + \text{p} \\ \quad\hookrightarrow \alpha_1 + \alpha_2 \\ \quad\rightharpoonup \alpha_2 + \text{p} \\ {}^{5}\text{Li} + \alpha_1 \end{array} \right. \tag{11-33}$$

initiated by bombarding a ^{6}Li target with ^{3}He particles of energy 2.7 MeV. Coincidences were registered between counts from a proton detector fixed at 60 degrees to the incident beam and a solid-state α detector variable from -150 degrees to $+150$ degrees. The coincidence yield displayed a strong maximum

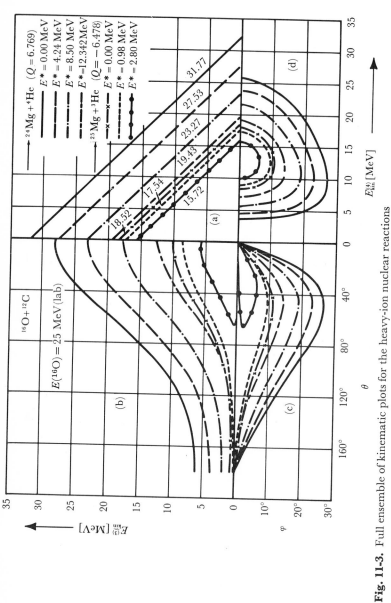

Fig. 11-3. Full ensemble of kinematic plots for the heavy-ion nuclear reactions

$$^{16}\text{O} + {}^{12}\text{C} \nearrow \searrow \begin{array}{l} {}^{24}\text{Mg}^* + {}^{4}\text{He} \\ {}^{25}\text{Mg}^* + {}^{3}\text{He} \end{array}$$

at an incident ^{16}O energy of 25 MeV. The plots feature the following kinematic relationships for each of the various exit channels to different excited states: (a) $E^{(3)}_{\text{kin}}$ vs $E^{(4)}_{\text{kin}}$; (b) $E^{(3)}_{\text{kin}}$ vs θ; (c) θ vs φ; (d) $E^{(4)}_{\text{kin}}$ vs φ.

at the −100 degree setting, corresponding to α-α coincidences between those α's originating via highly excited states ($E^* = 16.6$–16.9 MeV) of ⁸Be and their counterparts produced through the competing reaction via the ground state of ⁵Li. Two-dimensional kinematic spectra can to a considerable extent resolve ambiguities as to the origins of the detected particles, including identification of the various possible excited intermediate states, through

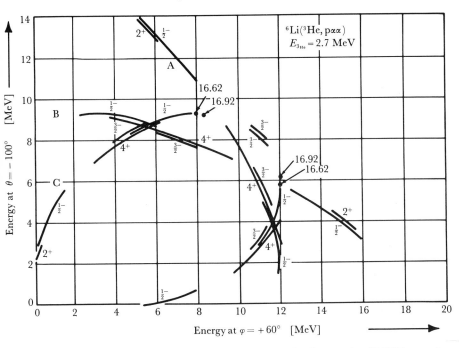

Fig. 11-4. Kinematic plots of α-α or p-α coincidences for the reaction ⁶Li(³He, pαα) at 2.7 MeV with the counters arranged at +60° and −100° to the incident beam. There are three sets of diagonally symmetric curves linking regions which correspond to discrete states in the intermediate ⁸Be and ⁵Li system having the spins shown. The points marked 16.92 refer to the 16.92-MeV level in ⁸Be (from [You 64]).

coincidence measurements of the energies of two out of the three emitted particles in the final ααp system as a function of emission angle θ and φ. The two-dimensional energy spectrum of $E_{kin}^{(3)}$ vs $E_{kin}^{(4)}$ is confined to discrete regions. Under these conditions at fixed θ and φ, the energy plot takes the form of discrete curves, certain points in which correspond to emission via given discrete states of the intermediate system. Figure 11-4 shows such kinematic plots in which curve A represents the situation for $\theta_\alpha = +60°$, $\varphi_p = −100°$, while curve B refers to $\theta_p = +60°$, $\varphi_\alpha = −100°$ and curve C to $\theta_\alpha = +60°$, $\varphi_\alpha = −100°$.

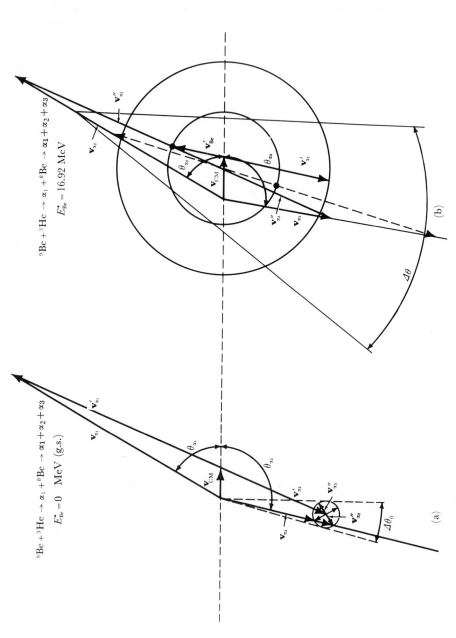

Fig. 11-5. Velocity vector diagrams for the reaction ^{9}Be (^{3}He, $\alpha\alpha\alpha$), at $E_{^3\text{He}} = 3.0$ MeV; (a) proceeding by way of the ground state in ^{8}Be, (b) proceeding by way of the 16.92 MeV...

Another similar instance is provided by the reaction $^9\text{Be}(^3\text{He}, \alpha\alpha\alpha)$ in which again two-body energy measurements in coincidence provide information on the intermediate transition sequence [Mo 64a]. The kinematics of the process are such that for the process $^9\text{Be}(^3\text{He}, \alpha_1)^8\text{Be(g.s.)} \rightarrow \alpha_2 + \alpha_3$, if α_1 is detected at an angle θ_{α_1}, the angle φ at which α_2 or α_3 are emitted must fall within a narrow cone (of half-angle $\Delta\varphi = 7.2°$ when $\theta_{\alpha_1} = 60°$) about the

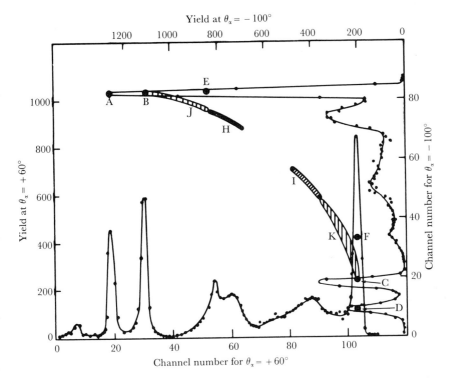

Fig. 11-6. Projections of two-dimensional spectra for the ^9Be $(^3\text{He}, \alpha\alpha\alpha)$ reaction at an energy $E_{^3\text{He}} = 3.0$ MeV, shown together with the reconstructed curve for $+60°$ and $-100°$ counter settings. The positions of the points and segments corresponding to various levels in ^8Be, as well as resonances in the α-α system, are indicated in accordance with the original lettering scheme [Mo64a], wherein G was not employed.

recoil direction of the intermediate ^8Be system. The velocity vector diagram of this process is shown in Fig. 11-5(a). Therein, $\mathbf{v}_{CM}$ represents the total center-of-mass velocity while the velocities (in the total center-of-mass system) of the intermediate products are $\mathbf{v}'_{\alpha_1}$ and $\mathbf{v}'_{^8\text{Be}}$, with $\mathbf{v}''_{\alpha_2}$ and $\mathbf{v}''_{\alpha_3}$ being the velocities of the two breakup α particles in the rest system of $^8\text{Be(g.s.)}$. This diagram shows the breakup emission cone and the symmetric dichotomy in the energy $E_{kin}^{(4)}$ for every φ within this cone. A permissible reversal of this diagram then indicates that, in all, *four* points appear on the kinematic curve, which represents reactions proceeding via the ^{8}Be ground state, namely, A, B, C, and D in Fig. 11-6. Furthermore, when φ (or θ) corresponds to the laboratory recoil direction of the ^{8}Be(g.s.) system, either detector may give pulses due to registering *both* breakup α's. These are denoted by E and F, which are not confined to the kinematic curve, since all three α's are captured in such events.

If, on the other hand, the reaction proceeds via a highly excited state (e.g., $E^* = 16.92$ MeV) of ^{8}Be, coincidences between α_1 and one of the breakup α's can be observed at all angles, giving unique kinematic solutions evinced as but two points along the kinematic curve, e.g., for θ_{α_1} and φ_{α_1}. However, the breakup emission cone is now wider ($\Delta\varphi \simeq 66°$), as can be seen from Fig. 11-5(b) in which $\mathbf{v}'_{s_{Be}}$ depicts the velocity vector of ^{8}Be(16.92) in the total center-of-mass system and $\mathbf{v}''_{\alpha_2}$, $\mathbf{v}''_{\alpha_3}$ the vectors for the two breakup α's in the rest system of ^{8}Be(16.92). This doubles the number of corresponding kinematic points to *four*, which are found to coincide almost exactly with those for the ^{8}Be(g.s.) sequence over a wide range of incident energies when θ and φ correspond with the ^{8}Be(g.s.) recoil directions (e.g., $\theta = +60°$, $\varphi = -100°$). The kinematic spectrum obtained by applying the signals from two solid-state α detectors to the x and y axes of an xy oscilloscope clearly shows the kinematic curve with points corresponding to various states of ^{8}Be and segments for the resonances in the α-α scattering system, as indicated in Fig. 11-7(a). By taking similar observations with φ displaced away from the ^{8}Be(g.s.) recoil axis, the main contribution is seen to stem from the sequence via ^{8}Be(16.92), as can be inferred from the trace in Fig. 11-7(b), taken with $\varphi = -120°$, outside the ^{8}Be(g.s.) breakup cone. The regions H and I correspond to an ($L=2$) α-α resonance. From an empirical projection of the kinematic curve on to the x and y axes separately, shown in Fig. 11-6, the kinematic curve can be reconstructed, with points and segments which correspond to various states of ^{8}Be and α-α scattering resonances. The lengths of the segments roughly correspond to the widths of the resonances, and therefore furnish information on Γ. The potentialities of kinematic methods of this type for the study of resonances are currently being exploited to an ever-increasing extent in nuclear physics (see, e.g., [Cha 67]) following their success in the analysis of strange-particle decay processes with the aid of DALITZ DIAGRAMS [Da 57a, Da 62a, Da 63a].

Another very interesting recent application of energy-correlation techniques in SEQUENTIAL DECAY to elucidate the lifetimes of short-lived intermediate nuclear states in a direct manner (i.e., without invoking the relation $\Gamma = \hbar/\tau$) is to be found in PROXIMITY SCATTERING studies [Fo 62, La 65, La 66]. These, together with bremsstrahlung interference methods [Eis 60, Eis 63, Ha 63b] constitute the only alternative direct way of determining lifetimes of nuclear states in the vicinity of 10^{-20} sec (cf. Fig. 6-2) which constitute an

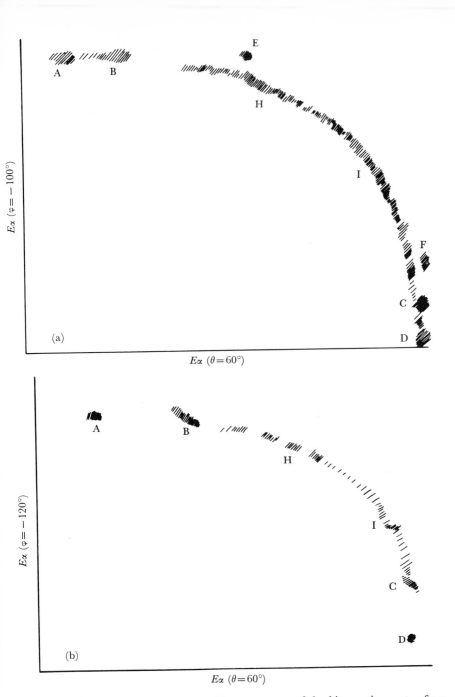

Fig. 11-7. Reconstruction of *x-y* oscilloscope traces of the kinematic spectra from two α-particle detectors, in which the shaded areas correspond to discrete excited states in ^{8}Be from the ^{9}Be(^{3}He, α) reaction at a bombarding energy of 3.0 MeV, followed by breakup of ^{8}Be into two α particles (from [Mo 64a]).

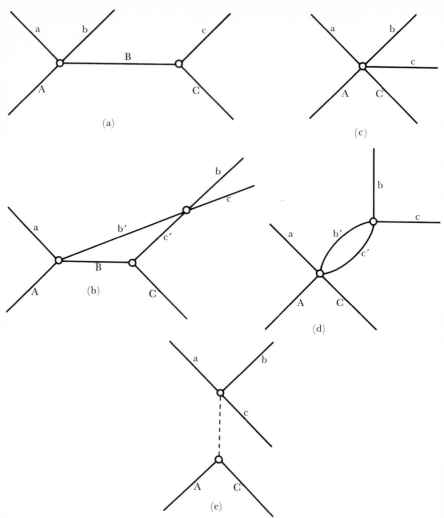

Fig. 11-8. Feynman graphs depicting various interaction processes associated with proximity scattering: (a) sequential decay; (b) sequential decay + proximity scattering; (c) "direct" interaction; (d) "direct" interaction + final-state interaction of particles 1 and 2; (e) peripheral or quasi-free processes (from [La 66]). (Used by permission of North-Holland Publishing Co.)

intermediate system in a two-stage sequential decay process of the following form:

$$a + A \longrightarrow \begin{array}{l} B^* + b(+Q_1) \\ \tau \bigg\lfloor_{\longrightarrow} C + c(+Q_2) \end{array}$$

$$(11\text{-}34)$$

In certain instances, if b and c emerge in the same direction and the velocity of the subsequently emitted particle c exceeds that of its predecessor b, it may overtake b and give rise to an interaction which can take the form of elastic scattering in the vicinity of the product nucleus ("proximity scattering") or some other final-state interaction. From the correlation between the particles after scattering, the lifetime of the unstable short-lived state B^* can be derived. To indicate this, we substitute simple classical considerations which lead to essentially the same result as more sophisticated quantum-mechanical

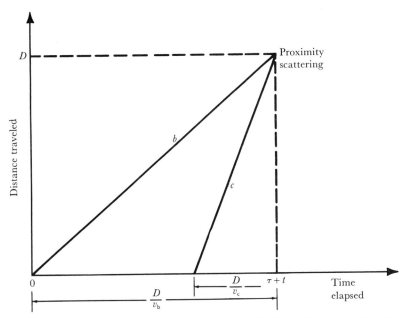

Fig. 11-9. Space-time diagram indicating the world lines of particles b and c up to the moment of collision in proximity scattering.

calculations in which processes represented by the Feynman diagrams of Fig. 11-8 have to be evaluated.

If the encounter between b (emitted with velocity v_b) and c (emitted in the same direction with velocity $v_c > v_b$ from essentially the same origin but a time τ later) takes place within a distance D after a time t following the decay of B^*, as indicated by the world lines in Fig. 11-9, it is evident that

$$\tau = (D/v_b) - (D/v_c) \tag{11-35}$$

The magnitudes of v_b and v_c are determined by the kinematics of the decay, which are known. The value of D can be deduced from a knowledge of the probability W for such a scattering process to occur. The theoretical expression for this is, in zeroth order approximation,

$$W \cong \frac{\sigma}{4\pi D^2} \tag{11-36}$$

where σ, which denotes the integrated scattering cross section at the particular relative scattering energy, is a known quantity (from measurements or from effective-range theory). Experimentally, W is obtained from the shape of

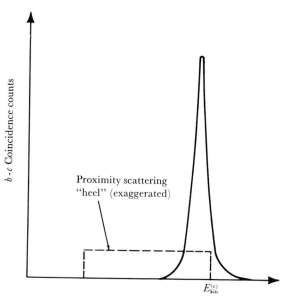

Fig. 11-10. Idealized energy spectrum of b-c coincidences, showing the rectangular spectral contribution to the principal peak if proximity scattering occurs.

the line depicting b-c coincidence intensity as a function of the kinetic energy of c for various relative angles $\varphi_{bc} = \varphi_b - \varphi_c$. The influence of proximity scattering on the symmetric natural (unperturbed) line shape is to add a "heel" to the low-energy side when φ_{bc} is smaller than a critical angle, fixed by kinematics, which also determine the heel width (Fig. 11-10). In practice, the distortion is not a step function, but has the form depicted in Fig. 11-11, which was calculated for the case of the reaction

$$d + {}^{12}C \longrightarrow {}^{13}N^* + n \qquad (Q_1 = -3.82 \text{ MeV})$$
$$\tau_{{}^{13}N^*} \longrightarrow {}^{12}C + p \qquad (Q_2 = +1.59 \text{ MeV}) \tag{11-37}$$

as studied experimentally [La 65, La 66] at an incident energy $E_d = 5.39$

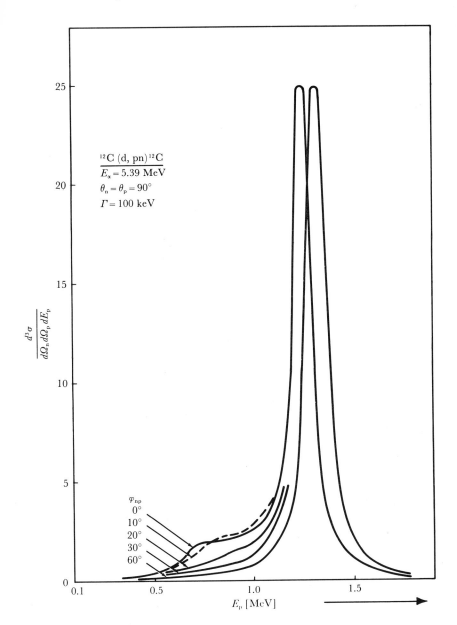

Fig. 11-11. Calculated actual energy spectra for various n-p angles φ_{np} when proximity scattering takes place in the reaction $^{12}C(d, pn)^{12}C$ at $E_d = 5.39$ MeV (from [La 66]). (Used by permission of North-Holland Publishing Co.)

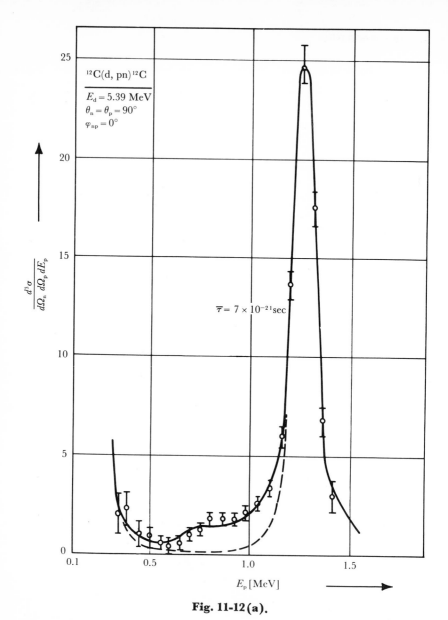

Fig. 11-12(a).

Fig. 11-12. Coincidence spectra between protons and neutrons from the reaction $^{12}C(d, pn)$ at 5.39 MeV for various azimuths: (a) $\varphi = 0°$; (b) $\varphi = 10°$; (c) $\varphi = 30°$, while $\theta_p = \theta_n = 90°$. The progressive disappearance of the "heel" due to proximity scattering vindicates the theoretical expectation. The dashed curves represent the line shape in absence of proximity scattering, while the solid curves are theoretical fits to the data, indicating a mean lifetime for the excited intermediate state in $^{13}N*$ of $\bar{\tau} = 7 \times 10^{-21}$ sec (from [La 66]). (Used by permission of North-Holland Publishing Co.)

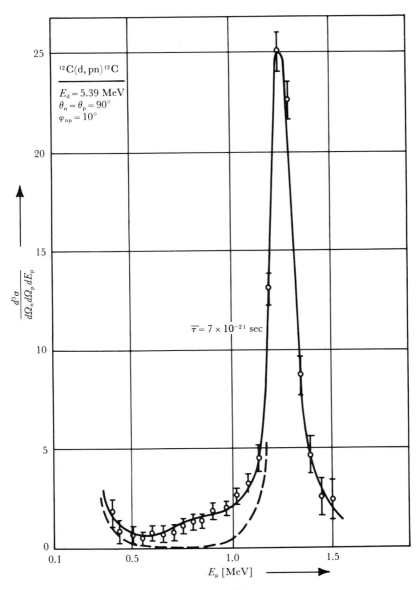

Fig. 11-12(b).

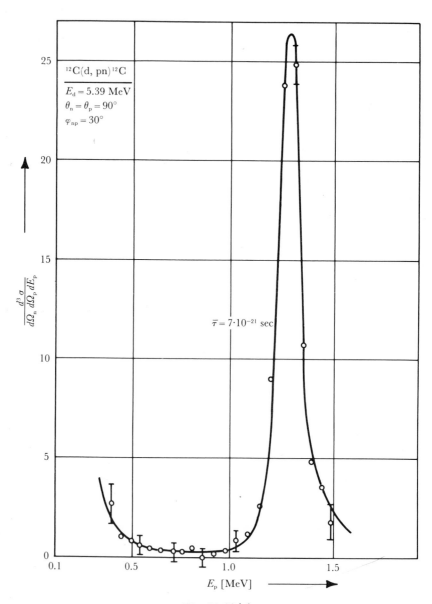

Fig. 11-12(c).

MeV (lab). Coincidences between protons detected in a silicon semiconductor counter (arranged to give a measure of E_p through pulse-height analysis) and neutrons detected in a plastic scintillator (using a time-of-flight technique to determine E_n) were measured at various relative azimuths φ_{np}. The results, as shown in Figs. 11-12(a, b), clearly indicate the occurrence of proximity scattering and permit an estimate of the lifetime of the $^{13}N^*$ system to be made. The empirical finding that (approximately) $W \approx 9 \times 10^{-4}$ yields a surprisingly large value for D, namely, $D \approx 340$ fm (about 50 nuclear diameters) and thence the result

$$\tau_{^{13}N^*} = (0.6^{+0.3}_{-0.2}) \times 10^{-20} \text{ sec} \tag{11-38}$$

for the excited intermediate state of $^{13}N^*$, which has been identified with two adjacent levels at an excitation energy of 3.51 and 3.56 MeV. The error limits represent estimated theoretical uncertainties. Theoretical curves giving best fits to the data result on taking $\bar{\tau} = (0.7 \pm 0.1) \times 10^{-20}$ sec, which compares well with the most recent indirect determination from level widths [Ar 66] of the lifetimes of these two levels as 1.05×10^{-20} sec and 0.89×10^{-20} sec, respectively.

11.2. General Features of Reaction Cross Sections

11.2.1. PROBABILITY CONSIDERATIONS

By extending the consideration of transition probability to nuclear reactions, building upon the foundations already laid down in Sections 3.4, 3.5, and 8.3, one can derive rather general information as to the overall features evinced by nuclear reaction cross sections.

We commence by applying the formalism to the reaction $A(a, b)B$ in which a particle a impinging with velocity v_a upon a stationary target nucleus which has a spin J_A gives rise to an outgoing particle b with velocity v_b and total angular momentum j_b, the residual nucleus B of spin J_B being assumed to be left practically at rest. The reaction is supposed to follow a statistical law and *not* to involve resonances (which will be considered separately in due course).

The transition probability per unit time, given by Fermi's "Golden Rule No. 2", Eq. (3-67),

$$W_{fi} = \frac{2\pi}{\hbar} |H'_{fi}|^2 \frac{dn}{dE_f} \tag{11-39}$$

depends directly upon the accessible final-state density dn/dE_f. For a continuum of final states,

$$\frac{dn}{dE_f} \to \infty \quad \text{and} \quad |H'_{fi}| \to 0 \tag{11-40}$$

so that it becomes necessary to resort to box normalization (see [Schi 55, p. 199] or [La 58a, pp. 15 and 144 ff.]) to obtain a definite, nonzero transition probability.

Just as in Section 8.3.3, one can express the number of states dn available to a particle b which has a momentum within the range p_b to $p_b + dp_b$ confined to a spatial volume V as

$$dn = \frac{4\pi p_b^2 \, dp_b}{h^3} \, V \, (2j_b + 1) \, (2J_B + 1) \tag{11-41}$$

The extra factors in brackets have been included to take cognizance of final-state multiplicity occasioned by the $(2J + 1)$-fold degeneracy of states of spin J. Each of these contributes individually to dn/dE_f. Using the relativistically invariant energy-momentum relation

$$dE_f = v_b \, dp_b \tag{11-42}$$

we can write

$$\frac{dn}{dE_f} = \frac{4\pi}{h^3} \frac{p_b^2}{v_b} \, V \, (2j_b + 1) \, (2J_B + 1) \tag{11-43}$$

which, substituted in (11-39), gives

$$W_{fi} = \frac{1}{\pi \hbar^4} \frac{p_b^2}{v_b} \, V \, |H_{fi}'|^2 \, (2j_b + 1) \, (2J_B + 1) \tag{11-44}$$

From this expression we can derive a general formula for the transition cross section. In terms of n_a, the density of incident particles a, the incident flux is $n_a v_a$ and hence the cross section is simply

$$\sigma = \frac{W_{fi}}{n_a v_a} \tag{11-45}$$

Assuming there to be but *one* particle of type a in the volume V we can set

$$n_a = \frac{1}{V} \tag{11-46}$$

and thus arrive at the final relation

$$\sigma = \frac{1}{\pi \hbar^4} \frac{p_b^2}{v_a v_b} \, |VH_{fi}'|^2 \, (2j_b + 1) \, (2J_B + 1) \tag{11-47}$$

Strictly speaking, this refers to the center-of-mass system, but if the nuclei A and B are sufficiently massive, the CM and laboratory system will essentially coincide and the formula also holds for laboratory velocities and momenta. The terms $|VH_{fi}'|^2$ and $(2j_b + 1) \, (2J_B + 1)$ may be unknown in certain instances, but it can be assumed that $|H_{fi}'|^2$ does not vary appreciably for the different accessible states within a fairly narrow energy range at low incident energy, and therefore such unknown terms can be treated as constants, e.g.,

$$\sigma \sim \frac{p_b^2}{v_a v_b} \tag{11-48}$$

More explicitly, $|H_{fi}'|^2$ contains nuclear Coulomb barrier penetrabilities which can modify σ. The matrix element $|H_{fi}'|$ involves a product of initial and final state wave functions. These are, if the incident particle a carries a charge, each reduced in amplitude at the nucleus by the transparency $(e^{-G_a})^{1/2}$, where G_a is the Gamow factor (cf. Sections 7.4.1 to 7.4.3) for particle a at an energy E_a (leaving orbital momentum considerations aside in order to avoid complicating the Gamow factor—i.e., we essentially confine the treatment to S waves). Hence in the case of absorption and emission of positively charged particles, we find that

$$|H_{fi}'|^2 \sim e^{-(G_a+G_b)} \tag{11-49}$$

These general considerations will now be applied to various types of nuclear reaction.

11.2.2. General Cross-Sectional Trend for Elastic Neutron Scattering

Since the (n, n) scattering is elastic it follows that

$$v_a \cong v_b \tag{11-50}$$

when the nuclear recoil energy is negligible. Hence

$$\frac{p_b^2}{v_a v_b} \to \frac{p_n^2}{v_n^2} = m_n^2 = \text{const} \tag{11-51}$$

and, accordingly, at low energies where $|H_{fi}'|^2$ can be taken as constant, it follows that

$$\sigma_{(n,n)} \approx \text{const} \tag{11-52}$$

as confirmed by experimental findings at energies up to about 5 eV (Fig. 11-13).

11.2.3. Characteristic Cross Section for Exothermic Reactions Induced by Low-Energy Neutrons

These embrace reactions of the type (n, γ), [i.e., radiative capture], (n, p), (n, α), and (n, f), [i.e., neutron-induced fission], which have fairly large positive Q values. Thus for moderate incident neutron energies ($\sim$ eV) v_b is sensibly constant, and hence so is p_b, since small variations in $v_a \equiv v_n$ cannot appreciably affect v_b. The G dependence in $|H_{fi}'|^2 \sim e^{-(G_a+G_b)}$ also vanishes, since G_a is zero for a neutral particle and e^{-G_b} is roughly constant, as the outgoing particle energy hardly varies at all. We can therefore set

$$\frac{p_b^2}{v_a v_b} \rightarrow \frac{1}{v_n}\frac{p_b^2}{v_b} = \text{const} \times \frac{1}{v_n} \tag{11-53}$$

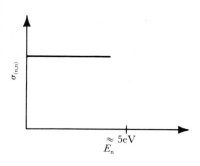

Fig. 11-13. Constant excitation function for elastic scattering of low-energy neutrons.

Fig. 11-14. The "$1/v$ law" for the excitation function for exothermic (n, b) reactions induced by low-energy neutrons.

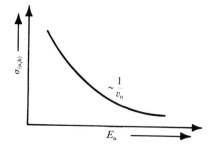

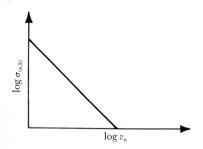

Fig. 11-15. Linear log-log plot of cross section *vs* neutron velocity for low-energy neutron-induced exothermic reactions (n, b).

and thence obtain the result known as the "$1/v$ LAW" (Fig. 11-14):

$$\sigma_{(n, b)} \sim \frac{1}{v_n} \tag{11-54}$$

Since this can be written in the form

$$\log \sigma_{(n, b)} = \text{const} - \log v_n \tag{11-55}$$

it follows that the intercepts of the log-log plot depicted in Fig. 11-15 furnish the value of the proportionality constant for any given exothermic reaction.

11.2.4. CHARACTERISTIC CROSS SECTION FOR INELASTIC NEUTRON SCATTERING

This (n, n') process is an instance of an endothermic reaction in which v_a exceeds v_b and the nucleus A is left with an excitation energy

$$E^* = -Q \approx 1 \text{ MeV} \tag{11-56}$$

This betokens a comparatively high reaction threshold. Above this threshold, a fairly small change in the incident energy E_n manifests itself by a large change in $E_{n'}$. Accordingly, even though v_n can be regarded as constant in the immediate vicinity of the threshold, the velocity of the scattered neutron obeys the relation

$$(v_{n'})^2 = \frac{\text{excess of energy above threshold}}{\frac{1}{2}m_n} \tag{11-57}$$

i.e.,

$$v_{n'} \sim (E_n - E^*)^{1/2} \tag{11-58}$$

and therefore

$$\frac{p_b^2}{v_a v_b} \rightarrow \frac{p_{n'}^2}{v_n v_{n'}} \sim v_{n'} \sim (E_n - E^*)^{1/2} \tag{11-59}$$

whence near threshold

$$\sigma_{(n, n')} \sim (E_n - E^*)^{1/2} \tag{11-60}$$

as illustrated in Fig. 11-16.

11.2.5. CHARACTERISTIC CROSS SECTION FOR ENDOTHERMIC NEUTRON-INDUCED REACTIONS LEADING TO EMISSION OF CHARGED PARTICLES

The fact that *charged* particles are emitted in such reactions as (n, p) or (n, α) necessitates a slight modification to the considerations of Section 11.2.4 in that a transparency factor e^{-G_b} has to be included, viz.

$$\sigma_{(n, b)} \sim e^{-G_b}(E_n - E^*)^{1/2} \tag{11-61}$$

with

$$G_b = \left(\frac{8m_b}{\hbar^2}\right)^{1/2} \int (U_b - E_b)^{1/2} \, dr \xrightarrow{\text{high Coulomb barrier}} \frac{2\pi Z_B z_b e^2}{\hbar v_b} \tag{11-62}$$

where

$$U_b \equiv (\text{charge of b}) \times (\text{Coulomb potential of residual nucleus } B) \tag{11-63}$$

This causes the cross-sectional curve to become *convex* to the energy axis, as in Fig. 11-17.

11.2.6. CROSS-SECTION TREND FOR EXOTHERMIC REACTIONS INVOLVING CHARGED INCOMING AND UNCHARGED OUTGOING PARTICLES

Typical examples of such processes are radiative proton or α capture, (p, γ) or (α, γ). Assuming that

$$E_a \ll Q \tag{11-64}$$

it follows that

$$\frac{p_b^2}{v_a v_b} \sim \frac{1}{v_a} \qquad \frac{v_b}{v_a} > > 1 \qquad i.e.\ v_a\ is\ very\ small \tag{11-65}$$

while, moreover, a factor e^{-G_a} has to be included to take account of the transparency of the Coulomb barrier to the incident charged particle:

$$\sigma_{(a, n)} \sim \frac{1}{v_a} e^{-G_a} \tag{11-66}$$

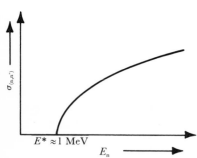

Fig. 11-16. Excitation function for low-energy inelastic neutron scattering.

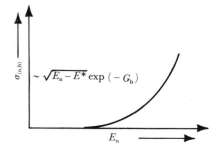

Fig. 11-17. Characteristic trend of the cross section σ with incident neutron energy E_n for endothermic reactions (n, b) in which charged particles b are emitted.

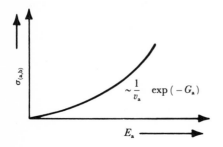

Fig. 11-18. Cross-section trend for exothermic reactions (a, b) which feature charged incident and uncharged emergent particles.

This cross section, depicted in Fig. 11-18, evinces no threshold, but is otherwise qualitatively similar in form to that of the previous section.

11.2.7. CHARACTERISTIC CROSS SECTION FOR EXOTHERMIC REACTIONS WITH CHARGED INCIDENT AND EMERGENT PARTICLES

For exothermic reactions of the type (α, p), there is once again no threshold, and the factor

$$\frac{p_b^2}{v_a v_b} \sim \frac{1}{v_a} \tag{11-67}$$

is now multiplied by the full penetrability factor $e^{-(G_a + G_b)}$. Hence the cross section, given by the relation

$$\sigma_{(a,b)} \sim \frac{1}{v_a} e^{-(G_a + G_b)} \tag{11-68}$$

displays a somewhat stronger dependence upon energy than in the previous case, as shown in Fig. 11-19.

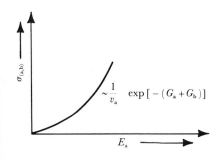

Fig. 11-19. Behavior of the cross section in terms of incident energy for exothermic reactions (a, b) involving charged incident and emergent particles. The energy-dependence is somewhat stronger than would be the case with outgoing *uncharged* particles in consequence of the additional Gamow penetrability factor.

11.3. Detailed Balance Predictions for Inverse Reaction Cross Sections

The ratio of cross sections for mutually inverse processes of the type

$$a + A \rightleftharpoons B + b \tag{11-69}$$

can be adduced from the general expression (11-47).

Noting that for the forward reaction $(i \to f)$

$$\sigma_{\mathrm{fi}} \sim \frac{p_b^2}{v_a v_b} (2j_b + 1)(2J_B + 1) \tag{11-70}$$

while for the reverse reaction $(f \rightarrow i)$

$$\sigma_{\mathrm{if}} \sim \frac{p_{\mathrm{a}}^2}{v_{\mathrm{a}} v_{\mathrm{b}}} (2j_{\mathrm{a}} + 1)(2J_{\mathrm{A}} + 1) \tag{11-71}$$

we see that, since the perturbation Hamiltonian is Hermitian $(|H'_{\mathrm{fi}}| = |H'_{\mathrm{if}}|)$,

$$\frac{\sigma_{\mathrm{fi}}}{\sigma_{\mathrm{if}}} = \frac{p_{\mathrm{b}}^2 (2j_{\mathrm{b}} + 1)(2J_{\mathrm{B}} + 1)}{p_{\mathrm{a}}^2 (2j_{\mathrm{a}} + 1)(2J_{\mathrm{A}} + 1)} \tag{11-72}$$

In this very important relation, the cross sections represent *averages* over the various possible spin states and *sums* of partial cross sections for the various

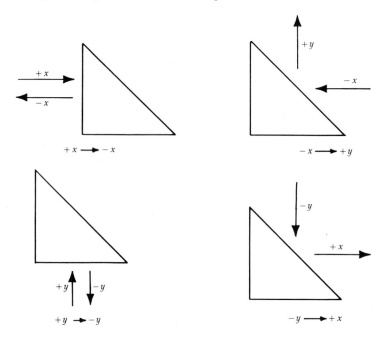

Fig. 11-20. Diagram to assist the visualization of probabilities for scattering on triangular obstacles in x and y directions as an illustration of probability inequality on reversal of the initial and final states of motion.

accessible final states. The formula can be considered to betoken a restricted form of the PRINCIPLE OF DETAILED BALANCE. This principle may be taken to pertain to total or to differential cross sections, provided that in the latter instance the CM angles of emergence are the same for the forward and reverse reactions. The energies also have to be the same in the CM system, taking the Q value into account. Of course, the cross sections and momenta refer to the CM system.

Detailed balance requires the probabilities of a process and its inverse under comparable conditions to be equal, viz.

$$W_{fi} = W_{if} \tag{11-73}$$

This implies that they must be suitably weighted by their respective momentum factors (but not by their statistical spin factors). In place of this rather sweeping principle, Heitler [He 56] has proposed a more limited version which applies to situations in which several spin states enter into consideration, namely, the PRINCIPLE OF SEMIDETAILED BALANCE, according to which the relation

$$\overline{W}_{fi} = \overline{W}_{if} \tag{11-74}$$

holds for the probabilities $\overline{W}$ *averaged over all spin states*.

To render this somewhat clearer, we make use of the following conceptual analogy: Assuming, for simplicity, a system in which the movement of particles is confined to only *two* dimensions, namely only to the $\pm x$, $\pm y$ directions, at uniform speeds, the alteration in their direction of motion when they encounter fixed obstacles depends upon the relative situation, as depicted in Fig. 11-20. We might consider the obstacles to be rigid, perfectly elastic, right-angled isosceles triangles as shown. Clearly, the respective probabilities for each of the four possible transitions under consideration are *nonzero and equal* (since the target cross section appears the same to particles traveling in the $+x$, $-x$, $+y$, or $-y$ directions), while the probabilities of all other transitions (such as, for example, $W_{-x \to +x}$) vanish identically. It therefore follows that

$$W_{+x \to -x} \neq W_{-x \to +x} \tag{11-75}$$

A more restricted form of equality can nevertheless be obtained on noting that if one commences with a beam of particles which impinge, along the $+x$ direction for instance, upon an array of such obstacles as we have considered, they will experience multiple collisions which will eventually yield a state of dynamic equilibrium between the numbers of particles moving in the $\pm x$, $\pm y$ directions finally. Accordingly, the *sum* of transition probabilities from a given initial state i to all possible final states f is equal to the sum of transition probabilities from all states f to that one particular state i, viz.

$$\sum_f W_{fi} = \sum_f W_{if} \tag{11-76}$$

Application of this rule to the above example would furnish the prediction

$$W_{+x \to -x} = W_{-y \to +x} \tag{11-77}$$

For a situation in which particle motion is restricted to a line only (e.g., along the $\pm x$ direction) it is clear that although the principle of detailed balance will hold when the obstacles have a fully symmetric form (e.g., square) it immediately breaks down for asymmetrical obstacles in various possible orientations (Fig. 11-21). Whereas, however, for each individual orientation detailed balance loses its validity, when we build the *average* over the four orientations which can occur, the resultant transition probabilities become equal once more:

$$\overline{W}_{fi} = \overline{W}_{if} \tag{11-78}$$

For example, if we designate the orientations as (1), (2), (3), and (4) and the individual distinct nonzero transition probabilities of Fig. 11-21 as W, we can write

$$\overline{W}_{+z\to-z} \equiv \tfrac{1}{4}(W^{(1)}_{+z\to-z} + W^{(2)}_{+z\to-z} + W^{(3)}_{+z\to-z} + W^{(4)}_{+z\to-z}) \tag{11-79}$$

$$= \tfrac{1}{4}(W^{(1)} + 0 + W^{(3)} + 0)_{+z\to-z} \tag{11-80}$$

$$= \tfrac{1}{2}W_{+z\to-z} \tag{11-81}$$

while for the inverse process we have

$$\overline{W}_{-z\to+z} \equiv \tfrac{1}{4}(W^{(1)}_{-z\to+z} + W^{(2)}_{+z\to+z} + W^{(3)}_{-z\to+z} + W^{(4)}_{-z\to+z}) \tag{11-82}$$

$$= \tfrac{1}{4}(0 + W^{(2)} + 0 + W^{(4)})_{-z\to+z} \tag{11-83}$$

$$= \tfrac{1}{2}W_{-z\to+z} \tag{11-84}$$

and therefore

$$\overline{W}_{+z\to-z} = \overline{W}_{-z\to+z} \tag{11-85}$$

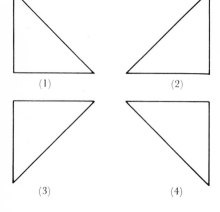

Fig. 11-21. Different orientations of asymmetrical obstacles.

11.3.1. Experimental Investigation of Inverse Reactions

The foregoing considerations have very wide-ranging applicability. To take a simple example from particle physics, the mean lifetime of the π^0 meson can be deduced from detailed-balance considerations for the conjugate processes

$$\pi^0 \rightleftharpoons \gamma + \gamma \tag{11-86}$$

The reverse (creation) process, or its equivalent, is derived by allowing high-energy γ radiation to interact with the electric field of a heavy nucleus and produce π^0's. The measurable creation cross section is related directly to the pions' decay constant, which is thereby found experimentally to correspond to a mean life of $\tau_{\pi^0} = 1/\lambda_{\pi^0} \approx 2 \times 10^{-16}$ sec.

Another instance is furnished by the direct determination of the pion spin.

The principle of detailed balance applied to the conjugate high-energy reactions

$$p + p \rightleftharpoons \pi^+ + d - 137 \text{ MeV} \tag{11-87}$$

when unpolarized beams impinge upon unpolarized targets yields the relation

$$\frac{\sigma_{(p,\pi)}}{\sigma_{(\pi,p)}} = \frac{(2J_d + 1)(2J_\pi + 1)}{(2J_p + 1)(2J_p + 1)} \frac{p_\pi^2}{p_p^2} \tag{11-88}$$

wherein cross sections and momenta are expressed in the CM system. Because of the indistinguishability of the identical protons emerging from the inverse reaction it is customary practice, as in [Sa 58a], to incorporate a factor $\frac{1}{2}$ into

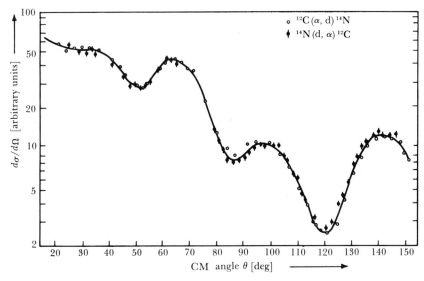

Fig. 11-22. Angular distributions in the center-of-mass system for the mutually inverse reactions $^{12}\text{C} + \alpha \rightleftharpoons {}^{14}\text{N} + d$ at the commensurate energies $E_\alpha = 41.7$ MeV, $E_d = 20.0$ MeV (lab) (from [Bo 59]).

the ensuing measured cross section; e.g., the net proton counting rate is halved in the evaluation of the cross section which is cited in the literature.

Substituting the intrinsic spin values $J_d \triangleq s_d = 1$, $J_p \triangleq s_p = \frac{1}{2}$ we derive the cross-section ratio

$$\frac{\sigma_{(p,\pi)}}{\sigma_{(\pi,p)}} = \frac{3(2J_\pi + 1)}{2} \frac{p_\pi^2}{p_p^2} \tag{11-89}$$

which can be measured at known energies (p_π, p_p known) and thence provide the result

$$J_{\pi^+} \triangleq s_{\pi^+} = 0 \tag{11-90}$$

From indirect evidence, one can confirm this finding, not only for π^+ mesons, but also for π^0's and π^-'s, thereby identifying pions as *spinless bosons*.

Turning now to nuclear physics, we remark that frequent use is made of the principle of detailed balance to assess reaction cross sections from those measured for the reverse process. The self-consistency in the values so deduced provides strong support for the validity of the underlying concept.

An investigation aimed explicitly at examining the extent to which the principle might be supposed to hold in a direct nuclear process was undertaken by Bodansky *et al.* [Bo 59], who measured the relative angular distributions of reaction products from the forward and inverse direct interactions

$$^{12}\text{C} + \alpha \rightleftharpoons {}^{14}\text{N} + \text{d} \tag{11-91}$$

at matched energies ($E_\alpha = 41.7$ MeV$_{\text{lab}}$, $E_\text{d} = 20.0$ MeV$_{\text{lab}}$) and obtained

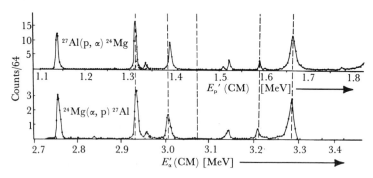

Fig. 11-23. Excitation functions measured by [Ka 52b] at matched energies for the mutually inverse reactions $^{27}\text{Al} + \text{p} \rightleftharpoons {}^{28}\text{Si}^* \rightleftharpoons {}^{24}\text{Mg} + \alpha$.

quite striking agreement between the arbitrarily normalized results (Fig. 11-22). The outcome can be interpreted to vindicate detailed balance to at least 96 percent.

At lower energies, reactions may not involve a direct interaction of the type considered above, but may proceed via an unstable intermediate state in a so-called COMPOUND NUCLEUS [Bo 36]. In order to establish the extent to which detailed balance holds for such processes, studies have been carried out at matched energies for the reactions

$$^{27}\text{Al} + \text{p} \rightleftharpoons {}^{28}\text{Si}^* \rightleftharpoons {}^{24}\text{Mg} + \alpha + 1.613 \text{ MeV} \tag{11-92}$$

Kaufmann *et al.* [Ka 52b] have measured yields in the forward and inverse reaction which, as can be seen from Fig. 11-23, agree remarkably well when allowance is made for the nuclear spins and for the energy matching ($Q = 1.613$ MeV) in the CM system.

A still more striking measure of agreement has been observed more recently [von W 66] with the same forward and reverse reactions at rather higher energies around $E_p = 10.3 \text{ MeV}_{\text{lab}}$ and $E_\alpha = 13.4 \text{ MeV}_{\text{lab}}$. These were chosen to correspond with a highly structured region in the excitation function (Fig. 11-24) for the forward reaction which, together with its inverse counterpart, was examined in fine detail (5-keV steps). The two sets of results, normalized to each other at the peak, are reproduced in Fig. 11-25; they yield a peak-to-valley ratio of 144 ± 4 in the forward (p, α) reaction and 140 ± 10 in the reverse (α, p) reaction. A theoretical analysis of these preliminary results, as undertaken by Ericson [Er 66], furnished an upper limit to possible violation of detailed balance of about 0.1–0.2 percent, but

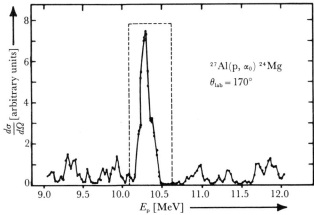

Fig. 11-24. Excitation function taken at 170° by [von W 66] for the reaction $^{27}\text{Al}(p, \alpha_0)^{24}\text{Mg}$ between 9 and 12 MeV. The framed central region was selected for high-precision measurement in forward and inverse sense. (Used by permission of North-Holland Publishing Co.)

it may be [Ma 66] that this is too low by a factor of 10 and that a more realistic figure would appear to be about 1 percent. Currently, further experimental and theoretical investigations are in progress (see, e.g., [Th 68]).

Heavy-ion reactions would appear to offer favorable possibilities for investigations of this type. An unpublished survey [Le 64] of processes that might be accessible to experiments with the same kind of Van de Graaff accelerator as has been employed in the above studies indicated that the following reactions could possibly be used:

$$^{12}\text{C} + {}^{16}\text{O} \rightleftharpoons {}^{14}\text{N} + {}^{14}\text{N} \tag{11-93}$$

$$^{12}\text{C} + {}^{20}\text{Ne} \rightleftharpoons {}^{16}\text{O} + {}^{16}\text{O} \tag{11-94}$$

$$^{14}\text{N} + {}^{9}\text{Be} \rightleftharpoons {}^{16}\text{O} + {}^{7}\text{Li} \tag{11-95}$$

$$^{14}\text{N} + {}^{11}\text{B} \rightleftharpoons {}^{16}\text{O} + {}^{9}\text{Be} \tag{11-96}$$

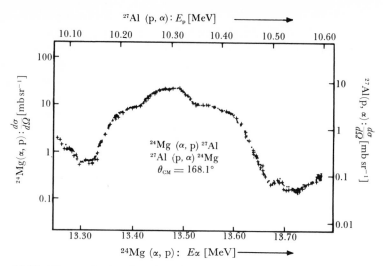

Fig. 11-25. The excitation functions for the reactions ^{27}Al + p $\rightleftharpoons$ ^{24}Mg + ^{4}He at matched energies measured in 5 keV steps in a highly structured region. The perfect match establishes the validity of the principle of detailed balance to a high degree of confidence (from [von W 66]). (Used by permission of North-Holland Publishing Co.)

and that in particular the cycle

$$^{11}B + {}^{12}C \rightleftharpoons {}^{16}O + {}^7Li$$
$$^{14}N + {}^9Be$$
$$(11\text{–}97)$$

could provide a valuable means of counterchecking results.

When the particle beams or the targets are polarized, the principle of (semi-) detailed balance still operates, but no longer furnishes so simple a relation as (11-72). Nevertheless, a valuable way of studying detailed balance or its closely related principle of time-reversal invariance in strong interactions [He 59, Ja 59] is to undertake experiments on oriented nuclei with polarized beams. Such measurements have been performed by directing a polarized neutron beam on to an aligned ^{49}Ti target [Ka 66]. They reduced the upper limit on possible violation from a previous estimate [Fu 64] of 4 percent to a maximum of 2 percent.

Rather than branching off to discuss time-reversal invariance further, we defer this for consideration in Volume II and instead go on to examine resonance reactions, which fall directly within the context of this concluding chapter and shed light upon the compound nucleus concept, which, briefly introduced in the present section, will also be given more attention in Volume II.

11.4. Resonance Reactions

11.4.1. RESONANCE ANOMALIES IN EXCITATION FUNCTIONS

The occurrence of resonances introduces discontinuities into the otherwise smooth trend of reaction cross sections as a function of energy. Such irregularities indicate that instead of being roughly constant, the matrix element

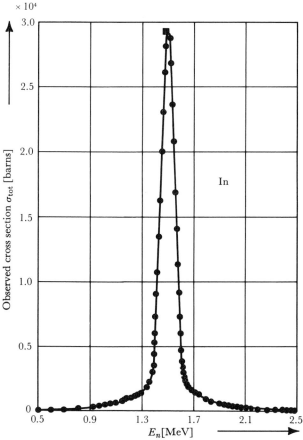

Fig. 11-26. Sharp resonance in the slow-neutron capture cross section evinced by natural indium (containing 95.8 percent of the isotope [115]In responsible for the photo-capture resonance at $E_0 = 1.457$ MeV). A Lorentzian curve has been drawn through the measured points, each of which constitutes an average over at least three runs (the square point has been determined with especially high statistical accuracy). The data were obtained by Landon and Sailor [La 55c] with a crystal spectrometer and analyzed using a modified form of Breit-Wigner single-level theory [Be 37a, Be 37b] designed to take the motion of the target atoms (treated as free atoms having a Maxwellian velocity distribution) into account. This yielded a still higher peak value and narrower width than that observed experimentally (for example, whereas the observed full width at half maximum (FWHM) was $\Gamma_{obs} = 0.109$ eV, the theoretical resonance parameters [Hu 58] for [115]In were established to be: $E_0 = 1.457$ eV, $\Gamma_n = 0.00304$ eV, $\Gamma_\gamma = 0.072$ eV, $J_0 = \frac{9}{2}$, $J_1 = 5$, whence $\sigma_0 = 3.81 \times 10^4$ b and $\Gamma_{th} = 0.075$ eV).

$|H_{fi}|^2$ fluctuates with energy—a phenomenon which may be observed in all reactions that proceed via an intermediate compound nucleus state, but which is particularly pronounced in such processes as radiative capture of slow neutrons. Thus the In(n, γ) reaction evinces a sharp resonance at $E_n = E_0 = 1.457$ eV, as shown in Fig. 11-26. This may be ascribed to S-wave neutron capture by the isotope ^{115}In (relative abundance $= 95.8$ percent; g.s. spin $J_0 = \frac{9}{2}$) to form an excited state (spin $J_1 = 5$) of ^{116}In. The data are fitted by a curve which is symmetrical in the neighborhood of the resonance and has the form given by the expression

$$\sigma_R \sim 1/(E - E_0)^2 \tag{11-98}$$

An explanation for these features is offered by postulating that a compound nucleus C is formed in an excited state as an intermediate step of the overall transition, rendering it a *second-order* process,

$$a + A \rightarrow C^* \rightarrow B + b \tag{11-99}$$

Applying second-order perturbation theory, the transition probability per unit time can be derived along analogous lines to those indicated in Section 3.4. For reactions proceeding from an initial state i by way of an intermediate "compound" state C to a final state f, this probability is given by the Fermi "GOLDEN RULE NO. 1" [Fe 50] as

$$W_{fi} = \frac{2\pi}{\hbar} \left| \frac{H'_{i \rightarrow C} H'_{C \rightarrow f}}{E_i - E_C} \right|^2 \frac{dn}{dE_f} \tag{11-100}$$

Thence, setting the volume V to unity in (11-47) we derive the cross section for the overall process,

$$\sigma_{i \rightarrow C \rightarrow f} = \frac{1}{\pi \hbar^4} \frac{p_b^2}{v_a v_b} \left| \frac{H'_{i \rightarrow C} H'_{C \rightarrow f}}{E_i - E_C} \right|^2 (2j_b + 1)(2J_B + 1) \tag{11-101}$$

and clearly this gives a sharp resonance when $E_i \rightarrow E_C$. In practice, instead of becoming infinite, the cross section rises abruptly to a large magnitude. (We failed to make allowance in the above simplified treatment for the short lifetime $\tau_C \approx 10^{-17}$ sec of the compound state.)

11.4.2. BREIT-WIGNER FORMULA

The shape of the resonance peak can be described by a *Lorentzian distribution*,

$$f(E) = K/[(E - E_0)^2 + (\Gamma/2)^2] \tag{11-102}$$

where K is a constant which will be determined later, E is the incident energy (in the vicinity of the resonance energy E_0), and Γ is the total width (FWHM) of the compound state, made up of the sum of the partial widths Γ_i (see

Section 6.2),

$$\Gamma = \sum_i \Gamma_i \qquad (11\text{-}103)$$

The inclusion of this term, which corresponds to a net lifetime $\tau = \hbar/\Gamma$ of the compound state, prevents the amplitude of $f(E)$ becoming infinite when $E \to E_0$. At resonance, σ attains its maximum value as given by the absolute cross section for incident radiation having a single sharp energy E_0 (rationalized de Broglie wavelength $\lambdabar$) and orbital angular momentum l (cf. Section 3.5.5),

$$\sigma_l = \pi \lambdabar^2 (2l + 1) \qquad (11\text{-}104)$$

Then, in the neighborhood of resonance the cross section for formation of a compound-nucleus state is given by the expression

$$\sigma_{CN} = \sigma_l\, T f(E) \qquad (11\text{-}105)$$

where the transparency T (cf. Section 7.4.1) has been inserted to take account of CN barrier penetration at the incident energy E. It can in principle range from 0 to 1 and hence, since $\sigma_{CN} \leqslant \sigma_l$, it follows that $f(E) \leqslant 1$. The function $f(E)$ can be normalized through the introduction of the mean level separation D in the compound system:

$$(1/D) \int_{-\infty}^{\infty} f(E)\, dE = 1 \qquad (11\text{-}106)$$

which corresponds to setting the *mean* value of $f(E)$ to unity after averaging over several resonances whose energy spacing is taken to have a constant value D. Using (11-106), we can eliminate the hitherto unknown constant K from (11-102), since

$$1 = \frac{1}{D}\frac{K}{\tfrac{1}{2}\Gamma}\left[\arctan\left(\frac{E - E_0}{\tfrac{1}{2}\Gamma}\right)\right]_{E=-\infty}^{E=+\infty} = \frac{2K}{D\Gamma}\left(\frac{\pi}{2} + \frac{\pi}{2}\right) \qquad (11\text{-}107)$$

i.e.,

$$K = D\Gamma/2\pi \qquad (11\text{-}108)$$

Also, we can use the theory presented in Section 7.4.3 to derive an expression for the transparency T in terms of a partial width Γ_a for formation (or decay) of the state C^* through the channel corresponding to particle a. This states that

$$\lambda_a \equiv \frac{1}{\tau_a} = \frac{\Gamma_a}{\hbar} = T \times (\text{repetition rate}) \qquad (11\text{-}109)$$

To derive the repetition rate, we proceed from the net CN wave function Ψ, which can be expressed as a sum of individual wave functions ψ describing the respective state of each energy niveau in the vicinity of the excitation energy E^* at, or near, resonance:

$$\Psi = \sum_{n=1}^{N} a_n \psi_n \exp\left(-\frac{iE_n t}{\hbar}\right) \qquad (11\text{-}110)$$

The exponential term in the above expression describes its behavior with time, while the remaining spatial part consists of the stationary energy eigenfunctions ψ_n of each unperturbed state of energy E_n multiplied by the respective wave-function amplitude (weighting) a_n. The summation, which extends over an appropriate number (N) of (spin-independent) states n, yields the eigenfunction Ψ for the perturbed state which, assuming

$$E_n = \epsilon_0 + nD \tag{11-111}$$

can be written as

$$\Psi = \left[\exp\left(-\frac{i\epsilon_0 t}{\hbar}\right) \right] \left[\sum_{n=1}^{N} a_n \psi_n \exp\left(-\frac{it}{\hbar} nD\right) \right] \tag{11-112}$$

in which the second exponential term is periodic, of period

$$\Delta t = \frac{2\pi\hbar}{D} = \frac{h}{D} \tag{11-113}$$

indicating that at times t, $t + (h/D)$, $t + (2h/D)$, ..., etc. the configuration of the state Ψ is reproduced. In other words, the repetition rate is D/h.

Substituting this result in (11-109), we find

$$\frac{\Gamma_a}{\hbar} = T\left(\frac{D}{2\pi\hbar}\right) \tag{11-114}$$

i.e.,

$$T = \frac{2\pi\Gamma_a}{D} \tag{11-115}$$

Hence, from (11-102), (11-104), (11-105), (11-108), and (11-115) we derive the cross section for *formation* of a compound nucleus near resonance,

$$\sigma_{CN} = \pi\lambda^2(2l+1) \frac{\Gamma_a \Gamma}{(E - E_0)^2 + (\Gamma/2)^2} \tag{11-116}$$

Of the various exit channels from the compound nucleus to all accessible final states, that going to the state f (composed of the reaction products $B + b$) will occur with a probability given by the ratio Γ_b/Γ. Hence we may set

$$\sigma_{(a,b)} = \sigma_{CN} \frac{\Gamma_b}{\Gamma} = \pi\lambda^2(2l+1) \frac{\Gamma_a \Gamma_b}{(E - E_0)^2 + (\Gamma/2)^2} \tag{11-117}$$

for the (a, b) reaction cross section near resonance.

Through a very considerably simplified approach, we have thus obtained the Breit-Wigner single-level formula, a more rigorous derivation of which entails the use of dispersion relations. The label "single-level" takes cognizance of the inherent assumption that only *one* resonance level of the compound nucleus is involved in the net reaction (a, b). One should note that $\sigma_{(a,b)}$ in the above form does *not* include contributions to the total cross section arising from *non*resonant reactions. The formula applies exclusively to resonance processes, and, e.g., it can also be used to describe inelastic resonance scattering (a, a'). However, it cannot be used as it stands for the special case of elastic resonance scattering (a, a), but has to be modified as indicated in Section 11.4.4.

11.4.3. RESONANCE CROSS SECTION NOMENCLATURE

The cross section which is derived from selecting one particular entrance channel and summing over all possible resonance exit channels *excepting the elastic outgoing a-channel* is termed the REACTION CROSS SECTION σ_r. This has the value

$$\sigma_r = \sum_{\substack{b \\ \text{including } b = a' \\ \text{excluding } b = a}} \sigma_{(a,b)} = \sigma_{CN} \sum_b \frac{\Gamma_b}{\Gamma} = \sigma_{CN}\left(1 - \frac{\Gamma_a}{\Gamma}\right) \qquad (11\text{-}118)$$

On the other hand, one can include the ELASTIC CROSS SECTION

$$\sigma_{el} \equiv \sigma_{(a,a)} \qquad (11\text{-}119)$$

in the summation, and thereby obtain the TOTAL CROSS SECTION

$$\sigma_{tot} = \sigma_r + \sigma_{el} \qquad (11\text{-}120)$$

(we assume throughout that the respective cross sections have been summed over the appropriate orbital momenta l).

It is important to note that of these *only σ_{tot} permits an unambiguous comparison between theory and experiment.* Its components σ_r and σ_{el} are in the current theoretical framework not directly calculable.

On the one hand, the experimental elastic cross section σ_{el} actually comprises two terms, namely,

(i) the COMPOUND ELASTIC CROSS SECTION $\sigma_{compound\,el}$ which describes the probability of *compound-elastic scattering,* also called *resonance scattering,* proceeding via an excited compound state, and

(ii) the SHAPE ELASTIC CROSS SECTION $\sigma_{shape\,el}$, which refers to so-called *shape-elastic scattering,* also termed *potential scattering,* in which no appreciable penetration of the target or formation of an intermediate complex occurs. Expressed in terms of *differential* cross sections for a scattering angle θ, the relation reads

$$\sigma_{el}(\theta) = \sigma_{compound\,el}(\theta) + \sigma_{shape\,el}(\theta) \qquad (11\text{-}121)$$

While $\sigma_{shape\,el}(\theta)$ can be calculated directly from a given scattering potential, $\sigma_{compound\,el}(\theta)$ cannot. The *total* cross sections $\sigma_{shape\,el}$ and $\sigma_{compound\,el}$ are, however, calculable from a modified form of Breit-Wigner theory, as indicated in the next section. Before going on to discuss this, we note that it is also not possible to calculate the *reaction* cross section σ_r, although this quantity too is measurable. Only the so-called ABSORPTION CROSS SECTION, defined as

$$\sigma_{\text{absorption}} \equiv \sigma_r + \sigma_{\text{compound el}} \qquad (11\text{-}122)$$

is calculable. Thus whereas σ_{el} and σ_r are measurable, only $\sigma_{\text{shape el}}$ and $\sigma_{\text{absorption}}$ are directly calculable, whence

$$\sigma_{\text{tot}} = \sigma_{\text{el}} + \sigma_r = \sigma_{\text{shape el}} + \sigma_{\text{absorption}} \qquad (11\text{-}123)$$

$$\diagdown \quad \diagup \qquad \diagdown \quad \diagup$$

measurable   calculable

One way to proceed in this situation is to derive the value of $\sigma_{\text{compound el}}$ by a difference method applicable to a given scattering potential (omitting spin-orbit effects):

$$\sigma_{\text{compound el}} = \sigma_{\text{absorption}} - \sigma_r \qquad (11\text{-}124)$$

with $\sigma_{\text{absorption}}$ calculated and σ_r measured. Then, on the assumption that the compound-elastic scattering intensity has an isotropic distribution it is permissible to express the differential cross section as

$$\sigma_{\text{compound el}}(\theta) = \frac{1}{4\pi} \sigma_{\text{compound el}} \qquad (11\text{-}125)$$

and thence build the "theoretical" elastic cross section,

$$\sigma_{\text{el}}(\theta) \overset{\triangle}{=} \sigma_{\text{shape el}}(\theta) + \left(\frac{1}{4\pi} \sigma_{\text{compound el}} \right) \qquad (11\text{-}126)$$

for comparison with the experimental value. Though consistent, and to a good approximation compatible with experimental data, this approach admittedly falls short of being altogether satisfactory. A more detailed discussion of this point, including alternative means of circumventing the difficulty, is presented in Ch. 14, Vol. II. At least in the vicinity of an elastic scattering resonance the Breit-Wigner treatment, suitably modified to take account of a shape-elastic contribution, might be expected to furnish theoretical predictions for the net elastic cross section σ_{el} in reasonably good agreement with experimental observations. That this is indeed the case, we demonstrate in the next section.

11.4.4. MODIFIED BREIT-WIGNER THEORY FOR ELASTIC SCATTERING

If shape-elastic scattering were completely suppressed, and, moreover, if the remaining compound-elastic scattering proceeded at resonance by way of a single intermediate state, the Breit-Wigner formula would apply, viz.

$$\sigma_{\text{el}} \overset{\triangle}{=} \sigma_{(a,a)\,\text{res}} = \pi \lambdabar^2 \frac{\Gamma_a^2}{(E - E_0)^2 + (\Gamma/2)^2} \qquad (11\text{-}127)$$

Thus for pure compound-elastic neutron scattering one would expect

$$\sigma_{(n,n)\,res} = \pi \lambdabar^2 \frac{\Gamma_n^2}{(E - E_0)^2 + (\Gamma/2)^2} \qquad (11\text{-}128)$$

in which provision is made through Γ for other *nonelastic* outgoing channels. Of course, when the compound nucleus is *highly damped*, i.e., if $\Gamma_n \ll \Gamma$, the above cross section will be small compared with that for a competing process such as radiative capture, for instance, in which thermal neutrons ($l = 0$:

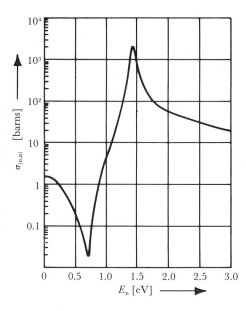

Fig. 11-27. Resonance anomaly in the slow-neutron elastic scattering cross section for ^{115}In, illustrating the occurrence of a dip due to interference between a negative resonance amplitude and a positive shape-elastic amplitude. The structure should be compared with that for *radiative capture* of slow neutrons by ^{115}In, shown in Fig. 11-26.

central collision, cf. Section 3.5.5) are resonantly captured to form a compound nucleus that subsequently experiences γ decay. The cross section for the latter process

$$\sigma_{(n,\gamma)} = \sigma_{capture} = \pi \lambdabar^2 \frac{\Gamma_n \Gamma_\gamma}{(E - E_0)^2 + \frac{1}{4}(\Gamma_n + \Gamma_\gamma)^2} \qquad (11\text{-}129)$$

is frequently much larger than that for elastic scattering, as reflected in the appreciably larger γ-decay partial width Γ_γ: for instance, the resonance parameters for the ^{115}In resonance depicted in Fig. 11-26, which can be derived from the shape and position of the cross section, are [Hu 58]

$$E_0 = 1.457 \text{ eV} \qquad \Gamma_\gamma = 0.072 \text{ eV} \qquad \Gamma_n = 0.00304 \text{ eV} \quad (11\text{-}130)$$

i.e.,

$$\Gamma_\gamma \gg \Gamma_n \qquad (11\text{-}131)$$

The occurrence of a coherent indistinguishable admixture of *shape*-elastic scattering entails a modification of the Breit-Wigner formula to include a nonresonant contribution to the amplitude. Dispersion theory [Bl 52b] indicates that for the net elastic scattering of S-wave neutrons, omitting spin considerations, the requisite formula for the ELASTIC CROSS SECTION becomes

$$\sigma_{(n,n)\,el} = \frac{\pi}{k^2} \left| \frac{\Gamma_n}{(E - E_0) + i\,(\Gamma/2)} + 2e^{ikR} \sin kR \right|^2 \qquad (11\text{-}132)$$

where $k \equiv 1/\lambda$ is the wave number of the incident neutron and R is the radius

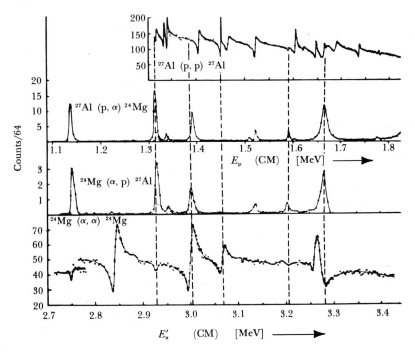

Fig. 11-28. Excitation functions for the forward and inverse reactions $^{27}\text{Al} + \text{p} \rightleftharpoons$ $^{28}\text{Si}^* \rightleftharpoons {}^{24}\text{Mg} + {}^4\text{He}$ displaying the resonances in the intermediate $^{28}\text{Si}^*$ system together with matching elastic-scattering resonance anomalies (cf. Fig. 11-23 and Section 11.3.1) (from [Ka 52b]).

of the target nucleus. The second term betokens the shape-elastic contribution, which can interfere with the resonant amplitude and give rise to characteristic dips in the cross section at energies just below resonance—an effect illustrated in Fig. 11-27, in which the logarithmic vertical scale serves to show up this phenomenon more clearly. The distinction between this Figure, which depicts elastic scattering only, and Fig. 11-26, which shows the *capture* cross section in the same energy region, should be noted.

Extensive experimental evidence for such resonance anomalies exists—an instructive example is reproduced in Fig. 11-28, which complements the results shown earlier in Fig. 11-23 by including two sets of elastic scattering data at matching energies for comparison with the reaction data. Not only do the plots show interference dips in the elastic data, but they indicate the extent of correlation between resonances attributed to levels in the compound nucleus ^{28}Si at excitation energies of about 12.7 to 13.4 MeV.

The shape-elastic scattering term in (11-132) merits further attention. If shape-elastic scattering were the sole reaction process, the cross section would take the form

$$\sigma_{(n,n)\,\text{shape}} = 4\pi\lambdabar^2 \sin^2 kR \qquad (11\text{-}133)$$

which, in the limit of small energies ($k \to 0$, whence $\sin kR \to kR$), reduces to

$$\sigma_{(n,n)\,\text{shape}} \xrightarrow{kR \ll 1} 4\pi R^2 \qquad (11\text{-}134)$$

an expression which corresponds to surface scattering from an impenetrable sphere, with $R \ll \lambdabar$. It is interesting to note that its magnitude is four times that of $\sigma_g = \pi R^2$, the geometric cross section (cf. Section 3.5.5). When the incident energy is increased, the number of open exit channels available to the compound nucleus increases commensurately, and the compound-elastic scattering cross section diminishes. Accordingly, with increasing energy the cross section would be expected to evince progressively less resonance structure and eventually to display a fairly smooth trend, as indicated in Fig. 11-29, a characteristic which it is usual to interpret in terms of a transition from a compound-nucleus mechanism to a direct interaction. (The relevant aspects will be discussed in more detail in Volume II.)

The asymptotic low-energy limit of $\sigma_{(n,n)\,\text{shape}}$ is more aptly described in terms of the SCATTERING LENGTH a (discussed further in Chapter 13),

$$\sigma_{(n,n)\,\text{shape}} \xrightarrow{kR \ll 1} 4\pi a^2 \qquad (11\text{-}135)$$

rather than of R, the nuclear radius. The concept of scattering length a originates in the attempt to describe neutron scattering in terms of a set of PARTIAL SCATTERING AMPLITUDES c_l which are functions of the orbital momentum l. The wave function at large distances from the scatterer is then written (asymptotically) as

$$\psi \simeq e^{ikz} + \sum_{l=0}^{\infty} \frac{c_l}{r} e^{ikr} P_l(\cos\theta) \qquad (11\text{-}136)$$

In the special case of large wavelength $\lambdabar = 1/k \gg R$, which corresponds to low energy, the summation can be restricted to just the first term ($l = 0$), whence the wave function can be written in the form of a difference between two terms,

$$\psi \approx e^{ikz} - \frac{a}{r}\, e^{ikr} \tag{11-137}$$

the first of which represents the contribution from the incident beam and the second that from the scattered beam. The scattering length a can be positive or negative. It is a complex quantity when the interaction involves some absorption of the incident particles.

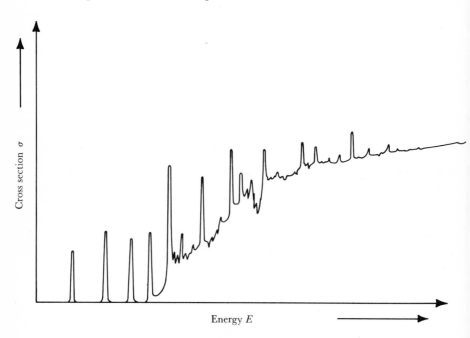

Fig. 11-29. Cross-sectional trend with rising incident energy.

This enables us to deduce the shape-elastic cross section, i.e.,

$$\sigma_{(n,\,n)\,\text{shape}} = \frac{\text{emission rate of scattered neutrons}}{\text{incident neutron flux}} \tag{11-138}$$

$$= 4\pi r^2\, v\, \frac{|(a/r)\, e^{ikr}|^2}{v\,|e^{ikz}|^2} \tag{11-139}$$

$$= 4\pi a^2 \tag{11-140}$$

where v is the incident neutron velocity. Schematically, the relationship between this and the compound-elastic cross section is depicted in Fig. 11-30.

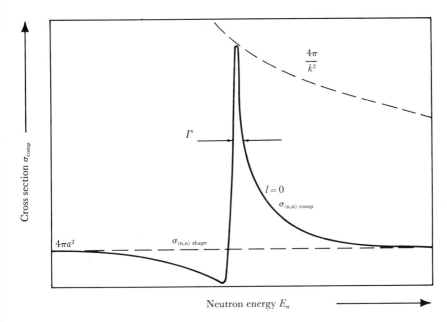

Fig. 11-30. Elastic scattering cross sections for S-wave neutrons near a resonance in the compound nucleus. The figure shows the relationship between the shape-elastic and the compound-elastic cross sections for a spinless target nucleus.

Collecting the various concepts together, we list the asymptotic expressions for elastic S-wave neutron scattering, writing these afresh in a form which introduces the mean energy-level separation D:

$$\sigma_{(n,n)\,shape} = 4\pi a^2 \qquad (11\text{-}141)$$

$$\sigma_{(n,n)\,res} = 2\pi^2 \lambdabar^2 \frac{\Gamma_n \Gamma_n}{\Gamma D} \qquad (11\text{-}142)$$

$$\sigma_{(n,n)\,el} \equiv \sigma_{el} = 4\pi a^2 + 2\pi^2 \lambdabar^2 \frac{\Gamma_n \Gamma_n}{\Gamma D} \qquad (11\text{-}143)$$

$$\sigma_{absorption} = 2\pi^2 \lambdabar^2 \frac{\Gamma_n}{D} \qquad (11\text{-}144)$$

$$\sigma_r = 2\pi^2 \lambdabar^2 \frac{\Gamma_n}{D}\left(1 - \frac{\Gamma_n}{\Gamma}\right) \qquad (11\text{-}145)$$

$$\sigma_{tot} = 4\pi a^2 + 2\pi^2 \lambdabar^2 \frac{\Gamma_n}{D} \qquad (11\text{-}146)$$

In these formulae, Γ_n, Γ, and D represent averages over many resonances.

The ratio Γ_n/D is termed the STRENGTH FUNCTION and is, as can be seen, closely related to the absorption. When plotted as a function of mass number A, it displays maxima wherever the scattering length a evinces discontinuities (cf. optical absorption and dispersion curves—a connection which can be exploited through setting up an optical model to describe scattering behavior, as will be discussed in Volume II). Between resonances a is of the same order of magnitude as the nuclear radius R.

11.4.5. STATISTICAL SPIN FACTOR

In deriving the basic form of the Breit-Wigner formula, we have so far taken no account of spin considerations. However, these can play a significant role in, for instance, neutron-induced reactions, which necessitate the inclusion of a STATISTICAL SPIN FACTOR

$$g = \frac{2J_C + 1}{(2J_A + 1)(2j_a + 1)(2l_a + 1)} \qquad (11\text{-}147)$$

multiplying the Breit-Wigner expression, which accordingly becomes a function of the orbital and total angular momenta l_a and j_a of the incident particle, and the nuclear spins J_A, J_C of the target and compound nuclei respectively.

In the specific case of *radiative capture of slow neutrons* $(l_a = 0)$, the (n, γ) process evinces a resonance when

$$J_C = J_A \pm \tfrac{1}{2} \qquad (11\text{-}148)$$

and therefore, with $j_a = \tfrac{1}{2}$ for slow neutrons, the "correction factor" becomes

$$g = \frac{1}{2}\left(1 \pm \frac{1}{2J_A + 1}\right) \qquad (11\text{-}149)$$

This can easily be verified as follows: The multiplicative factor g represents the probability that, for given nuclear spins J_A and J_C differing by $\tfrac{1}{2}$ unit, the incident neutron has the right orientation for capture. Now, the number of initial states with

parallel spin orientation [i.e., net spin $(J_A + \tfrac{1}{2})$] is $2(J_A + \tfrac{1}{2}) + 1$

$$(11\text{-}150)$$

and antiparallel spin orientation [i.e., net spin $(J_A - \tfrac{1}{2})$] is $2(J_A - \tfrac{1}{2}) + 1$

$$(11\text{-}151)$$

and hence the total number of initial states is $4J_A + 2$. Moreover, the probability that an incident neutron will form

$$\text{a parallel-spin state is} \qquad \frac{2(J_A + \frac{1}{2}) + 1}{4J_A + 2} \qquad (11\text{-}152)$$

$$\text{an antiparallel-spin state is} \; \frac{2(J_A - \frac{1}{2}) + 1}{4J_A + 2} \qquad (11\text{-}153)$$

Hence the total probability of the incident neutron's having the correct orientation for resonance capture is either

$$g = \frac{J_A + 1}{2J_A + 1} \qquad \text{(parallel spin)} \qquad (11\text{-}154)$$

or

$$g = \frac{J_A}{2J_A + 1} \qquad \text{(antiparallel spin)} \qquad (11\text{-}155)$$

which can be written collectively as

$$g = \frac{1}{2}\left(1 \pm \frac{1}{2J_A + 1}\right) \qquad \text{or} \qquad g = \frac{1}{2}\left(\frac{2J_C + 1}{2J_A + 1}\right) \qquad (11\text{-}156)$$

Unless $J_C = J_A \pm \frac{1}{2}$, there is no likelihood of a perceptible resonance for slow neutrons.

It should be noted that when fitting the Breit-Wigner formula to experimental resonance data it is customary to treat E_0 and Γ as if they were absolute constants of the particular resonance level. This is not altogether justified. A more rigorous mathematical treatment indicates that for charged-particle resonances and higher partial waves ($l_a \neq 0$), there can be significant variation in E_0 and Γ with incident energy E.

11.5. Formal Reaction Theory

Anyone whose enthusiasm has carried him this far should not be daunted by the fact that reaction theory involves a comprehension of R-, S-, T-, and U-matrix formalisms, some aspects of which we shall discuss in this section and later chapters. The formalisms are closely interrelated, but each is of value in its own right in that it is uniquely suited to the treatment of some particular situation (the field is still open to the development of ingenious V- to Z-matrix formalisms). In this section we shall explore, in part, at greater length some of the ideas introduced earlier, but at a more sophisticated level.

11.5.1. PARTIAL-WAVE APPROACH TO SCATTERING OF SPINLESS PARTICLES

The elastic scattering of spinless particles on a spinless scattering center constitutes one of the simplest problems to which the wave-interaction approach can be applied. The wave equation which describes this process takes the form

$$\nabla^2 \psi + k^2 \psi = 0 \qquad (11\text{-}157)$$

where k is the CM momentum of the scattered particle and the wave function ψ is, asymptotically (i.e., well outside the interaction region, where it refers to essentially free particles), of the form

$$\psi(\mathbf{r}) \xrightarrow{r \to \infty} \frac{1}{v^{1/2}}\left[e^{ikz} + f(\theta)\,\frac{e^{ikr}}{r}\right] \qquad (11\text{-}158)$$

when the z direction is taken to be that of the incoming beam, assumed to be a plane wave ($\sim e^{ikz}$) traveling with a velocity v in the CM system. The outgoing scattered wave is a spherical wave ($\sim e^{ikr}/r$) resulting from an inverse-square interaction, which emerges at the scattering angle θ to the z axis. This is the argument of the SCATTERING AMPLITUDE $f(\theta)$, which essentially determines the flux of particles scattered in the direction (θ, φ) into an element of area $dA = r^2\, d\Omega$ (cf. Section D.2)

$$j\, dA = \frac{\hbar}{2im}\left(\psi_{\text{sc}}^* \frac{\partial \psi_{\text{sc}}}{\partial r} - \psi_{\text{sc}} \frac{\partial \psi_{\text{sc}}^*}{\partial r}\right) r^2\, d\Omega = v|\psi_{\text{sc}}|^2 r^2\, d\Omega \qquad (11\text{-}159)$$

$$= |f(\theta)|^2\, d\Omega \qquad (11\text{-}160)$$

where

$$\psi_{\text{sc}} \equiv \frac{1}{v^{1/2}} f(\theta)\,\frac{e^{ikr}}{r} \qquad (11\text{-}161)$$

The factor $1/v^{1/2}$ serves to normalize the flux of the incoming beam to unity (i.e., 1 particle cm^{-2} sec^{-1}).

For the description of normalized unperturbed incident plane waves, the SPHERICAL BESSEL FUNCTION $j_l(kr)$ proves useful. This is related to the ordinary BESSEL FUNCTION $J_{l+(1/2)}(kr)$, as tabulated, e.g., in Jahnke et al. [Ja 60] through the formula

$$j_l(kr) = \left(\frac{\pi}{2kr}\right)^{1/2} J_{l+(1/2)}(kr) \qquad (11\text{-}162)$$

and can be represented as

$$j_l(kr) = (-kr)^l \left[\frac{1}{kr}\frac{d}{d(kr)}\right]^l \left(\frac{\sin kr}{kr}\right) \qquad (11\text{-}163)$$

whence, asymptotically,

$$j_l(kr) = \begin{cases} \dfrac{(kr)^l}{(2l+1)!!} & |kr| \ll l \\[2mm] \dfrac{1}{kr}\sin\left(kr - l\dfrac{\pi}{2}\right) & |kr| \gg l \end{cases} \qquad \begin{matrix}(11\text{-}164)\\[4mm](11\text{-}165)\end{matrix}$$

with $(2l+1)!! \equiv (2l+1)\,(2l-1)\ldots 5\times 3\times 1$. A plot of the lower-order $(l=0, 1, 2, 3)$ spherical Bessel functions against kr has been drawn in Fig. 14-11. In terms of these, the plane-wave representation runs

$$\frac{1}{v^{1/2}}e^{ikz} = \frac{1}{v^{1/2}}\sum_{l=0}^{\infty}[4\pi(2l+1)]^{1/2}i^l j_l(kr)\,Y_{l0}(\Omega) \qquad (11\text{-}166)$$

$$\xrightarrow{r\to\infty} \frac{1}{v^{1/2}}\frac{1}{kr}\sum_{l=0}^{\infty}[4\pi(2l+1)]^{1/2}i^l\sin\left(kr - l\frac{\pi}{2}\right)Y_{l0}(\Omega) \quad (11\text{-}167)$$

$$\to \left(\frac{4\pi}{v}\right)^{1/2}\frac{1}{kr}\sum_{l=0}^{\infty}(2l+1)^{1/2}\frac{i^{l+1}}{2}$$

$$\times \left\{\exp\left[-i\left(kr - \frac{l\pi}{2}\right)\right] - \exp\left[i\left(kr - \frac{l\pi}{2}\right)\right]\right\}Y_{l0}(\Omega) \quad (11\text{-}168)$$

where l is the orbital momentum quantum number which designates each incoming partial wave, and the z component is zero since

$$m_l = xk_y - yk_x = x(0) - y(0) = 0 \qquad (11\text{-}169)$$

The first of the exponential terms in (11-168) represents a wave impinging on the point $z = 0$, which we take to be the scattering origin, and the second term represents the complementary outgoing wave in an undisturbed situation:

$$\mathscr{I}_l \sim \exp\left[-i\left(kr - \frac{l\pi}{2}\right)\right] \qquad (11\text{-}170)$$

$$\mathscr{O}_l \sim \exp\left[i\left(kr - \frac{l\pi}{2}\right)\right] \qquad (11\text{-}171)$$

To make provision for the *disturbance* of the outgoing wave following scattering, we modify the wave by introducing a complex multiplicative SCATTERING

COEFFICIENT η_l, and write in asymptotic approximation

$$\psi \simeq \left(\frac{4\pi}{v}\right)^{1/2} \frac{1}{kr} \sum_{l=0}^{\infty} (2l+1)^{1/2} \frac{i^{l+1}}{2} \left\{ \exp\left[-i\left(kr - \frac{l\pi}{2}\right)\right] \right.$$

$$\left. - \eta_l \exp\left[i\left(kr - \frac{l\pi}{2}\right)\right] \right\} Y_{l0}(\Omega) \tag{11-172}$$

$$= \psi_{\text{in}} + \psi_{\text{out}} \equiv \psi_{\text{in}} + \psi_{\text{sc}} \tag{11-173}$$

$$\sim \sum_l (2l+1)^{1/2} \frac{i^{l+1}}{2} [\mathscr{I}_l - \eta_l \mathscr{O}_l] Y_{l0}(\Omega) \tag{11-174}$$

The coefficient η_l figures prominently in reaction theory, as may be appreciated on noting (e.g., from the equation of continuity) that

> $|\eta_l| > 1$ corresponds to generation of particles of the same species, momentum, and energy as the incident wave at the interaction center;
>
> $|\eta_l| < 1$ corresponds to absorption of (incident-type) particles at the interaction center;
>
> $|\eta_l| = 1$ corresponds to scattering of the incident particle flux;
>
> $|\eta_l| = 0$ corresponds to complete absorption of the incident particle flux (e.g., by a "black nucleus").

It is easy to ascertain the form taken asymptotically by the scattered wave, since from (11-168), (11-172), and (11-173),

$$\psi_{\text{sc}} = \psi - \psi_{\text{in}} = \psi - \frac{1}{v^{1/2}} e^{ikz} \tag{11-175}$$

$$= \left(\frac{4\pi}{v}\right)^{1/2} \frac{1}{kr} \sum_{l=0}^{\infty} (2l+1)^{1/2} \left(\frac{\eta_l - 1}{2i}\right) e^{ikr} Y_{l0}(\Omega) \tag{11-176}$$

on making use of the fact that $i^l e^{-i(l\pi/2)} = 1$. Comparison of this with (11-161) gives the value of the scattering amplitude

$$f(\theta) = (4\pi)^{1/2} \sum_{l=0}^{\infty} (2l+1)^{1/2} \left(\frac{\eta_l - 1}{2ik}\right) Y_{l0}(\Omega) \tag{11-177}$$

$$= \sum_{l=0}^{\infty} (2l+1) \left(\frac{\eta_l - 1}{2ik}\right) P_l(\cos\theta) \tag{11-178}$$

The second factor has been termed the SCATTERING FUNCTION and denoted by f_l (not to be confused with the scattering amplitude f):

$$f_l \equiv \frac{\eta_l - 1}{2ik} \tag{11-179}$$

As the incident flux has already been normalized to unity, the DIFFERENTIAL ELASTIC SCATTERING CROSS SECTION is simply

$$\sigma_{sc}(\theta) = \frac{\text{outgoing scattered flux}}{\text{incident flux } (=1)} \qquad (11\text{-}180)$$

$$= |f(\theta)|^2 \qquad (11\text{-}181)$$

$$= \frac{\pi}{k^2} \left| \sum_{l=0}^{\infty} (2l+1)^{1/2} (\eta_l - 1) Y_{l0}(\Omega) \right|^2 \qquad (11\text{-}182)$$

and the TOTAL ELASTIC CROSS SECTION is accordingly

$$\sigma_{sc} \equiv \sigma_{el} = \int \sigma_{sc}(\theta) \, d\Omega = \frac{\pi}{k^2} \sum_{l=0}^{\infty} (2l+1) |\eta_l - 1|^2 \qquad (11\text{-}183)$$

since the Y_{l0} harmonics are orthonormal. We can also build the REACTION CROSS SECTION by taking account of *particle loss* ($|\eta_l| < 1$) through taking the difference between incoming and outgoing beam intensities,

$$\sigma_r = \int vr^2 (|\psi_{in}|^2 - |\psi_{sc}|^2) \, d\Omega \qquad (11\text{-}184)$$

i.e.,

$$\sigma_r = \frac{\pi}{k^2} \sum_{l=0}^{\infty} (2l+1) (1 - |\eta_l|^2) \qquad (11\text{-}185)$$

We observe that σ_{el} is maximal (and $\sigma_r = 0$) when $\eta_l = -1$ (for S waves, $\sigma_{el} = 4\pi\lambda^2$), while σ_r is maximal when $\eta_l = 0$, and at this value of l (with $k \equiv 1/\lambda$),

$$\sigma_{r,l} = \sigma_{el,l} = \pi\lambda^2(2l+1) \qquad (11\text{-}186)$$

The physical regions of σ_{el} and σ_r are bounded by a curve, as shown in Fig. 11-31. The TOTAL CROSS SECTION is the sum of the elastic and reaction cross sections, i.e.,

$$\sigma_{tot} = \sigma_{el} + \sigma_r = \frac{\pi}{k^2} \sum_{l=0}^{\infty} (2l+1) (|\eta_l - 1|^2 + 1 - |\eta_l|^2) \qquad (11\text{-}187)$$

Evaluating this (e.g., by setting $\eta_l = a + ib$), we obtain

$$\sigma_{tot} = \frac{2\pi}{k^2} \sum_{l=0}^{\infty} (2l+1) (1 - \text{Re } \eta_l) \qquad (11\text{-}188)$$

where $\text{Re } \eta_l$ denotes the real part of the complex scattering coefficient.

There is a noteworthy relationship between the total cross section and the quantity $\text{Im} f(0)$, which denotes the imaginary part of the forward-scattering amplitude; namely, it follows from (11-178) and (11-188) that

$$\frac{4\pi}{k}\,\mathrm{Im}\,f(0) = \frac{2\pi}{k^2}\sum_{l=0}^{\infty}(2l+1)\,(1-\mathrm{Re}\,\eta_l) = \sigma_{\mathrm{tot}} \qquad (11\text{-}189)$$

This is known as the OPTICAL THEOREM, which remains valid also when particles with spin are considered.

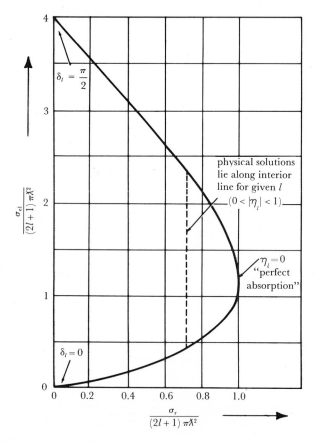

Fig. 11-31. Bounds upon the physical region of the elastic-scattering and resonance cross sections σ_{el} and σ_{r}, which are of necessity confined to the interior region of the limiting curve (adapted from [Bl 52b]).

We can apply the foregoing considerations to the simple example of SHADOW SCATTERING, in which we assume that a particle wave impinges on a perfectly absorbent "black" nucleus of radius R such that $R \gg \lambda$. The shadow effect corresponds to

complete absorption $(\eta_l = 0)$ of all partial waves with $l \leqslant \dfrac{R}{\lambda}$

perfect transmission $(\eta_l = 1)$ of all partial waves with $l > \dfrac{R}{\lambda}$

whence

$$\sigma_{\text{tot}} = 2\pi\lambda^2 \sum_{l=0}^{\infty} (2l + 1) = 2\pi\lambda^2 \left(\frac{R}{\lambda} + 1\right)^2 \qquad (11\text{-}190)$$

$$= 2\pi(R + \lambda)^2 \qquad (11\text{-}191)$$

$$\cong 2\pi R^2 \qquad (11\text{-}192)$$

which is *twice* the geometrical cross section, the doubling being caused by the scattering. The elastic scattering distributions for this situation follow on noting that the scattering amplitude is pure imaginary,

$$f(\theta) = \frac{i}{k} \sum_{l=0}^{l_{\max}=R/\lambda} (l + \tfrac{1}{2}) P_l(\cos\theta) \qquad (11\text{-}193)$$

as

$$\sigma_{\text{sc}}(\theta) = |f(\theta)|^2 = \frac{1}{k^2} \left| \sum_{l=0}^{l_{\max}} (l + \tfrac{1}{2}) P_l(\cos\theta) \right|^2 \qquad (11\text{-}194)$$

If many partial waves are involved, the sum can be replaced by an integral and, moreover, the scattering can be treated as confined to small angles, which allows one to make the approximations

$$l + \tfrac{1}{2} \to kb, \qquad P_l(\cos\theta) \to J_0(kb\sin\theta) \qquad (11\text{-}195)$$

where J_0 is a Bessel function of zeroth order and b is an impact parameter (viz. distance of closest approach). Introducing these approximations, we find

$$\sigma_{\text{sc}}(\theta) = \frac{1}{k^2} \left| \int_0^R k^2 \, b J_0(kb\sin\theta) \, db \right|^2 = R^2 \left[\frac{J_1(kR\sin\theta)}{\sin\theta} \right]^2 \qquad (11\text{-}196)$$

which corresponds to the classical formula for diffraction scattering from a "black" sphere of radius R. This approximation, which has a minimum at

$$\theta = \arcsin\left(3.8 \frac{\lambda}{R}\right) \qquad (11\text{-}197)$$

proves useful in the limit of scattering at high energies, and holds roughly even for particles with spin.

11.5.2. Phase Shifts

The interaction influences not only the amplitudes of the outgoing waves, but also their phases. The PHASE SHIFT δ_l (which differs for each partial wave and which is further elucidated in Exercise 11-18) can either be expressed as a complex quantity, defined by

$$\eta_l = e^{2i\delta_l} \quad \text{with} \quad \delta_l \equiv \alpha_l + i\beta_l \tag{11-198}$$

or, as in the more conventional physically equivalent representation which we also adopt, as a *real* quantity conjoined with a real INELASTICITY PARAMETER ρ_l, which lies between zero and unity,

$$\eta_l = \rho_l e^{2i\delta_l} \quad \text{with} \quad \delta_l \text{ and } \rho_l \text{ real} \tag{11-199}$$

This enables us to express the various cross sections in terms of δ_l and ρ_l; for instance, (11-185) can be rewritten as

$$\sigma_r = \pi\lambda^2 \sum_{l=0}^{\infty} (2l+1)(1-\rho_l^2) \tag{11-200}$$

and this vanishes when ρ_l takes on its extreme value $\rho_l = 1$, in which case pure elastic scattering occurs and

$$\eta_l \to e^{2i\delta_l} = \cos 2\delta_l + i \sin 2\delta_l = 1 + 2i e^{i\delta_l} \sin \delta_l \tag{11-201}$$

The scattering function $f_l \equiv (\eta_l - 1)/(2ik)$ can also be rewritten in terms of phase shifts for elastic scattering,

$$f_l = \frac{e^{i\delta_l} \sin \delta_l}{k} \tag{11-202}$$

and the scattering amplitude (11-178) then assumes the form

$$f(\theta) = \sum_{l=0}^{\infty} (2l+1)\left(\frac{e^{i\delta_l}\sin\delta_l}{k}\right) P_l(\cos\theta) \tag{11-203}$$

Substituting this in (11-181), we get the elastic scattering cross section

$$\sigma_{sc} \equiv \sigma_{el} = \int |f(\theta)|^2 d\Omega = 4\pi\lambda^2 \sum_{l=0}^{\infty} (2l+1)\sin^2\delta_l \tag{11-204}$$

which takes on maximal values at *resonance*, when

$$\delta_l = n\frac{\pi}{2} \quad \text{with} \quad n = 1, 3, 5, \ldots \tag{11-205}$$

The total cross section accordingly becomes

$$\sigma_{tot} = \sigma_{el} + \sigma_r = \pi\lambda^2 \sum_{l=0}^{\infty} (2l+1)(4\sin^2\delta_l + 1 - \rho_l^2) \tag{11-206}$$

11.5.3. RESONANCE CROSS SECTIONS

The scattering coefficient η_l which determines the absolute values of the various cross sections is itself determined by the logarithmic derivative of the (modified) wave function at the surface of the nucleus. The latter is assumed to have a well-defined surface of radius R. For simplicity, we consider the scattering of S-wave neutrons, which avoids complication through Coulomb or centrifugal barriers. We also refrain from incorporating the effects of spin, which would otherwise have necessitated the introduction of a statistical spin factor.

Assuming the wave function of the net system comprising a neutron and a scattering nucleus to be determined by the radial distance r from the interaction center, we define a modified wave function (for $l = 0$)

$$u_{\text{in}} \equiv r\psi_{\text{in}} \qquad \text{for} \qquad r < R \tag{11-207}$$

and, with (11-172),

$$u_{\text{out}} \equiv r\psi_{\text{out}} \equiv r\psi_{\text{sc}} = \left(\frac{4\pi}{v}\right)^{1/2} \frac{i}{2k} (e^{-ikr} - \eta_0 e^{ikr}) \tag{11-208}$$

requiring that at the nuclear surface $(r = R)$ these and their first derivatives match one another,

$$u_{\text{in}} = u_{\text{out}} \qquad \text{and} \qquad u'_{\text{in}} = u'_{\text{out}} \qquad \text{at} \qquad r = R \tag{11-209}$$

i.e.,

$$\left(\frac{u'}{u}\right)_{\text{in}} = \left(\frac{u'}{u}\right)_{\text{out}} \tag{11-210}$$

On introducing a dimensionless quantity

$$f \equiv \lim_{r \to R} \left(\frac{R}{u}\frac{du}{dr}\right) \tag{11-211}$$

which can be complex, we can use the condition of regularity of the logarithmic derivative to obtain the relation between f and η_0:

$$\left(\frac{u'}{u}\right)_{r=R} = \frac{f}{R} = \frac{k}{i}\frac{e^{-ikR} + \eta_0 e^{ikR}}{e^{-ikR} - \eta_0 e^{ikR}} \tag{11-212}$$

i.e.,

$$\eta_0 = \frac{f + ikR}{f - ikR} e^{-2ikR} \tag{11-213}$$

We abide by the usual convention of using the symbol f for this complex quantity, and writing

$$f = f_{\text{Re}} + if_{\text{Im}} \tag{11-214}$$

though it must be stressed that these f's are altogether different entities from the f_l and $f(\theta)$. When f is *real*, pure elastic scattering occurs ($|\eta_0| = 1$ and the reaction cross section vanishes), while in order that $|\eta_0| < 1$, the imaginary part f_{Im} must be negative. In terms of these quantities, we find that

$$\sigma_r^{(l=0)} = \pi \lambda^2 (1 - |\eta_0|^2) = \pi \lambda^2 \left[\frac{-4 f_{Im} kR}{(f_{Re})^2 + (kR - f_{Im})^2} \right] \tag{11-215}$$

and

$$\sigma_{sc}^{(l=0)} = \pi \lambda^2 |1 - \eta_0|^2 = 4\pi \lambda^2 \left| e^{ikR} \sin kR + \frac{kR}{i(kR - f_{Im}) - f_{Re}} \right|^2 \tag{11-216}$$

Explicitly, the squared modulus comprises three terms, the first of which represents POTENTIAL SCATTERING e.g.,

$$4\pi \lambda^2 \sin^2 kR \xrightarrow{kR \ll 1} 4\pi R^2 = 4 \times \text{geometrical cross section} \tag{11-217}$$

the second of which is a cross product representing an INTERFERENCE TERM, and the third of which corresponds to RESONANCE SCATTERING

$$\sigma_{res}^{(l=0)} = 4\pi R^2 \frac{1}{(kR - f_{Im})^2 + f_{Re}^2} \tag{11-218}$$

In the vicinity of a resonance at an energy E_0 we can expand f in a Taylor series,

$$f = f|_{E=E_0} + \left. \frac{\partial f}{\partial E} \right|_{E=E_0} (E - E_0) + \cdots \tag{11-219}$$

and neglect all derivatives of higher order than the first, while noting that $f|_{E=E_0}$ vanishes identically. If we now define a width Γ through the relation

$$\left. \frac{\partial f}{\partial E} \right|_{E=E_0} = -\frac{2kR}{\Gamma} \tag{11-220}$$

we can write

$$f = -\frac{2kR}{\Gamma} (E - E_0) \tag{11-221}$$

and therefore in this approximation

$$f = f_{Re} = -\frac{2kR}{\Gamma} (E - E_0) \qquad \text{while} \qquad f_{Im} = 0 \tag{11-222}$$

Substituting this in (11-218) we obtain

$$\sigma_{res}^{(l=0)} \simeq \pi \lambda^2 \frac{\Gamma^2}{(E - E_0)^2 + (\Gamma/2)^2} \tag{11-223}$$

which corresponds to the Lorentzian form employed in Section 11.4.4.

11.5.4. Reaction Theory in Matrix Formalism

The most general and elegant presentation of reaction theory involves the use of matrix formalisms in which the behavior of interacting systems is described in terms of the (asymptotic) interaction of particle-wave fields. The matrix elements in an appropriate representation provide a measure of the interaction probability and enable each of the various cross sections to be derived explicitly. Thus, in the course of time several formalisms have been developed to deal with different applications, e.g., M-matrix polarization theory [Wo 52], R-matrix ("reactance" or "reaction" matrix) reaction theory [La 58b], S-matrix ("scattering" matrix) interaction theory [He 43], T-matrix ("transition" matrix) transition theory, and U-matrix ("collision" matrix) collision theory.

Basically, they all stem from the S-matrix approach due to Heisenberg [He 43], in which an attempt was made to set up an "exact" field theory capable of predicting certain basic observable quantities such as free-particle momenta and energies or the energy levels of stationary bound systems without invoking an interaction Hamiltonian. Dealing with the asymptotic behavior of wave functions in collision, absorption and emission processes, the theory has been developed to furnish interaction rates for a wide range of processes and to describe the physical characteristics of interacting systems. Excellent reviews of the theory of low-energy nuclear reactions are to be found in survey articles [Mo 60c, Vo 62] and textbooks [Bl 52b, Vo 59, Pr 62].

11.5.5. Transition Amplitude in S-Matrix Theory

The foregoing sections have indicated how the cross sections for scattering processes are determined by the asymptotic wave functions of the system before and after interaction. The net (stationary) solution, as expressed by Eq. (11-173), is

$$\lim_{r \to \infty} \psi = \psi_{\text{in}} + \psi_{\text{out}} \tag{11-224}$$

The function of the S matrix, viewed as an operator, is to relate the magnitudes and phases of the incoming and outgoing waves described by ψ_{in} and ψ_{out}. Explicitly, the S matrix serves to transform the asymptotic form of the incoming wave (corresponding to a free-particle state) into that of the outgoing wave,

$$\psi_{\text{out}} = \underline{S}\psi_{\text{in}} \tag{11-225}$$

Herein ψ_{in} essentially refers to the system in the remote past and ψ_{out} to that in the distant future, with the $\underline{S}$ matrix operator extending over the entire time interval $-\infty < t < +\infty$ embracing the region around $t = 0$ which is taken to correspond with the occurrence of an interaction. With the aid of

$\underline{S}$, all possible final states ψ_{out} can be built from a given initial state ψ_{in}, the matrix element

$$S_{\text{fi}} \equiv \langle f|\underline{S}|i\rangle = \int \psi_{\text{out}}^* \underline{S}\psi_{\text{in}}\, d\Omega \tag{11-226}$$

giving the *transition probability amplitude* from a given initial state to a *particular final state*.

Starting out from these considerations, one can elucidate the basic properties of the S matrix which have far-reaching physical significance and use these to develop interaction theories along quantitative lines, as, for instance, Feynman has done in his formulation of quantum electrodynamics (Appendix F).

11.5.6. Basic Properties of the S Matrix (Probability Conservation and Time-Reversal Invariance)

The elements of the S matrix are complex quantities which are subject to the UNITARITY RELATION

$$\sum_c S_{cf}^* S_{ci} = \delta_{\text{if}} \tag{11-227}$$

or, in matrix notation,

$$S^\dagger S = SS^\dagger = S^{-1}S = \mathbf{1} \tag{11-228}$$

This condition corresponds to the physical requirement that probability be conserved in the transition, as follows from the relation

$$\langle f|f\rangle = \langle i|S^\dagger S|i\rangle \tag{11-229}$$

$$= \langle i|i\rangle \tag{11-230}$$

or, equivalently, from (11-227) since the condition

$$\sum_c |S_{cf}|^2 = 1 \tag{11-231}$$

implies that the sum of the probabilities for the interaction to proceed to some final state f is unity, i.e., if the incoming flux is unity then the net outgoing flux must also be unity.

Since the S matrix provides a formal link between incoming and outgoing wave functions according to Eq. (11-225), its SYMMETRY CHARACTER has a direct bearing upon the detailed-balanced and time-reversal properties of the system. It should be noted that the property of *reversibility* implied by detailed balance is not necessarily equivalent to the (stronger) condition of *reciprocity*, as implied by time-reversal invariance, though the inverse chain of dependence holds. The implication of (a) detailed balance, viz.

$$W_{\text{fi}} = W_{\text{if}} \tag{11-232}$$

or (b) semidetailed balance, in which an average over spins is built, viz.

$$\bar{W}_{fi} = \bar{W}_{if} \tag{11-233}$$

can in S-matrix symbolism be written as

$$|S_{fi}| = |S_{if}| \tag{11-234}$$

whereas reciprocity corresponds to

$$S_{fi} = S_{if} \tag{11-235}$$

Thus *symmetry of the S matrix implies time-reversal invariance* of the system [Coe 53]. When the S matrix is nonsymmetric, reversibility may nevertheless obtain irrespective of reciprocity, e.g., when:

(i) the system can exist only in one unique state i and another unique state f, for then the general relation [St 52]

$$\sum_f W_{fi} = \sum_f W_{if} \tag{11-236}$$

reduces to the detailed-balance condition

$$W_{fi} = W_{if} \tag{11-237}$$

or (ii) a first-order Born approximation can be applied (e.g., in weak interactions), for then the S matrix elements are proportional to the Hamiltonian (even if many channels are open):

$$S_{fi} \sim H'_{fi} \tag{11-238}$$

and since the Hamiltonian is Hermitian,

$$H'_{fi} = (H'_{if})^* \tag{11-239}$$

it follows that

$$S_{fi} = S_{if}^* \tag{11-240}$$

The same reasoning applies to other formalisms in which the S matrix can be reduced to a Hermitian matrix element.

It should be noted that there is more to the time-reversal operation than simply inversion of the time direction ($t \rightarrow -t$) and therefore of the sign of such dynamic quantities as $\mathbf{v}$, $\mathbf{p}$, $\mathbf{j}$, $\boldsymbol{\sigma}$, $\mathbf{i}$, $\mathbf{H}$, and $\mathbf{A}$ (while $\mathbf{r}$, m, E, ρ, $\mathbf{E}$, and ϕ remain unchanged). In order that a wave function $\psi(r, t)$ and its time conjugate ψ^T satisfy the *same* (Schrödinger) wave equation,

$$i\hbar \frac{\partial}{\partial t} \psi(\mathbf{r}, t) = \underset{\sim}{H}\psi(\mathbf{r}, t) \tag{11-241}$$

which, on time inversion becomes

$$\underline{H}\psi(\mathbf{r}, -t) = i\hbar \frac{\partial}{\partial(-t)}\, \psi(\mathbf{r}, -t) = -i\hbar \frac{\partial}{\partial t}\, \psi(\mathbf{r}, -t) \qquad (11\text{-}242)$$

it is necessary to obviate the disturbing minus sign by setting

$$\psi^{T} \equiv \psi^{*}(\mathbf{r}, -t) \qquad (11\text{-}243)$$

and thereby introducing complex conjugation into the time-reversal formalism. Explicitly, we can distinguish between

(1) interchange of initial and final states ⎫
                                                                              ⎬ time ⎫ detailed
                                                                              ⎭ inversion ⎬ balance
(2) change of sign of dynamic quantities ⎬ time ⎫
                                                                       ⎭ reversal ⎭

(3) complex conjugation

and note that the TIME-REVERSAL OPERATOR, defined by the relation

$$\underline{T}\psi = \psi^{T} \qquad (11\text{-}244)$$

is *unitary*,

$$\underline{T}^{\dagger}\,\underline{T} = \mathbf{1} \qquad (11\text{-}245)$$

but *anti*linear (i.e., involves complex conjugation),

$$\underline{T}(a_1\psi_1 + a_2\psi_2) = a_1^{*}\,\underline{T}\psi_1 + a_2^{*}\,\underline{T}\psi_2 \qquad (11\text{-}246)$$

$$= a_1^{*}\,\psi_1^{T} + a_2^{*}\,\psi_2^{T} \qquad (11\text{-}247)$$

It is therefore frequently termed *antiunitary*.

11.5.7. Cross Sections in Matrix Formalism

Though we shall continually refer back to the S matrix, it is more usual to express formal reaction theory in terms of the equivalent COLLISION MATRIX $\mathcal{U}$, whose elements are defined as follows: We start by considering ELASTIC SCATTERING, writing the modified asymptotic radial wave function $(u_l \equiv r\psi_l)$ as a superposition of incoming and outgoing waves $\mathcal{I}_l$ and $\mathcal{O}_l$ with amplitudes x_l, y_l respectively,

$$u_l = x_l\mathcal{I}_l + y_l\mathcal{O}_l \qquad (11\text{-}248)$$

and requiring continuity of the logarithmic derivative at a radius $r = a$ taken to represent the boundary of the interaction field,

$$u_l(a) = x_l\mathcal{I}_l(a) + y_l\mathcal{O}_l(a) \qquad (11\text{-}249)$$

and

$$u_l'(a) = x_l\mathcal{I}_l'(a) + y_l\mathcal{O}_l'(a) \qquad (11\text{-}250)$$

where the primes denote radial differentiation. In terms of the Wronskian, defined for two functions F and G of the same variable(s) by

$$W(F, G) \equiv FG' - F'G \tag{11-251}$$

one can derive the amplitudes as

$$x_l = \frac{1}{2ik} W(u_l, \mathcal{O}_l) \quad \text{and} \quad y_l = -\frac{1}{2ik} W(u_l, \mathcal{I}_l) \tag{11-252}$$

since

$$W(\mathcal{O}_l, \mathcal{I}_l) = -2ik \tag{11-253}$$

For different values of l, the ratio

$$\mathcal{U}_l \equiv -\frac{y_l}{x_l} = \frac{W(u_l, \mathcal{I}_l)}{W(u_l, \mathcal{O}_l)} \tag{11-254}$$

then defines the elements of the $\mathcal{U}$ matrix, which in turn determine the respective cross sections.

To normalize the incoming flux to unity we set

$$x_l = \frac{i}{k} [\pi(2l + 1)]^{1/2} \tag{11-255}$$

and thence derive

$$\sigma_{el}(\theta) = \frac{1}{4k^2} \left| \sum_l (2l + 1)(1 - \mathcal{U}_l)^2 P_l(\cos\theta) \right|^2 \tag{11-256}$$

This is equivalent to (11-182) with η_l replaced by the element $\mathcal{U}_l$, i.e.,

$$\eta_l = \rho_l e^{2i\delta_l} \rightarrow \mathcal{U}_l \tag{11-257}$$

For the total elastic scattering cross section we similarly obtain

$$\sigma_{el} = \sum_l \sigma_{el}^{(l)} = \frac{\pi}{k^2} \sum_l (2l + 1) |1 - \mathcal{U}_l|^2 \tag{11-258}$$

with the partial cross section given by the relation

$$\sigma_{el}^{(l)} = \frac{2\pi}{k^2} (2l + 1) \operatorname{Re}[1 - \mathcal{U}_l] \tag{11-259}$$

The condition for a *resonance* in the elastic scattering is [Hu 61]

$$W(u_l, \mathcal{O}_l)_{r=a} = 0 \tag{11-260}$$

Strictly speaking, the $\mathcal{U}_l$ are the *diagonal* elements of the unitary $\mathcal{U}$ matrix, e.g., for elastic scattering in channel c,

$$\mathcal{U}_l \hat{=} \mathcal{U}_{cc} \tag{11-261}$$

The generalization to *nuclear reactions* is effected by considering off-diagonal elements which correspond to different exit channels c', viz., $\mathscr{U}_{c'c}$ for a single entrance channel c. Then, by definition,

$$x_{c'} = 0 \qquad \text{for} \qquad c' \neq c \tag{11-262}$$

and

$$y_{c'} = -\mathscr{U}_{c'c}x_c \tag{11-263}$$

which linearly relates the outgoing amplitudes to the incoming amplitude, again normalized according to (11-255).

In terms of these elements, the reaction cross section can be written

$$\sigma_r(\theta) = \frac{\pi}{k^2} \left| \sum_l (2l+1)^{1/2} \, \mathscr{U}_{c'c} \, Y_{l0}(\Omega) \right|^2 \tag{11-264}$$

$$= \frac{1}{4k^2} \left| \sum_l (2l+1) \, \mathscr{U}_{c'c} \, P_l(\cos\theta) \right|^2 \tag{11-265}$$

whence

$$\sigma_r = \sum_l \sigma_r^{(l)} = \frac{\pi}{k^2} \sum_{\substack{l \\ c' \neq c}} (2l+1) \, |\mathscr{U}_{c'c}|^2 \tag{11-266}$$

which, because of the unitary condition (cf. (11-231))

$$\sum_{c'} |\mathscr{U}_{c'c}|^2 = 1 \tag{11-267}$$

reduces to

$$\sigma_r = \frac{\pi}{k^2} \sum_l (2l+1)\,(1 - |\mathscr{U}_{cc}|^2) \tag{11-268}$$

This gives the total reaction cross section in terms of the diagonal elements of the $\mathscr{U}$ matrix. The wave number k refers to the incident particle in all cases; if we explicitly label this according to the entrance channel, we obtain the RECIPROCITY RELATION from (11-265):

$$\frac{\sigma_r^{(c'c)}(\theta)}{\sigma_r^{(cc')}(\theta)} = \left(\frac{k_{c'}}{k_c}\right)^2 = \frac{p_{c'}^2}{p_c^2} \tag{11-269}$$

which is equivalent to (11-72) in absence of spin factors. The net total cross section is furnished by (11-258) and (11-268) as

$$\sigma_{\text{tot}} \equiv \sigma_{\text{el}} + \sigma_r = \frac{2\pi}{k^2} \sum_l (2l+1)\,(1 - \text{Re}\,\mathscr{U}_{cc}) \tag{11-270}$$

When the formalism is extended to include particle spins, as in [Bl 52b,

Pr 62], the expressions become more complicated, but their basic form remains unchanged.

Although the inherent physics of the collision interaction is contained within the $\mathscr{U}$ matrix, one can level the criticism against the formalism that it is somewhat remotely linked to the actual interaction since it is expressed through the asymptotic behavior of the wave functions. The quest for a formalism which, while still unconcerned with the complex situation *within* the interaction region, nevertheless "zooms in" from asymptotic remoteness to the very edge of the interaction region led to important new formulations of nuclear reaction theory. In particular, it gave rise to the Kapur-Peierls [Ka 38, Br 59] and Wigner-Eisenbud [Eis 41, Wi 46, Wi 47a, La 55a, La 58b] frameworks of $\mathscr{R}$ matrix theory, based upon the realization that the wave function throughout the entire external region is determined by the characteristics of the wave function on or just outside the boundary of the interaction region. The $\mathscr{U}$ matrix (i.e., essentially the S matrix) is expressed through the $\mathscr{R}$ matrix which constitutes a more immediate link between the interaction and the outside environment. Only the outcome of the collision process is subjected to scrutiny, while the effect of the actual interaction is simulated by building a hypothetical set of states inside the collision region. The $\mathscr{R}$ matrix can thereby be connected more readily than the $\mathscr{U}$ matrix with a physical model interaction. Nevertheless, it lacks the physical essence of the $\mathscr{U}$ matrix, and its sole justification is that of formalistic convenience, particularly in the case of resonance processes. From this point of view, the development of Humblet-Rosenfeld reaction theory [Hu 61, Hu 62, Ma 63a, Ma 63b, Hu 67] constitutes an appreciable stride toward a reversion to physical principles.

To illustrate the application of the $\mathscr{R}$ matrix approach, we start by considering the *potential scattering of spinless particles* in which the interaction is considered to be confined within the region $r \leqslant b$. For the initial stages of the treatment, we can retain a parallel between the element $\mathscr{R}_l$ and the corresponding collision-matrix element $\mathscr{U}_l$; e.g., we define $\mathscr{R}_l$ through the boundary condition

$$\mathscr{R}_l \equiv \frac{1}{b}\frac{u_l}{u_l'}\bigg|_{r=b} \qquad (11\text{-}271)$$

and write the external wave function as

$$\psi_{r>b} = \sum_l i\pi^{1/2}(2l+1)^{1/2}\,(\mathscr{I}_l - \mathscr{U}_l\mathscr{O}_l) \qquad (11\text{-}272)$$

while in the interior region

$$\psi_{r\leqslant b} = \sum_l u_l(r)\,Y_{lm}(\Omega) \qquad (11\text{-}273)$$

The requirement of continuity at $r = b$ then gives

$$\mathscr{R}_l = \left(\frac{1}{r}\frac{u_l}{u_l'}\right)_{r=b} = \left(\frac{1}{r}\frac{\mathscr{I}_l - \mathscr{U}_l\mathscr{O}_l}{\mathscr{I}_l' - \mathscr{U}_l\mathscr{O}_l'}\right)_{r=b} \tag{11-274}$$

A distinction between the Kapur-Peierls and Wigner-Eisenbud formalisms is already evident in the interpretation of the interaction radius cutoff b. It is found that a small value of b (roughly equivalent to the nuclear radius R) yields the most satisfactory results, but whereas in the Kapur-Peierls theory it is set to

$$b_{\mathrm{KP}} = -ikR \tag{11-275}$$

the Wigner-Eisenbud theory treats it as simply a dimensionless constant. Pursuing the latter course, we introduce an L function proportional to the logarithmic derivative,

$$L_l \equiv \left(\frac{r\mathscr{O}_l'}{\mathscr{O}_l}\right)_{r=b} \qquad \text{and} \qquad L_l^* \equiv \left(\frac{r\mathscr{I}_l'}{\mathscr{I}_l}\right)_{r=b} \tag{11-276}$$

and use this to express $\mathscr{U}_l$ in terms of $\mathscr{R}_l$, viz.

$$\mathscr{U}_l = \frac{\mathscr{I}_l}{\mathscr{O}_l}\bigg|_{r=b} \frac{1 - L_l^*\mathscr{R}_l}{1 - L_l\mathscr{R}_l} \tag{11-277}$$

The essence of the theory now lies in taking $\mathscr{R}_l$ to be expressible in terms of certain (hypothetical) states of the system in the interior region $r < b$, characterized by energy eigenvalues E_λ and orthonormal eigenfunctions u_λ (with $\lambda = 1, 2, 3, \ldots$). The actual wave function u_E of the scattering state with some energy E can be expressed as a linear combination of these,

$$u_E = \sum_\lambda A_\lambda u_\lambda(r) \tag{11-278}$$

the expansion coefficients being

$$A_\lambda = \int_0^b u_\lambda u_E \, dr \tag{11-279}$$

To obtain an expression for $\mathscr{R}_l$ by way of the definition (11-274), we integrate (11-279) by parts and substitute in (11-278) to obtain

$$A_\lambda = \frac{\hbar^2}{2m}\frac{1}{E_\lambda - E}(u_\lambda u_E')_{r=b} \tag{11-280}$$

and

$$u_E(r) = \frac{\hbar^2}{2mb}\sum_\lambda \frac{u_\lambda(r)\,u_\lambda(b)}{E_\lambda - E}(ru_E')_{r=b} \tag{11-281}$$

whence

$$\mathscr{R}_l = \left(\frac{1}{r}\frac{u_E}{u_E'}\right)_{r=b} = \frac{\hbar^2}{2mb}\sum_\lambda \frac{u_\lambda^2(b)}{E_\lambda - E} \qquad (11\text{-}282)$$

The evident resonance structure of this formula suggests the introduction of a REDUCED WIDTH

$$\gamma_\lambda \equiv \left(\frac{\hbar^2}{2mb}\right)^{1/2} u_\lambda(b) \qquad (11\text{-}283)$$

in order that (11-282) might be written as

$$\mathscr{R}_l = \sum_\lambda \frac{\gamma_\lambda^2}{E_\lambda - E} \qquad (11\text{-}284)$$

If we assume that just *one* state λ is dominant among the ensemble of states, we can reduce $\mathscr{R}_l$ to

$$\mathscr{R}_l = \frac{\gamma_\lambda^2}{E_\lambda - E} \qquad (11\text{-}285)$$

and in this SINGLE-LEVEL APPROXIMATION deduce a scattering formula of the Breit-Wigner type on writing

$$\Gamma_\lambda \equiv 2kR\gamma_\lambda^2 \qquad (11\text{-}286)$$

We proceed from the formula (11-258)

$$\sigma_{\text{el}} = \frac{\pi}{k^2}\sum_l (2l+1)\,|1 - \mathscr{U}_l|^2 \qquad (11\text{-}287)$$

and use (11-274) to write

$$\mathscr{U}_l = \left.\frac{\mathscr{I}_l - \mathscr{I}_l'b\mathscr{R}_l}{\mathscr{O}_l - \mathscr{O}_l'b\mathscr{R}_l}\right|_{r=b} \qquad (11\text{-}288)$$

whence, with

$$\mathscr{I}_l = N_l e^{-ikr} \qquad \text{and} \qquad \mathscr{O}_l = N_l e^{ikr} \qquad (11\text{-}289)$$

where the N_l are normalization factors, we get

$$\mathscr{U}_l = e^{-2ikR}\frac{1 + ikb\mathscr{R}_l}{1 - ikb\mathscr{R}_l} \qquad (11\text{-}290)$$

Hence, noting that $e^{2ikR} = \text{mod}\,(1)$, we can write

$$|1 - \mathscr{U}_l|^2 = \left|(e^{2ikR} - 1) + \left(1 - \frac{1 + ikb\mathscr{R}_l}{1 - ikb\mathscr{R}_l}\right)\right|^2 \qquad (11\text{-}291)$$

which on substituting (11-285) for $\mathscr{R}_l$ and introducing a LEVEL SHIFT

$$\Delta_\lambda \equiv b\gamma_\lambda^2 \qquad (11\text{-}292)$$

enables us to express (11-287) in the SINGLE-LEVEL RESONANCE form

$$\sigma_{\mathrm{el}} = \frac{\pi}{k^2} \sum_{l} (2l + 1) \left| 2e^{ikR} \sin{(kR)} - \frac{\Gamma_\lambda}{(E_\lambda - E + \Delta_\lambda) - \frac{1}{2}i\Gamma_\lambda} \right|^2 \quad (11\text{-}293)$$

This evinces a close affinity with (11-223), and contains terms corresponding to potential scattering, resonance scattering and an interference contribution. The occurrence of the level shift Δ_λ is significant in that it indicates that the resonance cross section peaks at an energy $E = E_\lambda + \Delta_\lambda$ rather than at E_λ itself and has a full width Γ_λ at the above energy. Clearly, Γ_λ must be appreciably smaller than the separation between successive resonances λ. In deriving the formula (11-293), we have striven to indicate the essence of the method without complicating the treatment by using more rigorous matrix formalism. Even in this simplified approach, it will be evident that as soon as we depart from the one-level approximation, $\mathcal{R}_l$ becomes appreciably more complicated and in deriving $\mathcal{U}_l$ from (11-290) one becomes faced with a difficult problem of matrix inversion. Thomas has shown [Tho 55] that this can be avoided by reformulating the Wigner-Eisenbud approach to exclude explicit occurrence of the $\mathcal{R}$ matrix (see also [Bl 57]). This modified treatment forms the basis of much recent work in reaction theory.

The matrix-inversion problem is also avoided in Kapur-Peierls formalism, as may be seen on redefining $\mathcal{R}_l$ through a different boundary condition (cf. (11-271)) in which b is set to $b \to -ikR$ and the denominator of the expression which corresponds to (11-290) is reduced to unity. This gives a result for the scattering cross section which differs somewhat from (11-293) and renders it difficult to relate corresponding parameters in either formalism to one another. Vogt [Vo 59, Vo 62] has endeavored to extract appropriate numerical values of the parameters from a comparison with experimental data. Although this has yielded striking success, one can level several basic criticisms against the $\mathcal{R}$-matrix approach. As already explained, the $\mathcal{R}$ matrix is simply an unphysical artifice introduced in order to permit the physical $\mathcal{U}$ matrix to be evaluated. Its characteristics are not to be confused with physically meaningful features of the $\mathcal{U}$ matrix itself. Thus the $\mathcal{U}$ matrix contains no radius dependence, whereas the $\mathcal{R}$ matrix does, which leads to difficulties in the selection of an appropriate radius at which to match interior and exterior logarithmic derivatives. Already in the simple case of S-wave neutron scattering described by a diffuse potential of the form shown in Fig. 11-32, instead of obtaining the same cross section irrespective of whether the match is effected at the radius r_1 or r_2, it is found that the value is quite sensitive to the choice of matching radius. Even more so is this the case in the scattering of charged particles having nonzero orbital momenta, for which the potential assumes a more complicated appearance owing to the introduction of Coulomb and centrifugal components.

In an attempt to obviate these shortcomings, Humblet and Rosenfeld [Hu 61, Ro 61, Hu 62, Ma 63a, Ma 63b, Hu 64a, Hu 64b, Hu 64c] have reverted to S-matrix theory and employ the $\mathcal{U}$-matrix formalism, but with a

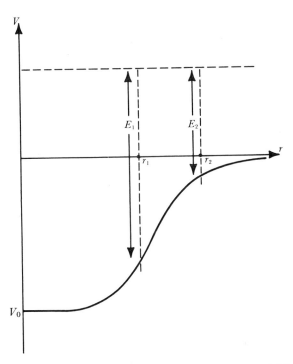

Fig. 11-32. Schematic representation of a diffuse nuclear potential in the scattering of S-wave neutrons. The sketch indicates the real central part of a typical potential, such as that due to Woods and Saxon [Woo 54] of the form $V(r) = V_0\{1 + \exp\,[(r-R)/a]\}^{-1}$ where a is a measure of the diffuseness, in general taken to be $a = 0.65$ fm, R is the nuclear radius which typically is given by $R = 1.3A^{1/3}$ fm, while V_0 is approximately $V_0 = -42$ MeV, depending upon the energy (and nature) of the incoming particles. As the nuclear radius is not clear-cut for such a diffuse surface, the relative neutron energy (e.g., E_1, E_2) and hence also the scattering cross section depends sensitively upon the choice of matching radius (e.g., r_1, r_2).

complex energy $\mathcal{E}_\lambda = E_\lambda - \frac{1}{2}i\Gamma_\lambda$ together with corresponding eigenfunctions ψ_λ. We have already presented the basic considerations in Eqs. (11-248) to (11-256) at the start of this section. The Humblet-Rosenfeld approach sets about parametrising the collision matrix $\mathcal{U}_l$ within the provisions of certain physical requirements, e.g., the expansion should take proper cognisance of (1) the boundary radius $r = b$; (2) time-reversal invariance; (3) threshold behavior of the cross section; (4) unitarity of the $\mathcal{U}$ matrix (flux conservation).

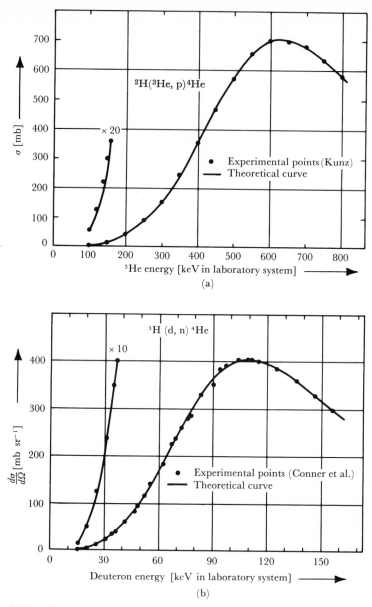

Fig. 11-33. Theoretical fits derived by Mahaux [Ma 63a] from the Humblet-Rosenfeld $\mathcal{U}$-matrix theory [Hu 61, Ro 61, Hu 62] in single-level approximation as a fit to experimental data: (a) total cross section for the ^{2}H(^{3}He, p)^{4}He reaction, fitted to resonance values measured by Kunz [Ku 55]; (b) differential cross section for the ^{3}H(d, n)^{4}He reaction, fitted to the experimental data of Conner *et al.* [Co 52].

One rather upsetting feature of this approach is the fact that so far no expansion has been found which can satisfy all four requirements simultaneously. For this reason, condition (4) has been relaxed and replaced by a weaker requirement, which can be fulfilled, namely, that the infinite expansion series be unitary even though the individual terms are not. The general expression which then ensues for $\mathscr{U}_{c'c}$ takes the form

$$\mathscr{U}_{c'c} = \delta_{c'c} + K_{c'c} - i \sum_\lambda \mathscr{P}_c^{1/2} \frac{A_{c'\lambda} A_{c\lambda}}{\mathscr{E} - \mathscr{E}_\lambda} \mathscr{P}_c^{1/2} e^{i\delta_\lambda} \qquad (11\text{-}294)$$

where the $\mathscr{P}_c^{1/2} \equiv \mathscr{E}_c k_c^{l+(1/2)}$ are energy-dependent factors, the $A_{c\lambda}$ are energy-dependent quantities proportional to widths $\Gamma_{c\lambda}^{1/2}$, the exponent δ_λ is a constant phase shift, and $K_{c'c}$ represents an energy-dependent *background contribution* whose value cannot be derived analytically. Since it has proved possible to fit experimental data well in absence of this term, it is customarily set to zero (or to a small constant value determined by least-squares analysis). Thus, for example, in the *single-level approximation without background*, in which only the first term in the summation (11-294) is taken, the cross section for a reaction proceeding from channel c to channel c' ($\neq c$) can be written

$$\sigma_{\rm r} \equiv \sigma_{c'c} = \frac{\pi}{k_c^2} g \mathscr{P}_c \frac{A_{c'\lambda}^2 A_{c\lambda}^2}{(E - E_\lambda)^2 + \frac{1}{4}\Gamma_\lambda^2} \mathscr{P}_{c'} \qquad (11\text{-}295)$$

in which g is known, and the number of adjustable parameters is reduced to but *three* (as against *seven* in $\mathscr{R}$-matrix theory), namely the resonance energy E_λ, the width Γ_λ, and the unknown $A_\lambda \equiv A_{c'\lambda}^2 A_{c\lambda}^2$. The choice of matching radius does not affect the penetrabilities and has no influence upon the value of the cross section. Even in this extremely simple version of the theory it is possible to attain remarkably good fits to experimental data with the three-parameter single-level formula, as exemplified by Fig. 11-33. The value of A_λ determined by fitting the resonance can be retained even away from the resonance. To illustrate this we show in Fig. 11-34 the type of fit that can be achieved with a slightly more complicated form of the single-level theory in which six free parameters exist [Ma 65a, Ma 65b]. The formalism can readily be extended beyond the single-level approximation and even then remains simple to apply.

Along somewhat different lines, the S- and $\mathscr{R}$-matrix formalism is also currently being developed by Danos and Greiner [Da 65b, Da 66c]. As we have seen, the theoretical formalisms are capable of providing a very good basic description of reaction behavior and in particular can account admirably for resonance effects without the need to invoke detailed models of the interaction process itself. Alternatively, explicit models of the reaction mechanism can be used to furnish theoretical predictions of interaction behavior, as will be demonstrated in Volume II of this work.

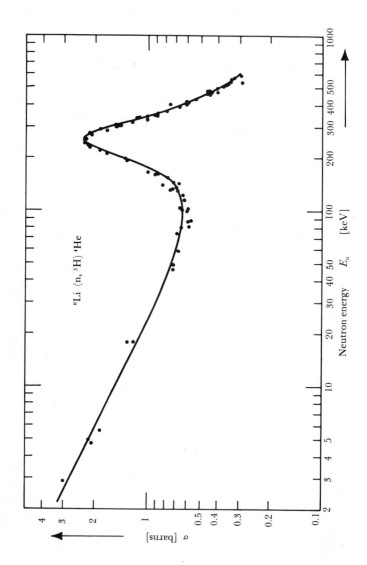

Fig. 11-34. Theoretical single-level $\mathscr{U}$-matrix fit to experimental data [Schw 65] for the $^6\mathrm{Li}(n,t)^4\mathrm{He}$ reaction on and off resonance (from [Ma 65b]). (Used by permission of North-Holland Publishing Co.)

EXERCISES

11-1   Calculate the absolute cross section σ_{abs} for capture of a 5-MeV neutron (treated as a point particle) which impinges upon a ^{82}Kr nucleus with an orbital momentum such that the collision can be regarded as a grazing collision.

$$\text{(Nuclear radius } R = r_0 A^{1/3} = 1.4 A^{1/3} \text{ fm.)}$$

11-2   A ^{12}C target is bombarded with neutrons, leading to formation of a compound nucleus state of spin-parity $\frac{1}{2}^+$ and excitation energy $E^* = 10.76$ MeV. If the ground state of the nucleus ^{13}C lies 4.947 MeV lower than that of the (^{12}C + n) system, calculate the absolute and the geometrical cross sections for the neutron capture reaction.

11-3   Calculate the resonance cross sections for the (n, n) and (n, γ) processes on ^{55}Mn (ground-state spin $J = \frac{5}{2}$) at the resonance energy $E_0 = 280$ eV, assuming that $\Gamma_n = 100\Gamma_\gamma$.

11-4   On an average, how many unvailing attempts to escape does a neutron trapped within a ^{56}Mn compound nucleus (with a mean kinetic energy of about 8 MeV) make before it succeeds? (The resonance width is $\Gamma_n = 20$ eV).

11-5   Calculate the excitation energy imparted to the compound nucleus in each of the following reactions when the energy of the incident particles or heavy ions is 6 MeV per nucleon in each case. Establish the identity of the compound nucleus and the residual nucleus in each case:

(a) ^{26}Mg (p, n)   (b) ^{23}Na (α, n)
(c) ^{13}C (^{14}N, n)   (d) ^{7}Li (^{20}Ne, n)

(Atomic masses: $M(^{26}\text{Mg}) = 25.982\ 593$ u; $M(^1\text{H}) = 1.007\ 825$ u; $M(^{23}\text{Na}) = 22.989\ 771$ u; $M(^4\text{He}) = 4.002\ 603$ u; $M(^{13}\text{C}) = 13.003\ 354$ u; $M(^{14}\text{N}) = 14.003\ 074$ u; $M(^7\text{Li}) = 7.016\ 004$ u; $M(^{20}\text{Ne}) = 19.992\ 441$ u. Atomic mass of compound nucleus: $M = 26.981\ 539$ u.)

11-6   Evaluate the maximum (n, n) and (n, γ) cross sections for ^{59}Co, which has a ground-state spin $J_0 = \frac{7}{2}$, for the resonance at a neutron energy $E_0 = 123$ eV, given that $\Gamma_n = 16\Gamma_\gamma$ and that the excited intermediate state has a spin $J_1 = 4$.
Compare your results with the geometrical cross section.

11-7 The radiative capture reaction $^{135}Xe(n, \gamma)^{136}Xe$ displays a marked resonance at an energy $E_0 = 0.084$ eV. The partial widths are: $\Gamma_\gamma = 0.0907$ eV, $\Gamma_n = 0.0257$ eV. (a) What is the magnitude of the capture cross section $\sigma_{(n,\gamma)}$ at resonance? (b) How large is the total cross section σ_{tot} at this energy? (c) Using the "compromise" value $g = \frac{1}{2}$ for the statistical spin factor, compare your result (b) with the experimental value [Hu 60] $\sigma_{exp} \simeq 3.5 \times 10^6$ b. (^{135}Xe has a ground-state spin $J = \frac{3}{2}$; the spin of the intermediate $^{136}Xe^*$ state is not known.)

11-8 Suppose that some other nucleus with a ground-state spin $J = \frac{3}{2}$ displays a neutron-capture resonance at $E_0 = 0.084$ eV. What is the largest value that the resonance cross section can assume? Compare this value with the geometrical cross section, taking the radius to be given by $R = 1.4A^{1/3}$ fm and assuming that $A = 135$. Also compare the value with that found for ^{135}Xe in the previous exercise.

11-9 As is well known, cadmium readily absorbs low-energy neutrons. The resonance occurs in ^{113}Cd and is characterized by the parameters ([Ra 47], updated in [Hu 60]) $E_0 = 0.178$ eV, $\Gamma_n = 0.00065$ eV, $\Gamma_\gamma = 0.113$ eV.

(a) Calculate the value of the resonance cross section for ^{113}Cd, and compare your result with that cited in the caption to Fig. 3-6.

(b) Show that since the neutron width Γ_n is inversely proportional to λ_n, i.e., directly proportional to $(E_n)^{1/2}$ while the total width Γ (and hence also the γ width Γ_γ) is essentially constant with energy, the cross section σ at energies E off resonance can be expressed in terms of the resonance cross section σ_0 as

$$\sigma_{(n,\gamma)}(E) = \sigma_0 \left(\frac{E_0}{E}\right)^{1/2} \frac{\Gamma^2}{4(E - E_0)^2 + \Gamma^2}$$

and calculate the capture cross section for *thermal* neutrons ($E_n = \frac{1}{40}$ eV) in *natural* cadmium.

(c) Demonstrate that for a broad resonance having $\Gamma \gg (E - E_0)$ the above formula expresses the $1/v$ law, e.g., when $E \ll E_0$.

(d) How thick a slab of natural cadmium would one need in order to effect a 10-percent reduction in the intensity of a (i) 0.025-eV (ii) 0.2-eV (iii) 0.3-eV neutron beam passing through it? (Isotopic abundance of $^{113}Cd = 12.26$ percent; density of natural cadmium: $\rho = 8.5$ g cm^{-3}. Ground-state spin of ^{113}Cd: $J_0 = \frac{1}{2}$; compound nucleus spin: $J_1 = 1$.)

11-10 A target of natural cadmium 0.5 mg cm^{-2} thick is bombarded by a flux of slow neutrons (10^8 neutrons per sec). On either side of the target, perpendicular to the incident beam, a neutron counter and a γ counter are set up. For the (n, n'γ) reaction in ^{113}Cd at the resonance energy $E_0 = 0.178$ MeV, calculate the counting rate in:

(a) the neutron counter, of diameter $d = 20$ cm and response efficiency to neutrons of this particular energy $\epsilon_n = 1$ percent, arranged 1 m away from the target;

(b) the γ counter, comprising an NaI(Tl) crystal of effective surface area 5 cm^2, γ-ray counting efficiency $\epsilon_\gamma = 25$ percent, also arranged 1 m away from the target.

(Isotopic abundance of ^{113}Cd = 12.26 percent; ground-state spin of ^{113}Cd, $J_0 = \frac{1}{2}$; compound nucleus spin, $J_1 = 1$; ratio of partial widths, $\Gamma_\gamma/\Gamma_n = 174$. Assume the distribution of the emitted radiation to be isotropic.)

11-11 Over the neutron energy range from 1 to 20 eV, measurements of the total cross section for ^{197}Au gave the results tabulated below. Plot these on log-log paper and endeavor to substantiate them on the assumption that the cross section may entirely be attributed to resonance absorption of S-wave neutrons to form a single level in the compound nucleus. In particular, determine (a) the resonance energy E_0; (b) the γ width Γ_γ; (c) the neutron width Γ_n, setting the spin factor equal to unity.

Plot the theoretical Breit-Wigner curve for comparison with the data and discuss the discrepancies you may observe in the upper energy region.

E_n [eV]	σ [barn]	E_n [eV]	σ [barn]
1	30	4.9	26,000
2	35	4.95	13,300
2.5	43	5.0	7500
3	60	5.1	2900
3.5	92	5.3	960
4	220	5.5	480
4.2	370	5.7	290
4.4	500	6	170
4.6	2200	7	56
4.7	5000	8	33
4.75	11,000	10	22
4.8	17,000	15	14
4.85	29,000	20	12

11-12 On integrating the Breit-Wigner formula over several adjoining resonances, the following expression is obtained [Fe 47] for the average (n, γ) cross section $\bar{\sigma}_l$ for neutrons with a broad energy distribution compared to the mean separation D between adjacent levels,

$$\bar{\sigma}_l = \frac{1}{D} \int \pi \lambda^2 (2l + 1) \frac{\Gamma_n \Gamma_\gamma}{(E - E_0)^2 + (\Gamma/2)^2} dE$$

$$\bar{\sigma}_l = \pi \lambda^2 (2l + 1) \frac{2\pi \overline{\Gamma_n \Gamma_\gamma}}{D(\overline{\Gamma}_n + \overline{\Gamma}_\gamma)}$$

where the bar over parameters indicates an average over the given energy range. By summing over all permitted l values, evaluate the total (n, γ) cross section

$$\sigma_{(n,\gamma)} = \sum_l \bar{\sigma}_l$$

and simplify the resultant expression for sufficiently high neutron energies $(E_n \gtrsim 10 \text{ keV})$ at which one can set $\overline{\Gamma}_n \gg \overline{\Gamma}_\gamma$. Measurements around $E_n \approx 1$ MeV on ^{65}Cu give an (n, γ) cross section $\sigma_{(n,\gamma)} = 6$ mb. From this result, calculate the mean level separation D at the appropriate excitation energy in ^{66}Cu if $\Gamma_\gamma = 0.1$ eV. (Cf. the experimental result [Hu 53], $D_{\exp} = 240$ eV.)

11-13 Imagine that you are planning an experiment to test the principle of detailed balance with the reactions

$$^{14}\text{N} + \text{d} \rightleftharpoons {}^{12}\text{C} + \alpha + 13.5 \text{ MeV}$$

and that you have at your disposal an accelerator giving 20-MeV deuterons or α particles of variable energy. What α-particle energy would be needed for the inverse reaction (in the laboratory system)? At what angle would the deuterons have to be observed in the inverse reaction if the α particles in the forward process are to be observed at $20°$ (lab)?

11-14 The cross section for the D-D reaction $^2\text{H}(\text{d}, \text{p})^3\text{H}$ with S-wave deuterons of energy $E_d = 100$ keV (lab) is $\sigma_{D-D} = 28$ mb and the Q value is $Q = +4.0329$ MeV. Calculate the cross section for the inverse reaction $^3\text{H}(\text{p}, \text{d})^2\text{H}$ with S-wave protons which have an energy such that the same excitation energy of the intermediate state is attained.

11-15 The total cross section for the deuteron photodisintegration process $^2\text{H}(\gamma, \text{n})^1\text{H}$ $(Q = -2.23$ MeV$)$ is $\sigma = 1.4$ mb for an incident γ energy $E_\gamma = 10$ MeV. What energy must neutrons possess if they are to give rise to γ radiation of the above energy when they participate in the inverse reaction ^{1}H$(n, \gamma)^2$H? What is the magnitude of the cross section for the inverse reaction under these conditions?

11-16 The total cross section of the reaction ^{3}He$(\gamma, p)^2$H $(Q = -5.49$ MeV$)$ is, theoretically, $\sigma_{tot} = 0.76$ mb for an incident γ energy $E_\gamma = 10$ MeV (lab). One would like to check this value experimentally, making use of the inverse reaction ^{2}H$(p, \gamma)^3$He. To this end one employs the following arrangement: A γ detector in the form of an NaI(Tl) crystal having the dimensions $L = 2R = 7.5$ cm is set up at right angles to the incident beam at a distance $a = 10$ cm from the target, which comprises deuterium gas within a container which is 3 cm long and under a pressure of 1 atm. The proton beam current is 0.1 μA. At what proton energy (in the laboratory system) should the experiment be undertaken?

What counting rate can be expected if the angular distribution of the emitted γ radiation assumes the theoretically predicted form

$$\frac{d\sigma}{d\Omega} \sim \sin^2 \theta$$

(The absorption coefficient for 10-MeV γ radiation in the crystal is $\mu = 0.14$ cm^{-1}.)

11-17 Use detailed-balance considerations to determine the spin of the π^+ meson, given that for the reaction

$$p + p \to \pi^+ + d$$

at $E_p = 341$ MeV (lab) the differential cross section in the forward direction $(\theta = 0°)$ is, according to [Ca 53], $d\sigma_{(p,\pi)} = 0.031$ mb sr^{-1} (CM), while for the reaction

$$\pi^+ + d \to p + p$$

in the forward direction at a roughly commensurate energy it is, according to [Sa 58a], $d\sigma_{(\pi,p)} = 0.85$ mb sr^{-1} (CM).

11-18 The reason for calling the quantity δ_l in Eqs. (11-198) and (11-199) a "phase shift" is that the lth term of the asymptotic wave function behaves essentially like the lth term of a free wave shifted by a phase angle δ_l. Justify this statement by showing that with the aid of Eq. (11-198) the lth term of the asymptotic wave function (11-172) can be cast into the form

$$\psi_l \simeq \frac{1}{v^{1/2}} \frac{1}{kr} [4\pi(2l + 1)]^{1/2} i^l e^{i\delta_l} \sin (kr - \tfrac{1}{2}l\pi + \delta_l) \, Y_{l0}(\Omega)$$

KINEMATICS OF RELATIVISTIC PARTICLES

A.1. Lorentz Transformation

The purpose of this Appendix is to serve to recall some relations and features inherent in the special theory of relativity which have a direct bearing upon nuclear physics. In the interests of simplicity and practicality we make no attempt at completeness or rigor but rather sketch out a selection of the salient points encountered in the applications to nuclear and particle behavior at relativistic velocities. For a more exhaustive treatment, the reader should refer to such authoritative texts as [Ei 23, Ei 51, Pa 58, Mø 66b].

When particles move with speeds approaching that of light, classical relations between length and time intervals as measured in the moving system intrinsic to the particle and those measured by a stationary observer (i.e., in the laboratory system) no longer hold. Thus the space-time coordinates (x, y, z, t) of a point in the stationary system Σ are no longer related by a classical Galilei transform to the coordinates (x', y', z', t') of that same point in space-time when referred to the particle's system of reference Σ' moving with a constant velocity $\mathbf{v}$ with respect to Σ in the x direction. If the origins of the two systems are so chosen that

$$x' = y' = z' = t' = 0 \qquad \text{for} \qquad x = y = z = t = 0 \tag{A-1}$$

the coordinates are related by the LORENTZ TRANSFORMATION

$$\left. \begin{aligned} x' &= \gamma(x - \beta ct) \\ y' &= y \\ z' &= z \\ t' &= \gamma\left(t - \frac{\beta}{c}x\right) \end{aligned} \right\} \tag{A-2}$$

where

$$\beta \equiv \frac{v}{c} \leqslant 1 \qquad \text{and} \qquad \gamma \equiv (1 - \beta^2)^{-1/2} \geqslant 1 \qquad \text{(A-3)}$$

with c representing the velocity of light *in vacuo*. It is instructive to note the graphical interrelation between these entities, as plotted in Fig. A-1. The Lorentz relations reduce to the classical Galilei transformation

$$\left. \begin{array}{l} x' = x - vt \\ t' = t \end{array} \right\} \qquad \text{(A-4)}$$

in the low-velocity limit $v \ll c$ (i.e., $\beta \to 0$ and $\gamma \to 1$).

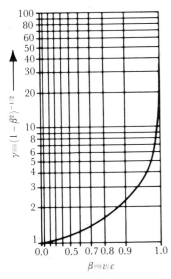

Fig. A-1. Log-log plot of the relativistic Einstein factor $\gamma \equiv (1 - \beta^2)^{-1/2}$ *vs* the reduced velocity $\beta \equiv v/c$.

The inverse Lorentz transformation, i.e., that from the (primed) coordinates in Σ' to those in the rest system Σ, is derived by simply reversing the sign of v, viz.

$$\left. \begin{array}{l} x = \gamma(x' + \beta ct') \\ y = y' \\ z = z' \\ t = \gamma \left(t' + \dfrac{\beta}{c} x' \right) \end{array} \right\} \qquad \text{(A-5)}$$

Figure A-2 depicts the situation schematically. A convenient way to bring to mind the form of the transformation in the first instance is to recall the classical spatial relation $x' = x - vt$ and simply multiply this by the Einstein factor γ to get

$$x' = \gamma(x - vt) = \gamma(x - \beta ct) \qquad \text{(A-6)}$$

Inherent in the nature of these linear relationships is the constancy of c as an ultimate terminal velocity,† conjoined with the relativity of simultaneity (events simultaneous in Σ' are not necessarily simultaneous in Σ), the Lorentz-Fitzgerald contraction (a spatial interval in Σ' can be different from that in Σ), and time dilation (a time interval in Σ' is perceived as a different time interval in Σ).

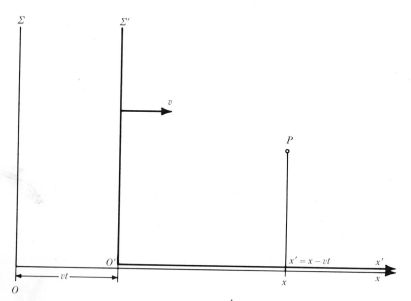

Fig. A-2. Diagram to illustrate the classical Galilei transformation of spatial coordinates. The abscissa in the system Σ', which has a velocity v in the x direction with respect to the observer's rest system Σ, is related to the Σ abscissa through the coordinate transformation for an arbitrary point P by the expression $x' = x - vt$. The relativistic generalization is represented by the Lorentz transformation $x' = \gamma(x - \beta ct)$.

A.1.1. LORENTZ CONTRACTION AND TIME DILATION

We consider two events which are separated in Σ' by a spatial x' interval $\Delta x'$. Then their spatial separation in Σ as measured instantaneously (i.e., $\Delta t = 0$: The events are simultaneous in Σ, but not in Σ', for there $\Delta t' \neq 0$) may be obtained from the relation

$$\Delta x' = \gamma(\Delta x - \beta c \Delta t) \tag{A-7}$$

† Cf. Exercise A-7 for a consideration of the possibility that certain particles may possess speeds in excess of c. In a recent theoretical model to account for the red shifts of quasars, Bludman and Ruderman also conceive of possible "superluminal" velocities.

as

$$\Delta x' = \gamma \Delta x \qquad \text{(A-8)}$$

Since $\gamma \geqslant 1$ it follows that

$$\Delta x \leqslant \Delta x' \qquad \text{(A-9)}$$

which implies that to an observer at rest, a moving object appears shortened along its direction of motion, as compared with the length of that same object when at rest relative to the observer. The LORENTZ CONTRACTION OF LENGTH is expressed by the factor γ and hence becomes of increasing importance as v tends to c. In the limit $v = c$, the moving object evinces vanishing spatial extension along its direction of motion.

By a similar argument applied to events which though equilocal in Σ' (i.e., $\Delta x' = 0$) are not simultaneous ($\Delta t' \neq 0$) one observes that the relation

$$\Delta t = \gamma \left(\Delta t' + \frac{\beta}{c} \Delta x' \right) \qquad \text{(A-10)}$$

reduces to

$$\Delta t = \gamma \Delta t' \geqslant \Delta t' \qquad \text{(A-11)}$$

and therefore to an observer at rest, time progression appears slowed down by a factor γ compared with the time intervals between the same events as measured in the moving system Σ': time is dilated.

An instance of such an effect as observed in practice is provided by the fact that the μ mesons' mean lifetime $\tau_\mu = 2.26 \times 10^{-6}$ sec (referred to stationary muons) is too small to be compatible in classical theory with their penetration of the Earth's atmosphere. Classical kinematics would restrict their mean free path, even if their velocity were roughly that of light, to

$$s = c\tau_\mu = 6.78 \times 10^4 \text{ cm} \simeq 700 \text{ m,}$$

whereas balloon observations indicate the mean penetrability to be about 30 km, corresponding to an apparent lifetime of approximately 10^{-4} sec. The explanation furnished by the special theory of relativity is that the intrinsic lifetime τ_μ is dilated to a value $\approx 10^{-4}$ sec when determined in a laboratory (rest) system of reference relative to which the muon moves with a relativistic velocity ($\rightarrow c$).

A.1.2. GEOMETRICAL REPRESENTATION OF THE LORENTZ TRANSFORMATION

By introducing the concept of a 4-space having coordinates

$$(x_1, x_2, x_3, x_4) \equiv (x, y, z, ict) \qquad \text{(A-12)}$$

in which the fourth component of the space-time coordinate vector is proportional to time t (since c is a constant), has the dimensions of space $[ct] = [L]$, and is pure imaginary $(i \equiv \sqrt{-1})$, Minkowski could interpret the invariant quantity

$$x^2 + y^2 + z^2 - c^2 t^2 = x'^2 + y'^2 + z'^2 - c^2 t'^2 \qquad (A\text{-}13)$$

as the square of a vector in a four-dimensional orthogonal coordinate system. This enabled the Lorentz transformation to be regarded analogously to a normal coordinate transformation between two simple Cartesian systems of which one has been rotated about the origin through an angle θ with respect

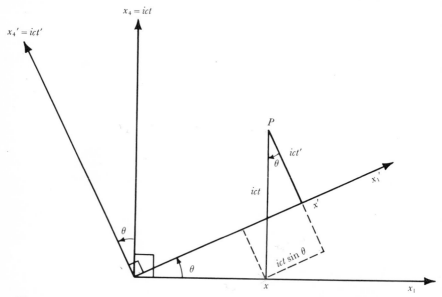

Fig. A-3. Minkowski representation of the Lorentz transformation as an axial rotation in terms of 4-component vector coordinates $(x, y, z, ict) \rightarrow (x', y', z', ict')$. The untransformed y and z axes are suppressed.

to the other (Fig. A-3 depicts this situation, omitting the untransformed y and z axes):

$$x_1' = x_1 \cos \theta + x_4 \sin \theta \qquad (A\text{-}14)$$

$$x_4' = -x_1 \sin \theta + x_4 \cos \theta \qquad (A\text{-}15)$$

Then, regarding the Lorentz transformation as a rotation in the (x_1, x_4) plane, we obtain the relations which connect θ with β:

$$\cos \theta = \gamma \qquad \sin \theta = i\gamma\beta \qquad \tan \theta = i\beta \qquad (A\text{-}16)$$

A useful application of this representation arises in the treatment of velocity addition.

A.1.3. COMPOSITION OF COLLINEAR VELOCITIES

If a body, instead of being at rest in Σ', moves in Σ' along the x' direction with a uniform velocity $\mathbf{u}'$, and Σ' itself has a translational velocity $\mathbf{v}$ with respect to Σ, the Minkowski approach renders it simple to calculate the body's velocity when observed in the rest system Σ (it is *not* just $\mathbf{u}' + \mathbf{v}$). If one writes

$$\frac{u'}{c} \equiv \beta' = -i \tan \theta' \tag{A-17}$$

and

$$\frac{v}{c} \equiv \beta = -i \tan \theta \tag{A-18}$$

one can derive the resultant combined velocity u'' of the body as measured in Σ from the relation

$$\frac{u''}{c} \equiv \beta'' = -i \tan (\theta + \theta') = -i \frac{\tan \theta + \tan \theta'}{1 - \tan \theta \tan \theta'} \tag{A-19}$$

viz.

$$\beta'' = \frac{\beta + \beta'}{1 + \beta \beta'} \tag{A-20}$$

or, explicitly,

$$u'' = \frac{v + u'}{1 + (vu'/c^2)} \tag{A-21}$$

A.1.4. RELATIVISTIC ADDITION OF NONCOLLINEAR VELOCITIES

We consider next three-dimensional spatial motion in Σ' as expressed functionally by

$$x' = x'(t') \qquad y' = y'(t') \qquad z' = z'(t') \tag{A-22}$$

and seek to deduce the connection between velocity components $(u_{x'}', u_{y'}', u_{z'}')$ in Σ' and those in Σ namely (u_x, u_y, u_z), as indicated in Fig. A-4. We can write the following expressions for the respective velocity components:

(a) in Σ:

$$\left. \begin{array}{l} u_x = \dfrac{dx}{dt} = u \cos \varphi \\[2ex] u_y = \dfrac{dy}{dt} \\[2ex] u_z = \dfrac{dz}{dt} \end{array} \right\} \tag{A-23}$$

(b) in Σ':

$$u'_{x'} = \frac{dx'}{dt'} = u' \cos \varphi'$$

$$u'_{y'} = \frac{dy'}{dt'}$$

(A-24)

$$u'_{z'} = \frac{dz'}{dt'}$$

and

$$u = (u_x^2 + u_y^2 + u_z^2)^{1/2} \qquad u' = (u_{x'}'^2 + u_{y'}'^2 + u_{z'}'^2)^{1/2} \qquad \text{(A-25)}$$

with

$$\varphi = \measuredangle\,(\mathbf{u}, \mathbf{x}) \qquad \varphi' = \measuredangle\,(\mathbf{u}', \mathbf{x}') \qquad \text{(A-26)}$$

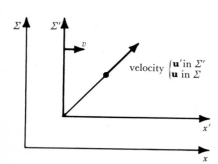

velocity $\begin{cases} \mathbf{u}' \text{ in } \Sigma' \\ \mathbf{u} \text{ in } \Sigma \end{cases}$

Fig. A-4. Diagram to illustrate the relativistic addition of noncollinear velocities.

Then, making use of the Lorentz transformation, with $\beta \equiv v/c$,

$$\left.\begin{array}{l} dx = \gamma(dx' + \beta c\, dt') \\ dy = dy' \\ dz = dz' \\ dt = \gamma\left(dt' + \dfrac{\beta}{c}\, dx'\right) \end{array}\right\} \qquad \text{(A-27)}$$

we can substitute throughout in order to relate unprimed to primed quantities,

$$u_x = \frac{dx' + \beta c\, dt'}{dt' + (\beta/c)\, dx'} = \frac{u'_{x'} + \beta c}{1 + (\beta/c)\, u'_{x'}} \qquad \text{(A-28)}$$

$$u_y = \frac{u'_{y'}}{\gamma[1 + (\beta/c)\, u'_{x'}]} \qquad \text{(A-29)}$$

$$u_z = \frac{u'_{z'}}{\gamma[1 + (\beta/c)\, u'_{x'}]} \qquad \text{(A-30)}$$

and

$$u^2 = u_x^2 + u_y^2 + u_z^2 = \left(1 + \frac{\beta}{c}\, u'_{x'}\right)^{-2}\left[(u'_{x'} + \beta c)^2 + \frac{1}{\gamma^2}\,(u_{y'}'^2 + u_{z'}'^2)\right] \qquad \text{(A-31)}$$

$$\frac{a^2}{t^2 + u' v \cos \varphi'}$$

i.e.,

$$u^2 = \left(1 + \frac{u' v}{c^2} \cos \varphi'\right)^{-2} \left[u'^2 + v^2 + 2u'_{x'} v - \beta^2 (u'^2_{y'} + u'^2_{z'})\right] \qquad \text{(A-32)}$$

viz.

$$u^2 = \left(1 + \frac{u' v}{c^2} \cos \varphi'\right)^{-2} \left[u'^2 + v^2 + 2u' v \cos \varphi' - \left(\frac{u' v}{c} \sin \varphi'\right)^2\right] \qquad \text{(A-33)}$$

Hence one can write

$$\frac{1}{\gamma} = \left(1 - \frac{u^2}{c^2}\right)^{1/2} = \left(1 + \frac{u' v}{c^2} \cos \varphi'\right)^{-1} \left(1 - \frac{u'^2}{c^2}\right)^{1/2} \left(1 - \frac{v^2}{c^2}\right)^{1/2} \qquad \text{(A-34)}$$

and

$$\tan \varphi = \frac{u' \sin \varphi'}{v + u' \cos \varphi'} \left(1 - \frac{v^2}{c^2}\right)^{1/2} \qquad \text{(A-35)}$$

The inverse formulae result from replacing v by $-v$.

A.1.5. DOPPLER EFFECTS (FREQUENCY SHIFTS)

It is usual to distinguish between two Doppler effects which result in frequency shifts, viz.

(1) the first-order Doppler effect in longitudinal motion ($\sim v/c$),

(2) the second-order Doppler effect in transverse motion ($\sim v^2/c^2$). The expressions for these can be derived in unified form through a considera- tion of the wave propagation or wave-number vector when viewed as having four components which are subject to the same kind of transformation as we have already applied to space-time coordinates. We consider the case of a stationary observer in system Σ who receives radiation emitted from a source which he perceives at an angle θ to his x direction (Fig. A-5). The emitter (system Σ') moves in the x-direction with a uniform velocity $\mathbf{v}$ as it emits the radiation of frequency $\nu' \equiv \omega'/2\pi = c/\lambda' = k' c/2\pi$. The spatial components of the wave vector in the rest system Σ are

$$\mathbf{k} = \left(k_x = -\frac{2\pi}{|\lambda|} \cos \theta, k_y, k_z\right) \qquad \text{(A-36)}$$

while the fourth component can be deduced from the invariance of the phase,

$$\omega t - (\mathbf{k} \cdot \mathbf{r}) = \omega t - (k_x x + k_y y + k_z z) = -\left(\frac{i\omega}{c}\right)(ict) - (\mathbf{k} \cdot \mathbf{r}) \qquad \text{(A-37)}$$

as

$$k_t \equiv \frac{k_4}{ic} = \frac{\omega}{c^2} \qquad \text{(A-38)}$$

which yields the wave-number 4-vector

$$k \equiv (k_1, k_2, k_3, k_4) = \left(k_x, k_y, k_z, \frac{i\omega}{c} \right) \tag{A-39}$$

The components of this may be subjected to a Lorentz transformation of the same sort as in Section A.1, namely,

$$k_x = \gamma(k'_{x'} + \beta c k'_{t'}) \tag{A-40}$$

$$k_t = \gamma \left(k'_{t'} + \frac{\beta}{c} k'_{x'} \right) \tag{A-41}$$

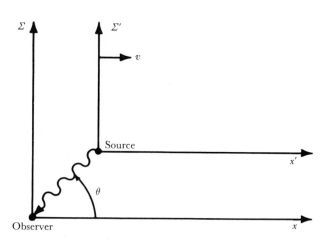

Fig. A-5. Representation of the emission of radiation in a system Σ' which moves with velocity v in the x direction relative to an observer's rest system Σ, in which the signal is received at an angle θ and a frequency $\nu = \omega/2\pi$.

Substituting in (A-40) for $k'_{x'}$ from (A-41), we find

$$k_x = \frac{c}{\beta} k_t - \gamma \frac{c}{\beta} (1 - \beta^2) k'_{t'} \tag{A-42}$$

which, on substitution for the k's yields the general Doppler relation

$$\omega = \frac{\omega'}{\gamma(1 + \beta \cos \theta)} \tag{A-43}$$

i.e.,

$$\nu = \frac{\nu'}{\gamma(1 + \beta \cos \theta)} \tag{A-44}$$

This reduces to the usual formulae for the first- and second-order Doppler effects on substitution:

(a) DOPPLER-I:

longitudinal motion $(\theta = 0)$:

$$\nu = \nu'\left(\frac{1 - \beta}{1 + \beta}\right)^{1/2} \tag{A-45}$$

hence at recession $(\beta > 0)$:

$$\nu < \nu' \quad \text{(i.e., a red shift)} \tag{A-46}$$

and at approach $(\beta < 0)$:

$$\nu > \nu' \quad \text{(i.e., a violet shift)} \tag{A-47}$$

(b) DOPPLER-II:

transverse motion $(\theta = \pi/2)$:

$$\nu = \nu'/\gamma \tag{A-48}$$

which corresponds to a red shift, independent of the sign of β (a manifestation of time dilation, $\Delta t = \gamma \Delta t'$).

A.2. Relativistic Mass, Momentum, and Energy

By the REST MASS m_0 of a particle, one means the mass determined in a system with respect to which the particle is at rest. One could conceive of this as being the intrinsic mass of a particle. However, when the particle (and therefore its eigensystem Σ') is moving linearly with a constant velocity relative to another system Σ in which an observer is at rest, the mass that this observer will measure is larger by a factor γ. Thus the RELATIVISTIC MASS m is related to the rest mass m_0 through the expression

$$m = \gamma m_0 \tag{A-49}$$

which tends to infinity as the particle's speed tends to the terminal velocity c. This tendency of particle masses to increase markedly with speed in the relativistic region necessitates a modification in the normal cyclotron phase to keep in step with the relativistic mass increase, which has led to the construction of synchro-cyclotrons.

If we apply the 4-vector concept to velocity, we find the spatial components to be†

$$\mathbf{V} = (V_1, V_2, V_3) = \gamma(v_x, v_y, v_z) \tag{A-50}$$

and, by analogy, we derive the timelike component as

$$V_4 = \gamma \frac{d}{dt}(x_4) = \gamma \frac{d}{dt}(ict) = \gamma ic \tag{A-51}$$

so that

$$V \equiv (V_1, V_2, V_3, V_4) \equiv \gamma(v_x, v_y, v_z, ic) \tag{A-52}$$

and

$$p = m_0 v = (p_1, p_2, p_3, p_4) \equiv \gamma(m_0 v_x, m_0 v_y, m_0 v_z, im_0 c) \tag{A-53}$$

To p_4 there corresponds a "time-component"

$$p_t \equiv \frac{p_4}{ic} = \gamma m_0 \tag{A-54}$$

from which one can build a product which corresponds to the TOTAL ENERGY E_{tot} of a moving body:

$$E_{\text{tot}} = c^2 p_t = \gamma m_0 c^2 = m_0 c^2 (1 + \tfrac{1}{2}\beta^2 + \cdots) \tag{A-55}$$

i.e.,

$$E_{\text{tot}} = m_0 c^2 + \tfrac{1}{2} m_0 v^2 + \cdots \tag{A-56}$$

We can now identify the first term on the right side as the energy of a body at rest $(E_{\text{tot}} \xrightarrow{v=0} m_0 c^2)$, and call this the REST ENERGY E_0, given by the famous mass-energy relation [Ei 05b]

$$E_0 = m_0 c^2 \tag{A-57}$$

The second term in the series corresponds to the kinetic energy E_{kin}, and

---

† In common with most other treatments (e.g. [Ro 64c]), we use a definition of 4-vector velocity components in which the differentiation of spacial coordinates is undertaken with respect to PROPER TIME τ rather than to the time t in the observer's rest system of reference. We accordingly consider the spacial components $V_k \equiv dx_k/d\tau$ rather than $v_k \equiv dx_k/dt$. This consequently gives rise to a factor γ, since the proper time corresponds to time intervals in the moving system Σ', which are related to intervals dt in the rest system Σ through (A-11), viz.

$$\delta\tau \triangleq \delta t' = \frac{1}{\gamma}\delta t$$

so that

$$V_k \equiv \frac{dx_k}{d\tau} = \gamma \frac{dx_k}{dt} = \gamma v_k$$

and similarly for the fourth component V_4.

the higher terms represent higher-order relativistic corrections to E_{kin}, so that

$$E_{tot} = E_0 + E_{kin} = \gamma m_0 c^2 = \gamma E_0 \tag{A-58}$$

whence

$$E_{kin} = E_{tot} - E_0 = (\gamma - 1) E_0 \tag{A-59}$$

The relation (A-58) can also be invoked to derive the more usual term for p_4, namely,

$$p_4 = i\gamma m_0 c = i\frac{E_{tot}}{c} \tag{A-60}$$

whence

$$p = (p_x, p_y, p_z, icp_t) = \left(mv_x, mv_y, mv_z, i\frac{E}{c}\right) \tag{A-61}$$

a link between energy and momentum which relativity theory exploits and develops further. Before dealing with this point, however, we explicitly write down the (Lorentzian) ENERGY-MOMENTUM TRANSFORMATION RELATIONS connecting quantities in a system Σ' with those in the observer's (rest) system Σ:

$$\left.\begin{aligned}
p'_{x'} &= \gamma\left(p_x - \frac{\beta}{c}E_{tot}\right) \\
p'_{y'} &= p_y \\
p'_{z'} &= p_z \\
E'_{tot} &= \gamma(E_{tot} - \beta c p_x)
\end{aligned}\right\} \tag{A-62}$$

and note in passing that relativistically, force is defined as the rate of change of relativistic momentum:

$$\mathbf{F} = \frac{d\mathbf{p}}{dt} = \frac{d}{dt}(\gamma m_0 \mathbf{v}) \tag{A-63}$$

We can readily derive the ENERGY-MOMENTUM RELATION for motion in the x direction, noting that

$$p^2 c^2 = \gamma^2 m_0^2 \beta^2 c^4 \qquad \text{and} \qquad E_0^2 = (m_0 c^2)^2 \tag{A-64}$$

viz.

$$p^2 c^2 + E_0^2 = (m_0 c^2)^2 (1 + \gamma^2 \beta^2) = (\gamma m_0 c^2)^2 = E_{tot}^2 \tag{A-65}$$

This second-order relation

$$E_{tot}^2 = E_0^2 + p^2 c^2 \tag{A-66}$$

holds simultaneously with the first-order formula

$$E_{tot} = E_0 + E_{kin} \tag{A-67}$$

in which

$$E_{tot} = mc^2 = \gamma m_0 c^2 \tag{A-68}$$

and

$$E_{\text{kin}} = (\gamma - 1) E_0 = (m - m_0) c^2 \tag{A-69}$$

It is important to note that the CONSERVATION OF ENERGY in nuclear and particle reactions applies to the *total* relativistic energy rather than to just some constituent thereof. This point should be borne in mind when dealing with high-energy kinematics, since particle energies are usually expressed as *kinetic* energies; for example, a 10-GeV proton has a total energy

$$E_{\text{tot}} = E_0 + E_{\text{kin}} = 0.938 + 10 = 10.938 \,\text{GeV} \tag{A-70}$$

Moreover, one may note that for particles of *vanishing rest mass* (photons,† neutrinos), as indeed for all bodies in the extreme-relativistic limit ($v \to c$ and $\gamma \to \infty$), it follows from (A-67), (A-68), and (A-69) that

$$E_{\text{tot}} = E_{\text{kin}} = pc \tag{A-71}$$

as in classical electrodynamic theory, and

$$m = E/c^2 = p/c \tag{A-72}$$

A.3. "Relativistic" Particles

Particles of *nonvanishing rest mass* behave "relativistically" (i.e., show deviations from classical behavior) beyond kinetic energies whose values stand in direct relationship to E_0. The rest mass of electrons is so comparatively small that for each 5-keV increase in kinetic energy there is roughly a 1 percent relativistic increase in the mass. Already at around 50 keV there is a quite perceptible deviation from classical conditions, and it is consequently almost always necessary to treat electrons as relativistic particles in nuclear physics.

Table A-1. VALUES OF PARTICLE KINETIC ENERGIES AT WHICH THE CLASSICAL MAGNITUDE DIFFERS BY 1 PERCENT OR BY 50 PERCENT FROM THAT CALCULATED RELATIVISTICALLY

Particle	$E_{1\%}$ [MeV]	$E_{50\%}$ [MeV]
e	0.003 44	0.316
μ	0.711	65.3
π	0.939	86.3
p	6.31	580
n	6.32	581
d	12.6	1159
α	25.1	2303

† The upper limit of possible mass for a photon has been determined [Go 68] from recent precise geomagnetic data to be $m_\gamma < 4.0 \times 10^{-48}$ g.

In showing how the electron kinetic energy varies with momentum (Fig. A-6), we also depict the rate of increase of β to its asymptotic value of unity.

As an index to the onset of relativistic characteristics, we list in Table A-1 the relativistic kinetic energies $E_{1\%}$ and $E_{50\%}$ at which a 1 percent or 50 percent

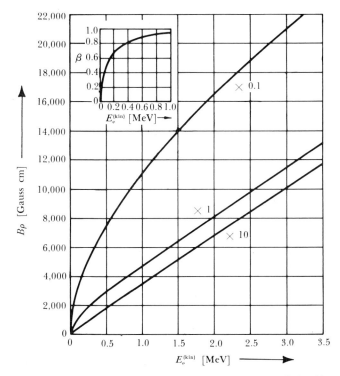

Fig. A-6. The relationship between magnetic rigidity $B\rho$ and the kinetic energy $E_e^{(kin)}$ of electrons over the range 0 to 35 MeV. The abscissa and ordinate scale units apply to the curve marked "×1". At lower energies, the ×0.1 curve applies, for which both scales are to be multiplied by this scale factor, while at higher energies the virtually linear relationship is represented by the ×10 curve, for which both scales should be multiplied by 10. Shown inset is the variation of β with kinetic energy $E_e^{(kin)}$ for electrons in the range below 1 MeV (from [Si 65]). (Used by permission of North-Holland Publishing Co.)

deviation is found from the relativistic value $E_{kin}^{(rel)} = (\gamma - 1)\, m_0 c^2$ when compared with the classical counterpart $E_{kin}^{(class)} = \frac{1}{2} m v^2$ [e.g., the value of $E_{kin}^{(rel)}$ for which $(E_{kin}^{(class)} - E_{kin}^{(rel)})/E_{kin}^{(rel)}$ is 1 percent or 50 percent).

By arbitrarily characterizing particles as "relativistic" when their kinetic energy starts to exceed a value corresponding to $\gamma = 1.1$, we derive the tabulation A-2.

Table A-2. Critical Kinetic Energies Corresponding to the Arbitrary Criterion ($\gamma = 1.1$)
for Onset of Relativistic Characteristics

Particle	Critical E_{kin} [MeV]
γ, ν	0
e	0.051
μ	11
π	14
p, n	94
d	188
α	373

10% E_0

A.4. Lifetimes of Relativistic Particles

Time dilation prolongs the intrinsic lifetime of swiftly moving particles by
a factor γ:

$$\tau_{lab} = \gamma \tau \qquad (A\text{-}73)$$

Thus the intrinsic lifetime of a particle which has a kinetic energy E_{kin} can be
determined from the measured lifetime if the rest mass is known, e.g.,

$$\tau = \tau_{lab} \frac{m_0 c^2}{E_{kin} + m_0 c^2} \qquad (A\text{-}74)$$

It is easy to check, for example, that the measured value $\tau_{lab} = 2.3 \times 10^{-5}$ sec
for the lifetime of 1-GeV muons indicates that they have an intrinsic life at
rest of $\tau = 2.2 \times 10^{-6}$ sec.

A.5. Speeds of Relativistic Charged Particles

The kinetic energy of particles which carry a charge q can be expressed in
terms of the electric potential difference U which would serve to accelerate
such particles from an initial state at rest to a final state of energy E_{kin}. Then

$$E_{kin} = |q| \, U \qquad (A\text{-}75)$$

and

$$\beta = \frac{v}{c} = \left[1 - \left(\frac{|qU|}{m_0 c^2} + 1 \right)^{-2} \right]^{1/2} \qquad (A\text{-}76)$$

In Table A-3 are listed the values of β and γ for various particles at different energies, as calculated according to the formula (A-76). The results are depicted graphically in Figs. A-7 and A-8.

Table A-3. VALUES OF β AND γ FOR VARIOUS PARTICLES OVER THE KINETIC ENERGY RANGE 0.1 eV TO 10^{12} eV

Kinetic energy E_{kin}	β (upper number) and γ (lower number) for particles of energy E_{kin}			
	e	p	d	α
0.1	0.0006	0.0000	0.0000	0.0000
	1.00	1.00	1.00	1.00
1.0	0.0020	0.0000	0.0000	0.0000
	1.00	1.00	1.00	1.00
10	0.0063	0.0001	0.0001	0.0001
	1.00	1.00	1.00	1.00
10^2	0.020	0.0005	0.0003	0.0002
	1.00	1.00	1.00	1.00
10^3	0.063	0.0015	0.0010	0.0007
	1.00	1.00	1.00	1.00
10^4	0.20	0.0046	0.0033	0.0023
	1.02	1.00	1.00	1.00
10^5	0.55	0.015	0.010	0.0073
	1.20	1.00	1.00	1.00
10^6	0.94	0.046	0.033	0.0232
	2.96	1.00	1.00	1.00
10^7	0.99	0.15	0.10	0.073
	20.6	1.01	1.01	1.00
10^8	1.00	0.43	0.31	0.227
	198	1.11	1.05	1.03
10^9	1.000	0.88	0.76	0.615
	> 1000	2.07	1.53	1.27
10^{10}	1.0000	0.996	0.99	0.962
	> 1000	11.7	6.28	3.70
10^{11}	1.0000	1.000	1.00	0.9994
	$\rightarrow \infty$	108	54.5	27.9
10^{12}	1.0000	1.0000	1.000	1.0000
	$\rightarrow \infty$	1067	535	270

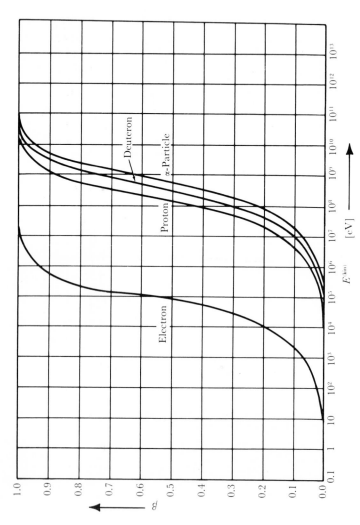

Fig. A-7. Variation of $\beta \equiv v/c$ with kinetic energy $E^{(\mathrm{kin})}$ for various particles (e, p, d, α) on a semilogarithmic scale.

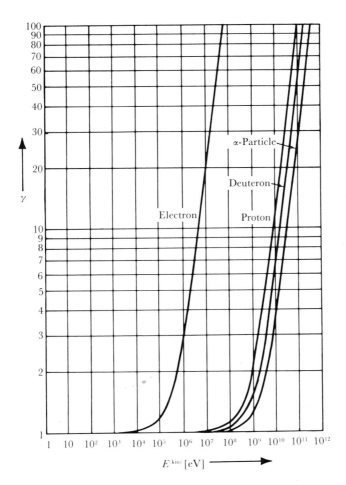

Fig. A-8. Log-log plots of the Einstein factor $\gamma \equiv (1 - \beta^2)^{-1/2}$ *vs* kinetic energy $E^{(\mathrm{kin})}$ for various particles (e, p, d, α).

A-1 An appealingly simple way of "deriving" the mass-energy relation $E = mc^2$ is by way of the following *Gedankenexperiment*: A perfectly isolated cylinder has a device at each end capable of emitting or absorbing pulses of electromagnetic radiation. When device A emits radiation of energy E (i.e., momentum $p = E/c$ according to *classical* electrodynamics) the system acquires a recoil momentum p and moves until the instant of reception of the radiation by device B. *Classically*, one would regard the radiation as massless, and would therefore be entitled to regard the masses of A and B as unchanged after this process—they could therefore be interchanged without any expenditure of work. This, however, leads to the paradox that an arbitrary displacement of the center of mass of the system could be undertaken without any work being done. The paradox can be resolved by ascribing a mass m to the radiation and M to each of the devices, mounted in a tube of negligible mass a distance $2L$ apart. Assume that the tube moves a distance during the flight time $t \cong 2L/c$ of the radiation and build moments about the center of mass to derive the relation $E = mc^2$.

A-2 A particle of rest mass m_0 moving at a relativistic velocity $\beta = v/c$ impinges upon a stationary particle of mass M_0 and is captured in the collision. What will be the speed of the ensuing composite particle?

A-3 Suppose that in a colliding-beam arrangement particles, each having a speed $0.99\ c$ relative to an outside observer, make a head-on collision. What is the speed of one particle with respect to the other?

A-4 The mean lifetime of μ mesons at rest is $\tau_0 = 2.26$ μsec. Their rest mass is $m_\mu = 207\ m_e$. What kinetic energy must μ mesons acquire upon creation at an altitude $h = 30$ km above the Earth's surface if, when they approach the earth perpendicularly, they just manage to reach sea level? (For simplicity, only the time dilation effect is to be considered.)

A-5 Using the Mössbauer technique with resonant absorption of 14.4-keV γ rays from ^{57}Fe, the Harwell group studied not only the gravitational shift in energy as photons move vertically [Cr 60b], but also the red shift between an emitter coated on to the spindle of a rotating disk and an absorber on the periphery [Ha 60c] (see also [She 60, Cha 63]). Calculate the specific energy shift in terms of the rotational frequency ω for a spindle radius $R_1 = 0.4$ cm and a peripheral radius $R_2 = 6.64$ cm. What would the specific energy shift be if source and absorber were mounted on diametrically opposite sides of the rim?

(The shift can be interpreted as a second-order Doppler effect due to transverse motion—i.e., a time dilation—or, equivalently, one could treat the acceleration as an effective gravitational field and calculate the difference in potential between source and absorber. The reader may care to examine this alternative approach.)

A-6 Check the numerical values collated in Tables A-1 and A-2.

A-7 It has been suggested [Bil 62, Fei 67] that it is not inconsistent within the framework of special relativity theory for particles ["tachyons"; $\tau\alpha\chi\iota s$ = fast] to exist whose velocity v lies in the range $c \leqslant v \leqslant \infty$. They would have an *imaginary rest mass* $(m_0 = i\mu$, where μ is a real quantity), but their "relativistic" mass, momentum and energy would be real, positive quantities. Moreover, as v increases from the lower limit c, *the momentum and energy both diminish*; at $v = \infty$ the momentum retains the non-zero value $p = \mu c$, however, while the energy becomes zero. Use "super-relativistic kinematics" to check these conclusions.

TRANSFORMATION RELATIONS BETWEEN THE LABORATORY AND CENTER-OF-MASS SYSTEMS FOR ELASTIC COLLISIONS

B.1. Characteristics of the Center-of-Mass System

By using the center-of-mass (CM) system of reference, defined as the system Σ' in which the CM of colliding particles is at rest, it is possible to reduce the two-body collision problem to a one-body problem of far lesser mathematical complexity. The calculation simplifies to one involving merely the interaction of a single particle, taken to have a reduced mass $\mu \equiv (m_1 m_2)/(m_1 + m_2)$ and velocity u_1 (which corresponds to the velocity with which a particle of mass m_1 impinges upon a stationary partner of mass m_2 in the laboratory system) with a potential field which can always be regarded as centered at the CM.

Because the CM system shares with the laboratory system the property of being an *inertial system,* any kinematic quantity referred to the one system can be transformed into the corresponding quantity in the other by application of the relativistic Lorentz relations.

An important feature of the CM system is the fact that the vector sum of momenta before and after collision vanishes,

$$\sum \mathbf{p}' = 0 \qquad (\text{B-1})$$

so that in a two-body collision not only are the respective *initial* momenta mutually antiparallel (in the incident direction), but the same holds (in the direction θ', ϕ') for the *final* momenta.

B.2. Nonrelativistic Elastic Collision
of a Moving Particle with a Stationary Target

We contrast the situation depicted in Fig. B-1(a) for the laboratory system with the corresponding elastic collision process as represented in the CM

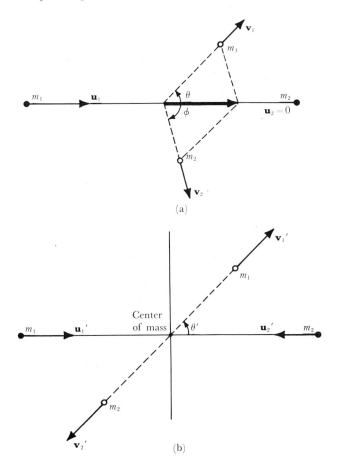

(a)

(b)

Fig. B-1. Kinematic quantities in an elastic collision process between a bombarding particle m_1 and a target particle m_2 (at rest in the laboratory system) as represented (a) in the laboratory system and (b) in the center-of-mass system.

system in Fig. B-1(b). In the following subsections, we set out to relate quantities in the lab system Σ (unprimed) with their counterparts in the CM system Σ' (primed).

For brevity, we use the following notation:

reduced mass: $\mu \equiv \dfrac{m_1 m_2}{m_1 + m_2}$ (B-2)

specific mass: $\mu_1 \equiv \dfrac{m_1}{m_1 + m_2}$ $\mu_2 \equiv \dfrac{m_2}{m_1 + m_2}$ (B-3)

mass ratio: $k \equiv \dfrac{m_1}{m_2} = \dfrac{\mu_1}{\mu_2}$ (B-4)

and note the interrelations

$$\left. \begin{array}{ll} \mu_1 = 1 - \mu_2 & \mu_2 = 1 - \mu_1 \\[2mm] m_1 \mu_2^2 = \mu \mu_2 & m_2 \mu_1^2 = \mu \mu_1 \end{array} \right\} \qquad \text{(B-5)}$$

Table B-1. KINETIC ENERGIES BEFORE AND AFTER A NONRELATIVISTIC ELASTIC
COLLISION

State	Particle	Kinetic energy in lab system	Kinetic energy in CM system
Initial	1	$\frac{1}{2}m_1 u_1^2$	$\frac{1}{2}m_1 u_1'^2 = \frac{1}{2}\mu\mu_2 u_1^2$
	2	0	$\frac{1}{2}m_2 u_2'^2 = \frac{1}{2}\mu\mu_1 u_1^2$
Final	1	$\frac{1}{2}m_1 v_1^2$	$\frac{1}{2}m_1 v_1'^2 = \frac{1}{2}\mu\mu_2 u_1^2$
	2	$\frac{1}{2}m_2 v_2^2$	$\frac{1}{2}m_2 v_2'^2 = \frac{1}{2}\mu\mu_1 u_1^2$
	CM	$\frac{1}{2}(m_1 + m_2) v_{CM}^2 = \frac{1}{2}m_1 \mu_1 u_1^2$	0

B.2.1. VELOCITY RELATIONS

The velocity of the CM in the lab system (e.g., with respect to the stationary
target m_2) is, in accordance with momentum conservation,

$$\mathbf{v}_{CM} = \frac{m_1}{m_1 + m_2}\, \mathbf{u}_1 = \mu_1 \mathbf{u}_1 \qquad \text{(B-6)}$$

In the CM system, by definition

$$v_{CM}' = 0 \qquad \text{(B-7)}$$

and hence the respective initial particle velocities are

$$\mathbf{u}_1' = \mathbf{u}_1 - \mathbf{v}_{CM} = (1 - \mu_1)\, \mathbf{u}_1 = \mu_2 \mathbf{u}_1 \qquad \text{(B-8)}$$

$$\mathbf{u}_2' = -\mathbf{v}_{CM} = -\mu_1 \mathbf{u}_1 \qquad \text{(B-9)}$$

Since the net kinetic energy is a conserved quantity in nonrelativistic processes, the particles recede from the collision center after their interaction with the same respective speeds (but in a new direction),

$$u_1' = v_1' = \mu_2 u_1 \quad \text{and} \quad u_2' = v_2' = \mu_1 u_1 \tag{B-10}$$

B.2.2. KINETIC ENERGY RELATIONS

Table B-1 collects the various quantities in a concise and self-explanatory form.

B.2.3. ANGULAR RELATIONS

The vector diagram in Fig. B-2 shows the geometrical relationship of the various velocities which are involved in elastic scattering. The various angular relations can readily be derived from this as follows:

Since $u_2' = v_2'$, the lengths $\overline{AB} = \overline{SC} = \overline{ED} = \overline{SE}$ and hence

$$\measuredangle\, ESD = \measuredangle\, SDE = \measuredangle\, DSC = \phi = \tfrac{1}{2}\phi' \tag{B-11}$$

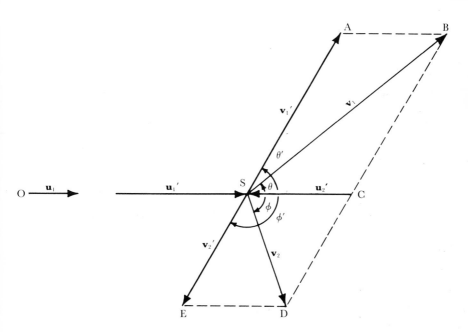

Fig. B-2. Vector diagram to illustrate the relationship between kinematic quantities in the laboratory and center-of-mass systems for an elastic scattering process.

so that

$$\text{\ding{73}} \, \text{ASC} \equiv \theta' = \pi - 2\phi \tag{B-12}$$

Thus the recoil angle ϕ, taken as negative by convention, is

$$\phi = \tfrac{1}{2}(\pi - \theta') \tag{B-13}$$

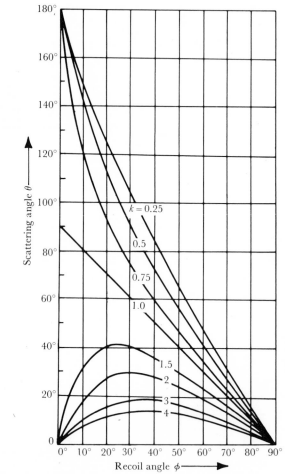

Fig. B-3. Graphical representation of the $\theta - \phi$ relation in the laboratory system for various values of the mass ratio $k \equiv m_1/m_2$ (from [Ma 58b]). (Used by permission of Cambridge University Press.)

Also, from the sine rule for triangle BSC,

$$\frac{v_1'}{\sin \theta} = \frac{u_2'}{\sin (\theta' - \theta)} \tag{B-14}$$

it follows that

$$\sin(\theta' - \theta) = \frac{m_1}{m_2} \sin\theta = k\sin\theta \qquad \text{(B-15)}$$

and

$$\tan\theta = \frac{v_1 \sin\theta}{u_2' + u_1' \cos\theta} = \frac{v_1 \sin\theta'}{\mu_1 u_1 + \mu_2 u_1 \cos\theta'}$$

i.e.,

$$\tan\theta = \frac{\sin\theta'}{k + \cos\theta'} \qquad \text{(B-16)}$$

The following relation can be shown to hold in the lab system between the scattering angle θ and the recoil angle ϕ:

$$\tan\theta = \frac{\sin 2\phi}{k - \cos 2\phi} \qquad \text{(B-17)}$$

This is represented graphically for various values of the mass ratio k in Fig. B-3.

B.2.4. Relations between Velocities, Energies, and Scattering Angles in the Laboratory System

The following results can be derived from the vector diagram (Fig. B-2):

(a) Since momentum is conserved, the perpendicular distance of the point P from $\overline{SD}$ must equal that of O from $\overline{SD}$, i.e.,

$$v_1 \sin(\theta + \phi) = u_1 \sin\phi \qquad \text{(B-18)}$$

whence

$$\frac{v_1}{u_1} = \frac{\sin\phi}{\sin(\theta + \phi)} \qquad \text{(B-19)}$$

(b) Also, because $\overline{SC} = \overline{CD}$ it follows that

$$v_2 = 2(\overline{SC})\cos\phi = 2u_2'\cos\phi = 2\mu_1 u_1 \cos\phi \qquad \text{(B-20)}$$

whence

$$\frac{v_2}{u_1} = 2\mu_1 \cos\phi \qquad \text{(B-21)}$$

(c) Further, the conservation of momentum along $\overline{OC}$ leads to the equality

$$m_1 u_1 = m_1 v_1 \cos\theta + m_2 v_2 \cos\phi \qquad \text{(B-22)}$$

i.e.,

$$\frac{v_2}{u_1} = \frac{m_1}{m_2} \frac{1}{\cos\phi}\left(1 - \frac{v_1}{u_1}\cos\theta\right) \qquad \text{(B-23)}$$

which, on substituting for v_1/u_1 from (B-19), gives

$$\frac{v_2}{u_1} = \frac{k \sin \theta}{\sin (\theta + \phi)} \tag{B-24}$$

(d) These relations yield the ratio of final to initial kinetic energy for particle 1:

$$\frac{(E_1)_f}{(E_1)_i} = \frac{\sin^2 \phi}{\sin^2 (\theta + \phi)} \tag{B-25}$$

and that of particle 2 referred to the initial kinetic energy of particle 1,

$$\frac{(E_2)_f}{(E_1)_i} = \frac{k \sin^2 \theta}{\sin^2 (\theta + \phi)} = 4\mu_1 \mu_2 \cos^2 \phi \tag{B-26}$$

(e) It follows that in terms of the maximum kinetic energy which can be transferred to the target body, namely

$$E_{\max} = 4\mu_1 \mu_2 (E_1)_i \tag{B-27}$$

the final recoil energy is

$$(E_2)_f = E_{\max} \cos^2 \phi \tag{B-28}$$

B.2.5. SOLID ANGLE RELATIONS

As shown in Fig. B-4, the solid angle $d\Omega$ in the lab system between two emergent cones of semiangle θ and $\theta + d\theta$ respectively, with a common apex at the target is

$$d\Omega = 2\pi \sin \theta \, d\theta \tag{B-29}$$

In the CM system, the corresponding solid angle is of the same form, viz.

$$d\Omega' = 2\pi \sin \theta' \, d\theta' \tag{B-30}$$

Hence from the relation (B-16) between θ and θ' it follows that the solid angles are connected by the formula

$$d\Omega = \frac{1 + k \cos \theta'}{(1 + 2k \cos \theta' + k^2)^{3/2}} d\Omega' \tag{B-31}$$

B.2.6. ANGULAR INTENSITY RELATIONS

We assume the intensity (i.e., the angular distribution) of emergent particles to be $I(\theta)$ in the lab system and $I(\theta')$ in the CM system. Furthermore, we make the simplifying assumption that the *CM distribution is isotropic, i.e.,*

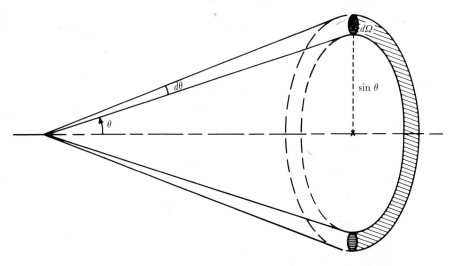

Fig. B-4. Diagram to illustrate the formula $d\Omega = 2\pi \sin \theta \, d\theta$.

$I(\theta') = \text{const.}$ Then we can write

$$I(\theta) \, d\Omega = I(\theta') \, d\Omega' \tag{B-32}$$

i.e.,

$$\frac{I(\theta)}{I(\theta')} = \frac{d\Omega'}{d\Omega} = \frac{2\pi \sin \theta' \, d\theta'}{2\pi \sin \theta \, d\theta} = \frac{d(\cos \theta')}{d(\cos \theta)} \tag{B-33}$$

But, from (B-16),

$$\cos \theta' = -k \sin^2 \theta \pm (1 - k^2 \sin^2 \theta)^{1/2} \cos \theta \tag{B-34}$$

whence

$$\frac{d(\cos \theta')}{d(\cos \theta)} = 2k \cos \theta \mp \frac{1 - k^2 + 2k^2 \cos^2 \theta}{(1 - k^2 \sin^2 \theta)^{1/2}} \tag{B-35}$$

Since physical reasons dictate that the ratio of intensities must be *positive*, we reject the solution with the minus sign and set

$$\frac{I(\theta)}{I(\theta')} = \frac{\sigma(\theta)}{\sigma(\theta')} = \frac{[(1 - k^2 \sin^2 \theta)^{1/2} + k \cos \theta]^2}{(1 - k^2 \sin^2 \theta)^{1/2}} \tag{B-36}$$

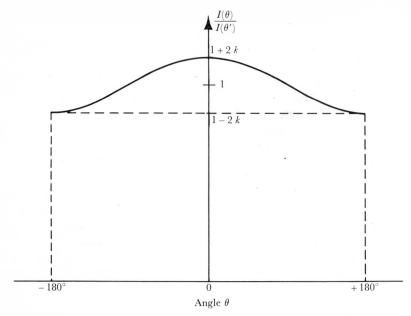

Fig. B-5. Variation of the ratio of intensities $I(\theta)/I(\theta')$ in the laboratory and center-of-mass systems over the range $-180°$ to $+180°$ of the scattering angle θ for the elastic scattering of a light particle on a massive target (mass ratio $k \equiv m_1/m_2 \ll 1$).

Three special cases now remain to be considered, namely:

(a) $k \ll 1$: scattering of a light particle on a heavy target:

$$I(\theta) \cong (1 + 2k \cos \theta)\, I(\theta') \qquad (\text{B-37})$$

which yields the relation shown in Fig. B-5.

(b) $k \gg 1$: scattering of a heavy particle on a light target: the maximum scattering angle is then

$$\sin \theta_{\max} \cong \theta_{\max} = \frac{1}{k} \qquad (\text{B-38})$$

and

$$I(\theta) = \left\{ 2k + \frac{1 + k^2}{[1 - (\theta/\theta_{\max})^2]^{1/2}} \right\} I(\theta') \qquad (\text{B-39})$$

a situation illustrated in Fig. B-6.

(c) $k = 1$: a collision of particles of equal mass:

$$I(\theta) = 4 \cos \theta\, I(\theta') \qquad (\text{B-40})$$

a relation depicted in Fig. B-7.

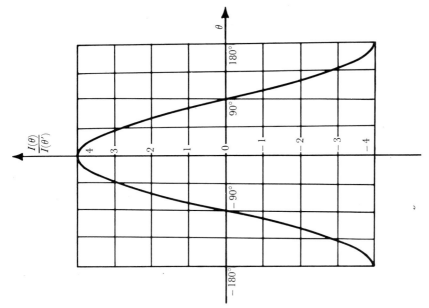

Fig. B-7. As Fig. B-5, but for particles of equal mass ($k = 1$).

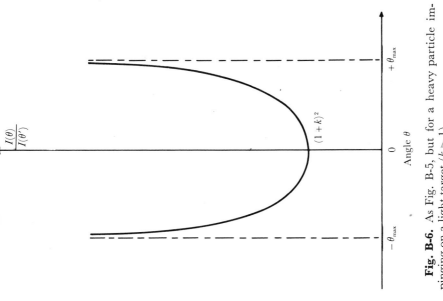

Fig. B-6. As Fig. B-5, but for a heavy particle impinging on a light target ($k \gg 1$).

**B.3. Relativistic Elastic Collision
of a Fast-Moving Particle with a Stationary Target**

The physical situation and vector collision diagrams remain unchanged, but we now have to introduce relativistic expressions from Appendix A for the velocities, momenta, and energies featured in the process, as well as observing the more complicated rules for, e.g., addition of velocities. The formulae therefore become rather involved. For a more detailed treatment, the reader is referred to Blaton [Bl 50a], Mather and Swan [Ma 58b], and Baldin *et al.* [Ba 61a].

We make use of the following 4-vector notation for the *initial* momenta:

$$
\left.
\begin{aligned}
\text{particle 1 in the lab system:} \quad & p_1 \equiv \left(p_{1x}, p_{1y}, p_{1z}, \frac{i E_{\text{tot}}^{(1)}}{c} \right) \\[2ex]
\text{particle 1 in the CM system:} \quad & p_1' \equiv \left(p_{1x}', p_{1y}', p_{1z}', \frac{i E_{\text{tot}}^{(1)'}}{c} \right) \\[2ex]
\text{particle 2 in the lab system:} \quad & p_2 = 0 \\[2ex]
\text{particle 2 in the CM system:} \quad & p_2' \equiv \left(p_{2x}', p_{2y}', p_{2z}', \frac{i E_{\text{tot}}^{(2)'}}{c} \right)
\end{aligned}
\right\}
\qquad \text{(B-41)}
$$

The fact that the respective initial and final speeds in the CM system are equal renders their initial and final CM momenta also equal (in magnitude, but not in direction), i.e.,

$$
\left.
\begin{aligned}
\text{final momentum of particle 1 in CM system} &= |\mathbf{p}_1'| \\
\text{final momentum of particle 2 in CM system} &= |\mathbf{p}_2'|
\end{aligned}
\right\}
\qquad \text{(B-42)}
$$

The modulus has been introduced to take account of equality in *magnitude*; if we take the absolute magnitude to be an amount p_0 and regard momentum components as *positive* if they lie in the positive x or y direction, we can write, explicitly,

$$
\left.
\begin{aligned}
p_{1x}' &= p_0 \cos \theta' & p_{2x}' &= -p_{1x}' = -p_0 \cos \theta' \\
p_{1y}' &= p_0 \sin \theta' & p_{2y}' &= -p_{1y}' = -p_0 \sin \theta' \\
p_{1z}' &= 0 & p_{2z}' &= 0
\end{aligned}
\right\}
\qquad \text{(B-43)}
$$

No
=

We also employ the usual relativistic notation,

$$\beta_i' \equiv \frac{u_i'}{c} \quad\text{and}\quad \gamma_i' = [1 - (\beta_i')^2]^{-1/2} \quad\text{with}\quad i = 1,2 \quad\text{(B-44)}$$

and use the abbreviation

$$\delta' \equiv \frac{\beta_2'}{\beta_1'} \tag{B-45}$$

For the respective energies, we use the expressions featured in Appendix A.

B.3.1. Velocity Relations

A simple relation connects *velocities within the CM system*. For the CM momenta before scattering we can use the definition (A-53) to write

$$p_1' = \gamma_1' m_1 u_1' = m_1 c \beta_1' \gamma_1' = m_1 c[(\gamma_1')^2 - 1]^{1/2} \tag{B-46}$$

and

$$p_2' = m_2 c[(\gamma_2')^2 - 1]^{1/2} = p_1' \tag{B-47}$$

whence

$$(m_1)^2[(\gamma_1')^2 - 1] = (m_2)^2[(\gamma_2')^2 - 1] \tag{B-48}$$

To derive *expressions relating velocities in the lab system to those in the CM system*, we note that the CM system moves with a velocity $v_{\mathrm{CM}} = u_2'$ with respect to the x direction of the lab system, and write the momentum-energy transformation in accordance with (A-62) as

$$p_1' = \gamma_2'\left(p_1 - \beta_2' \frac{E_{\mathrm{tot}}^{(1)}}{c}\right) \tag{B-49}$$

with

$$p_1 = \gamma_1 m_1 u_1 = m_1 c \beta_1 \gamma_1 = m_1 c[(\gamma_1)^2 - 1]^{1/2} \tag{B-50}$$

and

$$E_{\mathrm{tot}}^{(1)} = \gamma_1 m_1 c^2 \tag{B-51}$$

Since, by definition,

$$p_1' = m_1 c \beta_1' \gamma_1' = m_1 c[(\gamma_1')^2 - 1]^{1/2} \tag{B-52}$$

it follows from (B-49), (B-50), (B-51), and (B-52) that

$$\gamma_2'[(\gamma_1)^2 - 1]^{1/2} - \gamma_1[(\gamma_2')^2 - 1]^{1/2} = [(\gamma_1')^2 - 1]^{1/2} \tag{B-53}$$

Using the CM velocity relation (B-48) to substitute for the right side, we find

$$\gamma_2'[(\gamma_1)^2 - 1]^{1/2} = \left(\frac{1}{k} + \gamma_1\right)[(\gamma_2')^2 - 1]^{1/2} \tag{B-54}$$

i.e.,

$$\frac{(\gamma_1)^2 - 1}{[(1/k) + \gamma_1]^2} = 1 - \frac{1}{(\gamma_2')^2} \tag{B-55}$$

whence

$$\gamma_2' = \frac{(1/k) + \gamma_1}{[1 + (1/k^2) + 2(\gamma_1/k)]^{1/2}} = \gamma_{CM} \tag{B-56}$$

and, analogously,

$$\gamma_1' = \frac{k + \gamma_1}{(1 + k^2 + 2\gamma_1 k)^{1/2}} \tag{B-57}$$

B.3.2. EXPRESSIONS FOR ENERGY IN THE CENTER-OF-MASS SYSTEM

As stressed in Appendix A, the usual convention when referring to the energy of a particle is to imply its *kinetic* energy thereby. To obtain the *total* energy of that particle, its rest energy $E_0 = m_0 c^2$ must be added on:

$$E_{tot} = E_{kin} + E_0 \tag{B-58}$$

In the case of an elastic collision process the respective rest energies of the participants are each conserved, and therefore have no influence upon the energetics of the process. The COLLISION ENERGY, namely, the energy available for exciting a nucleus or initiating a nuclear reaction, is then composed of the sum of the individual kinetic energies, together possibly with such potential energies as may feature when the particles happen to be in an excited state. The magnitude of this total collision energy is equal to the CM ENERGY, which is the KINETIC ENERGY OF IMPACT in the rest system of the CM. The problem accordingly resolves itself into determining the net kinetic energy in the CM system when the corresponding kinetic energies in the lab system are specified.

We start out from the relativistic energy-momentum transformation (A-62),

$$E_{tot}^{(1)'} = \gamma_2'(E_{tot}^{(1)} - \beta_2' c p_1) = \gamma_1' m_1 c^2 \tag{B-59}$$

and substitute for $E_{tot}^{(1)}$ and p_1 from (A-58) and from (A-53) to get

$$\gamma_2'(\gamma_1 m_1 c^2 - \beta_2' m_1 c^2 \beta_1 \gamma_1) = \gamma_1' m_1 c^2 \tag{B-60}$$

i.e.,

$$\gamma_2' \gamma_1 - [(\gamma_2')^2 - 1]^{1/2} [(\gamma_1)^2 - 1]^{1/2} = \gamma_1' \tag{B-61}$$

To derive the total reaction energy in the CM system, defined as

$$E_{CM}' = E_{kin}^{(1)'} + E_{kin}^{(2)'} \tag{B-62}$$

we note, from (A-59), that

$$E_{\text{kin}}^{(1)\prime} = E_{\text{tot}}^{(1)\prime} - E_0^{(1)} = m_1 c^2 (\gamma_1' - 1) \tag{B-63}$$

$$E_{\text{kin}}^{(2)\prime} = E_{\text{tot}}^{(2)\prime} - E_0^{(2)} = m_2 c^2 (\gamma_2' - 1) \tag{B-64}$$

and thereby obtain, on substitution,

$$E_{\text{CM}}' = m_1 c^2 \{\gamma_2' \gamma_1 - [(\gamma_2')^2 - 1]^{1/2} [(\gamma_1)^2 - 1]^{1/2} - 1\} + m_2 c^2 (\gamma_2' - 1) \tag{B-65}$$

This can be cast into a more practical form when energies are expressed in electron-volts. Setting

$$E_{\text{kin}}^{(i)} / m_i c^2 \equiv \alpha_i \equiv e U_i / m_i c^2 \tag{B-66}$$

where U_i denotes the accelerating potential difference in the laboratory system, we obtain the formula

$$
\begin{aligned}
E_{\text{CM}}'{}^{\text{[eV]}} &\equiv e(U_1' + U_2') \\
&= \{\gamma_2' \gamma_1 - [(\gamma_2')^2 - 1]^{1/2} [(\gamma_1)^2 - 1]^{1/2} - 1\} (m_1 c^2)^{\text{[eV]}} \\
&\quad + (\gamma_2' - 1)(m_2 c^2)^{\text{[eV]}}
\end{aligned} \tag{B-67}
$$

If we deal with *particles of equal mass* $(m_1 = m_2 = m)$, we can reduce this to a much simpler expression for the net kinetic energy in the CM system,

$$E_{\text{CM}}'{}^{\text{[eV]}} \equiv e U' = 2(mc^2)^{\text{[eV]}} \left[\left(1 + \frac{\alpha}{2}\right)^{1/2} - 1 \right] \tag{B-68}$$

Approximate formulae can readily be deduced from this in the nonrelativistic and extreme-relativistic limits, viz.

nonrelativistic formula ($\alpha \ll 1$):

$$E_{\text{CM}}'{}^{\text{[eV]}} \hat{\triangleq} U' = 2mc^2 \frac{\alpha}{4} \doteq \tfrac{1}{2} U \tag{B-69}$$

extreme-relativistic formula ($\alpha \gg 1$):

$$E_{\text{CM}}'{}^{\text{[eV]}} \hat{\triangleq} U' = 2mc^2 \left[\left(\frac{\alpha}{2}\right)^{1/2} - 1 \right] \tag{B-70}$$

To illustrate the very striking difference between the net kinetic energy in the lab system and the very much smaller corresponding energy available in the CM system, we show in Fig. B-8 the variation in U' with U in the case of a collision between two protons. It will be seen that a 30-GeV proton beam impinging upon a stationary proton target (as in the CERN and Brookhaven experiments) can furnish a *reaction energy* in the CM system of only 5.8 GeV (this can be compared with an energy of 1.87 GeV necessary for nucleon pair creation, requiring a lab kinetic energy of about 5 GeV). To attain a CM energy of 30 GeV, one would have to accelerate the incident proton beam to 540 GeV, or, alternatively, to employ colliding beams of 15 GeV each, using a magnetic storage ring. (Incidentally, both of these projects are now in course of realization.) When identical particles of the same energy and mass collide, the CM system coincides with the lab system, and the reaction energy in the CM system is therefore equal to the sum of the kinetic energies in the laboratory system.

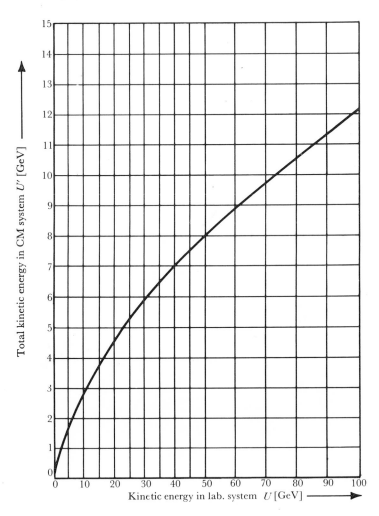

Fig. B-8. Relation between the total kinetic energy U in the laboratory system and the corresponding quantity U' in the center-of-mass system in the elastic collision of two particles of equal mass.

B.3.3. ANGULAR RELATIONS

For particle 1 after collision, the inverse momentum-energy transformation takes the form

$$p_{1x} = \gamma_2'\left(p_{1x}' + \beta_2'\frac{E_{\text{tot}}^{(1)\prime}}{c} \right) = \gamma_2'(p_{1x}' + \beta_2'\gamma_1' m_1 c) \qquad \text{(B-71)}$$

while

$$p_{1y} = p_{1y}' \qquad \text{and} \qquad p_{1z} = p_{1z}' = 0 \qquad \text{(B-72)}$$

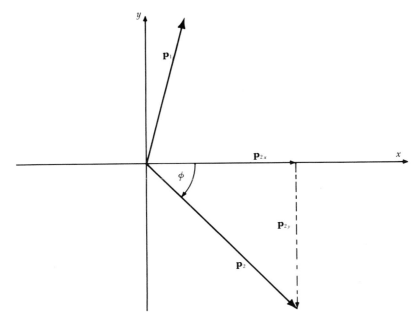

Fig. B-9. Resolution of the laboratory recoil momentum $\mathbf{p}_2$ into components p_{2x} and p_{2y}, showing that p_{2y} is negative and hence $p_{2y}/p_{2x} \equiv \tan\phi$ is also a negative quantity by convention.

Now, in the CM system one has

$$p_{1x}' = p_1' \cos\theta' \qquad \text{(B-73)}$$

$$p_{1y}' = p_1' \sin\theta' = p_{1y} \qquad \text{(B-74)}$$

whence, substituting, and writing $\delta' \equiv \beta_2'/\beta_1'$, we find

$$p_{1x} = \gamma_2' p_1'(\cos\theta' + \delta') \qquad \text{(B-75)}$$

i.e.,

$$\tan\theta \equiv \frac{p_{1y}}{p_{1x}} = \frac{\sin\theta'}{\gamma_2'(\cos\theta' + \delta')} \qquad \text{(B-76)}$$

The similarity to the corresponding nonrelativistic expression (B-16) will be noted. There is generally a large distinction between the lab and CM scattering angles in the case of fast-moving particles. For instance, when 500-MeV electrons are scattered through a lab angle $\theta = 2\frac{1}{2}°$, the corresponding CM angle is $\theta' = 90°$. The scattered energy is itself very sensitive to the lab scattering angle. In elastic e-e scattering it changes at the rate of about 100 MeV per degree (lab).

The dependence of ϕ upon the CM scattering angle θ' can be derived from an analogous treatment applied to particle 2. However, in this instance the sign convention plays an important role. By defining $\tan \phi$ as

$$\tan \phi \equiv \frac{p_{2y}}{p_{2x}} \tag{B-77}$$

we automatically make $\tan \phi$ *negative* by convention, since p_{2y} is in the negative y direction (see Fig. B-9). Of course, in actual magnitude ϕ is less than $\pi/2$ and $\tan \phi$ represents a positive quantity.

The relativistic inverse momentum-energy transformation is

$$p_{2x} = \gamma_2' \left(p_{2x}' + \beta_2' \frac{E_{tot}^{(2)'}}{c} \right) \tag{B-78}$$

(sign convention: $\oplus$ $\ominus$ $\oplus$)

i.e.,

$$p_{2x} = \gamma_2'(p_{2x}' + \beta_2' \gamma_2' m_2 c) \tag{B-79}$$

Since

$$p_{2x}' = -p_0 \cos \theta' \tag{B-80}$$

and

$$m_2 c \beta_2' \gamma_2' = +|\mathbf{p}_2'| = +p_0 \tag{B-81}$$

it follows that

$$p_{2x} = \gamma_2' p_0 (1 - \cos \theta') \tag{B-82}$$

Moreover, as

$$p_{2y} = p_{2y}' = -p_0 \sin \theta' \tag{B-83}$$

we arrive at the result

$$\tan \phi \equiv \frac{p_{2y}}{p_{2x}} = -\frac{1}{\gamma_2'} \frac{\sin \theta'}{1 - \cos \theta'} = -\frac{1}{\gamma_2'} \cot \frac{\theta'}{2} \tag{B-84}$$

where the minus sign indicates the angle convention (and will therefore be retained throughout), but is to be suppressed when only the magnitude of the angle is in question.

B.3.4. Solid Angle Relation

The same form of expression for the solid angle between emergent cones obtains in the relativistic case as in Section B.2.5, but the ensuing relation between $d\Omega$ and $d\Omega'$ becomes more complicated as a result of the greater complexity of the angular relations, viz.

$$\frac{d\Omega}{d\Omega'} \equiv \frac{2\pi \sin\theta \, d\theta}{2\pi \sin\theta' \, d\theta'} = \frac{d(\cos\theta)}{d(\cos\theta')} \tag{B-85}$$

which, making use of the intensity relations derived in the next section, may be written

$$\frac{d\Omega}{d\Omega'} = \frac{\gamma_2'(1 + \delta'\cos\theta')}{[\sin^2\theta' + (\gamma_2')^2(\cos\theta' + \delta')^2]^{3/2}} \tag{B-86}$$

B.3.5. Intensity Relation

Analogously to the nonrelativistic approach in Section B.2.6 we have

$$\frac{I(\theta)}{I(\theta')} = \frac{\sigma(\theta)}{\sigma(\theta')} = \frac{d\Omega'}{d\Omega} = \frac{d(\cos\theta')}{d(\cos\theta)} \tag{B-87}$$

However, we are here confronted with the more complicated angular relation (B-76)

$$\tan\theta = \frac{\sin\theta'}{\gamma_2'(\cos\theta' + \delta')} \tag{B-88}$$

which prompts us to evaluate the reciprocal differential $d(\cos\theta)/d(\cos\theta')$ for greater convenience. We first rearrange (B-88) to read

$$(1 - \cos^2\theta)(\gamma_2')^2(\cos\theta' + \delta')^2 = \sin^2\theta'\cos^2\theta \tag{B-89}$$

i.e.,

$$(\gamma_2')^2(\cos\theta' + \delta')^2 = [\sin^2\theta' + (\gamma_2')^2(\cos\theta' + \delta')^2]\cos^2\theta \tag{B-90}$$

whence

$$\cos\theta = \frac{\gamma_2'(\cos\theta' + \delta')}{[\sin^2\theta' + (\gamma_2')^2(\cos\theta' + \delta')^2]^{1/2}} \tag{B-91}$$

Differentiating this, we find

$$\begin{aligned}
\frac{d(\cos\theta)}{d(\cos\theta')} &= \gamma_2'(\cos\theta' + \delta')\left\{-\tfrac{1}{2}[\sin^2\theta' + (\gamma_2')^2(\cos\theta' + \delta')^2]^{-3/2}\right. \\
&\quad \left. \times [-2\cos\theta + 2(\gamma_2')^2(\cos\theta' + \delta')]\right\} \\
&\quad + \gamma_2'[\sin^2\theta' + (\gamma_2')^2(\cos\theta' + \delta')^2]^{-1/2} \tag{B-92}
\end{aligned}$$

i.e.,

$$\frac{d(\cos\theta)}{d(\cos\theta')} = \frac{\gamma_2'(1 + \delta'\cos\theta')}{[\sin^2\theta' + (\gamma_2')^2(\cos\theta' + \delta')^2]^{3/2}} \tag{B-93}$$

Insertion of this result into (B-87) yields

$$\frac{I(\theta')}{I(\theta)} = \frac{\sigma(\theta')}{\sigma(\theta)} = \frac{\gamma_2'(1 + \delta'\cos\theta')}{[\sin^2\theta' + (\gamma_2')^2(\cos\theta' + \delta')^2]^{3/2}} \tag{B-94}$$

The intensity ratio can also be expressed in terms of angles in the *laboratory* system. We first derive the ratio in terms of the recoil angle ϕ, and thereafter progress to the more complicated result as a function of the scattering angle θ.

With the aid of (B-84) we find that

$$\cos\theta' = 1 - \frac{2}{1 + (\gamma_2')^2\tan^2\phi} \tag{B-95}$$

which, upon differentiation and substitution in the ϕ-dependent intensity ratio

$$\frac{I(\theta')}{I(\phi)} = \frac{\sigma(\theta')}{\sigma(\phi)} = \frac{2\pi\sin\phi\,d\phi}{2\pi\sin\theta'\,d\theta'} = \frac{d(\cos\phi)}{d(\cos\theta')} \tag{B-96}$$

gives

$$\frac{I(\theta')}{I(\phi)} = \frac{\cos^4\phi + 2(\gamma_2')^2\sin\phi\cos\phi + (\gamma_2')^4\sin^4\phi}{4(\gamma_2')^2\cos\phi} \tag{B-97}$$

whence

$$\frac{I(\theta')}{I(\phi)} = \frac{(\gamma_2')^2[1 - (\beta_2')^2\cos^2\phi]^2}{4\cos\phi} \tag{B-98}$$

Finally, we embark upon deriving the intensity ratio $I(\theta')/I(\theta)$ in terms of θ rather than of θ' (cf. (B-94)). This involves differentiating the expression

$$\cos\theta' = \frac{-(\gamma_1')^2\delta'\sin^2\theta \pm [(\gamma_1)^2(1 - \delta'^2)\sin^2\theta + \cos^2\theta]^{1/2}\cos\theta}{(\gamma_1')^2\sin^2\theta + \cos^2\theta} \tag{B-99}$$

to obtain the result

$$\frac{I(\theta')}{I(\theta)} = \frac{\sigma(\theta')}{\sigma(\theta)} = \frac{d(\cos\theta)}{d(\cos\theta')}$$

$$= \frac{[(\gamma_1)^2(1 - \delta'^2)\sin^2\theta + \cos^2\theta]^{1/2}[(\gamma_1)^2\sin^2\theta + \cos^2\theta]^2}{K} \tag{B-100}$$

where

$$K \equiv \{\pm(\gamma_1)^4(1 - \delta'^2)\sin^4\theta + (\pm\gamma_1^4 \pm \gamma_1^2 \pm \gamma_1^2\delta'^2 \mp 2\gamma_1^4\delta'^2)\sin^2\theta\cos^2\theta$$

$$+ \gamma_1^2\delta'[\gamma_1^2(1 - \delta'^2)\sin^2\theta + \cos^2\theta]^{1/2}\cos\theta$$

$$+ (\mp\gamma_1^2 \pm \gamma_1^2\delta'^2 - 2\gamma_1^2 + 2 \pm 2)\cos^4\theta\} \tag{B-101}$$

EXERCISES

B-1 Slow neutrons can basically experience S-wave scattering only. This has an isotropic distribution in the CM system. On the assumption that the orbital momentum l (in units of $\hbar$) is quantized and that scattering can therefore occur only when the collision parameter d is less than the range of nuclear forces, calculate the minimum energy E_{min} at which the CM angular distribution begins to deviate from isotropy.

(Assume that the range of nuclear force is $\approx r_0 = 1.4$ fm. Note that quantum-mechanically $(l\hbar)^2 \rightarrow l(l+1)\hbar^2$.)

B-2 The threshold energy for the reaction

$$^4\text{He} + {}^{16}\text{O} \rightarrow {}^{19}\text{Ne} + \text{n}$$

lies at $E_{thresh} = 15.2$ MeV.

(a) Calculate the Q value of the reaction and the mass of ^{19}Ne. (Atomic mass of ^{16}O, $M(^{16}\text{O}) = 15.994\ 915$ u.)

(b) Determine the energy of the emitted neutrons in the center-of-mass system if the energy of the incident α particles is 20 MeV in the laboratory system.

B-3 A particle of mass m and nonrelativistic kinetic energy E makes a noncentral elastic collision with a stationary particle of the same mass and is scattered through 45° (lab). Compare the wavelengths of the particles in the laboratory and the CM systems after collision. Calculate the net combined energy of the two particles in the CM system and compare this with (a) the energy of the center-of-mass in the laboratory system, (b) the initial energy E in the laboratory system, and (c) the energy of each particle in the CM system.

B-4 By way of distinguishing between the characteristics of noncentral and central elastic collisions evaluate the following exercises:

(a) Suppose N_0 similar spherical projectiles having a mass m and radius r are directed with a uniformly distributed impact parameter at a stationary target sphere also of mass m and radius r. Calculate the energy distribution $E(\theta)$ as a function of the scattering angle θ and determine the angular distribution $dN/d\theta$ of the elastically scattered particles.

(b) For certain experiments in high-energy physics involving the bombardment of a stationary hydrogen target with high-speed protons it would be desirable to have an accelerator capable of accelerating protons to a laboratory energy E which is 50 times higher than the center-of-mass energy E'_{CM} upon collision. With the aid of Eq. (B-68) deduce the values of E and E'_{CM}. If a colliding-beam technique could be employed, by what factor could the laboratory energy of the accelerated protons be reduced?

B-5 (a) Currently, the highest-energy accelerators are (i) for light particles, the SLAC 20-GeV electron accelerator at Stanford; (ii) for heavy particles, the 33-GeV proton synchroton at Brookhaven and (iii) the 74-GeV proton synchroton at Serpukhov.

Calculate the total energy available in the CM system when the respective beams from these machines impinge upon protons at rest.

(b) Near in energy to the Brookhaven PS is the 28-GeV proton synchrotron at CERN which is now being equipped with a storage ring to enable experiments with colliding proton beams to be undertaken. What CM energy will be available with this set-up?

(c) The following proton accelerators are currently being projected:

    (i) a 200-GeV alternating gradient synchroton at Weston, Illinois, possibly extensible to 400 GeV;

    (ii) a 300-GeV alternating gradient synchroton in Europe, to be operated in conjunction with CERN;

    (iii) a 1000-GeV accelerator in the USSR.

For each of these, calculate the available CM energy in p-p collisions and estimate their probable diameters on the basis of reasonable magnetic field strengths and currently conventional electromagnetic engineering techniques. Will any of them yield a higher CM energy *without* a storage ring than the CERN PS *with* a storage ring?

(d) Discuss the advantages and shortcomings of "colliding-beam" experiments compared with those involving stationary targets.

B-6 What maximal energy can protons acquire in an accelerator of the synchrotron type set up around the Earth's equator, if a magnetic field of 10 kG is used to guide them?

B-7 Beyond what energy must a proton be accelerated in order that when it impinges upon a proton at rest it can give rise to a proton-antiproton pair?

B-8 The muon decay process

$$\mu \rightarrow e + \nu_\mu + \nu_e$$

produces electrons whose energy spectrum is continuous from zero up to a definite end-point. Calculate the value of this end-point energy. Supposing the decay of a muon at rest led to the production of an electron and a *single* neutrino, what form of energy spectrum would result and in particular what relationship would exist in first approximation between the electron energy and the muon rest energy?

THE DYNAMICS OF DECAY AND
REACTION PROCESSES

C.1. Decay and Reaction Kinematics

The kinematics of all interaction processes are fundamentally determined by the principle of energy and momentum conservation. In Appendix B, we proceeded to derive kinematic formulae for elastic scattering from this basic foundation. In the present appendix we amplify and supplement this treatment by listing some of the main results for decay and reaction dynamics. Since their derivation, which presupposes a knowledge of the material presented in Appendices A and B, follows along similar lines, we refrain from going into detail except in certain instances, preferring to collect end formulae together in a concise comprehensive form.

Some indication of the usefulness and importance of kinematic analysis is given in Chapter 11. Especially in high-energy physics it can play a vital role, and in many instances even at much lower energies it can serve as an invaluable adjunct to the identification of otherwise indistinguishable competing processes. For a more detailed discussion of decay and reaction kinematics and useful ancillary material, the reader is referred to [Ha 49, Bl 50a, Mo 53d, Fa 54, St 55, Ja 57b, Lei 58, Ma 58b, Ba 61a, De 62, De 63, Ha 63a, St 63, Ny 66]. Convenient tables exist [Ma X] for numerical transformation of distribution data, and of several fast computer programs which deal with reaction kinematics, that by Ball [Ba 62c] has been utilized most extensively.

C.2. Energetics and Kinematics for Two-Particle Decay

The consideration of the two-body decay of an unstable primary particle either at rest or in flight can be embraced within one formalism when referred to the *rest system* Σ' *of the primary particle*. This is yet one further instance of the inherent advantages in selecting an appropriate system of reference, which plays such a significant role in mechanics. According to a recent textbook [Ki 65], "The most powerful single intellectual device known in physics is the transformation of the reference frame from which we observe a process."

For the decay exemplified by the rest-mass schematic

$$m_0 \rightarrow m_1 + m_2 \tag{C-1}$$

the rest system of the primary particle is essentially analogous to the center-of-mass system, and all quantities referred to this system will be labeled with a prime. Thus momentum conservation requires that

$$\mathbf{p}_1' + \mathbf{p}_2' \equiv \sum_{i=1,2} \mathbf{p}_i' = 0 \tag{C-2}$$

since $\mathbf{p}_0' \equiv 0$, i.e., just as in the CM system, the vector momenta of the decay products sum to zero. If the primary particle decays while in flight (velocity v_0 relative to the lab system), the energy-momentum transformation relation considered in Appendix A.2 serves to express the total energy of each of the product particles in the lab system through primed quantities which refer to the primary rest system:

$$E_i = \gamma_0 (E_i' + \beta_0 \, c p_i' \cos \theta_i') \qquad (i = 1, 2) \tag{C-3}$$

where

$$\gamma_0 \equiv (1 - \beta_0^2)^{-1/2} \qquad \text{with} \qquad \beta_0 \equiv \frac{v_0}{c} \tag{C-4}$$

and θ_i' represents the emission angle in the rest system. The extrema, corresponding to $\theta_i' = 0, \pi$ are

$$(E_i)_{\substack{\text{max} \\ \text{min}}} = \gamma_0 (E_i' \pm \beta_0 \, c p_i') \tag{C-5}$$

To derive the energy distribution of the product particles, we assume the decay to be *spatially isotropic in the rest system*, viz.

$$\frac{dn}{d(\cos \theta_i')} = \frac{1}{2} = \text{const.} \tag{C-6}$$

and note that the number of decays per unit energy interval dE_i is, with (C-3) and (C-6),

$$\frac{dn}{dE_i} = \frac{dn}{d(\cos \theta_i')} \frac{d(\cos \theta_i')}{dE_i} = \frac{1}{2} \frac{1}{\gamma_0 \beta_0 \, c p_i'} = \frac{1}{E_{i,\text{max}} - E_{f,\text{max}}} \tag{C-7}$$

which is constant between the two extreme values of energy. This expression can also be written equivalently as

$$\frac{dn}{dE_i} = \frac{m_0}{2p_0 p_1'} \tag{C-8}$$

The angular distribution of the product particles entails a rather more complicated treatment, which has been given, e.g., by Baldin *et al.* [Ba 61a]. The distribution of the separation angle α between the two particles (in the laboratory system) can be derived [St 55] on setting

$$\frac{dn}{d\alpha} = \frac{dn}{dE_i} \frac{dE_i}{d\alpha} \tag{C-9}$$

and making use of the conservation equations for energy (see Appendix A.2),

$$E_0^2 = p_0^2 c^2 + (m_0 c^2)^2 = (E_1 + E_2)^2 \tag{C-10}$$

and for momentum,

$$p_1 \sin \theta_1 - p_2 \sin \theta_2 = 0 \tag{C-11}$$

$$p_1 \cos \theta_1 + p_2 \cos \theta_2 = p_0 \tag{C-12}$$

where θ_1 and θ_2 are the emission angles of the products in the laboratory system with respect to $\mathbf{p}_0$ and

$$p_i = \frac{1}{c} [E_i - (m_i c^2)^2]^{1/2} \tag{C-13}$$

Solving these, with $\alpha = \theta_1 + \theta_2$, we get the separation angle,

$$\cos \alpha = \frac{E_1 E_2 - K}{[E_1^2 - (m_1 c^2)^2]^{1/2} [E_2^2 - (m_2 c^2)^2]^{1/2}} \tag{C-14}$$

with

$$K \equiv \tfrac{1}{2} (m_0^2 - m_1^2 - m_2^2) c^4 \tag{C-15}$$

Differentiating (C-14) and noting that $\partial E_2 / \partial E_1 = -1$, one finds

$$-\sin \alpha \, d\alpha = \left\{ \frac{E_2 - E_1}{E_1 E_2 - K} - \frac{E_1}{[E_1^2 - (m_1 c^2)^2]^{1/2}} + \frac{E_2}{[E_2^2 - (m_2 c^2)^2]^{1/2}} \right\} \cos \alpha \, dE_1 \tag{C-16}$$

which, inserted in (C-9), gives the result

$$\frac{dn}{d\alpha} = \frac{m_0}{2p_0 p_1'} \left| \frac{\tan \alpha}{\dfrac{E_2 - E_1}{E_1 E_2 - K} - \dfrac{E_1}{[E_1^2 - (m_1 c^2)^2]^{1/2}} + \dfrac{E_2}{[E_2^2 - (m_2 c^2)^2]^{1/2}}} \right| \tag{C-17}$$

with α given by (C-14) as a function of the total energy E_1.

The kinematics of interactions involving three secondary particles introduce additional complications, which have been considered in some detail by Baldin *et al.* [Ba 61a, pp. 78ff]. Kinematic tables for two-body decays and two-body reactions in particle physics have been compiled by Leipuner [Lei 58].

We now proceed to a consideration of nonrelativistic and relativistic two-body interactions.

C.3. Scattering Kinematics

For convenience, results derived in Appendix B are collected here and extended, using the same notation to describe the elastic scattering of a projectile of mass m_1 on a stationary target m_2. Unprimed quantities refer to the laboratory system and primed quantities to the CM system.

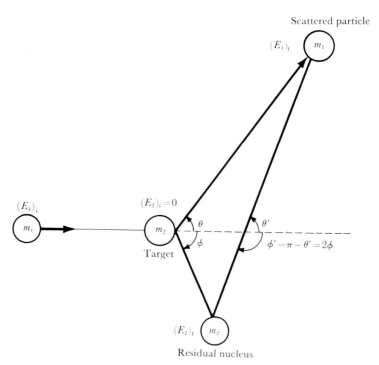

Fig. C-1. Vector diagram to illustrate the classical elastic scattering of a particle of mass m_1 and initial kinetic energy $(E_1)_i$ upon a stationary target of mass m_2. After scattering through an angle θ in the laboratory system (or θ' in the CM system) the particle leaves with a kinetic energy $(E_1)_f$, which is smaller than $(E_1)_i$ since a recoil energy $(E_2)_f$ is conferred upon the residual nucleus that recoils at an angle ϕ in the laboratory system (or ϕ' in the CM system) with respect to the incident particle direction.

C.3.1. NONRELATIVISTIC ELASTIC SCATTERING KINEMATICS

For the elastic scattering process depicted in vector-diagram form as Fig. C-1, we use the notation indicated in Table C-1 to express in Table C-2 some of the results derived in Appendix B.2.

Table C-1. NOTATION USED TO DESCRIBE ELASTIC SCATTERING ON A STATIONARY TARGET

Quantity	Lab system	CM system
Initial velocity of m_1	u_1	u_1'
Initial velocity of m_2	$u_2 = 0$	u_2'
Final velocity of m_1	v_1	$\lvert v_1' \rvert = \lvert u_1' \rvert$
Final velocity of m_2	v_2	$\lvert v_2' \rvert = \lvert u_2' \rvert$
Velocity of CM	$v_{\mathrm{CM}} = \lvert u_2' \rvert$	$v_{\mathrm{CM}}' \equiv 0$
Initial kinetic energy of m_1	$(E_1)_i$	$(E_1')_i$
Initial kinetic energy of m_2	$(E_2)_i = 0$	$(E_2')_i$
Final kinetic energy of m_1	$(E_1)_f$	$(E_1')_f$
Final kinetic energy of m_2	$(E_2)_f$	$(E_2')_f$
Scattering angle of m_1	θ	θ'
Recoil angle of m_2	ϕ	ϕ'
Specific mass of m_1	$\mu_1 \equiv \dfrac{m_1}{m_1 + m_2}$	
Specific mass of m_2	$\mu_2 \equiv \dfrac{m_2}{m_1 + m_2}$	
Reduced mass	$\mu \equiv \dfrac{m_1 m_2}{m_1 + m_2}$	
Mass ratio	$k \equiv \dfrac{m_1}{m_2} = \dfrac{\mu_1}{\mu_2}$	

C.3.2. RELATIVISTIC ELASTIC SCATTERING KINEMATICS

As a full relativistic treatment of scattering kinematics would be out of place in the present text, we confine ourselves to a presentation of some of the main considerations and indicate graphical methods of solving certain scattering problems. Blaton [Bl 50a] has presented relativistic methods appropriate to a wide range of situations; his treatment is in part complemented by Baldin *et al.* [Ba 61a] and Hagedorn [Ha 63a].

We consider first the *transfer of energy in a relativistic elastic scattering collision.* Application of the energy-momentum transformation from Appendix A, viz.

$$E_{\mathrm{tot}} = \gamma'(E_{\mathrm{tot}}' + \beta c p') \tag{C-18}$$

Table C-2. NONRELATIVISTIC ELASTIC SCATTERING FORMULAE

Quantity	Formula
Reduced CM kinetic energy of projectile	$\dfrac{(E_1')_i}{(E_1)_i} = \dfrac{(E_1')_f}{(E_1)_i} = \mu_2^2$
Reduced CM kinetic energy of recoil nucleus	$\dfrac{(E_2')_f}{(E_1)_i} = \mu_1 \mu_2$
Reduced lab kinetic energy of scattered particle	$\dfrac{(E_1)_f}{(E_1)_i} = 1 - 2\mu_1 \mu_2(1 - \cos\theta')$ $= \mu_1^2\left[\cos\theta \pm \left(\dfrac{1}{k^2} - \sin^2\theta\right)^{1/2}\right]^2$ wherein the plus sign is to be used except when $m_1 > m_2$, in which case $\theta_{max} = \arcsin\dfrac{1}{k}$
Reduced lab kinetic energy of recoil nucleus	$\dfrac{(E_2)_f}{(E_1)_i} = 1 - \dfrac{(E_1)_f}{(E_1)_i} = 4\mu_1 \mu_2 \cos^2\phi$ (where $\phi \leqslant \pi/2$)
Lab recoil angle	$\phi = \tfrac{1}{2}\phi' = \tfrac{1}{2}(\pi - \theta')$ $\sin\phi = \left[\dfrac{k\,(E_1)_f}{(E_2)_f}\right]^{1/2}\sin\theta$
Lab scattering angle	$\tan\theta = \dfrac{\sin\theta'}{k + \cos\theta'} = \dfrac{\sin 2\phi}{k - \cos 2\phi}$
CM scattering angle	$\theta' = \pi - \phi' = \pi - 2\phi$ $\theta' = \theta + \arcsin(k\sin\theta)$ $\cos\theta' = 1 - 2\cos^2\phi$
CM recoil angle	$\phi' = 2\phi = \pi - \theta'$
Intensity or solid-angle ratio for scattered particle	$\dfrac{\sigma(\theta')}{\sigma(\theta)} = \dfrac{I(\theta')}{I(\theta)} = \dfrac{d\Omega}{d\Omega'} = \dfrac{\sin\theta\,d\theta}{\sin\theta'\,d\theta'}$ $= \dfrac{\sin^2\theta}{\sin^2\theta'}\cos(\theta' - \theta)$ $= \mu_1 \mu_2\dfrac{(E_1)_i}{(E_1)_f}\left(\dfrac{1}{k^2} - \sin^2\theta\right)^{1/2}$
Intensity or solid-angle ratio for recoil nucleus	$\dfrac{\sigma(\phi')}{\sigma(\phi)} = \dfrac{I(\phi')}{I(\phi)} = \dfrac{\sin\phi\,d\phi}{\sin\phi'\,d\phi'}$ $= \dfrac{1}{4\cos\phi}$

and substitution of appropriate quantities from (B-80) and (B-81) yields an expression for the recoil energy in the laboratory system,

$$E_{\text{kin}}^{(2)} = E_{\text{tot}}^{(2)} - E_0^{(2)} = \gamma_2'(\gamma_2' - \gamma_2' \beta_2'^2 \cos \theta') E_0^{(2)} - E_0^{(2)} \qquad \text{(C-19)}$$

i.e.,

$$E_{\text{kin}}^{(2)} = m_2 c^2 [(\gamma_2')^2 - 1] [1 - \cos \theta'] \qquad \text{(C-20)}$$

or, in terms of the recoil angle ϕ,

$$E_{\text{kin}}^{(2)} = 2m_2 c^2 \frac{(\gamma_2')^2 - 1}{(\gamma_2')^2 \tan^2 \phi + 1} \qquad \text{(C-21)}$$

on using (B-84) to make the substitution

$$\cos \theta' = \frac{(\gamma_2')^2 - (\gamma_2')^2 \cos^2 \phi - \cos^2 \phi}{(\gamma_2')^2 - (\gamma_2')^2 \cos^2 \phi + \cos^2 \phi} \qquad \text{(C-22)}$$

It may be mentioned in passing that (C-21) indicates that, just as in the non-relativistic situation, the recoil energy attains its maximum value when $\phi = 0$.

On eliminating γ_2' from (C-21) with the aid of (B-56) we arrive at the relation (with $k \equiv m_1/m_2$)

$$E_{\text{kin}}^{(2)} = 2m_2 c^2 \frac{k^2(\gamma_1^2 - 1) \cos^2 \phi}{(1 + k\gamma_1)^2 + (1 - \gamma_1^2) k^2 \cos^2 \phi} \qquad \text{(C-23)}$$

which, using (B-50) and (B-51), can be written as

$$E_{\text{kin}}^{(2)} = 2m_2 c^2 \frac{p_1^2 c^2 \cos^2 \phi}{(E_{\text{tot}}^{(1)} + m_2 c^2)^2 - p_1^2 c^2 \cos^2 \phi} \qquad \text{(C-24)}$$

In terms of the incident kinetic energy $E_{\text{kin}}^{(1)} = m_1 c^2(\gamma_1 - 1)$ the maximum fraction of kinetic energy which can be transferred from the incident particle to the target is given on substituting $\phi = 0$ in (C-23) as

$$(E_{\text{kin}}^{(2)})_{\text{max}} = E_{\text{kin}}^{(1)} \frac{2k(\gamma_1 + 1)}{1 + 2k\gamma_1 + k^2} \qquad \text{(C-25)}$$

The ratio $(E_{\text{kin}}^{(2)})_{\text{max}}/E_{\text{kin}}^{(1)}$ increases with γ_1 from the nonrelativistic limit of $4\mu_1\mu_2$ (cf. Table C-2) obtained by setting γ_1 to unity. Its upper limit, in the extreme-relativistic region ($\gamma_1 \gg 1$) is 1, i.e., all the kinetic energy is transferred from the incident particle to the struck target.

C.3.3. Graphical Treatment of Elastic Scattering

It follows from the conservation of net momentum and total energy in the laboratory system that

$$(p_1)_i = (p_1)_f + (p_2)_f \qquad \text{(C-26)}$$

and

$$(E_{\text{tot}}^{(1)})_i + E_0^{(2)} = (E_{\text{tot}}^{(1)})_f + (E_{\text{tot}}^{(2)})_f \tag{C-27}$$

The scattering and recoil angles θ and ϕ are shown in the graphical construction depicted in Fig. C-2; the above conditions are satisfied for all values of incident momentum $(\mathbf{p}_1)_i$ when the possible end-points of $(\mathbf{p}_2)_f$ lie on an ellipse which passes through A, has its center at the point B along $(\mathbf{p}_1)_i$ and has the following dimensions:

semi-major axis $\overline{AB}$: $a = (p_1)_i \dfrac{E_0^{(2)}[(E_{\text{tot}}^{(1)})_i + E_0^{(2)}]}{(E_0^{(1)})^2 + 2E_0^{(2)}(E_{\text{tot}}^{(1)})_i + (E_0^{(2)})^2}$ (C-28)

semi-minor axis $\overline{BE}$: $b = (p_1)_i \dfrac{E_0^{(2)}}{[(E_0^{(1)})^2 + 2E_0^{(2)}(E_{\text{tot}}^{(1)})_i + (E_0^{(2)})^2]^{1/2}}$ (C-29)

with $a > b$ except in the nonrelativistic limit (when $a \simeq b$).

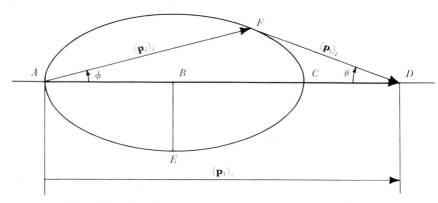

Fig. C-2. Ellipse diagram appropriate to elastic scattering.

In the case of an *elastic collision between identical particles* ($m_1 = m_2 = m$), the point D coincides with C and

$$a = \tfrac{1}{2}(p_1)_i \tag{C-30}$$

Generally, with m_1 and m_2 not necessarily the same, the maximum momentum transfer from particle 1 to the target 2 is

$$(p_2)_{f,\,\text{max}} = 2a \tag{C-31}$$

and the maximum energy transfer is

$$(E_{\text{kin}}^{(2)})_{\text{max}} = (E_{\text{tot}}^{(2)})_{f,\,\text{max}} - E_0^{(2)} \tag{C-32}$$

that is

$$(E_{\text{kin}}^{(2)})_{\text{max}} = 2a \frac{(p_1)_i c^2}{(E_{\text{tot}}^{(1)})_i + E_0^{(2)}} \tag{C-33}$$

$$= \frac{2E_0^{(2)} (p_1)_i^2 c^2}{(E_0^{(1)})^2 + 2E_0^{(2)} (E_{\text{tot}}^{(1)})_i + (E_0^{(2)})^2} \tag{C-34}$$

Depending upon the relationship of the participant masses, the maximum scattering angle is

(a) $\qquad \theta_{\text{max}} = \pi \qquad\qquad$ when $\qquad m_1 < m_2 \qquad$ (C-35)

(b) $\qquad \theta_{\text{max}} = \pi/2 \qquad\qquad$ when $\qquad m_1 = m_2 \qquad$ (C-36)

(c) $\qquad \theta_{\text{max}} = \arcsin\left(\dfrac{m_2}{m_1}\right) \qquad$ when $\qquad m_1 > m_2 \qquad$ (C-37)

The angle between the outgoing particles, $(\theta + \phi)$, can be derived from the formula

$$\tan(\theta + \phi) = \frac{\tan \phi}{1 - K} \tag{C-38}$$

where

$$K \equiv \frac{2E_0^{(2)}[(E_{\text{tot}}^{(1)})_i + E_0^{(2)}]}{[(E_{\text{tot}}^{(1)})_i + E_0^{(2)}]^2 - (p_1)^2 c^2 \cos^2 \phi} \tag{C-39}$$

C.3.4. Graphical Treatment of Inelastic Scattering

A similar approach to that of the previous section can be applied to deduce the kinematics of inelastic scattering, in which the incident particle additionally transfers an intrinsic excitation energy E^* to the target, so that

$$(E_{\text{tot}}^{(2)})_f = (E_{\text{kin}}^{(2)})_f + E_0^{(2)} + E^* \tag{C-40}$$

while the momentum and energy conservation equations (C-26) and (C-27) remain formally the same.

The appropriate diagram now takes the form shown in Fig. C-3. The endpoints F of (p_2) for given values of $(p_1)_i$ and E^* lie on an ellipse whose dimensions are such that

$$a < \bar{a} < (p_1)_i \tag{C-41}$$

with

$$\bar{a} \equiv \tfrac{1}{2}(p_1)_i \frac{[2E_0^{(2)}(E_{\text{tot}}^{(1)})_i + (E_0^{(2)})^2 + (E_0^{(2)} + E^*)^2]}{[2E_0^{(2)}(E_{\text{tot}}^{(1)})_i + (E_0^{(2)})^2 + (E_0^{(1)})^2]} \tag{C-42}$$

semi-major axis:

$$a = m_2 \frac{[(E_{\text{tot}}^{(1)})_i + E_0^{(2)}]\left\{\left[(E_{\text{tot}}^{(1)})_i - E^* - \dfrac{(E^*)^2}{2E_0^{(2)}}\right]^2 - [E_0^{(1)} + kE^*]^2\right\}^{1/2}}{2E_0^{(2)}(E_{\text{tot}}^{(1)})_i + (E_0^{(2)})^2 + (E_0^{(1)})^2} \tag{C-43}$$

semi-minor axis:

$$b = a \frac{[2E_0^{(2)}(E_{tot}^{(1)})_i + (E_0^{(1)})^2 + (E_0^{(2)})^2]^{1/2}}{E_0^{(2)} + (E_{tot}^{(1)})_i} \qquad \text{(C-44)}$$

The following condition must be fulfilled if inelastic scattering is to occur:

$$(E_{kin}^{(1)})_i \geqslant \frac{E^*}{E_0^{(2)}} (E_0^{(1)} + E_0^{(2)} + \tfrac{1}{2}E^*) \qquad \text{(C-45)}$$

In the special case of rectilinear propagation, when $\theta = \phi = 0$, the speeds of the outgoing particles are equal, being given by the expression

$$v_1 = v_2 = c\left\{1 - \left[\frac{E_0^{(1)} + E_0^{(2)} + E^*}{(E_{tot}^{(1)})_i + E_0^{(2)}}\right]^2\right\}^{1/2} \qquad \text{(C-46)}$$

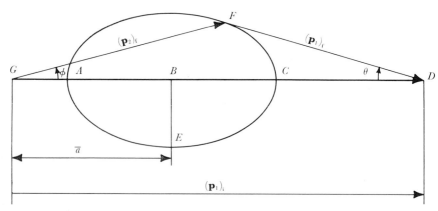

Fig. C-3. Modified ellipse diagram pertaining to inelastic scattering.

In the nonrelativistic limit,

$$a = b = \mu_2[(p_1)_i^2 - 2kE^*(m_1 + m_2)]^{1/2} \qquad \text{(C-47)}$$

$$\bar{a} = \mu_2(p_1)_i > a \qquad \text{(C-48)}$$

and the condition for inelastic scattering reduces to

$$(E_{tot}^{(1)})_i \geqslant \frac{E^*}{\mu_2} \qquad \text{(C-49)}$$

C.3.5. INELASTIC SCATTERING AT HIGH INCIDENT ENERGIES LEADING TO THE THRESHOLD OF PARTICLE CREATION

The foregoing treatment can be extended to include a special situation in inelastic scattering, namely, the creation of one or more particles whose total rest mass is M. Hence initially the rest mass of the participants is $(m_1 + m_2)$

and finally it is $(m_1 + m_2 + M)$. We assume that the target is not left in an excited state, and consider the incident energy to be such that it just reaches the threshold of the creation process, for then the CM *kinetic* energies are zero, and the total CM energy before and after collision remains unchanged, viz.

$$E' = (E_{\text{tot}}^{(1)\prime})_i + (E_{\text{tot}}^{(2)\prime})_i = (E_{\text{tot}}^{(1)\prime})_f + (E_{\text{tot}}^{(2)\prime})_f + Mc^2 \tag{C-50}$$

i.e.,

$$\gamma_1' m_1 c^2 + \gamma_2' m_2 c^2 = m_1 c^2 + m_2 c^2 + Mc^2 \tag{C-51}$$

e.g.,

$$c^2 (m_1^2 + 2\gamma_1 m_1 m_2 + m_2^2)^{1/2} = c^2 (m_1 + m_2 + M) \tag{C-52}$$

Since we are dealing with a threshold process,

$$(E_{\text{kin}}^{(1)})_i = m_1 c^2 (\gamma_1 - 1) = (E_{\text{kin}}^{(1)})_{\text{thresh}} \tag{C-53}$$

which, substituting for γ_1 from (C-52), yields the result

$$(E_{\text{kin}}^{(1)})_{\text{thresh}} = \frac{m_1 + m_2 + \frac{1}{2}M}{m_2} Mc^2 \tag{C-54}$$

Accordingly, the creation of a nucleon pair through a proton-proton collision (stationary target) requires an incident energy in the lab system of at least

$$(E_{\text{kin}}^{(1)})_{\text{thresh}} = 6m_p c^2 \cong 5.6 \text{ GeV} \tag{C-55}$$

whereas the commensurate threshold energy for a lighter projectile (e.g., photon or electron, with $m_1 \approx 0$) reduces to

$$(E_{\text{kin}}^{(1)})_{\text{thresh}} \cong 4m_p c^2 \cong 3.8 \text{ GeV} \tag{C-56}$$

Sternheimer [St 63] has presented similar considerations aimed at evaluating the thresholds for associated production of unstable particles in which two or more particles are emitted from reactions of the type

$$\mathcal{N} + \mathcal{N} \nearrow \begin{matrix} Y + K + \mathcal{N} \\ K^+ + K^- + 2\mathcal{N} \end{matrix} \tag{C-57}$$

$$\pi + \mathcal{N} \nearrow \begin{matrix} Y + K \\ K^+ + K^- + \mathcal{N} \end{matrix} \tag{C-58}$$

$$\gamma + \mathcal{N} \nearrow \begin{matrix} Y + K \\ K^+ + K^- + \mathcal{N} \end{matrix} \tag{C-59}$$

with the hyperon in each case designated by the symbol Y. A tabulation of numerical results for $\mathcal{N}$-, π-, and γ-induced associated production processes shows the threshold energies when the target particle is at rest to range from about 0.7 GeV to 7 GeV.

C.4. Nonrelativistic Reaction Kinematic Formulae

Instead of proceeding to a detailed derivation of the rather complicated kinematic relations which apply to nuclear reaction processes, we confine ourselves to presenting a summary of the main results in the convenient form adopted by Jarmie and Seagrave [Ja 57b].

For the reaction

$$m_1 + m_2 \rightarrow m_3 + m_4 + Q \qquad\qquad (\text{C-60})$$

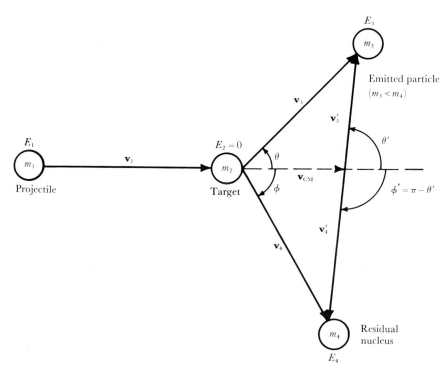

Fig. C-4. Vector diagram showing the relationship of kinematic quantities in the reaction process $m_1 + m_2 \rightarrow m_3 + m_4$.

the total reaction energy in the lab system is

$$E = E_{\text{kin}}^{(1)} + Q = E_{\text{kin}}^{(3)} + E_{\text{kin}}^{(4)} \qquad\qquad (\text{C-61})$$

The process is depicted in Fig. C-4. It is easy to include inelastic scattering within this scheme by setting $m_1 = m_3$, $m_2 = m_4$, $Q = -E^*$, while elastic scattering corresponds to the above substitution with $Q = 0$.

It is convenient for the sake of conciseness to express the kinematic formulae in terms of four parameters which are defined as follows:

$$A_{13} \equiv \frac{m_1 m_3}{(m_1 + m_2)(m_3 + m_4)} \frac{E_{\text{kin}}^{(1)}}{E} \tag{C-62}$$

$$A_{14} \equiv \frac{m_1 m_4}{(m_1 + m_2)(m_3 + m_4)} \frac{E_{\text{kin}}^{(1)}}{E} \tag{C-63}$$

$$A_{23} \equiv \frac{m_2 m_3}{(m_1 + m_2)(m_3 + m_4)} \left(1 + \frac{m_1}{m_2} \frac{Q}{E}\right) = \frac{E_{\text{kin}}^{(4)\prime}}{E} \tag{C-64}$$

$$A_{24} \equiv \frac{m_2 m_4}{(m_1 + m_2)(m_3 + m_4)} \left(1 + \frac{m_1}{m_2} \frac{Q}{E}\right) = \frac{E_{\text{kin}}^{(3)\prime}}{E} \tag{C-65}$$

These sum to unity,

$$A_{13} + A_{14} + A_{23} + A_{24} = 1 \tag{C-66}$$

The ensuing kinematic formulae have been collated in Table C-3.

Table C-3. NONRELATIVISTIC REACTION KINEMATIC FORMULAE

Quantity	Formula
Reduced lab kinetic energy of outgoing particle	$\dfrac{E_{\text{kin}}^{(3)}}{E} = A_{13} + A_{24} + 2(A_{14}A_{23})^{1/2} \cos\theta'$ $= A_{13}\left[\cos\theta \pm \left(\dfrac{A_{24}}{A_{13}} - \sin^2\theta\right)^{1/2}\right]^2$ wherein the plus sign is to be used except when $A_{13} > A_{24}$, in which case $\theta_{\max} = \arcsin\left(\dfrac{A_{24}}{A_{13}}\right)^{1/2}$
Reduced lab kinetic energy of residual nucleus	$\dfrac{E_{\text{kin}}^{(4)}}{E} = A_{14} + A_{23} + 2(A_{14}A_{23})^{1/2} \cos\phi'$ $= A_{14}\left[\cos\phi \pm \left(\dfrac{A_{23}}{A_{14}} - \sin^2\phi\right)^{1/2}\right]^2$ wherein the plus sign is to be used except when $A_{14} > A_{23}$, in which case $\phi_{\max} = \arcsin\left(\dfrac{A_{23}}{A_{14}}\right)^{1/2}$
CM angle of outgoing particle	$\sin\theta' = \left(\dfrac{1}{A_{24}} \dfrac{E_{\text{kin}}^{(3)}}{E}\right)^{1/2} \sin\theta$

Table C-3.—*Continued*

Quantity	Formula
Lab angle of residual nucleus	$\sin \phi = \left(\dfrac{m_3 E_{\text{kin}}^{(3)}}{m_4 E_{\text{kin}}^{(4)}} \right)^{1/2} \sin \theta$
Intensity ratio for outgoing particle	$\dfrac{I(\theta')}{I(\theta)} = \dfrac{\sigma(\theta')}{\sigma(\theta)} \dfrac{\sin^2 \theta}{\sin^2 \theta'} \cos(\theta' - \theta)$
	$= \dfrac{E}{E_{\text{kin}}^{(3)}} (A_{14} A_{23})^{1/2} \left(\dfrac{A_{24}}{A_{13}} - \sin^2 \theta \right)^{1/2}$
Intensity ratio for residual nucleus	$\dfrac{I(\phi')}{I(\phi)} = \dfrac{\sigma(\phi')}{\sigma(\phi)} \dfrac{\sin^2 \phi}{\sin^2 \phi'} \cos(\phi' - \phi)$
	$= \dfrac{E}{E_{\text{kin}}^{(4)}} (A_{14} A_{23})^{1/2} \left(\dfrac{A_{23}}{A_{14}} - \sin^2 \phi \right)^{1/2}$
Intensity ratio for both emergent particles in the lab system	$\dfrac{I(\phi)}{I(\theta)} = \dfrac{\sigma(\phi)}{\sigma(\theta)} \dfrac{\sin^2 \theta}{\sin^2 \phi} \dfrac{\cos(\theta' - \theta)}{\cos(\phi' - \phi)}$

EXERCISES

C-1   What is the proton "threshold" energy for the reaction $^7\text{Li}(\text{p},\text{n})^7\text{Be}$ ($Q = -1.6433$ MeV) leading to the production of neutrons at rest in the *laboratory* system?

C-2   For the elastic scattering of protons and deuterons on hydrogen, calculate the energies (in nonrelativistic approximation) of the two outgoing particles as a function of the angles θ and ϕ with respect to the incident direction.

C-3   Only in the case of *nonrelativistic* elastic scattering of two particles having equal mass is the relative angle between the outgoing particles a right angle. At higher incident energies the angle for elastic collision with a stationary target becomes less than 90° and can vary between the two extremes α_{min} and α_{max}. Calculate these limiting angles for incident particles of equal rest mass m when the incident relativistic kinetic energy is E_{kin}.

C-4   The reaction $^7\text{Li}(\text{p},\text{n})^7\text{Be}$ ($Q = -1.6433$ MeV) is often employed to produce monoenergetic neutrons in the energy range of order 0.1 MeV to several MeV.

   Calculate

   (a)   the threshold energy $E_{\text{thresh}}^{(\text{p})}$ and the corresponding neutron energy E_{n};

(b) the threshold energy $E_{\text{thresh}}^{(p)}$ $(\theta \geqslant 90°)$ for the production of neutrons in the backward direction $(\theta \geqslant 90°)$;

(c) the energy E_n of the neutrons emitted directly forward $(\theta = 0°)$ at the proton energy E_p.

C-5 One method for the detection of thermal neutrons $(E_n = \frac{1}{40}\text{ eV})$ employs the $^{10}_5\text{B}(n, \alpha)^7_3\text{Li}$ reaction $(Q = +2.790\text{ MeV})$, wherein the α particles are registered. What are the kinetic energies of the α particle and of the recoil nucleus?

C-6 For the "d-d" reaction $^2\text{H}(d,n)^3\text{He}$ $(Q = +3.268\text{ MeV})$ used for the production of low-energy neutrons, calculate the largest and the least neutron energy attainable using a fixed-energy accelerator which delivers 800-keV deuterons to a thin deuterium gas target. Also calculate the angular distribution of the neutrons in the laboratory system on the assumption that the CM angular distribution is isotropic.

C-7 How much energy is lost by 500-MeV electrons which undergo elastic scattering through 90° on stationary protons? (Compare the exact relativistic solution to that obtained by assuming the process to be akin to Compton scattering.)

C-8 A particle of mass m_1 incident on a target nucleus of mass m_2 is inelastically scattered. Calculate the energy E_3 (lab) of the scattered particle as a function of the incident energy E_1 (lab) and the scattering angle θ (lab), assuming the target nucleus receives an excitation energy $E*$.

For the scattering of 4.2-MeV neutrons on ^{24}Mg compare the laboratory energies $E_{3,\text{el}}$ and $E_{3,\text{inel}}$ of the elastically and inelastically scattered neutrons which are observed perpendicular to the incident beam.

(The lowest-lying levels of ^{24}Mg have excitation energies 1.368, 4.12, 4.24, and 5.22 MeV.)

C-9 The binding energy between a proton and a neutron in a deuterium nucleus is so low $(E_B = 2.225\text{ MeV})$ that the deuteron is quite easily disrupted, e.g., by the "nuclear photo-effect" initiated by γ radiation. In an investigation [Bö 63c; see also Bö 65b] of the nuclear photo-effect, a deuterium target was bombarded by a parallel beam of 9-MeV γ rays and the angular distribution of the photoneutrons emerging from the reaction $^2\text{H}(\gamma,n)^1\text{H}$ was studied. The emission probability $W(\theta)$ as a function of the laboratory angle of emergence θ referred to the incident γ direction was determined over the range $15° \leqslant \theta \leqslant 150°$. The results could be fitted by a curve of the form

$$W(\theta) = 1 + b\sin^2\theta$$

in which least-squares-fit analysis indicated that the coefficient had the value $b = 10$.

A theoretical treatment of the situation gives results in the *center-of-mass* system and indicates an emission probability of the form

$$W(\theta') = 1 + b' \sin^2 \theta' - c' \cos \theta' - d' \cos \theta' \sin^2 \theta'$$

By transforming the measured distribution into the center-of-mass system and comparing theory with experiment, determine the values of the coefficients b', c', and d', noting that $\theta \simeq \theta'$.

(The energy in the center-of-mass system should be taken as $E_\gamma - E_B$, thereby neglecting the center-of-mass energy. Furthermore, whenever the ratio v_{CM}/v_n' of the center-of-mass velocity to the neutron velocity in the CM system occurs to higher than *first* order, it should be neglected. The outgoing particles are to be treated nonrelativistically.)

Appendix D

WAVE MECHANICS

D.1. Schrödinger Equations

The state and behavior of a quantum-mechanical system is most readily described in terms of wave functions which are solutions of wave equations for the given system under specific conditions. Such wave equations may be time-dependent or in certain cases time-independent, corresponding to stationary waves. They assume different forms according to the spin composition of the system or to the relativistic character of the intrinsic motion. In general, the equations are expressed in the *center-of-mass* system of reference, and feature *reduced* masses.

We commence by considering the simplest conceivable system, composed of a single nonrelativistic free particle of mass m, and leave spin considerations aside for the present. The derivation of the SCHRÖDINGER WAVE EQUATION can then, in a nonrigorous way, be sketched as follows: In classical wave mechanics, the equation for a one-dimensional periodic disturbance ξ has the form

$$\frac{\partial^2 \xi}{\partial t^2} = c^2 \frac{\partial^2 \xi}{\partial x^2} \tag{D-1}$$

with a general solution

$$\xi = \xi_0 e^{-i(\omega t - kx)} \tag{D-2}$$

where $\omega = 2\pi\nu = c/\lambda$ is the frequency and $k \equiv 1/\lambda$ is the propagation vector of the wave traveling in the x direction with a phase velocity c. In the special case of a *standing wave* which has an amplitude a_0, the wave equation becomes independent of time and takes the form

$$\frac{\partial^2 a}{\partial x^2} + k^2 a = 0 \tag{D-3}$$

wherein

$$\xi = a \cos \omega t = a_0 \sin kx \cos \omega t \tag{D-4}$$

To make the requisite transition to quantum mechanics, we replace the disturbance ξ by a wave function ψ,

$$\xi \to \psi(x, y, z, t) \tag{D-5}$$

which describes, e.g., the momentum and energy state of the system (through a probability distribution). For example, a *free* electron (not under the influence of an external potential, so that $E_{pot} = V = 0$) can have its state represented by a *stationary* wave function which corresponds to nonrelativistic motion along, say, the x direction. The kinetic energy of this "matter wave" is, according to de Broglie,

$$E_{kin} = \frac{1}{2} mv^2 = \frac{1}{2} m \left(\frac{p}{m}\right)^2 = \frac{1}{2} m \left(\frac{\hbar}{m\lambda}\right)^2 \tag{D-6}$$

whence

$$k^2 = \frac{1}{\lambda^2} = \frac{2m}{\hbar^2} E_{kin} \tag{D-7}$$

which yields the TIME-INDEPENDENT SCHRÖDINGER EQUATION $(a \to \psi)$

$$\frac{\partial^2 \psi}{\partial x^2} + \frac{2m}{\hbar^2} E_{kin} \psi = 0 \tag{D-8}$$

This can readily be extended to the three-dimensional form,

$$\nabla^2 \psi + \frac{2m}{\hbar^2} E_{kin} \psi = 0 \tag{D-9}$$

with

$$\nabla^2 \equiv \frac{\partial^2}{\partial x^2} + \frac{\partial^2}{\partial y^2} + \frac{\partial^2}{\partial z^2} \tag{D-10}$$

It can be generalized further when the system is subject to a potential field V, for then

$$E_{kin} = E - V \tag{D-11}$$

where $E \equiv E_{tot}$ is the *total* energy of the system:

$$\nabla^2 \psi + \frac{2m}{\hbar^2} (E - V) \psi = 0 \tag{D-12}$$

The still more general TIME-DEPENDENT SCHRÖDINGER EQUATION can most simply be deduced upon introducing the *operator formalism* for momentum and energy. Writing

$$\psi = \psi_0 e^{-i(\omega t - kx)} \tag{D-13}$$

we note that on building the *spatial* partial derivative,

$$\frac{\partial \psi}{\partial x} = ik\psi = \frac{i}{\hbar}p_x\,\psi \qquad (D\text{-}14)$$

we can resolve this to obtain an expression for the momentum operator in one dimension as

$$\underline{p}_x = \frac{\hbar}{i}\frac{\partial}{\partial x} \qquad (D\text{-}15)$$

and therefore in three dimensions as

$$\underline{p} = \frac{\hbar}{i}\nabla \qquad (D\text{-}16)$$

Analogously, the time-derivative, with $\omega = c/\lambda = pc/\hbar$, i.e., $\omega = E/\hbar$, takes the form

$$\frac{\partial \psi}{\partial t} = -i\omega\psi = -\frac{i}{\hbar}E\psi \qquad (D\text{-}17)$$

and thereby yields the operational energy definition

$$\underline{E} = -\frac{\hbar}{i}\frac{\partial}{\partial t} = i\hbar\frac{\partial}{\partial t} \qquad (D\text{-}18)$$

By substituting the operator definitions into the energy-balance equation, viz.

$$E = E_{\text{kin}} + E_{\text{pot}} = \frac{p^2}{2m} + V(x,y,z,t) \qquad (D\text{-}19)$$

i.e.,

$$\underline{E}\psi = \left(\frac{1}{2m}\underline{p}^2 + \underline{V}\right)\psi \qquad (D\text{-}20)$$

we arrive at the time-dependent Schrödinger equation

$$i\hbar\frac{\partial \psi}{\partial t} = \left(-\frac{\hbar^2}{2m}\nabla^2 + \underline{V}\right)\psi \qquad (D\text{-}21)$$

The system is thus represented by a progressive wave which, in terms of the Hamiltonian energy operator $\underline{H}$, equivalent to

$$\underline{H} = -\frac{\hbar^2}{2m}\nabla^2 + \underline{V} \qquad (D\text{-}22)$$

has discrete energy states E_n which constitute the *eigenvalues* of the wave equation

$$\underline{H}\psi_n = E_n\,\psi_n \qquad (D\text{-}23)$$

in which the ψ_n are energy *eigenfunctions* that also satisfy the equation

$$i\hbar \frac{\partial \psi_n}{\partial t} = H\psi_n = E_n \psi_n \tag{D-24}$$

The physically observable energy states E_n, such as the energy levels of a nucleus, can accordingly in principle be determined from a knowledge of the appropriate wave function, which is normalized to

$$\int |\psi|^2 \, d\Omega \equiv \int \psi^* \, \psi \, d\Omega = 1 \tag{D-25}$$

where the volume integral over $d\Omega \equiv dx \, dy \, dz$ extends over all space. The physical interpretation of the wave function then follows when the square of the latter is taken to give a measure of the probability of locating the system at a time instant t within the volume element $d\Omega$. Integration over all space corresponds to *certainty* of locating the system and therefore to *unit probability*.

As the wave function describes the physical behavior of a system, it is reasonable to expect that, together with its first derivative, it will be *unique* and *regular* (continuous) over its entire range. This is equivalent to requiring simply that its logarithmic derivative

$$\frac{\partial(\ln \psi)}{\partial x} = \frac{1}{\psi} \frac{\partial \psi}{\partial x}$$

be continuous. Further, to avoid the possibility of ψ becoming infinite at any point, we require that

$$|\psi|^2 \, d\Omega \to 0 \qquad \text{as} \qquad d\Omega \to 0 \tag{D-26}$$

In the case of a bound particle, which has a discrete spectrum of energy states, the wave function must vanish to zero at $\pm\infty$, whereas in the case of a free particle, described by a plane wave, the wave amplitude can be nonzero even at infinity.

D.2. Probability Density and Electron Probability Distribution

The PROBABILITY DENSITY ρ is positive definite throughout, being given by the expression

$$\rho = |\psi(x,y, z,t)|^2 \equiv \psi^* \, \psi \tag{D-27}$$

and the PROBABILITY CURRENT DENSITY then becomes

$$\mathbf{j} = \frac{\hbar}{2im} \, (\psi^* \, \nabla\psi - \psi \, \nabla\psi^*) \tag{D-28}$$

satisfying the EQUATION OF CONTINUITY which expresses ρ conservation,

$$\frac{\partial \rho}{\partial t} + \operatorname{div} \mathbf{j} = 0 \tag{D-29}$$

To solve the three-dimensional wave equation analytically it is as a rule necessary to separate it into total differential equations in each of the three spatial coordinates, and in certain special cases, this may be achieved more readily by introducing polar coordinates (r, θ, φ) in place of the Cartesian coordinates, writing

$$\left.\begin{array}{l} x = r \sin \theta \cos \varphi \\ y = r \sin \theta \sin \varphi \\ z = r \cos \theta \end{array}\right\} \tag{D-30}$$

and

$$\nabla^2 = \frac{1}{r^2} \frac{\partial}{\partial r} \left(r^2 \frac{\partial}{\partial r} \right) + \frac{1}{r^2 \sin \theta} \frac{\partial}{\partial \theta} \left(\sin \theta \frac{\partial}{\partial \theta} \right) + \frac{1}{r^2 \sin \theta} \frac{\partial^2}{\partial \varphi^2} \tag{D-31}$$

The wave function can be decomposed into radial and angular parts,

$$\psi(x,y,z) \rightarrow \psi(r, \theta, \varphi) = f(r)\, Y(\theta, \varphi) = f(r)\, \Theta(\theta)\, \Phi(\varphi) \tag{D-32}$$

as in Section 5.3.10.

A frequently encountered such case is that of a system under the influence of a *spherically-symmetric* potential, e.g., the attractive Coulomb potential in a hydrogen-like atom of atomic number Z,

$$V = -\frac{Ze^2}{r} \tag{D-33}$$

The *radial* part of the wave equation then furnishes the bound energy levels corresponding to the Bohr theory, viz.

$$E_n = -\frac{1}{n^2} \frac{Z^2 e^4 m}{2\hbar^2} = -\frac{1}{n^2} \frac{Z^2 e^2}{2a_0} \tag{D-34}$$

in terms of the principal quantum number n (see Section 5.3.7), while the *angular* parts involve the orbital and magnetic quantum numbers l and m_l (Sections 5.3.1 and 5.3.2)—a detailed treatment of this situation is given by Schiff [Schi 55, pp. 69ff], in which separation into parabolic coordinates is also considered. The radial part of the energy eigenfunction is a complicated expression involving associated Laguerre polynomials which depend on n and l; the lowest of these reduce to give the following normalized radial functions $f_{nl}(r)$:

$$f_{10}(r) = \left(\frac{Z}{a_0}\right)^{3/2} 2e^{-Zr/a_0} \tag{D-35}$$

$$f_{20}(r) = \left(\frac{Z}{2a_0}\right)^{3/2} \left(2 - \frac{Zr}{a_0}\right) e^{-Zr/2a_0} \qquad \text{(D-36)}$$

$$f_{21}(r) = \left(\frac{Z}{2a_0}\right)^{3/2} \frac{Zr}{a_0\sqrt{3}} e^{-Zr/2a_0} \qquad \text{(D-37)}$$

Functions of this type enable us to visualize the wave functions in a physical way. (A good discussion can be found in Pauling and Wilson [Pau 35, pp.

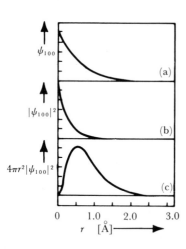

Fig. D-1. Radial dependence of the energy eigenfunction ψ_{nlm_l} of the hydrogen atom in its normal (1S) state with $n = 1$, $l = 0$, $m_l = 0$. (a) The normalized wave function ψ_{100}; (b) the absolute square $|\psi_{100}|^2 \equiv \psi^*_{100}\psi_{100}$; (c) the probability distribution function $4\pi r^2 |\psi_{100}|^2$, as functions of radial distance r from the center of the atom. The vertical scale is expressed in arbitrary units (adapted from [Pau 35]). (Used by permission of McGraw-Hill Book Co.)

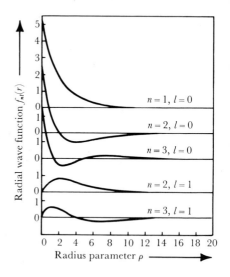

Fig. D-2. Plots of the ρ-dependence of the radial wave functions $f_{nl}(r)$ for various values of n and l in the case of a hydrogen atom. To convert the horizontal scale into a direct representation of electron-nucleus separation r, the units should in each case be multiplied by a factor n (from [Pau 35]). (Used by permission of McGraw-Hill Book Co.)

132ff].) The wave function ψ_{nlm_l} which describes the properties of the hydrogen atom in its normal (1S) state, in which $n = 1$, $l = 0$, $m_l = 0$, is

$$\psi_{100} = \frac{1}{(\pi a_0^3)^{1/2}} e^{-r/a_0} \qquad \text{(D-38)}$$

and this, together with $|\psi_{100}|^2$ and the PROBABILITY DISTRIBUTION FUNCTION $4\pi r^2 |\psi_{100}|^2$ is shown in Figs. D-1(a–c). The large probability of location within about 1 Å of the nucleus indicates the "size" of the hydrogen atom to correspond closely with that predicted by Bohr theory—indeed, the most probable electron-nucleus separation is given by the peak of the distribution function as 0.529 Å, which exactly corresponds to the Bohr radius a_0.

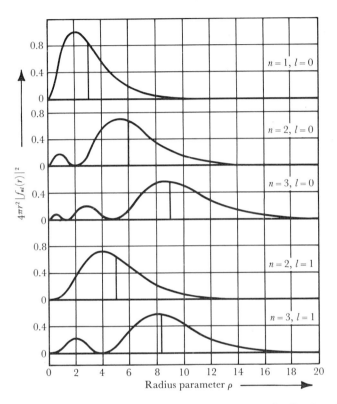

Fig. D-3. Corresponding plots of the radial electron distribution functions $4\pi r^2 |f_{nl}(r)|^2$ vs ρ for the hydrogen atom. The vertical lines denote the average distance of the electron from the nucleus, showing that the size of the atom increases roughly as n^2 (from [Pau 35]). (Used by permission of McGraw-Hill Book Co.)

The radial wave functions $f_{nl}(r)$ for various n and l are plotted in terms of a convenient dimensionless parameter,

$$\rho = \frac{2Z}{na_0} r \tag{D-39}$$

in Fig. D-2, and the electron radial distribution functions $4\pi r^2 |f_{nl}(r)|^2$ are plotted against ρ in Fig. D-3. To picture these in terms of the electron-nucleus separation, the horizontal scale would have to be imagined expanded by the factor n in each case. The vertical lines in Fig. D-3 indicate the average electron-nucleus separation, as given by

$$\bar{r} = \int \int \int \psi_{nlm_l}^* \, r \psi_{nlm_l} \, r^2 \, dr \sin \theta \, d\theta \, d\varphi = \frac{n^2 a_0}{Z} \left\{ 1 + \frac{1}{2} \left[1 - \frac{l(l+1)}{n^2} \right] \right\} \quad \text{(D-40)}$$

The "central-field approximation" can be extended to embrace atoms containing several electrons, as in the Thomas-Fermi [Tho 27, Fe 28] and Hartree-Fock [Ha 28, Fo 30a] approximations. Thus for atoms of alkali elements which consist of spherically symmetric closed shells plus one orbital (valence) electron, the distribution can be deduced. For the typical case of

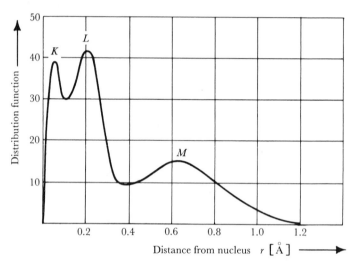

Fig. D-4. Electron density distribution in a singly ionized potassium atom, calculated in central-field approximation. The peaks distinguish the K, L, and M shells (from [To 49]).

singly ionized potassium (from which the valence electron lying outside closed K, L, and M shells has been removed) the distribution function takes the form shown in Fig. D-4, in which the peaks corresponding to the K, L, and M electrons can clearly be distinguished.

D.3. Heisenberg Uncertainty Relations

To give some measure of substantiation for the principle of indeterminacy [He 27] which has been utilized throughout the text, we indicate a derivation of the momentum-location uncertainty relation in which we define the uncertainty $\varDelta q$ in a quantity q as the *root-mean-square deviation of q from its* *expectation value* $\langle q \rangle \equiv \int \psi^* q\psi\, d\Omega$:

$$\Delta q \equiv \{\langle (q - \langle q \rangle)^2 \rangle\}^{1/2} = (\langle q^2 \rangle - \langle q \rangle^2)^{1/2} \tag{D-41}$$

Thus, the deviation in the x component of momentum is

$$\delta p_x \equiv p_x - \langle p_x \rangle \tag{D-42}$$

and in the x coordinate is

$$\delta x \equiv x - \langle x \rangle \tag{D-43}$$

It can easily be shown that the commutator bracket of these deviations gives the same value as the commutator bracket of p_x and x, namely,

$$[p_x, x] = [\delta p_x, \delta x] = \frac{\hbar}{i} \tag{D-44}$$

which shows that δp_x and δx differ in phase. By writing these latter as complex quantities, viz.

$$\left. \begin{array}{l} \delta p_x\, \delta x = A + iB \\ \delta x\, \delta p_x = A - iB \end{array} \right\} \tag{D-45}$$

we may set

$$[\delta p_x, \delta x] \equiv \delta p_x\, \delta x - \delta x\, \delta p_x = 2iB = \frac{\hbar}{i} \tag{D-46}$$

whence

$$B = -\frac{1}{2}\hbar \tag{D-47}$$

Noting that

$$|\delta p_x\, \delta x| = |\delta x\, \delta p_x| = (A^2 + B^2)^{1/2} \tag{D-48}$$

and that in accordance with the above definition we must set the uncertainty equal to the rms deviation,

$$\Delta p_x\, \Delta x = (|\delta p_x\, \delta x|^2)^{1/2} = (A^2 + B^2)^{1/2} = (A^2 + \tfrac{1}{4}\hbar^2)^{1/2} \tag{D-49}$$

it is evident that the positive quantity A^2 must be at least zero and therefore

$$\Delta p_x\, \Delta x \geqslant \frac{\hbar}{2} \tag{D-50}$$

It normally suffices to take a less rigorous definition of the uncertainties and write

$$\Delta p_x \Delta x \gtrsim \hbar \tag{D-51}$$

or, correspondingly,

$$\Delta E\, \Delta t \gtrsim \hbar \tag{D-52}$$

D.4. Klein-Gordon Equation for Spin-0 Particles

At the same time as he formulated his nonrelativistic wave equations, Schrödinger [Schrö 26] also put forward the *relativistic* equivalent, which can be obtained by substituting operational definitions into the relativistic relation

$$E^2 = p^2 c^2 + m^2 c^4 \tag{D-53}$$

as

$$\frac{\partial^2 \psi}{\partial t^2} = \left(c^2 \nabla^2 - \frac{m^2 c^4}{\hbar^2} \right) \psi \tag{D-54}$$

an expression now termed the KLEIN-GORDON EQUATION after Klein and Gordon who independently proposed a similar result very slightly later [Go 26, Kl 27].

An originally troublesome feature of this equation was the fact that the probability density

$$\rho = \frac{i\hbar}{2mc^2} \left(\psi^* \frac{\partial \psi}{\partial t} - \frac{\partial \psi^*}{\partial t} \psi \right) \tag{D-55}$$

defined in accordance with the relations (D-28) and (D-29) is able to assume *negative* as well as positive values, i.e., it could change with ψ and $\partial \psi / \partial t$ from positive to negative in different regions. This dilemma was resolved by Pauli and Weisskopf [Pa 34], who reinterpreted it as a field equation which was subjected to second quantization, with ρ viewed as a *charge* density which can take on positive or negative values according to the sign of the charge. The wave function ψ which satisfies the Klein-Gordon equation transforms as a *scalar* under proper Lorentz transformations. Thus the only degree of freedom which it permits is that of *translation*, but not of spin, and accordingly the Klein-Gordon equation is restricted to spin-0 particles. For example it should be appropriate to the description of π-meson orbits in π-mesic atoms.

D.4.1. RELATIVISTICALLY INVARIANT NOTATION

The Klein-Gordon equation can be written in a more elegant relativistically invariant form when one notes that for a momentum-energy 4-vector written as

$$p^\mu \equiv \left(\mathbf{p}, \frac{iE}{c} \right) = \left\{ \frac{E}{c}, \mathbf{p} \right\} \tag{D-56}$$

where now

$$\mu = 0, 1, 2, 3 \tag{D-57}$$

with

$$p^0 = \frac{E}{c} = \text{real} \tag{D-58}$$

and

$$p^1, p^2, p^3 = p_x, p_y, p_z \qquad \text{(D-59)}$$

the ENERGY-MOMENTUM RELATION (D-53) takes the form

$$\sum_\mu p^\mu p_\mu = m^2 c^2 \qquad \text{(D-60)}$$

and the operators become simply

$$p_\mu = i\hbar \frac{\partial}{\partial x^\mu} \equiv i\hbar\, \partial_\mu \qquad \text{(D-61)}$$

We observe that the CONTRAVARIANT vector p^μ (which transforms as the coordinates x, y, z, t) can be converted into the corresponding COVARIANT vector p_μ (which transforms as a gradient) through the relativistic metric transformation

$$v^\mu = \sum_\nu g^{\mu\nu} v_\nu \qquad \text{(D-62)}$$

and, inversely,

$$v_\mu = \sum_\nu g_{\mu\nu} v^\nu \qquad \text{(D-63)}$$

Therein, $g_{\mu\nu}$ is the METRIC TENSOR which has components $(+1, -1, -1, -1)$ and possesses the property that

$$g^{\mu\nu} = g^{\nu\mu}, \qquad g_{\mu\nu} = g_{\nu\mu} \qquad \text{(D-64)}$$

while

$$\sum_\nu g^{\mu\nu} g_{\nu\lambda} = \delta^\mu_\lambda = \begin{cases} 1 \text{ if } \mu = \lambda \\ 0 \text{ otherwise} \end{cases} \qquad \text{(D-65)}$$

With the aid of this notation, the Klein-Gordon equation can be written in Lorentz-invariant form as

$$\left\{ \Box + \left(\frac{mc}{\hbar}\right)^2 \right\} \psi(x^\mu) = 0 \qquad \text{(D-66)}$$

for the field-free case, introducing the d'Alembertian

$$\Box \equiv \frac{1}{c^2} \frac{\partial^2}{\partial t^2} - \nabla^2 \qquad \text{(D-67)}$$

In the presence of external fields the equation (D-66) becomes modified to

$$\left\{ \sum_\mu \left(p^\mu - \frac{e}{c} A^\mu \right)^2 - m^2 c^2 \right\} \psi(x^\mu) = 0 \qquad \text{(D-68)}$$

upon setting

$$p^\mu \to p^\mu - \frac{e}{c} A^\mu \qquad \text{(D-69)}$$

in terms of the 4-potential

$$A^\mu \equiv \{\phi, \mathbf{A}\} \qquad \text{(D-70)}$$

D.5. Dirac Equation for Spin-$\frac{1}{2}$ Relativistic Particles

One means of preventing the probability density from taking on negative values is to eliminate time derivatives from the expression for ρ—which can be attained by not permitting any time derivatives of higher than *first* order to appear in the wave equation. Space-time symmetry then imposes a similar restriction upon the *spatial* derivatives. A final requirement, imposed by the quantum-mechanical principle of superposition, is that the first-order differential wave equation must also be linear.

A wave equation with these properties was first formulated by Dirac [Di 28] in the field-free case. Starting out from the Hamiltonian form (D-24), which is of first order in the time derivative, he set out to derive a linear Hamilton operator containing only first-order derivatives by introducing parameters $\boldsymbol{\alpha}$ and β as a means of linearizing the energy-momentum relation

$$E^2 = (pc)^2 + (mc^2)^2 \tag{D-71}$$

to

$$E = (\boldsymbol{\alpha} \cdot \mathbf{p})\, c + \beta(mc^2) \tag{D-72}$$

The corresponding wave equation

$$[-E + (\boldsymbol{\alpha} \cdot \mathbf{p})\, c + \beta(mc^2)]\, \psi = 0 \tag{D-73}$$

can then, on substituting the operator definitions

$$E = i\hbar \frac{\partial}{\partial t}, \qquad p = \frac{\hbar}{i}\nabla \tag{D-74}$$

be written as

$$\frac{\hbar}{i}\frac{\partial \psi}{\partial t} + \frac{\hbar c}{i}\,(\boldsymbol{\alpha} \cdot \nabla)\, \psi + \beta mc^2\, \psi = 0 \tag{D-75}$$

which is the DIRAC EQUATION in terms of the vector and scalar parameters $\boldsymbol{\alpha}$ and β respectively. These can be shown to satisfy the relations

$$\alpha_x^2 = \alpha_y^2 = \alpha_z^2 = \beta^2 = 1 \tag{D-76}$$

$$\alpha_x \alpha_y + \alpha_y \alpha_x = \alpha_y \alpha_z + \alpha_z \alpha_y = \alpha_z \alpha_x + \alpha_x \alpha_z = 0 \tag{D-77}$$

$$\alpha_x \beta + \beta \alpha_x = \alpha_y \beta + \beta \alpha_y = \alpha_z \beta + \beta \alpha_z = 0 \tag{D-78}$$

i.e., their squares are unity and they anticommute in pairs. An appropriate representation through 4×4 matrices can be depicted schematically as

$$\alpha = \begin{pmatrix} 0 & \sigma \\ \sigma & 0 \end{pmatrix}, \qquad \beta = \begin{pmatrix} 1 & 0 \\ 0 & -1 \end{pmatrix} \tag{D-79}$$

where **0**, **1**, and **σ** are, respectively, zero, unit, and Pauli spin matrices of the second rank,

$$\sigma_x = \begin{pmatrix} 0 & 1 \\ 1 & 0 \end{pmatrix}, \qquad \sigma_y = \begin{pmatrix} 0 & -i \\ i & 0 \end{pmatrix}, \qquad \sigma_z = \begin{pmatrix} 1 & 0 \\ 0 & -1 \end{pmatrix} \qquad \text{(D-80)}$$

Written out explicitly, the **α**'s and β are Hermitian 4×4 matrices,

$$= \begin{pmatrix} 0 & 0 & 0 & 1 \\ 0 & 0 & 1 & 0 \\ 0 & 1 & 0 & 0 \\ 1 & 0 & 0 & 0 \end{pmatrix}, \quad \alpha_y = \begin{pmatrix} 0 & 0 & 0 & -i \\ 0 & 0 & i & 0 \\ 0 & -i & 0 & 0 \\ i & 0 & 0 & 0 \end{pmatrix}, \quad \alpha_z = \begin{pmatrix} 0 & 0 & 1 & 0 \\ 0 & 0 & 0 & -1 \\ 1 & 0 & 0 & 0 \\ 0 & -1 & 0 & 0 \end{pmatrix}, \quad \beta = \begin{pmatrix} 1 & 0 & 0 & 0 \\ 0 & 1 & 0 & 0 \\ 0 & 0 & -1 & 0 \\ 0 & 0 & 0 & -1 \end{pmatrix}$$

$$\text{(D-81)}$$

and the wave function is, accordingly, a column matrix with four components,

$$\psi = \begin{pmatrix} \psi_1(x, y, z, t) \\ \psi_2(x, y, z, t) \\ \psi_3(x, y, z, t) \\ \psi_4(x, y, z, t) \end{pmatrix} \qquad \text{(D-82)}$$

In this representation, (D-75) is equivalent to four simultaneous partial differential equations of first order that are linear and homogeneous in the four ψ components. Such products as, for instance, $\alpha_x \psi$ acquire a simple physical interpretation. Since

$$\alpha_x \psi = \begin{pmatrix} 0 & 0 & 0 & 1 \\ 0 & 0 & 1 & 0 \\ 0 & 1 & 0 & 0 \\ 1 & 0 & 0 & 0 \end{pmatrix} \begin{pmatrix} \psi_1 \\ \psi_2 \\ \psi_3 \\ \psi_4 \end{pmatrix} = \begin{pmatrix} \psi_4 \\ \psi_3 \\ \psi_2 \\ \psi_1 \end{pmatrix} \qquad \text{(D-83)}$$

this term corresponds simply to reversing the spin of the free particle.

The PROBABILITY DENSITY in matrix notation, viz.

$$\rho = \sum_{\mu} |\psi_{\mu}|^2 = \psi^{\dagger} \psi \qquad \text{(D-84)}$$

involves the Hermitian conjugate wave function $\psi^{\dagger}$ derived by transposing matrix rows and columns throughout and building the complex conjugate,

$$\psi^{\dagger} \equiv \overbrace{\psi_1^* \quad \psi_2^* \quad \psi_3^* \quad \psi_4^*} \qquad \text{(D-85)}$$

With this definition, the probability density is always positive *definite*.

The CURRENT DENSITY takes the form

$$\mathbf{j} = c\psi^{\dagger} \boldsymbol{\alpha} \psi \qquad \text{(D-86)}$$

satisfying the equation of continuity

$$\left(\psi^\dagger \frac{\partial \psi}{\partial t} + \frac{\partial \psi^\dagger}{\partial t} \psi \right) + \operatorname{div} \mathbf{j} = 0 \tag{D-87}$$

Here, the Hermiticity of the $\boldsymbol{\alpha}$ and β matrices has implicitly been assumed, viz.

$$\boldsymbol{\alpha}^\dagger = \boldsymbol{\alpha}, \qquad \beta^\dagger = \beta \tag{D-88}$$

This property can be shown to follow from a consideration of the adjoint Dirac equation and the Hermiticity of the Hamiltonian.

The requisite modification of the Dirac equation to take account of an external electromagnetic field $A^\mu = \{\phi, \mathbf{A}\}$ results on setting

$$E \to E - e\phi, \qquad \mathbf{p} \to \mathbf{p} - \frac{e}{c}\mathbf{A} \tag{D-89}$$

in (D-73) to derive the DIRAC EQUATION FOR A BOUND PARTICLE,

$$\left\{ (E - e\phi) + c\left(\boldsymbol{\alpha}\cdot\left[\mathbf{p} - \frac{e}{c}\mathbf{A}\right]\right) + \beta mc^2 \right\} \psi = 0 \tag{D-90}$$

The energy levels of a system comprising a relativistic spin-$\frac{1}{2}$ particle situated in a Coulomb field, for which

$$e\phi = -\frac{Ze^2}{r}, \qquad \mathbf{A} = 0 \tag{D-91}$$

are (upon including the rest energy mc^2) given by the formula

$$E_n = mc^2 \left\{ 1 + \frac{1/137}{(n - k + [k^2 - (1/137)^2]^{1/2})^2} \right\}^{-1/2} \tag{D-92}$$

with

$$n = 1, 2, \dots \tag{D-93}$$

and

$$k = 1, 2, \dots, n \tag{D-94}$$

This formula is equivalent to the Sommerfeld fine-structure formula [So 16]. Except for small, but significant, discrepancies [La 51] it furnishes good agreement with experimental data.

In the nonrelativistic limit, with $\phi = 0$ and $\partial \mathbf{A}/\partial t = 0$ (i.e., $\mathbf{A}$ stationary), the Dirac equation reduces to the Schrödinger equation with an extra term

$$-\frac{e\hbar}{2mc}(\boldsymbol{\sigma}\cdot\mathbf{H}) \tag{D-95}$$

in the Hamiltonian, which can physically be interpreted as the energy due to the particle's magnetic moment $e\hbar/2mc$ when situated in a magnetic field of strength **H**. The **σ**'s are the Pauli spin matrices and the wave function then has only two components.

In the nonrelativistic *central field* limit $(e\phi = V(r)$ and $\mathbf{A} = 0)$, one derives the Schrödinger equation with a spin-orbit energy term

$$\frac{1}{2mc^2}\frac{1}{r}\frac{\partial V(r)}{\partial r}\,(\mathbf{S}\cdot\mathbf{L}) \tag{D-96}$$

where $\mathbf{S} = \tfrac{1}{2}\hbar\boldsymbol{\sigma}$ represents the spin angular momentum and $\mathbf{L} = [\mathbf{r}\times\mathbf{p}]$ the orbital angular momentum of the bound particle.

D.5.1. COVARIANT FORM OF THE DIRAC EQUATION (GAMMA MATRICES)

To arrive at a concise, Lorentz-invariant representation of the Dirac equation, we introduce the γ MATRICES (also called CLIFFORD NUMBERS) which we define as

$$\gamma^\mu \equiv \{\gamma^0, \boldsymbol{\gamma}\} \tag{D-97}$$

where

$$\gamma^0 \equiv \beta \triangleq \begin{pmatrix} 1 & 0 \\ 0 & -1 \end{pmatrix} \tag{D-98}$$

and

$$\boldsymbol{\gamma} = (\gamma^1, \gamma^2, \gamma^3) \equiv \beta\boldsymbol{\alpha} \triangleq \begin{pmatrix} 0 & \boldsymbol{\sigma} \\ -\boldsymbol{\sigma} & 0 \end{pmatrix} \tag{D-99}$$

It should be borne in mind that this convention differs from other alternative but essentially similar representations, in which, e.g., $\boldsymbol{\gamma} = -i\beta\boldsymbol{\alpha}$, and which lead to the result

$$(\gamma_5)^2 \equiv (\gamma^0\gamma^1\gamma^2\gamma^3)^2 = +\mathbf{1}$$

while in our notation $(\gamma_5)^2 = -\mathbf{1}$, as will be shown later.

These γ matrices play an important role in the formulation of β-decay theory, as presented in Section 8.6, for which reason their properties will here be investigated in some detail. They satisfy the anticommutation rules

$$\gamma^\mu\gamma^\nu + \gamma^\nu\gamma^\mu = 2g^{\mu\nu}\mathbf{1} = 2g^{\mu\nu}\delta^{\mu\nu} \tag{D-100}$$

$$\gamma_\mu\gamma_\nu + \gamma_\nu\gamma_\mu = 2g_{\mu\nu}\mathbf{1} = 2g_{\mu\nu}\delta_{\mu\nu} \tag{D-101}$$

While γ^0 is Hermitian, one finds that the $\boldsymbol{\gamma} \equiv \gamma^1$, γ^2, γ^3 are anti-Hermitian. Before we examine these more closely, we show how they can be incorporated within the Dirac equation to yield a concise covariant representation.

On multiplying the field-free Dirac equation (D-75) from the left by β, dividing by $\hbar c$ and noting that

$$\frac{\partial}{\partial t} = c\frac{\partial}{\partial(ct)} = c\frac{\partial}{\partial x^0} \tag{D-102}$$

we find, with the notation (D-61) that

$$\left[\frac{\beta}{i}\partial_0 + \left(\sum_{k=1,2,3}\frac{\beta}{i}\alpha^k\partial_k\right) + \frac{mc}{\hbar}\right]\psi = 0 \tag{D-103}$$

which can be written in covariant form as

$$\left[\left(\frac{1}{i}\sum_\mu \gamma^\mu \partial_\mu\right) + \frac{mc}{\hbar}\right]\psi = 0 \tag{D-104}$$

One can abbreviate this still further by making use of the Feynman "dagger notation" [Fe 49a]:

$$\rlap{/}A \equiv \sum_\mu \gamma^\mu A_\mu = \sum_\mu \gamma_\mu A^\mu = \gamma^0 A^0 - (\boldsymbol{\gamma \cdot A}) \tag{D-105}$$

$$\rlap{/}\nabla \equiv \sum_\mu \gamma^\mu \partial_\mu = \gamma^0 \partial_0 + (\boldsymbol{\gamma \cdot \nabla}) \tag{D-106}$$

This yields the concise form of the COVARIANT FIELD-FREE DIRAC EQUATION

$$\left(\frac{1}{i}\rlap{/}\nabla + \frac{mc}{\hbar}\right)\psi = 0 \tag{D-107}$$

From this, with (D-61), we can derive the equivalent form in *momentum representation*,

$$\left(\rlap{/}p - mc\right)\psi = 0 \tag{D-108}$$

where

$$\rlap{/}p \equiv \sum_\mu \gamma^\mu p_\mu = \sum_\mu \gamma_\mu p^\mu \tag{D-109}$$

The probability and current densities ρ and $\mathbf{j}$ can be expressed in terms of the γ's and combined into a 4-vector

$$\frac{j^\mu}{c} \equiv \left\{\rho, \frac{\mathbf{j}}{c}\right\} \tag{D-110}$$

as ensues from considering the function

$$\bar{\psi} \equiv \psi^\dagger \beta = \psi^\dagger \gamma^0 \tag{D-111}$$

which is frequently called the "adjoint" wave function, although the true Hermitian adjoint wave function is $\psi^\dagger$. The above function is "adjoint" in the sense that

$$\bar{\psi}^* = \gamma^0 \psi \tag{D-112}$$

Making use of the definition of the γ's to write

$$\beta\gamma^k = \alpha^k \qquad (k = 1, 2, 3) \tag{D-113}$$

we can express the current density in terms of $\bar{\psi}$ as

$$j^k = c\psi^\dagger \alpha^k \psi = c\psi^\dagger \beta\gamma^k \psi = c\bar{\psi}\gamma^k \psi \tag{D-114}$$

and the probability charge density as

$$\rho \equiv \frac{j^0}{c} = \psi^\dagger \gamma^0 \gamma^0 \psi = \bar{\psi} \gamma^0 \psi \tag{D-115}$$

One can combine the above two equations into a single expression,

$$\frac{j^\mu}{c} = \bar{\psi} \gamma^\mu \psi \tag{D-116}$$

which is Hermitian. The 4-vector j^μ/c is divergence-free,

$$\partial_\mu \left(\frac{j^\mu}{c} \right) = 0 \tag{D-117}$$

and can be interpreted physically as a quantum-mechanical charge-density/current-density 4-vector. A comparison with the corresponding classical 4-vector,

$$i^\mu \equiv \left\{ \rho, \frac{\rho}{c} \mathbf{v} \right\} \tag{D-118}$$

immediately suggests that the role of the classical particle velocity $\mathbf{v}$ is here taken over by the operator $c\boldsymbol{\alpha}$, since

$$\mathbf{v} = \frac{\mathbf{j}}{\rho} \to \frac{c\psi^\dagger \boldsymbol{\alpha} \psi}{\psi^\dagger \psi} \to c\boldsymbol{\alpha} \tag{D-119}$$

As $\alpha^1 \equiv \alpha_x$ has eigenvalues ± 1, the x component of velocity would appear to have eigenvalues $\pm c$. This finds its explanation in the fast irregular motion, termed ZITTERBEWEGUNG, of the electron which is associated with the spin motion, whereas the actual mean velocity is dictated by the momentum,

$$\mathbf{v} = \frac{\mathbf{p}}{m} \tag{D-120}$$

D.5.2. PROPERTIES OF THE GAMMA MATRICES

When two or more γ matrices are multiplied together, new matrices ensue. On examining the structure of such product matrices, one finds that in all only 16 independent combinations of the γ^μ can be built, and these can be classified according to their transformation behavior as in Table 8-4. In terms of the more rigorous nomenclature introduced for the foregoing considerations, the respective types of product should strictly have been written as

$$\mathbf{1} \equiv \gamma^\mu \gamma_\mu, \qquad \gamma^\mu, \qquad i\gamma^\mu \gamma^\nu, \qquad i\gamma^\mu \gamma^\nu \gamma^\lambda, \qquad \gamma_5 \equiv \gamma^0 \gamma^1 \gamma^2 \gamma^3 \tag{D-121}$$

and the transformation property related to the matrix element formed on interposing each of the above between $\bar{\psi}$ and ψ. The i's have been included in order to make the resultant matrices Hermitian. The γ_5 matrix constitutes a rather special case. The index "5" has been retained from an alternative nomenclature which is still in use elsewhere and in which our matrices γ^0, γ^1, γ^2, γ^3 have been labeled respectively as γ^1, γ^2, γ^3, γ^4. A noteworthy feature of γ_5 is that it exists only in covariant form—the index cannot be raised. It is a convenient mathematical entity, since it has the value

$$\gamma_5 \equiv \gamma^0 \gamma^1 \gamma^2 \gamma^3 = \begin{pmatrix} 1 & \cdot & \cdot & \cdot \\ \cdot & 1 & \cdot & \cdot \\ \cdot & \cdot & -1 & \cdot \\ \cdot & \cdot & \cdot & -1 \end{pmatrix} \begin{pmatrix} \cdot & \cdot & \cdot & 1 \\ \cdot & \cdot & 1 & \cdot \\ \cdot & -1 & \cdot & \cdot \\ -1 & \cdot & \cdot & \cdot \end{pmatrix}$$

$$\times \begin{pmatrix} \cdot & \cdot & \cdot & -i \\ \cdot & \cdot & i & \cdot \\ \cdot & i & \cdot & \cdot \\ -i & \cdot & \cdot & \cdot \end{pmatrix} \begin{pmatrix} \cdot & \cdot & 1 & \cdot \\ \cdot & \cdot & \cdot & -1 \\ -1 & \cdot & \cdot & \cdot \\ \cdot & 1 & \cdot & \cdot \end{pmatrix} \quad \text{(D-122)}$$

$$= \begin{pmatrix} \cdot & \cdot & -i & \cdot \\ \cdot & \cdot & \cdot & -i \\ -i & \cdot & \cdot & \cdot \\ \cdot & -i & \cdot & \cdot \end{pmatrix} \quad \text{(D-123)}$$

or, symbolically,

$$\gamma_5 = -i \begin{pmatrix} \mathbf{0} & \mathbf{1} \\ \mathbf{1} & \mathbf{0} \end{pmatrix} \quad \text{(D-124)}$$

It anticommutes with each of the other covariant γ's,

$$\gamma_5 \gamma_\mu + \gamma_\mu \gamma_5 = 0 \quad \text{(D-125)}$$

and its square has the value

$$(\gamma_5)^2 = -1 \quad \text{(D-126)}$$

The quantity

$$\tfrac{1}{2}(1 \pm \gamma_5) \quad \text{(D-127)}$$

constitutes a projection operator which, when applied to the wave function ψ_ν of the neutrino field in the extended Fermi theory of β decay (Section 8.6) gives the neutrino eigenfunctions in a two-component theory. The plus sign yields left-handed neutrinos, whereas the minus sign would give right-handed neutrinos and can therefore be excluded on the basis of experimental findings. Since γ_5 thus essentially determines the "handedness" or CHIRALITY (from the Greek *chir*, hand) it is sometimes termed the chirality operator. The restriction to

$$\tfrac{1}{2}(1 + \gamma_5) \quad \text{(D-128)}$$

is reflected in the relative phases of the β-decay coupling constants (cf. the conclusions in Section 8.8):

$$C_\mathrm{A} = +C_\mathrm{A}', \qquad C_\mathrm{V} = +C_\mathrm{V}' \quad \text{(D-129)}$$

We might now test some of the contentions made hitherto by demonstrating the transformation properties of γ-matrix products and relating these to the formulation of β-decay theory. To this end, we make use of the Lorentz invariance of the field-free Dirac equation (D-104),

$$\left(\frac{1}{i}\gamma^\mu \partial_\mu + \frac{mc}{\hbar}\right)\psi = 0 \tag{D-130}$$

and the so-called "adjoint" equation which features $\bar{\psi} \equiv \psi^\dagger \gamma^0$, viz.

$$\left(i\gamma^\mu \partial_\mu + \frac{mc}{\hbar}\right)\bar{\psi} = 0 \tag{D-131}$$

For brevity, we have dropped the summation sign on the tacit understanding that summation is implied over any or all repeated ("dummy") indices.

The Lorentz transformation can be written as

$$\left.\begin{array}{l} x^{\mu'} = a_{\mu\nu} x^\nu \\ x^\nu = a_{\mu\nu} x^\mu \end{array}\right\} \tag{D-132}$$

with

$$a_{\mu\lambda} a_{\nu\lambda} = \delta_{\mu\nu} \tag{D-133}$$

Transforming the wave function

$$\psi' = \tau\psi \tag{D-134}$$

and substituting, we get

$$\left(\frac{1}{i}\gamma^\mu \partial_{\mu'} + \frac{mc}{\hbar}\right)\psi' = \left(\frac{1}{i}\gamma^\mu a_{\mu\nu} \partial_\nu \tau + \frac{mc}{\hbar}\tau\right)\psi \tag{D-135}$$

where

$$\partial_{\mu'} \equiv \frac{\partial}{\partial x^{\mu'}} \tag{D-136}$$

Then the transformed Dirac equation acquires the form

$$\left(\frac{1}{i}\tau^{-1}\gamma^\mu \tau a_{\mu\nu} \partial_\nu + \frac{mc}{\hbar}\right)\psi = 0 \tag{D-137}$$

Since the first term should be $(1/i)\gamma^\nu \partial_\nu$, comparison indicates that

$$\tau^{-1}\gamma^\mu \tau a_{\mu\nu} = \gamma^\nu \tag{D-138}$$

and, further,

$$\bar{\psi}' = \bar{\psi}\tau^{-1} \tag{D-139}$$

and

$$\psi' = \tau\psi \tag{D-140}$$

The character of τ is rendered evident on considering a rotation through an angle θ about the z axis, for then

$$\tau = \exp\left(\tfrac{1}{2}\theta\gamma^1\gamma^2\right) = \cos\frac{\theta}{2} + \gamma^1\gamma^2\sin\frac{\theta}{2} \tag{D-141}$$

or, on space inversion $(\mathbf{x}' = -\mathbf{x}, t' = t)$,

$$\tau = \gamma^4 \tag{D-142}$$

with

$$a_{\mu\nu} = \begin{pmatrix} -1 & \cdot & \cdot & \cdot \\ \cdot & -1 & \cdot & \cdot \\ \cdot & \cdot & -1 & \cdot \\ \cdot & \cdot & \cdot & -1 \end{pmatrix} \tag{D-143}$$

We are now in a position to identify the transformations, namely,

scalar: $\qquad \bar{\psi}\psi = \bar{\psi}'\,\tau^{-1}\,\tau\psi' = \bar{\psi}'\,\psi'$ $\qquad\qquad$ (D-144)

vector: $\qquad \bar{\psi}_1\gamma^\mu\psi_2 = \bar{\psi}_1\tau^{-1}\gamma^\mu\tau\psi_2 = \bar{\psi}_1'\gamma^\nu\psi_2'\,a_{\mu\nu}$ $\qquad$ (D-145)

tensor: $\qquad \bar{\psi}_1 i\gamma^\mu\gamma^\nu\psi_2 = \bar{\psi}_1' i\gamma^\mu\gamma^\nu\psi_2'$ $\qquad\qquad$ (D-146)

axial vector: $\quad \bar{\psi}_1 i\gamma^\mu\gamma_5\psi_2 = \bar{\psi}_1' i\gamma^\mu\gamma^0\gamma^1\gamma^2\gamma^3\psi_2'$ $\qquad$ (D-147)

pseudoscalar: $\;\; \bar{\psi}_1\gamma_5\psi_2 = -\bar{\psi}_1'\gamma_5\psi_2'$ $\qquad\qquad$ (D-148)

In connection with the axial vector transformation it should be noted that a triple γ product, i.e., $\gamma^\mu\gamma^\nu\gamma^\lambda$, can always be brought to the form $\gamma^\mu\gamma_5$.

When applied to the theory of β decay, these products are used to construct the general Hamiltonian density, namely,

scalar term $\qquad \mathcal{H} = C_S(\bar{\psi}_p\psi_n)(\bar{\psi}_e\psi_\nu)$

vector term $\qquad\qquad +C_V(\bar{\psi}_p\gamma^\mu\psi_n)(\bar{\psi}_e\gamma^\mu\psi_\nu)$

tensor term $\qquad\qquad + C_T(\bar{\psi}_p i\gamma^\mu\gamma^\nu\psi_n)(\bar{\psi}_e i\gamma^\mu\gamma^\nu\psi_\nu)$

axial vector term $\qquad + C_A(\bar{\psi}_p i\gamma^\mu\gamma_5\psi_n)(\bar{\psi}_e i\gamma^\mu\gamma_5\psi_\nu)$

pseudoscalar term $\qquad + C_P(\bar{\psi}_p\gamma_5\psi_n)(\bar{\psi}_e\gamma_5\psi_\nu)$

$\qquad\qquad\qquad\qquad + $ Hermitian conjugate $\qquad$ (D-149)

This is the most general conceivable expression, since interchange of ψ_n with ψ_ν introduces no new type of Fermi interaction. The part written out explicitly is that which refers to β decay of the *neutron*, while the Hermitian conjugate stands for the inverse β decay of the proton. The remainder of the development in β-decay theory is to be found in Section 8.6 *et seq.* Instead of pursuing this further, we return to the Dirac equation and indicate its application to electron-positron theory, as a supplement to the considerations presented in Section 4.6.1.

D.6. Dirac Electron-Positron Theory

The wave function ψ has four components ψ_μ and hence for any given value of p_μ one obtains *four* eigensolutions, one pair of which refers to a spin-$\frac{1}{2}$ particle having *positive* energy and either of two possible spin settings, and the other pair of which corresponds to *negative* energy, with either of two spin settings. This pairing suggests that one might simplify the formalism through the use of two-component spinors,

$$\psi = \begin{pmatrix} u \\ v \end{pmatrix} \tag{D-150}$$

where

$$u \equiv \begin{pmatrix} \psi_1 \\ \psi_2 \end{pmatrix} \quad \text{and} \quad v \equiv \begin{pmatrix} \psi_3 \\ \psi_4 \end{pmatrix} \tag{D-151}$$

which can be written in the form

$$u = \begin{pmatrix} u_{+1} \\ u_{-1} \end{pmatrix} \quad \text{and} \quad v = \begin{pmatrix} v_{+1} \\ v_{-1} \end{pmatrix} \tag{D-152}$$

On this basis, the field-free Dirac equation (D-73) can be expressed as

$$E \begin{pmatrix} u \\ v \end{pmatrix} = c(\boldsymbol{\sigma} \cdot \mathbf{p}) \begin{pmatrix} v \\ u \end{pmatrix} + mc^2 \begin{pmatrix} u \\ -v \end{pmatrix} \tag{D-153}$$

Taking the z direction to be the distinguished direction, we can make the substitution

$$(\boldsymbol{\sigma} \cdot \mathbf{p}) \to \sigma_z p_z = \begin{pmatrix} 1 & 0 \\ 0 & -1 \end{pmatrix} p_z \tag{D-154}$$

and thence decompose (D-153) into four linear equations, of which the first two correspond to spin "up" and the second pair to spin "down":

$$\text{spin "up":} \quad \begin{cases} (E - mc^2)\, u_{+1} = p_z v_{+1} & \text{(D-155)} \\ (E + mc^2)\, v_{+1} = p_z u_{+1} & \text{(D-156)} \end{cases}$$

$$\text{spin "down":} \quad \begin{cases} (E - mc^2)\, u_{-1} = -p_z v_{-1} & \text{(D-157)} \\ (E + mc^2)\, v_{-1} = -p_z u_{-1} & \text{(D-158)} \end{cases}$$

We first consider the *spin "up," positive energy* situation ($\Uparrow+$) for which, in arbitrary normalization, one has

$$\left. \begin{array}{l} u_{+1} = 1 \\ u_{-1} = v_{-1} = 0 \end{array} \right\} \tag{D-159}$$

Then, from (D-155) we derive

$$v_{+1} = \frac{E - mc^2}{p_z} = \frac{(E - mc^2)\,(E + mc^2)}{p_z\,(E + mc^2)} = \frac{p_z}{E + mc^2} \tag{D-160}$$

which is also given by (D-156), while (D-157) and (D-158) are compatible
with

$$u_{-1} = v_{-1} = 0 \tag{D-161}$$

Analogously, for the *spin* "*down,*" *negative energy* situation ($\Downarrow$–) one can set

$$u_{+1} = v_{+1} = 0 \quad \text{and} \quad v_{-1} = 1 \tag{D-162}$$

to find from (D-157) or (D-158) that

$$u_{-1} = \frac{p_z}{|E| + mc^2} \tag{D-163}$$

since now $|E| = -E$.

Similar procedures yield the other parameters, which can be depicted
collectively as

$$
(\Uparrow +) \qquad (\Downarrow +) \qquad (\Uparrow -) \qquad (\Downarrow -)
$$

$$
\begin{pmatrix} 1 \\ 0 \\ \dfrac{p_z}{E + mc^2} \\ 0 \end{pmatrix}
\begin{pmatrix} 0 \\ 1 \\ 0 \\ \dfrac{-p_z}{E + mc^2} \end{pmatrix}
\begin{pmatrix} \dfrac{-p_z}{|E| + mc^2} \\ 0 \\ 1 \\ 0 \end{pmatrix}
\begin{pmatrix} 0 \\ \dfrac{-p_z}{|E| + mc^2} \\ 0 \\ 1 \end{pmatrix}
\tag{D-164}
$$

In order to normalize to unit volume, one seeks to have the squares of the
wave-function components summing to unity, viz.

$$u^2 + v^2 = 1 \tag{D-165}$$

Since the squares of the above spinors sum to

$$\left[1 + \frac{p_z^2}{(|E| + mc^2)^2} \right] = \frac{2|E|}{|E| + mc^2} \tag{D-166}$$

the appropriate normalization constant for each of the above is

$$\left[\frac{|E| + mc^2}{2|E|} \right]^{1/2} \tag{D-167}$$

The physical interpretation is elucidated on causing first the σ_z function
and then the Hamiltonian H to operate in turn on the spinors, for one finds

$$\sigma_z \psi = \pm \psi \tag{D-168}$$

with the signs betokening respectively $\Uparrow$ or $\Downarrow$, and

$$H\psi = \pm |E| \psi \tag{D-169}$$

in which the signs betoken $E > 0$ and $E < 0$ respectively.

There are no solutions for which any other spin component than that in the direction of motion (e.g., σ_z, but not σ_x or σ_y) is sharp. The negative-energy solutions correspond physically to "Dirac hole" particles, namely to positrons.

Some further characteristics of the spinor equation (D-153) deserve to be given prominence. One should note that a proper Lorentz transformation serves to transform the v pair contragradiently to the u pair, and consequently

$$\psi_1^* \psi_3 + \psi_2^* \psi_4 = \text{invariant} \tag{D-170}$$

Writing the spinor form of the field-free equation explicitly as

$$\left[\begin{pmatrix} \boldsymbol{\sigma}_i & \mathbf{0} \\ \mathbf{0} & \boldsymbol{\sigma}_i \end{pmatrix} cp_i + mc^2 \begin{pmatrix} 1 & 0 \\ 0 & -1 \end{pmatrix}\right]\begin{pmatrix} u \\ v \end{pmatrix} = E\begin{pmatrix} u \\ v \end{pmatrix} \tag{D-171}$$

one sees that this implies a pair of two-component spinor equations, namely,

$$\left.\begin{array}{l} c\sigma_i p_i\, v + mc^2\, u = Eu \\ c\sigma_i p_i\, u - mc^2\, v = Ev \end{array}\right\} \tag{D-172}$$

It follows from (D-172) that

$$v = \frac{c\sigma_i p_i}{E + mc^2}\, u \tag{D-173}$$

On going over to the *nonrelativistic limit*

$$E \approx mc^2, \qquad p_i \approx mV_i \tag{D-174}$$

one finds that

$$v \cong \frac{c\sigma_i p_i}{2mc^2}\, u = \frac{c\sigma_i V_i}{2c^2}\, u = \frac{1}{2}\frac{V_i}{c}\,\sigma_i u \tag{D-175}$$

Consequently, the spinor wave function v is *smaller* than u by a factor $\sim V_i/c$, and one might interpret u and v, respectively, as the large positive-energy component and the small negative-energy component of the overall wave function ψ, describing electron and positron states (which are neglected in a nonrelativistic treatment). It should be noted, though, that the above considerations depend directly upon the particular Dirac representation chosen.

D.7. Weyl Equation for Massless Particles (Two-Component Neutrino Theory)

The two-component spinors u and v which satisfy the field-free Dirac equation (D-153) and from which current and charge densities can be built as in (D-114) and (D-115) can also be used to construct the quantities

$$\mathbf{j} = cu^\dagger \boldsymbol{\sigma} u, \qquad j^0 = cu^\dagger u \tag{D-176}$$

and

$$\mathbf{j} = cv^\dagger \boldsymbol{\sigma} v, \qquad j^0 = cv^\dagger v \tag{D-177}$$

which fulfil the identity

$$(j^0)^2 = \sum_{n=1,2,3} (j^n)^2 \tag{D-178}$$

It therefore follows that the matrix β which weights the rest-energy term in the Dirac equation cannot enter into covariant wave equations built from the components u and v. Thus in a two-component wave equation the term involving the rest mass would have to vanish identically. A *two-component wave equation* accordingly describes *massless particles* which move with the speed of light. Such an equation takes the form

$$\frac{\partial \phi}{\partial t} = c \sum_{n=1,2,3} \sigma_n \frac{\partial \phi}{\partial x^n} = 0 \qquad (\phi \equiv u, v) \tag{D-179}$$

or, with summation implied over repeated indices,

$$(\partial_0 + \sigma^n \partial_n) \phi = 0 \tag{D-180}$$

This WEYL EQUATION is relativistically invariant and the energy solutions that it yields correspond to negative as well as positive eigenstates for spin-$\frac{1}{2}$ massless particles. It leads to a particularly simple *equation of continuity*,

$$\sum_{\mu=1}^{4} \frac{\partial j^\mu}{\partial x^\mu} = 0 \tag{D-181}$$

and possesses the property of being *parity-violating* in character, for which reason it was long set aside [Pa 58] until its revival by Lee and Yang [Le 57] for the formulation of a *two-component theory of the neutrino*.

Its interpretation can be indicated along the following lines: The spinor equation (D-180) in fact stands for the pair of simultaneous wave equations

$$(\partial_0 + \sigma^n \partial_n) u = 0 \tag{D-182}$$
and
$$(\partial_0 + \sigma^n \partial_n) v = 0 \tag{D-183}$$

which, in momentum representation with

$$p_\mu \equiv \{p_0, \mathbf{p}\} = \left\{\frac{E}{c}, \mathbf{p}\right\} \text{ can be written as}$$

$$(p_0 + \sigma^n \partial_n) u = 0$$
$$(p_0 + \sigma^n \partial_n) v = 0 \tag{D-184}$$

When one takes the wave function $\phi \equiv u, v$ to be that of a plane wave with an energy eigenvalue E and a momentum eigenvalue expressed by the wave vector $\mathbf{k} = \mathbf{p}/\hbar$, one obtains the eigenfunction

$$\phi = \phi_0\, e^{i(k_n x_n + Et)} \tag{D-185}$$

which satisfies the equations

$$p_0 \phi = \frac{E}{c} \phi \qquad \text{and} \qquad p_n \phi = k_n \phi \qquad \text{(D-186)}$$

This permits (D-180) to be rewritten as

$$\left(\frac{E}{c} \pm \sigma^n k_n \right) \phi_0 = 0 \qquad \text{(D-187)}$$

and if one assumes that the particle described by this wave equation is travelling in the z direction with a momentum given by $\mathbf{k} = (0,0,k)$, one can simplify the equation to

$$\left(\frac{E}{c} \pm \sigma_z k \right) \phi_0 = 0 \qquad \text{(D-188)}$$

where σ_z is the Pauli spin operator in the z direction, which has eigenvalues ± 1 corresponding to antiparallel and parallel spin orientations respectively, e.g., to left and right helicity:

$$\frac{E}{c} = \mp \sigma_z k \qquad \text{(D-189)}$$

whence

$$\sigma_z k \phi = \mp \frac{E}{c} \phi \qquad \text{(D-190)}$$

i.e.,

$$\sigma_z \phi = \mp \phi \qquad \text{(D-191)}$$

The assignment of left helicity to neutrinos and right helicity to antineutrinos [Fe 58] tallies with experimental findings. (Lee and Yang's original choice of the converse assignment [Le 57] has since been proved wrong.)

D.8. Wave Equations for Bosons

The classic examples of wave equations for relativistic spin-1 particles are the Maxwell equations. However, their form is basically different from that employed for the present considerations. A relativistic wave equation of the desired structure for spin-1 particles was proposed by Proca [Pr 36] and thereafter subjected to second quantization by Bhabha [Bha 38] and Kemmer [Ke 38a, Ke 38b], who derived a set of four tensor equations for relativistic bosons of spin 0 or 1 by linearizing the Klein-Gordon equation. A consideration of these lies beyond the scope of the present account. For details the reader is referred to Roman [Ro 64a] and Muirhead [Mui 65]. We confine ourselves merely to pointing out some of the underlying considerations, emphasizing the value of the Lagrangian method for deducing field equations. The Lagrangian has the advantage over the Hamiltonian of leading naturally to Lorentz covariance, since it features time symmetrically, whereas in the Hamiltonian a distinction is made between space and time coordinates.

In classical theory the LAGRANGIAN FUNCTION L betokens the difference between the kinetic and potential energy of a system expressed in terms of position coordinates q_i (with $i = 1, 2, 3$). HAMILTON'S PRINCIPLE OF LEAST ACTION, expressed as a variational principle,

$$\delta \int_{t_1}^{t_2} L \, dt = 0 \tag{D-192}$$

then leads directly to the EULER-LAGRANGE EQUATION OF MOTION

$$\frac{\partial L}{\partial q_i} - \frac{d}{dt} \frac{\partial L}{\partial \dot{q}_i} = 0 \tag{D-193}$$

where the dot represents the time derivative. The corresponding equations of motion for a continuous field, i.e., the *field equations*, follow from a generalization of this formalism. In place of L we write a continuous LAGRANGIAN DENSITY $\mathscr{L}$,

$$L \to \mathscr{L} \tag{D-194}$$

while in place of the discrete coordinates q_i we introduce FIELD FUNCTIONS $q(x)$ in which the variable x is a continuous index,

$$q_i \to q^\mu(x) \qquad (\mu = 0, 1, 2, 3) \tag{D-195}$$

The variational principle (whose invariance under continuous Lorentz transformations forms the basis of NOETHER'S THEOREM [Noe 18] concerning the relationship between invariance principles and conservation laws)

$$\delta \int_\Omega \mathscr{L} \, d^4x = 0 \tag{D-196}$$

then yields the EULER-LAGRANGE FIELD EQUATION

$$\sum_\mu \left[\frac{\partial \mathscr{L}}{\partial q^\mu} - \partial_\nu \left(\frac{\partial \mathscr{L}}{\partial (\partial_\nu q^\mu)} \right) \right] = 0 \tag{D-197}$$

with the aid of which one can derive boson field equations. Thus considering first an electromagnetic (free photon) field, we can rewrite the conventional Maxwell equations

$$c[\mathbf{V} \times \mathbf{H}] = \dot{\mathbf{E}} + \mathbf{j} \tag{D-198}$$

$$(\mathbf{V} \cdot \mathbf{H}) \quad = 0 \tag{D-199}$$

$$c[\mathbf{V} \times \mathbf{E}] = -\dot{\mathbf{H}} \tag{D-200}$$

$$(\mathbf{V} \cdot \mathbf{E}) \quad = \rho \tag{D-201}$$

in terms of vector and scalar potentials **A** and ϕ. We set

$$c\mathbf{E} = -\dot{\mathbf{A}} - \nabla\phi \tag{D-202}$$

$$\mathbf{H} = [\nabla \times \mathbf{A}] \tag{D-203}$$

which satisfies the two middle equations (D-199) and (D-200), while (D-198) and (D-201) give

$$c\nabla^2\mathbf{A} - \frac{1}{c}\ddot{\mathbf{A}} = -\mathbf{j} \tag{D-204}$$

$$\nabla^2\phi - \frac{1}{c^2}\ddot{\phi} = -\rho \tag{D-205}$$

In 4-vector notation,

$$A^\mu \equiv \{\phi, \mathbf{A}\} \qquad \text{and} \qquad j^\mu = \{\rho c, \mathbf{j}\} \tag{D-206}$$

these reduce to the single MAXWELL FIELD EQUATION

$$c\,\square A^\mu = -j^\mu \tag{D-207}$$

the right side of which vanishes in the absence of electric charge. Implicit use has been made of the Lorentz condition

$$(\nabla \cdot \mathbf{A}) + \dot{\phi} \equiv \partial_\mu A^\mu = 0 \tag{D-208}$$

To derive other boson field equations, we take A^μ to be the 4-vector field operator for the particular field under consideration. The Lagrangian density has to be composed of *scalar* terms (so that it is invariant under proper Lorentz transformations) forming real quantities which have the dimensions of an energy density. The simplest form it might accordingly take is

$$\mathscr{L} = k_1\left(\frac{\partial A^\mu}{\partial x^\nu}\frac{\partial A_\mu}{\partial x_\nu}\right) + k_2(A^\mu A^\mu) \tag{D-209}$$

where k_1 and k_2 are weighting constants for which we shall make an *Ansatz* in a moment. An alternative, equally acceptable form is

$$\mathscr{L} = \tfrac{1}{2}k_1(\partial_\nu A^\mu - \partial_\mu A^\nu) + k_2(A^\mu A^\mu) \tag{D-210}$$

The A^μ take on the role of the q^μ here, and the Euler-Lagrange equation (D-197) can therefore be written as

$$\sum_\mu\left[\frac{\partial\mathscr{L}}{\partial A^\mu} - \partial_\nu\left(\frac{\partial\mathscr{L}}{\partial(\partial_\nu A^\mu)}\right)\right] = 0 \tag{D-211}$$

from which, on substituting (D-209), we get the field equation

$$-k_1(\partial_\nu A^\mu)^2 + k_2 A^\mu = 0 \tag{D-212}$$

i.e.,

$$\left(\Box - \frac{k_2}{k_1}\right) A^\mu = 0 \tag{D-213}$$

while with the alternative (D-210) we obtain

$$-k_1 \, \partial_\nu(\partial_\nu A^\mu - \partial_\mu A^\nu) + k_2 A^\mu = 0 \tag{D-214}$$

i.e.,

$$\partial_\nu(\partial_\nu A^\mu - \partial_\mu A^\nu) + \frac{k_2}{k_1} A^\mu = 0 \tag{D-215}$$

With the *Ansatz*

$$\frac{k_2}{k_1} = m^2 c^2 \tag{D-216}$$

a comparison with Eq. (D-68) shows that (D-213) is the vector equivalent of the KLEIN-GORDON EQUATION for the field due to a spinless particle. On the other hand, Eq. (D-215) then represents the PROCA EQUATION for spin-1 particles.

Moreover, on setting

$$\frac{k_2}{k_1} = m^2 c^2 = 0 \tag{D-217}$$

both (D-213) and (D-215) reduce to the MAXWELL EQUATION in absence of electric charge [cf. (D-207)],

$$\Box A^\mu = 0 \tag{D-218}$$

The Lagrangian density for a free photon field can therefore be derived from (D-209) on setting $k_2 = 0$ and choosing an appropriate value for k_1. A consideration of energy-momentum operators for a Maxwell field indicates this to be

$$k_1 = -\tfrac{1}{2} \tag{D-219}$$

whence, explicitly,

$$\mathscr{L} = -\frac{1}{2}\left(\frac{\partial A^\mu}{\partial x^\nu}\frac{\partial A_\mu}{\partial x_\nu}\right) \tag{D-220}$$

We shall use this formula in Appendix F when considering the interaction of particle fields with an electromagnetic field.

It remains only to mention that although spin-2 bosons have so far not been observed, quantized wave equations for gravitons, e.g., relativistic particles considered to propagate the gravitational field, have been derived [Li 55b, We 61] and used extensively.

EXERCISES

D-1 Why is it that when the potential V does not depend explicitly on time the Schrödinger equation can be written in either of two forms, viz.

$$\pm i\hbar \frac{\partial \psi}{\partial t} = \left(-\frac{\hbar^2}{2m}\nabla^2 + V\right)\psi$$

D-2 A particle of mass m moving under the action of an elastic force directed toward a fixed centre constitutes a one-dimensional harmonic oscillator. Derive the Schrödinger equation for such a system and show that it can be reduced to the form

$$\frac{d^2\psi}{d\xi^2} + (\lambda - \xi^2)\psi = 0$$

on introducing a new variable $\xi = \alpha x$. Find the values of the constants α and λ and determine the energy eigenvalues.

(The parameter λ can take on only the following values: $\lambda = 2n + 1$, $(n = 0, 1, 2, \ldots)$.)

D-3 Construct the symmetric and antisymmetric spin eigenfunctions for a system consisting of two identical spin-1 particles. How many functions of each kind are there? Discuss the significance of these functions.

ANGULAR MOMENTUM IN QUANTUM MECHANICS (RACAH ALGEBRA)

E.1. Angular Momentum Operators

In Appendix D.1 we presented the operator notation for *linear* momentum,

$$p = \frac{\hbar}{i}\nabla \tag{E-1}$$

For the *angular* momentum operator we might accordingly expect that

$$\mathbf{j} = [\mathbf{r} \times \mathbf{p}] = \frac{\hbar}{i}[\mathbf{r} \times \nabla] \tag{E-2}$$

(at this stage considering spinless particles, so that j represents *orbital* momentum). From elementary quantum mechanics, it is known that the angular momentum operators acting on the state function for a single (non-relativistic) particle, i.e., on its angular momentum eigenfunction $\psi(j,m)$ yield the eigenvalues

$$j^2\psi(j,m) = j(j+1)\,\hbar^2\,\psi(j,m) \tag{E-3}$$

$$j_z\psi(j,m) = m\hbar\,\psi(j,m) \tag{E-4}$$

and

$$(j_x + ij_y)\,\psi(j,m) = \hbar[(j-m)(j+m+1)]^{1/2}\psi(j,m+1) \tag{E-5}$$

$$(j_x - ij_y)\,\psi(j,m) = \hbar[(j+m)(j-m+1)]^{1/2}\psi(j,m-1) \tag{E-6}$$

where j_x, j_y, and j_z are components of the total angular momentum operator j, with

$$j^2 = j_x^2 + j_y^2 + j_z^2 \tag{E-7}$$

The components satisfy commutation relations,

$$j_x j_y - j_y j_x = i\hbar j_z \quad \text{(and cycl)} \tag{E-8}$$

$$j^2 j_x - j_x j^2 = 0 \quad \text{(and cycl)} \tag{E-9}$$

Whereas in classical theory the specification of the total angular momentum of a single independent system requires three parameters, j_x, j_y, j_z, the quantum-mechanical description involves only *two*, namely the total angular momentum j and its projection m along a distinguished direction taken as z axis. The angular momentum eigenfunction is completely specified by the values of j and m with respect to a particular origin, such as the center of mass in a collision system.

E.2. Composition of Angular Momentum Wave Functions (Clebsch-Gordan Coefficients)

When two particles, whose separate momenta are j_1, m_1 and j_2, m_2, respectively, combine to form a single composite system whose momentum is j, m, the classical composition formulae are

$$\mathbf{j} = \mathbf{j}_1 + \mathbf{j}_2 \tag{E-10}$$

$$m = m_1 + m_2 \tag{E-11}$$

Similar combination rules apply in quantum mechanics, except that the j's have magnitude $\hbar[j(j+1)]^{1/2}$. The problem that one frequently encounters in the quantum theory of angular momentum is how to combine two state functions $\psi(j_1, m_1)$ and $\psi(j_2, m_2)$ in order to derive the state function of the overall system $\Psi(j, m)$. It does not suffice merely to build the product of the constituent functions. The net state function of the compound system Ψ is derived by a unitary transformation of the product function $\psi_1 \psi_2$ effected with the aid of a unitary matrix having the dimension $(2j_1 + 1)(2j_2 + 1)$ and composed of numerical elements each of which we represent as

$$\langle jm | j_1 j_2 m_1 m_2 \rangle \tag{E-12}$$

in a notation resembling that of the Dirac bra-ket scheme. By convention, the phases of the wave functions are so chosen as to make the element *real*. This representation denotes the particular element of the matrix corresponding to the vector addition of $\mathbf{j}_1$ and $\mathbf{j}_2$ to a resultant $\mathbf{j}$, and the algebraic addition of m_1 and m_2 to a resultant m. In terms of these elements, we can write

$$\Psi(j, m) = \sum_{\substack{m_1 = -j_1 \\ m_2 = -j_2}}^{\substack{m_1 = +j_1 \\ m_2 = +j_2}} \langle jm | j_1 j_2 m_1 m_2 \rangle \, \psi(j_1, m_1) \, \psi(j_2, m_2) \tag{E-13}$$

and interpret the respective matrix elements as VECTOR ADDITION COEFFICIENTS, or "CLEBSCH-GORDAN COEFFICIENTS." They arise whenever vector coupling of momenta is undertaken in a quantum-mechanical calculation, be it in j–j coupling representation as illustrated, or in the equivalent L–S coupling representation.

Many different ways of writing essentially the same coupling coefficients have been adopted. We use the notation most generally employed, though some texts reverse or modify the sequence, as i.e., $\langle j_1 j_2 m_1 m_2 | jm \rangle$ or $\langle j_1 m_1, j_2 m_2 | jm \rangle$, while the following alternative designations have in the main been superseded in the meantime:

Blatt and Weisskopf [Bl 52b]:

$$C_{jj'}(J, M; m, m') \equiv C_{j_1 j_2}(j, m; m_1, m_2) \tag{E-14}$$

Condon and Shortley [Co 51]:

$$(jj' \, mm' | jj' \, JM) \equiv (j_1 j_2 \, m_1 \, m_2 | j_1 j_2 jm) \tag{E-15}$$

Wigner [Wi 31b]:

$$S^{jj'}_{Jmm'} \equiv S^{j_1 j_2}_{jm_1 m_2} \tag{E-16}$$

Rose [Ro 57]:

$$C(j_1 j_2 j; m_1 m_2 m) \tag{E-17}$$

Edmonds [Ed 57]:

$$(j_1 j_2 jm | j_1 m_1 j_2 m_2) \tag{E-18}$$

The numerical values of Clebsch-Gordan coefficients may be derived from algebraic expressions furnished by group-theoretical calculations. A general formula so obtained can be written as

$$\langle jm | j_1 j_2 m_1 m_2 \rangle$$

$$= (2j+1)^{1/2} [(j_1+m_1)! \, (j_1-m_1)! \, (j_2+m_2)! \, (j_2-m_2)! \, (j+m)! \, (j-m)!]^{1/2} \, \delta(j_1 j_2 j)$$

$$\times \sum_k \frac{(-1)^k}{k! \, (j_1+j_2-j-k)! \, (j_1-m_1-k)! \, (j_2+m_2-k)! \, (j-j_2+m_1+k)! \, (j-j_1-m_2+k)!} \tag{E-19}$$

with

$$\delta(j_1 j_2 j) \equiv \left[\frac{(j_1+j_2-j)! \, (j_1-j_2+j)! \, (-j_1+j_2+j)!}{(j_1+j_2+j+1)!} \right]^{1/2} \tag{E-20}$$

While many tables of Clebsch-Gordan and related coefficients exist [Si 54, Sea 54, Ro 59, Chi 62, In 66], it is customary nowadays when dealing with complicated problems to incorporate automatic routines for their evaluation into fast computer programs, based upon the formula (E-19) or upon recursion relations (cited, e.g., by Edmonds [Ed 57]), such as

$$[(j \mp m)(j \pm m + 1)]^{1/2} \langle j_1 \ m{\pm}1 | j_1 j_2 m_1 m_2 \rangle$$
$$= [(j_1 \pm m_1)(j_1 \mp m_1 + 1)]^{1/2} \langle jm | j_1 j_2, m_1 \mp 1, m_2 \rangle$$
$$+ [(j_2 \pm m_2)(j_2 \mp m_2 + 1)]^{1/2} \langle jm | j_1 j_2 m_1, m_2 \mp 1 \rangle \qquad \text{(E-21)}$$

A simple derivation of the Clebsch-Gordan coefficients has been given by Sharp [Sha 60], and more detailed discussions are to be found in the various textbooks already cited, as well as in [Sha 58, Br 62, Bie 65]. We confine ourselves to considering next some properties of Clebsch-Gordan coefficients and exploring their relationship to the WIGNER 3-*j* SYMBOLS.

E.3. Properties of Clebsch-Gordan Coefficients and Wigner 3-*j* Symbols

Clebsch-Gordan coefficients are real numbers which vanish unless $m = m_1 + m_2$ and the triad $(jj_1 j_2)$ satisfies a triangle relation,

$$\Delta j \equiv |j_1 - j_2| \leqslant j \leqslant j_1 + j_2 \equiv \Sigma j \qquad \text{(E-22)}$$

(a consequence of angular momentum conservation, which also decrees that $j_1 + j_2 - j$ is always an integer).

Since the underlying transformation from which the coefficients stem is *unitary*, it follows that any two rows or two columns of the transformation matrix must be orthogonal to one another. We accordingly have UNITARITY and COMPLETENESS RELATIONS:

$$\sum_{m_1, m_2} \langle jm | j_1 j_2 m_1 m_2 \rangle \langle j'm' | j_1 j_2 m_1 m_2 \rangle = \delta_{jj'} \delta_{mm'} \qquad \text{(E-23)}$$

$$\sum_{j,m} \langle jm | j_1 j_2 m_1 m_2 \rangle \langle jm | j_1 j_2 m_1' m_2' \rangle = \delta_{m_1 m_1'} \delta_{m_2 m_2'} \qquad \text{(E-24)}$$

The unitarity of the transformation matrix also specifies the INVERSE TRANS-FORMATION, viz.

$$\psi(j_1, m_1) \, \psi(j_2, m_2) = \sum_{j,m} \langle jm | j_1 j_2 m_1 m_2 \rangle \, \Psi(j, m) \qquad \text{(E-25)}$$

Also, the coefficients can be applied to VECTOR SUBTRACTION, e.g.,

$$\mathbf{j}_1 - \mathbf{j} = \mathbf{j}_2 \qquad \text{(E-26)}$$

which is characterized by $\langle j_2 m_2 | jj_1 m, -m_1 \rangle$.

The phase convention stipulates that when $j = m = j_1 + j_2$ the coefficient has the value $+1$:

$$\langle j_1 + j_2, j_1 + j_2 | j_1 j_2 j_1 j_2 \rangle = +1 \qquad \text{(E-27)}$$

Useful SYMMETRY PROPERTIES apply to interchange of two variables; with the above phase convention, as adopted from Condon and Shortley [Co 51], it follows that

$$\langle jm|j_1j_2\,m_1\,m_2\rangle = (-1)^{j_1+j_2-j}\langle jm|j_2j_1\,m_2\,m_1\rangle \qquad (E\text{-}28)$$

and upon sign reversal of the m's,

$$\langle jm|j_1j_2\,m_1\,m_2\rangle = (-1)^{j_1+j_2-j}\langle j,-m|j_1j_2,-m_1,-m_2\rangle \qquad (E\text{-}29)$$

while finally,

$$\langle jm|j_1j_2\,m_1\,m_2\rangle = (-1)^{j_2+m_2}\left(\frac{2j+1}{2j_1+1}\right)^{1/2}\langle j_1\,m_1|j_2j,-m_2\,m\rangle \qquad (E\text{-}30)$$

which betokens a change from the scheme $\mathbf{j}_1+\mathbf{j}_2=\mathbf{j}$ to $\mathbf{j}-\mathbf{j}_2=\mathbf{j}_1$, and involves a statistical weighting of final states that gives rise to the square-root factor.

The notation can be simplified and symmetry properties rendered more evident by changing over to WIGNER 3-j SYMBOLS, defined as

$$\begin{pmatrix} j_1 & j_2 & j_3 \\ m_1 & m_2 & m_3 \end{pmatrix} \equiv (-1)^{j_1-j_2-m_3}(2j_3+1)^{-1/2}\langle j_3,-m_3|j_1j_2\,m_1\,m_2\rangle \qquad (E\text{-}31)$$

where we replace j by j_3 and m by $-m_3$.

In terms of these entities, the coupling of vector momentum eigenfunctions reads

$$\Psi(j_3,-m_3) = (-1)^{j_1-j_2}(2j_3+1)^{1/2}$$
$$\times \sum_{m_1,m_2}(-1)^{m_3}\begin{pmatrix} j_1 & j_2 & j_3 \\ m_1 & m_2 & m_3 \end{pmatrix}\psi(j_1,m_1)\,\psi(j_2,m_2)$$
$$(E\text{-}32)$$

wherein we make use of the fact that $(-1)^{-m_3} \equiv (-1)^{m_3}$ since j_3 is an integer and therefore m_3 is also integral.

The 3-j symbols have the following properties:

(a) the momenta $(j_1j_2j_3)$ are subject to a triangle relation

$$\Delta j \equiv |j_1-j_2| \leqslant j_3 \leqslant j_1+j_2 \equiv \Sigma j \qquad (E\text{-}33)$$

(b) the magnetic quantum numbers sum to zero,

$$m_1+m_2+m_3 = 0 \qquad (E\text{-}34)$$

(c) an *even* permutation of columns leaves the 3-j symbol unchanged in value (i.e., introduces a phase factor equal to +1),

$$\begin{pmatrix} j_1 & j_2 & j_3 \\ m_1 & m_2 & m_3 \end{pmatrix} = \begin{pmatrix} j_3 & j_1 & j_2 \\ m_3 & m_1 & m_2 \end{pmatrix} \qquad (E\text{-}35)$$

(d) an *odd* permutation of columns introduces a phase factor $(-1)^{j_1+j_2+j_3}$,

$$\begin{pmatrix} j_1 & j_2 & j_3 \\ m_1 & m_2 & m_3 \end{pmatrix} = (-1)^{j_1+j_2+j_3} \begin{pmatrix} j_1 & j_3 & j_2 \\ m_1 & m_3 & m_2 \end{pmatrix} \tag{E-36}$$

(e) a sign reversal of *all* the m's introduces the same phase factor $(-1)^{j_1+j_2+j_3}$,

$$\begin{pmatrix} j_1 & j_2 & j_3 \\ m_1 & m_2 & m_3 \end{pmatrix} = (-1)^{j_1+j_2+j_3} \begin{pmatrix} j_1 & j_2 & j_3 \\ -m_1 & -m_2 & -m_3 \end{pmatrix} \tag{E-37}$$

(*N.B.*: All three m's must be subjected to sign reversal, since (b) would be violated if they were not *all* changed simultaneously.)

(f) simultaneous change of the signs of the m's combined with an *odd* permutation of columns accordingly leaves the 3-j symbol unaffected, since $(-1)^{2(j_1+j_2+j_3)} = +1$,

$$\begin{pmatrix} j_1 & j_2 & j_3 \\ m_1 & m_2 & m_3 \end{pmatrix} = \begin{pmatrix} j_1 & j_3 & j_2 \\ -m_1 & -m_3 & -m_2 \end{pmatrix} \tag{E-38}$$

E.4. Values of Simple 3-j Symbols

An extensive listing of 3-j symbol formulae and recursion relations has been compiled by Edmonds [Ed 57]. Some of the simpler expressions are presented in Table E-1; they are complemented by the following special results, in which for brevity we write $j_1 + j_2 + j_3 \equiv J$:

$$\begin{pmatrix} j & j & 0 \\ m & -m & 0 \end{pmatrix} = (-1)^{j-m}(2j+1)^{-1/2} \tag{E-39}$$

$$\begin{pmatrix} j_1 & j_2 & j_3 \\ 0 & 0 & 0 \end{pmatrix} = \begin{cases} 0 & J=\text{odd} \tag{E-40} \\ (-1)^{J/2} \left[\dfrac{(J-2j_1)!\,(J-2j_2)!\,(J-2j_3)!\,(J/2)!}{(J+1)!\,[(J/2)-j_1]!\,[(J/2)-j_2]!\,[(J/2)-j_3]!} \right]^{1/2} & J=\text{even} \end{cases} \tag{E-41}$$

$$\begin{pmatrix} J & J & 1 \\ m & -m & 0 \end{pmatrix} = (-1)^{J-m}\, m[J(J+1)(2J+1)]^{-1/2} \tag{E-42}$$

$$\begin{pmatrix} J & J & 1 \\ m & -m-1 & 1 \end{pmatrix} = (-1)^{J-m}\left[\frac{(J-m)(J+m+1)}{J(2J+1)(2J+2)} \right]^{1/2} \tag{E-43}$$

E.5. Examples of Wave-Function Coupling

As an example of wave-function coupling on the basis of the foregoing considerations, we construct the wave function for a system composed of two particles each of spin $s = \frac{1}{2}$ and zero relative orbital momentum $l = 0$. Their

Table E-1. EXPRESSIONS FOR SIMPLE 3-j SYMBOLS

$$\begin{pmatrix} j+\frac{1}{2} & j & \frac{1}{2} \\ m & -m-\frac{1}{2} & \frac{1}{2} \end{pmatrix} = (-1)^{j-m-1/2}\left[\frac{(j-m+\frac{1}{2})}{(2j+1)(2j+2)}\right]^{1/2}$$

$$\begin{pmatrix} j+1 & j & 1 \\ m & -m-1 & 1 \end{pmatrix} = (-1)^{j-m-1}\left[\frac{(j-m)(j-m+1)}{(2j+1)(2j+2)(2j+3)}\right]^{1/2}$$

$$\begin{pmatrix} j+1 & j & 1 \\ m & -m & 0 \end{pmatrix} = (-1)^{j-m-1}\left[\frac{(j-m+1)(j+m+1)}{(j+1)(2j+1)(2j+3)}\right]^{1/2}$$

$$\begin{pmatrix} j+\frac{3}{2} & j & \frac{3}{2} \\ m & -m-\frac{3}{2} & \frac{3}{2} \end{pmatrix} = (-1)^{j-m+1/2}\left[\frac{(j-m-\frac{1}{2})(j-m+\frac{1}{2})(j-m+\frac{3}{2})}{(2j+1)(2j+2)(2j+3)(2j+4)}\right]^{1/2}$$

$$\begin{pmatrix} j+\frac{3}{2} & j & \frac{3}{2} \\ m & -m-\frac{1}{2} & \frac{1}{2} \end{pmatrix} = (-1)^{j-m+1/2}\left[\frac{3(j-m+\frac{1}{2})(j-m+\frac{3}{2})(j+m+\frac{3}{2})}{(2j+1)(2j+2)(2j+3)(2j+4)}\right]^{1/2}$$

$$\begin{pmatrix} j+\frac{1}{2} & j & \frac{3}{2} \\ m & -m-\frac{3}{2} & \frac{3}{2} \end{pmatrix} = (-1)^{j-m-1/2}\left[\frac{3(j-m-\frac{1}{2})(j-m+\frac{1}{2})(j+m+\frac{3}{2})}{2j(2j+1)(2j+2)(2j+3)}\right]^{1/2}$$

$$\begin{pmatrix} j+\frac{1}{2} & j & \frac{3}{2} \\ m & -m-\frac{1}{2} & \frac{1}{2} \end{pmatrix} = (-1)^{j-m-1/2}\left[\frac{j-m+\frac{1}{2}}{2j(2j+1)(2j+2)(2j+3)}\right]^{1/2}(j+3m+\tfrac{3}{2})$$

$$\begin{pmatrix} j & j & 2 \\ m & -m & 0 \end{pmatrix} = (-1)^{j-m}\frac{6m^2-2j(j+1)}{[(2j-1)\,2j(2j+1)(2j+2)(2j+3)]^{1/2}}$$

$$\begin{pmatrix} j & j & 2 \\ m & -m-1 & 1 \end{pmatrix} = (-1)^{j-m}\left[\frac{6(j-m)(j+m+1)}{(2j-1)\,2j(2j+1)(2j+2)(2j+3)}\right]^{1/2}(1+2m)$$

$$\begin{pmatrix} j & j & 2 \\ m & -m-2 & 2 \end{pmatrix} = (-1)^{j-m}\left[\frac{6(j-m-1)(j-m)(j+m+1)(j+m+2)}{(2j-1)\,2j(2j+1)(2j+2)(2j+3)}\right]^{1/2}$$

spins either lie parallel or antiparallel to the quantization direction, whence the possible states of particle 1 are described by wave functions of the form $\psi_i(j,s)$, viz. $\psi_1(\frac{1}{2},\frac{1}{2})$ and $\psi_1(\frac{1}{2},-\frac{1}{2})$, while those of particle 2 are $\psi_2(\frac{1}{2},-\frac{1}{2})$ and $\psi_2(\frac{1}{2},\frac{1}{2})$, giving a singlet combination for the resultant two-particle system, whose wave function in accordance with the combination rules takes the form

$$\Psi(0,0) = \begin{pmatrix} \frac{1}{2} & \frac{1}{2} & 0 \\ \frac{1}{2} & -\frac{1}{2} & 0 \end{pmatrix}\psi_1(\tfrac{1}{2},\tfrac{1}{2})\,\psi_2(\tfrac{1}{2},-\tfrac{1}{2}) + \begin{pmatrix} \frac{1}{2} & \frac{1}{2} & 0 \\ -\frac{1}{2} & \frac{1}{2} & 0 \end{pmatrix}\psi_1(\tfrac{1}{2},-\tfrac{1}{2})\,\psi_2(\tfrac{1}{2},\tfrac{1}{2}) \tag{E-44}$$

But, from the standard form (E-39), we note that

$$\begin{pmatrix} \frac{1}{2} & \frac{1}{2} & 0 \\ \frac{1}{2} & -\frac{1}{2} & 0 \end{pmatrix} = -\begin{pmatrix} \frac{1}{2} & \frac{1}{2} & 0 \\ -\frac{1}{2} & \frac{1}{2} & 0 \end{pmatrix} = \frac{(-1)^0}{(1+1)^{1/2}} = \frac{1}{(2)^{1/2}} \tag{E-45}$$

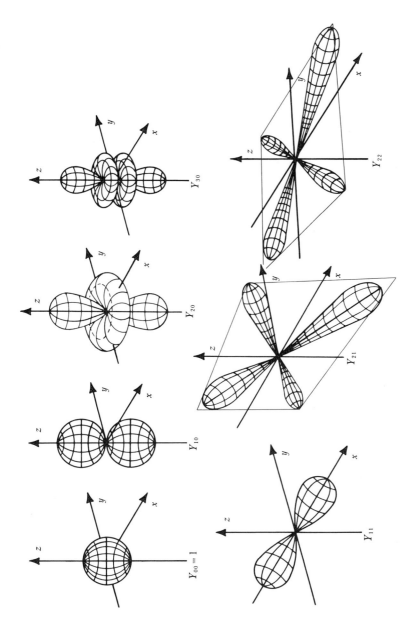

Fig. E-1. Polar representation of some spherical harmonics Y_{lm} of low order.

whence the singlet two-body wave function becomes

$$\Psi(0,0) = \frac{1}{(2)^{1/2}} [\psi_1(\tfrac{1}{2}, \tfrac{1}{2}) \, \psi_2(\tfrac{1}{2}, -\tfrac{1}{2}) - \psi_1(\tfrac{1}{2}, -\tfrac{1}{2}) \, \psi_2(\tfrac{1}{2}, \tfrac{1}{2})] \qquad \text{(E-46)}$$

$$\text{spin:} \qquad\qquad \Uparrow \qquad\quad \Downarrow \qquad\qquad \Downarrow \qquad\qquad \Uparrow$$

in appropriately normalized and antisymmetrized form (in conformity with the Pauli exclusion principle for fermions).

An alternative instance is that of *coupling of momentum eigenfunctions*. The eigenfunction which gives integer eigenvalues l and m_l when acted upon by the orbital momentum operators $\underline{l}$ and $\underline{l}_z$ (cf. Section E.1) is the SPHERICAL HARMONIC, which can be written in the form

$$Y_{lm}(\theta, \phi) = \frac{(-1)^l}{2^l \, l!} \left[\frac{2l+1}{4\pi} \frac{(l+m)!}{(l-m)!} \right]^{1/2} \sin^{-m} \theta e^{im\phi} \left(\frac{d}{du} \right)^{l-m} (1-u^2)^l \Big|_{u=\cos\theta}$$

$$\text{(E-47)}$$

in which u is set equal to $\cos\theta$ *after* differentiation. Fig. E-1 shows the form of some spherical harmonics in Cartesian representation. This formula can be simplified to that quoted in Section 5.3.10, viz.

$$Y_{lm}(\theta, \phi) = (-1)^m \left[\frac{2l+1}{4\pi} \frac{(l-|m|)!}{(l+|m|)!} \right]^{1/2} P_l^{|m|}(\cos\theta) \, e^{im\phi} \qquad \text{(E-48)}$$

with the phase factor $(-1)^m$ *omitted for negative values of m*, and the $P_l^m(\cos\theta)$ denoting ASSOCIATED LEGENDRE FUNCTIONS, given by the formula (for $m \geqslant 0$)

$$P_l^m(u) = \frac{(2l)!}{2^l \, l! \, (l-m)!} (1-u^2)^{m/2} \left[u^{l-m} - \frac{(l-m)(l-m-1)}{2(2l-1)} u^{l-m-2} \right.$$

$$\left. + \frac{(l-m)(l-m-1)(l-m-2)(l-m-3)}{2 \times 4 \times (2l-1)(2l-3)} u^{l-m-4} + \cdots \right]$$

$$\text{(E-49)}$$

and having the property that when $m = 0$ they reduce to LEGENDRE POLYNOMIALS

$$P_l^0(u) \equiv P_l(u)$$

$$= \frac{(2l-1)!!}{l!} \left[u^l - \frac{l(l-1)}{2(2l-1)} u^{l-2} + \frac{l(l-1)(l-2)(l-3)}{2 \times 4 \times (2l-1)(2l-3)} u^{l-4} - \cdots \right]$$

$$\text{(E-50)}$$

where $\qquad\qquad (2l-1)!! \equiv (2l-1)(2l-3)\cdots \begin{cases} 4 \times 2 \\ \text{or } 3 \times 1 \end{cases}$

and $P_l(0) \equiv 1$ $\qquad\qquad\qquad\qquad\qquad\qquad\qquad\qquad\qquad\qquad\qquad$ (E-51)

A graphical representation of these in function of u and of θ is given in Figs.

E-2(a,b). Some values of frequently used spherical harmonics, associated Legendre functions and Legendre polynomials are listed in Table E-2, from which it will be noted that when $m = 0$,

$$Y_{l0}(\theta, \phi) = \left[\frac{2l + 1}{4\pi}\right]^{1/2} P_l(\cos\theta) \tag{E-52}$$

The spherical harmonics, with $\Omega \equiv (\theta, \phi)$, satisfy the relations

$$\int Y_{lm}^*(\Omega) Y_{l'm'}(\Omega) \, d\Omega = \delta_{ll'}\delta_{mm'} \tag{E-53}$$

$$Y_{lm}^*(\Omega) = (-1)^m Y_{l,-m}(\Omega) \tag{E-54}$$

$$Y_{lm}(\Omega)|_{\theta=0} = \delta_{m0}\left[\frac{2l + 1}{4\pi}\right]^{1/2} \tag{E-55}$$

and

$$\int Y_{l_1 m_1}(\Omega) Y_{l_2 m_2}(\Omega) \, d\Omega$$
$$= \left[\frac{(2l_1 + 1)(2l_2 + 1)}{4\pi}\right]^{1/2} \sum_{\lambda,\mu} (2\lambda + 1)^{1/2} \begin{pmatrix} l_1 & l_2 & \lambda \\ m_1 & m_2 & \mu \end{pmatrix} \begin{pmatrix} l_1 & l_2 & \lambda \\ 0 & 0 & 0 \end{pmatrix} Y_{\lambda\mu}^*(\Omega) \tag{E-56}$$

$$\int Y_{l_1 m_1}(\Omega) Y_{l_2 m_2}(\Omega) Y_{l_3 m_3}(\Omega) \, d\Omega$$
$$= \left[\frac{(2l_1 + 1)(2l_2 + 1)(2l_3 + 1)}{4\pi}\right]^{1/2} \begin{pmatrix} l_1 & l_2 & l_3 \\ m_1 & m_2 & m_3 \end{pmatrix} \begin{pmatrix} l_1 & l_2 & l_3 \\ 0 & 0 & 0 \end{pmatrix} \tag{E-57}$$

The state function for a system composed of two particles with angular coordinates Ω_1 and Ω_2 which, having the same orbital momentum l and mutually opposite magnetic quantum numbers $m_1 = -m_2 = |m|$ combine to an antisymmetric system in its S state, is

$$\Psi = \sum_{m_1,m_2} \begin{pmatrix} l & l & 0 \\ m_1 & m_2 & 0 \end{pmatrix} Y_{lm_1}(\Omega_1) Y_{lm_2}(\Omega_2) \tag{E-58}$$

$$= \sum_{m_1,m_2} \langle 00|llm_1 m_2\rangle Y_{lm_1}(\Omega_1) Y_{lm_2}(\Omega_2) \tag{E-59}$$

$$= \frac{(-1)^l}{(2l + 1)^{1/2}} \sum_m Y_{lm}^*(\Omega_1) Y_{lm}(\Omega_2) \tag{E-60}$$

This is invariant under rotations and can accordingly be evaluated for the special case corresponding to choice of z axis along the direction of motion of particle 1:

$$\Psi = \left[\frac{2l + 1}{4\pi}\right]^{1/2} Y_{l0}(\Omega_{12}) = \frac{2l + 1}{4\pi} P_l(\cos\theta_{12}) \tag{E-61}$$

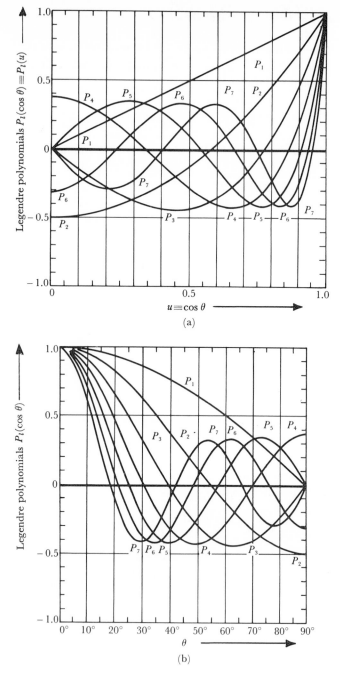

Fig. E-2. Angular dependence of Legendre polynomials: $P_l(\cos \theta)$: (a) in terms of $u \equiv \cos \theta$; (b) in terms of θ, for $0 \leqslant l \leqslant 7$ (from [Ja 60]).

Table E-2. Expressions for Low-Order Spherical Harmonics, Associated Legendre Functions, and Legendre Polynomials

l	m	$Y_{lm}(\Omega)$	$P_l^m(u)$	$P_l(u)$
0	0	$(4\pi)^{-1/2}$	1	1
1	0	$(3/4\pi)^{1/2}\cos\theta$	u	u
	± 1	$\mp (3/8\pi)^{1/2}\sin\theta\, e^{\pm i\phi}$	$(1-u^2)^{1/2}$	
2	0	$(5/16\pi)^{1/2}(3\cos^2\theta - 1)$	$\frac{1}{2}(3u^2 - 1)$	$\frac{1}{2}(3u^2 - 1)$
	± 1	$\mp (15/8\pi)^{1/2}\cos\theta\sin\theta\, e^{\pm i\phi}$	$3(1-u^2)^{1/2}\, u$	
	± 2	$(15/32\pi)^{1/2}\sin^2\theta\, e^{\pm 2i\phi}$	$3(1-u^2)$	
3	0	$(7/16\pi)^{1/2}(5\cos^3\theta - 3\cos\theta)$	$\frac{1}{2}(5u^3 - 3u)$	$\frac{1}{2}(5u^3 - 3u)$
	± 1	$\mp (21/64\pi)^{1/2}(5\cos^2\theta - 1)\sin\theta\, e^{\pm i\phi}$	$\frac{3}{2}(1-u^2)^{1/2}(5u^2 - 1)$	
	± 2	$(105/32\pi)^{1/2}\sin^2\theta\cos\theta\, e^{\pm 2i\phi}$	$15(1-u^2)\, u$	
	± 3	$\mp (35/64\pi)^{1/2}\sin^3\theta\, e^{\pm 3i\phi}$	$15(1-u^2)^{3/2}$	
4	0	$(3/16\pi^{1/2})\,(35\cos^4\theta - 30\cos^2\theta + 3)$	$\frac{1}{8}(35u^4 - 30u^2 + 3)$	$\frac{1}{8}(35u^4 - 30u^2 + 3)$

E.6. Recoupling of Angular Momenta
(Racah Coefficients and Wigner 6-j Symbols)

The momentum-coupling calculus as presented up to this point is capable of extension to the coupling of three or more momenta, which are subject to the complication that more than one way now exists for pairwise combination of momenta, e.g., to derive the resultant net momentum

$$\mathbf{j} = \mathbf{j}_1 + \mathbf{j}_2 + \mathbf{j}_3 \tag{E-62}$$

one may proceed in the sequence

(a) $\mathbf{j}_1 + \mathbf{j}_2 = \mathbf{j}_{12}$ $\quad$ followed by $\mathbf{j}_{12} + \mathbf{j}_3 = \mathbf{j}$ $\tag{E-63}$

or

(b) $\mathbf{j}_2 + \mathbf{j}_3 = \mathbf{j}_{23}$ $\quad$ followed by $\mathbf{j}_1 + \mathbf{j}_{23} = \mathbf{j}$ $\tag{E-64}$

The Racah algebra associated with the coupling of individual wave functions $\psi(j_1, m_1)$, $\psi(j_2, m_2)$, $\psi(j_3, m_3)$ to a net eigenfunction $\Psi(j, m)$ involves suitable RECOUPLING COEFFICIENTS when the angular momenta j_1, j_2, and j_3 are independent, i.e., when they commute. The coupling scheme (a) leads to

$$\Psi(j)_{12,3} = \langle j_{12}|j_1 j_2 \rangle \langle j|j_{12} j_3 \rangle \, \psi(j_1) \, \psi(j_2) \, \psi(j_3) \tag{E-65}$$

while (b) leads to

$$\Psi(j)_{23,1} = \langle j_{23}|j_2 j_3 \rangle \langle j|j_1 j_{23} \rangle \, \psi(j_1) \, \psi(j_2) \, \psi(j_3) \tag{E-66}$$

in a representation in which, for mathematical simplicity, the respective m's have been suppressed.

The fact that the $\Psi(j)_{12,3}$ must be expressible as a linear combination of the $\Psi(j)_{23,1}$ with coefficients which are independent of the m's (since the m's are unaffected by the addition sequence) prompts one to introduce expansion coefficients W and write

$$\Psi(j)_{12,3} = \sum_{j_{23}} [(2j_{12} + 1)(2j_{23} + 1)]^{1/2} \, W(j_1 j_2 j j_3 ; j_{12} j_{23}) \, \Psi(j)_{23,1} \tag{E-67}$$

Substituting for the $\Psi(j)$ from above, we see that

$$\langle j_{12}|j_1 j_2 \rangle \langle j|j_{12} j_3 \rangle = \sum_{j_{23}} [(2j_{12} + 1)(2j_{23} + 1)]^{1/2} \langle j_{23}|j_2 j_3 \rangle$$
$$\times \langle j|j_1 j_{23} \rangle \, W(j_1 j_2 j j_3 ; j_{12} j_{23}) \tag{E-68}$$

or, written in a more straightforward symbolic form,

$$\langle e|ab \rangle \langle c|ed \rangle = \sum_f [(2e + 1)(2f + 1)]^{1/2} \langle c|af \rangle \langle f|bd \rangle \times W(abcd; ef) \tag{E-69}$$

It is this relation which defines the RACAH COEFFICIENT that enters in the recoupling of three momenta. It is defined in magnitude and phase by the products of Clebsch-Gordan coefficients, and is independent of the magnetic quantum numbers m. Its value is given by the formula

$$W(abcd;ef) = \delta(abe)\,\delta(cde)\,\delta(acf)\,\delta(bdf) \sum_{k'} \frac{N(k')}{D(k')} \tag{E-70}$$

where

$$N(k') \equiv (-1)^{k'}(a+b+c+d+1-k')!$$
$$D(k') \equiv k'!\,(a+b-e-k')!\,(c+d-e-k')!\,(a+c-f-k')!$$
$$\times (b+d-f-k')!\,(-a-d+e+f+k')!\,(-b-c+e+f+k')!$$

where the δ functions $\delta(abe)$ etc. take the form indicated in Eq. (E-20). It vanishes identically unless triangle relations are satisfied by the triads

$$(abe),\,(cde),\,(acf),\,\text{and}\,(bdf) \tag{E-71}$$

e.g.,

$$|a-b| \leqslant e \leqslant a+b,\,\text{etc.} \tag{E-72}$$

Indicating the coupling scheme by a symbolic diagram having the form of Fig. E-3, we note that the symmetry between the two diagonals implies the following *orthogonality property*:

$$\sum_e (2e+1)(2f+1)\,W(abcd;ef)\,W(abcd;eg) = \delta_{fg} \tag{E-73}$$

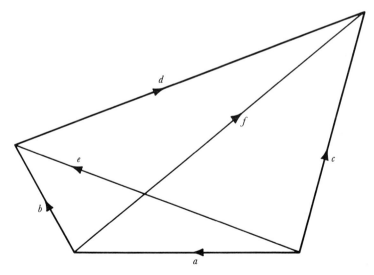

Fig. E-3. Symbolic representation of the coupling of three angular momenta.

A consideration of the respective Clebsch-Gordan coefficients reveals that the Racah coefficients have essentially the symmetries inherent in the diagram, namely

$$W(abcd;ef) = W(badc;ef) = W(cdab;ef) = W(acbd;fe) \qquad \text{(E-74)}$$

$$= (-1)^{e+f-a-d} W(ebcf;ad) \qquad \text{(E-75)}$$

$$= (-1)^{e+f-b-c} W(aefd;bc) \qquad \text{(E-76)}$$

and other combinations derived therefrom.

The symmetry properties are evinced more clearly by the equivalent WIGNER 6-j SYMBOLS, which differ from the Racah coefficients solely in the phase, being defined as

$$\begin{Bmatrix} a & b & e \\ d & c & f \end{Bmatrix} \equiv (-1)^{a+b+c+d} W(abcd;ef) \qquad \text{(E-77)}$$

and given by a product of four 3-j symbols,

$$\begin{Bmatrix} a & b & e \\ d & c & f \end{Bmatrix} = \sum_{g,h} (-1)^{a+b+c+d+e+f+h_1+h_2+h_3}$$

$$\times \begin{pmatrix} a & b & e \\ g_1 & g_2 & g_3 \end{pmatrix} \begin{pmatrix} a & c & f \\ -g_1 & h_2 & h_3 \end{pmatrix} \begin{pmatrix} d & b & f \\ -h_1 & -g_2 & h_3 \end{pmatrix} \begin{pmatrix} d & c & e \\ h_1 & -h_2 & -g_3 \end{pmatrix}$$

$$\text{(E-78)}$$

The curly brackets {} serve to distinguish 6-j symbols from their 3-j counterparts.

Interchange of a vertical pair leaves the 6-j symbol unaltered, e.g.,

$$\begin{Bmatrix} a & b & e \\ d & c & f \end{Bmatrix} = \begin{Bmatrix} b & a & e \\ c & d & f \end{Bmatrix}, \text{ etc.} \qquad \text{(E-79)}$$

as does also the *vertical interchange of a horizontal pair*,

$$\begin{Bmatrix} a & b & e \\ d & c & f \end{Bmatrix} = \begin{Bmatrix} d & c & e \\ a & b & f \end{Bmatrix}, \text{ etc.} \qquad \text{(E-80)}$$

The upper triad of a 6-j symbol fulfils a triangle relation,

$$|a - b| \leqslant e \leqslant a + b \qquad \text{(E-81)}$$

but this is *not* the case for the lower triad. *If one of the terms in the upper or lower triplet is zero*, the 6-j symbol reduces to

$$\begin{Bmatrix} a & a & 0 \\ d & d & f \end{Bmatrix} = (-1)^{a+d+f} [(2a+1)(2d+1)]^{-1/2} \qquad \text{(E-82)}$$

which, when expressed in Racah-coefficient form, reads

$$W(abcd;e0) = (-1)^{a+d-e} [(2a+1)(2d+1)]^{-1/2} \delta_{ac} \delta_{bd} \qquad \text{(E-83)}$$

Expressions for the coupling of more than three momenta can always be reduced to a set of 6-j symbols and, possibly, 3-j symbols.

Some values of rather simple 6-j symbols have been collected in Table E-3, setting $j_1 + j_2 + j_3 = J$.

Table E-3. VALUES OF SPECIAL 6-j SYMBOLS (WITH $J \equiv j_1 + j_2 + j_3$)

$$\begin{Bmatrix} j_1 & j_2 & j_3 \\ 0 & j_3 & j_2 \end{Bmatrix} = \frac{(-1)^J}{[(2j_2+1)(2j_3+1)]^{1/2}}$$

$$\begin{Bmatrix} j_1 & j_2 & j_3 \\ \tfrac{1}{2} & j_3-\tfrac{1}{2} & j_2+\tfrac{1}{2} \end{Bmatrix} = (-1)^J \left[\frac{(J-2j_2)(J-2j_3+1)}{(2j_2+1)(2j_2+2)\,2j_3(2j_3+1)} \right]^{1/2}$$

$$\begin{Bmatrix} j_1 & j_2 & j_3 \\ \tfrac{1}{2} & j_3-\tfrac{1}{2} & j_2-\tfrac{1}{2} \end{Bmatrix} = (-1)^J \left[\frac{(J+1)(J-2j_1)}{2j_2(2j_2+1)\,2j_3(2j_3+1)} \right]^{1/2}$$

$$\begin{Bmatrix} j_1 & j_2 & j_3 \\ 1 & j_3-1 & j_2-1 \end{Bmatrix} = (-1)^J \left[\frac{J(J+1)(J-2j_1)(J-2j_1-1)}{(2j_2-1)\,2j_2(2j_2+1)(2j_3-1)\,2j_3(2j_3+1)} \right]^{1/2}$$

$$\begin{Bmatrix} j_1 & j_2 & j_3 \\ 1 & j_3-1 & j_2 \end{Bmatrix} = (-1)^J \left[\frac{(J+1)(J-2j_1)(J-2j_2)(J-2j_3+1)}{j_2(2j_2+1)(2j_2+2)(2j_3-1)\,2j_3(2j_3+1)} \right]^{1/2}$$

$$\begin{Bmatrix} j_1 & j_2 & j_3 \\ 1 & j_3-1 & j_2+1 \end{Bmatrix} = (-1)^J \left[\frac{(J-2j_2)(J-2j_2-1)(J-2j_3+1)(J-2j_3+2)}{(2j_2+1)(2j_2+2)(2j_2+3)(2j_3-1)\,2j_3(2j_3+1)} \right]^{1/2}$$

$$\begin{Bmatrix} j_1 & j_2 & j_3 \\ 1 & j_3 & j_2 \end{Bmatrix} = (-1)^{J+1} \frac{(2)^{1/2}[j_2(j_2+1)+j_3(j_3+1)-j_1(j_1+1)]}{[j_2(2j_2+1)(2j_2+2)\,2j_3(2j_3+1)(2j_3+2)]^{1/2}}$$

E.7. Coupling of Four Angular Momenta (Wigner 9-j Symbols)

While problems involving the coupling of four angular momenta can always be split up into a sequence of simpler situations in which use can be made of 3-j and 6-j formalism, it is convenient to extend the Racah calculus to include an explicit symbol for the recoupling of all four momenta j_1, j_2, j_3, and j_4. One could conceive of these being coupled as follows:

$$\mathbf{j}_1 + \mathbf{j}_2 = \mathbf{j}_{12}, \text{ followed by } \mathbf{j}_3 + \mathbf{j}_4 = \mathbf{j}_{34} \text{ and finally } \mathbf{j}_{12} + \mathbf{j}_{34} = \mathbf{j} \quad \text{(E-84)}$$

or

$$\mathbf{j}_1 + \mathbf{j}_3 = \mathbf{j}_{13}, \text{ followed by } \mathbf{j}_2 + \mathbf{j}_4 = \mathbf{j}_{24} \text{ and finally } \mathbf{j}_{13} + \mathbf{j}_{24} = \mathbf{j} \quad \text{(E-85)}$$

The eigenfunctions in the two alternative coupling schemes must then be related by a unitary transformation. This enables us to write

$$\langle j_{12}|j_1 j_2\rangle\langle j_{34}|j_3 j_4\rangle\langle j|j_{12}j_{34}\rangle = \sum_{j_{13},j_{24}} [(2j_{12}+1)(2j_{34}+1)(2j_{13}+1)(2j_{24}+1)]^{1/2}$$

$$\times \langle j_{13}|j_1 j_3\rangle\langle j_{24}|j_2 j_4\rangle\langle j|j_{13}j_{24}\rangle \begin{Bmatrix} j_1 & j_2 & j_{12} \\ j_3 & j_4 & j_{34} \\ j_{13} & j_{24} & j \end{Bmatrix}$$

$$(E-86)$$

where the magnetic quantum numbers have again been suppressed. The recoupling coefficient $\{\}$ is termed the WIGNER 9-j SYMBOL or, in alternative notation for typographical convenience, the FANO X COEFFICIENT $X(j_1 j_2 j_{12}; j_3 j_4 j_{34}; j_{13} j_{24} j)$. It is *real and independent of magnetic quantum numbers*. The vector diagrams symbolic of these coupling schemes are depicted in Figs. E-4(a,b). They exhibit the six triangle relations implicit in the definition, which are also evinced by the constituent Racah coefficients:

$$\begin{Bmatrix} a & b & c \\ d & e & f \\ g & h & i \end{Bmatrix} = \sum_{\substack{K>0 \\ \text{half-integer}}} (2K+1)\, W(gibe;hK)\, W(cide;fK)\, W(gdbc;aK) \quad (E-87)$$

$$= \sum_{K} (2K+1)\, W(adih;gK)\, W(beKd;hf)\, W(cfaK;ib) \quad\quad (E-88)$$

or, in 6-j symbol notation,

$$\begin{Bmatrix} a & b & c \\ d & e & f \\ g & h & i \end{Bmatrix} = \sum_{K} (-1)^{2K}(2K+1) \begin{Bmatrix} a & d & g \\ h & i & K \end{Bmatrix}\begin{Bmatrix} b & e & h \\ d & K & f \end{Bmatrix}\begin{Bmatrix} c & f & i \\ K & a & b \end{Bmatrix}$$

$$(E-89)$$

Continuing the reduction further, one can express the 9-j symbol in terms of 3-j symbols, viz.

$$\begin{Bmatrix} j_{11} & j_{12} & j_{13} \\ j_{21} & j_{22} & j_{23} \\ j_{31} & j_{32} & j_{33} \end{Bmatrix} = \sum_{\text{all } m} \begin{pmatrix} j_{11} & j_{12} & j_{13} \\ m_{11} & m_{12} & m_{13} \end{pmatrix}\begin{pmatrix} j_{21} & j_{22} & j_{23} \\ m_{21} & m_{22} & m_{23} \end{pmatrix}$$

$$\times \begin{pmatrix} j_{31} & j_{32} & j_{33} \\ m_{31} & m_{32} & m_{33} \end{pmatrix}\begin{pmatrix} j_{11} & j_{21} & j_{31} \\ m_{11} & m_{21} & m_{31} \end{pmatrix}$$

$$\times \begin{pmatrix} j_{12} & j_{22} & j_{32} \\ m_{12} & m_{22} & m_{32} \end{pmatrix}\begin{pmatrix} j_{13} & j_{23} & j_{33} \\ m_{13} & m_{23} & m_{33} \end{pmatrix}$$

$$(E-90)$$

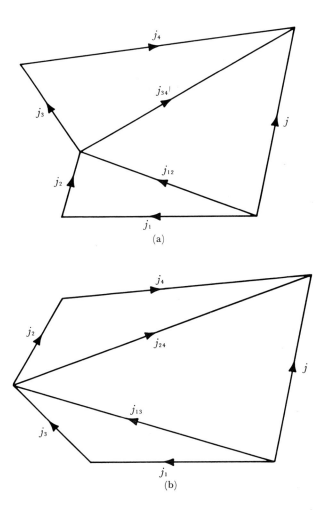

Fig. E.4. Symbolic representation of the coupling of four angular momenta according to two equivalent schemes, viz.

$$
\begin{array}{ll}
\text{(a)} \quad \mathbf{j}_1 + \mathbf{j}_2 \rightarrow \mathbf{j}_{12} & \text{(b)} \quad \mathbf{j}_1 + \mathbf{j}_3 \rightarrow \mathbf{j}_{13} \\
\qquad \mathbf{j}_3 + \mathbf{j}_4 \rightarrow \mathbf{j}_{34} & \qquad \mathbf{j}_2 + \mathbf{j}_4 \rightarrow \mathbf{j}_{24} \\
\qquad \mathbf{j}_{12} + \mathbf{j}_{34} \rightarrow \mathbf{j} & \qquad \mathbf{j}_{13} + \mathbf{j}_{24} \rightarrow \mathbf{j}
\end{array}
$$

An *interchange of the rows or columns* in a 9-j symbol leaves the value unaltered if the permutation is even, or introduces a phase factor $(-1)^{a+b+c+d+e+f+g+h+i}$ if the permutation is odd.

"*Stretched*" *X coefficients*, in which one element in a row or column is equal to the sum of the other two, have been discussed by Sharp [Sha 67].

When one of the terms in a 9-*j* symbol is zero, the recoupling coefficient reduces to a 6-*j* symbol,

$$
\begin{Bmatrix} a & b & c \\ d & e & c \\ f & f & 0 \end{Bmatrix} = (-1)^{b+c+d+f}[(2c+1)(2f+1)]^{-1/2} \begin{Bmatrix} a & b & c \\ e & d & f \end{Bmatrix} \tag{E-91}
$$

The 9-*j* symbol proves especially useful in angular correlation calculations or in the application to two-particle systems when one seeks to change the formalism from *jj*-coupling to *LS*-coupling or vice versa. Eigenfunctions of the angular momenta j_1 and j_2 of each of the particles can be combined to yield eigenfunctions of the total angular momentum J, viz. $\Psi(J)_{j_1 j_2}$, or alternatively eigenfunctions of the total *orbital* momentum L of the net two-particle system can be combined with eigenfunctions of the total *spin* momentum S to furnish eigenfunctions of J which are physically equivalent, but expressed in a different representation. The transformation from the one representation to the other, or vice versa, is expressed by the symmetric formula

$$
\Psi(J)_{j_1 j_2} = \sum_{L,S} [(2j_1+1)(2j_2+1)(2L+1)(2S+1)]^{1/2} \begin{Bmatrix} l_1 & s_1 & j_1 \\ l_2 & s_2 & j_2 \\ L & S & J \end{Bmatrix} \Psi(J)_{LS}
$$
$$\tag{E-92}$$

Although higher-order recoupling coefficients, e.g., 12-*j* symbols, etc. have been utilized for special applications (such as in the theory of fractional parentage coefficients for the description of nuclear shell-model configurations), the Clebsch-Gordan, Racah, and Fano coefficients suffice for most purposes, and in particular prove both adequate and useful in the formulation of angular distribution and correlation theory.

E.8. Racah Functions in
Angular Distribution and Correlation Theory

The development of the quantum theory of angular momentum in general, and of Racah algebra in particular, has been briefly surveyed by Biedenharn and Van Dam [Bie 65], who also reproduce some of the main papers in this field. Other review articles and textbooks (e.g., [Bie 53, De 57, Ed 57, Ro 57, Go 59, Bie 60, Li 61a, Br 62, Fr 65c]) deal with angular distribution and correlation theory for reactions which proceed by way of a definite sequence of levels. This sequence includes the formation and decay of an intermediate state and the theory accordingly refers to reactions which proceed via a "compound nucleus" mechanism rather than by a direct interaction.

(The theory for the latter type of process is given by Tobocman [To 61], and Bassel *et al.* [Ba 62d].) The coupling of angular momenta in each individual step of the overall sequence is responsible for the appearance of Racah functions in the formalism. The evaluation of angular distribution and correlation formulae is rendered much less complicated through the use of Racah algebra than would otherwise be the case. The most attractive feature of this approach is the fact that one can simply link characteristic functions for each transition step of a net sequence to derive a product which specifies the overall distribution or correlation. The way in which the quantum formalism relates to semi-classical considerations and permits of factorization into a product of individual "linking parameters" is very clearly shown by Biedenharn [Bie 60]. Having explained the concept of angular distributions and correlations in Sections 3.5.3 and 3.5.4, we set out from elementary principles in Section 9.4 to derive the expression for an anisotropic distribution $[\sim(1 + \cos^2\theta)]$ in a highly simplified case $(0 \to 1 \to 0$ transition). The calculation involved summation over each of the magnetic substates which can contribute to the process. Though quite easy in our application, this becomes exceedingly complicated in practical situations, but can be obviated altogether by the use of Racah methods which enable one in all cases to write down the explicit value of the numerical coefficients which weight the angular functions, e.g., the Legendre-polynomial weighting factor a_ν in the distribution formula [cf. Eq. (9–47)]

$$\frac{d\sigma}{d\Omega} = \sum_\nu a_\nu P_\nu(\cos\theta) \tag{E-93}$$

or the weighting coefficients $a_{\mu\nu\lambda}$ in the Legendre-hyperpolynomial expansion (which we shall define later) that constitutes the correlation function,

$$\frac{d^2\sigma}{d\Omega_1\,d\Omega_2} = \sum_{\mu,\nu,\lambda} a_{\mu\nu\lambda} S_{\mu\nu\lambda}(\theta_1, \theta_2, \phi) \tag{E-94}$$

The assembling of factors to build a_ν or $a_{\mu\nu\lambda}$ is still further facilitated by the fact that if any given transition is effected by some particle *other* than a photon, the requisite linking parameter is just the product of the appropriate "standard" parameter for γ radiation and a "particle parameter," such as already discussed in Section 10.3. Explicitly:

(a) The a_ν or $a_{\mu\nu\lambda}$ factorize into "linking parameters" which depend only upon the respective angular momenta that are involved in the individual steps, and are otherwise independent of one another; and

(b) the linking parameters depend upon the type of radiation effecting the transition, but this dependence is also factorizable; while

(c) it is immaterial whether a particular step represents an absorption or an emission transition, or whether the given sequence is undergone in the forward or the reverse sense.

The synthesis of the a_ν or $a_{\mu\nu\lambda}$ through linking parameters containing one or more Racah functions is described in detail by Biedenharn [Bie 60] (see also [Sh 64a]). We show next how to apply this procedure in cases of the type most frequently encountered.

E.8.1. COMPOSITION OF ANGULAR DISTRIBUTION FUNCTIONS

Taking γ transitions as our "standard basis," we start out by considering the simple case of a γ_1-γ_2 cascade (of pure multipolarity L_1 and L_2, respectively) between the states $J_0\pi_0$, $J_1\pi_1$ and $J_2\pi_2$ (Fig. E-5). The distribution

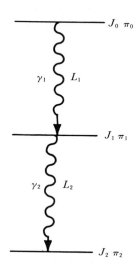

Fig. E-5. Nuclear spins and γ multipolarities involved in a γ_1-γ_2 cascade transition.

of γ_2 with respect to the γ_1 direction is expressed by the function a_ν which in this case comprises the product of two functions which are the respective linking parameters for γ transitions,

$$a_\nu = F_\nu(L_1 L_1 J_0 J_1) F_\nu(L_2 L_2 J_2 J_1) \tag{E-95}$$

The first factor applies to γ_1 decay and the second to γ_2 decay: note that the *intermediate* state in both terms is written last in the argument of F_ν. Since the F_ν are defined for the more general case of mixed multipolarity (L_1, L_1' and/or L_2, L_2') the multipolarity variable is featured twice in the argument. The definition of the F_ν (for a mixed γ transition L, L' between a state J and an intermediate state J_i) is

$$F_\nu(LL'JJ_i) = (-1)^{(J-J_i)-1}[(2L+1)(2L'+1)(2J_i+1)]^{1/2}\langle\nu 0|LL'1-1\rangle$$
$$\times W(LL'J_iJ_i; \nu J) \tag{E-96}$$

The range of even values that can be assumed by ν in the Legendre expansion is accordingly bounded by the properties of the constituent Racah functions:

$$|L - L'| \leqslant \nu \leqslant L + L' \quad \text{and} \quad 0 \leqslant \nu \leqslant 2J_i \quad \text{(E-97)}$$

Moreover, the F_ν are normalized to unity when $\nu = 0$,

$$F_0 = 1 \quad \text{(E-98)}$$

as immediately follows on setting $\nu = 0$ in the right-hand side of (E-96) and reducing. The F_ν functions were introduced by Biedenharn and Rose [Bie 53] and have been tabulated numerically by Ferentz and Rosenzweig [Fe 53]; from this tabulation (or from numerical values of the component Racah functions) it is easy to derive a_ν from (E-95) as a *numerical* Legendre weighting coefficient. For example, applying the method to the sequence

$$J_0 = 0 \xrightarrow[L_1 = 1]{\gamma_1} J_1 = 1 \xrightarrow[L_2 = 1]{\gamma_2} J_2 = 0 \quad \text{(E-99)}$$

one obtains

$$a_\nu = F_\nu(1101) \, F_\nu(1101) \quad \text{(E-100)}$$

whence the condition $0 \leqslant \nu \leqslant 2$ restricts ν to 0 and 2 only. For $\nu = 0$, Eq. (E-97) shows that $a_0 = 1$, while for $\nu = 2$ one finds from the tables that

$$F_2(1101) = 0.707107 = \frac{1}{\sqrt{2}} \quad \text{(E-101)}$$

which, substituted in (E-100) gives

$$a_2 = \tfrac{1}{2} \quad \text{(E-102)}$$

The normalized distribution $W(\theta)$, directly proportional to the differential cross section $d\sigma/d\Omega$, can accordingly be evaluated from (E-93), noting that $P_0 = 1$,

$$W(\theta) = 1 + \tfrac{1}{2}P_2(\cos \theta) \quad \text{(E-103)}$$

$$= 1 + \tfrac{1}{2}(\tfrac{3}{2}\cos^2 \theta - \tfrac{1}{2}) \quad \text{(E-104)}$$

$$\sim 1 + \cos^2 \theta \quad \text{(E-105)}$$

which tallies with the result derived in Section 9.4.

For the more general case of a γ_1-γ_2 *cascade* with either or both of the transitions having *mixed multipolarity* (L, L', with an amplitude mixing ratio δ), the linking parameter assumes the form

$$A_\nu(LL'; \delta; JJ_i) \equiv (1 + \delta^2)^{-1}[F_\nu(LLJJ_i) + 2\delta F_\nu(LL'JJ_i) \\ + \delta^2 F_\nu(L'L'JJ_i)] \quad \text{(E-106)}$$

and therefore the weighting coefficient is

$$a_\nu = A_\nu(L_1 L_1'; \delta_1; J_0 J_1) \, A_\nu(L_2 L_2'; \delta_2; J_2 J_1) \quad \text{(E-107)}$$

Generalizing further to the case in which instead of photons, *spinless particles* make the transition between either or both of the pair of levels, the linking coefficient may be expressed through $\bar{Z}$ functions (equivalent to Blatt-Biedenharn Z functions [Bl 52a] with a phase correction [Hu 54] which may or may not alter the sign). These are defined for a transition $J \to J_i$ effected by particles whose angular momentum can most generally be l or l' as

$$\bar{Z}(ll'J_iJ_i;\nu J) \equiv [(2l+1)(2l'+1)(2J_i+1)(2J_i+1)]^{1/2}$$
$$\times \langle \nu 0 | ll' 00 \rangle W(ll'J_iJ_i;\nu J) \tag{E-108}$$

The linking coefficient is, in terms of $\bar{Z}$,

$$(-1)^{J-J_i}(2J_i+1)^{-1/2}\,\bar{Z}(llJ_iJ_i;\nu J) \tag{E-109}$$

Dividing (E-109) by (E-96), we get the *particle parameter for spinless radiation* of type x (for equal angular momenta $l = L$),

$$b_\nu(ll;x) = -\frac{\langle \nu 0 | ll00 \rangle}{\langle \nu 0 | ll\ 1\ -1 \rangle} = \frac{2l(l+1)}{2l(l+1)-\nu(\nu+1)} \tag{E-110}$$

More generally, a transition between levels of spin J and J_i can be mediated by spinless-particle waves having either of two l values, e.g., the transition from $J_0 = 1$ to $J_1 = 2$ can be effected by α particles of momentum $l = 1$ *or* 3: making provision for this two-valuedness we write the *particle parameter for mixed momenta l,l'* as

$$b_\nu(ll';x) \sim \frac{2[l(l+1)\,l'(l'+1)]^{1/2}}{l(l+1)+l'(l'+1)-\nu(\nu+1)} \tag{E-111}$$

The constant of proportionality is a phase factor which depends upon the Coulomb functions in Wigner-Eisenbud reaction theory: it is governed by the incident particle energy and by the choice of target nucleus.

Before going on to illustrate this application, we mention that as soon as a transition $J_i \to J_f$ is effected by particles of spin s other than photons, the product of the linking parameters not only has to be multiplied by a STATISTICAL SPIN FACTOR (cf. Section 11.4.5)

$$g = \frac{2J_f+1}{(2s+1)(2J_i+1)} \tag{E-112}$$

but also by an appropriate PENETRABILITY TERM τ to take account of barrier tunneling. The angular distribution so derived can be transformed into an absolute differential cross section by multiplying by a factor $\lambda^2/4$, where λ is the rationalized wavelength of the incident particle in the center-of-mass system.

To take a specific example of a mixed particle-gamma transition sequence, we find for the radiative capture of α particles according to the scheme

$$J_0 = 1 \xrightarrow[\substack{l=1,\,l'=3 \\ (\text{mixing ratio }\delta_\alpha)}]{\alpha} J_1 = 2 \xrightarrow[\substack{L=1,\,L'=2 \\ (\text{mixing ratio }\delta_\gamma)}]{\gamma} J_2 = 1 \qquad (\text{E-113})$$

that the γ distribution referred to the incident α direction is

$$W(\theta) = g \sum_\nu [b_\nu(11;\alpha)F_\nu(1112) + 2\delta_\alpha b_\nu(13;\alpha)F_\nu(1312)$$
$$+ \delta_\alpha^2 b_\nu(33;\alpha)F_\nu(3312)][1 + \delta_\alpha^2]^{-1}$$
$$\times [A_\nu(12;\delta_\gamma;12)]\,\tau\,P_\nu(\cos\theta) \qquad (\text{E-114})$$

where

$$g = \frac{5}{3} \qquad (\text{E-115})$$

and

$$A_\nu(12;\delta_\gamma;12) = [1 + \delta_\gamma^2]^{-1}[F_\nu(1112) + 2\delta_\gamma F_\nu(1212) + \delta_\gamma^2 F_\nu(2212)] \qquad (\text{E-116})$$

When the transition is effected by *particles having a spin s* (other than photons), the appropriate procedure is to use the *spinless* particle parameter in the calculation, but at the same time replace the spin J in $F_\nu(LLJJ_i)$ or $A_\nu(LL';\delta;JJ_i)$ by a *channel spin*

$$\mathbf{J}_c = \mathbf{J} + \mathbf{s} \qquad (\text{E-117})$$

throughout, e.g., to form the linking parameter $b_\nu(ll';s=0)A_\nu(LL';\delta;J_cJ_i)$ for an *observed* transition which constitutes either the initial or final step of a sequence (when the observed particle transition does not correspond to bombarding or final radiation, the device employed consists of replacing J by a "fictitious" channel spin $\mathbf{J}_c' = \mathbf{J}_{\text{initial}} + \mathbf{s}$).

For a *nucleon transition* it is often more convenient to use the specific linking parameter introduced and tabulated by Satchler [Sa 53],

$$\eta_\nu(jj'JJ_i) \equiv b_\nu(jj';\mathcal{N})F_\nu(LL'JJ_i) \qquad (\text{E-118})$$

$$\equiv (-1)^{(J-J_1)-1/2}[(2j+1)(2j'+1)(2J_i+1)]^{1/2}$$
$$\times \langle\nu 0|jj'\tfrac{1}{2}-\tfrac{1}{2}\rangle W(jj'J_iJ_i;\nu J) \qquad (\text{E-119})$$

The form of this, for spin-$\frac{1}{2}$ particles, corresponds closely to that [Eq. (E-96)] for spin-1 photons.

The factorization also extends to the occurrence of *intermediate unobserved transitions* within the overall reaction sequence. For *any* radiation, irrespective of whether it be γ, nucleon, α, etc., the linking parameter for mixed momenta

L, L' (mixing ratio δ) is

$$U_\nu(LL'\,JJ_i) = (1 + \delta^2)^{-1}(-1)^{J+J_i}[(2J+1)(2J_i+1)]^{1/2}$$
$$\times\,[(-1)^L\,W(JJJ_iJ_i;\nu L) + \delta^2\,(-1)^{L'}\,W(JJJ_iJ_i;\nu L')]$$

$$(E\text{-}120)$$

This function is symmetrical in J and J_i, or L and L', and is normalized to

$$U_0 = 1 \qquad\qquad (E\text{-}121)$$

For particles with intrinsic spin s, the momenta L, L' are replaced by j, j'. An alternative, but unnormalized, function employed in some treatments [Sa 56] is

$$I_\nu \equiv \left(\frac{2J+1}{2J_i+1}\right)^{1/2} U_\nu \qquad\qquad (E\text{-}122)$$

The occurrence of unobserved radiation other than photons entails a modification in the penetrability term. The range of summation is prescribed by restrictions upon ν exercised by the Racah coefficients which feature in the resulting distribution formula. In particular, when a level of spin 0 or $\frac{1}{2}$ occurs at some stage of a transition sequence, the triangle relations confine ν to zero only and thereby render the distribution of the immediately subsequent radiation isotropic, since

$$F_0 = A_0 = b_0 = \eta_0 = U_0 = P_0 = 1 \qquad\qquad (E\text{-}123)$$

The application of these rules to a diversity of examples has been illustrated in [Bie 60, Sh 64a, Sh 66]. Including the penetrability term τ as indicated by Hauser-Feshbach theory [Ha 52], fast computer codes have been compiled for the numerical evaluation of distribution under various circumstances, many of which are listed in [Ra 65b] (see also [Sh 67]). The formalism can also be extended to furnish distributions from aligned nuclei, and computer programs have been set up to evaluate these automatically [Le 66c].

E.8.2. Composition of Angular Correlation Functions

Though more complicated in form, the correlation formulae can similarly be built up from a product of linking factors. The angular dependence is contained within the Legendre hyperpolynomial or "bipolar spherical harmonic" which is a function of *three* variables μ, ν, λ and three angles θ_1, θ_2, ϕ:

$$S_{\mu\nu\lambda}(\theta_1, \theta_2, \phi) = 4\pi \left(\frac{2\mu+1}{2\lambda+1}\right)^{1/2} \sum_{m=-\min(\nu,\lambda)}^{+\min(\nu,\lambda)} (-1)^m \langle \lambda m | \mu\nu 0m \rangle$$
$$\times\, Y_\nu^{-m}(\theta_1, 0)\, Y_\lambda^m(\theta_2, \phi) \qquad\qquad (E\text{-}124)$$

where the directions of emergence of the two correlated emergent radiations with respect to the incident direction are specified by θ_1, θ_2, and the azimuth ϕ. In terms of associated Legendre functions, this can be written as

$$S_{\mu\nu\lambda}(\theta_1, \theta_2, \phi) = [(2\mu + 1)(2\nu + 1)]^{1/2}\langle\lambda 0|\mu\nu 00\rangle P_\nu(\cos\theta_1) P_\lambda(\cos\theta_2)$$

$$+ 2 \sum_{m=1}^{+\min(\nu,\lambda)} [(2\mu + 1)(2\nu + 1)]^{1/2}\left[\frac{(\nu - m)!(\lambda - m)!}{(\nu + m)!(\lambda + m)!}\right]^{1/2}$$

$$\times \langle\lambda m|\mu\nu 0m\rangle P_\nu^m(\cos\theta_1) P_\lambda^m(\cos\theta_2) \cos m\phi \qquad \text{(E-125)}$$

The $S_{\mu\nu\lambda}$ are weighted by correlation coefficients $a_{\mu\nu\lambda}$ [cf. (E-94)] and, in the case of particle transitions, by a Hauser-Feshbach penetrability term τ. The overall correlation function can be expressed as an absolute double-differential cross section on multiplying it by the statistical spin factor g and an energy-dependent term $\lambda^2/16\pi$.

We now discuss the composition of the $a_{\mu\nu\lambda}$, once again referring to γ transitions as our "norm" which is readily capable of extension to other cases. The steps in the γ_1-γ_2-γ_3 cascade depicted in Fig. E-6 each contribute their characteristic linking parameter. For the initial and final steps, these take the same form as before, viz. $F_{(\mu \text{ or } \lambda)}$ or $A_{(\mu \text{ or } \lambda)}$, but for the *intermediate* step (observed together with the final step in a γ_2-γ_3 coincidence arrangement) one must employ the function [Sa 53]

$$R_{\mu\nu\lambda} \equiv \sum_{L,L'} r_\nu(LL')[(2L + 1)(2L' + 1)]^{1/2}[(2J_1 + 1)(2J_2 + 1)]^{1/2}\begin{pmatrix} J_1 & J_1 & \mu \\ L & L' & \nu \\ J_2 & J_2 & \lambda \end{pmatrix}$$

$$\text{(E-126)}$$

where for γ rays one has

$$r_\nu(LL') \xrightarrow{\ \gamma\ } (-1)^{L-1}\langle\nu 0|LL'1 - 1\rangle \qquad \text{(E-127)}$$

with appropriate weighting by the mixing ratio δ, while for a particle transition the expression has to be multiplied by the requisite particle parameter. Thus for nucleons (with $L = L' = j$), this assumes the form

$$r_\nu(LL') \xrightarrow{\ \mathcal{N}\ } |B(j)|^2(-1)^{j-\frac{1}{2}}\langle\nu 0|jj\tfrac{1}{2} -\tfrac{1}{2}\rangle \qquad \text{(E-128)}$$

where $B(j)$ is a reduced matrix element such that the fraction of the particle intensity having a momentum j is given by $|B(j)|^2$. It is in practice replaced by a penetrability factor τ.

The particle parameter has to be introduced whenever an *observed* transition is effected by particles other than photons; for an *unobserved* transition, the particle parameter is not required, and the appropriate linking factor is U as before.

For example, the γ_2-γ_3 correlation in a triple cascade in which all radiations are of mixed multipolarity (Fig. E-7) is given by the formula

$$\frac{d^2\sigma}{d\Omega_1\,d\Omega_2} = \frac{\lambda^2}{16\pi} \sum_{\mu,\nu,\lambda} A_\mu(L_1 L_1'; \delta_1; J_0 J_1)\, R_{\mu\nu\lambda}(L_2 L_2'; \delta_2; J_1 J_2)$$
$$\times A_\lambda(L_3 L_3'; \delta_3; J_3 J_2)\, S_{\mu\nu\lambda}(\theta_1, \theta_2, \phi) \qquad \text{(E-129)}$$

The formula reduces to an angular *distribution* when it is integrated over θ_1 or θ_2, as the case may be. It becomes equivalent to a distribution (though smaller

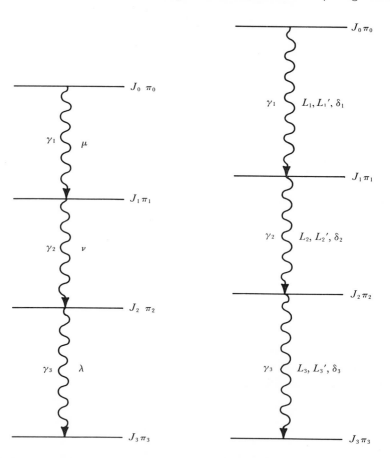

Fig. E-6. Nuclear spins and transition indices involved in a γ_1-γ_2-γ_3 cascade transition in which the γ_2-γ_3 angular correlation is evaluated.

Fig. E-7. Nuclear spin and multipolarity notation for a γ_1-γ_2-γ_3 transition sequence.

by a factor 4π) when J_1 or J_2 are 0 or $\frac{1}{2}$, for then the immediately subsequent radiation is emitted isotropically.

For a nucleon-induced $(\mathcal{N}, \gamma\text{-}\gamma)$ process, the term A_μ in (E-129) would have to be replaced by $g\eta_\mu(jjJ_0J_1)$. In the frequently encountered case of inelastic nucleon scattering $(\mathcal{N}, \mathcal{N}'\,\gamma)$, the $\mathcal{N}'$-γ correlation (Fig. E-8) takes the form

$$\frac{d^2\sigma}{d\Omega_{\mathcal{N}'} d\Omega_\gamma} = \frac{\lambda^2}{16\pi} \sum_{\mu,\nu,\lambda} g\,\eta_\mu(j_1j_1J_0J_1)\, R_{\mu\nu\lambda}(j_2j_2J_1J_2)$$
$$\times A_\lambda(LL';\delta;J_3J_2)\, S_{\mu\nu\lambda}(\theta_{\mathcal{N}'}, \theta_\gamma, \phi) \qquad \text{(E-130)}$$

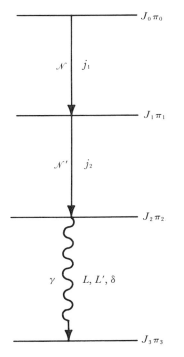

Fig. E-8. Momenta involved in an $\mathcal{N}$-$\mathcal{N}'$-γ sequence for the inelastic nucleon scattering process $(\mathcal{N}, \mathcal{N}'\,\gamma)$.

in which the penetrability term τ is contained within $R_{\mu\nu\lambda}$. The implicit Racah functions limit the range of the summation variables μ, ν, λ. Explicit reduced expressions are given by Biedenharn [Bie 60] (see also [Sh 63, Sh 64b]), useful numerical tables have been compiled by Ferguson and Rutledge [Fe 62] and Kaye *et al.* [Ka 63], and a computer program has been set up to evaluate the correlation problem numerically (see [Sh 67]).

EXERCISES

E-1 Equation (E-8) expresses the fact that components of angular momenta do not commute. Physically, this can be interpreted through an uncertainty relation. Write down the appropriate relation and discuss its physical significance.

E-2 Check the expressions in Table E-2 and determine the parity of the spherical harmonics Y_{22} and Y_{32}.

E-3 Show from the distribution function (E-105) for two cascade γ transitions in a $0 \to 1 \to 0$ sequence that the most probable angle of emergence between the two photons is about $55°$.

E-4 Discuss the physical implications of the statement that a determination of an angular correlation between two radiations is essentially equivalent to the measurement of the *distribution* of the second radiation from aligned nuclei (the observation of the first radiation serves to pick out those nuclei which happen to be aligned in a particular direction from among the total ensemble of randomly-oriented nuclei).

E-5 Show that in an $(\mathcal{N}, \mathcal{N}' \, \gamma_1\text{-}\gamma_2)$ process the differential cross section which gives the γ_1 distribution is in magnitude and angular dependence identical with that for the γ_2 distribution (γ_1 unobserved) when the two γ multipolarities are pure and equal, and the final nuclear spin J_4 is zero.

E-6 Verify the recursion-type identity

$$\frac{W(J-L, J-L, L, L; vJ)}{W(J-L, J-L, J, J; vL)} \equiv (-1)^{2J}(2J+1)\, W(JJLL; v, J+L)$$

and use this to show that the γ radiation making the transition $J_2 \to J_3$ has the same distribution as that from J_3 to J_4 (with $J_2 \to J_3$ unobserved) when

$$L_2 = L_2' = L_3 = L_3' = |J_2 - J_3| = |J_3 - J_4|$$

even for *nonzero* values of J_4.

E-7 Demonstrate, with the aid of Eqs. (E-83), (E-93), (E-112), (E-118) and (E-119), that the angular distribution of nucleons effecting a $0^+ \to J \to 0^+$ transition sequence is given by the expression

$$d\sigma/d\Omega = \tfrac{1}{8}\lambda^2 \sum_{v} (-1)^{2J-1}(2J+1)|\langle v0|JJ\tfrac{1}{2} - \tfrac{1}{2}\rangle|^2 \tau P_v(\cos\theta).$$

Appendix F

FEYNMAN INTERACTION THEORY

F.1. The Underlying Motivation
behind a Field-Interaction Approach

The problem of interacting particles, as considered, e.g., in Chapters 4, 8, and 11, lends itself readily to a treatment based upon the interaction of *fields*. In Section 4.6.3 we introduced Feynman graphs, pointing out that their value lies not only in the way that they could be used to depict an interaction process schematically but in the means they offer for the elucidation of transition amplitudes and cross sections. Next, in Sections 8.3, 8.6, and 8.8, the field-interaction calculus was employed in a formal derivation of the transition cross section for β decay in terms of appropriate matrix elements. Finally, in Section 11.5 the method was applied to reaction theory through the introduction of elements of the S matrix, whose properties were discussed in some detail. All these considerations, together with ancillary properties such as have been discussed in Appendix D, serve as a foundation for a presentation of the extremely versatile and convenient technique offered by Feynman interaction graphs.

The extreme usefulness of the Feynman approach comes from the fact that the matrix elements for given interactions can be built from a product of appropriate factors, each of which corresponds to a particular segment of the Feynman graph. In a similar way to the composition of a distribution or correlation cross section, the interaction probability can be constructed mathematically by simply linking appropriate terms for each step of the process in accordance with certain fundamental rules. It is then an easy matter to arrive at formal expressions for the cross section. The purpose of this Appendix is to illustrate this procedure through a few selected examples which indicate the power, elegance, and utility of the method.

F.2. Interaction Matrix Elements

Let us first consider the interaction of electromagnetic radiation described by the field 4-vector $A^\mu \equiv \{\varphi, \mathbf{A}\}$ with an electron field, as in Compton scattering. The net Lagrangian density (cf. Appendix D.8) for the interacting system is the sum of the Lagrangians for the participating fields together with an interaction Lagrangian whose form is determined by invariance principles (conjoined with simplicity considerations):

$$\mathscr{L} = \mathscr{L}_\gamma + \mathscr{L}_e + \mathscr{L}_{\text{int}} \tag{F-1}$$

The derivation of the free-photon Lagrangian density $\mathscr{L}_\gamma$ has been indicated in Appendix D.8, the result being expressed by Eq. (D-220),

$$\mathscr{L}_\gamma = -\frac{1}{2}\left(\frac{\partial A^\mu}{\partial x^\nu}\frac{\partial A_\mu}{\partial x_\nu}\right) \tag{F-2}$$

That for a free electron can be deduced along analogous lines. We find that if the basic form

$$\mathscr{L}_e = k_1 \frac{1}{i}\,\bar\psi\gamma^\mu\frac{\partial\psi}{\partial x^\mu} + k_2\,\bar\psi\psi = k_1\,\bar\psi\left(\frac{1}{i}\gamma^\mu\partial_\mu + \frac{k_2}{k_1}\right)\psi \tag{F-3}$$

is taken for the Lagrangian density of the electron field described by the wave function ψ, a comparison with the covariant Dirac equation (D-104) suggests that a suitable identification might be

$$\frac{k_2}{k_1} = \frac{mc}{\hbar} \tag{F-4}$$

and, in order to satisfy energy-momentum properties of the Dirac field, one should set

$$k_1 = -\hbar c \tag{F-5}$$

This gives us

$$\mathscr{L}_e = -\hbar c\,\bar\psi\left(\frac{1}{i}\gamma^\mu\partial_\mu + \frac{mc}{\hbar}\right)\psi \tag{F-6}$$

Since the interaction Lagrangian again has to be a scalar quantity but must contain the vector A_μ, the simplest *Ansatz* takes the form of a scalar product

$$\mathscr{L}_{\text{int}} = (j^\mu A_\mu) \tag{F-7}$$

of A_μ with the *electron* current 4-vector [cf. (D-116)] introducing the elementary charge e

$$j^\mu = e\bar\psi\gamma^\mu\psi \tag{F-8}$$

namely,

$$\mathscr{L}_{\text{int}} = e\bar\psi\gamma^\mu\psi A_\mu \tag{F-9}$$

Inserting (F-2), (F-6), and (F-9) in (F-1) we find

$$\mathscr{L} = -\frac{1}{2}\left(\frac{\partial A^{\mu}}{\partial x^{\nu}}\frac{\partial A_{\mu}}{\partial x_{\nu}}\right) - \hbar c\bar{\psi}\left(\frac{1}{i}\,\gamma^{\mu}\,\partial_{\mu} + \frac{mc}{\hbar}\right)\psi + e\bar{\psi}\gamma^{\mu}\,\psi A_{\mu} \qquad \text{(F-10)}$$

It is a simple matter to go over to the corresponding Hamiltonian form in interaction representation which is customarily used to express the S-matrix element. Specifically the interaction term constitutes the principal quantity of interest.

We proceed from the definition

$$p^{\mu} \equiv \frac{\partial \mathscr{L}}{\partial \dot{q}^{\mu}} \qquad \text{(F-11)}$$

of p^{μ} as canonically conjugate quantity to q^{μ} (it might, leaving dimensional considerations aside, be viewed as a sort of "generalized momentum density") and note that

$$H \equiv \int \mathscr{H}\,d^4x = \int\,(p^{\mu}\dot{q}^{\mu} - \mathscr{L})\,d^4x \qquad \text{(F-12)}$$

whence

$$\mathscr{H} = p^{\mu}\dot{q}^{\mu} - \mathscr{L} \qquad \text{(F-13)}$$

Accordingly, for $\mathscr{L}$ defined by (F-1), we could write

$$\mathscr{H}_{\gamma} + \mathscr{H}_{e} + \mathscr{H}_{int} = p^{\mu}\dot{q}^{\mu} - (\mathscr{L}_{\gamma} + \mathscr{L}_{e} + \mathscr{L}_{int}) \qquad \text{(F-14)}$$

and with justification assume this to be satisfied separately by the noninteracting (i.e., unperturbed) and interacting quantities. Equating the latter we obtain

$$\mathscr{H}_{int} = p^{\mu}\dot{q}^{\mu} - \mathscr{L}_{int} \qquad \text{(F-15)}$$

With (F-11) this gives us

$$\mathscr{H}_{int} = \frac{\partial \mathscr{L}}{\partial \dot{q}^{\mu}}\dot{q}^{\mu} - \mathscr{L}_{int} \qquad \text{(F-16)}$$

If we make the reasonable assumption in this case that $\mathscr{L}_{int}$ does not depend upon the time-derivative of the field $(\partial \mathscr{L}/\partial \dot{q}^{\mu} = 0)$, we obtain in conclusion

$$\mathscr{H}_{int} \cong -\mathscr{L}_{int} \qquad \text{(F-17)}$$

All that remains is to express the S matrix as a product of Hamiltonians in a time-ordered sequence. The equation of motion of a coupled interacting system can be written as a time-dependent Schrödinger equation in interaction representation

$$i\hbar\frac{\partial}{\partial t}\psi(t) = H(t)\,\psi(t) \qquad \text{(F-18)}$$

where the $\psi(t)$ represent state-vector eigenfunctions describing the interaction, whose asymptotic values are linked through the S matrix,

$$\psi(+\infty) = S\psi(-\infty) \tag{F-19}$$

The S matrix can accordingly be deduced by solving (F-18), which we write as an integral equation [for small $H(t)$]

$$\psi(t) = \psi(-\infty) + \frac{1}{i\hbar} \int_{-\infty}^{t} H(t_1)\,\psi(t_1)\,dt_1 \tag{F-20}$$

and solve this by iteration,

$$\psi(t) = \psi(-\infty) + \frac{1}{i\hbar} \int_{-\infty}^{t} dt_1\, H(t_1)\left[\psi(-\infty) + \frac{1}{i\hbar} \int_{-\infty}^{t_1} dt_2\, H(t_2)\,\psi(t_2)\right] = \cdots \tag{F-21}$$

$$= \psi(-\infty)\left[1 + \frac{1}{i\hbar} \int_{-\infty}^{t} dt_1\, H(t_1) + \frac{1}{(i\hbar)^2} \int_{-\infty}^{t} dt_1 \int_{-\infty}^{t_1} dt_2\, H(t_1)\,H(t_2) + \cdots\right] \tag{F-22}$$

Setting $t \to +\infty$ and using (F-19) we obtain the S matrix,

$$S = 1 + \frac{1}{i\hbar} \int_{-\infty}^{\infty} dt_1\, H(t_1) + \frac{1}{(i\hbar)^2} \int_{-\infty}^{\infty} dt_1 \int_{-\infty}^{t_1} dt_2\, H(t_1)\,H(t_2) + \cdots \tag{F-23}$$

$$= \sum_{n=0}^{\infty} \frac{1}{(i\hbar)^n} \int_{-\infty}^{\infty} dt_1 \int_{-\infty}^{t_1} dt_2 \cdots \int_{-\infty}^{t_{n-1}} dt_n [H(t_1)\,H(t_2)\cdots H(t_n)] \tag{F-24}$$

We introduce a Dyson [Dy 49] TIME-ORDERING OPERATOR P whose role consists of rearranging the factors in a product of time-specified operators so that they appear in chronological sequence, with operators for *earlier* times standing to the right of those for later, and therefore operating first. This enables us to write

$$S = \sum_{n=0}^{\infty} \frac{1}{n!}\frac{1}{(i\hbar)^n} \int_{-\infty}^{\infty} \int_{-\infty}^{\infty} \cdots \int_{-\infty}^{\infty} dt_1\cdots dt_n\, P\,[H(t_1)\cdots H(t_n)] \tag{F-25}$$

or, in terms of the interaction density defined by

$$H(t) = \int d^3x\, \mathscr{H}(x) \tag{F-26}$$

as

$$S = \sum_{n=0}^{\infty} \frac{1}{n!}\frac{1}{(i\hbar c)^n} \int_{-\infty}^{\infty} \cdots \int_{-\infty}^{\infty} d^4x_1 \cdots d^4x_n\, P\,[\mathscr{H}(x_1)\cdots \mathscr{H}(x_n)] \tag{F-27}$$

on noting that

$$\int dt \int d^3 x = \frac{1}{c} \int d^4 x \qquad \text{(F-28)}$$

When lowest-order perturbation theory is adequate and Eq. (F-17) is applicable, we obtain the particularly simple form

$$S = 1 - \frac{1}{i\hbar c} \int_{-\infty}^{\infty} d^4 x \, \mathscr{L}(x) \qquad \text{(F-29)}$$

Although this suffices in general for the evaluation of a weak-interaction matrix element, it is necessary to include higher terms in the S-matrix expansion when dealing with all but the simplest electromagnetic interactions, and a judicious choice of the appropriate order of approximation therefore has to be made for each individual problem. As we consider Compton scattering, we can avail ourselves of (F-17), (F-7), and (F-9) to write

$$\mathscr{H}_{\text{int}} = - \mathscr{L}_{\text{int}} = - (j^\mu A_\mu) = - e \bar{\psi} \gamma^\mu \psi A_\mu \qquad \text{(F-30)}$$

and allow the S-matrix expansion to extend formally to any given order while restricting our attention to at most the second order. Our sign convention, though in accord with that of Muirhead [Mui 65] and other authors, differs from that used in other treatments in which a different interpretation of the sign of e has been adopted.

The matrix elements describing the interaction of the photon-electron field are, explicitly,

zeroth order:

$$S_0 = 1 \qquad \text{(F-31)}$$

first order:

$$S_1 = -\frac{e}{i\hbar c} \int_{-\infty}^{\infty} d^4 x \, \bar{\psi}(x) \, \gamma^\mu A_\mu(x) \, \psi(x) \qquad \text{(F-32)}$$

second order:

$$S_2 = \frac{1}{2} \left(\frac{e}{i\hbar c} \right)^2 \int_{-\infty}^{\infty} d^4 x_1 \int_{-\infty}^{\infty} d^4 x_2 [\bar{\psi}(x_1) \, \gamma^\mu A_\mu(x_1) \, \psi(x_1)]$$
$$\times \, [\bar{\psi}(x_2) \, \gamma^\mu A_\mu(x_2) \, \psi(x_2)] \qquad \text{(F-33)}$$

bearing in mind that $\bar{\psi}$ and ψ both commute with A_μ. We note also that frequently Feynman dagger notation is employed:

$$\gamma^\mu A_\mu \equiv \mathcal{A} \qquad \text{(F-34)}$$

Each term in the power expansion can correspond to a variety of processes, virtual and real. Wick [Wi 50] and Dyson [Dy 51a,b] have shown how

$P[\mathscr{H}(x_1) \ldots \mathscr{H}(x_n)]$ can be expressed in a form which explicitly features all virtual processes through a decomposition of the chronological product into so-called NORMAL PRODUCTS. The latter are products of free-particle creation and destruction operators in which all the creation operators stand to the left of all destruction operators in second-quantization formalism. This procedure is discussed in detail by Schweber *et al.* [Schw 56], Mandl [Ma 59b], and Muirhead [Mui 65]. The essential point is that a Feynman graph then betokens a concise way of representing such a normal product in pictorial form. For our purposes it suffices to note that the Dyson time-ordering operator P has to be replaced by a more powerful Wick chronological operator T which in addition to time-ordering introduces a plus or minus sign according as the rearrangement of electron-positron factors involves an even or odd number of permutations. Furthermore, products of the type featured in (F-32) and (F-33) have to be reinterpreted as *normal products*, characterized by the notation $N[\cdots]$, in which creation operators, e.g.,

$\bar{\psi}^{(-)}(x)$, creation of an electron at the space-time point x

$\psi^{(-)}(x)$, creation of a positron at the space-time point x

stand to the left of destruction operators, viz.

$\psi^{(+)}(x)$, destruction of an electron at the space-time point x

$\bar{\psi}^{(+)}(x)$, destruction of a positron at the space-time point x

We accordingly rewrite the S matrix in our example as

$$S = \sum_{n=0}^{\infty} \frac{1}{n!} \left(-\frac{e}{i\hbar c}\right)^n \int_{-\infty}^{\infty} d^4 x_1 \ldots \int_{-\infty}^{\infty} d^4 x_n$$

$$\times \ T\{N[\bar{\psi}(x_1)\,A(x_1)\,\psi(x_1)] \ldots N[\bar{\psi}(x_n)\,A(x_n)\,\psi(x_n)]\} \quad \text{(F-35)}$$

and proceed in the next section to relate the corresponding segments of Feynman graphs to the terms appearing in this formula, hereafter closely following the treatment in Muirhead [Mui 65].

F.3. Feynman Graphs

F.3.1. RELATION TO MATRIX ELEMENTS

Feynman graphs consist of lines representing normal products of interaction operators, proceeding from or to a VERTEX which specifies the space-time point x_1, etc. Thus the creation and destruction operators are represented by the directed lines in Figs. F-1 (a–d) referred to the point x_1. Positrons are thereby visualized as time-reversed electrons in the Feynman picture.

Then the first-order term in the Compton scattering matrix element
[cf. (F-32)]

$$S_1 \sim \int d^4x_1 \, T\{N[\bar{\psi}(x_1) \, A(x_1) \, \psi(x_1)]\} \tag{F-36}$$

can be decomposed into a sum of normal products whose corresponding
Feynman graphs are depicted in Figs. F-2(a–d).

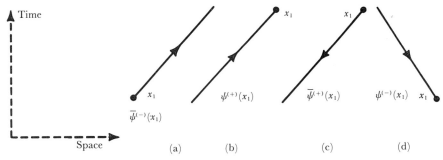

Fig. F-1. Segments of Feynman graphs and corresponding creation or destruction operators.

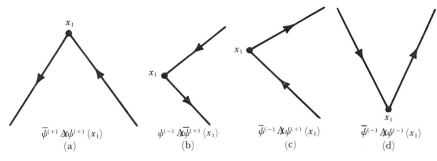

Fig. F-2. Feynman graphs corresponding to normal products of field operators.

When referred to a VACUUM STATE Ψ_0, the rules governing the construction
of Feynman graphs are:

(1) a solid line from x_1 to x_2 represents the factor pair (i.e., contraction)
$\langle \Psi_0 | \bar{\psi}(x_1) \, \psi(x_2) | \Psi_0 \rangle$;

(2) a solid line away from or into a vertex represents an unpaired operator
$\bar{\psi}$ or ψ;

(3) a wavy line between x_1 and x_2 represents the factor pair
$\langle \Psi_0 | A_\mu(x_1) A_\mu(x_2) | \Psi_0 \rangle$;

(4) a wavy line proceeding to or from a vertex represents an unpaired operator A_μ;

(5) lines joining a point to itself are not meaningful, since they represent a vanishing matrix element.

Omitting those matrix products which give either a zero or nonmeaningful contribution to physical observables, we are left with seven quantities when we consider the second-order S-matrix element

$$S_2 \sim \int d^4 x_1 \int d^4 x_2 \, T\{N[\bar{\psi}(x_1)\, A(x_1)\, \psi(x_1)]\, N[\bar{\psi}(x_2)\, A(x_2)\, \psi(x_2)]\}$$
$$\text{(F-37)}$$

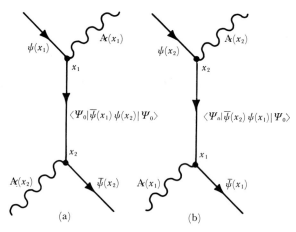

Fig. F-3. Feynman graphs which represent Compton scattering, labeled according to the corresponding contribution to the S-matrix element.

namely,

$$M_1 \sim \langle \Psi_0 | \bar{\psi}(x_1)\, \psi(x_2) | \Psi_0 \rangle \, N[A(x_1)\, \psi(x_1)\, \bar{\psi}(x_2)\, A(x_2)] \qquad \text{(F-38)}$$

which corresponds to COMPTON SCATTERING [Fig. F-3(a)], and the *topologically indistinguishable graph* [Fig. F-3(b)] with x_1 and x_2 interchanged, viz.

$$M_2 \sim \langle \Psi_0 | \bar{\psi}(x_2)\, \psi(x_1) | \Psi_0 \rangle \, N[\bar{\psi}(x_1)\, A(x_1)\, A(x_2)\, \psi(x_2)] \qquad \text{(F-39)}$$

together with the entity

$$M_3 \sim \langle \Psi_0 | A(x_1)\, A(x_2) | \Psi_0 \rangle \, N[\bar{\psi}(x_1)\, \psi(x_1)\, \bar{\psi}(x_2)\, \psi(x_2)] \qquad \text{(F-40)}$$

which represents MØLLER SCATTERING (elastic electron-electron scattering), shown in Fig. F-4,

$$M_4 \sim \langle \Psi_0|\bar{\psi}(x_1)\,\psi(x_2)|\Psi_0\rangle\langle\Psi_0|A(x_1)\,A(x_2)|\Psi_0\rangle\,N[\bar{\psi}(x_2)\,\psi(x_1)]$$
$$(F\text{-}41)$$

and the topologically indistinguishable

$$M_5 \sim \langle \Psi_0|\bar{\psi}(x_2)\,\psi(x_1)|\Psi_0\rangle\langle\Psi_0|A(x_2)\,A(x_1)|\Psi_0\rangle\,N[\bar{\psi}(x_1)\,\psi(x_2)]$$
$$(F\text{-}42)$$

which, shown in Figs. F-5(a, b), stand for the FERMION SELF-ENERGY, whereas

$$M_6 \sim \langle \Psi_0|\bar{\psi}(x_1)\,\psi(x_2)|\Psi_0\rangle\langle\Psi_0|\bar{\psi}(x_2)\,\psi(x_1)|\Psi_0\rangle\,N[A(x_1)\,A(x_2)]$$
$$(F\text{-}43)$$

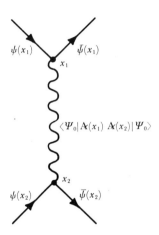

Fig. F-4. The Feynman graph representative of Møller scattering (elastic electron-electron scattering).

represents the PHOTON SELF-ENERGY (Fig. F-6), and

$$M_7 \sim \langle \Psi_0|\bar{\psi}(x_1)\,\psi(x_2)|\Psi_0\rangle\langle\Psi_0|\bar{\psi}(x_2)\,\psi(x_1)|\Psi_0\rangle\langle\Psi_0|A(x_1)\,A(x_2)|\Psi_0\rangle$$
$$(F\text{-}44)$$

corresponds to a VACUUM FLUCTUATION, depicted in Fig. F-7.

The method offers a way to consider higher-order effects, as for example the correction to the Compton scattering cross section when there is exchange of a virtual photon between electrons in initial and final states (Fig. F-8). The appropriate matrix-element correction can be written down straightway from the Feynman graph,

$$M_8 \sim \langle \Psi_0|\bar{\psi}(x_1)\,\psi(x_2)|\Psi_0\rangle\langle\Psi_0|\bar{\psi}(x_2)\,\psi(x_3)|\Psi_0\rangle$$
$$\times \langle \Psi_0|\bar{\psi}(x_3)\,\psi(x_4)|\Psi_0\rangle\langle\Psi_0|A(x_1)\,A(x_4)|\Psi_0\rangle\,N[\psi(x_1)\,\bar{\psi}(x_4)\,A(x_2)\,A(x_3)]$$
$$(F\text{-}45)$$

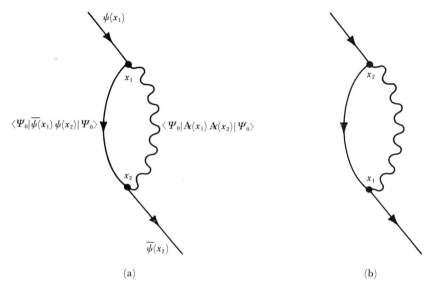

Fig. F-5. Feynman graphs representing the fermion self-energy.

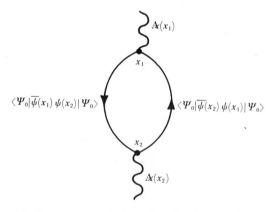

Fig. F-6. Feynman graph depicting the photon self-energy.

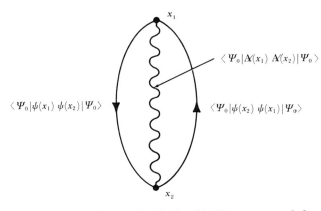

Fig. F-7. A vacuum fluctuation depicted in Feynman-graph form.

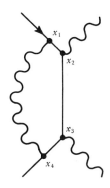

Fig. F-8. Third-order correction to Compton scattering represented as a Feynman diagram.

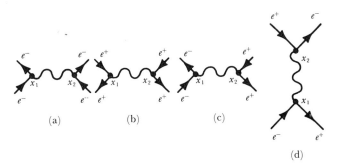

Fig. F-9. Feynman graphs depicting the scattering of electrons and positrons through a Coulomb interaction interpreted as exchange of a (virtual) photon.

Feynman's *space-time approach* also permits of extension to associated processes. Thus, for instance, Fig. F-4 inherently covers (a) electron-electron scattering [Fig. F-9(a)], (b) positron-positron scattering [Fig. F-9(b)], (c) electron-positron scattering through the intermediary of a virtual photon [Fig. F-9(c)], (d) electron-positron scattering represented by pair annihilation at x_1 followed by re-formation at x_2 [Fig. F-9(d)].

The interactions therefore include not only Møller scattering [Mø 32a] of electrons on electrons, but also Bhabha scattering [Bha 36] of electrons on positrons, for which the Bhabha equation (see Appendix D.8) was first assembled [Bha 38]. Pair annihilation constitutes an intrinsic part of the formalism.

F.3.2. FEYNMAN GRAPHS IN MOMENTUM SPACE

The theory can readily be reformulated to apply in momentum space, which is often more convenient for calculations than configuration space.

For Compton scattering, the two Feynman graphs, Figs. F-3(a,b) apply, and we can therefore write the second-order S-matrix element (inserting a factor 2 to take account of M_1 *and* M_2 being valid) as

$$S_2 = 2 \times \frac{1}{2}\left(-\frac{e}{i\hbar c}\right)^2 \int_{-\infty}^{\infty} d^4x_1 \int_{-\infty}^{\infty} d^4x_2 \langle \Psi_0 | T[\bar\psi(x_1)\,\psi(x_2)] | \Psi_0 \rangle$$
$$\times N[A(x_1)\,\psi(x_1)\,\bar\psi(x_2)\,A(x_2)] \tag{F-46}$$

Since we concern ourselves with incident and outgoing electrons we can set

$$N[\psi(x_1)\,\bar\psi(x_2)] = -\bar\psi^{(+)}(x_2)\,\psi^{(-)}(x_1) + \text{irrelevant terms} \tag{F-47}$$

and

$$N[A(x_1)\,A(x_2)] = A^{(+)}(x_1)\,A^{(-)}(x_2) + A^{(+)}(x_2)\,A^{(-)}(x_1) + \text{irrelevant terms} \tag{F-48}$$

so that (F-46) becomes

$$S_2 = -\left(-\frac{e}{i\hbar c}\right)^2 \int_{-\infty}^{\infty} d^4x_1 \int_{-\infty}^{\infty} d^4x_2 \langle \Psi_0 | T[\bar\psi(x_1)\,\psi(x_2)] | \Psi_0 \rangle$$
$$\times [A^{(+)}(x_1)\,\bar\psi^{(+)}(x_2)\,\psi^{(-)}(x_1)\,A^{(-)}(x_2) + A^{(+)}(x_2)\,\bar\psi^{(+)}(x_2)\,\psi^{(-)}(x_1)\,A^{(-)}(x_1)]$$
$$\tag{F-49}$$

This can be written as the sum of two terms,

$$S_2 = S_2^{(1)} + S_2^{(2)} \tag{F-50}$$

where

$$S_2^{(1)} \equiv -\left(-\frac{e}{i\hbar c}\right)^2 \int_{-\infty}^{\infty} d^4 x_1 \int_{-\infty}^{\infty} d^4 x_2 \{\bar{\psi}^{(+)}(x_2) \, A^{(-)}(x_2)$$

$$\times \langle \Psi_0 | T[\bar{\psi}(x_1) \, \psi(x_2) | \Psi_0 \rangle A^{(+)}(x_1) \, \psi^{(-)}(x_1)\} \qquad \text{(F-51)}$$

and

$$S_2^{(2)} \equiv -\left(-\frac{e}{i\hbar c}\right)^2 \int_{-\infty}^{\infty} d^4 x_1 \int_{-\infty}^{\infty} d^4 x_2 \{\bar{\psi}^{(+)}(x_2) \, A^{(+)}(x_2)$$

$$\times \langle \Psi_0 | T[\bar{\psi}(x_1) \, \psi(x_2)] | \Psi_0 \rangle A^{(-)}(x_1) \, \psi^{(-)}(x_1)\} \qquad \text{(F-52)}$$

The corresponding graphs are depicted in Figs. F-10(a, b), the 4-momenta of electron and photon, respectively, being denoted by p and k (since $p_\gamma = \hbar k$),

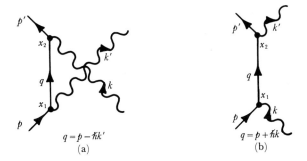

Fig. F-10. Feynman graphs in momentum space pertaining to Compton scattering in which there is a momentum transfer q to the scattered electron.

the outgoing terms carrying a prime. The intermediate electron line connecting x_1 and x_2 represents the ELECTRON PROPAGATOR which is associated with a momentum $q = p - \hbar k'$ or $q = p + \hbar k$ respectively. Other processes can also feature photon and/or meson propagators in the graph and the matrix element.

In order to derive explicit expressions for the various field operators which enter into (F-51) and (F-52) we must have recourse to the formalism of second quantization, briefly outlined in the next section.

F.3.3. SECOND QUANTIZATION

In Feynman's formulation of quantum electrodynamics, the interactions of fields are intimately bound up with the creation and destruction of particles, for which provision is made through the formalism of second quantization. Rather than get too deeply involved in so intricate and far-ranging a subject, however, we extract just a few of the results it provides, and refer the reader who desires further details to the standard textbooks on field quantization (e.g., [Schw 56, Ma 59b]).

The solution of the Klein-Gordon equation (D-68), representing a neutral, scalar, Hermitian field, can be written in the form

$$\varphi(x) = \frac{1}{V^{1/2}} \sum_k \frac{1}{(2\omega_k)^{1/2}} [e^{-ikx} a^\dagger(\mathbf{k}) + e^{ikx} a(\mathbf{k})] \qquad \text{(F-53)}$$

The factor $V^{-1/2}$ enters through box normalization in which V is conventionally assumed to be a large volume bounded by a rectangular box with sides of length l_i (where $i = 1, 2, 3$) such that

$$k_i l_i \equiv \frac{2\pi}{\lambda_i} l_i = 2\pi n_i \qquad \text{(F-54)}$$

Clearly, the total number n_{tot} of wave vectors $\mathbf{k}$ in the interval $d^3 k$ is

$$n_{\text{tot}} = \frac{1}{(2\pi)^3} V d^3 k \qquad \text{(F-55)}$$

and when V is sufficiently large a summation over all vectors $\mathbf{k}$ of a slowly-varying function $f(\mathbf{k})$ can be replaced by an integral,

$$\frac{1}{V^{1/2}} \sum_k f(\mathbf{k}) \rightarrow \frac{1}{(2\pi)^{3/2}} \int f(\mathbf{k}) \, d^3 k \qquad \text{(F-56)}$$

A further normalization in (F-53) is effected by the term $(2\omega_k)^{-1/2}$ in which $\omega_k \equiv E_k/\hbar$ represents the frequency corresponding to a wave-number $k = p/\hbar$. This enables the CREATION and DESTRUCTION OPERATORS $a^\dagger(\mathbf{k})$ and $a(\mathbf{k})$ to satisfy the commutation relations

$$\left. \begin{aligned} [a^\dagger(\mathbf{k}), a^\dagger(\mathbf{k}')] &= [a(\mathbf{k}), a(\mathbf{k}')] = 0 \\ [a^\dagger(\mathbf{k}), a(\mathbf{k}')] &= \delta_{\mathbf{kk'}} \end{aligned} \right\} \qquad \text{(F-57)}$$

Denoting a state comprising n particles, each of momentum $\mathbf{k}$, by $|n\rangle$, we may characterize these operators by the relations

$$\langle n + 1 | a^\dagger(\mathbf{k}) | n \rangle = (n + 1)^{1/2} \qquad \text{(F-58)}$$

$$\langle n | a(\mathbf{k}) | n + 1 \rangle = (n + 1)^{1/2} \qquad \text{(F-59)}$$

Thus $a^\dagger(\mathbf{k})$ augments the system by 1 particle of momentum $\mathbf{k}$, while leaving the number of particles with other momenta than $\mathbf{k}$ unchanged, while $a(\mathbf{k})$ brings about a corresponding diminution. All other matrix elements of $a^\dagger(\mathbf{k})$ and $a(\mathbf{k})$ vanish identically. In particular, (F-58) shows us that one can create a state of n particles with momentum $\mathbf{k}$ by operating n times with $a^\dagger(\mathbf{k})$ on the vacuum state comprising no particles, $|\Psi_0\rangle \equiv |0\rangle$:

$$|n_k\rangle = \frac{1}{(n!)^{1/2}} [a^\dagger(\mathbf{k})]^n |0\rangle \qquad \text{(F-60)}$$

The Hamiltonian for the system is simply

$$H = \sum_k \hbar\omega_k \, a^\dagger(\mathbf{k}) \, a(\mathbf{k}) \tag{F-61}$$

and $a^\dagger(\mathbf{k}) \, a(\mathbf{k})$ betokens the number of particles with momentum $\mathbf{k}$ present in the particular state.

Specifically, the scalar field operator φ (e.g., scalar electromagnetic potential) can be separated into positive- and negative-frequency Fourier components,

$$\varphi(x) = \varphi^{(+)}(x) + \varphi^{(-)}(x) \tag{F-62}$$

where, in interaction representation,

$$\varphi^{(+)}(x) \equiv \frac{1}{(2\pi)^{3/2}} \int_{\substack{-\infty \\ k_0>0}}^{\infty} \frac{d^3 k}{(2\omega_k)^{1/2}} \, a(\mathbf{k}) \, e^{-i(\mathbf{k}\cdot\mathbf{x})} \tag{F-63}$$

and

$$\varphi^{(-)}(x) \equiv \frac{1}{(2\pi)^{3/2}} \int_{\substack{-\infty \\ k_0>0}}^{\infty} \frac{d^3 k}{(2\omega_k)^{1/2}} \, a^\dagger(\mathbf{k}) \, e^{i(\mathbf{k}\cdot\mathbf{x})} \tag{F-64}$$

whence

$$[\varphi^{(-)}(x)]^\dagger = \varphi^{(+)}(x) \tag{F-65}$$

Here, $\varphi^{(+)}$ represents a destruction operator and $\varphi^{(-)}$ a creation field operator; thus if $\varphi^{(-)}$ acts upon the vacuum state it creates a particle of momentum $\mathbf{k}$ within a distance from $\mathbf{x}$ equal to the reciprocal of the Compton wavelength. This can be generalized to apply in slightly modified form to the 4-vector potential operator

$$A_\mu(x) = A_\mu^{(+)}(x) + A_\mu^{(-)}(x) \tag{F-66}$$

with

$$A_\mu^{(+)}(x) \equiv \frac{1}{2\pi} \left(\frac{\hbar c}{2\pi}\right)^{1/2} \int_{\substack{-\infty \\ k_0>0}}^{\infty} \frac{d^3 k}{(2\omega_k)^{1/2}} \sum_{\nu=0}^{3} a_\nu(\mathbf{k}) \, e^{-ikx} \, \epsilon_\mu^\nu(\mathbf{k}) \tag{F-67}$$

$$A_\mu^{(-)}(x) \equiv \frac{1}{2\pi} \left(\frac{\hbar c}{2\pi}\right)^{1/2} \int_{\substack{-\infty \\ k_0>0}}^{\infty} \frac{d^3 k}{(2\omega_k)^{1/2}} \sum_{\nu=0}^{3} a_\nu^\dagger(\mathbf{k}) \, e^{ikx} \, \epsilon_\mu^\nu(\mathbf{k}) \tag{F-68}$$

in which it has been necessary to incorporate unit POLARIZATION VECTORS $\epsilon_\mu^\nu(\mathbf{k})$ to describe the polarization state of the field. These can be represented as

$$\left.\begin{aligned}
\epsilon_\mu^0(\mathbf{k}) &= (1,0,0,0) \\
\epsilon_\mu^1(\mathbf{k}) &= (0,0,1,0) \\
\epsilon_\mu^2(\mathbf{k}) &= (0,0,0,1) \\
\epsilon_\mu^3(\mathbf{k}) &= (0,1,0,0)
\end{aligned}\right\} \tag{F-69}$$

and build the operators A_μ, $A_\mu^\dagger$,

$$A_\mu(\mathbf{k}) = \sum_{\nu=0}^{3} \epsilon_\mu^\nu(\mathbf{k})\, a^{(\nu)}(\mathbf{k}) \tag{F-70}$$

$$A_\mu^\dagger(\mathbf{k}) = \sum_{\nu=0}^{3} \epsilon_\mu^\nu(\mathbf{k})\, a^{(\nu)\dagger}(\mathbf{k}) \tag{F-71}$$

In classical theory, the transversality condition

$$k^\mu A_\mu(\mathbf{k}) = 0 \tag{F-72}$$

restricts the physically possible states to the two transverse ones. The present formalism indicates this also, but gives two additional results, viz.

$$k^\mu \epsilon_\mu^\nu = \begin{cases} 0 & \text{for } \nu = 0, 3: \text{transverse polarization} \\ k = \omega_k/c & \text{for } \nu = 1: \text{longitudinal polarization} \\ i\omega_k/c & \text{for } \nu = 2: \text{time-like, scalar photons} \end{cases} \tag{F-73}$$

The ϵ_μ^ν are orthonormal,

$$\epsilon_\mu^\nu \epsilon_\mu^{\nu'} = \delta_{\nu\nu'} \tag{F-74}$$

$$\epsilon_\mu^\nu \epsilon_{\mu'}^\nu = \delta_{\mu\mu'} \tag{F-75}$$

Having accounted for the (electromagnetic) field operators A_μ we now require the complementary expressions for the Dirac field operators $\bar\psi$, ψ which refer to fermions of momentum $\mathbf{p}$. We are thereby confronted with a *spinor* field described by the state vectors (in momentum space) $u_r(p)$ and $v_r(p)$ (cf. Appendix D.6). Since this is associated with *charged* particles, we are, moreover, obliged to introduce a further pair of creation and annihilation operators $b^\dagger(\mathbf{p})$, $b(\mathbf{p})$ in addition to $a^\dagger(\mathbf{p})$ and $a(\mathbf{p})$, which perform the same function as the a's but upon particles of opposite charge (same mass). The modification can most easily be seen through the example of a *charged* scalar Klein-Gordon field, whose solutions now have to be generalized from (F-53) to

$$\varphi^\dagger(x) = \frac{1}{V^{1/2}} \sum_k \frac{1}{(2\omega_k)^{1/2}} [e^{-ikx} a^\dagger(\mathbf{k}) + e^{ikx} b(\mathbf{k})] \tag{F-76}$$

$$\varphi(x) = \frac{1}{V^{1/2}} \sum_k \frac{1}{(2\omega_k)^{1/2}} [e^{-ikx} b^\dagger(\mathbf{k}) + e^{ikx} a(\mathbf{k})] \tag{F-77}$$

The a's and the b's have the commutation property

$$[a^\dagger(\mathbf{k}), a(\mathbf{k}')] = [b^\dagger(\mathbf{k}), b(\mathbf{k}')] = \delta_{\mathbf{kk'}} \tag{F-78}$$

while all other combinations of $a^\dagger$, a, $b^\dagger$, b do not commute. The Hamiltonian is modified from (F-61) to

$$H = \sum_k \hbar\omega_k [a^\dagger(\mathbf{k})\, a(\mathbf{k}) + b^\dagger(\mathbf{k})\, b(\mathbf{k})] \tag{F-79}$$

which corresponds to adding the energies of the two kinds of particle. The net charge is the difference between the individual charges

$$Q = e \sum_k [a^\dagger(\mathbf{k})\, a(\mathbf{k}) - b^\dagger(\mathbf{k})\, b(\mathbf{k})] \tag{F-80}$$

The quantization of the free *spinor Dirac field* can be effected in an analogous manner to yield solutions of the type

$$\psi_\alpha(x) \sim \frac{1}{V^{1/2}} \sum_{p,r} [e^{-(ipx)/\hbar}\, u_{r\alpha}^{(-)}(-\mathbf{p})\, b_r^\dagger(\mathbf{p}) + e^{(ipx)/\hbar}\, u_{r\alpha}^{(+)}(\mathbf{p})\, a_r(\mathbf{p})] \tag{F-81}$$

where the $a_r(\mathbf{p})$ are destruction operators for particles with momentum $\mathbf{p}$ and polarization (spin) r, while the $b_r(\mathbf{p})$ are destruction operators for *antiparticles* with momentum $\mathbf{p}$ and polarization r, which satisfy the *anticommutation* relations

$$a_r^\dagger(\mathbf{p})\, a_{r'}(\mathbf{p}') + a_{r'}(\mathbf{p}')\, a_r^\dagger(\mathbf{p}) = b_r^\dagger(\mathbf{p})\, b_{r'}(\mathbf{p}') + b_{r'}(\mathbf{p}')\, b_r^\dagger(\mathbf{p}) = \delta_{rr'}\, \delta_{pp'} \tag{F-82}$$

whereas all other anticommutators vanish. The energy is thereby rendered positive definite. The products $a_r^+(\mathbf{p})\, a_r(\mathbf{p})$ and $b_r^\dagger(\mathbf{p})\, b_r(\mathbf{p})$ which betoken occupation numbers can be either zero or unity, since the fermions are subject to the Pauli exclusion principle. When the expression (F-81) is taken as an equality, the spinors u and v are normalized to

$$u_r^\dagger u_s = \delta_{rs} \qquad \text{and} \qquad v_r^\dagger v_s = -\delta_{rs} \tag{F-83}$$

In practice, it is more usual and convenient to require that

$$\bar{u}_r u_s = \delta_{rs} \qquad \text{and} \qquad \bar{v}_r v_s = -\delta_{rs} \tag{F-84}$$

i.e.,

$$\sum_{\alpha=1}^{4} \bar{u}_{r\alpha}(\mathbf{p})\, u_{s\alpha}(\mathbf{p}) = \delta_{rs} \tag{F-85}$$

This is achieved by including a normalization factor $[mc^2/E(p)]^{1/2}$ in the right side of (F-81), where $E(p)$ is the energy corresponding to momentum p, viz.

$$E = + (\mathbf{p}^2 c^2 + m^2 c^4)^{1/2} \tag{F-86}$$

and writing

$$\psi_\alpha(x) = \frac{1}{V^{1/2}} \sum_{p,r} \left(\frac{mc^2}{E(p)}\right)^{1/2} [e^{-(ipx)/\hbar}\, v_{r\alpha}(\mathbf{p})\, b_r^\dagger(\mathbf{p}) + e^{(ipx)/\hbar}\, u_{r\alpha}(\mathbf{p})\, a_r(\mathbf{p})] \tag{F-87}$$

$$\bar{\psi}_\alpha(x) = \frac{1}{V^{1/2}} \sum_{p,r} \left(\frac{mc^2}{E(p)}\right)^{1/2} [e^{-(ipx)/\hbar}\, \bar{u}_{r\alpha}(\mathbf{p})\, a_r^\dagger(\mathbf{p}) + e^{(ipx)/\hbar}\, \bar{v}_{r\alpha}(\mathbf{p})\, b_r(\mathbf{p})] \tag{F-88}$$

The decomposition is then

$$\psi_\alpha(x) = \psi_\alpha^{(+)}(x) + \psi_\alpha^{(-)}(x) \tag{F-89}$$

$$\bar{\psi}_\alpha(x) = \bar{\psi}_\alpha^{(+)}(x) + \bar{\psi}_\alpha^{(-)}(x) \tag{F-90}$$

wherein

$$\bar{\psi}^{(+)} = \overline{\psi^{(-)}} \qquad \text{and} \qquad \bar{\psi}^{(-)} = \overline{\psi^{(+)}} \tag{F-91}$$

Explicitly,

$$\psi_\alpha^{(+)} = \frac{1}{V^{1/2}} \sum_{p,r} \left(\frac{mc^2}{E(p)} \right)^{1/2} [e^{-(ipx)/\hbar} v_{r\alpha}(\mathbf{p}) \, b_r^\dagger(\mathbf{p})] \tag{F-92}$$

$$\psi_\alpha^{(-)} = \frac{1}{V^{1/2}} \sum_{p,r} \left(\frac{mc^2}{E(p)} \right)^{1/2} [e^{(ipx)/\hbar} u_{r\alpha}(\mathbf{p}) \, a_r(\mathbf{p})] \tag{F-93}$$

$$\bar{\psi}_\alpha^{(+)} = \frac{1}{V^{1/2}} \sum_{p,r} \left(\frac{mc^2}{E(p)} \right)^{1/2} [e^{-(ipx)/\hbar} \bar{u}_{r\alpha}(\mathbf{p}) \, a_r^\dagger(\mathbf{p})] \tag{F-94}$$

$$\bar{\psi}_\alpha^{(-)} = \frac{1}{V^{1/2}} \sum_{p,r} \left(\frac{mc^2}{E(p)} \right)^{1/2} [e^{(ipx)/\hbar} \bar{v}_{r\alpha}(\mathbf{p}) \, b_r(\mathbf{p})] \tag{F-95}$$

The only remaining unknown in (F-49) is the operator CONTRACTION $\langle \Psi_0 | T[\bar{\psi}(x_1) \psi(x_2)] | \Psi_0 \rangle$. Its value, and that of other vacuum expectation values of operator products, can be derived through somewhat lengthy and complicated second-quantization calculations presented e.g., in Schweber *et al.* [Schw 56], Mandl [Ma 59b], and Muirhead [Mui 65]. We confine ourselves to quoting the results:

$$\langle \Psi_0 | T[A_\mu(x_1) A_\nu(x_2)] | \Psi_0 \rangle = -\frac{i}{(2\pi)^4} \delta_{\mu\nu} \int_{-\infty}^{\infty} d^4 k \, \frac{e^{ik(x_1 - x_2)}}{\hbar^2 k^2 - i\varepsilon} \tag{F-96}$$

and, for a product of Dirac field operators *without* time ordering,

$$\langle \Psi_0 | \bar{\psi}_\alpha(x_1) \psi_\beta(x_2) | \Psi_0 \rangle = \frac{1}{(2\pi)^3} \int_{-\infty}^{\infty} d^3 q \, \frac{(-c\gamma^\nu q_\nu + mc^2)_{\alpha\beta}}{2E(q)} \times \exp\left[\frac{iq(x_2 - x_1)}{\hbar} \right] \tag{F-97}$$

while *with* time ordering one obtains

$$\langle \Psi_0 | \bar{\psi}_\alpha(x_1) \psi_\beta(x_2) | \Psi_0 \rangle = -\frac{i}{(2\pi)^4} \int_{-\infty}^{\infty} d^4 q \, \frac{(c\gamma^\nu q_\nu - mc^2)_{\alpha\beta}}{q^2 c^2 + m^2 c^4 - i\varepsilon} \times \exp\left[\frac{iq(x_2 - x_1)}{\hbar} \right] \tag{F-98}$$

The infinitesimal energy contribution ε is a device introduced to prevent the occurrence of divergent integrals; it is set to zero after the integrations have been performed. The symbol q has been used for momentum in (F-97) and

(F-98) since it refers to the electron propagator (cf. Fig. F-10) of momentum $q = p - \hbar k'$ or $q = p + \hbar k$ connecting x_1 and x_2 in our present application.

F.3.4. FORMATION OF THE S-MATRIX ELEMENT

We are now in a position to assemble S-matrix elements from the components listed in the previous section. After simplification these furnish the appropriate interaction cross section directly.

The illustrative example of Compton scattering necessitates the formation of an S-matrix element in *second*-order perturbation theory, according to (F-37), (F-38), and (F-39), which we have written out explicitly in (F-49). With the decomposition in (F-50), we may write the requisite element as

$$S_{\mathrm{fi}} \equiv \langle \mathrm{f}|S_2|\mathrm{i}\rangle = \langle p'\,k'|S_2^{(1)}|pk\rangle + \langle p'\,k'|S_2^{(2)}|pk\rangle \qquad \text{(F-99)}$$

whose two terms have been represented as Feynman graphs in Fig. F-10. We examine the second of these in which an electron initially absorbs a photon (in state $\mathbf{k}$, $\epsilon(\mathbf{k})$) and later emits a photon (in state $\mathbf{k}'$, $\epsilon(\mathbf{k}')$)—in the first diagram, photon emission precedes absorption—and use (F-62) to write

$$
\langle p'\,k'|S_2^{(2)}|pk\rangle = \langle p'\,k'|(e/\hbar c)^2 \int_{-\infty}^{\infty} d^4 x_1 \int_{-\infty}^{\infty} d^4 x_2
$$
$$
\times \{\bar{\psi}^{(+)}(x_2)\,A^{(+)}(x_2)\langle \Psi_0|T[\bar{\psi}(x_1)\,\psi(x_2)|\Psi_0\rangle
$$
$$
\times A^{(-)}(x_1)\,\psi^{(-)}(x_1)\}|pk\rangle \qquad \text{(F-100)}
$$

into which we have to substitute (F-94), (F-67), (F-98), (F-68), and (F-93) for the respective terms. Using Feynman notation to write

$$\gamma^\mu \epsilon_\mu \equiv \not{\epsilon} \qquad \text{and} \qquad \gamma^\nu q_\nu \equiv \not{q} \qquad \text{(F-101)}$$

and noting that the various creation and destruction operators which enter into Eq. (F-100) simply reduce to unity,

$$\langle p'\,k'|a_r^\dagger(\mathbf{p}')\,a_\nu(\mathbf{k}')\,a_\nu^\dagger(\mathbf{k})\,a_r(\mathbf{p})|pk\rangle = \langle \Psi_0|\Psi_0\rangle = 1 \qquad \text{(F-102)}$$

we obtain

$$
\langle p'\,k'|S_2^{(2)}|pk\rangle = \left(\frac{e}{\hbar c}\right)^2 \int_{-\infty}^{\infty} d^4 x_1 \int_{-\infty}^{\infty} d^4 x_2
$$
$$
\times \left[\left(\frac{mc^2}{VE(p')}\right)^{1/2} e^{-(ip'x_2)/\hbar}\, u_{r'}(\mathbf{p}')\right]\left[\left(\frac{\hbar c}{V}\right)^{1/2} \frac{e^{-ik'x_2}}{(2\omega_{k'})^{1/2}}\, \not{\epsilon}\,(\mathbf{k}')\right]
$$
$$
\times \left[-\frac{i}{(2\pi)^4} \int_{-\infty}^{\infty} d^4 q\, \frac{c\not{q} - mc^2}{q^2 c^2 + m^2 c^4 - i\varepsilon}\, \exp\left(\frac{iq(x_2 - x_1)}{\hbar}\right)\right]
$$
$$
\times \left[\left(\frac{\hbar c}{V}\right)^{1/2} \frac{e^{ikx_1}}{(2\omega_k)^{1/2}}\, \not{\epsilon}\,(\mathbf{k})\right]\left[\left(\frac{mc^2}{VE(p)}\right)^{1/2} e^{(ipx_1)/\hbar}\, u_r(\mathbf{p})\right] \text{(F-103)}
$$

The integrations over x_1 and x_2 yield 4-vector δ functions,

$$\int_{-\infty}^{\infty} d^4 x_1 \int_{-\infty}^{\infty} d^4 x_2 \left[\exp\left(-\frac{iq}{\hbar} + ik + \frac{ip}{\hbar}\right)x_1\right]\left[\exp\left(-\frac{ip'}{\hbar} - ik' + \frac{iq}{\hbar}\right)x_2\right]$$

$$= [(2\pi)^4 \delta(p + \hbar k - q)][(2\pi)^4 \delta(q - p' - \hbar k')] \tag{F-104}$$

which, upon the remaining integration over q, reduce to an overall δ function expressing the conservation of energy and momentum in the process,

$$\int d^4 q\, \delta(p + \hbar k - q)\, \delta(q - p' - \hbar k')\, F(q) = \delta(p + \hbar k - p' - \hbar k')\, F(p + \hbar k) \tag{F-105}$$

where $F(q)$ stands for a given function of q. Thus (F-103) simplifies, upon setting $i\varepsilon \to 0$ and $q = p + \hbar k$, to

$$\langle p'\, k'|S_2^{(2)}|pk\rangle = - i(2\pi)^4 \left(\frac{e}{V}\right)^2 \delta(p + \hbar k - p' - \hbar k')$$

$$\times \left[\left(\frac{mc^2}{E(p')}\right)^{1/2} u_{r'}(\mathbf{p}')\right]\left[\frac{\boldsymbol{\varepsilon}(\mathbf{k}')}{(2\omega_{k'})^{1/2}}\right]$$

$$\times \left[\frac{cq - mc^2}{c^2 q^2 + m^2 c^4}\right]\left[\frac{\boldsymbol{\varepsilon}(\mathbf{k})}{(2\omega_k)^{1/2}}\right]\left[\left(\frac{mc^2}{E(p)}\right)^{1/2} u_r(\mathbf{p})\right] \tag{F-106}$$

The one-to-one correspondence between the terms in (F-106) and the segments of the Feynman graph depicted in Fig. F-11 is rendered evident. Leaving the numerical factor aside for the moment, we see that "Feynman graphology" enables one to write down the combination of physical variables comprising the S-matrix element immediately without the need for laborious intermediate calculation.

The technique can also be used for other interactions; e.g., the pion-nucleon interaction as described by

$$\mathcal{H}_{\text{int}} = -\mathcal{L}_{\text{int}} = -g\bar{\psi}\gamma^0\,\psi\varphi \tag{F-107}$$

where g is a constant, $\bar{\psi}$ and ψ refer to the nucleon field, and φ to the (pseudo-scalar) pion field. One finds by analogous methods that

(a) meson creation and destruction gives a factor

$$(2\omega_k)^{-1/2}$$

(b) a meson propagator gives a factor

$$(\hbar^2 c^2 k^2 + m^2 c^4)^{-1}$$

(c) a meson-nucleon vertex gives a factor

$$[\gamma^0 \delta(p - p' - \hbar k)]$$

Hence by drawing topologically distinguishable graphs to represent a given interaction process up to a given order n (there are $n!$ such graphs), the corresponding S-matrix product of factors can be built from the listing in Table F-1 and the weighting coefficient deduced from rules drawn up, e.g., by Schweber *et al.* [Schw 56] and Muirhead [Mui 65] or, for particles of arbitrary spin, by Weinberg [Wei 64].

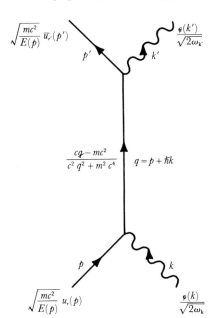

Fig. F-11. Feynman graph for Compton scattering in momentum space labeled according to matrix-element contribution.

Each component makes its own contribution to the resulting weighting coefficient, whose value is thus determined essentially by the number of internal and external lines, the number of vertices (equal to the order n) and of closed loops, together with the sign corresponding to the final electron state permutation. A 4-momentum δ function also enters through integration over momenta of internal and external lines. Closed loops correspond to an electron path which is inherently closed, and therefore involves no free electron functions, but only electron propagators. Normally, they arise only in higher-order graphs: a closed loop corresponds to a double sum over spinor indices and therefore constitutes a trace. Some graphical examples have been given in Fig. 4-36 of Section 4.6.3, and FURRY's THEOREM [Fu 37] that *diagrams containing a closed loop with an odd number of corners make no contribution to matrix*

Table F-1. FEYNMAN-GRAPH S-MATRIX FACTORS IN MOMENTUM SPACE†

Graph segment	Connotation	S-matrix factor
	Fermion (e⁻) creation	$\left(\dfrac{mc^2}{E(p)}\right)^{1/2}\bar{u}_r(p)$
	Fermion (e⁻) destruction	$\left(\dfrac{mc^2}{E(p)}\right)^{1/2}u_r(p)$
	Antifermion (e⁺) destruction	$\left(\dfrac{mc^2}{E(p)}\right)^{1/2}\bar{v}_r(p)$
	Antifermion (e⁺) creation	$\left(\dfrac{mc^2}{E(p)}\right)^{1/2}v_r(p)$
	Photon creation ⎱ Photon destruction ⎰	$\dfrac{\varepsilon_\mu^\nu(k)}{(2\omega_k)^{1/2}}$
	Meson creation	$\dfrac{1}{(2\omega_k)^{1/2}}$
	Meson destruction	$\dfrac{1}{(2\omega_k)^{1/2}}$
	Fermion propagator	$\dfrac{c\slashed{q}-mc^2}{c^2q^2+m^2c^4}$
	Photon propagator	$\dfrac{1}{\hbar^2k^2}\delta_{\mu\nu}$
	Fermion-photon interaction vertex	$\gamma^\mu\,\delta(p-p'-\hbar k)$

Table F-1. (*Continued*)

Graph segment	Connotation	S-matrix factor
	Nucleon-meson interaction vertex	$\gamma^0 \, \delta(p - p' - \hbar k)$

† From Muirhead [Mui 65].

elements is mentioned there, together with its generalized version. This simplifies the task of evaluating matrix elements very appreciably in certain instances.

When the rules are applied to Coulomb scattering in second-order perturbation, as depicted by Fig. F-10, one obtains precisely the numerical coefficient weighting the graph factors in Eq. (F-106). Thus with the Feynman technique this expression, and others of like character, could have been written down immediately without any preliminary lengthy calculation. This matrix element can then readily be transformed into one whose square gives the transition probability per unit volume and unit time. Fermi's "Golden Rule No. 2" can then be utilized to derive the differential cross section for the process. The various steps of the calculation have been presented by Schweber *et al.* [Schw 56], and Mandl [Ma 59b]. When performed for the scattering of initially unpolarized γ radiation, the evaluation leads directly to the KLEIN-NISHINA FORMULA [cf. Eq. (4-28)]

$$\sigma(\theta) = \frac{1}{2} r_e^2 \left(\frac{\omega_{k'}}{\omega_k} \right)^2 \left(\frac{\omega_{k'}}{\omega_k} + \frac{\omega_k}{\omega_{k'}} - \sin^2 \theta \right) \text{ cm}^2 \, \text{sr}^{-1} \text{ per electron} \quad \text{(F-108)}$$

where $r_e \equiv e^2/m_e c^2$ is the classical electron radius and θ is the γ-scattering angle. Moreover, in the low-energy limit one can make the substitution

$$\omega_{k'} \cong \omega_k \ll m_e c^2/\hbar \quad \text{(F-109)}$$

and immediately obtain the classical THOMSON FORMULA

$$\sigma(\theta) = \tfrac{1}{2} r_e^2 (1 + \cos^2 \theta) \quad \text{(F-110)}$$

On the other hand, the full Klein-Nishina formula (4-27) for the scattering of polarized γ radiation could equally well have been derived by the Feynman method. Its power, convenience, and elegance can be appreciated without any further commendation. It might, however, be appropriate to sound a note of warning against interpreting certain graphs too literally as physical processes, as in some instances this can lead to fallacious conclusions. Experience and discretion here, as elsewhere, can assist one to avoid pitfalls.

SOME MEASUREMENT TECHNIQUES
IN NUCLEAR PHYSICS

G.1. Introduction

In view of the extensive and rapidly advancing nature of nuclear instrumentation, we outline only a few of the many topics that would otherwise merit discussion. The reader who desires more detailed information is referred to the more specialized literature in this field, (e.g., [Pr 64, Dea 65, Kno 65, Nei 65, Si 65, Neu 66]). Our somewhat arbitrary selection accordingly does not include certain well-established detection techniques, such as those which involve ionization and Geiger counters, nor do we deal with the various acceleration methods which have received adequate coverage in other texts.

G.2. Beta Spectrometry

Measurements of the energy distribution of β rays emitted in the course of radioactive decay processes have played a most important role in the elucidation of the characteristics of β transitions from the shape and end-points of β spectra, not to mention the extremely valuable information concerning nuclear matrix elements they have provided in several cases. Moreover, β spectrometers also serve to measure the energies and intensities of conversion electrons and thereby provide the most powerful method to determine multipolarities of γ transitions, from which nuclear spins and parities of low-lying levels ($E^* \lesssim 3$ MeV) can be deduced.

Many noteworthy types of magnetic spectrometers have been designed, the guiding aim being to attain the best possible momentum resolution commensurate with high transmission and luminosity. Since high-resolution studies entail the use of very thin β-emitting sources, the figure of merit of any particular instrument also includes the *area* of the source which is associated with a given momentum resolution $\Delta p/p$. This consideration has, however, now in part become less important as very large specific activities have been made available through progress in accelerator and reactor design.

G.2.1. PRINCIPLE OF MAGNETIC SPECTROMETERS

The basic principle on which most β spectrometers operate is the deflection of the electrons in a magnetic field, which can be used to determine their momentum, and thence their kinetic energy (see Fig. G-1). The equation of

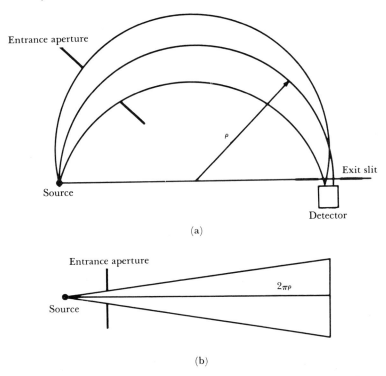

(a)

(b)

Fig. G-1. Principle of a flat β-spectrometer, showing the electron trajectories of mean curvature ρ from a radioactive source on the left to a thin-window Geiger counter on the right through a beam-defining slit. Between source and detector, lead shielding is interposed. (a) Top view, indicating the median and extreme trajectories on an exaggerated scale; (b) Side view, indicating the formation of a *line* focus.

motion (cf. Section 5.4.5) of a particle (i.e., electron) of charge e which moves with a velocity **v** perpendicular to a homogeneous magnetic field **B** along a circular path of radius ρ is

$$Bev = \frac{mv^2}{\rho} \tag{G-1}$$

whence

$$p = mv = e(B\rho) \qquad\qquad (G\text{-}2)$$

wherein $m = \gamma m_e$ is the *relativistic* mass. The radius of curvature ρ is therefore directly proportional to the momentum p, which can be expressed through the MAGNETIC RIGIDITY $B\rho$ in Gauss cm. The kinetic energy of the particles can then be related through the relativistic formula (see Appendix A.2)

$$E_{\text{kin}} = [p^2 c^2 + (m_e c^2)^2]^{1/2} - m_e c^2 \qquad\qquad (G\text{-}3)$$

i.e.,

$$E_{\text{kin}} = m_e c^2 \left\{ \left[\left(\frac{e}{m_e c} \right)^2 (B\rho)^2 + 1 \right]^{1/2} - 1 \right\} \qquad\qquad (G\text{-}4)$$

or, with E_{kin} in keV and $B\rho$ in Gauss cm,

$$E_{\text{kin}}^{[\text{keV}]} = 511.006 \left\{ [3441.8 \times 10^{-10} (B\rho)^2 + 1]^{1/2} - 1 \right\} \qquad\qquad (G\text{-}5)$$

Thus when $E_{\text{kin}} = m_e c^2$ the corresponding rigidity is approximately 3 kG cm. A plot of E_{kin} vs $B\rho$ has been given as Fig. A-6. When interpolating in tables of E_{kin} vs $B\rho$, such as have been assembled by Hagström *et al.* [Ha 65a], it is important to note that the specific energy change $\Delta E_{\text{kin}}/E_{\text{kin}}$ does not correspond numerically with that of momentum $\Delta p/p = \Delta(B\rho)/B\rho$, but is expressed by the relation

$$\frac{\Delta E_{\text{kin}}}{E_{\text{kin}}} = \left(1 + \frac{m_e c^2}{m_e c^2 + E_{\text{kin}}} \right) \frac{\Delta(B\rho)}{(B\rho)} \qquad\qquad (G\text{-}6)$$

The ratio of the specific momentum change to the specific energy change is plotted as a function of kinetic energy (or momentum) in Fig. G-2.

The basic arrangement of a flat 180° spectrometer depicted in Fig. G-1 allows a narrow set of rays from the source to come to a focus upon a photographic emulsion or an energy-defining slit placed in front of a detector, usually a Geiger counter equipped with a very thin window. The assembly has to be maintained under vacuum. In order to achieve high resolution, only a small fraction of the total radiation is utilized, determined by the size of the spectrometer entrance aperture. The TRANSMISSION T of a spectrometer is defined as the ratio of the number of particles reaching the detector to that emitted altogether by the source in the same momentum range. The LUMINOSITY of a particular instrument is defined as

$$L = T \cdot \sigma \qquad\qquad (G\text{-}7)$$

where σ represents the *effective* area of the source. Normally, the magnetic field B is varied in steps over the region of interest while the count rate of the detector is registered. It should be noted that this procedure does not furnish the actual momentum spectrum. For a given magnetic field strength B, the entrance slit permits a narrow band of orbital radii $\delta\rho$ to be accepted. These correspond to a momentum width δp given by

$$\delta p = \delta(B\rho) = B\delta\rho = p\,\frac{\delta\rho}{\rho} \qquad \text{(G-8)}$$

and therefore

$$\frac{\delta p}{p} = \frac{\delta\rho}{\rho} \qquad \text{(G-9)}$$

Since the constants $\delta\rho$ and ρ are predetermined by the setting and the geometrical position of the energy-defining slit, the relative momentum

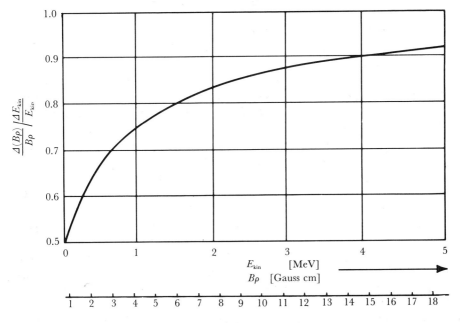

Fig. G-2. Ratio of specific momentum change to specific energy change for electrons as a function of their kinetic energy E_{kin} or corresponding magnetic rigidity $B\rho$ (from [Si 65]). (Used by permission of North-Holland Publishing Co.)

resolution is itself constant and the registered momentum interval δp is therefore proportional to p. For this reason, one first has to divide the count rate by p in order to arrive at the momentum spectrum $N(p)\,dp$. The energy spectrum $N(E)\,dE$ has its distinct shape in consequence of the fact that dE_{kin} and dp are related through the formula

$$dE_{\text{kin}} = \frac{c^2 p\,dp}{(E_{\text{kin}} + m_e c^2)} \qquad \text{(G-10)}$$

The various representations are contrasted in Fig. G-3. It should be noted that β^- *energy spectra* differ very markedly in shape from those for *positrons* in the-low energy region: the former, which are enhanced at small energies, do *not* become vanishingly small as the energy tends to zero. The effect of the

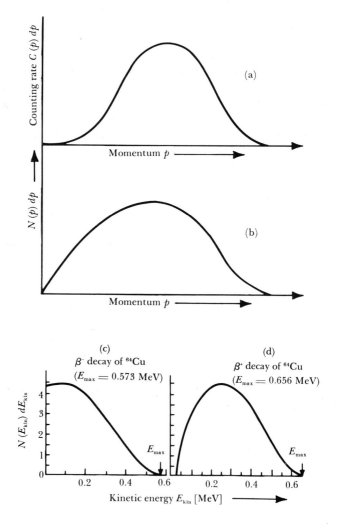

Fig. G-3. Various representations of β-ray spectra: (a) counting rate $C(p)\,dp$ *vs* momentum p; (b) momentum spectrum, $N(p)\,dp$ as a function of p; (c) kinetic energy spectrum for β^- decay (of ^{64}Cu to ^{64}Zn), displaying the finite ordinal intercept at $E_{\text{kin}} = 0$; (d) kinetic energy spectrum for β^+ decay (of ^{64}Cu to ^{64}Ni), which approaches the origin at zero energy.

nuclear charge on β^+ spectra, on the other hand, is to decrease the number of low-energy particles and the spectrum dwindles to zero at $E = 0$, i.e., approaches the origin.

When magnetic spectrometers are used for internal-conversion line measurements, the interpretation of the measured intensity calls for some attention. The natural energy spread of a conversion line, corresponding to the width of the appropriate atomic levels is very small and the "smearing-out" effect due to energy loss in the source material can usually be kept quite low. The dimensions of the source and the aberration of the spectrometer produce a finite focal line at the energy-defining slit. If the slit setting is such as to accept the entire set of monochromatic rays transmitted by the entrance aperture, the line intensity will be given by the *height* of the registered distribution. In the other extreme case of a very narrow slit, the *area* under the distribution will be proportional to the line intensity. If the slit is but slightly narrower than the focal line, the data have to be analyzed carefully. It should be noted that an optimal setting of the entrance aperture exists for each setting of the energy-defining slit.

Because of the difficulty of measuring with great precision the magnetic field along the path of the electrons, the calibration of β spectrometers is usually accomplished by reference to conversion lines of known energy. However, the line shape, which in some types of instrument is a rather involved fold, introduces complications; various schemes have been devised with the aim of introducing fiducial points such as the high-energy intercept of the line with the background. The difficulty of obtaining a single fiducial point on the β-ray line structure can, however, be circumvented in a simple manner for high-precision work. Since the momentum resolution is a constant, lines of any momentum plotted on a scale proportional to the logarithm of p (or of the current in the coils or the frequency of a nuclear magnetic resonance system) will have exactly the same profile when normalized to the same peak intensity. Thus, instead of utilizing one fiducial point, the entire profile may be matched by superposition to the profile of a calibration line.

G.2.2. Focusing Arrangements

The flat 180° semicircular spectrometer is capable of excellent momentum resolution, in favorable cases as good as 0.01 percent. Because focusing takes place in the median plane only, however, its transmission is very low. Double-focusing spectrometers, such as the $\pi\sqrt{2}$ Spectrometer whose principle is illustrated in Fig. G-4, overcome the disadvantage of having no space focusing while retaining some of the features and inherent qualities of one-dimensional arrangements. Trochoidal spectrometers, in which the rays spiral several times before being brought to a sharp focus, have recently

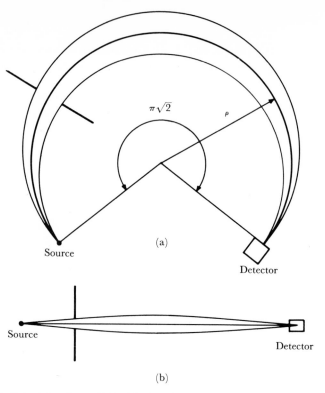

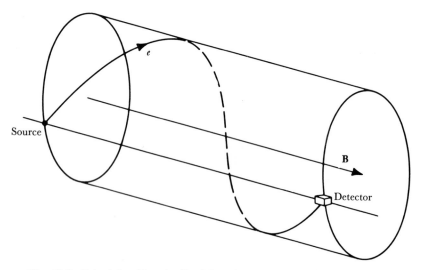

Fig. G-4. Principle of a $\pi\sqrt{2}$ double-focusing spectrometer: (a) Side view; (b) Top view (from [Si 65]). (Used by permission of North-Holland Publishing Co.)

Fig. G-5. Principle of longitudinal focusing in a helical β-spectrometer.

also come into use [Le 63; Ho 66a]; an interesting application of the principle has been utilized in the construction of a *β^+-β^- pair spectrometer* [Ba 67].

LENS SPECTROMETERS, in which the electrons describe helical paths in an *axial* magnetic field (Fig. G-5), usually provide much higher transmissions. With a homogeneous magnetic field, the trajectories roughly coincide at a single point independently of the emission angle (Fig. G-6). This represents

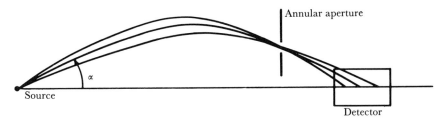

Fig. G-6. Coincidence of β-ray trajectories in a longitudinal lens spectrometer.

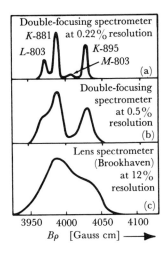

Fig. G-7. Momentum spectra of the conversion electrons from the transitions $^{206}\text{Bi} \rightarrow {}^{206}\text{Pb}$ at different resolutions, illustrating the emergence of fine detail at high resolution (from [Si 65]). (Used by permission of North-Holland Publishing Co.)

the most suitable position for incorporating an annular energy-defining slit. Momentum resolutions of 0.1 percent can be achieved at transmissions of a few percent. Lens spectrometers require additional baffles in order to distinguish between electrons and positrons.

The improvement in detail that ensues from an improvement in spectrometer resolution is illustrated by the example shown in Fig. G-7 for a portion of the conversion-electron energy spectrum obtained from the decay process $^{206}\text{Bi} \xrightarrow{EC} {}^{206}\text{Pb}$.

G.3.1. PRINCIPLE

The property whereby some solid or liquid materials emit quanta of visible light when traversed by fast charged particles has long been known and utilized in many non-nuclear applications. The mechanism of the scintillation process varies appreciably from one material to another. Thus, in an alkali-halide crystal such as sodium iodide or cesium iodide the incident charged radiation promotes an ionizing process in which electrons are raised out of the filled valence band into the conduction band. The electron and the hole in the valence band are free to move independently through the crystal. A different excitation process may also take place, in which the electron is raised into the forbidden zone. The electron-hole pair, also known as EXCITON, is a hydrogen-like structure which can wander through the lattice. Crystal imperfections (impurities, vacancies) give rise to energy levels in the forbidden band, which act as activator sites and traps at isolated points. These capture the excitons, also the free holes and free electrons. In the process of recombination taking place at such sites, a portion of the energy dissipated by the primary radiation in the crystal is converted to light. In order to increase the yield and to decrease the time required for the emission of light, small quantities of activators are added to the crystal, e.g., thallium to sodium iodide or cesium iodide, europium to lithium iodide. The light is emitted with a continuous spectrum, whose maximum lies around 4200 Å.

In organic-crystal scintillators like anthracene and trans-stilbene, the luminescence is produced by molecular excitation. In organic-liquid scintillators, the excitation energy given to the solvent (for example, xylene) is rapidly transferred to the solute (for example, p-terphenyl), which radiates part of it as fluorescent light.

The typical decay time within which 63 percent of the photons are emitted following the primary ionization is 250 nsec for NaI(Tl), or 30 nsec for anthracene. Plastic and liquid scintillators have decay times between 2 and 8 nsec. Yields of energy conversion extend from about 2 percent for the fast scintillators to 15 percent for the much slower NaI(Tl).

As shown in Fig. G-8, the scintillator is normally connected directly to a photomultiplier tube. The light emitted in the scintillation process is reflected by alumina or magnesium oxide on to the light-sensitive semitransparent SbCs PHOTOCATHODE, from which it ejects electrons through the photoelectric effect. Phototubes can give one electron at the cathode for about twenty incident light quanta, and then multiply this by a factor of 10^6 or more through secondary emission in a system of dynodes. The pulses delivered by the last dynode are then fed to an amplifier and analysed in an appropriate electronic system. For each particular species of particle producing the primary scintillation, the amplitude of the ensuing pulse is, to a good degree of approximation, directly proportional to the total energy dissipated in the scintillator. Conse-

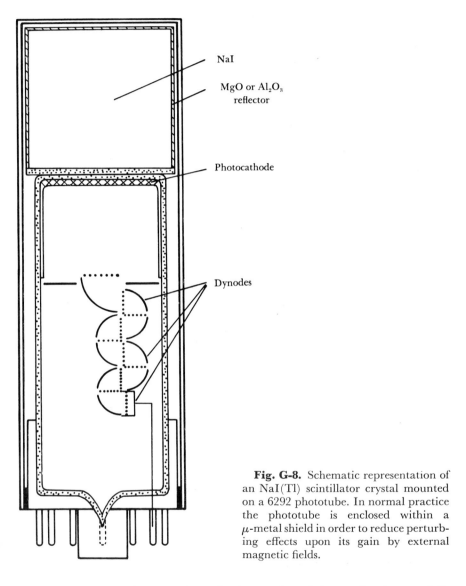

NaI

MgO or Al₂O₃ reflector

Photocathode

Dynodes

Fig. G-8. Schematic representation of an NaI(Tl) scintillator crystal mounted on a 6292 phototube. In normal practice the phototube is enclosed within a μ-metal shield in order to reduce perturbing effects upon its gain by external magnetic fields.

quently, a display of the number of pulses against their amplitude essentially corresponds to an energy spectrum of the primary radiation, and, in that sense, a scintillation counter acts as a type of spectrometer.

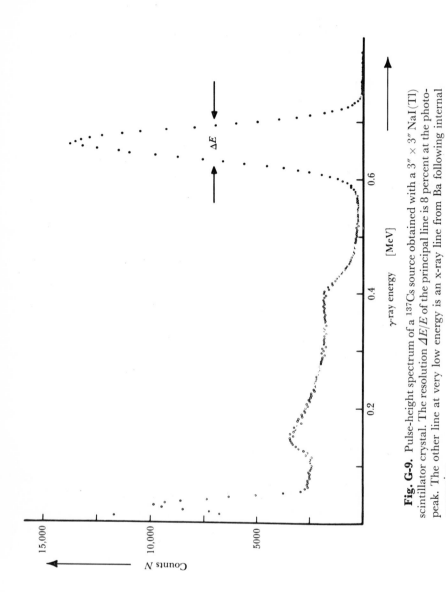

Fig. G-9. Pulse-height spectrum of a ^{137}Cs source obtained with a 3″ × 3″ NaI(Tl) scintillator crystal. The resolution $\Delta E/E$ of the principal line is 8 percent at the photopeak. The other line at very low energy is an x-ray line from Ba following internal conversion.

G.3.2. ENERGY RESOLUTION

The energy resolution which can be attained in scintillation spectrometry is limited primarily by the fact that the number of electrons ejected from the photocathode is rather small. As with all statistical processes, this number fluctuates about a mean value and is therefore not the perfectly sharp quantity requisite for ideal resolution. Thus, a 1-MeV photon absorbed in a scintillator with a light conversion efficiency of 10 percent produces 3.5×10^5 light quanta of 4200 Å wavelength and thereby causes 1800 ± 42 primary photoelectrons to be ejected at the photocathode for subsequent multiplication. Under these conditions the optimal attainable energy resolution would be approximately 5 percent. As the primary energy decreases, the resolution depreciates further, e.g., at 0.05 MeV the best obtainable resolution is only about 21 percent. A further broadening of the lines comes about through the energy spread associated with processes in the phototube itself and with the nonproportional response of the scintillator material.

Despite their limited energy resolution, scintillation counters represent an extremely useful and versatile tool. Their sensitive volume can be made very large, and their total counting efficiency can approach 100 percent. This feature, conjoined with their excellent time resolution, renders them particularly valuable for coincidence experiments. The fact that they exhibit different decay rates for different particle species makes it possible to discriminate between various particle species through pulse-shape analysis.

In Fig. G-9 we show the pulse-height spectrum obtained with a $3'' \times 3''$ NaI(Tl) crystal when irradiated with monoenergetic γ rays of energy 0.6616 MeV. The PHOTOPEAK incorporates all the events leading to the complete absorption of the incident γ energy, namely, (i) photoelectric absorption, and (ii) Compton scattering followed by photoelectric absorption of the scattered photon. The total line width at half maximum ($\Delta E =$ FWHM) gives an energy resolution $\Delta E/E \cong 8$ percent.

The broad COMPTON DISTRIBUTION extending from zero energy to the Compton edge at 0.476 MeV comprises all events in which Compton scattering of the primary photon occurs, but which is *not* followed by absorption of the scattered photon within the active volume of the detector. The Compton edge corresponds to back-scattering through 180°.

The broad peak at 0.185 MeV is due to radiation which has been back-scattered in the backing material on which the radioactive source had been deposited. It will be observed that its distance from the origin is the same as the interval between the Compton edge and the center of the photopeak.

Since the photopeak also contains pulses that involved Compton scattering, it would be wrong to deduce the photopeak efficiency simply from the photo-electric absorption coefficient of NaI depicted in Fig. G-10. In principle,

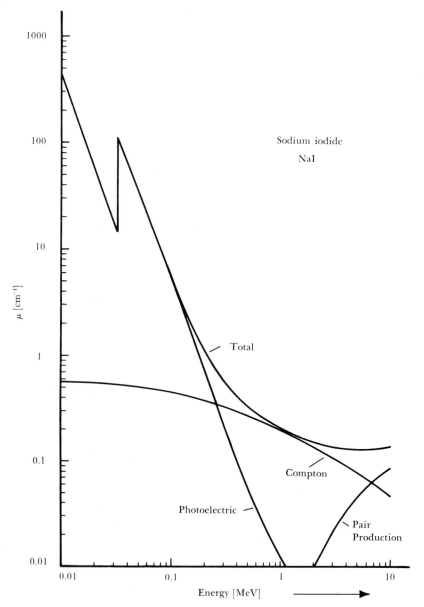

Fig. G-10. Linear attenuation coefficients for γ rays in NaI, showing the characteristic K edge. The corresponding mass attenuation coefficients may be obtained by dividing by the density $\rho = 3.67$ g cm^{-3} (i.e., by a shift of the vertical scale in the log-log representation). The data have been taken from tables of atomic cross sections compiled by White [Whi 52].

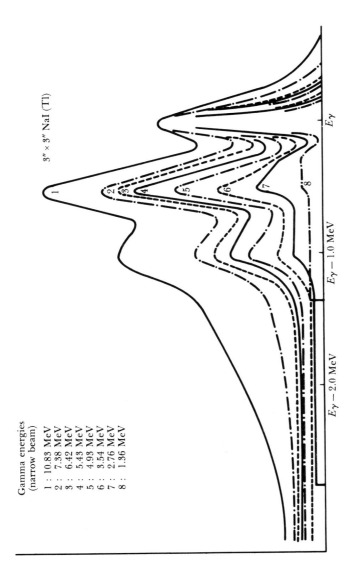

Gamma energies
(narrow beam)

1 : 10.83 MeV
2 : 7.38 MeV
3 : 6.42 MeV
4 : 5.43 MeV
5 : 4.93 MeV
6 : 3.54 MeV
7 : 2.76 MeV
8 : 1.36 MeV

$3'' \times 3''$ NaI (Tl)

$E\gamma$

$E\gamma - 1.0$ MeV

$E\gamma - 2.0$ MeV

Fig. G-11. The shapes of the photopeak and of the first and second escape lines for various γ-ray energies ranging from 1.36 to 10.83 MeV for a $3'' \times 3''$ NaI(Tl) scintillator crystal (from [Ja 62]). (Used by permission of North-Holland Publishing Co.)

one could evaluate this via a Monte-Carlo–type calculation, but in practice a direct calibration using a source of known intensity is much easier to perform. The calibration is, however, valid exclusively for one specific geometrical arrangement of the source, the scintillator and collimator.

An additional absorption mechanism operates when the incident γ energy exceeds $2m_e c^2 \simeq 1$ MeV, namely, PAIR PRODUCTION. The negative electron so formed is rapidly stopped in the crystal, and its kinetic energy is partially

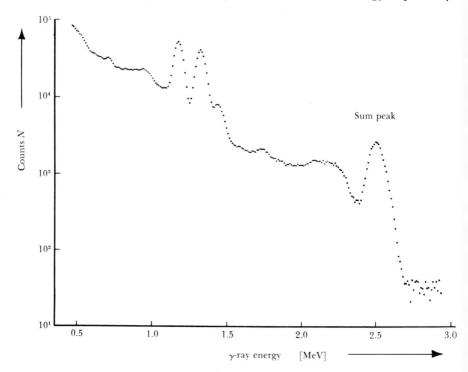

Fig. G-12. Upper portion of the ^{60}Co scintillation spectrum, showing the sum peak at 2.5 MeV.

converted to light. The positron, after it has been slowed down sufficiently, suffers annihilation. Consequently, the annihilation quanta, each of energy 0.511 MeV, are released within the crystal, which may absorb both, one, or none of them. If the scintillator absorbs both, a pulse is produced whose amplitude is equal to that of photoelectric absorption, whereas if it absorbs only one, there is an energy escape of 0.511 MeV out of the crystal; if it absorbs none, the energy escape is double this value, viz. 1.022 MeV. Hence two ESCAPE LINES are superposed upon the normal spectrum. Here again, the intensity ratios of the photopeak to the two escape peaks depend very much on the size and geometry of the scintillator. Jarczyk *et al.* [Ja 62] have investigated the photopeak efficiency for γ rays of energy ranging from 0.661 MeV to 10.83 MeV for NaI(Tl) crystals of various sizes. Fig. G-11 shows the spectral shapes obtained with a narrow beam impinging on a $3'' \times 3''$ crystal.

G.3.3. PILE-UP

When high incident intensities give rise to high counting rates, some of the pulses are added together in the scintillator or in the electronic system, with the result that spurious PILE-UP PULSES are discerned at amplitudes higher than that of the highest energy actually featured in the spectrum. Pile-up also causes a loss in resolution. Such undesirable effects can be eliminated by making provision for a reduction in the counting rate or, to some extent, by the incorporation of additional electronic safeguards. It should be noted that at high counting rates, the DEAD TIME of the system has to be taken into account.

A different type of pile-up effect can sometimes be employed to advantage. Namely, when sources which emit two or more γ rays in cascade are arranged either very close to, or actually within the counter, the scintillation due to the various cascade lines will in part add up and give rise to sum-peaks. This simple method on occasion serves to show that a nuclide emits γ rays in prompt cascade. The effect is illustrated for ^{60}Co decay in Fig. G-12. The intensity of a sum peak decreases with the fourth power of the distance between the source and the detector.

G.4. Semiconductor Detectors

As observed in Section G.3.2, the somewhat poor resolution of scintillation counters is to a large extent attributable to the statistical fluctuation in the rather small number of electrons ejected from the photocathode of the photo-tube. In semiconductor detectors, the conversion of the incident radiation's energy into an electrical pulse is achieved in a more direct manner, which results in greatly improved efficiency. In this conversion process, the primary energy is distributed among *many* carriers, and as a consequence, the relative statistical fluctuation is greatly reduced.

The operation of semiconductors, which are materials such as silicon or germanium crystals that behave as insulators at low temperatures and have a limited conduction at room temperature, depends upon electrons and holes to act as carriers of electricity. These can occupy only energy states within band regions; normally, the valence and deeper-lying bands are completely filled. The excitation produced by passage of a charged particle through a crystal raises the electrons into the conduction band or higher-lying unoccupied bands. In such *intrinsic semiconductors* as silicon or germanium, which have quite loosely bound valence electrons, comparatively little energy is needed to induce electron jumps over the rather narrow "energy gap" or "forbidden zone" (of width 1.1 eV for silicon and 0.7 eV for germanium). At room temperature, the vibration energy of the crystal gives rise to "spontaneous" jumps.

For each electron raised to a higher band, a hole is created in the valence (or deeper) band. An electron belonging to a nearby atom can thereupon fill the hole, but in so doing it leaves another hole at the point whence it came. In this way a hole can move in the lattice, behaving like a positive particle with a certain effective mass. The energy $\mathscr{E}$ needed to form one electron-hole pair

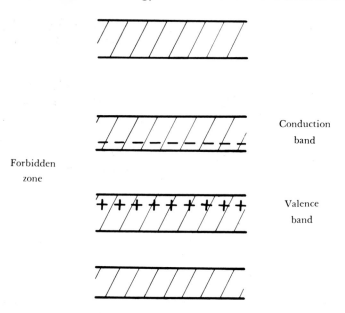

Fig. G-13. Schematic energy-level diagram for a semiconductor.

in passage of a charged particle proves effectively to be about three times larger than the energy gap itself: thus, measurements have furnished the values

$$\mathscr{E} = \begin{cases} 3.23 \text{ eV for silicon} \\ 2.84 \text{ eV for germanium} \end{cases}$$

For any given semiconductor, this energy does not appear to depend greatly upon the species of particle which initiates the ionization process. It is but one-tenth of the energy needed to produce an ion pair in a gas such as air. The fact that $\mathscr{E}$ is larger than the energy gap proves that electron-hole production does not take place only between the valence and conduction bands but that other bands are also involved in the primary process. After a very short time ($\approx 10^{-12}$ sec), however, the initial excited configuration goes over into a more stable one, depicted in Fig. G-13, in which the electrons are at the bottom of the conduction band and the holes at the top of the valence band. A voltage applied to the crystal causes electrons and holes to drift with their characteristic mobilities (in silicon at room temperature, approximately 500 cm^2 sec^{-1} V^{-1} for holes and 1500 cm^2 sec^{-1} V^{-1} for electrons) toward the electrodes. The charge collection time is usually in the range 10 to 50 nsec for a thin slab.

The principle of operation as described so far would impose extremely stringent requirements upon the material of the semiconductor, particularly

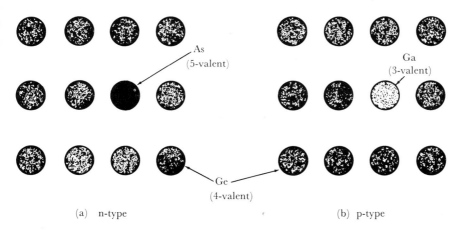

As
(5-valent)

Ga
(3-valent)

Ge
(4-valent)

(a) n-type   (b) p-type

Fig. G-14. Semiconductor doping to derive (a) an n-type; (b) a p-type material by introduction of impurity atoms having similar configurational properties but different valency from the host atoms.

as regards its resistivity. A more effective basis for the attainment of appropriate conditions of small carrier concentration is provided by the use of a P-N JUNCTION.

In *doped semiconductors*, a few of the lattice atoms are replaced by foreign atoms having a different valency. An intruder which has a higher number of valence electrons than the atom it has replaced acts as an electron DONOR and confers a net negative charge to the lattice cell, forming an *n-type* semiconductor. Conversely, an impurity atom having *fewer* valence electrons serves as an ACCEPTOR that effectively divests the cell of negative charge and therefore yields a *p-type* semiconductor. For example, by doping 4-valent germanium with 5-valent arsenic an n-type semiconductor can be made [Fig. G-14(a)], whereas if it is instead doped with 3-valent gallium, a p-type semiconductor results [Fig. G-14(b)].

At the boundary between a p- and an n-type semiconductor, namely, at the p-n junction, the energy structure has the form depicted in Fig. G-15. The electrons of the n-type material tend to move into the p-type material, because

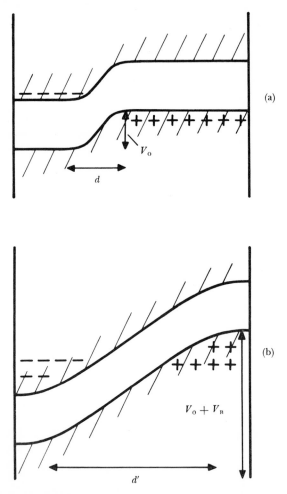

Fig. G-15. Energy diagram of a p-n semiconductor junction (a) in absence; (b) in presence of an applied bias potential V_B.

there are fewer electrons there. In a similar way, the positive carriers of the p material can diffuse across the junction into the n-type material. Under equilibrium conditions, the net diffusion current must be zero. This is brought about by the electric fields, which tend to draw the positive carriers back toward the p-type material and the electrons toward the n-type material. In the vicinity of the boundary a depletion zone *d* of very high resistivity is formed. It can be expanded by an external bias voltage V_{bias}, which also provides the charge collecting field. This space-charge region constitutes the active volume of a semiconductor radiation detector. Any ionizing particle giving up energy within the depletion layer will produce a current pulse in a circuit of the type shown in Fig. G-16.

In practice, silicon junction detectors are prepared by diffusing phosphorus or by drifting lithium (donors) into p-type silicon, thus "compensating" the material for a depth of a few mm. Increasing use is being made of SURFACE

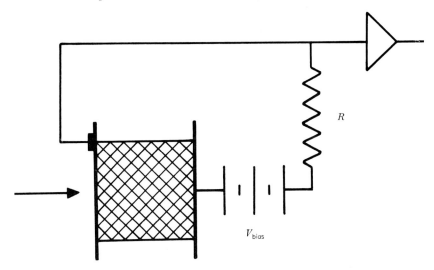

Fig. G-16. Semiconductor detector circuit in its simplest form. The amplifier is charge-sensitive.

BARRIER DETECTORS. These are prepared by etching the surface of n-type silicon, but the underlying principle is still not well understood. The effective thickness which it is possible to attain for such detectors is about 0.2 cm for a bias voltage of around 1000 V. A typical mounting arrangement for a silicon detector is shown in Fig. G-17.

Whereas silicon detectors have proved very valuable for heavy-particle spectrometry and for β spectrometry, in γ spectrometry wide use is made of lithium-drifted germanium counters, which provide a much larger photo-electric absorption. By drifting lithium from five sides of a crystal to a depth of 6 to 8 mm, sensitive volumes up to 100 cm^3 are currently being attained. Because of the narrowness of the energy gap in germanium (0.7 eV), it is necessary to keep Ge(Li) detectors at liquid nitrogen temperature in order to avoid thermal excitation.

The energy resolution of semiconductor detectors is limited by two main factors which can be of the same magnitude, namely, (i) the statistical fluctuation in the primary process of electron-hole production, and (ii) the fluctuation of the leakage current through the detector and the noise of the preamplifier loaded with the intrinsic capacity of the detector. The actual

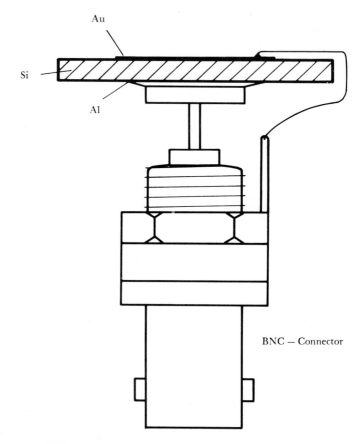

Fig. G-17. Typical mounting of a silicon surface-barrier detector.

line width (FWHM) is given to a good approximation by the formula:

$$\varDelta E = 2.3(E\mathscr{E} + \overline{\text{Noise}^2})^{1/2} \qquad \qquad \text{(G-11)}$$

where E is the particle or photon energy in eV and $\mathscr{E}$ the energy needed for the production of one electron-hole pair. The noise produced in the detector-amplifier combination varies between about 3 and 15 keV.

Typically, silicon counters yield line widths around $\varDelta E \cong 10$ keV for electrons and $\Delta E \simeq 25$ keV for heavy ions such as ^{12}C. By way of example, Fig. G-18 shows the K, L, and M conversion lines of the 1.0634-MeV γ ray emitted by a ^{207}Bi source.

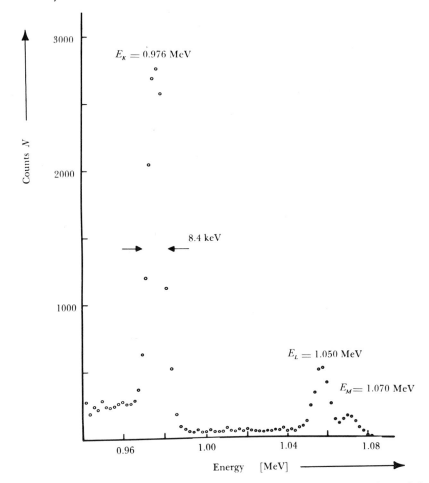

Fig. G-18. Conversion spectrum showing K-, L-, and M-conversion lines of the 1.0634-MeV γ radiation emitted from a ^{207}Bi source. The energies of the conversion lines are shown, together with the line-width $\Gamma = 8.4$ keV of the K line, which corresponds to a resolution of 0.86 percent (from [An 66]). (Used by permission of North-Holland Publishing Co.)

In Fig. G-19 we show the spectrum for the 2.753 and 1.368-MeV γ lines from the decay of ^{24}Na, obtained using a Ge(Li) detector having an active volume of 18 cm^3. The photopeak represents mainly photoelectric absorption. The FWHM of the lines is $\Delta E \simeq 6$ keV.

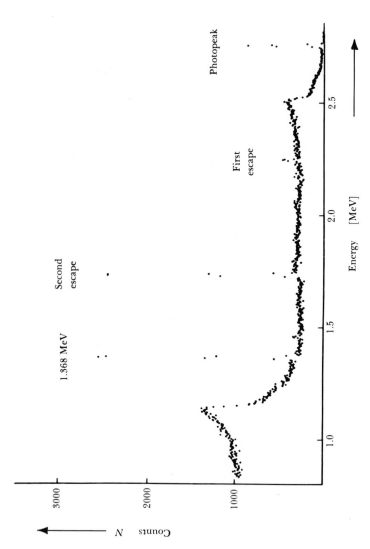

Fig. G-19. Pulse-height spectrum of the 2.753- and 1.368-MeV γ lines of ^{24}Na taken with a lithium-drifted Ge detecior.

G.5. Energy Scale in Low-Energy Nuclear Spectroscopy

Nuclear instruments such as scintillation or solid-state counters, as well as various types of β spectrometers, cannot of themselves give *absolute* energy values: it is necessary to calibrate them first with radiation of precisely known energy. The γ-energy scale has to a large extent been based on the work of DuMond and collaborators, who used the method of Bragg reflection in a bent crystal to derive high-precision values of the energies of numerous γ lines ranging up to 1.3 MeV. The diffraction method fundamentally relates the γ energies to the x-ray wavelength scale, which is itself ultimately based on a crystal density measurement.

An illustration of the basic geometry of the curved crystal spectrometer is presented in Fig. G-20. The crystal used to diffract the γ rays is shown

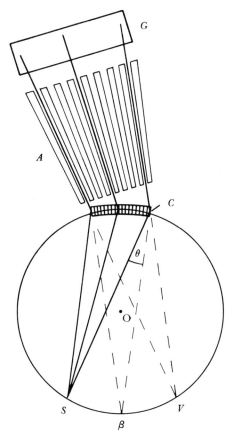

Fig. G-20. Basic geometry of the DuMond-type curved crystal spectrometer used in γ-ray spectrometry.

schematically at C. It consists of a 2-mm thick slab of quartz bent elastically in such a way that its [310] atomic planes, which are normal to the face of the crystal, converge toward the point β, some 2 m distant. The reason for bending the crystal in this manner is to enable all γ rays from a line-shaped source of γ rays at S to make the same angle θ with the crystal planes, thus enabling the crystal as a whole to serve as a grating. In this way, the useful solid angle is greatly improved over that of a flat crystal spectrometer. Simple geometrical considerations indicate that the source S has to be placed on a focal circle passing through C and having a radius equal to half the radius of curvature of the crystal. Reflection of γ rays from the lattice planes takes place when θ equals the Bragg angle for the particular radiation under study, viz.

$$\theta = \arcsin\left(\frac{n\lambda}{2d}\right) \qquad \text{(G-12)}$$

Since the angle of emergence of the radiation after Bragg reflection as it leaves the crystal is equal to the angle of incidence, the beam diverges as though it

Table G-1. STANDARD γ-TRANSITION ENERGY DATA AS MEASURED WITH DuMOND'S CURVED CRYSTAL SPECTOMETER

Parent nucleus	$T_{1/2}$	E_γ [MeV]	Reference
[131]I	8.08 d	0.08016; 0.28431; 0.36447	[Ho 53]
[177]Lu	6.75 d	0.11296; 0.20836	[Ma 55]
[198]Au	2.7 d	0.41177	[Mu 52]
[137]Cs	26.6 y	0.6616	[Mu 52]
[60]Co	5.24 y	1.1728; 1.3325	[Mu 52]

originated from a virtual focus V. This divergent beam is separated from the heterogeneous radiation passing directly through the crystal by a diverging collimator A made of tapered lead sheets. It is then registered in a NaI scintillation counter situated at G.

In an actual measurement the source S is moved in steps of 1/50 mm along the focal circle by means of a precision screw, and the number of counts in the detector is determined for each source position.

The overall efficiency of curved crystal spectrometers is low (10^{-8} at 0.5 MeV), and hence only strong sources (≈ 1 Ci) can be studied.

Because of the manifest desirability of referring measured γ energies to standard values from curved-crystal data, we list some useful calibration lines in Table G-1.

G.6. Coincidence Techniques

The decay sequence in many nuclei proceeds in a series of transitions which involve particle or γ emission in fairly rapid succession (e.g., within about 10^{-7} to 10^{-20} sec or less). One can utilize this strong time correlation as a means of selecting relative events or of deducing the parentage succession in a chain of decay steps. Essentially, the method consists of connecting two or more suitable radiation detectors to a common electronic "AND" gate which opens only if two (or more) channels have registered a count of the proper kind within a preset coincidence time 2τ. In some cases, one of the channels is gated by non-nuclear event, such as the pulsing system of an electrostatic accelerator. Coincidence equipment also proves extremely useful in time-of-flight studies or for lifetime measurements. It also serves to assist in reducing undesirable background or to effect the triggering of such devices as cloud chambers, etc.

As a simple example of the application of coincidence techniques, we consider the decay of ^{60}Co to ^{60}Ni, as observed with two scintillation counters 1 and 2 (Fig. G-21). Assuming the detectors to be insensitive to the β radiation, we seek to study the two γ-cascade transitions $4^+ \xrightarrow{\gamma_1} 2^+ \xrightarrow{\gamma_2} 0^+$ in ^{60}Ni. To this end, we set channel 1 to the photopeak of γ_1 (at 1.173 MeV) and channel 2 to the photopeak of γ_2 (at 1.332 MeV). If the rate of disintegration of the ^{60}Co sources is n_0 decays per sec, the actual counting rate of channel 1 is

$$n_1 = n_0\,\epsilon_1 \qquad\qquad \text{(G-13)}$$

where ϵ_1 represents the photopeak efficiency multiplied by the solid angle subtended by the detector 1. Similarly, the counting rate in channel 2 is

$$n_2 = n_0\,\epsilon_2 \qquad\qquad \text{(G-14)}$$

The lifetime of the intermediate 2^+ level in ^{60}Ni is sufficiently short (1.2×10^{-12} sec) to be essentially zero and therefore the coincidence rate is simply

$$n_c = n_0\,\epsilon_1\,\epsilon_2 \qquad\qquad \text{(G-15)}$$

In this formula, any anisotropy in the angular correlation between γ_1 and γ_2 has tacitly been neglected. It is evident that by measuring n_1, n_2, and n_c we can deduce the unknowns ϵ_1, ϵ_2, and n_0—hence the remark in Section 10.1.2 that by using two NaI(Tl) counters the γ-decay rate can be determined directly.

As a consequence of the finite resolving time 2τ, coincidences are automatically included which should in fact have been excluded as an unsought-for background contribution. It is possible to determine the number of such counts in order to subtract them and thereby arrive at the true coincidence rate. The following procedure can be used: Noting that whenever a pulse

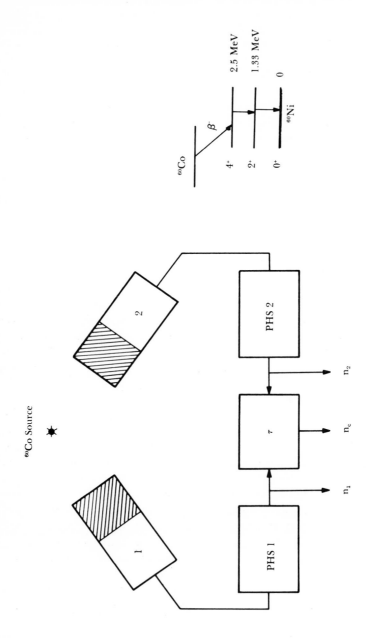

Fig. G-21. Schematic representation of a simple coincidence arrangement, as used, e.g., for measuring γ-γ pulse-height spectra singly (n_1, n_2) and in coincidence (n_c).

appears in channel 1 (and on an average, this will occur $n_0 \, \epsilon_1$ times/sec) the coincidence unit is activated for a time interval τ (during which $n_0 \, \epsilon_2 \tau$ pulses appear in channel 2, on an average, and each of these gives rise to a coincident pulse), we conclude that the rate of random coincidences is given by the expression

$$n_r^{(1-2)} = n_0^2 \, \epsilon_1 \, \epsilon_2 \, \tau \qquad \text{(G-16)}$$

However, similar reasoning can be applied to the situation in which a pulse first appears in channel 2 and gives rise to a spurious coincidence rate by activating the system for the same time τ while channel 1 counts are registered as coincidences,

$$n_r^{(2-1)} = n_0^2 \, \epsilon_1 \, \epsilon_2 \, \tau \qquad \text{(G-17)}$$

Hence the net overall rate of random coincidence is the commensurate expression with the "activation time" doubled, i.e.,

$$n_r = n_0^2 \, \epsilon_1 \, \epsilon_2 \, (2\tau) \qquad \text{(G-18)}$$

A simple way to measure the value of n_r is to introduce a time delay longer than 2τ into one of the channels, and thereby completely suppress the true coincidences.

It is important to note that whereas the true rate n_c is subject to a linear increase with n_0, the *random* rate n_r increases as n_0^2. Therefore in order to keep the number of random coincidences as low as possible one should aim to use a weak source combined with an arrangement which provides large solid angles, high counter efficiency and, of course, a small resolution time 2τ. The quadratic increase in the random coincidence rate sometimes gives rise to serious difficulties when coincidence measurements are taken with targets bombarded from a pulsed source, e.g., a cyclotron or some other pulsed accelerator which delivers brief bursts of particles. The pulse intensity is in such cases concentrated over only some small fraction of the time, which can lead to a very marked reduction in the actual efficiency of the machine. Expressing this reduction factor, to which the name DUTY CYCLE has been given, as the percentage ratio of the pulsed beam duration to the overall time, one finds that cyclotrons typically have a duty cycle of only 1 percent.

The extent to which the *resolving time* 2τ can be reduced is determined essentially by the "time jitter" occasioned by unavoidable system noise and the "walk" which occurs when the discriminators are handling pulses of different amplitude. Too narrow a setting of the coincidence width on the coincidence unit would then cause a loss of a fraction of the pulses. Resolving times down to $2\tau = 20$ nsec can easily be achieved; with special techniques using fast-slow electronics, plastic scintillators, and fast phototubes, resolving times can be reduced to as little as 0.1 nsec.

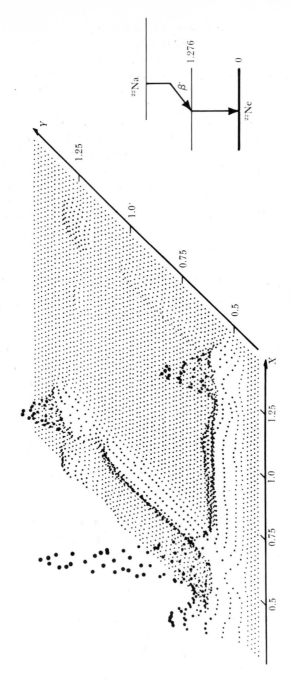

Fig. G-22. An *X-Y-Z* representation of the annihilation γ-quanta coincidences observed in the β^+ decay of ^{22}Na. The pulse-height spectra were obtained with two $3'' \times 3''$ NaI(Tl) scintillation counters; the resolving time was $2\tau = 500$ nsec. Random coincidences of the 1.277-MeV de-excitation γ line are discernible as symmetrical subsidiary peaks.

In some investigations it is of advantage to arrange for a three-dimensional display of two coincident spectra, in which the X coordinate corresponds to a particular energy interval in one of the spectra, the Y coordinate to a given energy interval in the complementary spectrum, and the Z coordinate represents the number of coincidences registered between both energy intervals.

An X-Y display obtained with an ^{22}Na β^+ emitter is depicted in Fig. G-22. This was derived using two $3'' \times 3''$ NaI(Tl) counters, whose coincident pulses were analyzed in a 64×64 channel array. The peak at $(0.511 \text{ MeV})_X$ $(0.511 \text{ MeV})_Y$ represents coincidences between the annihilation γ quanta from β^+ annihilation with atomic electrons. The peaks at $(0.511 \text{ MeV})_X$ $(1.277 \text{ MeV})_Y$ and $(1.277 \text{ MeV})_X(0.511 \text{ MeV})_Y$ correspond to coincidences between one of the annihilation quanta and the photopeak of the 1.277-MeV de-excitation γ ray from the (2^+) first level of ^{22}Ne populated by the β^+-decay process. The small hillock at $(1.277 \text{MeV})_X(1.277 \text{ MeV})_Y$ is produced by random coincidences between the photopeaks in the spectra. By using counter settings other than in the same straight line, the annihilation peak has been kept low. It will be observed that the display is symmetrical with respect to the diagonal $X = Y$, along which only those peaks due to annihilation quanta or random coincidences are to be found.

EXERCISES

G-1   A decay process in which two γ transitions occur in cascade is to be studied experimentally, using two detectors to register the γ quanta in coincidence, with a time resolution $\tau = 1$ μsec. What is the largest activity that the γ source may have in order that the random coincidence rate does not rise above 10 percent of the true coincidence rate, assuming that γ-ray emission takes place isotropically?

G-2   The dead time of a counter system is to be determined by taking measurements on two radioactive sources individually and collectively. If the pulse counts over a time interval t are, respectively, N_1, N_2, and N_{12}, what is the value of the dead time τ?

G-3   In an experiment the following counting rates were registered:
(effect under investigation) + (background) $= 100 \text{ sec}^{-1}$,
background alone $= 25 \text{ sec}^{-1}$.

The total measurement time was minimized.

Calculate the measurement time in the measurement of the effect alone if the mean statistical error fell below 3 percent.

G-4 Measurements in an experiment aimed at the investigation of a certain effect involve registering the count rate over 1 hour for (a) effect + background events, $N_1 = 8000$; (b) background events alone, $N_2 = 2000$.

What is the magnitude of the mean statistical error in the determination of the effect? Can the total available measurement period be reduced through a reallocation of relative counting duration without increasing the expected error limits? If so, by how much?

G-5 A 400-channel pulse-height analyser has a dead-time $\tau = (17 + 0.5 K)$ μsec when it registers counts in channel K. How large may the pulse frequencies become if in channel 100 the dead-time correction is not to exceed 10 percent? Repeat the calculation for channel 400.

G-6 In order to determine the half-life of ^{214}Po which forms a link in the decay chain

$$^{214}_{83}\text{Bi} \xrightarrow{\ \beta^-\ } {}^{214}_{84}\text{Po} \xrightarrow{\ \alpha\ } {}^{210}_{82}\text{Pb}$$

an arrangement to measure coincidences between the emitted α and β particles is set up, with the following characteristics:

The effective pulse duration in the α counter is $\tau_\alpha = 6$ μsec, whereas that in the β counter can be varied within extensive limits. For the counting rate of α-β coincidences we obtain $n = 48.6$ min^{-1} when a setting $\tau_\beta = 132$ μsec is used, whereas above $\tau_\beta = 1.2$ msec the coincidence counting rate essentially reaches its saturation value $n_{\max} = 109.8$ min^{-1}.

What value does this indicate for the half-life $T_{1/2}$ of ^{214}Po? Compare this with the currently accepted value $T_{1/2} = 164$ μsec and suggest a possible reason for any discrepancy.

G-7 In a semicircular β-spectrometer (Fig. G-1) the β particles emerge into a homogeneous magnetic field through a narrow slit with a divergence α, equal to half the opening angle. In the field their path is deflected through 180°, after which they impinge upon a photographic plate. Show that when α is small, there is first-order focusing in the plane of the film, and calculate the ratio of the image width to the slit-plate distance for divergences $\alpha = 2.5°$ and $\alpha = 8°$.

G-8 An axially focusing β-spectrometer with a longitudinal field B has a beam-limiting slit in the form of an annular ring arranged perpendicular to the beam trajectory between point source and detector so that only those rays which are emitted within an angular region $\alpha \pm \Delta\alpha$ with respect to the source-detector direction are allowed to pass through. The width of the slit is Δr_s and its radius with respect to the spectrometer axis is r_s.

For $\alpha = 45°$ determine the relationship between the momentum resolution $\delta p/p$ and the transmission T (defined as the ratio of the number of electrons transmitted through the slit to the total number of electrons emitted isotropically by the β source). It is necessary to calculate (a) the electron trajectories, viz. $r(z, \alpha)$, where z is the axial coordinate and r the radial; (b) the angular and momentum resolutions.

(Note the focusing condition $\partial r/\partial \alpha|_S = 0$ at the slit position S.)

G-9 A pair spectrometer with a field strength $B = 2000$ G is to be used to investigate the 15-MeV γ radiation emitted from nuclear reactions which occur in a point source S. The latter is mounted within a lead collimator and pair production occurs in a circular target, the geometry of the arrangement being as shown.

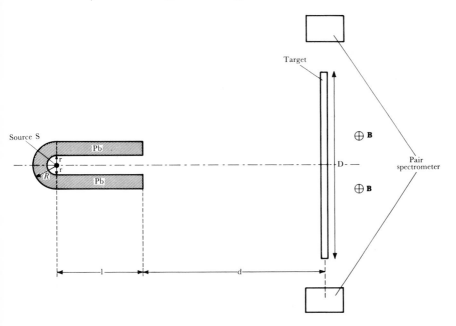

When $l = 8$ cm, $R = 8$ cm, $r = 0.1$ cm, and $D = 0.6$ cm, (a) what separation d is necessary in order that the γ intensity at the periphery of the circular target of diameter D be half that at the center? (b) How large must the separation between the two β detectors be? (c) What is the radius ρ of the β^+ trajectory in the pair spectrometer when the positrons have a total energy $E_{\beta^+} = 10$ MeV?

(Absorption coefficient of Pb for 15-MeV γ radiation: $\mu = 0.644$ cm^{-1}).

G-10   A counter arrangement is set up for the identification of particles which originate from bombardment of a target with 5.3-GeV protons. The negatively-charged particles are subjected to momentum analysis in a magnetic field, which permits only those having momenta $p = 1.19$ GeV/c(± 2 percent) to pass through into a telescope system comprising two scintillation counters in coincidence with a separation of 12 m between the two detectors.

   Identify the particles, whose time of flight in the telescope arrangement was determined as $t = 51 \pm 1$ nsec. What would have been the time of flight of π^- mesons?

G-11   Over what range does the Compton distribution extend in a scintillation spectrometer, and how high is the distribution (expressed as an intensity per unit energy interval and referred to the total number of absorbed γ rays) at the two extremes of this range in the case of a very small crystal (i.e., so small that absorption of secondary γ quanta can be neglected)? Qualitatively, what change would one expect for larger crystals?

G-12   In principle, plastic and liquid scintillators give a rectangular pulse-height distribution when monoenergetic neutrons of energy E_0 are incident upon them. The number of pulses per unit energy interval are constant between the energy limits 0 and E_0. The current pulses of the electron multiplier are shaped into rectangular pulses of width $\tau = 10$ μsec and height H, proportional to energy E. Because of the statistical distribution in time for occurrence of pulses, there is a certain probability for two arbitrary pulses out of the distribution to overlap. This is termed "pile-up."

   Calculate the pile-up spectrum one might expect for energies $E \geqslant E_0$. What is the largest permissible value of the total number of counts N if the pile-up at the energy $E = E_0$ is to be 1 percent of $N(E_0)$?

Appendix H

RADIATION DOSIMETRY

H.1. Biological Effects of Radiation

The ionizing effects of radiation upon living organisms constitute a biological hazard and necessitate the observance of precautions and tolerance limits. Primary and secondary radiation induces excitation or splitting of inorganic and organic molecules. Highly reactive radicals such as OH or O_2H and other cell poisons are produced within the intracellular and extracellular regions. It can be estimated that a lethal dose of radiation corresponds to no more than one molecule out of 10^7 being damaged in the initial process. This suggests the action of intensifying mechanisms leading to deleterious effects. Many important cell constituents, in particular enzymes and enzyme systems, perform *repetitive* functions in life-supporting processes. They might easily be inactivated by chemical bonding or reactions. In this way, the small-scale primary accident can give rise to multiplicative large-scale consequences. Radiation can also induce genetic modifications through its direct and indirect action upon DNA or RNA.

All types of radiation ultimately lead to ionization effects. For this reason, uncharged radiation constitutes as much of a hazard as charged-particle radiation, except that the density of ionization can be appreciably different. The action of uncharged radiation can be attributed to secondary charged particles. For example, γ *radiation* gives rise to fast electrons; *slow neutrons* give rise to reactions such as $H(n, \gamma) D$, from which 2.2-MeV γ's originate, and $^{14}N(n, p) \, ^{14}C$ from which 0.57-MeV p's originate; *fast neutrons* produce recoil ions in collisions with H, C, N, O atoms in living matter.

The various cells of the body evince considerable differences in their sensitivity to radiation. Lymphatic organs and bone marrow are among the highly radiosensitive; mucous membrane of the gastrointestinal tract, the lens of the eye, epithelium of the skin and cartilage are moderately sensitive;

the liver, the brain, the kidneys, striated muscles, and the central nervous system are quite resistant.

The extent of biological damage differs according to the time distribution of exposure. A partial recovery from somatic effects can be observed in most instances. This does not apply in the case of damage to genetic material.

H.2. Dosimetry Units

The biological effect is determined by the extent of energy absorption by an organ or by the body. Radiation dosimetry was first established in work with x and γ rays which was closely linked with the use of air ionization chambers as radiation detectors. As basic (but not very practical) dose-unit one chose the *röntgen* (r), the amount of x or γ radiation producing 2.08×10^9 ions pairs/ cm^3 of dry air at STP. Since an average energy of 32.5 eV must be expended to produce one pair of ions in air, the energy absorption corresponds to

$$1r \triangleq 32.5 \times 2.08 \times 10^9 \, eV \, cm^{-3} = 6.76 \times 10^4 \, MeV \, cm^{-3}$$

$$= 83.3 \text{ erg/g of air}$$

$$\simeq 93 \text{ erg/g of water or tissue}$$

The *dose rate* is the time derivative of the dose. Since the linear energy attenuation coefficient of air, μ_{air}, is fairly independent of energy between 0.07 and 2 MeV, we can write for the dose rate:

$$\text{dose rate} = \Phi \, E \, \mu_{air}$$

where Φ is the flux of photons (in units $cm^{-2} \, sec^{-1}$) having an energy E. The numerical mean value of μ is

$$\bar{\mu}_{air} = 3.35 \times 10^{-5} \, cm^{-1}$$

and thus the dose rate per hour is given by the formula:

$$\text{dose rate}^{[rh^{-1}]} = 1.78 \times 10^{-6} \, \Phi \, E^{[MeV]}.$$

The *energy dose* (radiation absorbed dose = rad) has been introduced as a more convenient general basis for expressing the radiation dose than the r (which specifically refers to *air* as an absorber). The rad is defined in terms of the amount of energy absorbed by 1 g of the particular absorber (e.g., water in the case of the human body),

$$1 \text{ rad} \equiv 100 \text{ erg/g of absorbing material.}$$

It will be observed that the rad is numerically close to the dose of 93 erg absorbed by 1 g of water subjected to an exposure of 1 r. The energy absorption by tissues exposed to γ irradiation can be deduced from Fig. 4-57(b).

The biological effect of radiation depends not only upon the absorbed energy dose, but varies considerably according to the kind of radiation to which the body is exposed. The *RBE factor, relative biological effectiveness,* is an empirical factor introduced to take account of the effects of specific ionization of different radiations and is based upon the action of 0.2-MeV x rays as a standard:

$$\text{RBE} = \frac{\left(\begin{array}{c}\text{Energy dose in rad of 0.2-MeV x rays producing a}\\ \text{certain biological effect}\end{array}\right)}{\left(\begin{array}{c}\text{Energy dose in rad of other given radiation producing}\\ \text{the same biological effect}\end{array}\right)}$$

Since all radiations are biologically more effective than γ rays or electrons, their RBE value is always larger than unity. On the basis of extensive experimental data, the United States National Committee on Radiation Protection has laid down a recommended RBE index which has been also adopted in many other countries (Table H-1).

Table H-1. RECOMMENDED VALUES OF THE RELATIVE BIOLOGICAL EFFECTIVENESS (RBE) OF DIFFERENT TYPES OF IONIZING RADIATION

Type of radiation	Mean RBE	Range of mean specific ionization [ion pairs per micron H_2O]	Range of RBE
x, γ, β	1	$\leqslant 100$	1
n_{th}	≈ 3	100–650	1–5
n, p (10 MeV)	≈ 10	650–1,500	5–10
d	$\gtrsim 10$	$\gtrsim 1,500$	10–15
α	≈ 15	1,500–5,000	10–20
Heavy ions	≈ 20	$\gtrsim 5,000$	$\gtrsim 20$

The *radiation dose* as expressed in rem *(röntgen equivalent man)* takes account of the magnitude of the energy dose and of the biological effectiveness of the particular radiation:

$$\text{rem} = \text{RBE} \times \text{energy dose in rad}$$

When an exposure to more than one kind of radiation takes place—simultaneously or otherwise—the individual radiation doses in rem can be summed.

A single *whole-body* irradiation of 25 rem is noticeable in the blood count, and 100 rem cause the typical symptoms of radiation sickness. A dose of 400 rem causes death within 1 month in 50 percent of the exposed persons, while 600 rem are almost invariably lethal. It should be noted that up to 50,000 rem are needed to kill bacteria and up to 10^6 rem to inactivate viruses.

A *tolerance dose* has been established for *occupational* exposure. Currently it stands at 5 rem/y or about 100 mrem/week (i.e., roughly 2.5 mrem/h of working time). This is approximately the dose delivered at a distance of 1 m by a 100-μCi source emitting a single γ line. When only part of the body is irradiated, the tolerance doses are, of course, larger [Eth 58].

By way of comparison we list the radiation doses received under normal conditions by the entire population from natural sources:

cosmic radiation	35 mrem/y
environmental radioactivity	70 mrem/y
^{40}K in the body	20 mrem/y
^{226}Ra in bones	50 mrem/y
Rn in air	25–200 mrem/y

The total natural dose (which varies somewhat with geographical position and altitude) is of the order of 200–400 mrem/y.

The exposure in medical diagnostic applications of radiation, in which an average of at least 2 rem/y is delivered to a large fraction of the Western population, at present constitutes by far the largest source of irradiation. The occupational exposure in most nuclear laboratories, on the other hand, can be kept to a small percentage of the tolerance limit.

The highest permissible amount of a specific radioisotope in the body, absorbed through ingestion or retention in the respiratory system, depends upon the kind and energy of the emitted radiation, upon the physical and the biological half-lives of the material, and finally, in some cases, upon physiological factors which might govern accumulation of the radioisotope in an organ. The lowest tolerance values appertain to α emitters. As an example, only 0.02 μCi of ^{239}Pu is permissible. The concentration of radioisotopes in air and water accordingly have to be kept very low.

EXERCISES

H-1 The biological effects of thermal-neutron irradiation are in large measure occasioned through the $^{14}_{7}N(n, p)^{14}_{6}C$ reaction which occurs in living tissue and produces radiocarbon. Determine the kinetic energies of the two product particles, given that the end-point energy of the β spectrum of $^{14}_{6}C$ lies at $E_0 = 0.156$ MeV.

H-2 At sea level, cosmic radiation interacting with air at STP produces 2.74 ion pairs/cm^3 in 1 sec at a latitude 41°N on an average. Express this ionization as a dose in röntgen per sec. What percentage of the tolerance dose does this represent?

H-3 Calculate the radiation dose in r/h, rad/h, and rem/h received from a ^{137}Cs source at a distance of 1 m if the activity is 1 Ci.

(Note that the branching ratio in β^- decay of ^{137}Cs to ^{137}Ba is 92 percent to the first excited state at 0.662-MeV excitation energy, and the rest to the stable ground state of ^{137}Ba. Assume that the β radiation is stopped in the source. Only the 0.662-MeV deexcitation γ rays constitute the radiation hazard, and for these the total energy absorption coefficient in air is $(\mu/\rho)_{air} = 0.029$ cm^2 g^{-1} and in water (tissue) is $(\mu/\rho)_{H_2O} = 0.032$ cm^2 g^{-1}.)

H-4 A Ra-Be source emits fast neutrons whose mean energy is 5 MeV. The neutron yield is 1.7×10^7 sec^{-1} per g Ra, in addition to which γ radiation is emitted whose total energy is 1.79 MeV per Ra decay. How close may one approach such a point source containing 0.1 g Ra without exceeding the tolerance dose for neutrons and for γ rays?

(For calculating the neutron dose, consider only the mean energy of the recoil protons in water, and use the n-p cross section $\sigma_{np} = 1.6$ b at this energy. The energy absorption coefficient for water is

$$(\mu/\rho)_{H_2O} = 0.032 \text{ cm}^2 \text{ g}^{-1}.)$$

H-5 A 1-kCi ^{60}Co source is to be transported in a hollow sphere of lead which has an inner radius $r = 1$ cm. How large must the outer radius R be if a radiation dose rate of 200 mrem/h is not to be exceeded?

(β^- decay of ^{60}Co occurs essentially to the 2.505-MeV level of ^{60}Ni, which de-excites by a two-stage γ cascade via the 1.332-MeV first excited state. For this energy the mass absorption coefficient (a) in air is $(\mu/\rho)_{air} = 0.026$ cm^2 g^{-1}, (b) in lead is $(\mu/\rho)_{Pb} = 0.033$ cm^2 g^{-1}, (c) in water (tissue) is $(\mu/\rho)_{H_2O} = 0.030$ cm^2 g^{-1}.)

H-6 The 230-MW nuclear reactor at Shippingport has a mean thermal neutron flux of $\Phi = 6 \times 10^{13}$ neutrons cm^{-2} sec^{-1}. Calculate the thickness of concrete (which has an absorption coefficient $\mu = 0.067$ cm^{-1}) needed in order to reduce the neutron flux to the tolerance level at the outer face of the shielding wall. (Assume the tolerance dose corresponds to 600 thermal neutrons cm^{-2} sec^{-1}.)

H-7 For an investigation in which a 20-mCi ^{198}Au source is mounted at a distance of 50 cm from the experimenter, lead shielding has to be interposed in order to reduce the radiation dose to tolerance limits. What thickness is necessary? If the source strength is doubled, in what proportion need the shielding thickness be increased?

(In ^{198}Au, 98.6 percent of the β^- decays lead to the first excited state of ^{198}Hg, which makes a 412-keV γ transition to the ground state.

Absorption coefficient of Pb at 412 keV: $\mu = 2.20$ cm^{-1}. Energy absorption coefficient of H$_2$O at 412 keV: $(\mu/\rho)_{H_2O} = 0.033$ cm^2 g^{-1}.)

H-8 Suppose that you have a "hot cell" at your disposal for handling highly radioactive substances. It is designed for a maximum activity $\mathscr{A} = 1$ kCi of ^{60}Co and its dimensions are $2 \times 2 \times 4$ m, bounded by 90-cm thick walls of reinforced concrete.

(a) What would be the mass of a ^{60}Co source having the maximum permissible activity?

(b) If the above source is situated at the very center of the cell, what is the radiation dose at the outer surface of the shielding wall?

(c) For how long would you be allowed to stand at the same distance directly exposed to the radiation through an open door without infringing a total tolerance dose set at 0.1 rem for the research week?

(d) What mass of extra lead shielding around the source would be needed if a 10-fold stronger ^{60}Co source were to be used?

(The external dose rate must be maintained unchanged.

$$\text{Mean density of concrete } \bar{\rho}_c = 3.5 \text{ g cm}^{-3}$$
$$\text{Mean density of lead } \quad \bar{\rho}_{Pb} = 11.35 \text{ g cm}^{-3}$$
$$\text{Mean density of cobalt } \bar{\rho}_{Co} = 8.8 \text{ g cm}^{-3}$$

Total energy absorption coefficient for the two γ's from ^{60}Ni produced by β decay of ^{60}Co, having a mean energy $\bar{E}_\gamma \cong 1.3$ MeV each:

$$\text{in water,} \quad (\mu/\rho)_{H_2O} = 0.030 \text{ cm}^2 \text{ g}^{-1}$$
$$\text{in concrete,} \quad (\mu/\rho)_c = 0.027 \text{ cm}^2 \text{ g}^{-1}$$
$$\text{in air,} \quad (\mu/\rho)_{air} = 0.026 \text{ cm}^2 \text{ g}^{-1}$$
$$\text{in lead,} \quad (\mu/\rho)_{Pb} = 0.033 \text{ cm}^2 \text{ g}^{-1}$$

The absorption in air may be neglected in the calculation.)

H-9 For the sterilization of medical preparations the following arrangement might be used: An electron beam (0.5 mA and 1.25 MeV) is directed on to a tungsten target in which it is stopped, emitting bremsstrahlung. The preparations are exposed to this at $0°$ for 100 sec at a distance of 10 cm. Their dimensions are $3 \times 3 \times 3$ cm and their density is that of water.

(a) How much bremsstrahlung energy is taken up by each preparation?

(b) How strong a ^{60}Co source would have to be used in place of the above arrangement under the above conditions?

(c) What thickness of lead shielding is needed in case (a) at a distance of 1 m from the tungsten target in order to prevent an observer at this distance from getting a higher radiation dose than the tolerance limit of 2.5 mrem/h?

(Confine your considerations to $E_\gamma > 1$ MeV.)

(From the form of the bremsstrahlung spectrum at 1 m from the target, one can deduce the maximum intensity to occur at an energy $E_\gamma = 0.5$ MeV and to be $I_{max} = 2500$ erg cm^{-2} sec^{-1} MeV^{-1}. It drops linearly to zero at 0.3 MeV and 1.3 MeV, being in first approximation a triangular spectrum.)

(Energy absorption coefficients:

$$(\mu/\rho)_{Pb} = 0.033 \text{ cm}^2 \text{ g}^{-1}; \quad (\mu/\rho)_{H_2O} = 0.030 \text{ cm}^2 \text{ g}^{-1}.)$$

CONSTANTS AND CONVERSION FACTORS IN ATOMIC, NUCLEAR, AND PARTICLE PHYSICS

Based upon the recommendations of the NAS-NRC, with the support of the International Union of Pure and Applied Physics, the following set of mutually consistent values has been compiled from listings in *Physics Today*, 1964, **17**, No. 2, 48 and Cohen and DuMond [Co 65a, especially p. 590]. The values are expressed in the CGS system of units employed throughout this book and also in the rationalized international system for ease of comparison. The symbols follow the system adopted by the I.U.P.A.P. in Document U.I.P. 11 (S.U.N. 65-3), 1965. Particle data is based on values tabulated by Rosenfeld *et al.* [Ro 64b, updated in Ro 68].

Table I-1. Rest Masses and Energies

Atomic mass unit (unified, $^{12}C = 12u$)	$1\ u$	$1.660\ 431 \times 10^{-24}\ g \simeq 931.478\ MeV$
Mass of electron	m_e	$9.1091 \times 10^{-28}\ g = 5.4865 \times 10^{-4}\ u \simeq 0.511\ 006\ MeV$
Mass of muon	m_μ	$206.767\ m_e = 1.8834 \times 10^{-25}\ g \simeq 105.659\ MeV$
Mass of pion	$m_{\pi^\pm}$	$273.19\ m_e = 2.4885 \times 10^{-25}\ g \simeq 139.60\ MeV$
	m_{π^0}	$264.20\ m_e = 2.4066 \times 10^{-25}\ g \simeq 135.01\ MeV$
Mass of proton	m_p	$1835.918\ m_e = 1.67252 \times 10^{-24}\ g = 1.007\ 2766\ u \simeq 938.256\ MeV$
Mass of neutron	m_n	$1838.628\ m_e = 1.67482 \times 10^{-24}\ g = 1.008\ 6654\ u \simeq 939.550\ MeV$
Mass of deuteron	m_d	$3.342\ 45 \times 10^{-24}\ g = 2.013\ 005\ u \simeq 1875.069\ MeV$
Mass of α particle	m_α	$6.642\ 397 \times 10^{-24}\ g = 4.000\ 408\ u \simeq 3726.292\ MeV$
Mass of H atom	M_H	$1.673\ 423 \times 10^{-24}\ g = 1.007\ 825\ u \simeq 938.767\ MeV$
Mass of D atom	M_D	$3.344\ 275 \times 10^{-24}\ g = 2.014\ 102\ u \simeq 1876.092\ MeV$
Mass of ^{4}He atom	M_{He}	$6.646\ 042 \times 10^{-24}\ g = 4.002\ 603\ u \simeq 3728.337\ MeV$
Mass-energy conversion factors	$1\ g$	$6.022\ 535 \times 10^{23}\ u \simeq 5.609\ 86 \times 10^{26}\ MeV$
	$1\ MeV$	$1.602\ 10 \times \begin{cases} 10^{-6}\ erg \\ 10^{-13}\ J \end{cases} \simeq 1.073\ 56 \times 10^{-3}\ u$
	$1\ J$	$10^7\ erg = 0.24\ cal$
	$1\ cal$	$4.1840 \times 10^7\ erg = 4.1840\ J$
Reduced mass of electron in H atom	$\mu_e = m_e\, m_p / M_H$	$9.104\ 99 \times 10^{-28}\ g = 5.483\ 51 \times 10^{-3}\ u \simeq 0.510\ 777\ MeV$
Schrödinger constant for H atom	$2\mu_e/\hbar^2$	$1.637\ 63 \times \begin{cases} 10^{27}\ J^{-1}\ m^{-2} \\ 10^{38}\ J^{-1}\ m^{-2} \end{cases}$
Schrödinger constant for massive nucleus	$2m_e/\hbar^2$	$1.638\ 37 \times \begin{cases} 10^{27}\ erg^{-1}\ cm^{-2} \\ 10^{38}\ J^{-1}\ m^{-2} \end{cases}$

Table I-2. PHYSICAL CONSTANTS

Speed of light in vacuum	c	$2.997\ 925 \times 10^{10}$ cm sec^{-1}
Planck constant of action	h	$6.6256 \times \begin{cases} 10^{-27} \text{ erg sec} \\ 10^{-34} \text{ J sec} \end{cases}$
Rationalized Planck constant	$\hbar \equiv h/2\pi$	$1.054\ 50 \times \begin{cases} 10^{-27} \text{ erg sec} \\ 10^{-34} \text{ J sec} \end{cases}$
Elementary charge	e	$4.802\ 98 \times 10^{-10}$ esu $= 1.602\ 10 \times \begin{cases} 10^{-20} \text{ emu} \\ 10^{-19} \text{ C} \end{cases}$
Specific charge of electron	e/m_e	$5.272\ 74 \times 10^{17}$ esu g$^{-1} = 1.758\ 796 \times \begin{cases} 10^{7} \text{ emu g}^{-1} \\ 10^{13} \text{ C kg}^{-1} \end{cases}$
Avogadro constant	N_A	$6.022\ 531 \times 10^{23}$ mol^{-1}
Faraday constant	$F = N_A e$	$2.892\ 61 \times 10^{14}$ esu mol$^{-1} = 9.648\ 70 \times \begin{cases} 10^{5} \text{ emu mol}^{-1} \\ 10^{4} \text{ C mol}^{-1} \end{cases}$
Boltzmann constant	k	$1.380\ 54 \times \begin{cases} 10^{-16} \text{ erg } {}^{\circ}\text{K}^{-1} \\ 10^{-23} \text{ J } {}^{\circ}\text{K}^{-1} \end{cases}$
Molar gas constant	$R = N_A k$	$8.314\ 3 \times \begin{cases} 10^{-7} \text{ erg } {}^{\circ}\text{K}^{-1} \text{ mol}^{-1} \\ 1 \text{ J } {}^{\circ}\text{K}^{-1} \text{ mol}^{-1} \end{cases}$

Gravitational constant	G	$6.670 \times \begin{cases} 10^{-8} \text{ dyn cm}^2 \text{ g}^{-2} \\ 10^{-11} \text{ N m}^2 \text{ kg}^{-2} \end{cases}$
Fine-structure constant	$\alpha = e^2/\hbar c$	$7.297\,20 \times 10^{-3} \approx 1/137$
Classical electron radius	$r_e = e^2/m_e c^2$	$2.817\,77 \times 10^{-13} \text{ cm}$
Bohr radius	$a_0 \equiv \hbar^2/m_e e^2 = r_e/\alpha^2$	$5.291\,67 \times 10^{-9} \text{ cm}$
Thomson cross section for electrons	$\sigma_{\text{Th}} = \tfrac{8}{3}\pi r_e^2$	$0.665\,16 \times 10^{-24} \text{ cm}^2$
Nuclear radius constant	$r_0 = R A^{-1/3}$	$\approx 1.4 \times 10^{-13} \text{ cm}$
Fermi weak-interaction constant	g_F	$\approx 1.4061 \times \begin{cases} 10^{-49} \text{ erg cm}^3 \\ 10^{-48} \text{ J m}^3 \end{cases}$
Rydberg constant for massive nucleus	$R_\infty = 2\pi^2 m_e e^4/h^3 c$	$1.097\,3731 \times 10^5 \text{ cm}^{-1}$
	$R_\infty hc = m_e e^4/2\hbar^2$	$2.179\,72 \times \begin{cases} 10^{-11} \text{ erg} \\ 10^{-18} \text{ J} \end{cases} = 13.61 \text{ eV}$
Rydberg constant for H atom	$R_H = R_\infty \left(1 + \dfrac{m_e}{m_p}\right)^{-1}$	$1.096\,7758 \times 10^5 \text{ cm}^{-1}$
	$R_H hc$	$2.178\,53 \times \begin{cases} 10^{-11} \text{ erg} \\ 10^{-18} \text{ J} \end{cases} = 13.60 \text{ eV}$
Influence constant	ϵ_0	$8.854\,3 \times 10^{-12} \text{ F m}^{-1}$
Induction constant	μ_0	$1.256\,64 \times 10^{-6} \text{ H m}^{-1}$

Table I-3. Magnetic Moments

Bohr magneton	$\mu_B \equiv e\hbar/2m_e c$	$9.2732 \times$	$\begin{cases} 10^{-21}\,\text{erg G}^{-1} \\ 10^{-24}\,\text{J Wb}^{-1}\,\text{m}^2 = 10^{-24}\,\text{J T}^{-1} \end{cases}$
Nuclear magneton	$\mu_N \equiv e\hbar/2m_p c$	$5.0505 \times$	$\begin{cases} 10^{-24}\,\text{erg G}^{-1} \\ 10^{-27}\,\text{J Wb}^{-1}\,\text{m}^2 = 10^{-27}\,\text{J T}^{-1} \end{cases}$
Magnetic moment of electron	μ_e	$1.001\,159\,\mu_B = 9.2840 \times$	$\begin{cases} 10^{-21}\,\text{erg G}^{-1} \\ 10^{-24}\,\text{J T}^{-1} \end{cases}$
Magnetic moment of proton	μ_p	$2.792\,76\,\mu_N = 1.410\,49 \times$	$\begin{cases} 10^{-23}\,\text{erg G}^{-1} \\ 10^{-26}\,\text{J T}^{-1} \end{cases}$
Magnetic moment of neutron	μ_n	$-1.913\,148\,\mu_N = -9.662\,354 \times$	$\begin{cases} 10^{-24}\,\text{erg G}^{-1} \\ 10^{-27}\,\text{J T}^{-1} \end{cases}$
Magnetic moment of deuteron	μ_d	$0.857\,393\,\mu_N = 4.330\,263 \times$	$\begin{cases} 10^{-24}\,\text{erg G}^{-1} \\ 10^{-27}\,\text{J T}^{-1} \end{cases}$
Gyromagnetic ratio of proton	γ_p	$2.675\,19 \times$	$\begin{cases} 10^4\,\text{sec}^{-1}\,\text{G}^{-1} \\ 10^8\,\text{sec}^{-1}\,\text{T}^{-1} \end{cases}$
Zeeman splitting constant	$e/4\pi m_e c^2 = \mu_B/hc$	$4.668\,58 \times$	$\begin{cases} 10^{-5}\,\text{cm}^{-1}\,\text{G}^{-1} \\ 10^1\,\text{m}^{-1}\,\text{T}^{-1} \end{cases}$

Table I-4. Lifetimes

Time conversion factors	$1\,\text{y}$	$3.15569 \times 10^7\,\text{sec} = 5.25948 \times 10^5\,\text{m}$
	$1\,\text{d}$	$8.640 \times 10^4\,\text{sec} = 1.440 \times 10^3\,\text{m}$
Half-life	$T_{1/2} = \tau \ln 2$	$0.693\,\tau$
Mean lifetime of neutron	τ_n	$1.01 \times 10^3\,\text{sec}$
Half-life of neutron	$T_{1/2}^{(n)}$	$7.02 \times 10^2\,\text{sec} = 11.7\,\text{m}$
ft value for neutron	$ft_n = f\,T_{1/2}^{(n)}$	$1.211 \times 10^3\,\text{sec}$
Mean lifetime of muon	τ_μ	$2.2001 \times 10^{-6}\,\text{sec}$
Mean lifetime of pion	$\tau_{\pi^\pm}$	$2.551 \times 10^{-8}\,\text{sec}$
	τ_{π^0}	$1.80 \times 10^{-16}\,\text{sec}$
Time-of-flight per cm	t	$7.192 \times 10^{-11}\,(A/E^{[\text{MeV}]})^{1/2}\,\text{sec}$

Table I-5. Velocities

Electron velocity (non-relativistic)	v	$5.931 \times 10^7\,(E_{\text{kin}}^{[\text{eV}]})^{1/2}\,\text{cm sec}^{-1}$
Velocity of thermal electrons ($T = 300\,°\text{K}$)	$v_{\text{th e}} = (3kT/m_e)^{1/2}$	$1.117 \times 10^8\,\text{cm sec}^{-1}$
Velocity of thermal neutrons ($E_{\text{kin}} = 1/40$ eV)	$v_{\text{th n}}$	$2.187 \times 10^5\,\text{cm sec}^{-1}$

Table I-6. Binding Energies

Total binding energy of Z electrons in (Thomas-Fermi) atom	$B_e(Z)$	$15.73\, Z^{7/3}$ eV
Binding energy per electron in K shell of H atom	$I_0 = \dfrac{m_e e^4}{2\hbar^2} = \dfrac{e^2}{2a_0}$	13.597 65 eV
Ionization energy of nth shell (neglecting screening)	$I = I_0 \dfrac{Z^2}{n^2}$	
Average binding energy per nucleon	E_B	≈ 8 MeV
Binding energy of deuteron	E_d	2.224 62 MeV
Binding energy of α particle	E_α	28.3 MeV

Table I-7. Wavelengths

Compton wavelength of electron	$\lambda_C^{(e)} = \dfrac{h}{m_e c}$	2.42621×10^{-10} cm
	$\lambda_C^{(e)} = \dfrac{\hbar}{m_e c} = \dfrac{r_0}{\alpha}$	3.86144×10^{-11} cm
Compton wavelength of proton	$\lambda_C^{(p)} = \dfrac{h}{m_p c}$	1.32140×10^{-13} cm
	$\lambda_C^{(p)} = \dfrac{\hbar}{m_p c}$	2.10307×10^{-14} cm
Compton wavelength of neutron	$\lambda_C^{(n)} = \dfrac{h}{m_n c}$	1.31958×10^{-13} cm
	$\lambda_C^{(n)} = \dfrac{\hbar}{m_n c}$	2.100186×10^{-14} cm
Compton wavelength of muon	$\lambda_C^{(\mu)} = \dfrac{h}{m_\mu c}$	1.1734×10^{-12} cm
	$\lambda_C^{(\mu)} = \dfrac{\hbar}{m_\mu c}$	1.8676×10^{-13} cm
Compton wavelength of pion	$\lambda_C^{(\pi)} = \dfrac{h}{m_\pi c}$	8.8811×10^{-13} cm
	$\lambda_C^{(\pi)} = \dfrac{\hbar}{m_\pi c}$	1.4135×10^{-13} cm
Wavelength of photon ($E_\gamma = h\nu$)	$\lambda = \dfrac{c}{\nu} = \dfrac{hc}{E_\gamma}$	$\dfrac{1.2397 \times 10^{-10}}{E_\gamma^{[MeV]}}$ cm
Wavelength of thermal neutrons	$\lambda_{th\,n} = \dfrac{h}{p_n}$	$\dfrac{2.8601 \times 10^{-9}}{(E_n^{[eV]})^{1/2}}$ cm $\approx 1.81 \times 10^{-8}$ cm
	$\lambda_{th\,n} = \dfrac{\hbar}{p_n}$	$\dfrac{4.552 \times 10^{-10}}{(E_n^{[eV]})^{1/2}}$ cm $\approx 2.88 \times 10^{-9}$ cm

Table I-7. *(Continued)*

Wavelength of nonrelativistic electrons	$\lambda_e = \dfrac{h}{p_e}$	$\dfrac{1.2264 \times 10^{-7}}{(E_e^{[\text{eV}]})^{1/2}}$ cm
	$\lambdabar_e = \dfrac{\hbar}{p_e}$	$\dfrac{1.9519 \times 10^{-8}}{(E_e^{[\text{eV}]})^{1/2}}$ cm
Wavelength of protons	$\lambda_p = \dfrac{h}{p_p}$	$\dfrac{2.8620 \times 10^{-9}}{(E_p^{[\text{eV}]})^{1/2}}$ cm
	$\lambdabar_p = \dfrac{\hbar}{p_p}$	$\dfrac{4.5550 \times 10^{-10}}{(E_p^{[\text{eV}]})^{1/2}}$ cm

REFERENCES

The numbers in brackets at the end of each reference are cross-references to the respective Sections, Figures and Tables within this volume.

Ab 39   ABELSON, P. H., *Phys. Rev.* 1939, **56**, 753.   [*8.5*]

Ab 64   ABOV, YU. G., KRUPCHITSKY, P. A., and ORATOVSKY, YU. A., *Phys. Letters* 1964, **12**, 25.   [*9.2.4*]

Ad 65   ADELBERGER, E., and BARNES, C. A., *Bull. Am. Phys. Soc.* [2] 1965, **10**, 1195.   [*5.4.4*]

Ad 66   ADAMS, J. L., THOMPSON, W. J., and ROBSON, D., *Nucl. Phys.* 1966, **89**, 377.   [*5.4.4, Fig. 5.17*]

Ah 55   AHRENS, L. H., *Geochim. Cosmochim. Acta* 1955, **8**, 1.   [*6.5*]

Ah 56   AHRENS, L. H., "Radioactive Methods for Determining Geological Age," *Rept. Progr. Phys.* 1956, **19**, 80.   [*6.5*]

Aj 52   AJZENBERG, F., and LAURITSEN, I., *Rev. Mod. Phys.* 1952, **24**, 321.   [*5.3.8, 9.1*]

Aj 60   AJZENBERG-SELOVE, F. (ed.), *Nuclear Spectroscopy*, Parts A and B. Academic Press, New York, 1960.   [*9.1*]

Aj 65   AJDAČIĆ, V., CERINEO, M., LALOVIĆ, B., PAIĆ, G., ŠLAUS, I., and TOMAŠ, P., *Phys. Rev. Letters* 1965, **14**, 444.   [*2.3.5*]

Ak 61   AKSENOV, S. I., ALFIMENKOV, V. P., LUSHCHIKOV, V. I., OSTANEVICH, Y. M., SHAPIRO, F. L., and WU-KUANG, Y., *Zh. Eksperim. i Teor. Fiz.* 1961, **40**, 88; *Soviet Phys. JETP (English Transl.)* 1961, **13**, 62; see also Dubna Rept. P-845 (1961).   [*9.5.2*]

Ak 66   AKERLOF, C. W., HIEBER, R. H., KRISCH, A. D., EDWARDS, K. W., RATNER, L. G., and RUDDICK, K., *Phys. Rev. Letters* 1966, **17**, 1105.   [*1.4*]

Al 31   ALLIS, W. P., and MORSE, P. M., *Z. Physik* 1931, **70**, 567.   [*4.9*]

Al 37   ALVAREZ, L. W., *Phys. Rev.* 1937, **52**, 134L.   [*8.5*]

Al 38   ALVAREZ, L. W., *Phys. Rev.* 1938, **54**, 486.   [*8.5*]

Al 40   ALVAREZ, L. W., and BLOCH, F., *Phys. Rev.* 1940, **57**, 111.   [*5.4.5*]

Al 48   ALBERT, R. D., and WU, C. S., *Phys. Rev.* 1948, **74**, 847.   [Fig. 8-18(c)]

Al 57   ALDER, K., STECH, B., and WINTHER, A., *Phys. Rev.* 1957, **107**, 728.   [*8.6.4*]

Al 58a ALDRICH, L. T., and WETHERILL, G. W. "Geochronology by Radioactive Decay," *Ann. Rev. Nucl. Sci.* 1958, **8**, 257.   [*6.5*]

Al 58b ALLEN, J. S., *The Neutrino*. Princeton Univ. Press, Princeton, New Jersey, 1958.   [*8.3.7*]

Al 59   ALLEN, J. S., BURMAN, R. L., HERRMANNSFELDT, W. B., STÄHELIN, P., and BRAID, T. H., *Phys. Rev.* 1959, **116**, 134.   [*8.6.4*]

Al 61   ALBURGER, D. E., PIXLEY, R. E., WILKINSON, D. H., and DONOVAN, P., *Phil. Mag.* 1961, **6**, 171.   [*Exercise 7-5*]

Am 63   AMALDI, E., BARONI, G., BRADNER, H., DE CARVALHO, H. G., HOFFMANN, L., MANFREDINI, A., and VANDERHAEGHE, G., "Search for Dirac Magnetic Poles," CERN Rept. 63-13 (April, 1963).   [*5.4.5*]

An 33   ANDERSON, C. D., *Phys. Rev.* 1933, **43**, 491.   [*1.2, 4.6*]

An 36   ANDERSON, C. D., and NEDDERMEYER, S. H., *Phys. Rev.* 1936, **50**, 263.
           [*1.2, 3.2.3*]

An 62a   ANDERS, E., *Rev. Mod. Phys.* 1962, **34**, 287.   [*6.5*]

An 62b   ANDERSON, J. D., WONG, C., and McCLURE, J. W., *Phys. Rev.* 1962, **126**,
           2170.   [*5.4.4*]

An 65   ANDERSON, J. W., WONG, C., and McCLURE, J. W., *Phys. Rev.* 1965, **138**,
           B 615.   [*5.4.4*]

An 66   ANDERSSON-LINDSTRÖM, G., and ZAUSIG, B., *Nucl. Instr. Methods* 1966,
           40, 277.   [Fig. G-18]

Ar 46   ARNOLD, W. R., and ROBERTS, A., *Phys. Rev.* 1946, **70**. 766.   [*5.4.5*]

Ar 47   ARNOLD, W. R., and ROBERTS, A., *Phys. Rev.* 1947, **71**, 878.   [*5.4.5*]

Ar 66   ARMSTRONG, J. C., BAGGETT, M. J., HARRIS, W. R., and LATORRE, V. A.,
           Phys. Rev. 1966, **144**, 823.   [*11.1.5*]

As 60   ASARO, F., STEPHENS, F. S., HOLLANDER, J. M., and PERLMAN, I., *Phys.
           Rev.* 1960, **117**, 492.   [*10.1.2*]

Au 25   AUGER, P., *Compt. Rend.* 1925, **180**, 65.   [*4.5*]

Au 61   AUFFRAY, E. P., *Phys. Rev. Letters* 1961, **6**, 120.   [*5.4.8*]

Aw 56   AWSCHOLOM, M., *Phys. Rev.* 1956, **101**, 1051.   [*8.2.2*]

Ay 62   AYRES, R., HORNYAK, W. F , CHAN, L., and FANN, H., *Nucl. Phys.* 1962,
           29, 212.   [*2.3.8*]

Ba 11   BARKLA, C. G., *Phil. Mag.* [6] 1911, **21**, 648.   [*1.2, 4.2.1*]

Ba 27   BACK, E., and GOUDSMIT, S., *Z. Physik* 1927, **43**, 321.   [*5.2*]

Ba 50   BARRETT, J. H., *Phys. Rev.* 1950, **79**, 907.   [*8.2.2*]

Ba 53   BAINBRIDGE, K. T., GOLDHABER, M., and WILSON, E. D., *Phys. Rev.* 1953,
           90, 430.   [*10.1.4*]

Ba 56   BARKAS, W. H., BIRNBAUM, W., and SMITH, F. M., *Phys. Rev.* 1965, **101**,
           778.   [*8.3.7*]

Ba 58   BARNES, C. A., FOWLER, W. A., GREENSTEIN, H. B., LAURITSEN, C. C.,
           and NORDBERG, M. E., *Phys. Rev. Letters* 1958, **1**, 328.   [*8.2.2*]

Ba 61a   BALDIN, A. M., GOL'DANSKII, V. I., and ROZENTHAL, I. L., *Kinematics of
           Nuclear Reactions.* Pergamon Press, Oxford, 1961; translated by W. E.
           Jones from the Russian, *Kinematica yadernykh reatskii.* Fitzmatgiz,
           Moscow, 1959.   [*B.3, C.1, C.2, C.3.2*]

Ba 61b   BARNARD, A. G. L., and KIM, C. C., *Nucl. Phys.* 1961, **28**, 428.   [*5.4.4*]

Ba 62a   BAHCALL, J. N., *Phys. Rev. Letters* 1962, **9**, 500.   [*8.5.3*]

Ba 62b   BATES, D. R. (ed.), *Quantum Theory*, Vol. III: *Radiation and High-Energy
           Physics.* Academic Press, New York, 1962.   [*4.2.1*]

Ba 62c   BALL, J. B., "Kinematics II: A Nonrelativistic Kinematics FORTRAN
           Program to Aid Analysis of Nuclear Reaction Angular Distribution
           Data," Oak Ridge Nat. Lab. Rept. ORNL-3251 (1962).   [*C.1*]

Ba 62d   BASSEL, R. H., DRISKO, R., and SATCHLER, G. R., "The Distorted Wave
           Theory of Direct Nuclear Reactions," Oak Ridge Nat. Lab. Rept.
           ORNL-3240 (1962).   [*E.8*]

Ba 63   BAHCALL, J. N., *Phys. Rev.* 1963, **129**, 2683.   [*8.5.3*]

Ba 64a   BARKAS, W. H., and BERGER, M. J., p. 103ff. of [NRC 64].   [*4.10*]

Ba 64b   BARTLETT, A. A., *Am. J. Phys.* 1964, **32**, 120.   [*4.2.2*]

Ba 64c   BAHCALL, J. N., *Phys. Rev. Letters* 1964, **12**, 300.   [*Fig. 8-11*]

Ba 65a   BAND, I. M., LISTENGARTEN, M. A., and SLIV, L. A., "Tables of the
           Conversion Matrix Elements and Phases," Appendix 6, on p. 1673ff. of
           [Si 65].   [*10.1.3*]

Ba 65b   BAND, I. M., LISTENGARTEN, M. A., SLIV, L. A., and THUN, J. E., "Particle Parameters for Angular Correlations of Conversion Electrons," Appendix 7, on p. 1683ff. of [Si 65].   [*10.3*]

Ba 65c   BAUER, M., and CANUTO, V., *Nucl. Phys.* 1965, **72**, 33.   [*2.3.8*]

Ba 66   BAHCALL, J. N., *Phys. Rev. Letters* 1966, **17**, 398.   [*8.2.2, Fig. 8-11*]

Ba 67   BALZER, R., BHARUCHA, D., HEINRICH, F., and HOFMANN, A., *Helv. Phys. Acta* 1967, **40**, 197.   [*G.2.2*]

Ba 68   BAHCALL, J. N., BAHCALL N. A., and SHAVIV, G., *Phys. Rev. Letters* 1968, **20**, 1209.   [*8.2.2*]

Be 96   BECQUEREL, H., *Compt. Rend.* 1896, **122**, 420 and 501.   [*1.2*]

Be 30   BETHE, H. A., *Ann. Physik* [5] 1930, **5**, 325.   [*4.9.1*]

Be 32   BETHE, H. A., *Z. Physik.* 1932, **76**, 293.   [*4.9.1*]

Be 34a   BETHE, H. A., and HEITLER, W., *Proc. Roy. Soc.* 1934, **A146**, 83.   [*4.6.4, 4.9.3, Fig. 4-79*]

Be 34b   BETHE, H. A., and HEITLER, W., *Proc. Cambridge Phil. Soc.* 1934, **30**, 524. [*4.9.3*]

Be 36   BETHE, H. A., and BACHER, R. F., *Rev. Mod. Phys.* 1936, **8**, 82.   [*2.3.1*]

Be 37a   BETHE, H. A., and PLACZEK, G., *Phys. Rev.* 1937, **51**, 42.   [*Fig. 11-26*]

Be 37b   BETHE, H. A., *Rev. Mod. Phys.* 1937, **9**, 140.   [*Fig. 11-26*]

Be 39   BETHE, H. A., *Phys. Rev.* 1939, **55**, 103 and 434.   [*Exercise 8.12*]

Be 49   BEYER, R. T. (ed.), *Foundations of Nuclear Physics.* Dover, New York, 1949. [*1.2*]

Be 52   BELL, P. R., JAUCH, J., and CASSIDY, J. M., *Science* 1952, **115**, 12.   [*Fig. 8-35(b)*]

Be 53   BETHE, H. A., and ASHKIN, J., "Passage of Radiations through Matter," in *Experimental Nuclear Physics* (E. Segrè, ed.), Vol. I, p. 166ff. Wiley, New York, 1953.   [*4.9, Figs. 4-47, 4-50, 4-58, 4-66, 4-67, 4-68, 4-69, 4-70, 4-72*]

Be 54   BETHE, H. A., and MAXIMON, L. C., *Phys. Rev.* 1954, **93**, 768.   [*4.6.4*]

Be 58   BELTRAMETTI, E. G., *Nuovo Cimento* [10] 1958, **8**, 445.   [*9.4*]

Be 59   BEL'SKII, S. A., and STARODUBTSEV, S. V., *Zh. Eksperim. i Teor. Fiz.* 1959, **37**, 983; *Soviet Phys. JETP* (*English Transl.*) 1960, **37** (10), 700. [*4.3*]

Be 62   BERENYI, D., *Phys. Letters* 1962, **2**, 332.   [*8.5.3*]

Be 63   BERENYI, D., *Nucl. Phys.* 1963, **48**, 121.   [*8.5.3*]

Be 64   BERNARDINI, G., BIENLEIN, J. K., VON DARNEL, G., FAISSNER, H., FERRERO, F., GAILLARD, J. M., GERBER, H. J., HAHN, B., KAFTANOV, V., FRIENEN, F., MANFREDOTTI, C., REINHARZ, M., and SALMERON, R. A., *Phys. Letters* 1964, **13**, 86.   [*3.2.3*]

Be 65   BERGSTRÖM, I., and NORDLING, C., "The Auger Effect," on p. 1523ff. of [Si 65].   [*4.5, Fig. 4-18*]

Be 66a   BENJAMIN, R. W., BUCHANAN, P. S., and MORGAN, I. L., *Nucl. Phys.* 1966, *79*, 241.   [*9.3, Table 9-4, Figs. 9-12, 9-13*]

Be 66b   BETHGE, K., SANDER, P., and SCHMIDT, H., *Z. Naturforsch.* 1966, **21a**, 1052.   [*4.10, Figs. 4-95, 4-96*]

Bei 52   BEISER, A., *Rev. Mod. Phys.* 1952, **24**, 273.   [*Fig. 4-61*]

Bha 36   BHABHA, H. J., *Proc. Roy. Soc.* 1936, **A154**, 195.   [*F.3.1*]

Bha 38   BHABHA, H. J., *Proc. Roy. Soc.* 1938, **A166**, 501.   [*D.8, F.3.1*]

Bi 62   BIAVATI, M. H., NASSIF, S. J., and WU, C. S., *Phys. Rev.* 1962, **125**, 1364. [*8.5.3*]

Bi 65   BISHOP, G. R., *Phys. Letters* 1965, **17**, 311.   [*9.2.4*]

Bie 53   BIEDENHARN, L. C., and ROSE, M. E., *Rev. Mod. Phys.* 1953, **25**, 729.
            [*9.4, 10.3, E.8, E.8.1*]
Bie 60   BIEDENHARN, L. C., "Angular Correlations in Nuclear Spectroscopy,"
            p. 732ff. of [Aj 60], Part B.   [*9.4, E.8, E.8.1, E.8.2*]
Biє 64   BIEDENHARN, L. C., and ROSE, M. E., *Phys. Rev.* 1964, **134**, B8.   [*10.3*]
Bie 65   BIEDENHARN, L. C., and VAN DAM, H. (eds.), *Quantum Theory of Angular
            Momentum.* Academic Press, New York, 1965.   [*E.2, E.8*]
Bil 62   BILANIUK, O. M. P., DESHPANDE, Y. K., and SUDARSHAN, E. C. G.,
            Am. J. Phys. 1962, **30**, 718.   [*Exercise A-7*]
Bl 33a   BLAKE, F. C., *Rev. Mod. Phys.* 1933, **5**, 169.   [*9.5.2*]
Bl 33b   BLOCH, F., *Ann. Physik* [5] 1933, **16**, 285.   [*4.9.1*]
Bl 46a   BLOCH, F., *Phys. Rev.* 1946, **70**, 460.   [*5.4.5*]
Bl 46b   BLOCH, F., HANSEN, W. W., and PACKARD, M., *Phys. Rev.* 1946, **69**, 127L;
            70, 474.   [*5.4.5*]
Bl 46c   BLEULER, E., and ZÜNTI, W., *Helv. Phys. Acta* 1946, **19**, 137.   [*4.9.2*]
Bl 48   BLOCH, F., NICODEMUS, D., and STAUB, H. H., *Phys. Rev.* 1948, **74**, 1025.
            [*5.4.5*]
Bl 50a   BLATON, J., *Kgl. Danske Videnskab. Selskab, Mat.-Fys. Medd.* 1950, **24**,
            No. 20.   [*B.3, C.1, C.3.2*]
Bl 50b   BLUNCK, O., and LEISEGANG, S., *Z. Physik* 1950, **128**, 500.   [*4.9.2*]
Bl 52a   BLATT, J. M., and BIEDENHARN, L. C., *Rev. Mod. Phys.* 1952, **24**, 258.
            [*9.4, E.8.1*]
Bl 52Bb   BLATT, J. M., and WEISSKOPF, V. F., *Theoretical Nuclear Physics.* Wiley,
            New York, 1952.   [*7.4.3, 8.3.9, 9.2, 9.4, 10.1.4, 11.4.4, 11.5.4,
            11.5.7, E.2, Fig. 11-31*]
Bl 57   BLOCH, C., *Nucl. Phys.* 1957, **4**, 503.   [*11.5.7*]
Bl 60a   BLIN-STOYLE, R. J., *Phys. Rev.* 1960, **118**, 1605.   [*9.2.4*]
Bl 60b   BLIN-STOYLE, R. J., *Phys. Rev.* 1960, **120**, 181.   [*9.2.4*]
Bl 62   BLOCHINZEW, D. I., *Grundlagen der Quantenmechanik*, 3rd ed. Deutsch,
            Frankfurt, 1962.   [*5.1*]
Bl 65   BLAKE, R. S., JACOBS, D. J., NEWTON, J. O., and SHAPIRA, J. P., *Phys.
            Letters* 1965, **14**, 219.   [*9.2.4*]
Bo 13a   BOHR, N., *Phil. Mag.* [6] 1913, **25**, 10.   [*4.9.1, 5.3.7*]
Bo 13b   BOHR, N., *Phil. Mag.* [6] 1913, **26**, 1 and 476.   [*1.2*]
Bo 26   BORN, M., *Z. Physik* 1926, **37**, 863.   [*1.2*]
Ƃo 27   BORN, M., *Nature* 1927, **119**, 354.   [*1.2*]
Bo 32   BOHR, N., *Chem. Soc. J.* 1932, p. 349.   [*8.1*]
Bo 36   BOHR, N., *Nature*, 1936, **137**, 344.   [*11.3.1*]
Bo 39   BOHR, N., and WHEELER, J. A., *Phys. Rev.* 1939, **56**, 426.   [*1.2*]
Bo 47   BORSELLINO, A., *Nuovo Cimento* [10] 1947, **4**, 112.   [*Fig. 4-50*]
Bo 48   BOHR, N., *Kgl. Danske Videnskab. Selskab, Mat.-Fys. Medd.* 1948, **18**, No. 8.
            [*4.9.1*]
Bo 49   BOUCHEZ, R., DAUDEL, P., DAUDEL, R., MUXART, R., and ROGOZINSKI,
            A., *J. Phys. Radium* 1949, **10**, 201.   [*8.5.3*]
Bo 52   BOUCHEZ, R., *Physica* 1952, **18**, 1171.   [*8.5.3*]
Bo 53   BOCKELMAN, C., BROWNE, C., BUECHNER, W., and SPERDUTO, A., *Phys.
            Rev.* 1953, **92**, 665.   [*Fig. 5-7*]
Bo 56   BOUCHEZ, R., TOBAILEM, J., ROBERT, J., MUXART, R., MELLET, R.,
            DAUDEL, P., and DAUDEL, R., *J. Phys. Radium* 1956, **17**, 363.   [*8.5.3*]
Bo 59   BODANSKY, D., ECCLES, S. F., FARWELL, G. W., RICKEY, M. E., and
            ROBISON, P. C., *Phys. Rev. Letters* 1959, **2**, 101.   [*11.3.1, Fig. 11-22*]

Bo 60   Bouchez, R., and Depommier, P., "Orbital Electron Capture by the
            Nucleus," *Rept. Progr. Phys.* 1960, **23**, 395.   [*8.5.3*]

Bo 62   Boyle, A. J. F., and Hall, H. E., "The Mössbauer Effect," *Rept. Progr.
            Phys.* 1962, **25**, 441.   [*9.5.2*]

Bö 63a   Bösch, R., Lang, J., Müller, R., and Wölfli, W., *J. Phys. Soc. Japan*
            1963, **18**, 1551.   [*2.3.5*]

Bö 63b   Bösch, R., Lang, J., Müller, R., and Wölfli, W., *Helv. Phys. Acta* 1963,
            36, 625.   [*4.7, Fig. 4-53*]

Bö 63c   Bösch, R., Lang, J., Müller, R., and Wölfli, W., *Helv. Phys. Acta*
            1963, **36**, 657.   [*Exercise C-9*]

Bo 64   Boccaletti, D., de Sabbata, V., and Gualdi, C., *Nuovo Cimento* [10]
            1964, **33**, 520 and 1216.   [*8.3.8*]

Bo 65a   Bock, P., and Schopper, H., *Phys. Letters* 1965, **16**, 284.   [*9.2.4*]

Bö 65b   Bösch, R., Lang, J., Müller, R., and Wölfli, W., *Helv. Phys. Acta*
            1965, **38**, 753.   [*Exercise C-9*]

Bo 66   Bocquet, J. P., Chu, Y. Y., Kistner, O. C., Perlman, M. L., and
            Emery, G. T., *Phys. Rev. Letters* 1966, **17**, 809.   [*9.5.2*]

Boe 64   Boehm, F., and Kankeleit, E., "Evidence for a Parity Impurity in a
            Nuclear Gamma Transition," *Proc. Intern. Conf. Nucl. Phys.*, *Paris*,
            Vol. 2, p. 1181. C.N.R.S., Paris, 1964.   [*9.2.4*]

Boe 65   Boehm, F., and Kankeleit, E., *Phys. Rev. Letters* 1965, **14**, 312.   [*9.2.4*]

Boe 68   Boehm, F., and Kankeleit, E., *Nucl. Phys.* 1968, A109, 457.   [*9.2.4*]

Boo 64   Booth, E. C., Chasen, B., and Wright, K. A., *Nucl. Phys.* 1964, **57**, 403.
            [*6.1.1*]

Br 13   van den Broek, A., *Physik. Z.* 1913, **14**, 32.   [*1.3*]

Br 26   Brillouin, L., *Compt. Rend.* 1926, **183**, 24.   [*5.1*]

Br 55   Brysk, H., and Rose, M. E., Oak Ridge Nat. Lab. Rept. ORNL-1830
            (1955).   [*8.5.3, Fig. 8-33*]

Br 57   Brown, G. E., and Mayers, D. F., *Proc. Roy. Soc.* 1957, A242, 89 (and
            earlier references therein).   [*4.3, Fig. 4-9*]

Br 58   Brysk, H., and Rose, M. E., *Rev. Mod. Phys.* 1958, **30**, 1169.   [*8.5.3*]

Br 59   Brown, G. E., *Rev. Mod. Phys.* 1959, **31**, 893.   [*11.5.7*]

Br 62   Brink, D. M., and Satchler, G. R., *Angular Momentum.* Oxford Univ.
            Press (Clarendon), London and New York, 1962.   [*E.2, E.8*]

Br 63   Broude, C., and Gove, H. E., *Ann. Phys. (N.Y.)* 1963, **23**, 71.   [*9.4*]

Bu 52   Burhop, E. H. S., *The Auger Effect.* Cambridge Univ. Press, London and
            New York, 1952.   [*4.5*]

Bu 55a   Burcham, W. E., *Phys. Rev.* 1962, **126**, 2170.   [*5.4.4*]

Bu 55b   Burhop, E. H. S., *J. Phys. Radium* 1955, **16**, 625.   [*4.5*]

Bu 58a   Burgy, M. T., Krohn, V. E., Novey, T. B., Ringo, G. R., and Telegdi,
            V. L., *Phys. Rev.* 1958, **110**, 1214.   [*8.8*]

Bu 58b   Burgov, N. A., and Terekhov, I. N. V., *Zh. Eksperim. i Teor. Fiz.* 1958,
            35, 932; *Soviet Phys. JETP (English Transl.)* 1959, *8*, 651.   [*8.2.2*]

Bu 60   Burgy, M. T., Krohn, V. E., Novey, T. B., Ringo, G. R., and Telegdi,
            V. L., *Phys. Rev.* 1960, **120**, 1829.   [*8.8*]

Bu 62   Burbidge, G., "Nuclear Astrophysics," *Ann. Rev. Nucl. Sci.* 1962, **12**, 507.
            [*6.5*]

Bü 65a   Bühring, W., and Schopper, H., Report Kernforschungszentrum
            Karlsruhe No. 307 (1965).   [*8.6.3*]

Bu 65b   BURNS, R., GOULIANOS, K., HYMAN, E., LEDERMAN, L., LEE, W., MISTRY, N., RETTBERG, J., SCHWARTZ, M., SUNDERLAND, J., and DANBY, G., *Phys. Rev. Letters* 1965, **15**, 42.   [*3.2.3*]

Bue 48   BUECHNER, W. W., VAN DE GRAAFF, R. J., BURRILL, E. A., and SPERDUTO, A., *Phys. Rev.* 1948, **74**, 1348.   [*Fig. 4-83*]

Ca 53   CARTWRIGHT, W. F., RICHMAN, C., WHITEHEAD, M. N., and WILCOX, H. A., *Phys. Rev.* 1953, **91**, 677.   [*Exercise 11-17*]

Ca 57   CAMERON, A. G. W., *Can. J. Phys.* 1957, **35**, 1021.   [*2.3.8*]

Ca 59   CARTER, R. E., REINES, F., WAGNER, M. E., and WYMAN, J. J., *Phys. Rev.* 1959, **113**, 280.   [*8.2.2*]

Ca 63   CARLSON, T. A., *Phys. Rev.* 1963, **132**, 2239.   [*8.6.4*]

Ca 68   CARLSON, T. A., ERMAN, P., and FRANSSON, K., *Nucl. Phys.* 1968, **A111**, 371.   [*9.5.2, 10.1.4*]

Ce 34   ČERENKOV, P. A., *Dokl. Akad. Nauk SSSR* 1934, **2**, 451.   [*4.9.4*]

Če 37a   ČERENKOV, P. A., *Dokl. Akad. Nauk SSSR* 1937, **14**, 105.   [*4.9.4*]

Če 37b   ČERENKOV, P. A., *Phys. Rev.* 1937, **52**, 378.   [*4.9.4*]

Ce 66   CERNY, J., PEHL, R. H., BUTLER, G., FLEMING, D. G., MAPLES, C., and DETRAZ, C., *Phys. Letters* 1966, **20**, 35.   [*5.4.4*]

Ch 30a   CHAO, C. Y., *Proc. Natl. Acad. Sci. U.S.* 1930, **16**, 431.   [*4.6*]

Ch 30b   CHAO, C. Y., *Phys. Rev.* 1930, **36**, 1519.   [*4.6*]

Ch 32   CHADWICK, J., *Proc. Roy. Soc.* 1936, **A136**, 692; reproduced on p. 5ff of [Be 49].   [*1.2, Exercise 1-9*]

Ch 34   CHADWICK, J., and LEA, E. A., *Proc. Cambridge Phil. Soc.* 1934, **30**, 59. [*8.2.2*]

Ch 55   CHAMBERLAIN, O., SEGRÈ, E., WIEGAND, C., and YPSILANTIS, T., *Phys. Rev.* 1955, **100**, 947.   [*4.9.4, Fig. 4-94*].

Cha 63   CHAMPENEY, D. C., ISAAK, G. R., and KHAN, A. M., *Nature* 1963, **198**, 1186.   [*Exercise A-5*]

Cha 67   CHAPMAN, R., and MACLEOD, A. M., *Nucl. Phys.* 1967, **A94**, 324. [*11.1.5*]

Chi 61   CHIU, H. Y., *Ann. Phys. (N.Y.)* 1961, **15**, 1.   [*8.2.2*]

Chi 62   CHI, B. E., *A Table of Clebsch-Gordan Coefficients.* Rensselaer Polytechnic Inst., Troy, New York, 1962.   [*E.2*]

Chi 65   CHIU, H. Y., *Neutrino Astrophysics.* Gordon and Breach, New York, 1965. [*8.3.8*]

Chi 66   CHIU, H. Y., "Neutrinos in Astrophysics and Cosmology," *Ann. Rev. Nucl. Sci.* 1966, **16**, 591.   [*8.3.8*]

Chr 67   CHRISTENSEN, C. J., NIELSEN, A., BAHNSEN, A., BROWN, W. K., and RUSTAD, B. M., *Phys. Letters* 1967, **26B**, 11.   [*1.4*]

Chu 56a   CHURCH, E. L., and WENESER, J., *Phys. Rev.* 1956, **103**, 1035.   [*10.1.4, Fig. 10-11*]

Chu 56b   CHURCH, E. L., and WENESER, J., *Phys. Rev.* 1956, **104**, 1382.   [*10.1.3*]

Chu 60   CHURCH, E. L., and WENESER, J., "Nuclear Structure Effects in Internal Conversion," *Ann. Rev. Nucl. Sci.* 1960, **10**, 193.   [*10.1.3*]

Chu 64   CHURCH, E. L., SCHWARZSCHILD, A., and WENESER, J., *Phys. Rev.* 1964, **133**, B35.   [*10.3*]

Ci 65   CIERJACKS, S., MARKUS, G., MICHAELIS, W., and PÖNITZ, W., *Phys. Rev.* 1965, **137**, B345.   [*2.3.5*]

Cl 63   CLEMENTEL, E., and VILLI, C. (eds.), *Direct Interactions and Nuclear Reaction Mechanisms.* Gordon and Breach, New York, 1963.

Co 22   COMPTON, A. H., *Bull. Natl. Res. Council (U.S.)* 1922, **4**, Part 2, No. 20, 10; see also *Phil. Mag.* [6] 1921, **41**, 749; *Phys. Rev.* 1921, **18**, 96. [*4.2.2*]

Co 23a   COMPTON, A. H., *Phys. Rev.* 1923, **21**, 207 and 483.   [*4.2.2*]

Co 23b   COMPTON, A. H., *Phys. Rev.* 1923, **22**, 409.   [*4.2.2*]

Co 24   COMPTON, A. H., *Phys. Rev.* 1924, **24**, 168.   [*4.2.2*]

Co 28a   COX, R. T., McILWRAITH, C. G., and KURRELMEYER, B., *Proc. Natl. Acad. Sci. U.S.* 1928, **14**, 544.   [*8.7*]

Co 28b   CONDON, E. U., and GURNEY, R. W., *Nature* 1928, **122**, 439.   [*7.3*]

Co 29   CONDON, E. U., and GURNEY, R. W., *Phys. Rev.* 1929, **33**, 127.   [*7.3*]

Co 32   COCKCROFT, J. D., and WALTON, E. T. S., *Proc. Roy. Soc.* 1932, **A137**, 229; reproduced on p. 23ff. of [Be 49].   [*1.2*]

Co 51   CONDON, E. U., and SHORTLEY, G. H., *The Theory of Atomic Spectra.* Cambridge Univ. Press, London and New York, 1951; corrected reprint of 1st ed. (1935).   [*5.3.10, E.2, E.3*]

Co 52   CONNER, J. P., BONNER, T. W., and SMITH, J. R., *Phys. Rev.* 1952, **88**, 468. [*Fig. 11-33(b)*]

Co 53   COOPER, L. N., and HENLEY, E. M., *Phys. Rev.* 1953, **92**, 801.   [*2.4.2*]

Co 56a   COHEN, V. W., CORNGOLD, N. R., and RAMSEY, N. F., *Phys. Rev.* 1956, **104**, 283.   [*5.4.5*]

Co 56b   COWAN, C. L., Jr., REINES, F., HARRISON, F. B., KRUSE, H. W., and McGUIRE, A. D., *Science* 1956, **124**, 103.   [*8.2.2, Fig. 8-8*]

Co 58   CONDON, E. U., and ODISHAW, H., (eds.), *Handbook of Physics.* McGraw-Hill, New York, 1958.   [*9.2, 9.5, Figs. 9-2, 9-3, 9-4*]

Co 59   CORNILLE, H., and CHAPDELAINE, M., *Nuovo Cimento* [10] 1959, **14**, 1386. [*4.3*]

Co 62   COMPTON, D. M. J., and SCHOEN, A. H., (eds.), *The Mössbauer Effect (Proc. 2nd Intern. Conf. Mössbauer Effect, Saclay, France, 13–15 Sept. 1961).* Wiley, New York, 1962.   [*9.5.2*]

Co 64   COWSIK, R., PAL, Y., and TANDON, S. N., *Phys. Letters* 1964, **13**, 265. [*8.3.8*]

Co 65a   COHEN, E. R., and DuMOND, J. W. M., *Rev. Mod. Phys.* 1965, **37**, 537.   [*I*]

Co 65b   COHEN, R. C., KANARIS, A. D., MARGULIES, S., and ROSEN, J. L., *Phys. Letters* 1965, **16**, 292.   [*2.3.5*]

Coe 53   COESTER, F., *Phys. Rev.* 1953, **89**, 619.   [*11.5.6*]

Cr 39   CRANE, H. R., *Phys. Rev.* 1939, **55**, 501.   [*8.2.2*]

Cr 59   CRAWFORD, F. S., CRESTI, M., DOUGLASS, R. L., GOOD, M. L., KALBFLEISCH, G. R., STEVENSON, M. L., and TICHO, H. K., *Phys. Rev. Letters* 1959, **2**, 266.   [*5.3.9*]

Cr 60a   CRAIG, P. P., NAGLE, D. E., and COCHRAN, D. R. F., *Phys. Rev. Letters* 1960, **4**, 561.   [*9.5.2*]

Cr 60b   CRANSHAW, T. E., SCHIFFER, J. P., and WHITEHEAD, A. B., *Phys. Rev. Letters* 1960, **4**, 163.   [*3.2.4, Exercises 9-8, A-5*]

Cr 64   CRANSHAW, T. E., and SCHIFFER, J. P., *Proc. Phys. Soc. (London)* 1964, **84**, 245.   [*3.2.4, Exercise 9-8*]

Cu 34   CURIE, I., and JOLIOT, M. F., *Compt. Rend.* 1934, **198**, 254; reproduced on p. 39ff. of [Be 49].   [*1.2, 6.3.2*]

Cu 49   CURRAN, S. C., ANGUS, J., and COCKROFT, A. L., *Phil. Mag.* [7] 1949, **40**, 53.   [*Fig. 8-18(b)*]

Cu 54   CUTKOSKY, R. E., *Phys. Rev.* 1954, **95**, 1222.   [*8.5.3*]

Da 08   DALTON, J., *A New System of Chemical Philosophy* (1808).   [*1.2*]
Da 39   DANCOFF, S. M., and MORRISON, P., *Phys. Rev.* 1939, **55**, 122.   [*10.1.3*]
Da 47   DAUDEL, R., *Rev. Scie.* 1947, **85**, 162.   [*8.5.3*]
Da 52a   DAVIS, R., Jr., *Phys. Rev.* 1952, **86**, 976.   [*8.2.2, Exercise 8-5*]
Da 52b   DAVISSON, C. M., and EVANS, R. D., *Rev. Mod. Phys.* 1952, **24**, 79.   [*4.1,
           Figs. 4-5, 4-15, 4-16*]
Da 53   DAVIDSON, J. P., and PEASLEE, D. C., *Phys. Rev.* 1953, **91**, 1232.   [*8.6.2*]
Da 54   DAVIES, H., BETHE, H. A., and MAXIMON, L. C., *Phys. Rev.* 1954, **93**, 788.
           [*4.6.4, Fig. 4-49*]
Da 55   DAVIS, R., Jr., *Phys. Rev.* 1955, **97**, 766.   [*3.5, 8.2.2*]
Da 56   DAVIS, R., Jr., *Bull. Am. Phys. Soc.* [2] 1956, **1**, 219.   [*8.2.2*]
Da 57a   DALITZ, R. H., "K Mesons and Hyperons: Their Strong and Weak
           Interactions," *Rept. Progr. Phys.* 1957, **20**, 163.   [*11.1.5*]
Da 57b   DAVIS, R., Jr., "An Attempt to Observe the Capture of Reactor Neutrinos
           in Chlorine-37," *Proc. Intern. Conf. Radioisotopes Sci. Res., Paris*, 1957
           p. 728ff. Pergamon Press, Oxford, 1958.   [*8.2.2*]
Da 58a   DANIEL, H., *Nucl. Phys.* 1958, **8**, 191.   [*8.6.2*]
Da 58b   DAVYDOV, A. S., and FILIPPOV, G. S., *Nucl. Phys.* 1958, **8**, 237.   [*5.4.8,
           9.2.4, Fig. 9-7*]
Da 59   DAVIDSON, J. P., *Am. J. Phys.* 1959, **27**, 457.   [*5.4.4*]
Da 62a   DALITZ, R. H., *Strong Interaction Physics and the Strange Particles*. Oxford
           Univ. Press, London and New York, 1962.   [*11.1.5*]
Da 62b   DANBY, G., GAILLARD, J. M., GOULIANOS, K., LEDERMAN, L. M., MISTRY,
           N., SCHWARTZ, M., and STEINBERGER, J., *Phys. Rev. Letters* 1962, **9**, 36.
           [*8.2.2, Fig. 8-13*]
Da 63a   DALITZ, R. H., "Strange-Particle Resonant States," *Ann. Rev. Nucl. Sci.*
           1963, **13**, 339.   [*11.1.5*]
Da 63b   DANIEL, H., and GENTNER, W., "Determination of Life-Time," on p.
           275ff. of *Nuclear Physics*, (L. C. L. Yuan and C. S. Wu, eds.), Vol. 5 of
           Methods of Experimental Physics, (L. Marton, ed.-in-chief). Academic
           Press, New York, 1963.   [*6.1*]
Da 64   DAVIS, R., Jr., *Phys. Rev. Letters* 1964, **12**, 302.   [*8.2.2*]
Da 65a   DABBS, J. W. T., HARVEY, J. A., PAYA, D., and HORSTMANN, H., *Phys.
           Rev.* 1965, **139**, B756.   [*3.2.4*]
Da 65b   DANOS, M., and GREINER, W., *Phys. Rev.* 1965, **138**, B93.   [*11.5.7*]
Da 65c   DAVISSON, C. M., "Interaction of γ-Radiation with Matter," on p. 37ff.
           of [Si 65].   [*Figs. 4-7, 4-8, 4-54*]
Da 66a   DAVIDSON, W., and NARLIKAR, J. V., "Cosmological Models and their
           Observational Validation," *Rept. Progr. Phys.* 1966, **29/II**, 539.   [*6.5*]
Da 66b   DAVIS, H. L., *Nucleonics* 1966, **24**, No. 11, 48.   [*1.4*]
Da 66c   DANOS, M., and GREINER, W., *Phys. Rev.* 1966, **146**, 708.   [*11.5.7*]
Da 68   DAVIS, R., Jr., HARMER, D. S., and HOFFMAN, K. C., *Phys. Rev. Letters*
           1968, **20**, 1205.   [*8.2.2*]
De 23   DEBYE, P., *Physik. Z.* 1923, **24**, 161.   [*4.2.2*]
De 27   DENNISON, D. M., *Proc. Roy. Soc.* 1927, **A115**, 483.   [*5.2*]
De 30   DEBYE, P., *Physik. Z.* 1930, **31**, 419.   [*4.3*]
De 57   DEVONS, S., and GOLDFARB, L. J. B., "Angular Correlations," in *Handbuch
           der Physik* (S. Flügge, ed.), Vol. 42, p. 362ff. Springer, Berlin, 1957.
           [*9.4, E.8*]
De 58   DEUTSCH, M., "Annihilation of Positrons," *Progr. Nucl. Phys.* 1958, **3**, 131.
           [*4.6.5*]

De 60a DELL'ANTONIO, G. F., and FIORINI, E., Nuovo Cimento Suppl. [10] 1960, **17**, 132.　[8.2.2]

De 60b DEPOMMIER, P., NGUYEN-KHAC, U., and BOUCHEZ, R., J. Phys. Radium 1960, **21**, 456.　[8.5.3]

De 60c DEVONS, S., "The Measurement of Very Short Lifetimes," on p. 512ff. of [Aj 60].　[6.1, 9.5.1]

De 62 DEDRICK, K. G., Rev. Mod. Phys. 1962, **34**, 429.　[C.1]

De 63 DEDRICK, K. G., Rev. Mod. Phys. 1963, **35**, 414.　[C.1]

De 68 DEBERTIN, K., and RÖSSLE, E., Nucl. Phys. 1968, **A107**, 693.　[2.3.5]

Dea 65 DEARNALEY, G., and NORTHROP, D. C., Semiconductor Counters for Nuclear Radiations. Spon, London, 1965.　[G.1]

DeBe 50 DEBENEDETTI, S., COWAN, C. E., KONNEKER, W. R., and PRIMAKOFF, H., Phys. Rev. 1950, **77**, 205.　[Exercise 8-13]

DeBe 54 DEBENEDETTI, S., and CORBEN, H. C., "Positronium," Ann. Rev. Nucl. Sci. 1954, **4**, 191.　[4.6.5]

DeBe 61 DEBENEDETTI, S., LANG, G., and INGALLS, R., Phys. Rev. Letters 1961, **6**, 60.　[9.5.2]

DeBe 66 DEBENEDETTI, S., BARROS, F. DE S., and HOY, G. R., "Chemical and Structural Effects on Nuclear Radiations," Ann. Rev. Nucl. Sci. 1966, **16**, 31.　[8.5.3]

de B 24 DE BROGLIE, L., Dissertation. Masson, Paris, 1924; Phil. Mag. [7] 1924, **47**, 446; Ann. Phys. (Paris) [10] 1925, **3**, 22.　[1.4]

Di 26 DIRAC, P. A. M., Proc. Roy. Soc. 1926, **A112**, 661.　[3.4.1]

Di 27 DIRAC, P. A. M., Proc. Roy. Soc. 1927, **A114**, 243.　[3.4.1]

Di 28 DIRAC, P. A. M., Proc. Roy. Soc. 1928, **A117**, 610.　[D.5]

Di 31 DIRAC, P. A. M., Proc. Roy. Soc. 1931, **A133**, 60.　[1.2, 5.4.5]

Di 48 DIRAC, P. A. M., Phys. Rev. 1948, **74**, 817.　[5.4.5]

Di 66 DICKE, R. H., Science J. 1966, **2**, No. 10, 95.　[Fig. 6-23]

Do 59 DOBROKHOTOV, E. I., LAZARENKO, V. R., and LUK'YANOV, S. YU., Zhur. Eksperim. i Teor. Fiz. 1959, **36**, 76; Soviet Phys. JETP (English Transl.) 1959, **36**, 54.　[8.2.2.]

Do 62 DOUGAN, P. W., LEDINGHAM, K. W. D., and DREVER, R. W. P., Phil. Mag. [8] 1962, **7**, 1223.　[8.5.3]

Dr 56 DREVER, R. W. P., MOLIJK, A., and SCOBIE, J., Phil. Mag. [8] 1956, **1**, 942.　[8.6.2]

Dr 60 DREVER, R. W. P., quoted by B. L. Robinson and R. W. Fink, in Rev. Mod. Phys. 1960, **32**, 117.　[8.5.3]

Du 15 DUANE, W., and HUNT, F. L., Phys. Rev. 1915, **6**, 166.　[4.9.3, Fig. 4-78]

Du 57 DUCKWORTH, H. E., "Masses of Atoms of A > 40," Progr. Nucl. Phys. 1957, **6**, 138.　[2.3.8]

Dy 49 DYSON, F. J., Phys. Rev. 1949, **75**, 486 and 1736.　[4.6.3, F.2]

Dy 51 DYSON, F. J., Phys. Rev. 1951, **82**, 428; ibid. 1951, **83**, 608.　[F.2]

Dz 56 DZHELEPOV, B. S., and ZYRIANOVA, L. J., Influence of the Atomic Electric Field on β-Decay. Acad. Sci. U.S.S.R., Moscow, 1956.　[8.3.9]

Ed 57 EDMONDS, A. R., Angular Momentum in Quantum Mechanics, Princeton Univ. Press, Princeton, New Jersey, 1957.　[E.2, E.4, E.8]

Ed 66 EDER, G., Nucl. Phys. 1966, **78**, 657.　[8.2.2]

Eh 64 EHLOTZKY, F., and SHEPPEY, G. C., Nuovo Cimento [10] 1964, **33**, 1185.　[4.7, Fig. 4-53]

Ei 05a   EINSTEIN, A., *Ann. Physik* [4] 1905, **17**, 891; English translation on p. 35ff. of [Ei 23].   [*1.4*]

Ei 05b   EINSTEIN, A., *Ann. Physik* [4] 1905, **18**, 639; English translation on p. 69ff. of [Ei 23]; see also *Ann. Physik* [4] 1906, **20**, 627.   [*A.2*]

Ei 23   EINSTEIN, A., LORENTZ, H. A., WEYL, H., and MINKOWSKI, H., *The Principle of Relativity:* Collection of original papers, with notes by A. Sommerfeld. Dover, New York; reprinted from Methuen, London, 1923.   [*A.1*]

Ei 51   EINSTEIN, A., *The Meaning of Relativity*, 5th ed. Methuen, London, 1951. [*A.1*]

Eis 41   EISENBUD, L., and WIGNER, E. P., *Proc. Natl. Acad. Sci. U.S.* 1941, **27**, 281. [*11.5.7*]

Eis 60   EISBERG, R. M., YENNIE, D. R., and WILKINSON, D. H., *Nucl. Phys.* 1960, **18**, 338.   [*11.1.5*]

Eis 63   EISBERG, R. M., "A Bremsstrahlung Technique for Measurement of Time Delay in Nuclear Reactions," on p. 352ff. of [Cl 63].   [*11.1.5*]

El 27   ELLIS, C. D., and WOOSTER, W. A., *Proc. Roy. Soc.* 1927, **A117**, 109. [*8.1.3*]

El 61   ELTON, L. J. B., *Nuclear Sizes*. Oxford Univ. Press, London and New York, 1961.   [*1.4*]

En 59   ENDT, P. M., and DEMEUR, M. (eds.), *Nuclear Reactions*, Vol. I. North-Holland Publ. Co., Amsterdam, 1959.   [*11.5.4, 11.5.7*]

Er 66   ERICSON, T. E. O., *Phys. Letters* 1966, **23**, 97.   [*11.3.1*]

Es 37   ESTERMANN, I., SIMPSON, O. C., and STERN, O., *Phys. Rev.* 1937, **52**, 535. [*5.4.5*]

Es 46   ESTERMANN, I., *Rev. Mod. Phys.* 1946, **18**, 300.   [*5.4.5*]

Es 47   ESTERMANN, I., SIMPSON, O. C., and STERN, O., *Phys. Rev.* 1947, **71**, 238. [*3.2.4*]

Et 64   ETTER, J. E., WAGGONER, M. A., MOAZED, C., HOLMGREN, H. D., and HAN, C., University of Maryland Tech. Rept. No. 427 (1964); contribution to the Gatlinburg Conference on Correlations of Particles Emitted in Nuclear Reactions (Oct., 1964) *Rev. Mod. Phys.* 1965, **37**, 444.   [*11.1.5*]

Eth 58   ETHRINGTON, H. (ed.), *Nuclear Engineering Handbook*. McGraw-Hill, New York, 1958.   [*H.2*]

Ev 55   EVANS, R. D., *The Atomic Nucleus*. McGraw-Hill, New York, 1955. [*2.3.7, 4.1, 4.8, 4.9, 6.3, Figs. 3-10, 4-14, 4-56, 4-57, 4-65, 4-76, 4-78, 4-82(a), 6-12, 7-7, 8-16, 8-25, 9-14*]

Ev 60   EVERLING, F., KOENIG, L. A., MATTAUCH, J. H. E., and WAPSTRA, A. H., *Nuclear Data Tables, Parts I & II*. Natl. Acad. Sci.–Natl. Res. Council, Washington, D.C., 1962.   [*2.1*]

Ew 57   EWAN, G. T., KNOWLES, J. W., and MACKENZIE, D. R., *Phys. Rev.* 1957, **108**, 1308.   [*10.1.2, Fig. 10-6*]

Ew 59   EWAN, G. T., GEIGER, J. S., GRAHAM, R. L., and MACKENZIE, D. R., *Phys. Rev.* 1959, **116**, 950.   [*10.1.2, Fig. 10-6*]

Ew 65   EWAN, G. T., and GRAHAM, R. L., "Internal Conversion Studies at Very High Resolution," on p. 951ff. of [Si 65].   [*10.1.2, Fig. 10-7*]

Fa 54   FARLEY, F. J. M., *Nucleonics* 1954, **12**, No. 10, 56; erratum, *Nucleonics* 1955, **13**, No. 7, 67.   [*C.1*]

Fa 59   FANO, U., SPENCER, L. V., and BERGER, M. J., "Penetration and Diffusion of X Rays," in *Handbuch der Physik*, (S. Flügge, ed.), Vol. 38, Part II, p. 660ff., Springer, Berlin, 1959.   [*4.1*]

Fa 62   FASOLI, U., MANDUCHI, C., and ZANNONI, G., *Nuovo Cimento* [10] 1962, **23**, 1126.   [*8.5.3*]

Fa X   FAISSNER, H., and FRANZINETTI, C., "Neutrino Interactions," *Ergeb. exakt. Naturw.* (to be published).   [*8.2.2*]

Fe 28   FERMI, E., *Z. Physik* 1928, **48**, 73.   [*D.2*]

Fe 33   FERMI, E., *La Ricerca Scientifica*, 1933, **2**, No. 12.   [*8.3.1*]

Fe 34a   FERMI, E., *Nature* 1934, **133**, 898; reproduced on p. 43ff. of [Be 49].   [*1.2, 6.3.2*]

Fe 34b   FERMI, E., *Z. Physik* 1934, **88**, 161; reproduced on p. 45ff. of [Be 49].   [*1.2, 8.3.1, 8.3.3*]

Fe 38   FEATHER, N., *Proc. Cambridge Phil. Soc.* 1938, **34**, 599.   [*4.9.2*]

Fe 47   FESHBACH, H., PEASLEE, D. C., and WEISSKOPF, V. F., *Phys. Rev.* 1947, **71**, 145 and 564 (erratum).   [*Exercise 11-12*]

Fe 49a   FEYNMAN, R. P., *Phys. Rev.* 1949, **76**, 749.   [*4.6.2, D.5.1*]

Fe 49b   FEYNMAN, R. P., *Phys. Rev.* 1949, **76**, 769.   [*4.6.2, 4.6.3*]

Fe 49c   FEYNMAN, R. P., *Phys. Rev.* 1949, **80**, 440.   [*4.6.3*]

Fe 50   FERMI, E., *Nuclear Physics*: Notes compiled by J. Orear, A. H. Rosenfeld, and R. A. Schluter. Univ. of Chicago Press, Chicago, Illinois, 1950 (rev. ed.).   [*2.3.5, 2.3.6, 3.4.2, 4.9.1, 5.4.8, 8.5.3, 10.1.3, 11.4.1*]

Fe 51   FEINGOLD, A. M., *Rev. Mod. Phys.* 1951, **23**, 10.   [*8.3.9*]

Fe 53   FERENTZ, M., and ROSENZWEIG, N., "Table of F-Coefficients," Argonne Natl. Lab. Rept. ANL-5234 (1953); reproduced as "Table of Angular Correlation Coefficients," Appendix 8, on p. 1687ff. of [Si 65].   [*E.8.1*]

Fe 58   FEYNMAN, R. P., and GELL-MANN, M., *Phys. Rev.* 1958, **109**, 193. [*8.8, D.7*]

Fe 62   FERGUSON, A. J., and RUTLEDGE A. R., "Coefficients for Triple Angular Correlation Analysis in Nuclear Bombardment Experiments," Chalk River Rept. CRP-615 (**AECL-420**), 1957 (revised Nov., 1962).   [*E.8.2*]

Fe 64   FEYNMAN, R. P., and GELL-MANN, M., *Phys. Rev.* 1964, **109**, 193.   [*9.2.4*]

Fee 50   FEENBERG, E., and TRIGG, G., *Rev. Mod. Phys.* 1950, **22**, 399.   [*8.3.9, Figs. 8-25, 8-34*]

Fei 50   FEISTER, I., *Phys. Rev.* 1950, **78**, 375.   [*8.3.9, Figs. 8-25, 8-34*]

Fei 67   FEINBERG, G., *Phys. Rev.* 1967, **159**, 1089.   [*Exercise A-7*]

Fie 37   FIERZ, M., *Z. Physik* 1937, **104**, 553.   [*8.6.2*]

Fl 39   FLAMMERSFELD, A., *Z. Physik* 1939, **114**, 227.   [*Fig. 10-2*]

Fl 64   FLÜGGE, S., *Lehrbuch der Theoretischen Physik*, Vol. IV: *Quantentheorie I.* Springer, Berlin, 1964.   [*3.6.4*]

Fo 30a   FOCK, V., *Z. Physik* 1930, **61**, 126.   [*D.2*]

Fo 30b   FOWLER, R. H., *Proc. Roy. Soc.* 1930, **A129**, 1.   [*10.2*]

Fo 62   FOX, R., *Phys. Rev.* 1962, **125**, 311.   [*11.1.5*]

Fo 63   FORD, K. W., *Sci. Am.* 1963, **209**, No. 6, 122.   [*5.4.5*]

Fo 64   FOX, J. D., MOORE, C. F., and ROBSON, D., *Phys. Rev. Letters* 1964, **12**, 198.   [*5.4.4, Fig. 5-9*]

Fo 66   Fox, J. D., and Robson, D. (eds.), *Isobaric Spin in Nuclear Physics*.
           Academic Press, New York, 1966.   [*5.4.4, Figs. 5-10, 5-11, 5-12,*
           5-13, 5-14, 5-15]

Fr 27   Friedrich, W., and Goldhaber, G., *Z. Physik* 1927, **44**, 700.   [*Fig.*
           4-6]

Fr 35   Franz, W., *Z. Physik* 1935, **95**, 652.   [*4.3*]

Fr 36   Franz, W., *Z. Physik* 1936, **98**, 314.   [*4.3*]

Fr 37   Frank, I., and Tamm, I., *Dokl. Akad. Nauk SSSR* 1937, **14** (3), 109.
           [*4.9.4*]

Fr 60   Frauenfelder, H., and Lustig, H. (eds.), "Mössbauer Effect," Univer-
           sity of Illinois Rept. AF-IN 60-698 (1960).   [*9.5.2*]

Fr 62   Frauenfelder, H., *The Mössbauer Effect*. Benjamin, New York, 1962.
           [*9.5.2, Fig. 9-25*]

Fr 63   Franzinetti, C. (ed.), "The Neutrino Experiment," CERN Rept. 63-37
           (1963).   [*8.2.2*]

Fr 65a   Franzinetti, C. (ed.), "Informal Conference on Experimental Neutrino
           Physics," CERN Rept. 65-32 (1965).   [*8.2.2*]

Fr 65b   Freeman, J. M., Murray, G., and Burcham, W. E., *Proc. Intern. Conf.*
           Nucl. Phys., (Paris, 1964), Vol. 2, p. 1178. C.N.R.S., Paris, 1965; *Phys.*
           Letters 1965, **17**, 317.   [*8.6.3*]

Fr 65c   Frauenfelder, H., Steffen, R. M., de Groot, S. R., Tolhoek, H. A.,
           and Huiskamp, W. J., "Angular Distribution of Nuclear Radiation,"
           on p. 997ff. of [Si 65], Vol. II.   [*E.8*]

Fr 66   Franzinetti, C., "Neutrino Interactions in the CERN Heavy Liquid
           Bubble Chamber," CERN Rept. 66-13 (1966).   [*8.2.2*]

Fu 37   Furry, W. H., *Phys. Rev.* 1937, **51**, 125.   [*4.6.3, F.3.4*]

Fu 64   Fuschini, E., Gadjokov, V., Maroni, G., and Varonesi, P., *Nuovo*
           Cimento [10] 1964, **33**, 709 and 1309.   [*11.3.1*]

Fu 65   Fuller, G. H., and Cohen, V. W., "Nuclear Moments," *Nucl. Data*
           Sheets, Appendix 1 (May, 1965).   [*5.4.5, 5.4.7, 5.4.8*]

Ga 28   Gamow, G., *Z. Physik* 1928, **51**, 204.   [*7.3*]

Ga 34a   Gamow, G., *Proc. Roy. Soc.* 1934, **A146**, 217.   [*8.4*]

Ga 34b   Gamow, G., *Physik. Z.* 1934, **35**, 533.   [*8.4*]

Ga 41   Gamow, G., and Schönberg, M., *Phys. Rev.* 1941, **109**, 193.   [*8.2.2*]

Ga 48   Gardner, E., and Lattes, C. M. G., *Science* 1948, **107**, 270.   [*1.2*]

Ga 51   Gardner, J. H., *Phys. Rev.* 1951, **83**, 996.   [*5.4.5*]

Ga 66   Garvey, G. T., and Kelson, I., *Phys. Rev. Letters* 1966, **16**, 197.   [*5.4.4*]

Ge 53   Gell-Mann, M., *Phys. Rev.* 1953, **92**, 833.   [*5.3.9*]

Ge 55   Gell-Mann, M., and Pais, A., "Theoretical views on the new particles,"
           Proc. Conf. Nucl. Meson Phys., (Glasgow, 1954), p. 342. Pergamon Press,
           London and New York, 1955.   [*5.3.9*]

Ge 57   Gell-Mann, M., and Rosenfeld, A. H., "Hyperons and Heavy Mesons
           (Systematics and Decay)," *Ann. Rev. Nucl. Sci.* 1957, **7**, 407.   [*5.3.9,*
           9.2.4]

Ge 58   Gerhart, J. B., *Phys. Rev.* 1958, **109**, 897.   [*8.6.2*]

Ge 59   Gell-Mann, M., *Rev. Mod. Phys.* 1959, **31**, 834.   [*9.2.4*]

Ge 61   Gerholm, T. R., Pettersson, B. G., van Nooijen, R., and Grabowski,
           Z., *Nucl. Phys.* 1961, **24**, 177, 196, and 251.   [*10.1.2*]

Ge 65   Gerholm, T. R., and Pettersson, B. G., "Some Consequences of the
           Finite Nuclear Size," on p. 981ff. of [Si 65].   [*10.1.2*]

Ge 67   Gentry, R. V., *Nature* 1967, **213**, 487.   [*6.5*]

Gei 09   GEIGER, H., and MARSDEN, E. *Proc. Roy. Soc.* 1909, **A82**, 495.   [*1.2, 3.5.7*]

Gei 11   GEIGER, H., and NUTTALL, J. M., *Phil. Mag.* [6] 1911, **22**, 613.   [*7.3, Fig. 7-2*]

Gei 12   GEIGER, H., and NUTTALL, J. M., *Phil. Mag.* [6] 1912, **23**, 439.   [*7.3*]

Gei 13   GEIGER, H., and MARSDEN, E., *Phil. Mag.* [6] 1913, **25**, 604.   [*1.2, 3.5.7, Fig. 3-10*]

Gei 22   GEIGER, H., *Z. Physik* 1922, **8**, 45.   [*Fig. 7-2*]

Gi 40a   GINZBURG, V. L., *J. Phys. (USSR)*, 1940, **2**, No. 6, 441.   [*4.9.4*]

Gi 40b   GINZBURG, V. L., *Zh. Eksperim. i Teor. Fiz.* 1940, **10** (6), 589.   [*4.9.4*]

Gi 65   GILBOY, W. B., and TOWLE, J. H., *Nucl. Phys.* 1965, **64**, 130.   [*9.3, Fig. 9-11*]

Gl 48   GLENDENIN, L. E., *Nucleonics* 1948, **2**, No. 1, 12.   [*4.9.2*]

Gl 54   GLAUBER, R. J., and MARTIN, P. C., *Phys. Rev.* 1954, **95**, 572.   [*8.5.3*]

Gl 56   GLAUBER, R. J., and MARTIN, P. C., *Phys. Rev.* 1956, **104**, 158.   [*8.5.3*]

Gl 60   GLASHOW, S., *Phys. Rev.* 1960, **118**, 316.   [*3.2.3*]

Gl 63   GLEIT, C. E., TANG, C. W., and CORYELL, C. D., "Beta-Decay Transition Probabilities," *Nucl. Data Sheets*, Appendix 5 (1963).   [*8.3.9, Fig. 8-26*]

Go 26   GORDON, W., *Z. Physik* 1926, **40**, 117.   [*D.4*]

Go 48   GORTER, C. J., *Physica* 1948, **14**, 504.   [*8.7.1*]

Go 51   GOLDHABER, M., and SUNYAR, A. W., *Phys. Rev.* 1951, **83**, 906.   [*9.2, Figs. 9-4, 9-6*]

Go 52   GOLDHABER, M., and HILL, R. D., *Rev. Mod. Phys.* 1952, **24**, 179.   [*9.2, Figs. 9-4, 9-6*]

Go 58   GOLDHABER, M., GRODZINS, L., and SUNYAR, A. W., *Phys. Rev.* 1958, **109**, 1015.   [*8.7.2, Figs. 8-48, 8-49*]

Go 59   GOLDFARB, L. J. B., "Angular Correlations and Polarization," on p. 159ff. of [En 59], Vol. I.   [*E.8*]

Go 61   GOUDSMIT, S. A., *Phys. Today* 1961, **14**, 18.   [*5.2*]

Go 63   GOTO, E., HOLM, H., and FORD, K., *Phys. Rev.* 1963, **132**, 387.   [*5.4.5*]

Go 65a   GOLDMAN, D. T., "Chart of the Nuclides," 8th ed. Prepared at the Knolls Atomic Power Laboratory and distributed by Educational Relations, General Electric Co., Schenectady, New York, 1965.   [*5.4.2, Fig. 2-1*]

Go 65b   GOUDSMIT, S. A., Plenarvortr. *Physikertag., Frankfurt*, 1965, p. 1ff. Teubner, Stuttgart, 1965.   [*5.2*]

Go 66   GOVE, N. B., and ROBINSON, R. L. (eds.), *Nuclear Spin-Parity Assignments.* Academic Press, New York, 1966.   [*9.2.4*]

Go 68   GOLDHABER, A. S., and NIETO, M. M., *Phys. Rev. Letters* 1968, **21**, 567.   [*1.4, A.2*]

Gr 54   GREEN, A. E. S., *Phys. Rev.* 1954, **95**, 1006.   [*2.4.1*]

Gr 55   GREEN, A. E. S., *Nuclear Physics.* McGraw-Hill, New York, 1955. [*Fig. 7-3*]

Gr 58   GREEN, T. A., and ROSE, M. E., *Phys. Rev.* 1958, **110**, 105.   [*10.1.3*]

Gr 60a   GRAHAM, R. L., EWAN, G. T., and GEIGER, J. S., *Nucl. Instr. Methods* 1960, **9**, 245.   [*10.1.2*]

Gr 60b   GREULING, E., and WHITTEN, R. C., *Ann. Phys. (N.Y.)* 1960, **11**, 510.   [*8.2.2*]

Gr 61   GRODZINS, L., and GENOVESE, F., *Phys. Rev.* 1961, **121**, 228.   [*9.5.2*]

Gü 65   GÜNTHER, C., BLUMBERG, H., ENGELS, W., STRUBE, G., VOSS, J., LIEDER, R. M., LUIG, H., and BODENSTEDT, E., *Nucl. Phys.* 1965, **61**, 65.   [*9.2.4*]

Ha 28   HARTREE, D. R., *Proc. Cambridge Phil. Soc.* 1928, **24**, 111.   [*D.2*]

Ha 39   HAHN, O., and STRASSMANN, F., *Naturwissenschaften* 1939, **27**, 11.   [*1.2*]

Ha 41   HAMILTON, D. R., *Am. J. Phys.* 1941, **9**, 319.   [*5.4.5*]

Ha 49   HANSON, A. O., TASCHEK, R. F., and WILLIAMS, J. H., *Rev. Mod. Phys.* 1949, **21**, 635.   [*C.1*]

Ha 52   HAUSER, W., and FESHBACH, H., *Phys. Rev.* 1952, **87**, 366.   [*9.4, E.8.1*]

Ha 53   HAMILTON, D. R., ALFORD, W. P., and GROSS, L., *Phys. Rev.* 1953, **92**, 1521.   [*8.3.7*]

Ha 59   HANNA, G. C., "Alpha-Radioactivity," in *Experimental Nuclear Physics* (E. Segrè, ed.), Vol. III, p. 54ff. Wiley, New York, 1959.   [*6.6, 7.4.3, Fig. 6-24*]

Ha 60a   HAGEDOORN, H. L., and WAPSTRA, A. H., *Nucl. Phys.* 1960, **15**, 146.   [*4.5*]

Ha 60b   HANNA, S. S., HEBERLE, J., LITTLEJOHN, C., PERLOW, G. J., PRESTON, R. S., and VINCENT, D. H., *Phys. Rev. Letters* 1960, **4**, 177.   [*9.5.2, Fig. 9-27*]

Ha 60c   HAY, H. J., SCHIFFER, J. P., CRANSHAW, T. E., and EGELSTAFF, P. A., *Phys. Rev. Letters* 1960, **4**, 165.   [*Exercise A-5*]

Ha 61   HAWKINGS, R. C., EDWARDS, W. J., and McLEOD, E. M., "Tables of Gamma Rays from the Decay of Radionuclides," Chalk River Rept. CRDC-1007, Parts I and II (1961).   [*6.4*]

Ha 63a   HAGEDORN, R., *Relativistic Kinematics*, Benjamin, New York, 1963.   [*C.1, C.3.2*]

Ha 63b   HANSEN, L. S., "Preliminary Results of Time Delay Measurements in Nuclear Reactions using the Bremsstrahlung Technique," on p. 367ff. of [Cl 63].   [*11.1.5*]

Ha 65a   HAGSTRÖM, S., NORDLING, C., and SIEGBAHN, K., "Tables of Electron Binding Energies and Kinetic Energy versus Magnetic Rigidity," Appendix 2, on p. 845ff. of [Si 65].   [*G.2.1*]

Ha 65b   HAMAMOTO, I., *Nucl. Phys.* 1965, **73**, 225.   [*9.2.1*]

Ha 65c   HARRIS, G. I., HENNECKE, H. J., and WATSON, D. D., *Phys. Rev.* 1965, **139**, B 1113; erratum *ibid.* 1966, **141**, B 1214.   [*9.4*]

Ha 66   HAMILTON, J. H. (ed.), *Internal Conversion Processes*. Academic Press, New York, 1966.   [*10.1.3, 10.2.1, 10.3*]

Ha 68   HANNON, J. P., and TRAMMEL, G. T., *Phys. Rev. Letters* 1968, **21**, 726. [*9.5.2*]

He 27   HEISENBERG, W., *Z. Physik* 1927, **43**, 172.   [*D.3*]

He 32   HEISENBERG, W., *Z. Physik* 1932, **77**, 1; *ibid.* 1932, **78**, 156; *ibid.* 1932, **80**, 587.   [*1.2, 1.4, 3.1, 5.3.8*]

He 33   HEITLER, W., and SAUTER, F., *Nature* 1933, **132**, 892.   [*4.6.4*]

He 36   HEITLER, W., *Proc. Cambridge Phil. Soc.* 1936, **32**, 112.   [*9.1*]

He 43   HEISENBERG, W., *Z. Physik* 1943, **120**, 513 and 673.   [*11.5.4*]

He 44   HEITLER, W., *The Quantum Theory of Radiation*, 2nd ed. Oxford Univ. Press, London and New York, 1944.   [*Fig. 4-82(b)*]

He 54   HEITLER, W., *The Quantum Theory of Radiation*, 3rd ed. Oxford Univ. Press (Clarendon), London and New York, 1954.   [*4.1, 4.2.3, 4.4.1, 4.6.4, 4.6.5, 4.9.3, 9.1, 9.5, Figs. 4-6, 4-13, 4-48, 4-51, 4-55, 4-80*]

He 56   HEITLER, W., *Ann. Inst. Henri Poincaré* 1956, **15**, 67.   [*11.3*]

He 59   HENLEY, E. M., and JACOBSOHN, B. A., *Phys. Rev.* 1959, **113**, 225.   [*11.3.1*]

He 61   HENDRIE, D., and GERHART, J., *Phys. Rev.* 1961, **121**, 846.   [*8.3.4*]

He 64   HELLER, L., "Some Low-Energy Neutrino Cross Sections," Los Alamos Sci. Lab. Rept. LA-3013 (1964).   [*8.2.2*]

He 65   HEBACH, H., and KÜMMEL, H., *Z. Physik* 1965, **186**, 452.   [*2.3.6*]

He 66   Heidelberg Conference, *Heidelberg Conf. Recent Progr. Nucl. Phys. with Tandems*, 1966.   [*5.4.4*]

Hi 50   HIPPLE, J. A., SOMMER, H., and THOMAS, H. A., *Phys. Rev.* 1950, **80**, 487. [*5.4.5*]

Hi 57a   HISDAL, E., *Phys. Rev.* 1957, **105**, 1821.   [*4.9.3*]

Hi 57b   HISDAL, E., *Arch. Math. Naturvidenskab* 1957, **53**, No. 3, 1.   [*4.9.3*]

Ho 48   HOUGH, P. V. C., *Phys. Rev.* 1948, **74**, 80.   [*4.9.3*]

Ho 53   HOYT, H. C., and DuMOND, J. W. M., *Phys. Rev.* 1953, **91**, 1027. [*Table G-I*]

Ho 54   HOLLADAY, W. G., and SACHS, R. G., *Phys. Rev.* 1954, **96**, 810.   [*5.4.6*]

Ho 57   HOFSTADTER, R., "Nuclear and Nucleon Scattering of High-Energy Electrons," *Ann. Rev. Nucl. Sci.* 1957, **7**, 231.   [*1.4, Fig. 1-4*]

Ho 63   HOFSTADTER, R., (ed.), *Nuclear and Nucleon Structure: a collection of reprints with an introduction.* Benjamin, New York, 1963.   [*1.4*]

Ho 64   HOFSTADTER, R., and SCHIFF, L. I. (eds.), *Nucleon Structure.* Stanford Univ. Press, Stanford, California, 1964.   [*1.4*]

Ho 66a   HOFMANN, A., *Nucl. Instr. Methods* 1966, **40**, 13.   [*G.2.2*]

Ho 66b   HOLLANDER, J. M., RASMUSSEN, J. O., and HAMILTON, J. H., *Phys. Today* 1966, **19**, No. 2, 58.   [*10.1.3, 10.1.4, 10.2.1, 10.3*]

Ho 66c   HOYLE, F., and NARLIKAR, J. V., *Proc. Roy. Soc.* 1966, **A290**, 143.   [*6.5*]

Ho 67   HORNSHØJ, P., DEUTCH, B. I., and MIRANDA, A., *Nucl. Phys.* 1967, **A95**, 65. [*10.3*]

Hou 54   HOUTERMANS, F. G., and THIRRING, W., *Helv. Phys. Acta* 1954, **27**, 81. [*8.2.2*]

Hu 27   HUND, F., *Z. Physik* 1927, **42**, 93.   [*5.2*]

Hu 53   HUGHES, D. J., GARTH, R. C., and LEVIN, J. S., *Phys. Rev.* 1953, **91**, 1423. [*Exercise 11-12*]

Hu 54   HUBY, R., *Proc. Phys. Soc. (London)* 1954, **67**, 1103.   [*E.8.1*]

Hu 58   HUGHES, D. J., and SCHWARTZ, R. B., "Neutron Cross Sections," 2nd ed. Brookhaven Natl. Lab. Rept. BNL-325 (1958).   [*11.4.4, Figs. 3-3, 3-6*]

Hu 60   HUGHES, D. J., MAGURNO, B. A., and BRUSSEL, M. K., "Supplement No. 1 to 'Neutron Cross Sections'," 2nd ed. Brookhaven Natl. Lab. Rept. BNL-325 (Suppl., 2nd ed.) (1960).   [*Ex. 11-7, 11-9*]

Hu 61   HUMBLET, J., and ROSENFELD, L., *Nucl. Phys.* 1961, **26**, 529 and 579. [*11.5.7, Fig. 11-33*]

Hu 62   HUMBLET, J., *Nucl. Phys.* 1962, **31**, 544.   [*11.5.7, Fig. 11-33*]

Hu 64a   HUMBLET, J., *Nucl. Phys.* 1964, **50**, 1.   [*11.5.7*]

Hu 64b   HUMBLET, J., *Nucl. Phys.* 1964, **57**, 386.   [*11.5.7*]

Hu 64c   HUMBLET, J., *Nucl. Phys.* 1964, **58**, 1.   [*11.5.7*]

Hu 67   HUMBLET, J., and LEBON, G., *Nucl. Phys.* 1967, **A96**, 593.   [*11.5.7*]

Im 66   IMBODEN, D., PAULI, C., and ALDER, K., *Helv. Phys. Acta* 1966, **39**, 600. [*10.1.3*]

In 53   INGLIS, D. R., *Rev. Mod. Phys.* 1953, **25**, 390.   [*5.3.8*]

In 66   INOUE, T., *Tables of the Clebsch-Gordan Coefficients.* Tokyo Tasho Co., Tokyo, 1966.   [*E.2*]

Iw 32   IWANENKO, D., *Nature* 1932, **129**, 798.   [*1.2*]

Ja 51   JAUCH, J., Oak Ridge Natl. Lab. Rept. ORNL-1102 (1951).   [*8.5.3*]

Ja 55   JAUCH, J. M., and ROHRLICH, F., *The Theory of Photons and Electrons.* Addison-Wesley, Reading, Massachusetts, 1955.   [*Fig. 4-32*]

Ja 57a   JACKSON, J. D., TREIMAN, S. B., and WYLD, H. W., Jr., *Nucl. Phys.* 1957, **4**, 206; *Phys. Rev.* 1957, **106**, 517.   [*8.6.4*]

Ja 57b   JARMIE, N., and SEAGRAVE, J. D. (eds.), "Charged Particle Cross Sections," Los Alamos Sci. Lab. Rept. LA-2014 (1957).   [*C.1, C.4*]

Ja 59   JACOBSOHN, B. A., and HENLEY, E. M., *Phys. Rev.* 1959, **113**, 234.
           [*11.3.1*]

Ja 60   JAHNKE, E., EMDE, F., and LÖSCH, F., *Tables of Higher Functions*, 6th ed.
           Teubner, Stuttgart, 1960.   [*11.5.1, Fig. E-2*]

Ja 61   JARCZYK, L., KNOEPFEL, H., LANG, J., MÜLLER, R., and WÖLFLI, W.,
           Nucl. Instr. Methods 1961, **13**, 287.   [*6.4*]

Ja 62   JARCZYK, L., KNOEPFEL, H., LANG, J., MÜLLER, R., and WÖLFLI, W.,
           Nucl. Instr. Methods 1962, **17**, 310.   [*G.3.2, Fig. G-11*]

Ja 65   JASTRZEBSKI, J., SUJKOWSKI, Z., and ZYLICZ, J., (eds.), *Role of Atomic
           Electrons in Nuclear Transformations (Proc. Intern. Conf.*, Warsaw, 24–28
           Sept., 1963.) Nucl. Energy Inform. Center, Palace of Culture and
           Science, Warsaw, 1965.   [*10.1.3, Fig. 10-8*]

Jae 35   JAEGER, J. C., and HULME, H. R., *Proc. Roy. Soc.* 1935, **A148**, 708.   [*10*]

Jae 60   JAENECKE, J., *Z. Physik* 1960, **160**, 171.   [*5.4.4*]

Je 50   JENKS, G. H., SWEETON, F. H., and GHORMLY, J. A., *Phys. Rev.* 1950,
           80, 990.   [*8.1.3*]

Je 58   JELLEY, J. V., *Čerenkov Radiation and its Applications.* Pergamon Press,
           Oxford, 1958.   [*4.9.4*]

Jo 58   JOSEPH, J., and ROHRLICH, F., *Rev. Mod. Phys.* 1958, **30**, 354.   [*4.9.3*]

Jo 60   JOSEPHSON, B. D., *Phys. Rev. Letters* 1960, **4**, 341.   [*9.5.2*]

Jo 63   JOHNSON, C. H., PLEASONTON, F., and CARLSON, T. A., *Phys. Rev.* 1963,
           132, 1149.   [*8.6.4*]

Ka 38   KAPUR, P. L., and PEIERLS, R. E., *Proc. Roy. Soc.* 1938, **A166**, 277.
           [*11.5.7*]

Ka 51   KAPLAN, I., *Phys. Rev.* 1951, **81**, 962.   [*7.6.2*]

Ka 52a   KATZ, L., and PENFOLD, A. S., *Rev. Mod. Phys.* 1952, **24**, 28.   [*4.9.2*]

Ka 52b   KAUFMANN, S. G., GOLDBERG, E., KOESTER, L. J., and MOORING, F. P.,
           Phys. Rev. 1952, **88**, 673.   [*11.3.1, Figs. 11-23, 11-28*]

Ka 63   KAYE, G., READ, E. J. C., and WILLMOTT, J. C., "Tables of Coefficients
           for the Analysis of Triple Angular Correlations from Aligned Nuclei,"
           Liverpool University Report (1963).   [*E.8.2*]

Ka 66   KAJFOSZ, J., KOPECKÝ, J., and HONZÁTKO, J., *Phys. Letters* 1966, **20**, 284.
           [*11.3.1*]

Ke 38a   KEMMER, N., *Proc. Roy. Soc.* 1938, **A166**, 127.   [*D.8*]

Ke 38b   KEMMER, N., *Proc. Cambridge Phil. Soc.* 1938, **34**, 354.   [*D.8*]

Ke 39   KELLOGG, J. M. B., RABI, I. I., RAMSEY, N. F., and ZACHARIAS, J. R.,
           Phys. Rev. 1939, **55**, 318L.   [*5.4.8*]

Ke 40   KELLOGG, J. M. B., RABI, I. I., RAMSEY, N. F., and ZACHARIAS, J. R.,
           Phys. Rev. 1940, **57**, 677.   [*5.4.8*]

Ke 46   KELLOGG, J. M. B., and MILLMAN, S., *Rev. Mod. Phys.* 1946, **18**, 323.
           [*5.4.5*]

Ke 66   KELSON, I., and GARVEY, G. T., *Phys. Letters* 1966, **23**, 689.   [*5.4.4*]

Ki 59   KISER, R. W., and JOHNSTON, W. H., *J. Am. Chem. Soc.* 1959, **81**, 1810.
           [*8.5.3*]

Ki 60   KISTNER, O. C., and SUNYAR, A. W., *Phys. Rev. Letters* 1960, **4**, 412.
           [*9.5.2*]

Ki 65   KITTLE, C., KNIGHT, W. D., and RUDERMAN, M. A., *Mechanics, Berkeley
           Physics Course*, Vol. I. McGraw-Hill, New York, 1965.   [*C.2*]

Kl 27   KLEIN, O., *Z. Physik* 1927, **41**, 407.   [*D.4*]

Kl 29   KLEIN, O., and NISHINA, Y., *Z. Physik* 1929, **52**, 853.   [*4.2.3*]

Kl 67   KLEINHEINZ, P., VUKANOVIĆ, R., ŽUPANČIĆ, M., SAMUELSSON, L., and
           LINDSTRÖM, H., *Nucl. Phys.* 1967, **A91**, 329.   [*9.2.4*]

Kno 65   KNOWLES, J. W., "Crystal Diffraction Spectroscopy of Nuclear γ-Rays," on p. 203ff. of [Si 65].   [*G.1*]

Ko 41   KONOPINSKI, E. J., and UHLENBECK, G. E., *Phys. Rev.* 1941, **60**, 308. [*8.4.4*]

Ko 43   KONOPINSKI, E. J., *Rev. Mod. Phys.* 1943, **15**, 209.   [*8.4.4*]

Ko 49   KOCKEL, B., *Ann. Physik* [6] 1949, **4**, 279.   [*4.6.3*]

Ko 52   KOLSKY, H. G., PHIPPS, T. E., Jr., RAMSEY, N. F., and SILSBEE, H. B., *Phys. Rev.* 1952, **87**, 395.   [*5.4.8*]

Ko 53   KONOPINSKI, E. J., and LANGER, L. M., "The Experimental Clarification of the Theory of β-Decay," *Ann. Rev. Nucl. Sci.* 1953, **2**, 261.   [*8.6*]

Ko 55   KONOPINSKI, E. J., "Theoretical Expectations Concerning Double β-Decay," USAEC Los Alamos Sci. Lab. Rept. LAMS-1949 (1955). [*8.2.2*]

Ko 56   KOPFERMANN, H., *Kernmomente*, 2nd ed. Akad. Verlagsges., Frankfurt, 1956.   [*5.2*]

Ko 57   KORSUNSKI, M. I., *Isomerie der Atomkerne*. Deut. Verlag Wiss., Berlin, 1957.   [*9.1*]

Ko 58a   KOFOED-HANSEN, O. M., *Rev. Mod. Phys.* 1958, **30**, 449.   [*2.4.2*]

Ko 58b   KONIJN, J., VAN NOOIJEN, B., HAGEDOORN, H. L., and WAPSTRA, A. H., *Nucl. Phys.* 1958, **9**, 296.   [*8.6.2*]

Ko 59   KOCH, H. W., and MOTZ, J. W., *Rev. Mod. Phys.* 1959, **31**, 920.   [*4.9.3*]

Ko 61   KOHMAN, T. P., *J. Chem. Educ.* 1961, **38**, 73.   [*6.5*]

Ko 62   KOFOED-HANSEN, O., and CHRISTENSEN, C. J., "Experiments on Beta Decay," in *Handbuch der Physik* (S. Flügge, ed.), Vol. 41, Part 2, p. 1ff. Springer, Berlin, 1962.   [*8.1, 8.2.2, Fig. 8-38*]

Ko 65   KOFOED-HANSEN, O. M., "Neutrino Recoil Experiments," on p. 1397ff. of [Si 65].   [*8.2.2*]

Ko 67   KOLM, H. H., *Phys. Today* 1967, **20**, No. 10, 69.   [*5.4.5*]

Ko 68   KOLM, H. H., *Science J.* 1968, **4**, No. 9, 60.   [*5.4.5*]

Kr 23   KRAMERS, H. A., *Phil. Mag.* [6] 1923, **46**, 836.   [*4.9.3*]

Kr 53   KRAUSHAAR, J. J., WILSON, E. D., and BAINBRIDGE, K. T., *Phys. Rev.* 1953, **90**, 610.   [*8.5.3*]

Ku 29   KUHN, W., *Phil. Mag.* [7] 1929, **8**, 625.   [*9.5*]

Ku 36   KURIE, F. N. D., RICHARDSON, J. R., and PAXTON, H. C., *Phys. Rev.* 1936, **49**, 368.   [*8.3.6*]

Ku 40   KUSCH, P., MILLMAN, S., and RABI, I. I., *Phys. Rev.* 1940, **57**, 765. [*5.4.5*]

Ku 47   KUSCH, P., and FOLEY, H. M., *Phys. Rev.* 1947, **72**, 1256L.   [*5.4.5*]

Ku 48   KUSCH, P., and FOLEY, H. M., *Phys. Rev.* 1948, **74**, 250.   [*5.4.5*]

Ku 55   KUNZ, W. E., *Phys. Rev.* 1955, **97**, 456.   [*Fig. 11-33(a)*]

Ku 65a   KUCHOWICZ, B., "New Proposals of Neutrino Experiments," Review Rept. No. 18, Nuclear Energy Information Center, Warsaw, 1965. [*8.2.2.*]

Ku 65b   KUZMIN, V. A., *Zh. Eksperim. i Teor. Fiz.* 1965, **49**, 1532; *Soviet. Phys. JETP (English Transl.)* 1966, **22**, 1051.   [*8.2.2*]

Ku 66   KUSCH, P., *Phys. Today* 1966, **19**, No. 2, 23.   [*5.4.5*]

Kü 66   KÜMMEL, H., MATTAUCH, J. H. E., THIELE, W., and WAPSTRA, A. H., *Nucl. Phys.* 1966, **81**, 129.   [*2.3.8*]

Ku X   KUCHOWICZ, B., *The Bibliography of the Neutrino*. Gordon and Breach, New York (undated, presumably 1967/68).   [*8.2.2*]

La 38   LANDAU, L., and RUMER, G., *Proc. Roy. Soc.* 1938, **A166**, 213.   [*4.9.3*]

La 39   LAMB, W. E., Jr., *Phys. Rev.* 1939, **55**, 190.   [*9.5.2*]

La 44   LANDAU, L., *J. Phys. (USSR)* 1944, **8**, No. 4, 201.   [*4.9.2*]

La 47a   LAMB, W. E., Jr., and RETHERFORD, R. C., *Phys. Rev.* 1967, **72**, 241.
            [*5.4.5*]

La 47b   LATTES, C. M. G., MUIRHEAD, H., OCCHIALINI, G. P. S., and POWELL,
            C. F., *Nature*, 1947, **159**, 694.   [*1.2*]

La 49a   LANGER,L.M.,and PRICE,H.C.,Jr.,*Phys.Rev.*1949,**75**,1109.[*8.4.4,Fig. 30*]

La 49b   LANGER,L. M.,and PRICE,H. C.,Jr.,*Phys.Rev.*1949,**76**,641.[*8.4.4,Fig. 30*]

La 51   LAMB, W. E., "Anomalous Fine Structure of Hydrogen and Singly
            Ionized Helium," *Rept. Progr. Phys.* 1951, **14**, 19.   [*D.5*]

La 52   LANGER, L. M., and MOFFATT, R. J. D., *Phys. Rev.* 1952, **88**, 689.
            [*8.3.7, Fig. 8-23*]

La 54   LANGEVIN, M., and RADVANYI, P., *Compt. Rend.* 1954, **238**, 77; see also
            ibid. 1955, **241**, 33.   [*8.5.3*]

La 55a   LANE, A. M., THOMAS, R. G., and WIGNER, E. P., *Phys. Rev.* 1955, **98**, 693.
            [*11.5.7*]

La 55b   LANGEVIN, M., and RADVANYI, P., *Compt. Rend.* 1955, **241**, 33.   [*8.5.3*]

La 55c   LANDON, H. H., and SAILOR, V.L., *Phys. Rev.* 1955, **98**, 1267.   [*Fig. 11-26*]

La 57   LANDAU, L., *Nucl. Phys.* 1957, **3**, 127.   [*8.7.1*]

La 58a   LANDAU, L. D., and LIFSHITZ, E. M., *Quantum Mechanics: Non-Relativistic
            Theory.* Pergamon Press, Oxford, 1958.   [*3.5.8, 11.2.1*]

La 58b   LANE, A. M., and THOMAS, R. G., *Rev. Mod. Phys.* 1958, **30**, 257.
            [*11.5.4, 11.5.7*]

La 59   LAUTERJUNG, K. H., SCHIMMER, B., SCHMIDT-ROHR, V., and MAIER-
            LEIBNITZ, H., *Z. Physik* 1959, **155**, 547.   [*8.2.2*]

La 65   LANG, J., MÜLLER, R., WÖLFLI, W., BÖSCH, R., and MARMIER, P.,
            Phys. Letters 1965, **15**, 248.   [*11.1.5*]

La 66   LANG, J., MÜLLER, R., WÖLFLI, W., BÖSCH, R., and MARMIER, P., *Nucl.
            Phys.* 1966, **88**, 576.   [*11.1.5, Figs. 11-8, 11-11, 11-12*]

Lau 58   LAURITSEN, T., BARNES, C. A., FOWLER, W. A., and LAURITSEN, C. C.,
            Phys. Rev. Letters 1958, **1**, 326.   [*8.2.2*]

Lau 66   LAURITSEN, T., and AJZENBERG-SELOVE, F., *Nucl. Phys.* 1966, **78**, 1.
            [*Fig. 5-8*]

Le 91   LEONARDO DA VINCI, *Codex Trivulzi.* Castello Sforcesco, Milan (205)
            (Luca Beltrami, Milan, 1891).   [*1.1*]

Le 47   LEA, D. E., *Actions of Radiations on Living Cells.* Macmillan, New York,
            1947.   [*4.2.3*]

Le 49   LEININGER, R. F., SEGRÈ, E., and WIEGAND, C. E., *Phys. Rev.* 1949, **76**,
            897.   [*8.5.3*]

Le 51   LEININGER, R. F., SEGRÈ, E., and WIEGAND, C. E., *Phys. Rev.* 1951, **81**,
            280.   [*8.5.3*]

Le 55   LEVY, H. B., "A New Empirical Mass Equation. I and II." University of
            California Radiation Lab. Repts. UCRL-4588 (1955) and UCRL-4713
            (1956); see also [Ri 56].   [*2.3.8*]

Le 56   LEE, T. D., and YANG, C. N., *Phys. Rev.* 1956, **104**, 254.   [*8.7*]

Le 57   LEE, T. D., and YANG, C. N., *Phys. Rev.* 1957, **105**, 1671.   [*8.2.2,
            8.7.1, D.7*]

Le 59   LEIGHTON, R. B., *Principles of Modern Physics.* McGraw-Hill, New York,
            1959.   [*8.5.2*]

Le 60   LEE, T. D., and YANG, C. N., *Phys. Rev.* 1960, **119**, 1410.   [*3.2.3*]

Le 61   LEUTZ, H., *Z. Physik* 1961, **164**, 78.   [*8.6.2*]

Le 63   LEE-WHITING, G. E., *Can. J. Phys.* 1963, **41**, 496.   [*G.2.2*]

Lé 64   LÉGER, J. J., and SHELDON, E., unpublished work (1964).   [*11.3.1*]

Le 65   LEIPUNSKII, O. I., NOVOZHILOV, B. V., and SAKHAROV, V. N., *The Propagation of Gamma Quanta in Matter* (translated from the Russian by Pra Senjit Basu). Pergamon Press, Oxford, 1965.   [*4.1*]

Le 66a   LEE, T. D. (ed.), *Weak Interactions and High-Energy Neutrino Physics.* Academic Press, New York, 1966.   [*8.2.2*]

Le 66b   LEDERMAN, L. M., "Recent Neutrino Experiments," on p. 176ff. of [Le 66a].   [*8.2.2*]

Le 66c   LEIGH, J. L., "An APEX Program for the Analysis of Gamma-Gamma Angular Correlation Data," Chalk River Rept. **AECL 2529** (1966). [*E.8.1*]

Lei 58   LEIPUNER, L. B., "Relativistic Two-Body Kinematics," Brookhaven Natl. Lab. Cosmotron Internal Rept. LBL-2 (1958).   [*C.1, C.2*]

Li 37   LIVINGSTON, M. S., and BETHE, H. A., *Rev. Mod. Phys.* 1937, **9**, 245. [*Fig. 4-58*]

Li 55a   LIBBY, W. F., *Radiocarbon Dating*, 2nd ed. Univ. of Chicago Press, Chicago, Illinois, 1955.   [*6.5*]

Li 55b   LICHNEROWICZ, A., *Théories relativistes de la gravitation et de l'électromagnetisme.* Masson, Paris, 1955.   [*D.8*]

Li 61a   LITHERLAND, A. E., and FERGUSON, A. J., *Can. J. Phys.* 1961, **39**, 788. [*9.4, E.8*]

Li 61b   LITTAUER, R. M., SCHOPPER, H. F., and WILSON, R. R., *Phys. Rev. Letters* 1961, **7**, 144; reproduced on p. 626ff of [Ho 63].   [*1.4, Fig. 1-5*]

Li 61c   LINDHARD, J., and SCHARFF, M., *Phys. Rev.* 1961, **124**, 128.   [*4.10*]

Li 63   LINDHARD, J., SCHARFF, M., and SCHIØTT, H. E., *Kgl. Danske Videnskab. Selskab, Mat.-Fys. Medd.* 1963, **33**, No. 14.   [*4.10, Fig. 4-97*]

Li 64   LIPNIK, P., PRALONG, G., and SUNIER, J. W., *Nucl. Phys.* 1964, **59**, 504. [*8.5.3, Fig. 8-36*]

Li 65   LICHTENBERG, D. B., *Meson and Baryon Spectroscopy.* Springer, New York, 1965; augmented and updated from *Ergeb. exakt. Naturw.* 1964, **36**, 83. [*5.3.10*]

Lo 55   LOUW, J. D., *Trans. Geol. Soc. S. Africa* 1955, **57**, 1.   [*6.5*]

Lo 66a   LOBASHOV, V. M., NAZARENKO, V. A., SAENKO, L. F., and SMOTRITSKII, L. M., *Zh. Eksperim. i Teor. Fiz. Pis'ma 1966*, **3**, No. 2, 76, *JETP Letters (English Transl.)* 1966, **3**, 47.   [*9.2.4*]

Lö 66b   LÖBNER, K. E. G., and MALMSKOG, S. G., *Nucl. Phys.* 1966, **80**, 505. [*9.2*]

Ma 51a   MATHER, R. L., *Phys. Rev.* 1951, **84**, 181.   [*4.9.4, Figs. 4-92, 4-93*]

Ma 51b   MARMIER, P., "Fonctions d'excitation de la réaction (p, n)," Doctoral Thesis, Eidg. Techn. Hochsch., Zürich, (1951).   [*Fig. 4-64*]

Ma 52a   MAHMOUD, H. M., and KONOPINSKI, E. J., *Phys. Rev.* 1952, **88**, 1266. [*8.6.2*]

Ma 52b   MAXIMON, L. C., and BETHE, H. A., *Phys. Rev.* 1952, **87**, 156.   [*4.6.4*]

Ma 54a   MAJOR, J. K., and BIEDENHARN, L. C., *Rev. Mod. Phys.* 1954, **26**, 321. [*8.5.2, Fig. 8-32*]

Ma 54b   MARTIN, C. N., *Tables numériques de physique nucléaire.* Gauthier-Villars, Paris, 1954.   [*7.43, 7.6.3, Fig. 7-13*]

Ma 55   MARMIER, P., and BOEHM, F., *Phys. Rev.* 1955, **97**, 103.   [*Table G-1*]

Ma 56a   MANN, A. K., *Phys. Rev.* 1956, **101**, 4.   [*4.3*]

Ma 56b   MARTON, L., MARTON, C., and HALL, W. G., "Electron Physics Tables,"
            Natl. Bur. Std. (U.S.), Circ. **571** (1956).   [*1.4*]

Ma 58a   MARTIN, P. C., and GLAUBER, R. J., *Phys. Rev.* 1958, **109**, 1037.   [*8.5.3*]

Ma 58b   MATHER, K. B., and SWAN, P., *Nuclear Scattering.* Cambridge Univ. Press,
            London and New York, 1958.   [*B.3, C.1, Fig. B-3*]

Ma 58c   MATZ, D., and KAEMPFFER, F. A., *Bull. Am. Phys. Soc.* [2] 1958, **3**, 317.
            [*3.2.4*]

Ma 59a   MALIK, S. S., POTNIS, V. R., and MANDEVILLE, C. E., *Nucl. Phys.* 1959,
            11, 691.   [*9.2.4, Fig. 9-8*]

Ma 59b   MANDL, F., *Introduction to Quantum Field Theory.* Wiley (Interscience),
            New York, 1959.   [*4.6.3, F.2, F.3.3, F.3.4*]

Ma 61   MANDUCHI, C., and ZANNONI, G., *Nuovo Cimento* [10] 1961, **22**, 462;
            see also *ibid.* 1962, **24**, 181; *Nucl. Phys.* 1962, **36**, 497.   [*8.5.3*]

Ma 63a   MAHAUX, C., *Bull. Soc. Roy. Sci. Liège* 1963, No. 1–2, 62 and 70.   [*11.5.7,*
            Fig. 11-33]

Ma 63b   MAHAUX, C., *Bull. Soc. Roy. Sci. Liège* 1963, No. 3–4, 240.   [*11.5.7*]

Ma 63c   MARX, G., *Nuovo Cimento* [10] 1963, **30**, 1555.   [*8.3.8*]

Ma 64   MANG, H. J., "Alpha Decay," *Ann. Rev. Nucl. Sci.* 1964, **14**, 1.   [*7.4.3*]

Ma 65a   MAHAUX, C., *Nucl. Phys.* 1965, **68**, 481.   [*11.5.7*]

Ma 65b   MAHAUX, C., and ROBAYE, G., *Nucl. Phys.* 1965, **74**, 161.   [*11.5.7,*
            Fig. 11-34]

Ma 65c   MALFORS, K. G., "Resonance Scattering of γ-Rays," on p. 1281ff. of
            [Si 65].   [*9.5.1*]

Ma 65d   MATTAUCH, J. H. E., THIELE, W., and WAPSTRA, A. H., *Nucl. Phys.*
            1965, **67**, 1.   [*2.1, 2.3.8*]

Ma 65e   MATTAUCH, J. H. E., THIELE, W., and WAPSTRA, A. H., *Nucl. Phys.*
            1965, **67**, 32.   [*2.1*]

Ma 66   MAHAUX, C., and WEIDENMÜLLER, H. A., *Phys. Letters* 1966, **23**, 100.
            [*11.3.1*]

Ma 67a   MACDONALD, J. R., and GRACE, M. A., *Nucl. Phys.* 1967, **A92**, 593.
            [*9.3*]

Ma 67b   MAQUEDA, E., and BLIN-STOYLE, R. J., *Nucl. Phys.* 1967, **A91**, 460.
            [*9.2.4*]

Ma X   MARION, J. B., and GINZBARG, A. S., "Tables for the Transformation of
            Angular Distribution Data from the Laboratory System to the Center
            of Mass System." Shell Development Co. Report (undated).   [*C.1*]

Mae 51   MAEDER, D., and PREISWERK, P., *Phys. Rev.* 1951, **84**, 595.   [*Fig. 8-35*]

McCa 55   McCARTHY, J. A., *Phys. Rev.* 1955, **97**, 1234.   [*8.2.2*]

McK 60   McKINLEY, W. A., *Am. J. Phys.* 1960, **28**, 129.   [*1.4*]

McKi 48   McKINLEY, W. A., and FESHBACH, H., *Phys. Rev.* 1948, **74**, 1759.
            [*4.9.2*]

McR 51   McREYNOLDS, A. W., *Phys. Rev.* 1951, **83**, 172 and 233.   [*3.2.4*]

Me 30   MEITNER, L., and ORTHMANN, W., *Z. Physik* 1930, **60**, 143.   [*8.1.3*]

Me 33   MEITNER, L., and KÖSTERS, H., *Z. Physik* 1933, **84**, 144.   [*4.7*]

Me 52   METZGER, F., *Phys. Rev.* 1952, **88**, 1360.   [*8.3.6, Fig. 8-20*]

Me 59   METZGER, F. R., "Resonance Fluorescence in Nuclei," *Progr. Nucl. Phys.*
            1959, **7**, 54.   [*9.5.1*]

Me 64   MEAD, C. A., *Phys. Rev.* 1964, **135**, B 849.   [*1.4*]

Me 68   MEAD, C. A., *Phys. Rev.* 1968, **143**, 990.   [*1.4*]

Mi 54   MILLER, W., MOTZ, J. W., and CIALELLA, C., *Phys. Rev.* 1954, **96**, 1344.
            [*Fig. 4-85*]

Mi 55   MITCHELL, A. C. G., "The Coincidence Method," on p. 201ff. of [Si 55].
            [*Fig. 8-4*]

Mi 56   MICHALOWICZ, A., *Compt. Rend.* 1956, **242**, 108.   [*8.5.3*]
Mi 64   MICHEL, F. C., *Phys. Rev.* 1964, **133**, B329.   [*9.2.4*]
Mi 66   MITRA, A. N., and BHASIN, V. S., *Phys. Rev. Letters* 1966, **16**, 523.   [*2.3.5*]
Mi 67   MILLER, P. D., DRESS, W. B., BAIRD, J. K., and RAMSEY, N. F., *Phys. Rev. Letters* 1967, **19**, 381.   [*5.4.5*]
Mi X   MINEHART, R. C., COULSON, L., GRUBB, W. F., and ZIOCK, K., Preprint (1968); *Phys. Rev.* (in press) 1968.   [*5.4.4*]
Mo 13   MOSELEY, H. G., *Phil. Mag.* [6] 1913, **26**, 1024.   [*1.2*]
Mo 14   MOSELEY, H. G., *Phil. Mag.* [6] 1914, **27**, 703.   [*1.2*]
Mo 29   MOTT, N. F., *Proc. Roy. Soc.* 1929, **A124**, 425.   [*4.9.2*]
Mo 30   MOTT, N. F., *Proc. Roy. Soc.* 1930, **A126**, 259.   [*3.5.8*]
Mø 32a MØLLER, C., *Ann. Physik* [5] 1932, **14**, 531.   [*4.9.2, F.3.1*]
Mo 32b MOTT, N. F., *Proc. Roy. Soc.* 1932, **A135**, 429.   [*4.9.2*]
Mo 40   MORRISON, P., and SCHIFF, L. I., *Phys. Rev.* 1940, **58**, 24.   [*8.5.3*]
Mo 46   MORETTE, C., *J. Phys. Radium* 1946, **7**, 135.   [*Fig. 4-16*]
Mo 47   MOLIÈRE, G., *Z. Naturforsch.* 1947, **2a**, 133.   [*4.9.2*]
Mo 49   MOTT, N. F., and MASSEY, H. S. W., *The Theory of Atomic Collisions*, 2nd ed. Oxford Univ. Press (Clarendon), London and New York, 1949.   [*3.5.8*]
Mo 50   MOON, P. B., *Proc. Phys. Soc. (London)* 1950, **A63**, 1189.   [*4.3, 9.5*]
Mo 51a MOON, P. B., *Proc. Phys. Soc. (London)* 1951, **A64**, 76.   [*9.5*]
Mo 51b MOSZKOWSKI, S. A., *Phys. Rev.* 1951, **82**, 35.   [*8.3.9*]
Mo 53a MOON, P. B., *Proc. Phys. Soc. (London)* 1953, **A66**, 585.   [*9.5*]
Mo 53b MOSZKOWSKI, S. A., *Phys. Rev.* 1953, **89**, 474.   [*9.2*]
Mo 53c MOTZ, H., and SCHIFF, L. I., *Am. J. Phys.* 1953, **21**, 258.   [*4.9.4*]
Mo 53d MORRISON, P., "A Survey of Nuclear Reactions," in *Experimental Nuclear Physics* (E. Segrè, ed.), Vol. II, p. 1ff. Wiley, New York, 1953   [*C.1*]
Mo 53e MOTZ, J. W., MILLER, W., and WYCKOFF, H. O., *Phys. Rev.* 1953, **89**, 968.   [*Fig. 4-86*]
Mo 58a MORPURGO, G., *Phys. Rev.* 1958, **110**, 721.   [*9.1*]
Mo 58b MORRISON, P., *Am. J. Phys.* 1958, **26**, 358.   [*3.2.4*]
Mö 58c MÖSSBAUER, R. L., *Z. Physik* 1958, **151**, 124.   [*9.5.2, Figs 9-20, 9-22*]
Mö 59a MÖSSBAUER, R. L., *Z. Naturforsch.* 1959, **14a**, 211.   [*9.5.2, Figs. 9-21, 9-22, 9-23*]
Mo 59b MOZER, F. S., *Phys. Rev.* 1959, **116**, 970.   [*2.3.8, 2.4.1, Figs. 2-7, 2-8, 2-9, 2-13*]
Mo 60a MOFFAT, J., and STRINGFELLOW, M. W., *Proc. Roy. Soc.* 1960, **A254**, 242.   [*4.7*]
Mö 60b MÖSSBAUER, R. L., and WIEDEMANN, W. H., *Z. Physik* 1960, **159**, 33.   [*9.5.2*]
Mo 60c MOORE, R. G., Jr., *Rev. Mod. Phys.* 1960, **32**, 101.   [*11.5.4*]
Mo 61   MOTZ, J. W., and MISSONI, G., *Phys. Rev.* 1961, **124**, 1458.   [*Fig. 4-9*]
Mo 63   MOLER, R. B., and FINK, R. W., *Phys. Rev.* 1963, **131**, 821.   [*8.5.3*]
Mo 64a MOAZED, C., ETTER, J. E., HOLMGREN, H. D., and WAGGONER, M. A., University of Maryland Tech. Rept. No. 427 (1964); contribution to the Gatlinburg Conference on Correlations of Particles Emitted in Nuclear Reactions (Oct. 1964); *Rev. Mod. Phys.* 1965, **37**, 354.   [*11.1.5, Figs. 11-5, 11-6, 11-7*]
Mo 64b MOAZED, C., ETTER, J. E., HOLMGREN, H. D., and WAGGONER, M. A., University of Maryland Tech. Rept. No. 427 (1964); contribution to the Gatlinburg Conference on Correlations of Particles Emitted in Nuclear Reactions (Oct. 1964); *Rev. Mod. Phys.* 1965, **37**, 441.   [*11.1.5*]

Mö 65   Mössbauer, R. L., "Recoilless Nuclear Resonance Absorption and its
             Applications," on p. 1293ff. of [Si 65].   [*9.5.2*]

Mo 66a   Moak, C. D., and Brown, M. D., *Phys. Rev.* 1966, **149**, 244.   [*4.10,
             Fig. 4-98*]

Mo 66b   Møller, C., *The Theory of Relativity*. Oxford Univ. Press (Clarendon),
             London and New York, 1966.   [*A.1*]

Mu 52   Muller, D. E., Hoyt, H. C., Klein, D. J., and DuMond, J. W. M.,
             Phys. Rev. 1952, **88**, 775.   [*Table G-1*]

Mü 63   Münnich, K. O., *Naturwissenschaften* 1963, **50**, 211.   [*6.5*]

Mui 65   Muirhead, H., *The Physics of Elementary Particles*. Pergamon Press,
             Oxford, 1965.   [*D.8, F.2, F.3.3, F.3.4, Table F-1*]

Mui 66   Muir, A. H., Jr., Ando, K. J., and Coogan, H. M., *Mössbauer
             Effect Data Index 1959–1965*, Wiley (Interscience), New York, 1966.
             [*9.5.2*]

Na 04   Nagaoka, H., *Phil. Mag.* [6] 1904, **7**, 445.   [*1.2*]

Na 53   Nakato, T., and Nishijima, K., *Progr. Theoret. Phys. (Kyoto)* 1953, **10**, 581.
             [*5.3.9*]

NBS 52   National Bureau of Standards, "Tables for the Analysis of Beta
             Spectra," N.B.S. Applied Mathematics Series No. 13 (2nd June,
             1952) (reissue of 1951 report.) Initiated at the request of L. F. Curtiss
             and I. Feister, under the supervision of I. A. Stegun, with an introduc-
             tion by U. Fano.   [*8.3.5, 8.3.9, Figs. 8-16, 8-17*]

Ne 56   Nelms, A. T., *Nat. Bur. Std. (U.S.) Circ.* **577** (1956).   [*4.9.2*]

Ne 58   Nelms, A. T., *Nat. Bur. Std. (U.S.), Circ.* **577**, Suppl. (1958).   [*4.9.2*]

Ne 59   Nelson, D. F., Schupp, A. A., Pidd, R. W., and Crane, H. R., *Phys. Rev.
             Letters* 1959, **2**, 492.   [*5.4.5*]

Ne 66   Nezrick, F. A., and Reines, F., *Phys. Rev.* 1966, **142**, 852.   [*3.5, 8.2.2*]

Nei 65   Neiler, J. H., and Bell, P. R., "The Scintillation Method," on p. 245ff.
             of [Si 65].   [*G.1*]

Nei 68   Neito, M. M., *Phys. Rev. Letters* 1968, **21**, 488.   [*3.2.3*]

Neu 66   Neuert, H., *Kernphysikalisches Messverfahren*. Braun, Karlsruhe, 1966.
             [*3.5, 8.2.2, [G.1] Fig. 8-10*]

Nie 61   Nielsen, L. P., *Kgl. Danske Videnskab. Selskab, Mat.-Fys. Medd.* 1961, **33**,
             No. 6.   [*Fig. 4-71(b)*]

Ni 65   Nix, J. R., and Swiatecki, W. J., *Nucl. Phys.* 1965, **71**, 1.   [*2.2*]

No 60   Northcliffe, L. C., *Phys. Rev.* 1960, **120**, 1744.   [*4.10*]

No 63   Northcliffe, L. C., "Passage of Heavy Ions through Matter," *Ann. Rev.
             Nucl. Sci.* 1963, **13**, 67; reproduced on p. 353ff. of [NRC 64].   [*4.10,
             Fig. 4-97*]

No 64   Northcliffe, L. C., "Passage of Heavy Ions through Matter: II. Range–
             Energy Curves," on p. 173ff. of [NRC 64].   [*4.10*]

No 65   Novakov, T., Hollander, J. M., and Graham, R. L., "Anomalous
             L-Subshell Ratios in Mixed MI-E2 Transitions," on p. 158ff. of
             [Ja 65].   [*10.1.3*]

Noe 18   Noether, E., *Nachr. Ges. Wiss. Goettingen, Math.-Phys. Kl.* 1918, p. 235.
             [*D.8*]

NRC 64   National Research Council, U.S. National Academy of Sciences, "Studies
             in Penetration of Charged Particles in Matter," *Natl. Acad. Sci.—
             Nat. Res. Council, Publ.* No. 1133 (1964).   [*4.9*]

Ny 66   Nyborg, P., *Am. J. Phys.* 1966, **34**, 932.   [*C.1*]

Ok 65   Okun, L. B., and Pomeranchuk, I. Ya., *Phys. Letters* 1965, **16**, 338.
             [*8.2.2*]

Op 30   Oppenheimer, J. R., *Phys. Rev.* 1930, **35**, 939.   [*1.2*]

Op 33   OPPENHEIMER, J. R., and PLESSET, M. S., *Phys. Rev.* 1933, **44**, 54L. [*4.6*]
Pa 24   PAULI, W., *Naturwissenschaften* 1924, **12**, 741.   [*5.2*]
Pa 27   PAULI, W., *Z. Physik* 1927, **43**, 601.   [*5.2*]
Pa 33   PAULI, W., *Proc. 7th Solvay Congr.*, (*Brussels*, 1933). Gauthier-Villars, Paris, 1934, p. 324.   [*8.1.3*]
Pa 34   PAULI, W., and WEISSKOPF, V., *Helv. Phys. Acta* 1934, **7**, 709.   [*D.4*]
Pa 50   PAKE, G. E., *Am. J. Phys.* 1950, **18**, 438 and 473.   [*5.4.5*]
Pa 52   PAIS, A., *Phys. Rev.* 1952, **86**, 663.   [*5.3.9*]
Pa 55   PATTERSON, C., *Geochim. Cosmochim. Acta* 1955, **7**, 151.   [*6.5*]
Pa 57   PAULI, W., *Nuovo Cimento* [10] 1957, **6**, 204.   [*8.6.1*]
Pa 58   PAULI, W., *Theory of Relativity*. Pergamon Press, London and New York 1958; translated from the article "Relativitätstheorie," in *Enzyklopädie der mathematischen Wissenschaften*, Vol. V 19. Teubner, Leipzig, 1921. [*A.1, D.7*]
Pa 64   PARKER, S., ANDERSON, H. L., and REY, C., *Phys. Rev.* 1964, **133**, B768. [*8.2.2*]
Pa 67   PAULI, H. C., "Tables of the Internal Conversion Coefficients of the *K*, *L*, and *M* Shell for Heavy Nuclei," unpublished compilation, University of Basel, Basel, 1967.   [*10.1.3, Fig. 10-1*]
Pau 35   PAULING, L., and WILSON, E. B., Jr., *Introduction to Quantum Mechanics.* McGraw-Hill, New York, 1935.   [*D.2, Figs. D-1, D-2, D-3*]
Pe 50   PERLMAN, I., GHIORSO, A., and SEABORG, G. T., *Phys. Rev.* 1950, **77**, 26. [*7.4.3*]
Pe 54   PEASLEE, D. C., *Phys. Rev.* 1954, **95**, 717.   [*2.3.4*]
Pe 57   PERLMAN, I., and RASMUSSEN, J. O., "Alpha Radioactivity," in *Handbuch der Physik* (S. Flügge, ed.), Vol. 42, p. 109. Springer, Berlin, 1957. [*7.4.3*]
Pe 58   PERLMAN, M. L., WELKER, J. P., and WOLFSBERG, M., *Phys. Rev.* 1958, **110**, 381.   [*8.5.3*]
Pe 65   PETTERSON, B. G., "The Internal Bremsstrahlung," on p. 1569ff. of [Si 65]. [8.5.3]
Pl 00   PLANCK, M., *Verhandl. Deut. Physik. Ges.* 1900, **2**, 237.   [*1.2*]
Po 49   PONTECORVO, B., KIRKWOOD, D. H. W., and HANNA, G. C., *Phys. Rev.* 1949, **75**, 982.   [*8.5.3*]
Po 52   POUND, R. V., "Nuclear Paramagnetic Resonance," *Progr. Nucl. Phys.* 1952, **2**, 21.   [*5.4.5*]
Po 60   POUND, R. V., and REBKA, G. A., Jr., *Phys. Rev. Letters* 1960, **4**, 274. [*3.2.4, 9.5.2*]
Po 63   POTNIS, V. R., and RAO, G. N., *Nucl. Phys.* 1963, **42**, 620.   [*9.2.4*]
Po 64   POUND, R. V., and SNIDER, J. L., *Phys. Rev. Letters* 1964, **13**, 539.   [*3.2.4*]
Po 65   POUND, R. V., and SNIDER, J. L., *Phys. Rev.* 1965, **140**, B788.   [*3.2.4*]
Pr 36   PROCA, A., *J. Phys. Radium* 1936, **7**, 347.   [*D.8*]
Pr 47   PRESTON, M. A., *Phys. Rev.* 1947, **71**, 865.   [*7.4.3*]
Pr 59   PRIMAKOFF, H., and ROSEN, S. P., "Double Beta Decay," *Rept. Progr. Phys.* 1959, **22**, 121; see also *Proc. Phys. Soc.* (*London*) 1961, **A78**, 464. [*8.2.2, 8.5.3, Fig. 8-37*]
Pr 60   PRATT, R. H., *Phys. Rev.* 1960, **117**, 1017.   [*4.4.1*]
Pr 61   PRIMAKOFF, H., and ROSEN, S. P., *Proc. Phys. Soc.* (*London*) 1961, **A78**, 464; see also [Pr 59].   [*8.2.2*]
Pr 62   PRESTON, M. A., *Physics of the Nucleus.* Addison-Wesley, Reading, Massachusetts, 1962.   [*10.1.3, 11.5.4, 11.5.7*]

Pr 64   PRICE, J., *Nuclear Radiation Detection*, McGraw-Hill Series in Nuclear
           Engineering. McGraw-Hill, New York, 1964.   [*G.1*]

Pu 46   PURCELL, E. M., TORREY, H. C., and POUND, R. V., *Phys. Rev.* 1949,
           69, 37L.   [*5.4.5*]

Pu 62   PUPPI, G., *Proc. Intern. Conf. High Energy Phys.*, CERN, 1962, p. 713ff.
           [*8.8*]

Pu 63   PURCELL, E. M., COLLINS, G. B., FUJII, T., HORNBOSTEL, J., and TURKOT,
           F., *Phys. Rev.* 1963, **129**, 2326.   [*5.4.5*]

Ra 36   RASETTI, E., *Elements of Nuclear Physics*, p. 134. Prentice-Hall, Englewood
           Cliffs, New Jersey, 1936.   [*10.1.3*]

Ra 38   RABI, I, I., ZACHARIAS, J. R., MILLMAN, S., and KUSCH, P., *Phys. Rev.*
           1938, **53**, 318L.   [*5.4.5*]

Ra 39   RABI, I. I., MILLMAN, S., KUSCH, P., and ZACHARIAS, J. R., *Phys. Rev.*
           1939, **55**, 526.   [*5.4.5*]

Ra 41   RARITA, W., and SCHWINGER, J., *Phys. Rev.* 1941, **59**, 436.   [*5.4.8*]

Ra 47   RAINWATER, L. J., HAVENS, W. W., Jr., WU, C. S., and DUNNING, J. R.,
           Phys. Rev. 1947, **71**, 65.   [*Exercise 11-9*]

Ra 52a   RADICATI, L. A., *Phys. Rev.* 1952, **87**, 521.   [*9.1*]

Ra 52b   RADVANYI, P., *Compt. Rend.* 1952, **235**, 289 and 428.   [*8.5.3*]

Ra 53   RAMSEY, N. F., *Nuclear Moments*. Wiley, New York, 1953.   [*5.2, 5.4.5*]

Ra 56   RAMSEY, N. F., *Molecular Beams*. Oxford Univ. Press, London and New
           York, 1956.   [*5.2, 5.4.5*]

Ra 61   RAMASWAMI, M. K., *Indian J. Phys.* 1961, **35**, 610.   [*8.6.2*]

Ra 65a   RASMUSSEN, J. O., "Alpha-Decay," on p. 701ff. of [Si 65].   [*7.4.3,
           Fig. 7-1*]

Ra 65b   RAZ, B. J., and BOCKELMAN, C. K., "Abstracts of Computer Codes,"
           Brookhaven Natl. Lab. Rept. BNL-9108 (1965).   [*E.8.1*]

Re 53a   REINES, F., and COWAN, C. L., Jr., *Phys. Rev.* 1953, **90**, 492.   [*8.2.2*]

Re 53b   REINES, F., and COWAN, C. L., Jr., *Phys. Rev.* 1953, **92**, 830.   [*8.2.2*]

Re 57   REINES, F., and COWAN, C. L., Jr., *Phys. Today* 1957, **10**, No. 8, 12.
           [*8.2.2, Fig. 8-8*]

Re 59   REINES, F., and COWAN, C. L., Jr., *Phys. Rev.* 1959, **113**, 273. [*8.2.2, Fig. 8*]

Re 60   REINES, F., COWAN, C. L., Jr., HARRISON, F. B., McGUIRE, A. D., and
           KRUSE, H. W., *Phys. Rev.* 1960, **117**, 159; see also *ibid.* 1959, **113**,
           273 and 280.   [*3.5, 8.2.2, Figs. 8-8, 8-9*]

Re 64   REINES, F., and KROPP, W. R., *Phys. Rev. Letters* 1964, **12**, 457.   [*8.2.2*]

Re 65a   REINES, F., CROUCH, M. F., JENKINS, T. L., KROPP, W. R., GURR, H. S.,
           SMITH, G. R., SELLSCHOP, J. P. F., and MEYER, B., *Phys. Rev. Letters*
           1965, **15**, 429.   [*8.2.2*]

Re 65b   REINES, F., and WOODS, R. M., Jr., *Phys. Rev. Letters* 1965, **14**, 20.
           [*8.2.2*]

Re 66a   REINES, F., and SELLSCHOP, J. P. F., *Sci. Am.* 1966, **214**, No. 2, 40 (see also
           references on pp. 138 and 139 of this journal).   [*8.2.2*]

Re 66b   REINES, F., *Science J.* 1966, **2**, No. 10, 84.   [*8.2.2*]

Ri 56   RIDDELL, J., "A Table of Levy's Empirical Atomic Masses," Chalk River
           Rept. CRP-654 (**AECL-339**) (1956).   ⌐*2.3.8*]

Rie 55   RIEZLER, W., and RUDLOFF, A., *Ann. Physik* [6] 1955, **15**, 224. [*Fig. 4-60*]

Rö 95   RÖNTGEN, W. K., *Sitzber. Physik.-Med. Ges.*, *Würzberg*, 1895; condensed
           English translation in W. F. Magie, *Source Book in Physics*. McGraw-
           Hill, New York, 1935.   [*1.2*]

Ro 48   Rose, M. E., "Polarization of Nuclear Spins," Oak Ridge Natl. Lab. Rept. ORNL-48 (**AECD-2119**), (1948).   [*8.7.1*]

Ro 49a   Rose, M. E., *Phys. Rev.* 1949, **75**, 213.   [*8.7.1*]

Ro 49b   Rose, M. E., and Jackson, J. L., *Phys. Rev.* 1949, **76**, 1540.   [*8.5.3*]

Ro 51a   Robson, J. M., *Phys. Rev.* 1951, **83**, 349.   [*Fig. 8-18(a)*]

Ro 51b   Rose, M. E., Goertzel, G. H., Spinrad, B. I., Harr, J., and Strong, P., *Phys. Rev.* 1951, **83**, 79.   [*10.1.3, Fig. 10-8*]

Ro 52a   Rodeback, G. W., and Allen, J. S., *Phys. Rev.* 1952, **86**, 446. [*8.2.2, Fig. 7*] *Fig. 8-7*]

Ro 52b   Rose, M. E., Dismuke, N. M., Perry, C. L., and Bell, P. R., "Fermi Functions for Allowed Beta Transitions," Oak Ridge Natl. Lab. Rept. ORNL-1222 (1952).   [*8.3.5, 8.3.9*]

Ro 53   Rose, M. E., Perry, C. L., and Dismuke, N. M., "Tables for the Analysis of Allowed and Forbidden Beta Transitions," Oak Ridge Natl. Lab. Rept. ORNL-1459 (1953).   [*8.3.5*]

Ro 55   Robinson, B. L., and Fink, R. W., *Rev. Mod. Phys.* 1955, **27**, 424. [*8.3.9*]

Ro 57   Rose, M. E., *Elementary Theory of Angular Momentum.* Wiley, New York, 1957.   [*E.2, E.8*]

Ro 58   Rose, M. E., *Internal Conversion Coefficients.* North-Holland Publ. Co., Amsterdam, 1958.   [*10.1.3*]

Ro 59   Rotenberg, M., Bivins, R., Metropolis, N., and Wooten, J. K., Jr. *The 3-j and 6-j Symbols.* M.I.T. Press, Cambridge, Massachusetts, 1959.   [*E.2*]

Ro 60   Robinson, B. L., and Fink, R. W., *Rev. Mod. Phys.* 1960, **32**, 117. [*8.3.9, 8.5.3*]

Ro 61   Rosenfeld, L., *Nucl. Phys.* 1961, **26**, 594.   [*11.5.7, Fig. 11-33*]

Ro 64a   Roman, P., *Theory of Elementary Particles*, 2nd ed. North-Holland Publ. Co., Amsterdam, 1964.   [*D.8*]

Ro 64b   Rosenfeld, A. H., Barbaro-Galtieri, A., Barkas, W. H., Bastien, P. L., and Kirz, J., *Rev. Mod. Phys.* 1964, **36**, 977.   [*I*]

Ro 64c   Rosser, W. G. V., *An Introduction to the Theory of Relativity.* Butterworth, London and Washington, D.C., 1964.   [*A.2*]

Ro 65a   Rose, M. E., "Theory of Internal Conversion," on p. 887ff. of [Si 65]. [*10.1.3, 10.1.4*]

Ro 65b   Rosen, S. P., and Primakoff, H., "Double β-Decay," on p. 1499ff. of [Si 65].   [*8.2.2*]

Ro 66   Robson, D., "Isobaric Spin in Nuclear Physics," *Ann. Rev. Nucl. Sci.* 1966, **16**, 119.   [*5.4.4*]

Ro 67   Rose, H. J., and Brink, D. M., *Rev. Mod. Phys.* 1967, **39**, 306.   [*9.2.4*]

Ro 68   Rosenfeld, A. H., Barash-Schmidt, N., Barbero-Galtieri, A., Price, L. R., Söding, P., Wohl, C. G., Roos, M., and Willis, W. J., *Rev. Mod. Phys.* 1968, **40**, 77.   [*I*]

Ru 11   Rutherford, E., *Phil. Mag.* [6] 1911, **21**, 669.   [*1.2, 3.5.7*].

Ru 65   Rudermann, M. A., "Astrophysical Neutrinos," *Rept. Progr. Phys.* 1965, **28**, 411.   [*8.2.2*]

Sa 31   Sauter, F., *Ann. Physik* [5] 1931, **11**, 454.   [*4.4.1, Fig. 4-15*]

Sa 33   Sargent, B. W., *Proc. Roy. Soc.* 1933, **A139**, 659.   [*8.4, Fig. 8-27*]

Sa 52   Sachs, R. G., *Phys. Rev.* 1952, **87**, 1100.   [*5.4.6*]

Sa 53   Satchler, G. R., *Proc. Roy. Soc.* 1953, **66A**, 1081.   [*E.8.1, E.8.2*]

Sa 54   Satchler, G. R., *Phys. Rev.* 1954, **94**, 1304.   [*9.4*]

Sa 56   Satchler, G. R., *Phys. Rev.* 1956, **104**, 1198; erratum, *ibid.* 1958, **111**, 1747.   [*9.4, E.8.1*]

Sa 57   SALAM, A., *Nuovo Cimento* [10] 1957, **5**, 299.   [*8.7.1*]

Sa 58a   SACHS, A. M., WINICK, H., and WOOTEN, B. A., *Phys. Rev.* 1958, **109**, 1733.
           [*11.3.1, Exercise 11-17*]

Sa 58b   SAKURAI, J. J., *Phys. Rev. Letters* 1958, **1**, 40.   [*8.3.7*]

Sa 60a   SAKURAI, J. J., *Ann. Phys.* (*N.Y.*) 1960, **11**, 1.   [*3.2.4*]

Sa 60b   SANTOS OCAMPO, A. G., and CONWAY, D. C., *Phys. Rev.* 1960, **120**, 2196.
           [*8.5.3*]

Sa 61   SAKISAKA, M., and TOMITA, M., *J. Phys. Soc. Japan* 1961, **16**, 2597.
           [*2.3.5*]

Sa 62   SANTOS OCAMPO, A. G., and CONWAY, D. C., *Phys. Rev.* 1962, **128**, 258.
           [*8.5.3*]

Sa 64   SAKAI, M., and YAMAZAKI, T., "Memorandum on Mixed Angular
           Correlations," Institute for Nuclear Study, Tokyo, Rept. INSJ-66
           (1964).   [*9.2.4*]

Sa 65   SANDAGE, A. R., *Astrophys. J.* 1965, **141**, 1560.   [*6.5*]

Schi 55   SCHIFF, L. I., *Quantum Mechanics*, 2nd ed. McGraw-Hill, New York, 1955.
           [*3.4.1, 3.6.1, 3.6.4, 7.4.3, 9.1, 11.2.1, D.2*]

Schi 58   SCHIFF, L. I., *Phys. Rev. Letters* 1958, **1**, 254.   [*3.2.4*]

Schi 59   SCHIFF, L. I., *Proc. Natl. Acad. Sci. U.S.* 1959, **45**, 69.   [*3.2.4*]

Schi 60   SCHINTLMEISTER, J., (ed.), *Der Isospin von Atomkernen*. Akademie-Verlag,
           Berlin, 1960.   [*5.4.4, Figs. 5-6, 5-7*]

Schm 37   SCHMIDT, T., *Z. Physik* 1937, **106**, 358.   [*5.4.7*]

Scho 59   SCHOPPER, H., and MÜLLER, H., *Nuovo Cimento* [10] 1959, **13**, 1028.
           [*8.8*]

Scho 60   SCHOPPER, H., *Fortschr. Physik* 1960, **8**, 327.   [*8.3.7*]

Scho 66   SCHOPPER, H. F., *Weak Interactions and Nuclear Beta-Decay*. North-Holland
           Publ. Co., Amsterdam, 1966.   [*8.1, 8.2.2, 8.3.5, 8.5.3, 8.6.1, 8.6.3,
           8.7.2, Figs. 8-24, 8-40*]

Scho 67   SCHOPPER, H., CERN Rept. 67-3 (February, 1967).   [*1.4*]

Scho 67   SCHOPPER, H., "Some Remarks concerning the Proton Structure,"
           CERN Rept. 67-3 (February, 1967).   [*1.4*]

Schr 55   SCHREINER, G. D. L., JAMIESON, R. T., and SCHONLAND, B. F. J., *Nature*
           1955, **175**, 464.   [*6.5*]

Schrö 26   SCHRÖDINGER, E., *Ann. Physik* [4] 1926, **81**, 109.   [*D.4*]

Schw 48   SCHWINGER, J., *Phys. Rev.* 1948, **73**, 416L.   [*5.4.5*]

Schw 49   SCHWINGER, J., *Phys. Rev.* 1949, **76**, 760.   [*5.4.5*]

Schw 56   SCHWEBER, S. S., BETHE, H. A., and DE HOFFMANN, F., *Mesons and Fields*,
           Vol. I: *Fields*. Row, Peterson and Co., Evanston, Illinois, 1956.
           [*4.6.3, F.2, F.3.3, F.3.4*]

Schw 58   SCHWINGER, J., (ed.), *Selected Papers on Quantum Electrodynamics*. Dover,
           New York, 1958.   [*4.6.2*]

Schw 65   SCHWARZ, S., STRÖMBERG, L. G., and BERGSTRÖM, A., *Nucl. Phys.* 1965,
           63, 593.   [*Fig. 11-34*]

Schw 66   SCHWARTZ, M., "Neutrino Experiments," on p. 161ff. of [Le 66a].
           [*8.2.2*]

Sco 59   SCOBIE, J., MOLER, R. B., and FINK, R. W., *Phys. Rev.* 1959, **116**, 657.
           [*8.5.3*]

Se 47   SEGRÈ, E., *Phys. Rev.* 1947, **71**, 247.   [*8.5.3*]

Se 49a   SEGRÈ, E., and HELMHOLZ, A. C., *Rev. Mod. Phys.* 1949, **21**, 271.
           [*10.1.3*]

Se 49b   SEGRÈ, E., and WIEGAND, C. E., *Phys. Rev.* 1949, **75**, 39.   [*8.5.3*]

Se 51   SEGRÈ, E. and WIEGAND, C. E., *Phys. Rev.* 1951, **81**, 284.   [*8.5.3*]

Se 55   SELDOWITSCH, J. B., *Usp. Fiz. Nauk* 1955, **56**, 165; *(German transl.)* *Fortschr. Physik* 1956, **4**, 33.   *[8.6]*

Se 59a   SEEGER, P. A., *Bull. Am. Phys. Soc.* [2] 1959, **4**, 461.   *[2.3.8]*

Se 59b   SEGRÈ, E., (ed.), *Experimental Nuclear Physics*, Vol. III. Wiley, New York, 1959.   *[8.3.5, Figs. 8-31, 8-39]*

Se 60   SENGUPTA, S., *Nucl. Phys.* 1960, **21**, 542.   *[5.4.4]*

Se 61a   SEEGER, P. A., *Nucl. Phys.* 1961, **25**, 1.   *[2.3.8]*

Se 61b   SEPPI, E. J., and BOEHM, F., *Bull. Am. Phys. Soc.* [2] 1961, **116**, 503. *[9.5.1]*

Se 61c   SEGEL, R. E., OLNESS, J. W., and SPRENKEL, E. L., *Phil. Mag.* [8] 1961, **6**, 163.   *[Exercise 7-5]*

Se 61d   SEGEL, R. E., OLNESS, J. W., and SPRENKEL, E. L., *Phys. Rev.* 1961, **123**, 1382.   *[Exercise 7-5]*

Se 64   SEGRÈ, E., *Nuclei and Particles*. Benjamin, New York, 1964.   *[3.4.1, 5.4.5, Fig. 6-22]*

Se 66   SEXL, R. U., *Phys. Letters* 1966, **20**, 376.   *[3.2.3]*

Sea 54   SEARS, B. J., and RADTKE, G., "Algebraic Tables of Clebsch-Gordan Coefficients," Chalk River Rept. TPI-75 (1954).   *[E.2]*

Sh 54   SHERR, R., and MILLER, R. H., *Phys. Rev.* 1954, **93**, 1076.   *[8.6.2]*

Sh 61   SHIRLEY, D. A., *Phys. Rev.* 1961, **124**, 354.   *[9.5.2]*

Sh 63   SHELDON, E., *Rev. Mod. Phys.* 1963, **35**, 795.   *[9.4]*

Sh 64a   SHELDON, E., "Interpretation of Angular Distributions and Correlations," on p. 155ff. of *Les mécanismes des réactions nucléaires*. Bur d'impression MRP, Lausanne, 1964.   *[9.4, E.8, E.8.1, Fig. 9-15]*

Sh 64b   SHELDON, E., *Phys. Rev.* 1964, **133**, B732.   *[E.8.2]*

Sh 65a   SHAPIRO, M. H., FRANKEL, S., KOICKI, S., WALES, W., and WOOD, G. T., *Bull. Am. Phys. Soc.* [2] 1965, **10** (4), Paper AB 3.   *[8.2.2]*

Sh 65b   SHAPIRO, M. H., *Bull. Am. Phys. Soc.* [2] 1965, **10** (5), Paper FA 2.   *[8.2.2]*

Sh 66   SHELDON, E., and VAN PATTER, D. M., *Rev. Mod. Phys.* 1966, **38**, 143. *[9.4, E.8.1, Fig. 9-17]*

Sh 67   SHELDON, E., and GANTENBEIN, P., *J. Appl. Math. Phys. (ZAMP)* 1967, **18**, 397.   *[9.4, E.8.1, E.8.2]*

Sha 58   SHARP, W. T., "The Quantum Theory of Angular Momentum," Chalk River Rept. **AECL 465** (1957) (revised 1958).   *[E.2]*

Sha 60   SHARP, W. T., *Am. J. Phys.* 1960, **28**, 116.   *[E.2]*

Sha 67   SHARP, R. T., *Nucl. Phys.* 1967, **A95**, 222.   *[E.7]*

She 60   SHERWIN, C. W., *Phys. Rev.* 1960, **120**, 17.   *[Exercise A-5]*

Shu 67   SHULL, C. G., and NATHANS, R., *Phys. Rev. Letters* 1967, **19**, 384.   *[5.4.5]*

Si 54   SIMON, A., "Numerical Table of the Clebsch-Gordan Coefficients," Oak Ridge Natl. Lab. Rept. ORNL-1718 (1954).   *[E.2]*

Si 55   SIEGBAHN, K., (ed.), *Beta- and Gamma-Ray Spectroscopy*. North-Holland Publ. Co., Amsterdam, 1955.   *[8.3.5]*

Si 58   SIMONS, L., "Positronium," in *Handbuch der Physik* (S. Flügge, ed.), Vol. 34, p. 139ff. Springer, Berlin, 1958.   *[4.6.5]*

Si 63   SINGHAL, N. C., and TREHAN, P. N., *Proc. Bombay Nucl. Phys. Symp.* 1963, p. 80.   *[9.2.4, Figs. 9-7, 9-9]*

Si 65   SIEGBAHN, K., *Alpha-, Beta- and Gamma-Ray Spectroscopy*, Vols. I and II. North-Holland Publ. Co., Amsterdam, 1965.   *[4.1, 4.3, 4.5, 4.8, 8.1, 8.3.5, 8.4.4, 8.7.2, 9.1, 9.2, 10, G.1, Figs. 9-24, 10-9, 10-10, A-6, G-2, G-4, G-7]*

Sl 59   SLACK, L., and WAY, K., "Radiations from Radioactive Atoms," *U.S. At. Energy Comm. Publ.* 1959.   [*6.4*]

Sl 61   SLIV, L. A., and BAND, I. M., *Gamma Rays.* Acad. Sci. U.S.S.R., Moscow-Leningrad, 1961.   [*10.1.3, Fig. 10-8*]

Sl 62   SLATER, D. N., *Gamma-Rays of Radionuclides in Order of Increasing Energy.* Butterworth, London and Washington, D. C., 1962.   [*6.4*]

Sl 65   SLIV, L. A., and BAND, I. M., "Tables of Internal Conversion Coefficients," reprinted as Appendix 5 on p. 1639ff. of [Si 65].   [*10.1.3*]

Sm 51   SMITH, A. M., *Phys. Rev.* 1951, **82**, 955L.   [*8.4.4*]

Sm 57   SMITH, J., PURCELL, E. M., and RAMSEY, N. F., quoted by Segrè on pp. 221–222 of [Se 64].   [*5.4.5*]

Sm 65   SMITH, C. M. H., *A Textbook of Nuclear Physics.* Pergamon Press, Oxford, 1965.   [*Figs. 4-3, 4-78, 6-13*]

Sn 55   SNELL, A. H., and PLEASANTON, F., *Phys. Rev.* 1955, **100**, 1396.   [*8.3.7*]

So 16   SOMMERFELD, A., *Ann. Physik* [4] 1916, **51**, 1.   [*1.2, 5.3.7, D.5*]

So 39   SOMMERFELD, A., *Atombau and Spektrallinien*, Vol. 2, p. 551. Vieweg, Braunschweig, 1939.   [*4.9.3*]

So 58   SOSNOVSKII, A. N., SPIVAK, P. E., PROKOFIEV, IU. A., KUTIKOV, I. E., and DOBRININ, IU. P., *Zh. Eksperim. i Teor. Fiz.* 1958, **35**, 1059; *Soviet Phys. JETP (English Transl.)* 1959, **8**, 739 (L).   [*1.4*]

So 59a   SOSNOVSKII, A. N., SPIVAK, P. E., PROKOFIEV, IU. A., KUTIKOV, I. E., and DOBRININ, IU. P., *Zh. Eksperim. i Teor. Fiz.* 1959, **36**, 1012; *Soviet Physics JETP (English Transl.)* 1959, **36**, 717.   [*1.4*]

So 59b   SOSNOVSKII, A. N., SPIVAK, P. E., PROKOFIEV, IU. A., KUTIKOV, I. E., and DOBRININ, IU. P., *Nucl. Phys.* 1959, **10**, 395.   [*1.4*]

St 22   STERN, O., and GERLACH, W., *Z. Physik* 1922, **8**, 110; **9**, 349   [*5.1, 5.4.5*]

St 41a   STUECKELBERG DE BREIDENBACH, E. C. G., *Helv. Phys. Acta* 1941, **14**, 51.   [*4.6.2*]

St 41b   STUECKELBERG DE BREIDENBACH, E. C. G., *Helv. Phys. Acta* 1941, **14**, 588.   [*4.6.3*]

St 52   STUECKELBERG DE BREIDENBACH, E. C. G., *Helv. Phys. Acta* 1952, **25**, 577.   [*11.5.6*]

St 55   STERNHEIMER, R. M., *Phys. Rev.* 1955, **98**, 205.   [*C.1, C.2*]

St 62   STIERLIN, U., SCHOLZ, W., and POVH, B., *Z. Physik* 1962, **170**, 47.   [*4.7*]

St 63   STERNHEIMER, R. M., "Kinematics," Appendix 2 on p. 821 ff. of *Nuclear Physics* (L. C. L. Yuan and C. S. Wu, eds.), Vol. 5, Part B of *Methods of Experimental Physics* (L. Marton, ed.-in-chief). Academic Press, New York, 1963.   [*C.1, C.3.5*]

St 65   STELSON, P. H., and GRODZINS, L., *Nucl. Data* 1965, **A1**, 21. [*9.2.1, Fig. 5] Fig. 9.5*]

Su 57   SULLIVAN, W. H., "Trilinear Chart of Nuclides," 2nd ed. (revised). Prepared at the Oak Ridge Natl. Lab. for the U.S. Atomic Energy Commission, 1957.   [*Fig. 2-1*]

Su 58   SUESS, H. E., "The Radioactivity of the Atmosphere and Hydrosphere," *Ann. Rev. Nucl. Sci.* 1958, **8**, 243.   [*6.5*]

Sw 12   SWINNE, R., *Physik. Z.* 1912, **13**, 14.   [*7.3*]

Sw 13   SWINNE, R., *Physik. Z.* 1913, **14**, 142.   [*7.3*]

Szy 66   SZYMANSKI, Z., *Nucl. Phys.* 1966, **76**, 539.   [*9.2.4*]

Ta 30   TARRANT, G. T. P., *Proc. Roy. Soc.* 1930, **A128**, 345.   [*4.6*]

Ta 32   TARRANT, G. T. P., *Proc. Roy. Soc.* 1932, **A135**, 223.   [*4.6*]

Ta 57   TANNER, N., *Phys. Rev.* 1957, **107**, 1203.   [*7.1, Exercise 7-5*]

Ta 61   TAAGEPERA, R., and NURMIA, M., *Ann. Acad. Sci. Fennicae: Ser. A. VI*, 1961, No. 78, 1.   [*7.3*]

Ta 65   TANG, Y. C., and BAYMAN, B. F., *Phys. Rev. Letters* 1965, **15**, 165.   [*2.3.5*]

Ta 66   TAYLER, R. J., "The Origin of the Elements," *Rept. Progr. Phys.* 1966, **29/II**, 489.   [*6.5*]

Te 66   TERRELL, G., and MOORE, C. F., *Phys. Rev. Letters* 1966, **16**, 804. [*5.4.4, Fig. 5-16*]

Th 97   THOMSON, J. J., *Phil. Mag.* [5] 1897, **44**, 293.   [*1.2*]

Th 04   THOMSON, J. J., *Phil. Mag.* [6] 1904, **7**, 237; *Proc. Cambridge Phil. Soc.* 1904, **13**, 49; see also *The Corpuscular Theory of Matter.* Constable, London, 1907.   [*1.2*]

Th 66   THORNTON, S. T., BAIR, J. K., JONES, C. M., and WILLARD, H. B., *Phys. Rev. Letters* 1966, **17**, 701.   [*2.3.5*]

Th 68   THORNTON, S. T., JONES, C. M., BAIR, J. K., MANCUSI, M. D., and WILLARD, H. B., *Phys. Rev. Letters* 1968, **21**, 447.   [*11.3.1*]

Tho 27   THOMAS, L. H., *Proc. Cambridge Phil. Soc.* 1927, **23**, 542.   [*D.2*]

Tho 55   THOMAS, R. G., *Phys. Rev.* 1955, **97**, 224.   [*11.5.7*]

To 49   TOLANSKY, S., and BRAGG, L., *Introduction to Atomic Physics*, 3rd ed. Longmans, Green, New York, 1949.   [*Fig. D-4*]

To 61   TOBOCMAN, W., *Theory of Direct Nuclear Reactions.* Oxford Univ. Press, London and New York, 1961.   [*E.8*]

Tr 51   TRALLI, N., and GOERTZEL, G., *Phys. Rev.* 1951, **83**, 399.   [*10.1.4*]

Tr 65   TRIPP, R. D., "Spin and Parity Determination of Elementary Particles," *Ann. Rev. Nucl. Sci.* 1965, **15**, 325.   [*5.3.10*]

Uh 25   UHLENBECK, G. E., and GOUDSMIT, S. *Naturwissenschaften* 1925, **13**, 953. [*1.2, 5.2*]

Uh 26   UHLENBECK, G. E., and GOUDSMIT, S., *Nature* 1926, **107**, 264.   [*1.2, 5.2*]

Ur 32a   UREY, H. C., BRICKWEDDE, F. G., and MURPHY, G. N., *Phys. Rev.* 1932, **39**, 164 (L).   [*1.2*]

Ur 32b   UREY, H. C., BRICKWEDDE, F. G., and MURPHY, G. N., *Phys. Rev.* 1932, **40**, 1.   [*1.2*]

Va 34   VAVILOV, S. I., *Dokl. Akad. Nauk SSSR* 1934, **2**, 457.   [*4.9.4*]

Van P 64   VAN PATTER, D. M., and MOHINDRA, R. K., *Phys. Letters* 1964, **12**, 223. [*Fig. 9-17*]

Vi 60   VISSHER, W. M., *Ann. Phys. (N.Y.)* 1960, **9**, 194.   [*9.5.2, Figs. 9-20, 9-23*]

Vl 59   VLASOV, N. A., and RUDAKOV, V. P., *Zh. Eksperim. i Teor. Fiz.* 1959, **36**, 24; *Soviet Phys. JETP (English Transl.)* 1959, **9**, 17.   [*8.8*]

Vo 59   VOGT, E., "Resonance Reactions, Theoretical," on p. 215ff. of [En 59]. [*11.5.4, 11.5.7*]

Vo 62   VOGT, E., *Rev. Mod. Phys.* 1962, **34**, 723.   [*11.5.4, 11.5.7*]

von B 64   VON BUTTLAR, H., *Einführung in die Grundlagen der Kernphysik.* Akad. Verlags-Ges. Frankfurt, 1964 (*English Transl.*) *Nuclear Physics: An Introduction.* Academic Press, New York, 1968.   [*Fig. 5-20*]

von W 35   VON WEIZSÄCKER, C. F., *Z. Physik* 1935, **96**, 431.   [*2.3.1, 2.3.5*]

von W 66   VON WITSCH, W., RICHTER, A., and VON BRENTANO, P., *Phys. Letters* 1966, **22**, 631.   [*11.3.1, Figs. 11-24, 11-25*]

Wa 52   WALSKE, M. C., *Phys. Rev.* 1952, **88**, 1283.   [*Fig. 4-58*]

Wa 53   WALCHLI, H. E., "A Table of Nuclear Moment Data," Oak Ridge Natl. Lab. Rept. ORNL-1469 (1953).   [*5.4.5*]

Wa 55 WASSERBURG, G. J., and HAYDEN, R. J., *Geochim. Cosmochim. Acta* 1955, **7**, 51. [*6.5*]

Wa 58a WAPSTRA, A. H., "Atomic Masses of Nuclides," in *Handbuch der Physik* (S. Flügge, ed.), Vol. 38, Part I, p. 1ff. Springer, Berlin, 1958. [*2.3.7, 2.3.8*]

Wa 58b WARBURTON, E. K., *Phys. Rev. Letters* 1958, **1**, 68. [*9.1, 9.2.4*]

Wa 59 WAPSTRA, A. H., NIJGH, G. J., and VAN LIESHOUT, R., *Nuclear Spectroscopy Tables*. North-Holland Publ. Co., Amsterdam, 1959. [*8.3.9, 8.5.3*]

Wa 60 WARBURTON, E. K., and PINKSTON, W. T., *Phys. Rev.* 1960, **118**, 733. [*9.2.4*]

Wa 61 WALKER, L. R., WERTHEIM, G. K., and JACCARINO, V., *Phys. Rev. Letters* 1961, **6**, 98. [*9.5.2*]

Wa 64 WAGGONER, M. A., ETTER, J. E., HOLMGREN, H. D., and MOAZED, C., University of Maryland Tech. Rept. No. 427 (1964); contribution to the Gatlinburg Conference on Correlations of Particles Emitted in Nuclear Reactions (Oct., 1964); published in *Rev. Mod. Phys.* 1965, **37**, 358; see also *Nucl. Phys.* 1966, **88**, 81. [*11.1.5*]

Wa 65 WAHLBORN, S., *Phys. Rev.* 1965, **138**, B530. [*9.2.4*]

We 26 WENTZEL, G., *Z. Physik* 1926, **38**, 518. [*5.1*]

We 61 WEBER, J., *General Relativity and Gravitational Waves*. Wiley (Interscience), New York, 1961. [*D.8*]

Wei 30 WEISSKOPF, V. F., and WIGNER, E., *Z. Physik* 1930, **63**, 54; **65**, 18. [*9.5*]

Wei 31 WEISSKOPF, V. F., *Ann. Physik*, 1931, **9**, 23. [*9.5*]

Wei 51 WEISSKOPF, V. F., *Phys. Rev.* 1951, **83**, 1073. [*9.2*]

Wei 62 WEINBERG, S., *Phys. Rev.* 1962, **128**, 1457. [*8.3.8*]

Wei 64 WEINBERG, S., *Phys. Rev.* 1964, **133**, B1318. [*F.3.4*]

Wei 65 WEISSKOPF, V. F., *Science* 1965, **149**, 1181; also appeared as "Quantum Theory and Elementary Particles," CERN Rept. No. 65–26 (July, (1965). [*1.1*]

Wh 39 WHEELER, J. A., and LAMB, W. E., *Phys. Rev.* 1939, **55**, 858. [*4.9.3*]

Wha 58 WHALING, W., "The Energy Loss of Charged Particles in Matter," in *Handbuch der Physik* (S. Flügge, ed.), Vol. 34, p. 193, Springer, Berlin, 1958. [*4.9, 4.10*]

Wha 60 WHALING, W., "The Interaction of Nuclear Particles with Matter," on p. 3ff. of [Aj 60]. [*Fig. 4-62*]

Whi 52 WHITE, G. R., "X-Ray Attenuation Coefficients from 10 keV to 100 MeV," *Natl. Bur. Std. (U.S.) Rept.* **1003** (1952). [*4.8, Figs. 4-57, G-10*]

Wi 31a WILLIAMS, E. J., *Proc. Roy. Soc.* 1931, **A130**, 310. [*Fig 4-75*]

Wi 31b WIGNER, E. P., *Gruppentheorie und ihre Anwendungen auf die Quantenmechanik der Atomspektren*. Vieweg, Braunschweig, 1931; Translated by J. J. Griffin, as *Group Theory and its Application to the Quantum Mechanics of Atomic Spectra*. Academic Press, New York, 1959 (with 3 extra chapters). [*E.2*]

Wi 37 WIGNER, E. P., *Phys. Rev.* 1937, **51**, 106. [*5.3.8*]

Wi 45 WILLIAMS, E. J., *Rev. Mod. Phys.* 1945, **17**, 217. [*3.6.4*]

Wi 46 WIGNER, E. P., *Phys. Rev.* 1946, **70**, 15 and 606. [*11.5.7*]

Wi 47a WIGNER, E. P., and EISENBUD, L., *Phys. Rev.* 1947, **72**, 29. [*11.5.7*]

Wi 47b WILSON, R. R., *Phys. Rev.* 1947, **71**, 385 (L). [*4.9.1, Exercise 4-10*]

Wi 50 WICK, G. C., *Phys. Rev.* 1950, **80**, 268. [*F.2*]

Wi 52   WINSLOW, G. H., and SIMPSON, O. C., "On the One-Body Model of Alpha Radioactivity: Part IIA," Argonne Natl. Lab. Rept. ANL-4901 (1952); see also Parts I and IIB, ANL-4841 and ANL-4910 (1952) and Part III, ANL-5277 (1954).   [*7.4.3*]

Wi 58   WILKINSON, D. H., *Phys. Rev.* 1958, **109**, 1603, 1610 and 1614. [*Ex. 7-5*]

Wi 53   WIMETT, T. F., *Phys. Rev.* 1953, **91**, 499A.   [*5.4.8*]

Wi 64   WILLARD, H. B., BAIR, J. K., and JONES, C. M., *Phys. Letters* 1964, **9**, 339. [*2.3.5*]

Wi 66   WINKLER, H., PIXLEY, R. E., and BLOCH, R., *Bull. Am. Phys. Soc.* [2] 1966, **10**, 204.   [*5.4.4*]

Wo 52   WOLFENSTEIN, L., and ASHKIN, J., *Phys. Rev.* 1952, **85**, 947.   [*11.5.4*]

Wo 67   WONG, C. Y., *Nucl. Phys.* 1967, **A103**, 625.   [*7.6.3*]

Woo 54   WOODS, R. D., and SAXON, D. S., *Phys. Rev.* 1954, **95**, 577.   [*Fig. 11-32*]

Wu 49   WU, C. S., and ALBERT, R. D., *Phys. Rev.* 1949, **75**, 315 and 1107. [*Fig. 8-18(d)*]

Wu 57   WU, C. S., AMBLER, E., HAYWARD, R. W., HOPPES, D. D., and HUDSON, R. P., *Phys. Rev.* 1957, **105**, 1413.   [*8.7.1, Figs. 8-43, 8-45*]

Wu 64   WU, C. S., *Rev. Mod. Phys.* 1964, **36**, 618.   [*8.6.3*]

Wu 65   WU, C. S., "The Shape of β-Spectra," on p. 1365ff. and "The Rest Mass of the Neutrino," on p. 1391ff. of [Si 65].   [*Fig. 8-22*]

Wu 66   WU, C. S., and MOSZKOWSKI, S. A., *Beta Decay*. Wiley (Interscience), New York, 1966.   [*8.1, 8.1.3, 8.2.2, 8.6.1, Fig. 8-12*]

Ya 50   YANG, C. N., and TIOMMO, J., *Phys. Rev.* 1950, **79**, 495.   [*8.7*]

Ya 59   YANOUKH, F., *Zh. Eksperim. i Teor. Fiz.* 1959, **36**, 335; *Soviet Phys. JETP (English Transl.)* 1959, **9**, 231.   [*8.8*]

You 64   YOUNG, F. C., JAYARAMAN, K. S., ETTER, J. E., HOLMGREN, H. D., and WAGGONER, M. A., University of Maryland Techn. Rept. No. 427 (1964); contribution to the Gatlinburg Conference on Correlations of Particles Emitted in Nuclear Reactions (Oct., 1964); *Rev. Mod. Phys.* 1965, **37**, 362.   [*11.1.5, Fig. 11-4*]

Yu 35a   YUKAWA, H., *Proc. Phys.-Math. Soc. Japan* [3] 1935, **17**, 48; reproduced on p. 139ff. of [Be 49].   [*1.2, 1.4, 3.1*]

Yu 35b   YUKAWA, H., and SAKATA, S., *Proc. Phys. Soc. Japan* 1935, **17**, 467.   [*8.5*]

Yu 35c   YUKAWA, H., and SAKATA, S., *Proc. Phys.-Math. Soc. Japan* [3] 1935, **17**, 397.   [*10.2*]

Yu 53   YUASA, T., and LABERRIGUE-FROLOW, J., *J. Phys. Radium* 1953, **14**, 95. [*8.3.5*]

Zi 68   ZIOCK, K., MINEHART, R., COULSON, L., and GRUBB, W., *Phys. Rev. Letters* 1968, **20**, 1386.   [*5.4.4*]

Zl 35   ZLOTOWSKI, I., *J. de Phys.* 1935, **6**, 242.   [*8.1.3*]

Zl 41   ZLOTOWSKI, I., *Phys. Rev.* 1941, **60**, 483.   [*8.1.3*]

Zu 48   ZUMWALT, L. R., CANNON, C. V., JENKS, G. H., PEACOCK, W. C., and GUNNING, L. M., *Science* 1948, **107**, 47.   [*8.1.3*]

Zu 65   ZUCKER, A., and BROMLEY, D. A., *Science* 1965, **149**, 1197.   [*1.1*]

Zw 57   ZWEIFEL, P. F., *Phys. Rev.* 1957, **107**, 329.   [*8.5.3*]

Zy 61   ZYRANOVA, L. N., and MIKHAILOV, V. M., *Izv. Akad. Nauk SSSR, Ser. Fiz.* 1961, **25**, 56; *Bull. Acad. Sci. USSR, Phys. Ser. (English Transl.)* 1961, **25**, 57.   [*8.3.9*]

Zy 63   ŻYLICZ, J., SUJKOWSKI, Z., JASTRZEBSKI, J., WOŁCZEK, O., CHOJNACKI, S., and YUTLANDOV, I., *Nucl. Phys.* 1963, **42**, 330.   [*8.5.3*]

SOLUTIONS TO EXERCISES

Chapter 1

1-1 (a) 63.8 cm;   (b) 6.38×10^{-4} cm.

1-2 $\approx 10^{-33}$ cm.

1-3 $\dfrac{m_{\text{blue}}}{m_{\text{red}}} = \dfrac{7}{5} = 1.4.$

1-4 $\theta = 30°$; $E_{\text{x}} = 4.34$ keV.

1-5 (a) 3.12×10^{15} p sec^{-1};   (b) 10 kW $= 2.4 \times 10^6$ cal sec^{-1};
(c) $F = 32.6$ dyn $= 3.26 \times 10^{-4}$ Newton;   (d) 2.3 percent.

1-6 1.36 cm^{-3} min^{-1}.

1-7 $s = 10.77$ m.

1-8 Initially, $\beta = 0.943$;   Finally, $\beta = 0.986$.

1-9 1.067 u. The presently accepted value, $m_{\text{n}} = 1.008\ 665$ u is 6 percent smaller than that derived by Chadwick.

1-10 $a_0 = \dfrac{\hbar^2}{m_{\text{e}}\,e^2} = 0.53$ Å $= 5.3 \times 10^{-9}$ cm;

$p = \dfrac{\hbar}{a_0} = \dfrac{m_{\text{e}}\,e^2}{\hbar} = 1.99 \times 10^{-19}$ g cm sec^{-1};

$\Delta a_0 \cdot \Delta p = \hbar$, which corresponds to the Heisenberg uncertainty relation.

1-11 (a) 5.09×10^3 g; (b) 1.51×10^{20} fissions sec^{-1}; (c) 14.2×10^3 tonne.

Chapter 2

2-1 The available energy of 20 MeV does *not* suffice to initiate the reaction, since an energy $E > E_{\text{thresh}} = 26.87$ MeV is needed.

2-2 (a) $E_{\text{kin}} = 212$ keV;   (b) 69.5 MeV.

2-3 1.66×10^9 kg $= 1660$ tonne.

2-4 16.7 percent.

2-5 (a) 1.69 fm;   (b) 1.41 fm.

2-6 (a) For a volume distribution of charge, the Coulomb energy is $E_C^{vol} = 844$ MeV.
(b) For a surface distribution of charge, the Coulomb energy is $E_C^{surf} = 704$ MeV. Hence $E_C^{vol} - E_C^{surf} = 140$ MeV $> \Delta E$.
(c) $R_1/R_2 \cong 0.322$.

2-8 $^{142}_{54}$Xe is β^--unstable because $Z_{Xe} < Z_{stable}$.

2-9 $r_0 = 1.59$ fm.

2-10 $E_0 = 3.44$ MeV. Hence $E_{\beta^+} = E_0 - 2m_e c^2 = 3.44 - 1.02 = 2.42$ MeV. This is 15.4 percent lower than the experimental value.

2-11 The Q-value for (symmetric) fission vanishes at around $Z = 35$, $A = 80$ (e.g., in the vicinity of Br). Above this, the Q-value is *positive* and fission is exothermic. In the case of *asymmetric* fission, the limiting values of Z and A lie somewhat higher.

Chapter 3

3-1 $\dfrac{F_G}{F_C} = \dfrac{\gamma m_e m_p}{e^2} = 4.4 \times 10^{-40}$;

$$\left[\text{cf. ratio of coupling constants} = \frac{2 \times 10^{-39}}{7 \times 10^{-3}} \approx 3 \times 10^{-37} \right].$$

3-2 $E_C/E_G = 2.55 \times 10^{36}$.

3-3 $E_{kin}^{(min)} = 3.2 \times 10^{-9}$ erg $= 2.0$ keV. Setting $E_{kin}^{(min)} = \frac{3}{2}kT$ yields $T = 1.55 \times 10^7$ °K.

3-4 The equilibrium value $\dfrac{M}{m_e} = \dfrac{1}{\gamma}\left(\dfrac{e}{m_e}\right)^2 = 4.16 \times 10^{42}$ indicates that when $M = 10^{10} m_e$ there would be a net repulsion. For equilibrium, one would require that $M = 3.8 \times 10^{15}$ g.

3-5 $\bar{\sigma} = 36$ mb.

3-6 Probability, expressed as N/N_0, is equal to 1.9×10^{-8} percent.

3-7 $x_1'(t) = \dfrac{\dfrac{x_1}{x_2} \exp\left(-\Phi\sigma t\right)}{1 + \dfrac{x_1}{x_2} \exp\left(-\Phi\sigma t\right)}$; $\quad x_2'(t) = \dfrac{1}{1 + \dfrac{x_1}{x_2} \exp\left(-\Phi\sigma t\right)}$.

3-8 $\sigma_f = \sigma(^{235}U) = 586$ b;
$N_f = \Phi n^{\square} \sigma d = 2 \times 10^7$ cm^{-2} sec^{-1}.

3-10 A classical estimate gives $R = r_0' A^{1/3} \leqslant b = 30.1$ fm, whence $r_0' \lesssim 5$ fm (cf. the accepted value $r_0 \simeq 1.4$ fm. See the discussion by Evans [Ev 55], pp. 47 ff.).

Chapter 4

4-1 Thickness of Pb: $x = 3.14 \times 10^{-2}$ cm.   For $E_\gamma = 90$ keV: $A_1 = e^{-\mu_1 x} = 0.071$;   For $E_\gamma = 85$ keV: $A_2 = e^{-\mu_2 x} = 0.71$.

4-2 Writing $\mathscr{E} \equiv h\nu/m_e c^2$ we find

(a) $\mathscr{E}_{(a)} = \mathscr{E}(1 + 2\mathscr{E})^{-1}$;   (b) $\mathscr{E}_{(b)} = \mathscr{E}(1 + 2\mathscr{E})^{-1}$;
(c) $\mathscr{E}_{(c)} = \mathscr{E}(1 + \tfrac{3}{2}\mathscr{E})^{-1}$.

Hence the energy loss is identical in cases (a) and (b), but lower in case (c).

4-3 For pure Be the transmitted intensity is

$$\mathbf{I} = \exp\left[-\left(\frac{\mu}{\rho}\right)_{\text{Be}} \rho d\right] = 0.95,$$

corresponding to an attenuation $A = 5$ percent. Hence the foil does *not* consist of pure Be. Assuming the impurity to be Fe, one finds that the percentage contamination is $f = 0.96$ percent.

4-4 $\theta = 0.309$ rad $= 17°\ 42'$.

4-6 $h\nu = E_{\text{kin}} + \tfrac{3}{2}m_e c^2$.

4-7 $\tau = 1.70 \times 10^{-10}$ sec.

4-8 $2(\rho_+ + \rho_-) = 2E_\gamma/Bec = $ constant.

4-9 $\left.\dfrac{dE}{dx}\right|_d (E_d) = \left.\dfrac{dE}{dx}\right|_p (E_p = \tfrac{1}{2}E_d)$; $\left.\dfrac{dE}{dx}\right|_\alpha (E_\alpha) = 4 \cdot \left.\dfrac{dE}{dx}\right|_p (E_p = \tfrac{1}{4}E_\alpha)$;

$R_d(E_d) = 2R_p(E_p = \tfrac{1}{2}E_d)$; $R_\alpha(E_\alpha) = R_p(E_p = \tfrac{1}{4}E_\alpha)$.

4-10 $R_\alpha^{[m]} = \left(\dfrac{E_\alpha^{[\text{MeV}]}}{37.2}\right)^{1.8}$.

4-11 Interchange has *no* effect upon the end energy or upon the energy straggling, viz. mean energy $\bar{E} = 3$ MeV; energy straggling $\alpha(\text{Cu/Au}) = \alpha(\text{Au/Cu}) = \sqrt{\alpha_1^2 + \alpha_2^2} = 72.5$ keV. The angle straggling changes, however, e.g.,

$$\bar{\theta}(\text{Cu/Au}) = 17°\ 42', \qquad \bar{\theta}(\text{Au/Cu}) = 14°\ 23'.$$

4-12 (a) $E_{p_0} = 11.4$ MeV;   (b) $\bar{E}_p = 10.9$ MeV;   (c) $\alpha = 36$ keV.

4-13 (a) $x_0 = 6.1 \times 10^{-3}$ cm $(\xi_0 \equiv \rho x_0 = 52.5$ mg cm$^{-2})$;

    (b) $\bar{E} = 100\alpha = 7.07$ MeV;   (c) $\bar{\theta} = 5°18'$.

4-14 $t \approx 10^{-10}$ sec.

4-15

(a) $E_\gamma = m_e c^2 + E_{kin}$;

   $\cos\theta =$

$$\frac{\sqrt{E_{kin}^2 + 2E_{kin} m_e c^2}}{m_e c^2 + E_{kin}}$$

(b) $E_\gamma = 2m_e c^2 = 1.022$ MeV.

Numerically,

	K	L	M
$B = E_{kin}$	117 keV	26.8 keV	9.86 keV
E_γ	628 keV	538 keV	521 keV
θ	54.5°	71.8°	78.8°

4-16 $\Phi = \dfrac{3}{(2\mu R)^3}\{\frac{1}{2}(2\mu R)^2 - 1 + \exp[-2\mu R(2\mu R + 1)]\}$;

$$\varphi(x, \Delta T) = 1 + 2\alpha\Delta T\frac{\frac{1}{2}x^2 - 3 + e^{-x}(3 + 3x + x^2)}{\frac{1}{2}x^2 - 1 + e^{-x}(x + 1)}.$$

Numerically, $\Phi = 0.728$; $\Delta\varphi = 1.05$ percent.

4-17 $\dfrac{m}{m_e} = \dfrac{E_{kin}}{m_e c^2(\gamma - 1)} = 281.$

Hence the particles are π-mesons $(m_\pi = 273\, m_e)$.

Chapter 5

5-1 $\pi = -$, whereas for an α particle with $j = l = 4$, the parity is $\pi = +$.

5-2 (a) $0^+ \to \frac{3}{2}^+, \frac{5}{2}^+$

    (b) $\frac{1}{2}^- \to 1^-, 2^-, 3^-$

    (c) $1^+ \to \frac{1}{2}^+, \frac{3}{2}^+, \frac{5}{2}^+, \frac{7}{2}^+$

    (d) $1^- \to \frac{1}{2}^-, \frac{3}{2}^-, \frac{5}{2}^-, \frac{7}{2}^-$

    (e) $\frac{5}{2}^+ \to 0^+, 1^+, 2^+, 3^+, 4^+, 5^+$

    (f) $\frac{5}{2}^- \to 0^-, 1^-, 2^-, 3^-, 4^-, 5^-$

    (g) $\frac{11}{2}^- \to 3^-, 4^-, 5^-, 6^-, 7^-, 8^-$

    (h) $l = 2$;

    (i) $l = 3$;

    (j) $l = 3, 5$;   (k) $l = 3, 5$.

A $1^- \to 4^-$ transition can occur for an α-particle with $j = l = 4$.

5-3 3.16×10^8 cm/sec.

5-4 $\pi_{ortho} = -$; $\pi_{para} = +$.   A 2-γ system would have *positive* parity, whereas $\pi_{ortho} = -$.

5-5 $\omega_L = \dfrac{2\mu_p}{\hbar e R}[E_{kin}(E_{kin} + 2m_p c^2)]^{1/2}.$

5-6 $\gamma = Z/A$.

5-7 $\gamma_E/\gamma_e = 6.5 \times 10^{-9}$.

5-8 $E = \dfrac{Ze^2}{r_c} + \underbrace{\dfrac{e}{r_c^2} \displaystyle\int \rho r_v \cos\theta\, dV}_{\substack{\text{dipole}\\\text{moment}}} + \underbrace{\dfrac{e}{2r_c^3} \displaystyle\int \rho r_v^2\, (3\cos^2\theta - 1)\, dV}_{\substack{\text{quadrupole}\\\text{moment}}};$

$Q = \tfrac{2}{5} Ze(a^2 - b^2).$

5-9 $\mu_p = 2.79\ \mu_N.$ $N(+\tfrac{1}{2})/N(-\tfrac{1}{2})$ 0.999 996 6, or $\Delta N/N = 1.7 \times 10^{-6}.$

The population of the substates tends to become equalized.

5-10 Permitted combination is $^3S_1 + {}^3D_1$ with 96.1 percent of $^3S_1$ and 3.9 percent of $^3D_1$.

Chapter 6

6-1 $V = 384\ \text{cm}^3.$

6-2 (a) $\mathscr{A}_i = 2.60\ \text{kCi}, \mathscr{A}_f = \tfrac{1}{2}\mathscr{A}_i = 1.30\ \text{kCi};$
   (b) $\mathscr{A}'_i = \mathscr{A}'_f = \mathscr{A}' = 9.72\ \text{kCi g}^{-1};$
   (c) $P_i = 1\ \text{atm}, P_f = \tfrac{3}{2}\ \text{atm}$, hence $\Delta P \equiv P_f - P_i = \tfrac{1}{2}\ \text{atm}.$

6-3 Particle emission (a *strong* interaction process) is energetically possible and thus takes preference over γ emission, which involves an electromagnetic interaction.

6-4 $\mathscr{A}' = 1.42\ \text{Ci g}^{-1}.$

6-5 12.3 kCi; $m = 728\ \text{g}.$

6-6 (a) 0.883 mg;   (b) $V = 0.042\ \text{cm}^3.$

6-7 (a) 15 h;   (b) 30 h.

6-8 1.082 kCi; 4.36×10^{-5} percent; $\mathscr{A}(t = nT_{1/2}) = \mathscr{A}_{\text{max}}\ [1 - (1/2^n)].$

6-9 $t = 2.75\ \text{d}.$

6-10 Rate of ^{235}U decay is 2.27×10^8 per hour, of which fission accounts for 0.973 per hour and α-decay for the balance.

6-11 $2.97\ \mu\text{Ci}.$

6-12 (a) 5 samples;   (b) 1.06 mCi;   Net activity is 1.20 mCi.

6-13 (a) 1.92 h;   (b) 6.37 d.

6-14 06.16 h.

6-15 7.59×10^{-9} percent.

6-16 $1:2.3$, i.e., 43.5 percent.

6-17 92.9 g ^{235}U and 214 g ^{238}U;   5.58 mg ^{235}U and 45.9 g ^{238}U.

6-18 (a) 7.5×10^7 y;   (b) 7.5×10^8 y;
(d) 1.33 percent and 10.4 percent respectively;   (e) 4.14×10^{-7} g.

Chapter 7

7-1 $\Gamma = 144$ eV.

7-2 $T = 6.6 \times 10^{-12}$.

7-3 $T = 2.31 \times 10^{-15}$.

7-4 Writing $\log_{10} T_{1/2} = c_1[(Z/\sqrt{E_\alpha}) - c_2 Z^{2/3}] - c_3$, one finds $c_1 = 1.72$, $c_2 = 1.07$, $c_3 = 28.6$.

7-5 (a) Decay to the 0^+ ground and first excited states of ^{16}O is parity-forbidden, to the second, third and fourth levels is parity-allowed (with $l = 3, 2$, and 1 respectively); to higher levels is energy-forbidden.
(b) Since $E_{CM}^{(\alpha)} = 8.88 - 7.15 = 1.73$ MeV it follows that $E_{lab}^{(\alpha)} = 12 \times 1.73/16 = 1.30$ MeV.
(c) The peak at $E_{lab}^{(\alpha)} = 1.73$ MeV in the α-particle spectrum corresponds to an allowed transition from a broad 1^- level at 9.58 MeV in $^{16}O^*$; the corresponding mean transition energy is accordingly 2.43 MeV (CM), that is, 1.82 MeV (lab).
(d) $\Gamma_\alpha < 6 \times 10^{-9}$ eV as against the estimated value for a parity-allowed transition, $\Gamma \approx 3 \times 10^{-3}$ eV, whence it was concluded that the fraction of a possible parity-odd admixture corresponds to an intensity ratio $\mathscr{F}^2 \lesssim 2 \times 10^{-12}$.

7-6 $T = [\cosh(qb/\hbar)]^{-2}$, instead of the approximate formula $T = \exp[-2qb/\hbar]$ to which it reduces when $qb/\hbar \gg 1$.

7-7 (a) $T = \exp\left\{-2b\left[\dfrac{2\mu}{\hbar^2}(U_0 - E)\right]^{1/2}\right\}$;

(b) $T = \left[1 + \dfrac{U_0^2 \sin^2 Kb}{4E(U_0 + E)}\right]^{-1}$, where $K = \left[\dfrac{2\mu}{\hbar^2}(U_0 + E)\right]^{1/2}$.

7-8 $l = 3$, $\hbar\sqrt{l(l+1)} = \hbar\sqrt{12} = 3.653$ erg sec.

7-9 $G_C = 120$ and $G_\alpha = 77$, whence $G_C - G_\alpha = 43$, showing that $T_C/T_\alpha = e^{-43} \simeq 10^{-19}$.

7-10 $l = 18$.

Chapter 8

8-1 $\bar{E} = \frac{1}{3}E_0$.

8-2 $\bar{E}_\nu = 12.07$ keV.

8-3 $\bar{E} = 0.345$ MeV and hence $\bar{E}/E_0 = 0.295$.

8-4 $E_0 = 2$ MeV; $\bar{E} = \frac{1}{2}E_0 = 1$ MeV.

8-5 $m_\nu < 0.26\ m_e$; $\Delta m_\nu/m_\nu = 56$ percent.

8-6 $N(p_e)\,dp_e = \text{const.} \times p_e^2(E_0 - E_e)^5\,dp_e$.

8-7 $E_0 = 1.98$ MeV: hence ^{114}In; $\bar{E}_\beta = 660$ keV; $\bar{E}_\nu = 1.32$ MeV.

8-8   n decay: F superallowed;
   ^{17}F decay: F superallowed;
   ^{35}S decay: F allowed;
   ^{75}Ge decay: $G\text{-}T$ allowed;
   ^{87}Rb decay: third forbidden (parity forbidden);
   ^{91}Y decay: first forbidden (unique forbidden).

8-9   n decay: mixed $F/G\text{-}T$;
   ^{3}H decay: mixed $F/G\text{-}T$;
   ^{6}He decay: pure $G\text{-}T$;
   ^{14}O decay: pure F;
   ^{15}O decay: mixed $F/G\text{-}T$;
   $g_F = 1.41 \times 10^{-49}$ erg cm^3;
   $|C_{GT}|^2/|C_F|^2 = 1.46$;
   $|M_{^3H}|^2 = 6.3$, $|M_{^6He}|^2 = 7.6$,
   $|M_{^{15}O}|^2 = 1.45$.

8-10 When $B_L(Z) < [M(Z) + B_{\text{tot}}(Z)] - [M(Z-1) + B_{\text{tot}}(Z-1)] -$
$$Q < B_K(Z),$$
where the B's are electron binding energies and Q is the sum of the neutrino energy and the nuclear energy, atomic masses being expressed in energy units.

8-11 $T = 0.006\ 88$ °K; $dT/dt = 0.151$ °K sec^{-1}.

8-12 $\Phi_\nu = 7.03 \times 10^{10}$ cm^{-2} sec^{-1}.

8-13 $E = 38$ eV.

Chapter 9

9-1 $\Gamma_\gamma/\Gamma_d = 2.36 \times 10^{-9}$.

9-2 $E^*(J^\pi) = 0(\frac{3}{2}+), 0.057(\frac{5}{2}+), 0.134(\frac{7}{2}+), 0.315(\frac{3}{2}+), 0.418(\frac{5}{2}-), 0.481(\frac{1}{2}-),$ 0.586(?). The first three spin-parity assignments tally with the accepted level scheme for ^{161}Tb; the other assignments are doubtful.

9-3 $E^*(J^\pi) = 0(\frac{1}{2}+), \quad 0.133(\frac{5}{2}+), \quad 0.137(\frac{3}{2}+), \quad 0.273(\frac{7}{2}+), \quad 0.619(\frac{7}{2}+),$ in disaccord with the presently accepted level scheme of ^{181}Ta.

9-4 $E^*(J^\pi) = 0(\frac{9}{2}+), 0.335(\frac{1}{2}-), 0.597(\frac{5}{2}-), 0.827(\frac{3}{2}-), 0.862(\frac{3}{2}-),$ which is consistent with the currently accepted level scheme of ^{115}In. The ft values are: $ft_1 = 10^7$, $ft_2 = 8 \times 10^7$, $ft_3 = 10^7$, $ft_4 = 3 \times 10^6$, which indicates that all the β transitions are first forbidden. Since the ground-state γ transition is pure M4, the Weisskopf estimate for the lifetime of the isomeric state is $\tau = 1.8 \times 10^{-5}$ sec.

9-5 (a) $E(\theta,t) = E_0\left(1 + \frac{v_i}{c}e^{-K\rho t/m}\cos\theta\right)$;

(b) $E(\theta) = E_0\left[1 + \left(1 + \frac{K\rho}{\lambda m}\right)^{-1}\frac{v_i}{c}\cos\theta\right]$;

(c) $\tau = \frac{M}{K\rho}\left(\frac{v_i}{c}\frac{1}{\Delta E/E_0} - 1\right) = 1.01 \times 10^{-12}$ sec.

9-6 (a) $\dfrac{\Delta E}{E} = \dfrac{E_\gamma - E_0}{E_0} = \left[\left(\dfrac{2E_1}{Mc^2}\right)^{1/2}\cos\theta - \dfrac{E_0}{2Mc^2}\right]$;

(b) $\theta_0 = \arccos\left[\dfrac{E_0}{2(2Mc^2 E_1)^{1/2}}\right]$;

(c) $\dfrac{\Delta E}{E}(\theta = 0) = \left[\left(\dfrac{2E_1}{Mc^2}\right)^{1/2} - \dfrac{E_0}{2Mc^2}\right]$,

whereas $\dfrac{\Delta E}{E}\bigg|_{\text{Doppler}} = \left(\dfrac{2E_1}{Mc^2}\right)^{1/2}$

The second term in square brackets represents the specific energy loss due to recoil.

(d) $E_1 = \dfrac{m}{m + M}\left[2\left(\dfrac{M^2}{(m+M)^2}E_n^2 - \dfrac{M}{m+M}E_0 E_n\right)^{1/2}\right.$

$\left. + \dfrac{2M}{m+M}E_n - E_0\right] = 0.925$;

$\dfrac{\Delta E}{E}(\theta = 0) = 1.69$ percent; $\theta_0 = 89° \, 52' \, 28''$.

9-7 $\Delta E = \dfrac{E_0^2}{Mc^2} = 3.90 \times 10^{-3}$ eV; $\quad v_R = \dfrac{E_0}{Mc} = 81.4$ m sec^{-1}.

For a recoil-free source, $v = \dfrac{E_0}{2Mc} = \frac{1}{2}v_R = 40.7$ m sec^{-1}.

$\Gamma = \dfrac{v}{c}E_0 = 4.80 \times 10^{-9}$ eV; $\quad \tau = 0.137$ μsec.

9-8 $\dfrac{\partial}{\partial T}\left(\dfrac{\Delta E}{E_0}\right) = \dfrac{C_{\mathrm{Fe}}}{2c^2} = 2.68 \times 10^{-15}$ °K^{-1};

$\dfrac{\Delta E}{E_\gamma} = \dfrac{gh}{c^2} = 1.36 \times 10^{-15}$; $\quad v = 4.08 \times 10^{-7}$ m sec^{-1}.

Chapter 10

10-1 $E_{\gamma_1} \approx E_{\beta_1} = 1.17$ MeV; $E_{\gamma_2} \approx E_{\beta_2} = 1.33$ MeV.

10-2 $E_\gamma = 25, 66, 81, 97, 121, 136, 199, 265, 280, 305$ and 402 keV. Self-consistent level scheme has levels at $E^* = 0, 0.199, 0.265, 0.280, 0.305,$ and 0.402 MeV.

10-3 Level scheme:
$E^*(J^\pi) = 0(\frac{3}{2}^+), 0.080(\frac{1}{2}^+), 0.164(\frac{11}{2}^-), 0.365(\frac{5}{2}^+),$
$0.638(\frac{5}{2}^+ \text{ or } \frac{7}{2}^+), 0.724(\frac{5}{2}^+ \text{ or } \frac{7}{2}^+).$ $\quad \tau = 898$ d.

Chapter 11

11-1 $\sigma_{\mathrm{abs}} = 0.915$ b.

11-2 $\sigma_{\mathrm{abs}} = (2l + 1)\,\pi\lambda^2 \xrightarrow{l=0} \pi\lambda^2 = 0.112$ b;
$\sigma_g = \pi r_0^2(A^{2/3} + 1) = 0.668$ b.

11-3 $\sigma_0(n, n) = (1 \pm \frac{1}{6}) \times 4640$ b;
$\sigma_0(n, \gamma) = (1 \pm \frac{1}{6}) \times 46.4$ b.

11-4 $N = 1.2 \times 10^5$.

11-5 ^{27}Al$^* \rightarrow n + {}^{26}$Al in all cases.
(a) $E^* = 14.05$ MeV; (b) $E^* = 30.54$ MeV;
(c) $E^* = 63.63$ MeV; (d) $E^* = 56.17$ MeV.

11-6 $\sigma(n, n)_{\mathrm{max}} = 10\,548$ b; $\quad \sigma(n, \gamma)_{\mathrm{max}} = 659$ b; $\quad \sigma_g = 0.933$ b.

11-7 (a) $\sigma(n, \gamma) = (1 \pm \frac{1}{4}) \times 2.67 \times 10^6$ b for $J_1 = \frac{3}{2} \pm \frac{1}{2}$;
(b) $\sigma_{\mathrm{tot}} = (1 \pm \frac{1}{4}) \times 3.42 \times 10^6$ b for $J_1 = \frac{3}{2} \pm \frac{1}{2}$;
(c) $\sigma_{\mathrm{tot}} = 0.978\,\sigma_{\mathrm{exp}}$.

11-8 $\sigma(n, \gamma)_{\mathrm{max}} = 19.4 \times 10^6$ b; $\quad \sigma_g = \pi r_0^2(A^{2/3} + 1) = 1.68$ b.

11-9 (a) $\sigma(n, \gamma) = 62\ 389$ b;   (b) $\sigma(n_{th}, \gamma) = 2474$ b;
(c) $\sigma(n, \gamma) \sim \sigma_0 (E_0/E)^{1/2} \sim E^{-1/2} \sim 1/v$;
(d) $d_{0.025} = 9.4 \times 10^{-4}$ cm; $d_{0.2} = 3.7 \times 10^{-4}$ cm; $d_{0.3} = 2.2 \times 10^{-3}$ cm.

11-10 (a) $N_n = 0.297$ sec^{-1};   (b) $N_\gamma = 20.6$ sec^{-1}.

11-11 (a) $E_0 = 4.87$ eV;   (b) $\Gamma_\gamma = 0.15$ eV;   (c) $\Gamma_n = 8.4 \times 10^{-3}$ eV.
Discrepancies with the Breit-Wigner values may be attributed to potential scattering and the influence of higher-lying levels.

11-12 $\sigma(n, \gamma) \rightarrow 2\pi^2 (R + \lambda)^2\ \Gamma_\gamma/D$; $D = 340$ eV.

11-13 $E_\gamma = 4.13$ MeV (lab); $\theta = 23° 13'$ (CM) $= 19° 6'$ (lab).

11-14 $\sigma(p, d) = 1.03$ mb.

11-15 $E_n = 3.885$ MeV (lab); $\sigma(n, \gamma) = 28.8$ μb.

11-16 $E_p = 6.76$ MeV; $\sigma(p, \gamma) = 8.99$ μb, whence $N = 16.4$ sec^{-1}.

11-17 $s_\pi = 0$.

Appendix A

A-2 $\beta' = \beta(\alpha^2 + \beta^2 + \alpha^2\beta^2)^{-1/2}$, with $\alpha \equiv (M_0 + m_0)/m_0$. In the classical limit $\beta \rightarrow 0$, the velocity reduces to $v' \rightarrow \dfrac{v}{\alpha} = \dfrac{m_0}{m_0 + M_0} v$.

A-3 $v = 0.999\ 949\ 5\ c$.

A-4 $E_{kin} = \left(\dfrac{h}{\tau_0 c} - 1\right) m_\mu c^2 = 3020$ MeV.

A-5 $\dfrac{\Delta E}{E_0} \cong \dfrac{\omega^2}{2c^2} (R_1^2 - R_2^2) = -2.44 \times 10^{-20}\ \omega^2$;
no net shift if diametrically mounted.

Appendix B

B-1 $E_{min} = \dfrac{\hbar^2}{mr_0^2} = 21$ MeV.

B-2 (a) $Q = -12.16$ MeV, $M_{Ne} = 19.0019$ u;
(b) $E_n' = 3.65$ MeV.

B-3 (a) $E' = E_{CM}$;   (b) $E' = \frac{1}{2}E$;   (c) $E' = 2E_1' = 2E_2'$.

B-4 (a) $E(\theta) = E\cos^2\theta$; $dN/d\theta = N_0\sin 2\theta$;
(b) $E = 4500$ GeV; $E_{CM}' = 90$ GeV; $E_{kin}/E_{kin}^{(CB)} = 100$.

B-5 (a) (i) $E'_{CM} = 142$ MeV;   (ii) $E'_{CM} = 6.20$ GeV;
         (iii) $E'_{CM} = 9.8$ GeV.
   (b) $E^{(CB)}_{kin} = 56$ GeV.
   (c) (i) $E'_{CM} = 17.5$ GeV, $d = 1.33$ km;
        (ii) $E'_{CM} = 22.0$ GeV, $d = 2.00$ km;
        (iii) $E'_{CM} = 41.6$ GeV, $d = 6.67$ km.

None can surpass the CM energy of $E_{CM} = E_{lab} = 56$ GeV attainable with the CERN PS coupled to a storage ring. However, colliding beam techniques entail the disadvantage of low intensity, poor definition of interaction region and the limitation to identical-particle interaction.

B-6 $E_{kin} = 1.91 \times 10^6$ GeV.

B-7 $E_{kin} = 6m_p c^2 = 563$ GeV.

B-8 $E^{(kin)}_{e\,max} = \frac{1}{2} m_\mu c^2 = 52.5$ MeV; $E_e = \frac{1}{2} m_\mu c^2$, that is, a line spectrum comprising a single line at an energy equal to the two-neutrino decay end-point energy.

Appendix C

C-1 $E_p = 1.92$ MeV.

C-2 For proton scattering, $(E_1)_f/(E_1)_i = \cos^2 \theta$, and $(E_2)_f/(E_1)_i = \cos^2 \phi = \sin^2 \theta$.
   For deuteron scattering, $(E_1)_f/(E_1)_i = \frac{4}{9}[\cos \theta - (\frac{1}{4} - \sin^2 \theta)^{1/2}]$, and $(E_2)_f/(E_1)_i = \frac{8}{9}\cos^2 \phi$.

C-3 $\alpha_{min} = \arccos\left(\dfrac{\gamma_1 - 1}{\gamma_1 + 3}\right) = \arccos\left(\dfrac{E_{kin}}{E_{kin} + 4mc^2}\right)$; $\alpha_{max} = \pi/2$.

C-4 (a) $E^{(p)}_{thresh} = 1.88$ MeV, $E_n = 0.0294$ MeV;
   (b) $E_p(\theta \geqslant 90°) = 1.92$ MeV;
   (c) $E_n(\theta = 0°) = 0.120$ MeV.

C-5 $E_\alpha = 1.015$ MeV, $E_{Li} = 1.775$ MeV.

C-6 $(E_n)_{max} = 3.92$ MeV, $(E_n)_{min} = 1.80$ MeV; $\sigma(\theta) \sim (\cos \theta + 5.25)^2$.

C-7 $\Delta E_e \simeq - \dfrac{E_e}{1 + \dfrac{m_p c^2}{E_e}} = -174$ MeV.

C-8 $x = v \pm (v^2 + w)^{1/2}$, where

$$x \equiv \sqrt{E_3}, \quad v \equiv \frac{m_1 E_1^{1/2} \cos \theta}{m_1 + m_2}, \quad w \equiv \frac{E_1(m_2 - m_1) - E^* m_2}{m_1 + m_2}.$$

$E_{3,\text{el}} = 3.86$ MeV, $E_{3,\text{inel}} = 2.56$ MeV. Inelastic scattering cannot take place to higher levels of ^{24}Mg than the first excited state at this energy.

C-9 $b' = 10,$ $c' = -0.113,$ $d' = -1.132.$

Appendix D

D-1   The $\pm$ sign refers respectively to incoming and outgoing waves.

D-2 $\left(-\dfrac{\hbar^2}{2m}\dfrac{d^2}{dx^2} + \tfrac{1}{2}m\omega_0^2 x^2\right)\psi = E\psi$, which reduces to the quoted form with

$\alpha \equiv \left(\dfrac{m\omega_0}{\hbar}\right)^{1/2}$ and $\lambda \equiv \dfrac{2E}{\hbar\omega_0}$.

D-3   Symbolizing spin "up" by $\alpha = \uparrow$, spin-0 by $\beta = \rightarrow$ and spin "down" by $\gamma = \downarrow$, one can construct the wave functions for net spin

$s = +2 : \psi_{+2} = \alpha(1)\,\alpha(2)$ symm. $\uparrow\uparrow$

$s = -2 : \psi_{-2} = \gamma(1)\,\gamma(2)$ symm. $\downarrow\downarrow$

$s = +1 : \psi_{+1} =$
$\dfrac{1}{\sqrt{2}}[\alpha(1)\,\beta(2) + \beta(1)\,\alpha(2)]$ symm. $\uparrow\rightarrow, \rightarrow\uparrow$

$\dfrac{1}{\sqrt{2}}[\beta(1)\,\alpha(2) - \alpha(1)\,\beta(2)]$ antisymm. $\rightarrow\uparrow, \uparrow\rightarrow$

$s = -1 : \psi_{-1} =$
$\dfrac{1}{\sqrt{2}}[\beta(1)\,\gamma(2) + \gamma(1)\,\beta(2)]$ symm. $\rightarrow\downarrow, \downarrow\rightarrow$

$\dfrac{1}{\sqrt{2}}[\beta(1)\,\gamma(2) - \gamma(1)\,\beta(2)]$ antisymm. $\rightarrow\downarrow, \downarrow\rightarrow$

$s = 0 : \psi_0 =$
$\dfrac{1}{\sqrt{2}}[\alpha(1)\,\gamma(2) + \gamma(1)\,\alpha(2)]$ symm. $\uparrow\downarrow, \downarrow\uparrow$

$-\dfrac{1}{\sqrt{2}}[\alpha(1)\,\gamma(2) - \gamma(1)\,\alpha(2)]$ antisymm. $\uparrow\downarrow, \downarrow\uparrow$

$\beta(1)\,\beta(2)$ $\rightarrow\rightarrow$

Appendix E

E-1 $\Delta j_x \Delta j_y \gtrsim \hbar.$

E-2 Y_{22} has *even* parity; Y_{32} has *odd* parity.

E-3   Setting the differential of Eq.(E-105) to zero yields the condition $\cos^2\theta = \tfrac{1}{3}$, that is $\theta = \arccos\left(\tfrac{1}{3}\right) = 54.7°$, namely the angle at which $P_2(\cos\theta)$ vanishes [cf. Eq. (E-103)].

Appendix G

G-1 $\mathscr{A} = 1.35 \ \mu\text{Ci}.$

G-2 $\tau = \dfrac{N_1 + N_2 - N_{12}}{2N_1 N_2} t$ when $N_{12}\tau \ll t.$

G-3 Net measurement time $t = 44.4$ sec, comprising $t\,(\text{effect} + \text{background}) = 29.6$ sec, and $t\,(\text{background}) = 14.8$ sec.

G-4 $\sigma_0 = 1.7$ percent;
If the measurement times are modified to
$t\,(\text{effect} + \text{background}) = 1.2$ h, $t\,(\text{background}) = 0.6$ h,
the total measurement time is reduced to $t = 1.8$ h from 2.0 h.

G-5 $n_{100} = 1.66 \times 10^3 \ \text{sec}^{-1}, \ n_{400} = 5.13 \times 10^3 \ \text{sec}^{-1}.$

G-6 $T_{1/2} = 157 \ \mu\text{sec}$; discrepancy may be attributed to differences in the time delay in the α and β channels or to random coincidences.

G-7 $\dfrac{\Delta d}{d} \ (\alpha = 2.5°) \simeq 0.1$ percent; $\dfrac{\Delta d}{d} \ (\alpha = 8°) \simeq 1$ percent.

G-8 $\dfrac{\delta p}{p} = \left[1 + \left(\dfrac{4}{\sin 2\alpha} + \dfrac{e^2 B^2 z_s^2}{4p^2 \cos^2 \alpha} \right) (1 + \cot \alpha)^{-1} \right] \dfrac{T^2}{2 \sin^2 \alpha} = 5.06 \ T^2.$

G-9 (a) $d = 12.75$ cm;   (b) $a = 500$ cm;   (c) $\rho = 167$ cm.

G-10 Since $m_0 c^2 = 940$ MeV, the particles are *antiprotons*; $t_{\pi^-} = 40.3$ nsec.

G-11 $0 \leqslant E_e \leqslant \dfrac{2\mathscr{E}}{1 + 2\mathscr{E}} h\nu$ for $0 \leqslant \theta \leqslant \pi.$

G-12 The pile-up spectrum has a triangular distribution; $N = 1000 \ \text{sec}^{-1}.$

Appendix H

H-1 $E_p = 0.585$ MeV; $E_C = 0.042$ MeV.

H-2 1.32×10^{-9} r/sec; 0.77 percent.

H-3 0.36 rad/h $\triangleq$ 0.30 rad/h for air $\triangleq$ 0.331 rad/h for water.
$\triangleq$ 0.331 rem/h.

H-4 $d = 91.3$ cm (with regard to the *neutron* flux; the γ flux would permit an approach to within $d = 623$ cm).

H-5 $R = 30.7$ cm (If, instead, one uses the linear attenuation coefficient $(\mu_0)_{\text{Pb}} = 0.64$ cm^{-1} and a dose build-up factor of 5 one obtains the result $R = 23$ cm).

H-6 $x = 380$ cm.

H-7 $x_1 = 0.9$ cm; $x_2 = 1.2$ cm.

H-8 (a) $m = 0.843$ g;
(b) 71.7 mrem/h $= 29 \times$ Tolerance Dose (when the linear absorption coefficient $(\mu_a)_{\text{C}} = 0.0945$ cm^{-1} is used and no correction for dose build-up is applied, whereas if one uses the attenuation coefficient $(\mu_0)_{\text{C}} = 0.198$ cm^{-1} and a build-up factor of 32.5 one calculates the dose rate to be 0.2 mrem/h $= 0.08 \times$ Tolerance Dose).
(c) $t = 1.02$ sec for a dose of 0.1 rem.
(d) $x = 6.1$ cm (using $(\mu_a)_{\text{Pb}} = 0.376$ cm^{-1}, as against $x = 3.6$ cm for $(\mu_0)_{\text{Pb}} = 0.645$ cm^{-1}).

H-9 (a) $W = 1.08 \times 10^7$ erg $= 1.08$ J;
(b) $\mathscr{A} = 1.06$ kCi;
(c) $x = 36$ cm (using $(\mu_a)_{\text{Pb}} = 0.370$ cm^{-1}, as against $x = 24$ cm when the attenuation coefficient $(\mu_0)_{\text{Pb}} = 0.632$ cm^{-1} and a build-up factor of 5 is used).

H-10 The magnitude of μ_d conjoined with parity considerations rule out all combinations other than ${}^3S_1 + {}^3D_1$, with 96.10 percent ${}^3S_1$ and 3.90 percent ${}^3D_1$, as discussed in Section 12.4.8.

SUBJECT INDEX

Page numbers printed in italic type refer to entries in footnotes, tables, and figure legends; those printed in boldface type refer to the main discussion of the entry.

A

Accelerators
  electron, 12
  proton, 5, 13
Acceptor, 715
Actinium decay chain, *274*
Activity, **267**ff
  analysis, **269, 283**ff
  daughter, **272**ff
  induced, *see* Radioactivity, induced
  maximum, 275
  mixture, **283**ff
  specific, **270**ff, *see also* Specific activity
  per unit time (yield), *see* Yield
Allowed transitions in Beta decay, *see* Beta decay, allowed transitions
Alpha decay
  application of Gamow formula, **321**ff
  lifetime, 265, **298**ff
  stability, 323
  systematics, *300,* 323
Alpha particle
  identified as helium nucleus, 4
  isospin, 229
  range, 301
Alpha rays
  discovery of, 2
  long-range, 320
  short-range, 320
Angle
  of emission, *see* Scattering kinematics, Decay kinematics, Reaction kinematics
  of scattering, *see* Scattering kinematics
  solid, *see* Solid angle
Angle straggling, 170
Angström unit, 8
Angular correlation, 78, **440**ff, **670**ff
  function, 78
  particle parameter, 482ff

Angular distribution, 77, **440**ff, **666**ff
  function 77, **667**
  of gamma radiation, **440**ff, **666**ff
  particle parameter, 482ff
  of photoelectrons, 119
Angular momentum
  conservation, 257, 259, **415**
  nuclear (spin), *see* Nuclear spin
  operators, **646**ff
  selection rule in gamma decay, *see* Gamma decay, momentum selection rule
    in internal conversion, *see* Internal conversion, momentum selection rule
  wave-function composition, **647**ff
Angular wave equation, *see* Wave equation, angular
Annihilation
  in flight, 144
  one-quantum, 144
  operator, *see* Second quantization
  radiation, 144
  at rest, 145
  three-quantum, 146
  two-quantum, 145
Antilinear operator, *258*, 544
Antilinear transformation, *258*
Antineutrino $\bar{\nu}$, 56, *see also* Neutrino
  helicity, **340**, 403, **404**ff
Antiparticle, 56, 691
  effect of gravitational field, 58
  time-reversal, 125
Antiunitary operator, *258*, 544
Antiunitary transformation, *258*, 544
Associated Legendre function, *see* Legendre associated function
Asymmetry energy due to neutron excess, *see* Nuclear liquid-drop model, asymmetry energy

791

Atomic beam
  effect of gravitational field on, 59
  technique for determination of magnetic
    moment, 242
Atomic Bohr radius, 29, 623
Atomic hypothesis, Dalton's, 2
Atomic mass
  parabolas, 46ff
  survey, 39
  tabulation, 28
  unit u, *15*
Atomic model
  Bohr, **3**
  Nagaoka, **3**
  Thomson, **3**
Atomic nucleus, *see* Nucleus
Atomic number Z, 15
Atomic physics, development of, **2**ff
Atomic shell capture ratios, *see* Electron
  capture ratios
Atomic weight, 24
Attenuation coefficient
  for Compton scattering, 148, 710
  linear, 76, 97, 120, 710
  mass, 76, 82, 120, **152**
  for pair production, 148, 710
  for photoelectric effect, 148, 710
  total linear, **148**, 710
Auger effect, **120**ff
Auger electron, **120**ff, 329
Auger lines, *332*

B

$B(EL)$ and $B(ML)$ reduced transition
  rates, *see* Gamma decay, multipole
  transition probability (reduced)
BeV, unit (GeV $\equiv 10^9$ eV), 11
Balance, detailed, *see* Detailed balance
Barn, unit ($\equiv 10^{-24}$ cm), 69
Barrier, *see* Nuclear barrier
Baryon, 210
  number B, **218**
Becquerel's discovery of natural radioactivity, *see* Radioactivity, natural, discovery of
Beta decay
  allowed transitions, 371, **375**ff, 391
  anisotropy parameter, **402**
  application of gamma-matrices, **634**ff

B-x diagram, 394
chain, 50, *274*
classification of transitions, **371**ff, **391**ff
comparative half-life, *see* Beta decay,
  ft value
Coulomb correction factor, **357**ff
coupling constants (strengths), **390**ff,
  408ff
coupling-strength restrictions, **391**
double, *see* Double beta decay
electron-neutrino angular correlation,
  396ff
energetics, **329**ff, 333
energy spectrum, *331*, *see also* Beta decay, momentum spectrum
Fermi function (Coulomb factor),
  358ff
Fermi integral function, **369**ff
modes, 264, **328**ff
Fermi transitions, **391**ff
Fermi weak-interaction constant, 56,
  355, **393**
Fierz interference term, *see* Beta decay,
  interference term
forbidden transitions, 371, **376**ff
ft histogram, *370*
ft value (comparative half-life $fT_{\frac{1}{2}}$),
  371ff, **393**ff
Gamow-Teller transitions, **391**ff
half-life, **368**ff, **370**ff
Hamiltonian density, **389**
interaction matrix element, **355**ff, 373,
  393ff
interference term (Fierz term), **392**
inverse, 60
Kurie plot, **360**ff
lifetime, **368**ff
mass-energy balance, **329**ff
momentum spectrum, 352, 364
of neutron, 60, 63, 137, 263, **351**, 389
parity violation, *see* Parity violation
probability function, **352**
Sargent diagram, 371
spectra, various representations contrasted, *702*
stability (valley of *or* line of), *35*, 46,
  294
statistical factor (final-state density),
  353ff
superallowed transitions, **372**, **374**ff, 391

theory, 5, **323ff**, 351ff
universal time constant τ_0, **369**
Beta rays, identified as electrons, 2, 4
Beta spectrometer
  double-focusing, **703ff**
  flat type, **699ff**
  lens (longitudinal) type, **705**
  luminosity, **700**
  $\pi\sqrt{2}$, **703ff**
  pair, **705**
  semicircular (180°), **703ff**
  transmission, **700**
  trochoidal, **703ff**
Beta spectrometry, **698ff**
Beta spectroscopy, *see* Beta spectrometry
Bethe equation, **158**
Bethe–Weizsäcker formula, *see* Nuclear
  liquid-drop model
Bessel function, 532, 537
  spherical, 532
Bhabha equation, 641, 686
Bhabha scattering, 686
Binding energy, *see also* Nuclear liquid-
  drop model
  K-shell, L-shell, M-shell, 114
  mean, per particle, 13, 24, 31
  nuclear, 24, 31
  total electron, *115*
  von Weizäcker's treatment, 31ff
Biological effects of radiation, **731ff**
Bipolar spherical harmonic, *see* Spherical
  harmonic
Bloch effect, 245
Bohr atomic model, *see* Atomic model,
  Bohr
Bohr magneton, *see* Magneton
Bohr radius, atomic, *see* Atomic Bohr
  radius
Bohr-Sommerfeld atomic model, *see*
  Atomic model, Bohr
Bohr and Wheeler's theory of fission, *see*
  Fission, Bohr-Wheeler theory
Boltzmann factor, 441
Born approximation, 93, 140, 181, 543
Born collision formula, 93
Bose-Einstein statistics, 210
Boson, 210
  intermediate vector (W-particle, *or*
    "schizon"), 56, 57, 350, 409
  wave equations, **641ff**

Bound state, *see* Nuclear bound state
Box normalization, 89
Bragg angle, 722
Bragg condition, 722
Bragg-Kleemann rule, 166
Bragg scattering (reflection), 722
Bragg spectrometer, *see* Gamma spectrom-
  eter
Branching ratio, 270
Bremsstrahlung, 144, **176ff**
  cross section, *see* Cross section, brems-
  strahlung
  external, 189, 388
  internal (inner), 189, 385
  spectral distribution, 187
Breit-Wigner resonance formula, **520ff**,
  549
Breit-Wigner theory, **520ff**
  modified for elastic scattering, **524ff**

C

Center-of-mass (CM) system, **26ff**, 532,
  580ff
  energy, **582ff**, **592ff**, *see also* Cross sec-
    tion, elastic; Scattering kinematics
Centrifugal barrier, *see* Nuclear barrier,
  centrifugal
Centrifugal force, *312*
Centrifugal potential, *312*
Čerenkov (-Vavilov) radiation, **190ff**
  angle, 192
  cone, 193
  detector (counter), 197
  dispersion, 196
  threshold, 192
Chadwick's discovery of neutrons, *see*
  Neutrons, discovery of
Charge
  conjugation, 56, *258*, 403
  distribution of heavy ions, *200*
  effective, 160
    of heavy ions, 199
  electric, conservation, *258*
  parity, *258*
Charge-density distribution, nuclear, *see*
  Nuclear charge distribution
Chemical binding, 135, **382ff**, 456, 459,
  479
Chirality, **634**

Classical cross section, *see* Cross section, Rutherford; Cross section, Thomson

Classical electron radius, *see* Electron radius

Clebsch-Gordan coefficient, **647ff**
symmetry properties, **650**

Clifford numbers, *see* Gamma matrices

Cockcroft-Walton experiment, 5

Coincidence spectroscopy, **723ff**
counter, **723ff**

Collision
diameter, 84
formula, see Born collision formula
matrix, *see* *U*-Matrix

Completeness relation, 649

Complex energy, 551

Compound elastic cross section, *see* Cross section, compound elastic

Compound elastic scattering (resonance scattering), *see* Scattering, compound elastic

Compound nucleus, *see* Nucleus, compound

Compton cross section, *see* Cross section, Compton

Compton effect, **103ff**

Compton electron, **104ff**

Compton electron distribution, 709

Compton energy transfer, 106

Compton recoil electron energy, 106

Compton scattering, *see* Scattering, Compton

Compton wavelength, *see* Wavelength, Compton

Compton X-ray line measurements, 104

Conjugate particle, *see* Antiparticle

Constants, physical (numerical), **738ff**

Continuity equation, **621**

Continuum state, *see* Nuclear continuum state

Contraction of operators, 692

Conversion, *see* Internal conversion

Conversion factors (numerical), **739ff**

Coulomb cross section, *see* Cross section, Rutherford; Mott formula; Cross section, Mott

Coulomb displacement energy, 232, *see also* Nuclear liquid-drop model, Coulomb energy

Coulomb energy, 232, 313

Coulomb energy term in liquid-drop model, *see* Nuclear liquid-drop model, Coulomb energy

Coulomb excitation, 155
multiple, 155

Coulomb factor (Fermi function), *see* Beta decay, Fermi function

Coulomb potential barrier, 313

Coulomb radius constant, *see* Nuclear radius parameter

Coupling constant, **55ff**
electromagnetic, 55
gravitational, 57
strong-interaction, 55
weak-interaction, 55ff

Coupling scheme (angular momentum), **225ff**
intermediate, 226
j-j (spin-orbit), **225**
L-S (Russell-Saunders), **225**

Cowan-Reines experiments, **338ff**

Crane experiment, 340

Creation operator, *see* Second quantization

Cross section, *see also* Attenuation coefficient
absolute, 79
absorption, **524, 529**
per atom, 82, 102
Born collision, **93**
bremsstrahlung, **179ff**
for charged-particle exothermic reactions, 510, 511
collision, 107, 109
compound-elastic (resonance), **523, 529, 540**
compound nucleus formation, **522**
Compton, **103, 107**
atomic, **110**
Klein-Nishina formula, **107, 697**
definition, **68ff**, 75
differential, 77, 535
double-differential, 78
elastic, **523, 529**, 535, 545
elastic neutron scattering, 507, 523, 529
per electron, 82, 102
energy-absorption, 109
energy-scattering, 109
geometrical, 68, 79, 540
inelastic neutron scattering, 509
Klein-Nishina formula, **107, 697**

Mott, **88**
for neutron-induced endothermic reactions, 509
for neutron-induced exothermic reactions ("$1/v$ law"), 507
pair production, **138**
partial, 77
photoelectric, total, **117**
Rayleigh, **113**
reaction, 71, **505**ff, 523, 529
resonance, **519**ff, 529, 539, *see also* Cross section, compound-elastic
resonance anomaly, *525, 526,* 527
Rutherford, **86,** 95
shape-elastic (potential), **523,** 529, 540
thin-target, 71
thick-target, 72
Thomson, **100,** 107, **697**
total, 77, **523,** 529, 535
units, 69
Crystal spectrometer, *see* DuMond curved crystal spectrometer
Cut-off radius, *see* Interaction cut-off radius
Curie, unit of activity, **270**
Cyclotron frequency, 243

D

d'Alembertian, 627
Dalitz diagram (plot), 496
Dalton's atomic hypothesis, *see* Atomic hypothesis, Dalton's
Dating (radioactive), *see* Radioactive dating methods
Daughter activity, *see* Activity, daughter
Davis experiments, 336, 341
Davydov model, *430, 431,* 432
Dead time, 713
de Broglie "matter waves," 8
de Broglie wavelength, *see* Wavelength, de Broglie
Debye temperature Θ, 453
Debye–Waller factor, 453, 454
Decay constant, **267,** 290, 315, 468
partial, **271**
Decay kinematics, **601**ff
Decay law (exponential), **269**
Decay schemes
radioactive, *264, 274*

of widely used radioactive sources, **286**ff
Delbrück scattering, *see* Scattering, Delbrück
Density of nuclear matter, *see* Nuclear matter, density of
Destruction (annihilation) operator, *see* Second quantization
Detailed balance, *258,* **511**ff, 542
experiments, **514**ff
principle of semi-detailed balance, **513,** 542
Deuteron isospin, 230
Dicke pulsating universe theory, 292
Dineutron, **34,** 230
Dirac bra-ket notation, 647
Dirac electron field, **676**
Dirac electron theory, **123**ff, 242, **637**ff, **676**
Dirac limits (magnetic moment), *251, 252, 253*
Dirac line, *see* Dirac limits
Dirac particles, *see* Positrons
Dirac perturbation treatment, *see* Perturbation theory
Dirac "sea," 124, 127
Dirac wave equation, **628**ff
Donor, 715
Doppler broadening of spectral lines, 447
Doppler effects, **567**ff
Doppler shift
attenuation, 265
of spectral lines, 448
Dose, *see* Radiation dosimetry
Dose rate, *see* Radiation dosimetry
Dosimetry, *see* Radiation dosimetry
Double beta decay, **345**ff
half-life, **346**
Double electron capture, **388**
Double-focusing spectrometer, *see* Beta spectrometer
Duane and Hunt's Law, *177,* 178
DuMond curved crystal spectrometer, 721
Dyson time-ordering operator, *see* Time-ordering operator

E

Effective charge, *see* Charge, effective

Effective nuclear charge, *see* Nuclear charge, effective
Eigenfunction, **620**
Eigenvalue, 619
Einstein factor γ, 9, **561**
Einstein mass-energy relation, 9, *16*, **570**
Einstein special relativity theory, **560**ff
Eisenbud–Wigner reaction formalism, *see* Wigner–Eisenbud reaction formalism
Electric multipole radiation, *see* Gamma decay, electric multipole radiation; Parity rule for multipole moments
Electromagnetic potential, 627
Electron
  Auger, *see* Auger electron
  cyclotron frequency, **243**
  density distribution in potassium atom, *624*
  Dirac theory, **123**ff
  electric dipole moment, **242**
  magnetic dipole moment, **242, 247**
  neutrino field, 351
  noncontainment by nuclei, 13
  radial distribution in hydrogen atom, *623*
  radius, 13, 29, 697
  rest energy, 10
  rest mass, 10
  scattering, 12, 18, 596, *685*
  self-energy, 132, **683**
  wavelength, 10
Electron capture, 60, 329, 363, **377**ff
  double, **388**
  Fermi integral function, **380**
  ft value, **381**
  lifetime, **378**ff
  neutrinos, 377
  ratios, **381**ff
  transition energy E_0, 385
Electron cross section, *see* Cross section, per electron
Electron-volt
  relation to β and γ, **574**ff, *575, 576, 577*
  unit of energy, **10**
Electrostatic separation energy, 54
Energetics
  decay, *see* Decay kinematics
  scattering, *see* Scattering kinematics

Energy
  conservation, 17, 257, 259, 327, 485ff, 490, 572, 603, 619
  correlation analysis, **490**ff
  dose, *see* Radiation dosimetry, energy dose
  level, *see* Nuclear energy level
  loss of electrons, **170**ff
    of heavy charged particles (protons), **156**ff, 171
      correction term, 158
    of heavy ions, **199**ff, *see also* Stopping power
  mass relation, *see* Mass-energy relation
  momentum relation (relativistic), 9, **571, 627**
  momentum transformation, **571**
  operator (Hamilton), *see* Hamilton operator
  resonance, 520
  scale in low-energy nuclear spectroscopy, **721**ff
  straggling of electrons, 176
    of heavy particles, 168
  threshold, *see* Nuclear reaction threshold
  total, 570
Energy-gap, 714
Energy level, shift, 550
Equation of continuity, **621**
Equilibrium
  ideal, *276,* **277**
  secular, **279**
  transient, **279,** *280*
Escape lines, *711, 720*
Eta meson decay mode, 62
Euler–Lagrange equation of motion, **642**
Euler–Lagrange field equation, **642**
Exchange forces, 17, 54, 135
Excitation energy, *see* Nuclear excitation energy
Exciton, 706, 714
External bremsstrahlung, *see* Bremsstrahlung, external

F

F-K plot, *see* Beta decay, Kurie plot
Fano X coefficient, *see* Wigner 9-*j* symbol
Femtometer (Fermi unit $\equiv 10^{-13}$ cm), 8

Fermi-Dirac statistics, 210
Fermi function in beta decay, *see* Beta decay, Fermi function
Fermi "Golden Rule No. 1", **520**
Fermi "Golden Rule No. 2", 68, 90, 91, 93, 179, 505, 697
Fermi integral function
  in beta decay, *see* beta decay, Fermi integral function
Fermi integral function
  in electron capture, *see* Electron capture, Fermi integral function
Fermi plot, *see* Beta decay, Kurie plot
Fermi theory of beta decay, *see* Beta decay, theory
Fermi transitions in beta decay, *see* Beta decay, Fermi transitions
Fermi unit (femtometer), 8
Fermi weak-interaction constant g_F, *see* Beta decay, Fermi weak-interaction constant
Fermion, 210
  self-energy, **683**
Feynman dagger notation, 679
Feynman diagram, *see* Feynman graph
Feynman graph (diagram), **127**ff, *498, 680*ff
  closed loop, **134**, 695
  vertex, **128, 680**ff
Feynman interaction theory, 675ff
Fierz interference term in beta decay, *see* Beta decay, interference term
Fine structure
  constant, 29, 55
  equation, 630
  splitting, 455
Fission
  Bohr-Wheeler theory, 5
  discovery of, 5
  energetics, 44ff
  of heavy nuclides
    Beta decay chain, 50
    energy release, 26
  symmetrical, 44
Fluorescence, nuclear resonance, *see* Nuclear resonance absorption
Forbidden transitions in beta decay, *see* Beta decay, forbidden transitions

in gamma decay, *see* Gamma decay, forbidden transitions
Forbidden zone of semiconductor, 714
Forces, nuclear, *see* Nuclear forces
Form factor, analysis, 18
Four-component neutrino theory, *see* Neutrino, four-component theory
Frequency shifts (relativistic), *see* Doppler effects
Furry theorem, 135, 695
  generalized, 135
Fusion, 26

G

g Factor, **246**, 248, 459
γ factor (Einstein), 9, **561**
γ matrices, **631**ff, **633**ff
Galilei transformation, **561**
Gamma decay
  application to compilation of nuclear energy-level schemes, *see* Nuclear energy level scheme compilation
  electric multipole radiation, **415**ff
  "forbidden" transitions, **427**
  lifetime, *see* Gamma decay, multipole transition probability
  magnetic multipole radiation, **415**ff
  momentum selection rule, **415**
  multipolarity, **414**
  multipole mixing, **428**ff, 445
  multipole mixing ratio, **430**, 445
  multipole order, **414**
  multipole transition matrix element, 425
  multipole transition probability, **418**ff
    reduced, **425**ff
  parity selection rule, *see* Parity rule for multipole moments
  spectroscopy, *see* Gamma spectroscopy; Angular distribution; Angular correlation
  Weisskopf estimate of multipole transition probability (lifetime), *see* Gamma decay, multipole transition probability
Gamma line width, *see* Width, natural line; Nuclear energy-level width
Gamma spectrometer, 721
  DuMond crystal, 721

Gamma spectroscopy, nuclear resonance absorption and fluorescence, see Nuclear resonance absorption; see also Angular distribution; Angular correlation

Gamow "big bang" cosmological theory, 292

Gamow factor, 309, 310, 317, 323, 507ff

Gamow formula, **309**

Gamow-Teller transitions in beta decay, see Beta decay, Gamow-Teller transitions

Ge(Li) detector, 717

GeV, unit (Giga electron-volts ≡ 10^9eV), 11

Geiger and Marsden's experiments, 3, **83**

Geiger–Nuttall plot, *302*

Geiger–Nuttall relation, 301

Geiger's rule, 301

Germanium counter, 717

Gravitational constant (Newton), 57

Gravitational coupling constant, 57

Gravitational energy, 54

Gravitational interactions, 57

Gravitational potential, 58

Graviton, 57, 210, 644
    Compton wavelength of, see Wavelength, Compton

Ground state, see Nuclear ground state

Gyromagnetic ratio, **246**

H

Hahn and Strassmann's discovery of fission, see Fission, discovery of

Half-life $T_{\frac{1}{2}}$, see Nuclear half-life
    numerical values, 742

Hamilton density, 389, 677

Hamilton-Jacobi theory, 209

Hamilton operator, 224, 257, 389, 543, 619, 630, 677

Hartree-Fock model, 624

Hauser-Feshbach theory, 670

Health hazards, **731**ff, 734, see also Radiation dosimetry

Heavy ion
    charge distributions, *200*
    energy loss of, see Energy loss of heavy ions
    reactions, 517

Heisenberg uncertainty relations, see Uncertainty relations

Helicity of neutrinos, see Neutrino helicity

Historical development of atomic and nuclear physics, see Atomic physics, development of; Nuclear physics, development of

Humblet-Rosenfeld reaction formalism, 547

Hydrogen atom, electron distribution, *623*

Hypercharge, see Quantum number, hypercharge

Hyperfine splitting, see Nuclear hyperfine structure

Hyperon decay modes, 63

I

Ideal equilibrium, see Equilibrium, ideal

Impact parameter, 80, 85, 537

Indeterminacy principle, see Uncertainty relations

Induced radioactivity, see Radioactivity, induced

Inelasticity parameter, **538**

Interaction cut-off radius b, 548

Interaction
    of charged particles with matter, **155**ff
    of electromagnetic radiation with matter, **98**ff, *99*
    particle response to, 61ff

Interference between compound-elastic (resonance) and shape-elastic (potential) scattering, *525*, *526*, 540

Interference (Fierz) term, see Beta decay, interference term

Intermediate coupling scheme, see Coupling scheme, intermediate

Internal bremsstrahlung, see Bremsstrahlung, internal

Internal characteristic x-rays, see X-rays, characteristic

Internal conversion, **464**ff
    coefficient, **467**ff
    decay constant, **468**
    mean lifetime, **479**ff
    momentum selection rule, **481**
    multipole mixing, **481**ff

Internal pair formation, *see* Pair production, internal
Internucleon forces, *see* Nuclear forces
Inverse beta decay, *see* Beta decay, inverse
Inverse pair production, *see* Annihilation
Ionization potential (excitation potential), 158
  specific, 160
Isoanalogue, *see* Isobaric analogue
Isobar, 23, 45
  "most stable," 50
Isobaric analogue (isoanalogue) state, **228**
Isobaric mass formula, **233**
Isobaric multiplet, **228**
Isobaric resonance, **232**ff
Isomer, 23, *see also* Nuclear isomerism
Isomeric state, *see* Nuclear isomerism
Isomerism, *see* Isomer; Nuclear isomerism
Isospin
  conservation, *258*
  nuclear, *see* Nuclear isospin
  particle classification scheme (table), *217*
  quantum number, *see* Quantum number, isospin
  rotation invariance, *258*
Isotone, 23
Isotope, 23
Isotropy of space, 260

J

j-j Coupling scheme, *see* Coupling scheme, *j-j*
Jauch plot, 386

K

Kapur-Peierls reaction formalism, 547, 548
Kemmer equation, 641
Kinematics, *see* Scattering kinematics; Reaction kinematics; Decay kinematics
Klein-Gordon equation, **626**ff, 644, 688
Klein-Nishina formula, *see* Cross section, Compton
Kramer's rule in quantum mechanics, 249
Kurie plot, *see* Beta decay, Kurie plot

L

L-Function, 548
L-S Coupling scheme, *see* Coupling scheme, *L-S*
Lagrangian density, **642, 676**ff
Lagrangian function, **642**
Lagrangian operator, *258*
Laguerre polynomial, **621**
Lamb-Mössbauer factor, 453, 454
Landé factor, *see* g Factor
Larmor precession frequency, 243
Legendre associated function, 223, **654,** *655, 656*
Legendre hyperpolynomial, **670**ff
Legendre polynomial, 443, 527, 545, **654,** *655, 656*
Lemaître "primeval atom" cosmological theory, 292
Length, contraction (relativistic), **562**ff
Lens spectrometer, *see* Beta spectrometer
Lepton, 210, 335
  gauge invariance, *258*
  number, *258*
Level density, *see* Nuclear energy level density
Level shift, *see* Energy level shift
Libby, radiocarbon dating technique, **291**
Lifetime, *see* Nuclear lifetime, mean
  alpha decay, *see* Alpha decay lifetime
  beta decay, *see* Beta decay lifetime
  gamma decay, *see* Gamma decay, multipole transition probability
  numerical, *742*
  of relativistic particles, *see* Relativistic particle lifetime
Line width (spectral), *see* Width
Linear attenuation coefficient, *see* Attenuation coefficient, linear
Liquid drop model, *see* Nuclear, liquid drop model
Lorentz curve, 265, *267, 519*
Lorentz distribution, **520,** 540
Lorentz-Fitzgerald contraction, **562**ff
Lorentz transformation, *258, 560*
Luminosity of spectrometer, **700**

M

M-Matrix, 541
Magnetic moment, *see* Moment, magnetic

Magnetic monopole, **242**
Magnetic multipole radiation, *see* Gamma decay, magnetic multipole radiation; Parity rule for multipole moments
Magnetic potential energy, 54
Magnetic quantum number, *see* Quantum number
Magnetic rigidity, *332, 573,* **700**
Magnetic spectrometer, *see* Beta spectrometer
Magneton unit
  Bohr, **246,** 742
  nuclear, **246,** 742
Mass
  balance in beta decay, **329**ff
  energy relation, 9, *16,* **570**
  of neutrino, 334, **363**ff
  reduced, *see* Reduced mass
Mass-attenuation coefficient, *see* Attenuation coefficient, mass
Mass defect (mass decrement), 24
Mass formula, *see* Nuclear liquid-drop model, semiempirical mass formula; Isobaric mass formula
Mass number *A,* 15
Mass parabolas, *see* Atomic mass parabolas
Mass stopping power, *see* Stopping power, mass
Matrix
  element, elastic scattering, 91
  forms of reaction theory, *see* Reaction theory, matrix formalism
  inversion, 550
Maxwell equations, **642**
  field, **643, 644**
Mean free path, 76
Mean range, *see* Range, mean
Meson
  classification, 210
  discovery of π, 5
  of μ, 5
  as exchange particle, *see* Nuclear meson model
  nuclear model, **247**ff
  prediction by Yukawa, 5, 16, 54
  pseudoscalar and pseudovector theory, 55
Meteorite ages, **291**ff
Metric tensor, **627**

Microscopic reversibility, *see* Reversibility, microscopic; Detailed balance
Mineral ages, **291**ff
Minkowski representation, **564**
Mirror
  nuclei, 23, 43
  transitions in beta decay, 394
Mixed multipoles
  in gamma decay, *see* Gamma decay, multipole mixing
  in internal conversion, *see* Internal conversion, multipole mixing
Model
  Davydov, *see* Davydov model
  nuclear liquid drop, *see* Nucleus, liquid drop model
Møller scattering, *see* Scattering, Møller
Mössbauer effect, **450**ff
  applications, 58, 457
    nuclear hyperfine structure, 458
    Pound-Rebka experiment, 58
    study of parity conservation, 459
  temperature dependence, 456
Mössbauer experiments, **450**ff
Mössbauer-Lamb factor, 453ff
Mössbauer line, **451**
Molecular beam technique for determination of magnetic moment, 243
Moment
  electric, **251**ff
    dipole, 255
    monopole, 255
    quadrupole, **256**
  magnetic, *see* respective particle
    numerical values, 742
  magnetic dipole
    orbital, 246
    spin, 246
  multipole, **242**
    determination methods, **243**
    of particles, *see* respective particle
  nuclear
    Schmidt formulae, **250**
    Schüler-Schmidt lines, **250**ff
  nuclear electric, **251**ff
  nuclear higher-order, **251**
  nuclear magnetic, **248**ff, **250**ff
  nuclear quadrupole, **256**ff
  order, 414
  quadrupole, **256**

Momentum
 conservation, 257, 259, **415**, 481, 485ff, 580, 602
 selection rule, *see* Momentum conservation; Angular momentum
Monopole, magnetic, *see* Magnetic monopole
Moon experiment, 448
Mott formula, **88**
Mott scattering, **88**ff, 171
Mozer's semiempirical mass formula, *see* Nuclear semiempirical mass formula, Mozer's
Mu meson, **55**ff
 decay, 138
  modes, 56, 62
 discovery of, *see* Meson, discovery of μ
 lifetime, 56, **563**, 742
Multiple scattering, *see* Scattering, multiple
Multiplicity, 115
Multipole mixing
 in gamma decay, *see* Gamma decay, multipole mixing
 ratio, *see* Gamma decay, multipole mixing
Multipole moment, *see* Moment

N

n-type semiconductor, *see* Semiconductor, n-type
Nagaoka's atomic model, *see* Atomic model, Nagaoka
NaI(Tl) counter, *see* Scintillaton counter
Neutretto (muon-type neutrino), **347**ff
 detection experiments, **349**ff
Neutrino ν, 56, 63
 bibliography, 351
 charge, 335
 continuous spectrum (in β^+ decay), 378
 electron-type *vs* muon-type, **347**
 four-component theory, 340
 helicity, **340**, 403, **404**ff, 641
 line spectrum in electron capture, **377**ff
 magnetic moment, 335
 mass, 334
 production, 63
 properties, **334**ff
 searches, **336**ff

 spectrum, solar, *345*
 spin, 335
 two-component theory, 340, 403, **639**ff
Neutron
 Beta decay of, *see* Beta decay of neutron
 cross section
  capture
   for Cd, 73, 74
   for In, *519*
   resonance, **519**ff
   for Xe, 70
  scattering, **542**ff, 529ff
   for In, *525*
   for S-waves, *529*
 discovery of, 4
 effect of gravitational field on, 59
 electric dipole moment, **242**
 excess, 34, 228
 existence of dineutron, 34, 230
  of tetraneutron, 34
  of trineutron, 34, 230
 lifetime, 335, 742
 magnetic dipole moment, **245, 247**
 number N, 15
 structure, *see* Nucleon structure
Noether's theorem, 642
Normal product, 680
Nuclear alpha decay, *see* Alpha decay
Nuclear angular momentum, *see* Nuclear spin
Nuclear barrier, 303, **310**
Nuclear beta decay, *see* Beta decay
Nuclear binding energy; *see also* Binding energy, nuclear
 deformation effects, 40ff
 numerical values, 743
 shell-model effects, 39ff
Nuclear bound state, 224
Nuclear centrifugal barrier, **311, 312,** *see also* Coulomb potential barrier
Nuclear charge (effective), 381
Nuclear charge distribution, 12
Nuclear composition
 meson model, 5, 16
 p-e model, 14
 p-n model, 15
Nuclear constitution
 p-n model, by Iwanenko and Heisenberg, 4
Nuclear continuum state, 224

Nuclear energy level (niveau), 92, **224**
  density, 92
  width Γ, **265**ff
  scheme compilation, **433**ff
Nuclear energy surface, 45
Nuclear excitation energy, 224
Nuclear fission, *see* Fission
Nuclear forces, 17, 53ff
Nuclear fusion, 26
Nuclear *g* factor, *see g* Factor
Nuclear gamma decay, *see* Gamma decay
Nuclear ground state, 224
  spin systematics (table) *226*
Nuclear half life, **269**, 499, 742
Nuclear hyperfine structure, 458, 459
Nuclear isomerism, 23, **427**
Nuclear isospin, **227**
Nuclear lifetime
  chemical effects on, 282
  mean, **263**, *266*, 382, 499ff, 742
  partial, **271**
  with respect to gamma decay (Weiss-
    kopf estimate), *see* Gamma decay,
    multiple transition probability
Nuclear liquid-drop model, 5, 31
  asymmetry energy, 34
  Bethe-Weizsäcker formula, 31ff
  Coulomb energy, 33
  stability limits, 294
  surface energy, 32
  volume energy, 32
  von Weizsäcker's treatment of nuclear
    binding energy, 31ff
Nuclear magneton, *see* Magneton
Nuclear mass, 24
Nuclear mass formula, *see* Nuclear liquid-
  drop model, Nuclear semiempirical
  mass formula
Nuclear mass parabolas, **46**ff
Nuclear matter density, **30**
Nuclear meson model, 5, 16
Nuclear multipole moment, *see* Moment,
  nuclear
Nuclear parity, *see* Parity, nuclear
Nuclear penetration, *see* Nuclear tunnel-
  ing
Nuclear penetration factor, 306
Nuclear photoelectric effect (photodisin-
  tegration), *see* Photoelectric effect

Nuclear physics, development of, **4**ff
Nuclear proximity scattering, *see* Scatter-
  ing, proximity
Nuclear quadrupole moment, *see* Moment
Nuclear radius, 28, 30ff
  parameter r_0, 28, **42**ff, 322
Nuclear reaction
  cross section, *see* Cross section, reac-
    tion
  kinematics (energetics), **485**ff
    threshold, 27, **488**ff
  sequential decay processes, **491**ff
Nuclear recoil energy, 336, 447
Nuclear reflectivity, **309**
Nuclear resonance absorption of gamma
  rays and fluorescence, **446**ff
Nuclear resonance technique for determination of magnetic moment, 243
Nuclear scattering state, 224
Nuclear semiempirical mass formula, 31ff
  application to alpha decay, **299**
  Mozer's, 39ff
  parameters, 37ff
  survey, 39
Nuclear single-particle model, Weisskopf
  prediction of multipole transition
  probability (lifetime), *see* Gamma
  decay, multipole transition probabil-
  ity
Nuclear size and constitution, 1, 8
Nuclear species, nomenclature, *23*
Nuclear spin, 211, 225
Nuclear stability
  application of mass formula, 42ff
  energy considerations, **22**ff, **42**ff, *294*
  systematics, 36
Nuclear states (Boltzmann population),
  441
Nuclear superfluid model, 42
Nuclear thermal motion, 447
Nuclear transmission coefficient (penetrability), **307**
Nuclear tunneling, 209, **306**ff
Nuclear Zeeman splitting, 459
Nuclei, asymmetric (Davydov model),
  432
Nucleon
  designation, 16
  electric dipole moments, **242**

magnetic dipole moments, **243, 245, 247**
structure, 18
Nucleor, **247**ff
Nucleus, compound, 516
Nuclide
definition of, 15
nomenclature, 23

O

Orbital quantum number, *see* Quantum number, orbital
Orthopositronium, 146

P

p-type semiconductor, *see* Semiconductor, *p*-type
$\pi\sqrt{2}$ spectrometer, *see* Beta spectrometer
Packing fraction, 24
Pair formation, *see* Pair production
Pair annihilation, 129ff
Pair internal conversion, *see* Pair production, internal
Pair production, 98, **122**ff, 138ff, *710, 712*
cross section, *see* Cross section, pair production
internal, **464**ff
Pair spectrometer, *see* Beta spectrometer
Parapositronium, 145
Parity, **220**
associated with orbital momentum, **221**ff
conservation, 257, 298, 459
intrinsic, **221**ff
nuclear, **227**
rule for multiple moments, **242, 416, 481**
selection rule in beta decay, 392
tests, **399**ff
violation, **397**
Partial conversion coefficient, *see* Internal conversion coefficient
Partial decay constant, *see* Decay constant, partial
Partial lifetime, *see* Nuclear lifetime, partial
Partial waves, 80, **532**
Particle

conjugation, 56
creation threshold, 611
response to electromagnetic interactions, 61
to strong interactions, 61
to weak interactions, 62
Yukawa, *see* Meson
Particle parameter, **482**ff, **668**
Path length
extrapolated, 173
mean, 173
Penetrability, 668, 671
Penetration of heavy ions through matter, **199**ff
Penetration factor, **306,** *see also* Penetrability
Perturbation theory, **64**ff, 697
Dirac treatment (variation of constants), 65
Phase shift, **538**
Photodisintegration (nuclear), *see* Photoelectric effect
Photo effect, *see* Photoelectric effect
Photoelectric cross section, *see* Cross section, photoelectric
Photoelectric effect (photodisintegration), 98, **113,** *710*
Photon
"falling" in gravitational field, 58
self-energy, 132, **683**
Photopeak, **709,** *711, 719, 720, 726,* 727
Physical constants, 740, 741
Pi meson
annihilation, 137
artificial production of, 5
capture by nucleons, 61
decay modes, 61
discovery of, *see* Meson, discovery of π
as exchange particle, 5, 16, 54, 136
lifetime, 514ff
radiative capture, 61, 137
Pile-up pulses, **713**
Pleochroic halos, 291
Plural scattering, *see* Scattering, plural
Polarization
vacuum, 132
vector, **689**
Population of nuclear states (Boltzmann), 441

Positron
  electron conjugation, **125**
  prediction by Dirac, 4
  as time-reversed electron, **125**
Positronium, 145
  annihilation, 146
  orthopositronium, 146
  parapositronium, 145
Potassium-argon ratio (dating), *291*
Potential barrier, *see* Nuclear barrier
Potential scattering, *see* Scattering, shape
  elastic
Pound-Rebka experiments, 58
Principal quantum number, *see* Quantum
  number, principal
Principle of detailed balance, *see* Detailed
  balance
Probability
  current density, **620, 629**
  density, **620, 629,** 632ff
  distribution function, **622**
Proca equation, 641, 644
Propagation vector, *see* Wave-number vec-
  tor
Proper time, 570
Proton
  magnetic dipole moment, *243, 247*
  structure, *see* Nucleon structure
Proximity scattering, *see* Scattering, prox-
  imity
Pulsating Universe, Dicke theory, 292
Puppi triangle, **409**

Q

Q-Equation
  nonrelativistic, **487**
  relativistic, **488**
Q-Value, 26, **487**ff
  for alpha decay, **299**
  tabulation, 28
Quadrupole moment, *see* Moment
Quantum electrodynamics, correction to
  magnetic dipole moment of electron,
  242, 247
Quantum mechanics, **617**ff, *see also* Racah
  algebra; Second quantization
Quantum number
  baryon, **218,** *258*
  hypercharge, **216**ff, *258*
  isospin, **215,** *258*
  magnetic orbital, **212**
  magnetic spin, **212**
  magnetic total, **214**
  orbital, 80, **212**
  parity, **220,** *see also* Parity
  principal, **214**
  radial, **214**
  spin, **210**ff, **212**
  strangeness, **216**ff, *258*
  total, **213**
Quantum properties of nuclei and particles, **208**ff

R

R-Matrix, 541
Racah algebra, **646**ff
Racah coefficient, **658**ff
Racah functions, 444
rad (radiation absorbed dose) unit, **732**
Radial quantum number, *see* Quantum
  number, radial
Radial wave equation, *see* Wave equation,
  radial
Radial wave function, **621**
Radiation dosimetry, **731**ff
  dose rate, **732**
  energy dose, **732**
  rad unit, **732**
  radiation dose, **733**
  RBE factor, **733**
  rem unit, **733**
  röntgen unit, **732**
  tolerance dose, **734**
Radiation hazard, *see* Radiation dosimetry; Health hazards
Radiative loss for bremsstrahlung, 183
Radioactive carbon, *see* Radiocarbon
Radioactive dating methods, **290**
Radioactive decay constant, *see* Decay
  constant
Radioactivity, **263**ff
  artificial, *see* Radioactivity, induced
  extinct, 291, *see also* Activity
  induced, **281**ff
  natural, **263**ff
  discovery of, **2**
Radiocarbon, **291**
Radius
  interaction cut-off, 548
  nuclear, *see* Nuclear radius

Radius parameter, r_0, *see* Nuclear radius parameter
Ramsauer effect, 155
Random coincidences, 725
Range, 163
  mean, **163ff**, 173
  of electrons, 174ff
  straggling, **167**
Rayleigh cross section, *see* Cross section, Rayleigh
Rayleigh scattering, *see* Scattering, Rayleigh
RBE factor, *see* Radiation dosimetry, RBE factor
Reactance matrix, *see* R-Matrix
Reaction
  cross section, *see* Cross section, reaction
  endothermic ($Q < 0$), 26
  energy, *see* Q-Value
  exothermic ($Q > 0$), 26
Reaction
  formalisms (Kapur-Peierls, Wigner-Eisenbud, Humblet-Rosenfeld), *see* respective authors
  nuclear, *see* Nuclear reaction
  probability and cross section, 68
Reaction kinematics, **487ff**, **612ff**
  in CM system, 27ff
Reaction matrix, *see* R-Matrix
Reaction theory, matrix formalism, **541ff**
Reciprocity, **542ff**
  relation, **546**
Recoil energy (nuclear), *see* Nuclear recoil energy
Recoil-free gamma-ray absorption, *see* Nuclear resonance absorption
Recoupling coefficients, **658**, *see also* Racah algebra; Wigner 3-*j*, 6-*j*, 9-*j* symbols
Rectangular-well potential, *see* Nuclear barrier
Red shift, *see* Doppler effects
Reduced mass, **28, 582**
Reduced transition probability in gamma decay, *see* Gamma decay, multipole transition probability (reduced)
Reduced width, γ_λ, **549**
Reflectivity, 309
Reines-Cowan experiments, **338ff**

Relative biological effectiveness (RBE), *see* Radiation dosimetry, RBE factor
Relative population of nuclear states, *see* Nuclear states (Boltzmann population)
Relative stopping power, *see* Stopping power, relative
Relativistic elastic scattering, **590ff**
Relativistic energy-momentum relation, 9, 571
Relativistic energy-momentum transformation, **571**
Relativistic mass, **569**
Relativistic particle, 572ff
  lifetime, **574**
  speed, 574
Relativistic scattering kinematics, **590ff, 605ff**
Relativistic space contraction, *see* Lorentz-Fitzgerald contraction
Relativistic space-time (Lorentz) transformation, *see* Lorentz transformation
Relativistic time dilation, *see* Time dilation
Relativity, check of special theory, 57ff
Rem (röntgen equivalent man) unit, **733**
Repetition rate, 315, **521ff**
Resonance
  absorption and fluorescence (nuclear), *see* Nuclear resonance absorption
  cross section, *see* Cross section, resonance
  energy, *see* Energy, resonance
Resonance anomaly, *see* Cross section, resonance anomaly
Resonance scattering, *see* Scattering, compound elastic
Rest energy, **570**
  of particles (numerical), **739**
Rest mass, **569**
  of electron, 10, 739
  of particles, **739**
  of proton, 13, 739
Rest system, 560ff
Reversibility, **542ff**
Rigidity (magnetic), *see* Magnetic rigidity
Rodeback-Allen experiment, **336ff**
Röntgen
  equivalent man (rem) unit, **733**
  unit of radiation, **732**

Röntgen's discovery of x rays, *see* X-Rays, discovery of
Russell-Saunders coupling scheme, *see* Coupling scheme, *L-S*
Rutherford-Bohr atomic model, *see* Atomic model, Bohr
Rutherford formula, 83ff
Rutherford scattering, 3, 83ff
Rydberg constant, 114

S

S-Matrix, **541ff, 678ff**
  properties, **542ff**
Sargent diagram, **371**, *372*
Saxon-Woods potential, *551*
Scattering
  amplitude, *525*, **532**
  coefficient, **533**
  coherent, 111, 526
  compound elastic (resonance scattering), **523**
  Compton, **98ff**, *see also* Cross section, Compton; Scattering, electron
    Feynman diagram, **131ff, 682ff**
  Delbrück, 135, **147**
  elastic, *see* Scattering, compound elastic; Scattering, shape elastic; Scattering kinematics; Neutron, elastic scattering; Cross section, elastic
    Born collision formula, **93**
    matrix element, 91
  electron, 12, 18, 171, 596, *685*, 686, *see also* Scattering, Compton; Scattering, Møller
  function, **534**
  incoherent, 111
  kinematics, **580ff, 604ff**
    relativistic, **590ff, 605ff**
  length *a,* **527**
  Møller, 129, **171,** 682ff, *see also* Scattering, electron
  Mott, **88ff**, 171, 172
  multiple, 155, 172
  p-p, 19
  partial, **527**
  plural, **172**
  proximity, **496ff**
  Rayleigh, 112
  resonance, 148, 545, *see also* Scattering, compound elastic
  shadow, 536
  shape elastic (potential scattering), **523**
  single, 172
  state (nuclear), *see* Nuclear scattering state
  Thomson, **98**
Scattering matrix, *see* *S*-matrix
Schrödinger equation, 224, 310, 543, 617, 631, 677
  time-dependent, **618**
  time-independent, **618**
Schüler-Schmidt lines, **250ff**
Schwinger reformulation of quantum elecrodynamics (correction factor to magnetic dipole moment), 242
Scintillation counter, **706ff**
  characteristics, *710, 711, 712*
  energy resolution, **709ff**
Second quantization, **687ff**
  creation operator, **688**
  destruction (annihilation) operator, **688**
Secular equilibrium, *see* Equilibrium, secular
Segrè chart, 23
Self-energy
  electron, *see* Electron self-energy
  fermion, *see* Fermion self-energy
  photon, *see* Photon self-energy
Semiconductor
  doped, 715
  intrinsic, 713
  n-type, 715
  p-type, 715
  p-n junction, 715
Semiconductor detector, **713ff**
  energy resolution, 718
Semiempirical mass formula, *see* Nuclear liquid-drop model, semiempirical mass formula
Sequential decay processes, *see* Nuclear reaction, sequential decay processes
Shadow scattering, 536
Shape elastic-cross section, *see* Cross section, shape elastic
Shape elastic scattering, *see* Scattering, shape elastic
Sigma hyperon, decay mode, 62, 63
Single-level approximation, 520ff, 549

Single scattering, *see* Scattering, single
Sodium iodide, *see also* Scintillation counter
NaI(Tl) scintillator crystal characteristics, *710, 711*
Solid angle, 586, 597
Sommerfeld fine-structure equation, 630, *see also* Fine-structure constant
Space
  inversion, 257
  isotropy of, *see* Isotropy
  rotation invariance, 257
  symmetry of, *see* Parity
  translation invariance, 257
Specific activity, **270ff**
Spectroscopic designation of partial waves, 80
Spherical Bessel function, *see* Bessel function, spherical
Spherical harmonic, 223, 311, 414, 425, 443, 533, 535, *653*, **654ff**
  bipolar, **670ff**
Spin
  electron, 210
  factor, *see* Statistical spin factor
  nuclear, *see* Nuclear spin
  quantum number, *see* Quantum number, spin
Spin-orbit coupling scheme, *see* Coupling scheme, *j-j*
Spin-quantum number, *see* Quantum number, spin
Spinor formalism, **637ff**
Square-well potential, *see* Nuclear barrier
Statistical factor
  in beta decay, *see* Beta decay, statistical factor; Statistical spin factor
  in gamma decay (Weisskopf), **420**, *see also* Statistical spin factor
Statistical spin factor, **668**
  in neutron-induced reactions, **530ff**
Steady state cosmological theory, 292
Stern-Gerlach experiment, 242
Stopping power, 97, **156ff**, 171
  of heavy ions, 203
  mass, **162**
  relative, **162ff**
Straggling
  angle, *see* Angle straggling
  energy, *see* Energy straggling

parameter, **169**
range, *see* Range straggling
Strangeness conservation, **219**
Strangeness quantum number, *see* Quantum number, strangeness
Strength function, 529, **530**
Surface barrier detector, **717**
Surface energy, *see* Nuclear liquid-drop model, surface energy
Superallowed transitions in beta decay, *see* Beta decay, superallowed transitions
Symmetrical fission, *see* Fission, symmetrical
Symmetry
  operation, 258
  of Racah functions (Wigner symbols), **650**
  of S-matrix, 542
  spatial, *see* Parity

T

T-Matrix, 541
Tetraneutron (^{4}n), 34
Thick-target cross section, *see* Cross section, thick-target
Thin-target cross section, *see* Cross section, thin-target
Thomas-Fermi model, 624
Thomson scattering, *see* Scattering, Thomson
Thomson's atomic model, *see* Atomic model, Thomson
Thorium decay chain, *274*
Threshold energy, 26, *see also* Nuclear reaction threshold
Time
  dilation, **563**, 574
  of flight, 742
  ordering operator, **678**
  proper, *see* Proper time
  reversal, 125, 459, 543, 551
    operator, 544
  translation invariance, 257
Time constant, universal in beta decay, *see* Beta decay, universal time constant
Time-dependent perturbation theory, *see* Perturbation theory

Tolerance dose, **734**
Transformation
 antilinear, *258*
 antiunitary, 258
 continuous and discrete, *258*
Transient equilibrium, *see* Equilibrium, transient
Transition energy, 263
Transition matrix, *see* *T*-Matrix
Transition probability, **64**ff
 per unit time, 67, 505
Transmission of beta spectrometer, **700**
Transmission coefficient, transparency *or* penetrability, **307**, 521
Transuranic elements, 323
Trineutron (^{3}n), 34, 230
Triproton, 230
Tritium activity (dating), 291
Trochoidal spectrometer, *see* Beta spectrometer
Tunneling, *see* Nuclear barrier tunneling
Two-component neutrino theory, *see* Neutrino, two-component theory

U

U-Matrix, 541, **544**
Uncertainty relations, 14, 208, 305, 353, 418, **624**ff
Unitarity
 relation, 542, 649
 of *S*-matrix, 542
 of *U*-matrix, 551
Universal time constant τ_0 in beta decay, *see* Beta decay, universal time constant
Universe
 age of, 293
 pulsating (Dicke theory), 292
Uranium
 decay chain, *274*
 fission, *see* Fission
 lead ratio (dating), **291**
Urca process, 344

V

Vacuum fluctuation, **683**

Vacuum polarization, *see* Polarization vacuum
Vacuum state, **681**
Vector model (of angular momentum), 213
Velocity, addition (relativistic), **565**ff
Vertex (in Feynman graph), **128**
Violet shift, *see* Doppler effects
Volume energy, *see* Nuclear liquid-drop model, volume energy
von Weizsäcker formula, *see* Nuclear liquid-drop model, Bethe-Weizsäcker formula

W

W-Particle, *see* Boson, intermediate charged vector
Wave equation, 523, 617, *see also* Klein-Gordon equation; Dirac equation; Weyl equation; Boson wave equations; Proca equation
 radial, **311**
 angular, **311**
 Schrödinger, *see* Schrödinger equation
Wave-function coupling, **651**ff
Wave mechanics, **617**ff, *see also* Schrödinger equation, Quantum number; Klein-Gordon equation; Uncertainty relations; Dirac equation; Weyl equation; Boson wave equations; Second quantization; Proca equation
Wave-number vector (propagation vector), 91
Wavelength
 Compton, 29, 689, 743
 of electrons, **105**, 743
 of gravitons, *57*
 shift, **105**
 de Broglie, 9ff, 521, **743**, *744*
 numerical, **743**, **744**
 rationalized, 10
 shift (red *or* violet), *see* Doppler effects
Weisskopf estimate of multipole transition probability (lifetime), **419**ff
Weizsäcker formula, *see* Nuclear liquid-drop model, Bethe-Weizsäcker formula
Weyl equation, **639**ff

Wick chronological operator, 680
Width
  natural line (Γ_0), **447**
  nuclear energy-level, *see* Nuclear energy-level width Γ
  reduced, *see* Reduced width
Wigner-Eisenbud reaction formalism, 547, 548
Wigner 3-*j* symbol, **650**ff
  properties, **650**ff
Wigner 6-*j* symbol, **660**ff
  properties, **660**ff
Wigner 9-*j* symbol, **662**ff
  properties, **663**ff
Woods-Saxon potential, *551*
World line, 127
Wronskian, 545
Wu experiment, **399**ff

X

X Rays
  characteristic, 189
  discovery of, **2**

Y

Yield (radioactive), 282
Yukawa particle, *see* Meson
Yukawa potential, 54
Yukawa's prediction of mesons, *see* Mesons, prediction by Yukawa

Z

Z-Function (and $\overline{Z}$-function), 668
Zeeman splitting (nuclear), *see* Nuclear Zeeman splitting
Zitterbewegung, 633